To Allison, Sam, Brooks, David, and Beverly.

Lydia with Anthony Louis in Ljubljana, Slovenia, 2005.

Lydia Mihelič Pulsipher is a cultural-historical geographer who studies the landscapes of ordinary people through the lens of archaeology, historical geography, and ethnography. She has contributed to several geography-related exhibits at the Smithsonian Museum of Natural History in Washington, D.C., including "Seeds of Change," which featured her research in the eastern Caribbean. Lydia Pulsipher has ongoing research projects in the eastern Caribbean (historical archaeology) and in central Europe, where her graduate students are now studying issues of national identity in several countries. She has taught European, North American, Mesoamerican, cultural, and gender geography at the University of Tennessee at Knoxville since 1980; through her research, she has given many students their first experience in fieldwork abroad. Previously, she taught at Hunter College and Dartmouth College. She received her B.A. from Macalester College, her M.A. from Tulane University, and her Ph.D. from Southern Illinois University.

Alex A. Pulsipher is a Ph.D. candidate in geography at Clark University, where he is studying the diffusion innovations related to real estate development, storm-water management, and sustainability in the United States. In the early 1990s, Alex spent some time in South Asia working for a sustainable development research center and then went on to do an undergraduate thesis at Wesleyan University on the history of Hindu nationalism. Alex has been a co-author on all editions of *World Regional Geography: Global Patterns, Local Lives.* He was the guiding force behind this first edition of *World Regional Geography Concepts.*

In the writing of *World Regional Geography,* Lydia and Alex Pulsipher were assisted in many ways by Lydia's husband, **Conrad "Mac" Goodwin,** a historical archaeologist who specializes in sites created during the European colonial era in North America, the Caribbean, and the Pacific. He has particular expertise in the archaeology of agricultural systems, gardens, domestic landscapes, and urban spaces. When not working on the textbook, Mac is a master organic gardener and slow-food chef. He holds a research appointment in the Department of Anthropology at the University of Tennessee.

(Left to right) Alex A. Pulsipher, Lydia Mihelič Pulsipher, Conrad "Mac" Goodwin.

Brief Contents

Contents

Preface

After eight years and four editions of *World Regional Geography,* we have decided to answer the call for a shorter version of our text that gives greater attention to geographic themes and concepts. Accordingly, we have created a new textbook that is 30% shorter than the long book and highlights eight thematic concepts to help students grapple with the amount of complex information contained in the book. We have also integrated the Web site more closely with the text. Two hundred sixty new web-based videos, with accompanying quizzes, are now keyed into the text with icons. Each region has a striking two-page photo-map feature, "Views of the Region," linking to web-based exercises that use three-dimensional mapping technology to illustrate concepts. We are also launching a new online social network aimed at helping instructors teach and manage this course, participate in related online discussions, and use the Internet to further explore the themes, concepts, and issues addressed in this book.

Main Aspects of the Book

Thematic Concepts. Teaching World Regional Geography is never easy. Many instructors have found that focusing their course on a few key ideas makes their job easier and helps students to retain information. We have developed eight thematic concepts that provide a few basic hooks on which students can hang their growing knowledge of the world. The concepts are the focus of major discussions in each chapter, and are related to each other and to other global patterns in the end of chapter "Reflections on the Region" section. Further, all the thematic concepts are reflected in the book's features, which are outlined below. The eight thematic concepts are:

- **Global Warming:** What human activities in the region contribute significant amounts of greenhouse gases? How are places, people, and ecosystems in the region vulnerable to the changes global warming may bring? How are people and governments in the region responding to the threats posed by global warming?

- **Water:** How do issues of water scarcity, water pollution, and water management affect people and environments in the region? Water issues are often discussed in the context of global warming or food production.

- **Food:** How do food production systems impact environments and societies in the region? How has the use of new agricultural technologies impacted farmers? How has it created pressure for urbanization?

- **Population:** What are the major forces driving population growth or decline in the region? How have changes in gender roles influenced population growth? How are changes in life expectancy, family size, and the age of the population influencing population growth?

- **Urbanization:** What forces are driving urbanization in the region? How have cities in the region responded to growth? How are the changes that accompany urbanization—for example in employment, education, and access to health care—affecting the region?

- **Globalization:** How has the region been impacted by globalization historically and in the present? How are people's lives changing as flows of people, ideas, products, and resources become more global?

- **Democratization:** What is the history and current status of democracy in the region? What factors are working for or against the political freedoms that make democracy possible?

- **Gender:** How do gender roles influence societies in this region? How do the livelihoods of men and women differ? How do women participate in politics?

Questions to Consider. This feature early in each chapter replaces the "Themes to Explore" of the larger book with questions that explore the major issues of each region. Each question is based on one or more of the eight thematic concepts and is designed to accomplish several goals:

- Give students a way into each chapter by highlighting some of the major questions they should be looking for answers to as they study the region.

- Show students how thematic concepts are related—for example, the connections between population growth and gender roles—in order to help guide their understanding of each region.

Questions to consider as you study East Asia

1. How is China's spectacular urban growth, and the massive increase in pollution that has come with it, linked to processes of globalization set in motion three decades ago? China's cities have grown by 450 million since 1980, when China's economy was opened to the global economy. Rapid urban economic growth has made China's cities some of the most polluted on the planet. How did globalization and China's response to it lead the country's cities to where they are today?

2. How might global warming result in changes in water availability? East Asia has long suffered from enormously destructive droughts and devastating floods. Now China's highest glaciers are melting. How might this effect the country's rivers? In what ways is China more vulnerable to global warming than Japan, the Koreas, or Taiwan?

3. In what ways has China's embrace of globalization created pressures that might eventually lead to democratization? The growing gap in wealth between China's globalizing cities and its much poorer rural areas has created a reservoir of discontent, yet the most recent shifts toward democracy are happening in urban areas. What changes in urban areas have led to these new developments?

4. How has East Asia's ability to feed itself been transformed by globalization? Why might the region's food systems be criticized for a lack of sustainability? East Asian countries have become more able to buy food on the global market and less able to produce the food they need at home. What kinds of problems might this create both for East Asia and for the rest of the world?

- Draw connections between world regions and highlight global patterns. For example, questions related to the impact of global warming or urbanization in all chapters will help students see how these are issues around the globe yet may be expressed differently in different regions.

- Give instructors ideas on how to provoke discussion in the classroom based on the reading. All of the "Questions to Consider" address issues covered at some length in the chapter.

Views of the Region. This striking two-page photo and text spread reflects the book's overall theme of "Global Patterns, Local Lives." By tying the eight thematic concepts visually to specific places and people in the region, it engages students, helps make geographic issues real for them, and answers the common question, "What does this have to do with me?" Photos are located on a composite satellite image of the region and captioned with explanations that relate them to the major issues of the chapter. The feature is complimented by web-based "fly throughs" with commentary that can be run on many different virtual globes or using videos accessible from the student Web site.

Emphasis on Global Warming and Other Environmental Issues. In response to demand from students and instructors, our environmental coverage is extensive and focuses on some of the most current global environmental problems—global warming (including major sources of greenhouse gases, efforts to reduce emissions, and vulnerability to the effects of global warming in each region), water, and food production, as well as many other environmental issues. Reflecting the importance of this topic, the "Environmental Issues" section comes early in each chapter, right after "Physical Patterns."

Ten Key Aspects of the Region. This feature provides a thumbnail sketch of each region by highlighting ten specific aspects of the region that echo the thematic concepts. Like

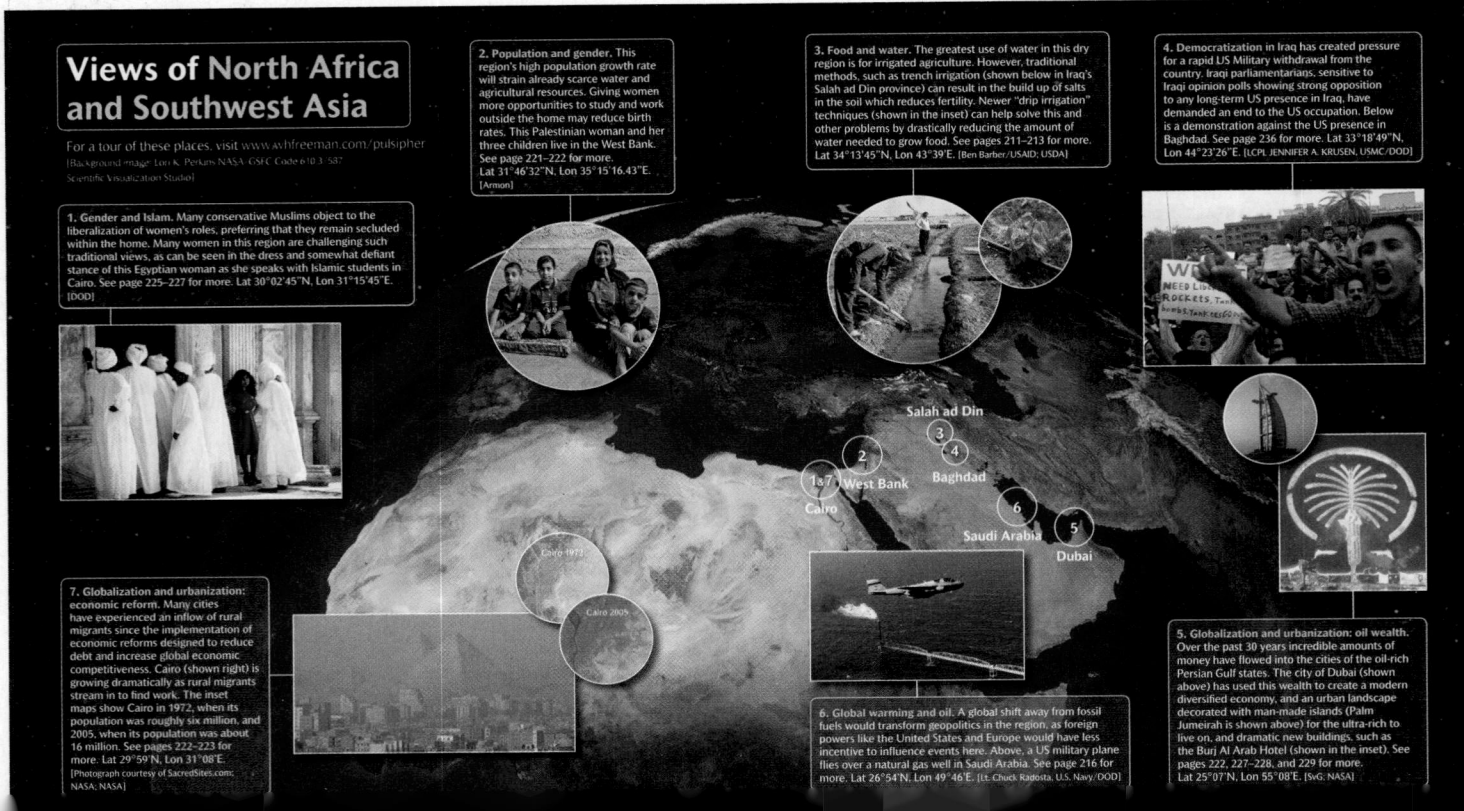

Views of North Africa and Southwest Asia

For a tour of these places, visit www.whfreeman.com/pulsipher
[Background image: Lori K. Perkins NASA-GSFC Code 610.3-587 Scientific Visualization Studio]

1. Gender and Islam. Many conservative Muslims object to the liberalization of women's roles, preferring that they remain secluded within the home. Many women in this region are challenging such traditional views, as can be seen in the dress and somewhat defiant stance of this Egyptian woman as she speaks with Islamic students in Cairo. See page 225–227 for more. Lat 30°02'45"N, Lon 31°15'45"E. [DOD]

2. Population and gender. This region's high population growth rate will strain already scarce water and agricultural resources. Giving women more opportunities to study and work outside the home may reduce birth rates. This Palestinian woman and her three children live in the West Bank. See pages 221–222 for more. Lat 31°46'32"N, Lon 35°15'16.43"E. [Armon]

3. Food and water. The greatest use of water in this dry region is for irrigated agriculture. However, traditional methods, such as trench irrigation (shown below in Iraq's Salah ad Din province) can result in the build up of salts in the soil which reduces fertility. Newer "drip irrigation" techniques (shown in the inset) can help solve this and other problems by drastically reducing the amount of water needed to grow food. See pages 211–213 for more. Lat 34°13'45"N, Lon 43°39'E. [Ben Barber/USAID; USDA]

4. Democratization in Iraq has created pressure for a rapid US Military withdrawal from the country. Iraqi parliamentarians, sensitive to Iraqi opinion polls showing strong opposition to any long-term US presence in Iraq, have demanded an end to the US occupation. Below is a demonstration against the US presence in Baghdad. See page 236 for more. Lat 33°18'49"N, Lon 44°23'26"E. [LCPL JENNIFER A. KRUSEN, USMC/DOD]

5. Globalization and urbanization: oil wealth. Over the past 30 years incredible amounts of money have flowed into the cities of the oil-rich Persian Gulf states. The city of Dubai (shown above) has used this wealth to create a modern diversified economy, and an urban landscape decorated with man-made islands (Palm Jumeirah is shown above) for the ultra-rich to live on, and dramatic new buildings, such as the Burj Al Arab Hotel (shown in the inset). See pages 222, 227–228, and 229 for more. Lat 25°07'N, Lon 55°08'E. [SvG: NASA]

6. Global warming and oil. A global shift away from fossil fuels would transform geopolitics in the region, as foreign powers like the United States and Europe would have less incentive to influence events here. Above, a US military plane flies over a natural gas well in Saudi Arabia. See page 216 for more. Lat 26°54'N, Lon 49°46'E. [Lt. Chuck Radosta, U.S. Navy/DOD]

7. Globalization and urbanization: economic reform. Many cities have experienced an inflow of rural migrants since the implementation of economic reforms designed to reduce debt and increase global economic competitiveness. Cairo (shown right) is growing dramatically as rural migrants stream in to find work. The inset maps show Cairo in 1972, when its population was roughly six million, and 2005, when its population was about 16 million. See pages 222–223 for more. Lat 29°59'N, Lon 31°08'E. [Photograph courtesy of SacredSites.com; NASA: NASA]

Ten Key Aspects of Southeast Asia

- High rates of deforestation and associated burning make Indonesia the world's third largest contributor to global warming after China and the United States.
- Over the last five centuries, several European countries, and later the United States and Japan, established colonies or quasi-colonies that covered almost all of Southeast Asia.
- From the 1960s to the 1990s, Southeast Asian countries pursued export-led growth strategies, investing heavily in industries that manufactured products for export, primarily to developed countries.
- About 39 percent of the region's population is urban, and this proportion is increasing, thanks to changes in agriculture and industry.
- By 2050, 766 million people will be packed into Southeast Asia's land area, which is roughly half the size of the United States.
- The economic crisis of the late 1990s exposed corruption in Thailand and Indonesia, leading to widespread protest movements that expanded democracy in both countries.
- Women's wages average only about one-half to two-thirds those of men in Southeast Asia.
- Glaciers that feed mainland Southeast Asia's four largest rivers are melting so rapidly that they may be gone in 50 years.
- With the exception of the animist belief systems of the indigenous peoples, the major religious traditions of Southeast Asia all originated outside the region.
- Repression by Burma's military government has resulted in at least 680,000 refugees, with 500,000 displaced within Burma and 180,000 having fled to Thailand.

"Questions to Consider," it is designed to help students focus on the main points of each chapter and avoid being overwhelmed by the amount of information.

Videos. Over 260 videos (an average of 24 per chapter) have been added to the existing video offering for *World Regional Geography Concepts*. All are 2–6 minutes long and cover key issues discussed in the text. Each video is keyed into the text with the icon at the point in the discussion where it is most relevant. These videos are available on our Web site and are accompanied with multiple-choice questions that can be automatically graded and entered into a grade book. Visit www.whfreeman.com/geographyvideos.

Current Content. One of the most significant aspects of the book is content that is as close to current news as possible. For example, some major news events included are:

- the 2008 Olympic games in Beijing
- the conflict between Russia and Georgia
- the ongoing conflicts in Iraq and Afghanistan
- political change in Zimbabwe
- mass resistance to the government of Burma
- political change in Nepal

In addition, many news stories are covered by the videos referenced in the text. New videos covering developing global news stories will also be made available via the Web site.

Less "news oriented" but extremely important content areas of the book include:

- The melting of glaciers in the Himalayas, and associated vulnerabilities to global warming in South, East, Southeast, and Central Asia.
- Vulnerability to global warming in Africa.
- Democracy and Islamism in North Africa and Southwest Asia.
- Aging populations in Russia and China.
- The growth of India as a global manufacturing power.
- Africa's partial but growing reorientation toward China and India as trade and investment partners.
- Growing pressures for democracy in China.
- Globalization, urbanization, and the rise of democracy in Europe.

The Enduring Vision: Global and Local Perspectives

The Global View Just as *World Regional Geography* has done through four editions, *World Regional Geography Concepts* emphasizes global trends and the interregional linkages that are changing lives throughout the world. The following linkages are explored throughout the book wherever appropriate:

- **The multifaceted economic linkages of globalization.** These include (1) the effects of colonialism; (2) trade; (3) the role of corporations in the world economy; and (4) the influence exercised by the World Bank and the International Monetary Fund in the form of structural adjustment programs.
- **Migration.** Migrants are changing economic and social relationships in virtually every part of the globe. The text explores the local and global effects of the European,

African, Chinese, and Indian diasporas; foreign workers in such places as Japan, Europe, the Americas, and Southwest Asia; and the increasing number of refugees resulting from conflicts around the world.

- Mass communications and marketing techniques are promoting **world popular culture** across regions. The text integrates coverage of popular culture and its effects throughout in discussions of topics such as TV viewing in North Africa and Southwest Asia and the blending of Western and traditional culture in Japan.

The Local View. Our approach pays special attention to the local scale—a town, a village, a household, an individual person. Our hope is, first, that stories of individual people and families will make geography interesting and real to students. We also hope that seeing the effects of abstract processes and trends on ordinary lives will make the effects of these developments clearer to students. Reviewers have told us that students particularly appreciate the personal vignettes, which are often stories of real people (with names disguised).

> **Vignette** Charity Kaluki Ngilu will never forget the day she became a professional politician. She was washing dishes in her kitchen in Kitui, east of Nairobi, Kenya, when she saw a group of women who had worked with her on community health projects approaching her back door. Mrs. Ngilu answered the knock on the door, drying her hands on an apron. The women said they wanted her to run for parliament in Kenya's first multiparty elections. She assumed they were joking.
>
> That was in 1992. Ngilu beat the governing party's incumbent, then became a major advocate for women's issues. In 1997, she was the first woman to run for president in Kenya. She didn't win, but men were among her strongest supporters because they believed she was capable of making bigger changes than a man could. By 2004, she was minister of health, in charge of Kenya's greatest challenge: responding to the HIV-AIDS epidemic.
>
> *Sources: James C. McKinley, Jr., "A woman to run in Kenya? One says, 'Why not?'" New York Times (August 3, 1997): 3; Kennedy Graham, ed., The Planetary Interest: A New Concept for the Global Age (London: Rutgers University Press, 1999); "Women taking control of power in Africa," "Talk of the Nation," National Public Radio, April 20, 2006.* ■

The following local responses are examined for each region, as appropriate:

- **Cultural change:** This topic explores changes in the family, gender roles, and social organization in response to urbanization, modernization, and the global economy.
- **Issues of identity:** Paradoxically, as the world becomes more tightly knit through global communications and media, ethnic and regional identities often become stronger. The text examines how modern developments such as the Internet are used to reinforce particular cultural identities.
- **Local attitudes toward globalization:** People often have ambivalent reactions to global forces: they are repelled by the seeming power of these forces, fearing effects on their own jobs and on local traditional cultural values, but they

are also attracted by the opportunities that may emerge from greater global integration. The text looks at what a region's people say in favor of or against cultural or economic globalization.

For the Instructor

Instructor's Web site: www.whfreeman.com/pulsipherconcepts (most of this material can also be provided on CD-ROM)

To help instructors create their own Web sites and orchestrate dynamic lectures, the discs contain:

- **Map Builder** available in PowerPoint and Flash format. Built by W. H. Freeman and Company and Maps.com, this is a revolutionary, first-of-its-kind program that allows instructors to create custom maps and then print them or display them as slides in the classroom. This program allows for in-class display of thousands of different custom maps; instructors can zoom in on any section they want to enlarge. There will be 15–25 quiz questions associated with each map.
- **An online social network for instructors** featuring blogs by the authors, your blogs, and numerous opportunities for peer-to-peer interaction all with the goal of helping instructors teach the course. (facultylounge.whfreeman. com/wrg website coming soon).
- **Online quizzing, powered by Questionmark,** by Jason Dittmer, University College London, and Toby Applegate, PhD candidate, Rutgers University. Instructors can easily and securely quiz students using the online multiple-choice Sample Tests (32 questions per chapter). Students receive instant feedback and can take the quizzes multiple times. Instructors can go into a protected Web site to view results by quiz, student, or question, or can get weekly results via e-mail in a simple spreadsheet with all quizzes compiled and graded.
- **Videos.** Please see page xv of the preface for more information about our video program.
- **All text images** in PowerPoint and JPEG formats with enlarged labels for better projection quality.
- **Test Bank,** by Jason Dittmer, University College London; Andy Walter, West Georgia University; and Toby Applegate, PhD candidate, Rutgers University. The *Test Bank* is designed to match the pedagogical intent of the text and offers more than 2000 test questions (multiple-choice, short answer, matching, true/false, and essay) in a Word format that makes it easy to edit, add, and resequence questions.
- **Clicker Questions,** by Jason Dittmer, University College London. Written in PowerPoint for easy integration into lecture presentations, *Clicker Questions* allow instructors to jump-start discussions, illuminate important points, and promote better conceptual understanding during lectures.

Course Management

All instructor and student resources are also available via WebCT/Blackboard to enhance your course. W. H. Freeman and Company offers a course cartridge that populates your site with content tied directly to the book.

Overhead Transparency Set (1-4292-0504-0)

Overhead transparencies showing every map from the text are available to adopters. All labels have been resized for easy readability.

Student Supplements

World Regional Geography Online at www.whfreeman.com/pulsipherconcepts

Tim Oakes and Chris McMorran,
University of Colorado, Boulder

- **Map Builder Assessment**—Between 15 and 25 questions per chapter ask students to explore the innovative Map Layering program and answer questions about the patterns they see.

- **Map Learning Exercises**—Students use this interactive feature to identify and locate countries, cities, and the major geographic features of each region. These exercises make learning place locations fun and are instructive for future work.

- **Thinking Geographically**—These activities allow students to explore a set of current issues—such as deforestation, human rights, or free trade—and to experience how geography helps clarify our understanding of them. Linked Web sites are matched with a series of questions, brief activities, or both that give students an opportunity to think about the ways they are connected to the places and people they read about in the text. This aid helps students focus on key geography concepts, such as scale, region, place, and interaction, by using these concepts to drive analysis of compelling issues.

- **Working with Maps**—This feature offers two sets of map-related exercises that develop students' analytical abilities:

  Thematic Maps: Students can place various maps from the text side-by-side to compare and contrast data. Associated questions accompany each option.

  Animated Population Maps: Animated maps show how regional populations have changed or fluctuated with time. Related questions ask students how and why the changes may have occurred.

- **Blank Outline Maps**—Printable maps of the world and of each region are available for use in note-taking, exam review, or both, as well as for preparing assigned exercises.

- **Online Quizzing**—A self-quizzing feature (25 questions per chapter) enables students to review key text concepts and sharpens their ability to analyze geographic material for exam preparation. Answers (correct or incorrect) prompt feedback referring students to the specific section in the text where the question is covered.

- **Flashcards**—Matching exercises teach vocabulary and definitions.

- **Audio Pronunciation Guide**—A spoken guide of place names, regional terms, and names of historical figures.

- **World Recipes and Cuisines**—From International Home Cooking, the United Nations International School Cookbook.

Rand McNally's Atlas of World Geography, 2007 Edition, paperback, 176 pages

This atlas contains:

- Fifty-two physical, political, and thematic maps of the world and continents; forty-nine regional, physical, political, and thematic maps; and dozens of metro area inset maps.

- Geographic facts and comparisons covering topics such as population, climate, and weather.

- A section on common geographic questions, a glossary of terms, and a comprehensive 25-page index.

Acknowledgments

First Edition World Regional Geography Concepts

Gillian Acheson
Southern Illinois University, Edwardsville

Tanya Allison
Montgomery College

Keshav Bhattarai
Indiana University, Bloomington

Leonhard Blesius
San Francisco State University

Jeffrey Brauer
Keystone College

Donald Buckwalter
Indiana University of Pennsylvania

Craig Campbell
Youngstown State University

John Comer
Oklahoma State University

Kevin Curtin
George Mason University

Ron Davidson
California State University, Northridge

Tina Delahunty
Texas Tech University

Dean Fairbanks
California State University, Chico

Allison Feeney
Shippensburg University of Pennsylvania

Eric Fournier
Samford University

Qian Guo
San Francisco State University

Carole Huber
University of Colorado at Colorado Springs

Paul Hudak
University of North Texas

Christine Jocoy
California State University, Long Beach

Ron Kalafsky
University of Tennessee, Knoxville

David Keefe
University of the Pacific

Mary Klein
Saddleback College

Max Lu
Kansas State University

Donald Lyons
University of North Texas

Barbara McDade
University of Florida

Victor Mote
University of Houston

Darrell Norris
SUNY Geneseo

Gabriel Popescu
Indiana University, South Bend

Claudia Radel
Utah State University

Donald Rallis
University of Mary Washington

Pamela Riddick
University of Memphis

Jennifer Rogalsky
SUNY Geneseo

Tobie Saad
University of Toledo

Charles Schmitz
Towson University

Sindi Sheers
George Mason University

Ira Sheskin
University of Miami

Dmitri Siderov
CSU Long Beach

Steven Silvern
Salem State College

Ray Sumner
Long Beach City College

Stan Toops
Miami University

Karen Trifonoff
Bloomsburg University of Pennsylvania

Jim Tyner
Kent State University

Michael Walegur
University of Delaware

Scott Walker
Northwest Vista College

Mark Welford
Georgia Southern University

First Edition World Regional Geography

Helen Ruth Aspaas
Virginia Commonwealth University

Brad Bays
Oklahoma State University

Stanley Brunn
University of Kentucky

Altha Cravey
University of North Carolina at Chapel Hill

David Daniels
Central Missouri State University

Dydia DeLyser
Louisiana State University

James Doerner
University of Northern Colorado

Bryan Dorsey
Weber State University

Lorraine Dowler
Penn State University

Hari Garbharran
Middle Tennessee State University

Baher Ghosheh
Edinboro University of Pennsylvania

Janet Halpin
Chicago State University

Peter Halvorson
University of Connecticut

Michael Handley
Emporia State University

Robert Hoffpauir
California State University, Northridge

Glenn G. Hyman
International Center for Tropical Agriculture

David Keeling
Western Kentucky University

Thomas Klak
Miami University of Ohio

Darrell Kruger
Northeast Louisiana University

David Lanegran
Macalester College

David Lee
Florida Atlantic University

Calvin Masilela
West Virginia University

Janice Monk
University of Arizona

Heidi Nast
De Paul University

Katherine Nashleanas
University of Nebraska

Tim Oakes
University of Colorado, Boulder

Darren Purcell
Florida State University

Susan Roberts
University of Kentucky

Dennis Satterlee
Northeast Louisiana University

Kathleen Schroeder
Appalachian State University

Dona Stewart
Georgia State University

Ingolf Vogeler
University of Wisconsin, Eau Claire

Susan Walcott
Georgia State University

Second Edition World Regional Geography

Helen Ruth Aspaas
Virginia Commonwealth University

Cynthia F. Atkins
Hopkinsville Community College

Timothy Bailey
Pittsburgh State University

Robert Maxwell Beavers
University of Northern Colorado

James E. Bell
University of Colorado, Boulder

Richard W. Benfield
Central Connecticut State University

John T. Bowen, Jr.
University of Wisconsin, Oshkosh

Stanley Brunn
University of Kentucky

Donald W. Buckwalter
Indiana University of Pennsylvania

Gary Cummisk
Dickinson State University

Roman Cybriwsky
Temple University

Cary W. de Wit
University of Alaska, Fairbanks

Ramesh Dhussa
Drake University

David M. Diggs
University of Northern Colorado

Jane H. Ehemann
Shippensburg University

Kim Elmore
University of North Carolina at Chapel Hill

Thomas Fogarty
University of Northern Iowa

James F. Fryman
University of Northern Iowa

Heidi Glaesel
Elon College

Ellen R. Hansen
Emporia State University

John E. Harmon
Central Connecticut State University

Michael Harrison
University of Southern Mississippi

Douglas Heffington
Middle Tennessee State University

Robert Hoffpauir
California State University, Northridge

Catherine Hooey
Pittsburgh State University

Doc Horsley
Southern Illinois University, Carbondale

David J. Keeling
Western Kentucky University

James Keese
California Polytechnic State University

Debra D. Kreitzer
Western Kentucky University

Jim LeBeau
Southern Illinois University, Carbondale

Howell C. Lloyd
Miami University of Ohio

Judith L. Meyer
Southwest Missouri State University

Judith C. Mimbs
University of Tennessee, Chattanooga

Monica Nyamwange
William Paterson University

Thomas Paradis
Northern Arizona University

Firooza Pauri
Emporia State University

Timothy C. Pitts
Edinboro University of Pennsylvania

William Preston
California Polytechnic State University

Gordon M. Riedesel
Syracuse University

Joella Robinson
Houston Community College

Steven M. Schnell
Northwest Missouri State University

Kathleen Schroeder
Appalachian State University

Dean Sinclair
Northwestern State University

Robert A. Sirk
Austin Peay State University

William D. Solecki
Montclair State University

Wei Song
University of Wisconsin, Parkside

William Reese Strong
University of North Alabama

Selima Sultana
Auburn University

Suzanne Traub-Metlay
Front Range Community College

David J. Truly
Central Connecticut State University

Alice L. Tym
University of Tennessee, Chattanooga

Third Edition World Regional Geography

Kathryn Alftine
California State University, Monterey Bay

Donna Arkowski
Pikes Peak Community College

Tim Bailey
Pittsburg State University

Brad Baltensperger
Michigan Technological University

Michele Barnaby
Pittsburg State University

Daniel Bedford
Weber State University

Richard Benfield
Central Connecticut State University

Sarah Brooks
University of Illinois at Chicago

Jeffrey Bury
University of Colorado, Boulder

Michael Busby
Murray State University

Norman Carter
California State University, Long Beach

Gary Cummisk
Dickinson State University

Cyrus Dawsey
Auburn University

Elizabeth Dunn
University of Colorado, Boulder

Margaret Foraker
Salisbury University

Robert Goodrich
University of Idaho

Steve Graves
California State University, Northridge

Ellen Hansen
Emporia State University

Sophia Harmes
Towson University

Mary Hayden
Pikes Peak Community College

R. D. K. Herman
Towson University

Samantha Kadar
California State University, Northridge

James Keese
California Polytechnic State University, San Luis Obispo

Phil Klein
University of Northern Colorado

Debra D. Kreitzer
Western Kentucky University

Soren Larsen
Georgia Southern University

Unna Lassiter
California State University, Long Beach

David Lee
Florida Atlantic University

Anthony Paul Mannion
Kansas State University

Leah Manos
Northwest Missouri State University

Susan Martin
Michigan Technological University

Luke Marzen
Auburn University

Chris Mayda
Eastern Michigan University

Michael Modica
San Jacinto College

Heather Nicol
State University of West Georgia

Ken Orvis
University of Tennessee

Thomas Paradis
Northern Arizona University

Amanda Rees
University of Wyoming

Arlene Rengert
West Chester University of Pennsylvania

B. F. Richason
St. Cloud State University

Deborah Salazar
Texas Tech University

Steven Schnell
Kutztown University

Kathleen Schroeder
Appalachian State University

Roger Selya
University of Cincinnati

Dean Sinclair
Northwestern State University

Garrett Smith
Kennesaw State University

Jeffrey Smith
Kansas State University

Dean Stone
Scott Community College

Selima Sultana
Auburn University

Ray Sumner
Long Beach City College

Christopher Sutton
Western Illinois University

Harry Trendell
Kennesaw State University

Karen Trifonoff
Bloomsburg University

David Truly
Central Connecticut State University

Kelly Victor
Eastern Michigan University

Mark Welford
Georgia Southern University

Wendy Wolford
University of North Carolina at Chapel Hill

Laura Zeeman
Red Rocks Community College

Fourth Edition World Regional Geography

Robert Acker
University of California, Berkeley

Joy Adams
Humboldt State University

John All
Western Kentucky University

Jeff Allender
University of Central Arkansas

David L. Anderson
Louisiana State University, Shreveport

Donna Arkowski
Pikes Peak Community College

Jeff Arnold
Southwestern Illinois College

Richard W. Benfield
Central Connecticut University

Sarah A. Blue
Northern Illinois University

Patricia Boudinot
George Mason University

Michael R. Busby
Murray State College

Norman Carter
California State University, Long Beach

Gabe Cherem
Eastern Michigan University

Brian L. Crawford
West Liberty State College

Phil Crossley
Western State College of Colorado

Gary Cummisk
Dickinson State University

Kevin M. Curtin
University of Texas at Dallas

Kenneth Dagel
Missouri Western State University

Jason Dittmer
University College London

Rupert Dobbin
University of West Georgia

James Doerner
University of Northern Colorado

Ralph Feese
Elmhurst College

Richard Grant
University of Miami

Ellen R. Hansen
Emporia State University

Holly Hapke
Eastern Carolina University

Mark L. Healy
Harper College

David Harms Holt
Miami University

Douglas A. Hurt
University of Central Oklahoma

Edward L. Jackiewicz
California State University, Northridge

Marti L. Klein
Saddleback College

Debra D. Kreitzer
Western Kentucky University

Jeff Lash
University of Houston, Clear Lake

Unna Lassiter
California State University, Long Beach

Max Lu
Kansas State University

Donald Lyons
University of North Texas

Shari L. MacLachlan
Palm Beach Community College

Chris Mayda
Eastern Michigan University

Armando V. Mendoza
Cypress College

Katherine Nashleanas
University of Nebraska, Lincoln

Joseph A. Naumann
University of Missouri at St. Louis

Jerry Nelson
Casper College

Michael G. Noll
Valdosta State University

Virginia Ochoa-Winemiller
Auburn University

Karl Offen
University of Oklahoma

Eileen O'Halloran
Foothill College

Ken Orvis
University of Tennessee

Manju Parikh
College of Saint Benedict and Saint John's University

Mark W. Patterson
Kennesaw State University

Paul E. Phillips
Fort Hays State University

Rosann T. Poltrone
Arapahoe Community College, Littleton, Colorado

Waverly Ray
MiraCosta College

Jennifer Rogalsky
SUNY-Geneseo

Gil Schmidt
University of Northern Colorado

Yda Schreuder
University of Delaware

Tim Schultz
Green River Community College, Auburn, Washington

Sinclair A. Sheers
George Mason University

D. James Siebert
North Harris Montgomery Community College, Kingwood

Dean Sinclair
Northwestern State University

Bonnie R. Sines
University of Northern Iowa

Vanessa Slinger-Friedman
Kennesaw State University

Andrew Sluyter
Louisiana State University

Kris Runberg Smith
Lindenwood University

Herschel Stern
MiraCosta College

William R. Strong
University of North Alabama

Ray Sumner
Long Beach City College

Rozemarijn Tarhule-Lips
University of Oklahoma

Alice L. Tym
University of Tennessee, Chattanooga

James A. Tyner
Kent State University

Robert Ulack
University of Kentucky

Jialing Wang
Slippery Rock University of Pennsylvania

Linda Q. Wang
University of South Carolina, Aiken

Keith Yearman
College of DuPage

Laura A. Zeeman
Red Rocks Community College

These books have been a family project many years in the making. Lydia Pulsipher came to the discipline of geography at the age of five, when her immigrant father, Joe Mihelič, hung a world map over the breakfast table in their home in Coal City, Illinois, where he was pastor of the New Hope Presbyterian Church. They soon moved to the Mississippi Valley of eastern Iowa, where Lydia's father, then a professor in the Presbyterian theological seminary in Dubuque, continued his geography lessons on the passing landscapes whenever Lydia accompanied him on Sunday trips to small country churches. Lydia's sons, Anthony and Alex, first traveled abroad and learned about the hard labor of field geography when as 12- and 8-year-olds they were expected to help with the archaeological and ethnographic research Lydia and her colleagues were conducting on the eastern Caribbean island of Montserrat. It was Lydia's brother John Mihelič, who first suggested that Lydia, Alex, and Mac write a book like this one, after he too came to appreciate geography. He has been a loyal cheerleader during the process, as have our extended family and friends in Knoxville, Montserrat, San Francisco, Slovenia, and beyond.

Alex's wife Allison has also been unfailingly supportive of this book with her penetrating insights about how best to write for students, and her keen eye for visual composition. Brooks and David Eggers and Beverly Owen have also helped make this book possible in more ways than they know.

Graduate students and faculty colleagues in the Geography Department at the University of Tennessee have been generous in their support, serving as helpful impromptu sounding boards for ideas. Ken Orvis, especially, has advised us on the physical geography sections of all editions.

Maps for this edition were conceived by Mac Goodwin with the help of Alex Pulsipher and produced by Will Fontanez and the University of Tennessee cartography shop staff; Joshua Calhoun, Tracy Pollock, and Steve Ahrens; and by Martha Bostwick and her staff at Maps.com; Deane Plaister, Brandi Webber, Mike Powers, Lisa Basanese, Lucy Pendl, and Jesse Wickizer.

Sara Tenney and Liz Widdicombe at W. H. Freeman and Company were the first to persuade us that together we could develop a new direction for *World Regional Geography*, one that included the latest thinking in geography written in an accessible style. In accomplishing this goal, we are especially indebted to our developmental editor, Susan Weisberg, who has proved remarkably efficient and helpful in shortening the text and in reconciling all the many details that make for a useful and elegant book. We are also indebted to the W. H. Freeman staff for all they have done to ensure that this book is well-written, beautifully designed, and well-presented to the public.

We would also like to gratefully acknowledge the efforts of the following people at W. H. Freeman: Marc Mazzoni, senior acquisitions editor for the first edition of concepts; Leigh Renhard, project editor; Scott Guile, senior marketing manager; Norma Roche, copy editor; Diana Blume, art director; Bill Page, senior illustration coordinator; Julia De Rosa, production manager; Sheridan Sellers, W. H. Freeman and Company Electronic Publishing Center; Kathy Bendo, photo editor, and Inge King, photo researcher for the first edition; Meg Kuhta, photo editor, and Bianca Moscatelli and Julie Tesser, photo researchers for the second edition; Bianca Moscatelli, photo editor, and Elyse Rieder, photo researcher for the third edition; Trish Marx and Ted Szczepanski, photo editors, and Elyse Rieder and Donna Ranieri, photo researchers for the fourth edition; Deepa Chungi, supplements editor; Eleanor Wedge and Martha Solonche, proofreaders for the text and maps; and Daniel Gonzalez, editorial assistant. We are also grateful to the supplements authors, who have created what we think are unusually useful, up-to-date, and labor-saving materials for instructors who use our book: Jason Dittmer, author of the Test Bank, PowerPoint lecture notes, Clicker questions, sample tests, and video guides, and Toby Applegate who updated Jason Dittmer's test bank for this edition.

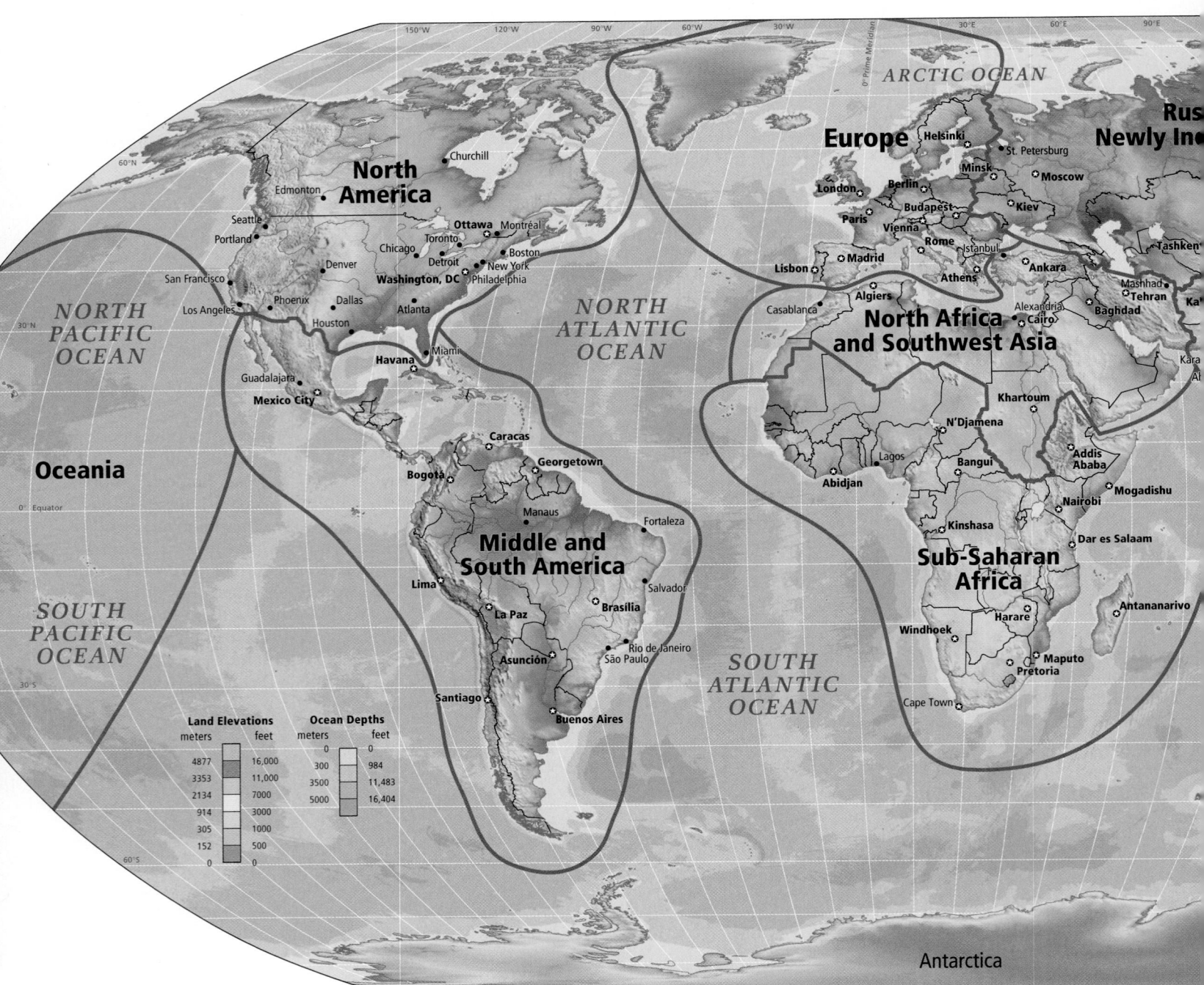

Figure 1.1 Regions of the world.

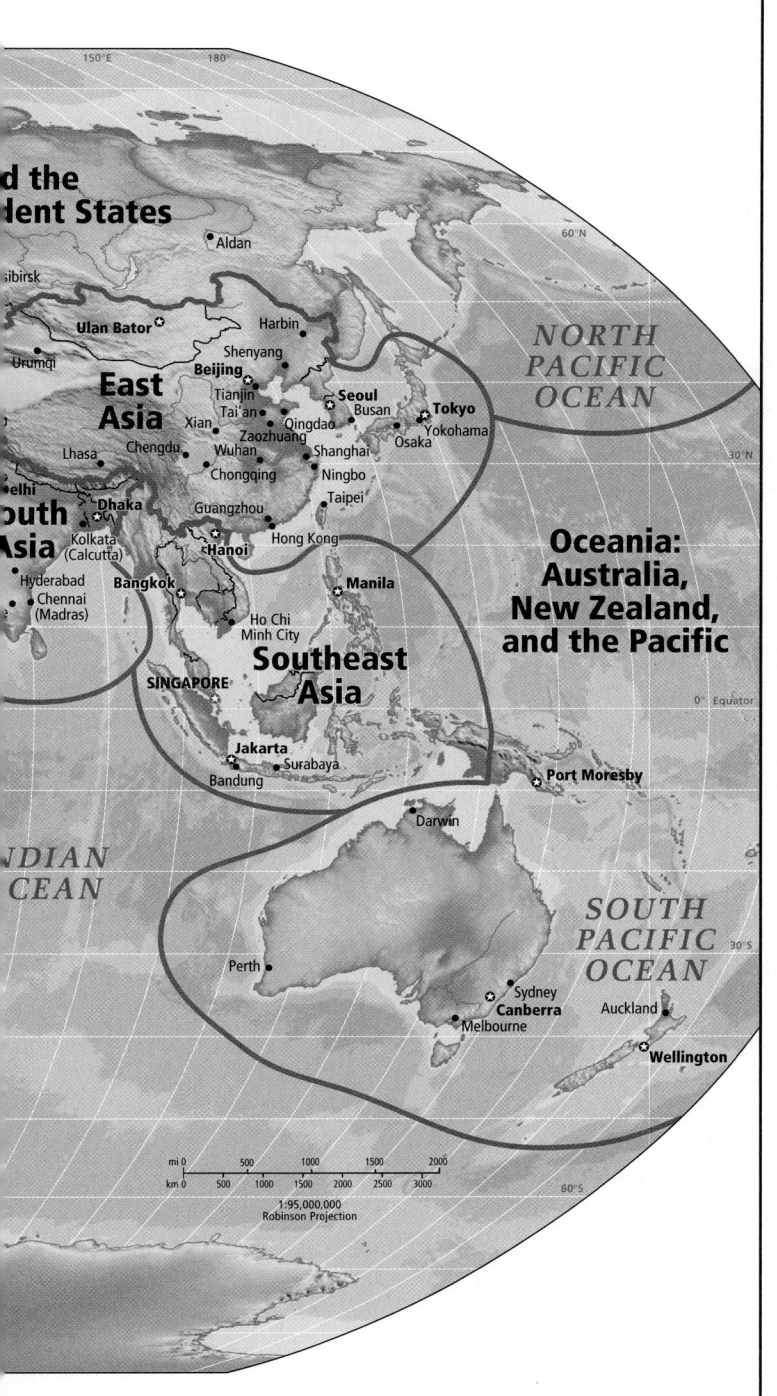

CHAPTER 1

Geography: An Exploration of Connections

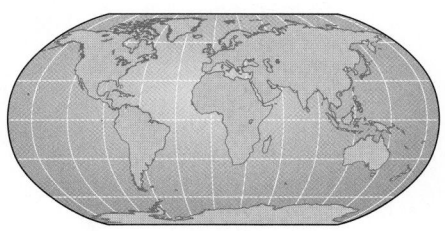

WHERE IS IT? WHY IS IT THERE? WHY DOES IT MATTER?

Where are you? You may be in a house or a library or sitting under a tree on a fine fall afternoon. You are probably in a community (perhaps a college or university), and you are in a country (perhaps the United States) and a region of the world (perhaps North America, Southeast Asia, or the Pacific). Why are you where you are? There are immediate answers, such as "I have an assignment to read." But there are also larger explanations, such as your belief in the value of an education, your career plans, or someone's willingness to sacrifice to pay your tuition. Even past social movements that opened up higher education to more than a fortunate few may help to explain why you are where you are.

The questions *where* and *why* are central to geography. Think about a time you had to find the site of a party on a Saturday night, the location of the best grocery store, or the fastest and safest route home. Like a geographer, you were interested in location, spatial relationships, and connections between the environment and people.

Geographers want to understand why different places have different sights, sounds, smells, and arrangements of features. They study what has contributed to the look and feel of a place, to the standard of living and customs of its people, and to the way people in one place relate to people in other places. Furthermore, geographers often think on several scales. For example, when choosing the best location for a new grocery store, a geographer might consider the socioeconomic circumstances of the neighborhood, traffic patterns for the broader area in which the neighborhood is located, as well as the store's location relative to the main population concentrations for the

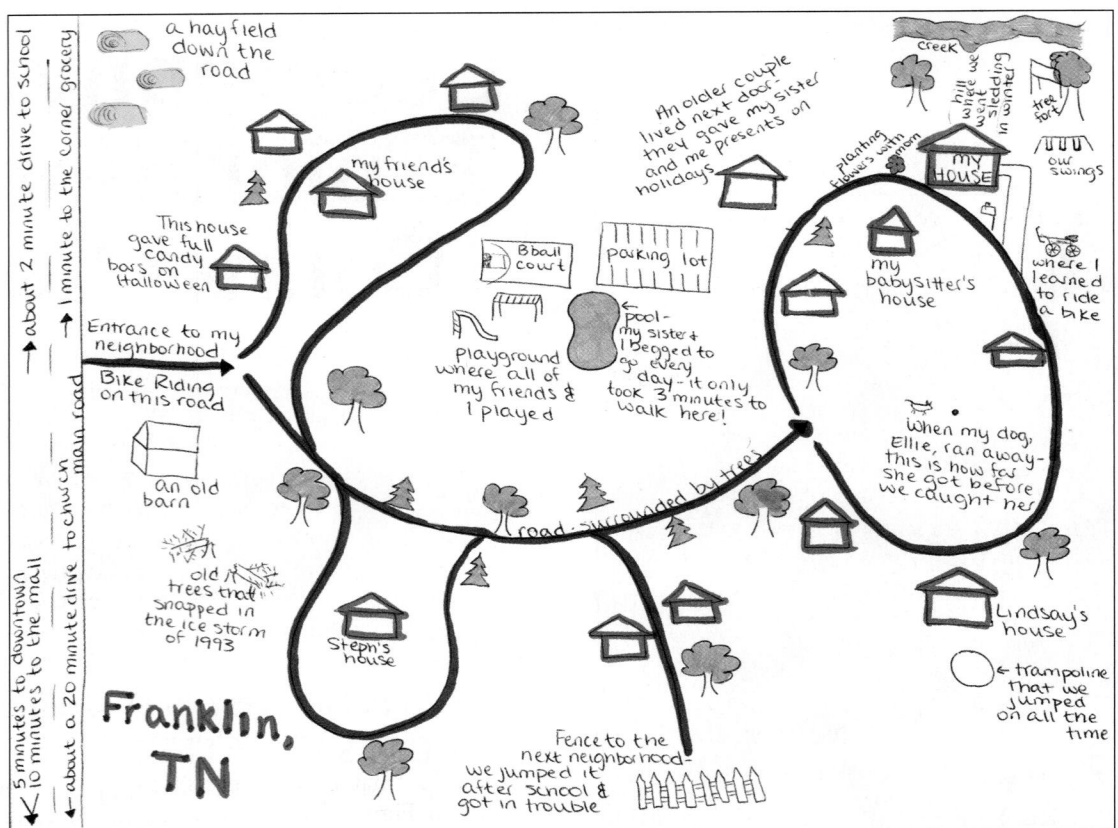

Figure 1.2 A childhood landscape map. Julia Stump drew this map of her childhood landscape in Franklin, Tennessee, as an exercise in Dr. Pulsipher's geography class in January 2006. [Courtesy of Julia Stump.]

whole city. She could also consider national or even international transportation routes, possibly to determine cost-efficient connections to suppliers.

To make it easier to understand a geographer's many interests, please try this exercise. On a piece of blank paper, draw a map of your favorite childhood landscape. Relax, and let your mind recall the objects and experiences that were most important to you in that place. If the place was your neighborhood, you might start by drawing and labeling your home. Then fill in other places you encountered regularly, such as your backyard, your best friend's home, or your school. Don't worry about creating a work of art—just make a map that reflects your experiences as you remember them. For example, Figure 1.2 shows the childhood landscape of Julia Stump in Franklin, Tennessee.

When you finish your map, think about how the map reveals the ways in which your life was structured by space. What is the **scale** of your map? That is, how much space did you decide to illustrate on the paper? The amount of space your map covers may represent the degree of freedom you had as a child, or how aware you were of the world around you. Were there places you were not supposed to go? Does your

map reveal, perhaps subtly, such emotions as fear, pleasure, or longing? Does it indicate your sex, your ethnicity, or the makeup of your family? Did you use symbols to show certain features? In making your map and analyzing it, you have engaged in several aspects of geography:

• Landscape observation

• Description of the earth's surface and consideration of the natural environment

• Spatial analysis (the study of how people, objects, or ideas are or are not related to one another across space)

• The use of different scales of analysis (your map probably shows the spatial features of your childhood at a detailed local scale)

• Cartography (the making of maps)

As you progress through this book, you will acquire geographic information and skills. Perhaps you are planning to travel or thinking about investing in East Asian timber stocks. Maybe you are searching for a good place to market an idea or trying to understand current local events in the context of world events. Knowing how to practice geography will make your task easier and more engaging.

WHAT IS GEOGRAPHY?

Geography can be defined as the study of our planet's surface and the processes that shape it. Yet this does not begin to convey the fascinating interactions of human and environmental forces that have given the earth its diverse landscapes and ways of life.

Geography, as an academic discipline, is unique in the extent to which it links the physical sciences—such as geology, physics, chemistry, biology, and botany—with the social sciences—such as anthropology, sociology, history, economics, and political science. **Physical geography** is the study of earth's physical processes. Physical geographers have generally focused on how they work independently of humans, but increasingly many are interested in how physical processes may affect humans, and how they are affected by humans in return. **Human geography** is the study of the various aspects of human life that create the distinctive landscapes and regions of the world. Physical and human geography are often tightly linked. For example, geographers might aim to understand

- How and why people came to occupy a particular place
- Gow people use the physical aspects of that place (climate, landforms, and resources) and then modify them to suit their particular needs
- How people may create environmental problems
- How people interact with other places, far and near

Geographers usually specialize in one or more fields of study, or *subdisciplines.* You will see some of these particular types of geography mentioned as we go along. Despite their individual specialties, geographers often cooperate in studying the interactions between people and places. For example, in the face of increasing global warming, climate geographers, cultural geographers, and economic geographers are working together. They want to understand the spatial distribution of carbon dioxide emissions, the cultural practices which might be changed to limit such emissions, and the economic effects of these potential changes.

Many geographers specialize in a particular region of the world, or even in one small part of a region. *Regional geography* is the analysis of the geographic characteristics of a particular place, the size and scale of which can vary radically. The study of a region can reveal connections among physical features and ways of life, as well as connections to other places. These links are a key to understanding the present and the past and are essential in planning for the future. This book follows a "world regional" approach, focusing on general knowledge about specific regions of the world. We will see just what geographers mean by *region* a little later in this chapter.

GEOGRAPHERS' VISUAL TOOLS

Among geographers' most important tools are maps, which they use to record, analyze, and explain spatial relationships, as you did on your childhood landscape map. Geographers who specialize in depicting geographic information graphically are called *cartographers.* The following discussion of map reading will help you understand the maps in this book and elsewhere.

When geographers want to show the spatial aspects of a particular place or concept on a map—say, the locations and relative sizes of islands in the Caribbean Sea—they begin by deciding at what scale to map the information. Earlier, we described *scale* as the amount of space you illustrated on your childhood map. Phrased another way, the scale of a map represents the relationship between the distance shown on the map and actual distance on the earth's surface. Sometimes a numerical ratio shows what one unit of measure on the map equals in the same units on the face of the earth. For example, 1:1,000,000, or 1/1,000,000, means that 1 inch on the map equals 1 million inches on the face of the earth. Alternatively, a simple bar may express the information visually:

0   10   20   30   40   50 km

Notice that each of the maps of the eastern Caribbean shown in Figure 1.3 is drawn using a different scale. As the area shown on a map increases, the amount of detail decreases.

Throughout this book, you will encounter different kinds of maps at different scales. Some will show physical features such as landforms or climate patterns at the regional or global scale. Others will show aspects of human activities at these same scales—for example, the routes taken by drug traders. Yet other maps will show settlement or cultural features at the scale of cities or countries.

The *title, caption,* and *legend* give basic information about the map. The title tells you the subject of the map, and the caption usually points out some features of the map that the cartographer wants you to notice. The legend is the box that explains what the symbols and colors on the map represent (see, for example, Figure 1.9 on page 10). The map's scale is usually placed in an open space or near the map's legend (see, for example, Figure 2.5 on page 57). Often whole world maps will not have scale information, though Figure 1.1, a map of world regions, does have a scale bar.

Longitude and Latitude

Most maps contain *lines of latitude* and *longitude,* which enable a person to establish a position on the map relative to other points on the globe. Lines of **longitude** (also called meridians) run from pole to pole; lines of **latitude** (also called parallels) run parallel to the equator (Figure 1.4).

Both latitude and longitude lines describe circles, so there are 360° in each circle of latitude and longitude. Each degree

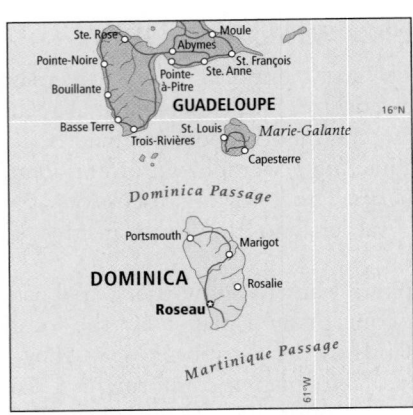

(a) **A map of Guadeloupe and Dominica,** in the eastern Caribbean, at a scale of 1:3,000,000. This scale makes it possible to show towns, a few roads, and a few landforms, but not much else.

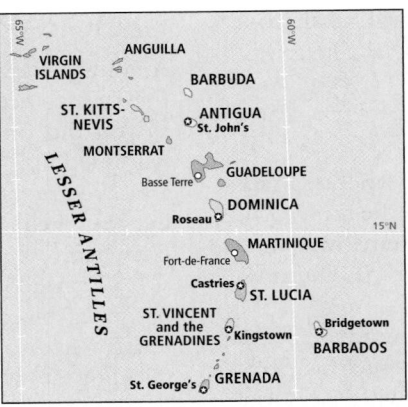

(b) **The same area at a scale of 1:15,000,000.** You can see much more of the eastern Caribbean, but the only detail that can be shown is the shape of the islands and the locations of some capital cities.

(c) **The map at a scale of 1:45,000,000.** It shows most of the Caribbean Sea and its general location between Central and South America, but now the eastern Caribbean islands are too small to identify clearly.

Figure 1.3 Examples of scale. For comparison, Figure 1.1, on pages 2–3, is at a scale of 1:95,000,000. [Adapted from *The Longman Atlas for Caribbean Examinations,* 2nd ed. (Essex, U.K.: Addison Wesley Longman, 1998), p. 4.]

spans 60 minutes (designated with the symbol ′), and each minute has 60 seconds (designated with the symbol ″). Keep in mind that these are measures of relative space on a circle, not time. They do not even represent real distance because the circles of latitude get successively smaller to the north and south of the equator, until they become a virtual dot at the poles.

The globe is also divided into hemispheres. The *prime meridian,* 0° longitude, runs from the North Pole through Greenwich, England, to the South Pole. The half of the globe's surface west of the prime meridian is called the Western Hemisphere; the half to the east is called the Eastern Hemisphere. The longitude lines both east and west of the prime meridian are labeled from 1° to 180° by their direction and distance in degrees from the prime meridian. For example, 20 degrees east longitude would be written as 20° E. The longitude line at 180° runs through the Pacific Ocean and is used as the *international date line,* where the calendar day officially begins.

The equator divides the globe into the Northern and Southern Hemispheres. Latitude is measured from 0° at the equator to 90° at the North Pole and South Pole, respectively.

Lines of longitude and latitude form a grid that can be used to designate the location of a place. In Figure 1.3a, you can see that the island of Marie-Galante lies just south of the parallel at 16° N and just about 18′ west of the 61st west meridian. Hence, the position of Marie-Galante's northernmost coast is 16° N, 61°18′ W.

Map Projections

Printed maps must solve the problem of showing the spherical earth on a flat piece of paper. Imagine drawing a map of the earth on an orange, peeling the orange, and then trying

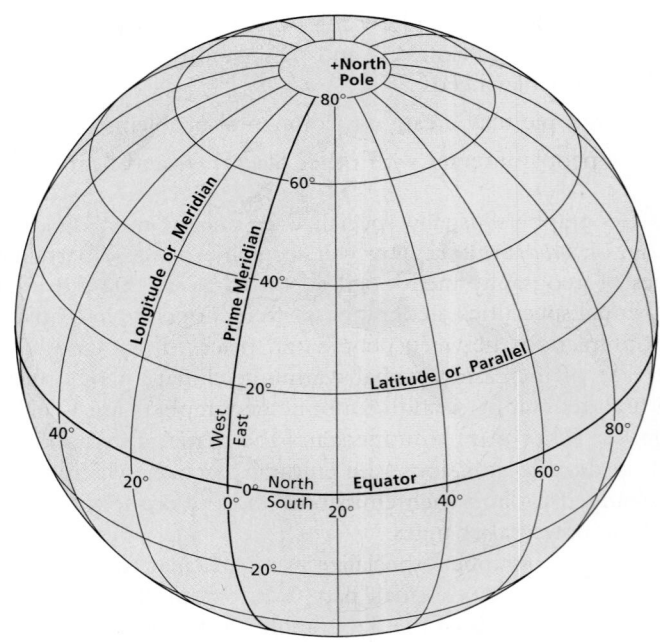

Figure 1.4 Summary of longitude and latitude. Lines of longitude (meridians) extend from pole to pole. The distance between them on the globe decreases steadily toward the poles, where they all meet. Lines of latitude (parallels) are equally spaced north and south of the equator and intersect the longitude lines at right angles. The only line of latitude that spans the complete circumference of the earth is the equator; all other lines of latitude describe ever smaller circles heading away from the equator. [Adapted from *The New Comparative World Atlas* (Maplewood, N.J.: Hammond, 1997), p. 6.]

to flatten out the orange-peel map and transfer it exactly to a flat piece of paper. The various ways of showing the spherical surface of the earth on flat paper are called *projections*. All projections create some distortion. For maps of small parts of the earth's surface, the distortion is minimal. Developing a projection for the whole surface of the earth that minimizes distortion is much more challenging.

The *Mercator projection* (Figure 1.5a) is popular, but geographers rarely use it because of its gross distortion near the poles. To make his flat map, the Flemish cartographer Gerhardus Mercator (1512–1594) stretched out the poles, depicting them as lines equal in length to the equator! Greenland, for example, appears about as large as Africa, even though it is only about one-fourteenth Africa's size. Nevertheless, Mercator's projection is still useful for navigation because it portrays the shapes of landmasses more or less accurately, and

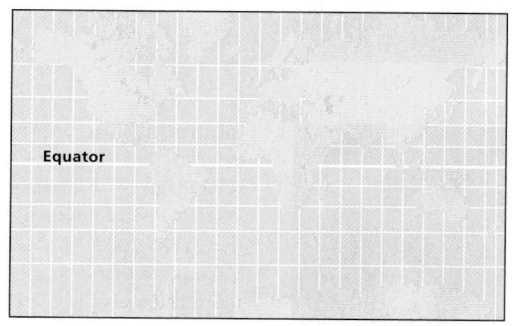

(a) Mercator Projection

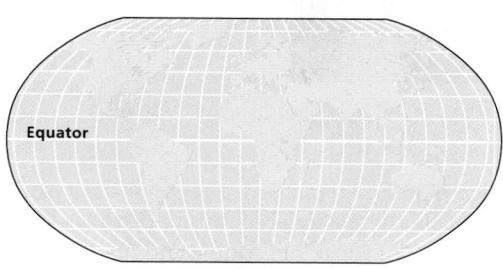

(b) Goode's Interrupted Homolosine Projection

(c) Robinson Projection

Figure 1.5 Three common map projections. [Mercator and Robinson projections adapted from *The New Comparative World Atlas* (Maplewood, N.J.: Hammond, 1997), pp. 6–7. Goode's interrupted homolosine projection adapted from *Goode's World Atlas,* 19th ed. (Chicago: Rand McNally, 1995), p. x.

because a straight line between two points on this map gives the compass direction between them.

Goode's interrupted homolosine projection (Figure 1.5b) flattens the earth rather like an orange peel, thus preserving some of the size and shape of the landmasses. In this projection, the oceans are split. The *Robinson projection* (Figure 1.5c) shows the longitude lines curving toward the poles to give an impression of the earth's curvature, and it shows an uninterrupted view of land and ocean; however, the shapes of landmasses are slightly distorted. In this book we often use the Robinson projection for world maps.

Maps are not unbiased. Most currently popular world projections reflect the European origins of modern cartography. Europe or North America is usually placed near the center of the map, where distortion is minimal; other population centers, such as East Asia, are placed at the highly distorted periphery. For a less-biased study of the modern world, we need world maps that center on different parts of the globe. For example, much of the world's economic activity is now taking place in and around Japan, Korea, China, Taiwan, and Southeast Asia. Discussions of the world economy require maps that focus on these regions but also include other parts of the world in the periphery.

The Detective Work of Photo Interpretation

Most geographers make use of photographs to help them understand or explain a geographic issue or depict the character of a place. Interpreting a photo to extract its geographical information can sometimes be like detective work. Below are some points to keep in mind as you look at pictures; try them out with the photos in Figures 1.6 and 1.7.

- *Landforms:* Notice the lay of the land and the landform features. How do landforms and humans influence each other? Is environmental stress visible?

- *Vegetation:* Notice whether the vegetation indicates a wet or dry, warm or cold environment. Does the vegetation appear to be natural or disturbed by human use?

- *Material culture:* Are there buildings, tools, clothing, foods, or vehicles that give clues about the wealth, values, or aesthetics of the people who live where the picture was taken?

- *The global economy:* Can you see evidence of the global economy, such as goods probably not produced locally (for example, the light bulb in Figure 1.7)?

- *Location:* From your observations, can you tell where the picture was taken or narrow down the possible locations that might be depicted?

Now, think of every possible statement that you can make about the picture, taking note of any doubts you have, as they can be useful in your detective work. You can use this system to analyze any of the photos in this book or elsewhere.

Figure 1.6 Interpreting a photo: Ecuador. Orlando Ayme, his wife Ermelinda (carrying some groceries), and daughter Livia (with her schoolbooks) are returning home from their weekly trip to market. They live in Tingo, Ecuador, a village in the central Andes. Notice the well-used path, the sparse trees, the cultivated fields and grazing animals, the relative dryness, and the high mountain peaks. Also note the warm clothing that the family is wearing to ward off the cold winds and temperatures at high altitudes. [Peter Menzel/http://www.menzelphoto.com.]

Figure 1.7 Interpreting a photo: Bhutan. The Namgay family of the village of Shingkhey, Bhutan, display a week's worth of food in the prayer room of their house. Take particular note of the number of people who are part of this family and of the kinds of foods that they eat regularly, especially the types of produce and the amount of rice. Notice also the chili peppers, potatoes, and tomatoes, all of which originated in Central and South America. [Peter Menzel/http://www.menzelphoto.com.]

THE REGION AS A CONCEPT

A **region** is a unit of the earth's surface that contains distinct

> A region is a unit of the earth's surface that contains distinct physical and human features.

physical and human features. We could speak of a desert region, a region that produces rice, or a region experiencing ethnic violence. In this book, it is rare for any two regions to be described by the same set of indicators. For example, the region of the southern United States might be defined by its distinctive vegetation, architecture, ways of speaking, foods, and historical experience. Meanwhile Siberia, in eastern Russia, could be defined primarily by its climate, remoteness, and population size.

Another problem in defining regions is that their boundaries are rarely crisp. The more closely we look at the border zones, the fuzzier the divisions appear. Take the case of the boundary between the United States and Mexico (Figure 1.8).

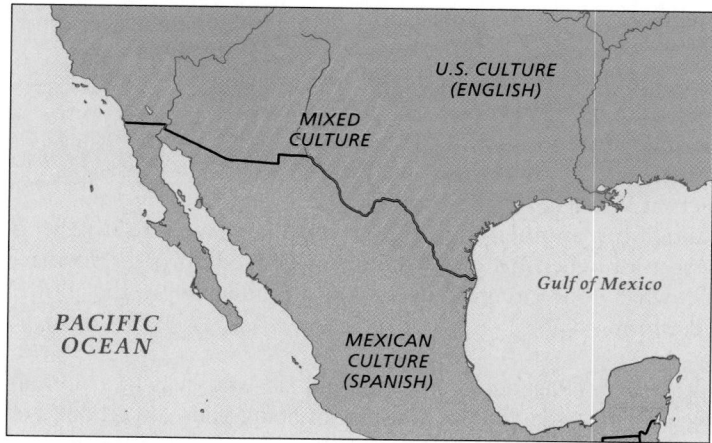

Figure 1.8 Cultural patterns along the border of the United States and Mexico. Along the border there is a wide band of cultural blending in which language, food, religion, and architecture show influences of both U.S. and Mexican cultures. Sometimes a phenomenon new to both cultures emerges. [Adapted from *The World Book Atlas* (Chicago: World Book, 1996), p. 89.]

The clearly delineated political border is not a marker of separation between cultures or economies. In a wide band extending over both sides of the border, there is a blend of Native American, Spanish colonial, Mexican, and Anglo-American cultural features. Languages, place-names, food customs, music, and family organization are only a few examples. And the economy of the border zone depends on interactions across a broad swath of territory.

In fact, the Mexican national economy is becoming more closely connected to the economies of the United States and Canada than to those of its close neighbors in Middle America (the countries of Central America and the Caribbean).

Why, then, does this book place Mexico in Middle America? Even though northern Mexico has much in common with parts of the southwestern United States, overall Mexico still has more in common with Middle America than with North America. The use of Spanish as its official language ties Mexico to Middle and South America and separates it from the United States and Canada, where English dominates (except in Québec). These language patterns are symbolic of the larger cultural and historical differences between the two regions, which will be discussed in Chapters 2 and 3.

If regions are so difficult to define and describe, why do geographers use them? It is impossible to discuss the whole world at once, and so we seek a reasonable way to divide it into manageable parts. There is nothing sacred about the criteria or the boundaries we use. They are here to help you learn. In defining the world regions for this book, we have considered such factors as physical features, political boundaries, cultural characteristics, and what the future may hold.

We have organized the material in this book into three scales: the global scale, the world regional scale, and the local

scale. At the global scale, explored in this chapter, the entire world is treated as a single area—a unity that is more and more relevant as our planet becomes a global system. We use the term *world region* for the largest divisions of the globe, such as East Asia, Southeast Asia, and North America (see Figure 1.1). We have defined ten world regions, each of which is covered in a separate chapter. Each regional chapter considers the interaction of human and physical geography in relation to cultural, social, economic, population, environmental, and political topics.

Most of us are familiar with geography at the local scale—where we live and work, whether in a city, town, or rural area. Local geography shapes our lifestyles and the culture we create together. Throughout this book, life at the local scale is illustrated with vignettes about people in a wide range of places. Often their lives reflect trends that can be tracked across several world regions or may even occur at the global scale.

The regions and local places discussed in this book vary dramatically in size and complexity. A region can be relatively small, such as Europe, or very large, as in the case of East Asia. At the local scale, a place can be a backyard in Polynesia, a neighborhood in Rome, or a town in Kenya.

In summary, regions have the following traits:

- A region is a unit of the earth's surface that contains distinct environmental or cultural patterns.

- Regions are defined by people to help them identify spaces for varying purposes, so regional definitions are fluid.

- No two regions are necessarily described by the same set of indicators.

- Regions can vary greatly in size (scale).

- The boundaries of regions are usually fuzzy and hard to agree upon.

INTERREGIONAL LINKAGES AND GLOBALIZATION

Due to economic, technological, social, and political changes, regions widely separated in space can now have interdependent economic relationships that used to be possible only between close neighbors. For example, much of the clothing and tropical hardwoods used by North Americans comes from Southeast Asia. Likewise, South Americans and Africans produce cash crops that end up on European tables.

These connections between distant regions are known as **interregional linkages.** They began to attain their present global reach during the early stages of **European colonialism,** the practice of taking over the human and natural resources of often distant places in order to produce wealth for Europe. Their voyages to America after 1492 led Spain and Portugal to establish colonies in Middle and South America. The British, Portuguese, Spanish, Dutch, and French founded colonies in North America and Asia in the sixteenth and seventeenth centuries and in the Pacific islands in the eighteenth century. By the end of the nineteenth century, most of Africa was divided into European colonies.

Although most colonies are now independent countries, European colonization led to transglobal ties that still remain. For example, there are Muslims of Indonesian descent who have lived for generations in Suriname on the northern coast of South America. Suriname and Indonesia were both colonies of the Dutch. Similarly, many of today's violent conflicts have their roots in the colonial era.

Today, because of rapid transportation and the speedy flow of electronic information, widely separated places can be intimately linked. Moreover, the details of their relationships can change from day to day. For example, construction workers from Pakistan who leave their homes to work in Saudi Arabia send substantial portions of their wages (**remittances**) back to support their families. But an outbreak of hostilities in and around Saudi Arabia can send hundreds of thousands of workers scurrying home. As a result, not only construction in Saudi Arabia but also tens of thousands of family budgets in Pakistan are disrupted.

The term **globalization** encompasses the changes brought about by many types of interregional linkages and flows (Figure 1.9). Individuals can be harmed or empowered by these changes, and sometimes both at once. As you read the many references to globalization in the following sections, reflect on how you as an individual are part of the picture. Perhaps you buy inexpensive products at Wal-Mart made by low-wage labor from another country. Or you may have a relative who lost a job when his or her company moved abroad. No matter what your circumstances, you are affected by the global economy.

> The term globalization encompasses the changes brought about by many types of interregional linkages and flows.

Vignette Rajinder is an 8-year-old kid who lives in a slum in New Delhi, India. He has a history of doing poorly in school. Yet in just a few minutes, he leapt across the digital divide after encountering one of Sugata Mitra's kiosk computers. The **digital divide** is the gap between the small percentage of the world's population who have access to computers and the huge majority who do not.

Mitra is the head of research and development for a computer software and training company with offices around the world and annual sales of $300 million. His company was interested in developing kiosk computers for the global market—computers that give passersby quick access to the Internet.

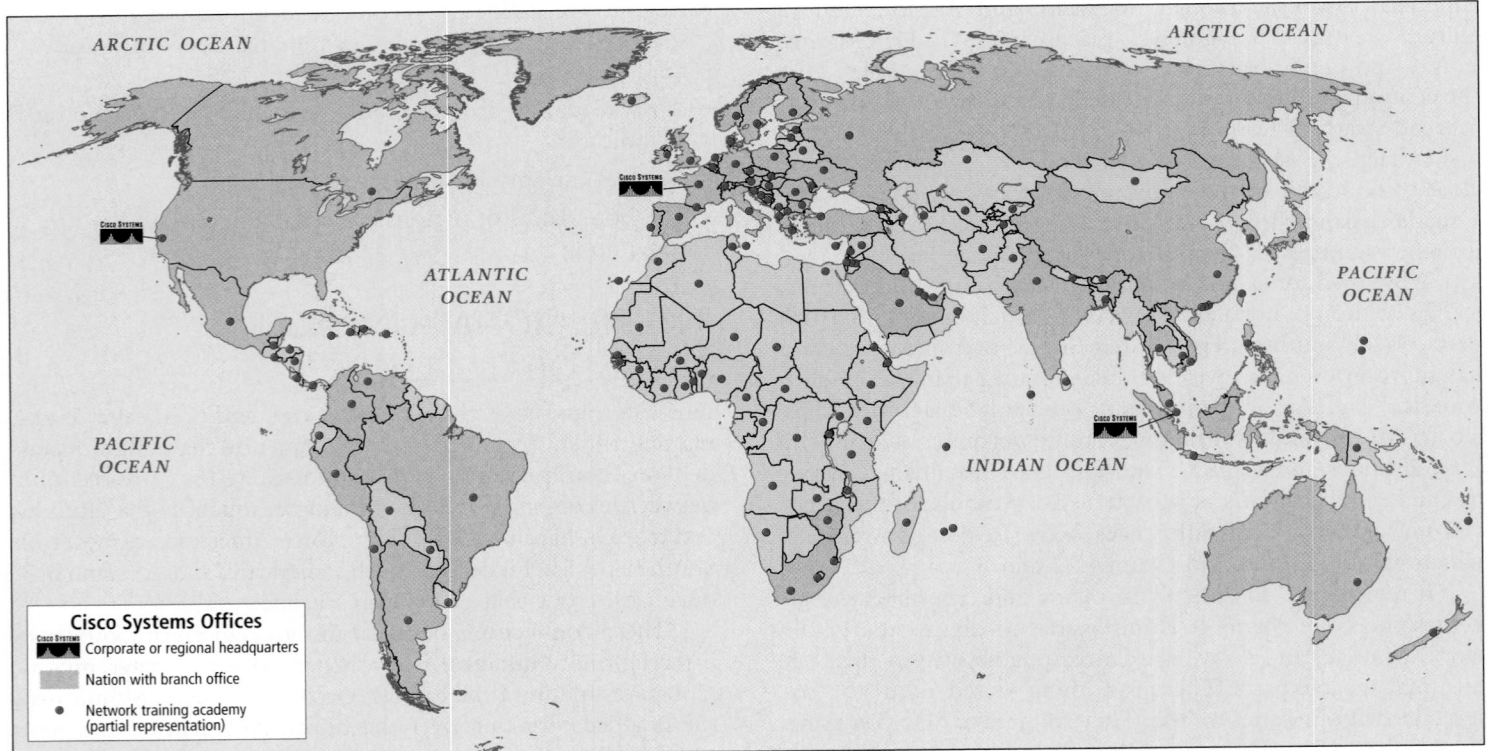

Figure 1.9 Globalization: Cisco Systems' global network. Cisco provides hardware and services for Internet networking. This map shows several levels of Cisco's activities: corporate and regional headquarters, countries with branch offices, and the locations of network training academies. Notice the uneven distribution of Cisco Systems offices.

Mitra cut a hole in the wall of his office compound in New Delhi and installed a high-speed computer with Internet access facing an adjacent slum. As he had guessed, within hours the local kids were browsing the Internet (Figure 1.10). There was "a spiral of self-instruction," says Mitra, with one kid making a discovery, three witnessing it and saying "Cool!" and then sharing three or four more discoveries while they explored together. In a matter of hours, Rajinder had visited the Disney site, learned to use a drawing tool, and read news stories about the Taliban in Afghanistan. A young girl found a graphics program to help her father, a tailor, design the clothes he sews.

"The hole in the wall gives us a method to create a door, if you like," Mitra says, "through which large numbers of children can rush into this new arena. When that happens, it will have changed our society forever."

There are now hundreds of such computers in slum neighborhoods across India. This is important because although India is a leader in technological development, it is plagued by the digital divide. Bangalore, located in the southern Indian state of Karnataka (known as India's "Silicon State"), is ranked by the United Nations as the world's fourth-best [not clear what "best" means] hub of technological innovation. Yet in Karnataka, 85 percent of the people still don't have access to a computer, and 100,000 school-age children don't go to school.

Find out more about Rajinder and the millions of Indian children who love the Internet at http://www.hole-in-the-wall.com/. ∎

Figure 1.10 Hole-in-the-wall computing. When Sugata Mitra first decided to address the "digital divide," he did so with one computer stuck in a wall facing a New Delhi slum. The experiment was immediately successful, and now hundreds of such computers, like that shown here, are available to poor children across India and beyond. [© Hole in the Wall Education, India/http://www.hole-in-the-wall.com/.]

Globalization and Economic Geography

Geographers have long been interested in economic interactions in different parts of the world. They consider how economies affect the resources that people use and the ways that people arrange themselves across the land. In recent decades, economic geographers have focused increasingly on the economic aspects of globalization, or the **global economy**—the ways in which goods, capital, labor, and resources are exchanged among distant and very different places. We begin by looking at a few examples of people who are part of that exchange and whose day-to-day lives are strongly affected by the global economy.

> In recent decades, economic geographers have focused increasingly on the economic aspects of globalization—the ways in which goods, capital, labor, and resources are exchanged among distant and very different places.

WORKERS IN THE GLOBAL ECONOMY

Vignette Olivia lives in Soufrière on St. Lucia, an island in the Caribbean. Soufrière was once a quiet fishing village, but it now hosts cruise ship passengers several times a week. Olivia is 60. She, her daughter Anna, and her three grandchildren live in a wooden house surrounded by a leafy green garden dotted with fruit trees. Anna has a tiny shop at the side of the house, from which she sells various small everyday items and preserves that she and her mother make from the garden fruits.

On days when the cruise ships dock, Olivia strolls down to the market shed on the beach with a basket of papayas, bags of roasted peanuts, and rolls of cocoa paste made from cacao beans picked in a neighbor's yard. She calls out to the passengers as they are rowed to shore, offering her spices and snacks for sale. In a good week she makes U.S.$50. Her daughter makes about U.S.$100 per week in the shop and is constantly looking for other ways to earn a few dollars. She often makes necklaces for tourists or takes in laundry.

Usually the family of five makes do with about U.S.$170 a week (U.S.$8840 per year). From this income they pay rent on the house, the electric bill, and school fees for the granddaughter who will go to high school in the capital next year and perhaps college if she succeeds. They also buy clothes for the children and whatever food (chiefly flour and sugar) they can't grow themselves. Their livelihood puts them at or above the standard of living of most of their neighbors.

In Riau, Indonesia, thirty-year-old Reyhan sits on the dock, anxiously waiting with several other male and female Indonesians. The arriving boat will take them across the Strait of Malacca to Malaysia, where they plan to work illegally. They will join more than a million fellow Indonesians in Malaysia, some working legally, some not. All are attracted by Malaysia's booming economy, where average wages are four times higher than at home.

This is Reyhan's second trip to Malaysia. His first was in 1996, when he signed up for an overseas employment program of the Malaysian government's Ministry of Manpower. He was promised a 2-year work visa and a contract to work legally on a Malaysian oil palm plantation. Upon arrival, however, Reyhan found that his first three months' wages would go toward paying off his boat fare and that his visa was valid for only two months. Not one to give in easily, he soon escaped to the city of Malacca, where he secured a construction job earning U.S.$10 a day (about U.S.$2600 a year). On this wage, he was able to send enough money to his wife and two children in Indonesia to pay for their food, school fees, and a new roof for their house.

In 1998, Reyhan was one of thousands of Indonesians deported by Malaysia (Figure 1.11). The country was suffering growing unemployment due to the Asian financial crisis, and its leaders wanted to create more jobs for locals by expelling foreign workers. The crisis affected Indonesia even more severely, and after several years of only part-time farm work, Reyhan is willing to give Malaysia another try.

Tanya is a 50-year-old grandmother with a son in high school and a married daughter. Tanya works at a fast-food restaurant in North Carolina, making less than U.S.$6 an hour. She had been earning U.S.$8 an hour sewing shirts at a textile plant until it closed and moved to Indonesia. Her husband is a delivery truck driver for a snack-food company.

Between them, Tanya and her husband make $27,000 a year, but from this income they must cover their mortgage and car loan and meet regular monthly expenses for food and utilities. In addition, they help their daughter, Rayna, who quit school after eleventh grade and married a man who is now out of work. They and a baby live in the old mobile home at the back of the lot that Tanya and her family once lived in.

With Tanya's lower wage ($5000 less a year), there will not be enough money to pay the college tuition for her son. He had hoped to be an engineer and would have been the first in the family to go to college. For now, he is working at the local gas station.

Adapted from Lydia Pulsipher's field notes, 1992–2000 (Olivia and Tanya), and Alex Pulsipher's field notes, 2000 (Reyhan). ■

These people, living worlds apart, are all part of the global economy. Workers around the world are paid startlingly different rates for jobs that require about the same skill level. Varying costs of living and varying local standards of wealth make a difference in how people live. Though Tanya's family has the highest income by far, they live in near poverty,

Figure 1.11 Illegal migrants. These construction workers, male and female, have fled the economic crisis in Indonesia to work illegally in Malaysia. They are under arrest on a Malaysian construction site and will be questioned by immigration officials before being deported. Notice their faces and try to imagine their worries at this point in their lives. [AP Photo/S. Thinakaran.]

and their hopes for the future are threatened. Olivia's family are not well off, but they do not think of themselves as poor because they have what they need and others around them live in similar circumstances. The growing tourist trade promises increased income, but it also means dependence on circumstances beyond their control—in an instant, the cruise-line companies can choose another port of call. Reyhan, by far the poorest, seems trapped by his status as an illegal worker, which robs him of many of his rights. Still, the higher pay that he can earn in Malaysia offers him a possible way out of poverty.

WHAT IS THE ECONOMY?

Economic geography focuses on how people interact across space and in different environments as they earn a living. The *economy* is the forum in which people make their living, and *resources* are what they use to do so.

Some resources are tangible materials, such as mineral ores, timber, plants, and soil. Because they must be mined from the earth's surface or grown from its soil, they are called **extractive resources.** There are also non-material human resources, such as skills and brainpower. Often resources must be transformed to produce new commodities (such as refrigerators or sugar) or bodies of knowledge (such as books or computer software). Extraction (mining and agriculture) and industrial production are two types of economic activities, or *sectors* of the economy.

A third economic sector is the **exchange** or **service sector:** the bartering and trading of resources, products, and services. Generally speaking, as people in a society shift from extractive activities, such as farming, to industrial and service activities, their material standards of living rise. However, their actual well-being may be compromised by dependence on cash, the need to migrate, urban crowding, or poor working conditions.

The **formal economy** includes all the activities that are officially recorded as part of a country's production. Examples from the vignettes at the beginning of this section include Olivia's daughter in her shopkeeper's role, and Tanya, her husband, and their son. All are registered workers who earn recorded wages and pay taxes to their governments. The activity of formal economies is measured by the **gross domestic product (GDP).** This number gives the total value, in monetary terms, of all goods and services officially recognized as produced in a country during a given year.

Many goods and services are produced outside formal markets, in the **informal economy.** Here, work is often traded for payment other than cash or for cash payment that is not reported to the government as taxable income. It is estimated that one-third or more of the world's work falls into this category. Examples of workers in the informal economy include Olivia when she sells her goods to the tourists and Reyhan when he works illegally in Malaysia. Any members of the three families who contribute to their own or someone else's well-being through such unpaid services as housework, gardening, and elder and child care are also part of the informal economy.

Remittances sent home from afar may be part of either economy. If they are sent through banks or similar financial institutions, they become part of the formal economy of the receiving society. If they are transmitted in pockets or via the mail as cash, or some other "off-the-books" (perhaps illegal) transaction, they are part of the informal economy.

WHAT IS THE GLOBAL ECONOMY?

The global economy includes the parts of any country's economy that are involved in global flows of resources: extracted materials, manufactured products, money, and people. Most of us participate in the global economy every day. For example, this book was manufactured using paper made from trees cut down in Southeast Asia, North America, or Siberia and shipped to a paper mill in Oregon. It was printed in the United States, though many books are now printed in Asia because labor costs are lower there. Such long-distance movement of resources and products has grown tremendously in the past 200 years. But it existed at least 2500 years ago, when silk and other goods were traded along Central Asian land routes that connected Rome with China.

Starting in about 1500, Europeans began extracting resources from their colonies and organizing systems to process those resources into higher-value goods to be traded wherever there was a market. Sugarcane, for example, was grown on Caribbean, Brazilian, and Asian plantations (Figure 1.12) and made into crude sugar and rum locally. These products were then sold in Europe and North America, where further refining took place. The global economy grew as each region produced goods for export, rather than just for local consumption, and became increasingly dependent on imported food, clothing, machinery, energy, and knowledge.

New wealth and ready access to global resources led to Europe's **Industrial Revolution,** a series of innovations and ideas that changed the way goods were produced. No longer was one woman producing the cotton or wool for cloth, spinning thread, weaving the thread into cloth, and sewing a shirt. Instead, tasks were spread out among many workers, often in distant places, with some people specializing in producing the fiber and others in spinning, weaving, or sewing. These innovations were followed by labor-saving improvements such as mechanized reaping, spinning, weaving, and sewing.

This larger-scale mechanized production accelerated globalization as it created a demand for raw materials and a need for markets in which to sell finished goods. European colonies in the Americas, Africa, and Asia provided both. For example, in the British Caribbean colonies, hundreds of thousands of African slaves wore garments made of cotton cloth woven in England from cotton grown in other colonies in North America, Africa, and Asia. The sugar they produced on British-owned plantations was transported to European markets in

Figure 1.12 European use of colonial resources. Among the first global economic institutions were Caribbean plantations like Old North Sound on Antigua, shown here in an old painting. In the eighteenth century, thousands of sugar plantations in the British West Indies, subsidized by the labor of slaves, provided huge sums of money for England and helped fund the Industrial Revolution. [Museum of Antigua and Barbuda.]

ships made in the British Isles of trees and resources from various parts of the world.

Until the early twentieth century, much of the activity of the global economy took place within the huge colonial empires ruled by a few European nations. Global economic and political changes brought an end to these empires by the 1960s, and almost all colonial territories are now independent countries. Nevertheless, the global economy continues to grow as the flow of resources and manufactured goods are sustained by private companies, many of which first developed during colonial times.

Multinational corporations such as Shell, Wal-Mart, Bechtel, and Cisco (see Figure 1.9 on page 10) operate across international borders. They extract resources from many places, make products in factories located in places where they can take advantage of cheap labor and transport facilities, and market their products wherever they can make the most profit. Their global influence, wealth, and importance to local economies enables the multinationals to influence the economic and political affairs of the countries in which they operate.

Multinationals are important conduits for the flow of **capital,** or investment money. For example, consider what was once a family-owned Mexico City cement company we will call MEXCRETE. It recently borrowed heavily from international banks to acquire controlling interests in cement industries in Texas, Mexico, the Philippines, and Thailand. MEXCRETE

also plans to use this borrowed money to build new cement plants that use cutting-edge technology in order to expand sales to Europe and especially to China.

Such international investment has advantages and disadvantages. In this case, high-quality cement may be delivered more efficiently and cheaply to all the markets served. Competition may spur technological advancement, and jobs will be created in Southeast Asia, Mexico, Texas, and in the markets the company serves. But to maximize its profits, MEXCRETE may not pay its workers enough to live on or provide a healthy workplace, and its cement plants may cause pollution. By taking the profits home to Mexico, MEXCRETE deprives local economies of capital needed for investment. Moreover, if demand for concrete slows and MEXCRETE misses payments on the huge debt that it amassed while expanding, its creditors could foreclose. Then thousands of jobs from Texas to Southeast Asia would be at risk.

THE DEBATE OVER FREE TRADE AND GLOBALIZATION

Free trade is the unrestricted international exchange of goods, services, and capital. Currently, all governments impose some restrictions on trade to protect their own national economies from foreign competition. Restrictions take two main forms: tariffs and import quotas. **Tariffs** are taxes imposed on imported goods that increase the cost of those goods to the consumer. **Import quotas** set limits on the amount of a given good that may be imported over a set period of time.

These and other forms of trade protection are a subject of contention, and views on the value of free trade and the globalization of markets vary widely. Proponents of free trade argue that it encourages efficiency, lowers prices, and gives consumers more choice. Companies can sell to larger markets and take advantage of mass-production systems that lower costs further. As a consequence, they can grow faster, thereby providing people with jobs and opportunities to raise their standard of living.

Proponents of free trade have been quite successful in recent years, and restrictions on trade imposed by individual countries are being reduced. Several **regional trade blocs**—associations of neighboring countries that agree to lower trade barriers for one another—have been formed. The main ones are the North American Free Trade Agreement (NAFTA), the European Union (EU), the Southern Common Market (Mercosur) in South America, and the Association of Southeast Asian Nations (ASEAN).

One of the main global institutions that supports free trade is the **World Trade Organization (WTO),** whose stated mission is to lower trade barriers and establish ground rules for international trade. Two other global institutions are the **World Bank** (officially named the International Bank for Reconstruction and Development) and the **International Monetary Fund (IMF).** Both of these organizations make loans to countries that need money to pay for economic development projects. Before approving a loan, the World Bank or the IMF may require a borrowing country to reduce and eventually remove tariffs and import quotas. These requirements are part of larger **structural adjustment policies (SAPs)** that the IMF imposes on countries seeking loans. SAPs have become highly influential and controversial in virtually every region of world. The most detailed explanation of SAPs is given in Chapter 3 (see page 119); please refer to that discussion as you read.

Those opposed to free trade argue that the gains of a less-regulated global economy can lead to rapid cycles of growth and decline that can wreak havoc on smaller national economies (Figure 1.13). Labor unions point out that as corporations relocate factories and services to poorer countries where wages are lower, jobs are lost in richer countries. In the poorer countries, multinational corporations often work with governments to prevent workers from organizing themselves into unions that could bargain for **living wages** (wages that support a minimum healthy life). Environmentalists argue that in newly industrializing countries, which often don't have effective environmental protection laws, multinational corporations tend to use highly polluting and unsafe production methods to lower costs. Many fear that a "race to the bottom" in wages, working conditions, government services, and environmental quality is underway as countries compete for profits and potential investors.

Figure 1.13 A demonstration against the World Trade Organization. Korean farmers demonstrate at the December 2005 WTO meeting in Hong Kong. When they were arrested, a hunger strike in their support ensued. [AP Photo/Kin Cheung.]

Fair trade, defined as trade that gives a fair price to producers and upholds environmental and safety standards in the workplace, is now proposed as an alternative to free trade. Some profits are sacrificed in order to provide markets for producers from developing countries. For example, "fair trade" coffee and chocolate are now marketed to North America and Europe. Prices are somewhat higher for consumers, but the extreme profits of middlemen are eliminated and growers of coffee and cocoa beans receive living wages and improved working conditions.

In developing an opinion about free trade and globalization, consider how many of the things you own or consume—computer, clothes, furniture, appliances, car, and foods—were produced in the global economy. These products are cheaper for you to buy, and your standard of living is higher as a result of lower production costs and competition among many global producers.

On the other hand, you or a relative may have lost a job because a company moved to another location where labor and resources are cheaper. You may be concerned that the products you buy so cheaply were made under harsh conditions by underpaid workers (even children), or that resources were used unsustainably. High levels of pollution may have occurred in the manufacturing and transport processes. Given all these considerations, what might be the advantages and drawbacks of free and fair trade?

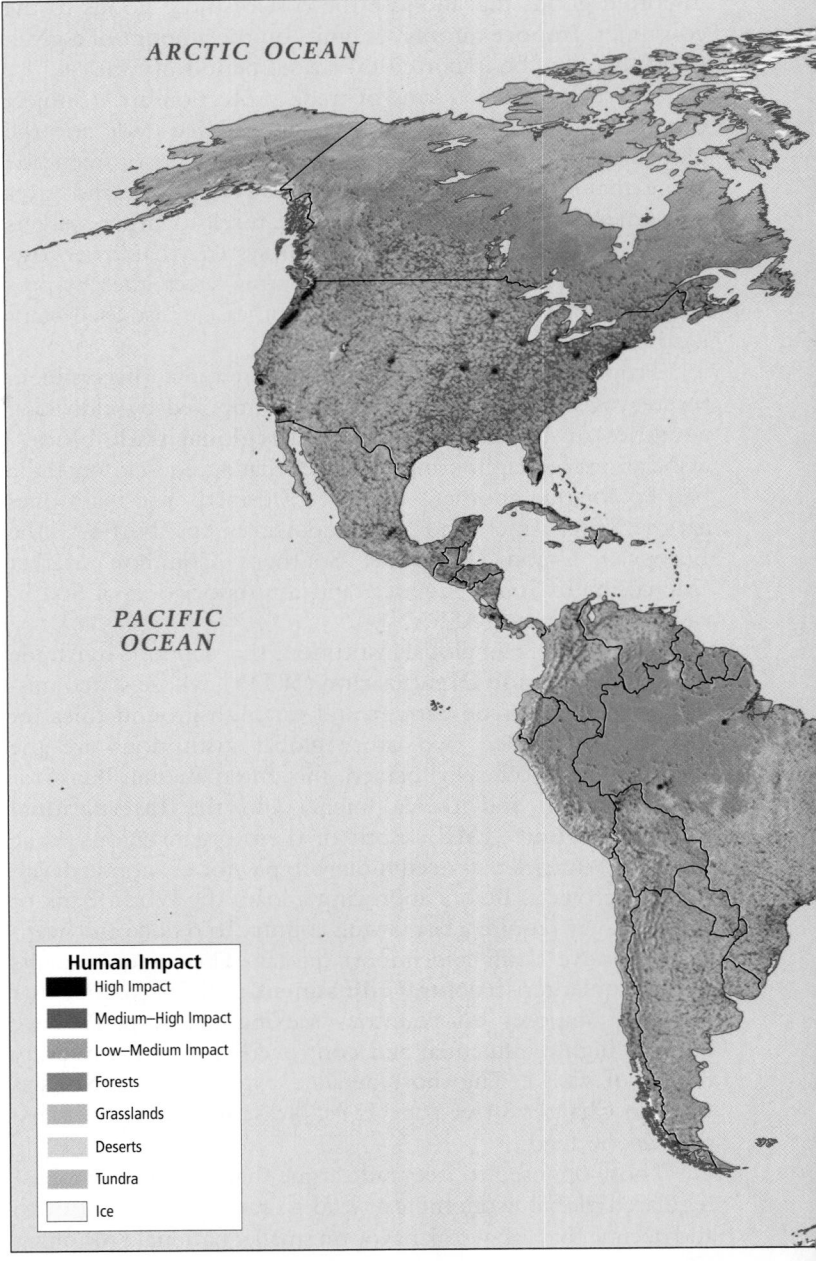

Figure 1.14 Human impact on the earth, 2002. The impacts depicted here are derived from a synthesis of hundreds of studies. High impact areas are associated with intense urbanization. Medium-High impact areas are associated with roads, railways, agriculture, or other intensive land uses. Low-Medium impact areas are experiencing disturbance to natural systems as a result of human activity. To see details of human impact, go to http://www.globio.info/region/world/, download the poster of World, Robinson projection pdf at the bottom of the page, and zoom in on the place of interest. [Adapted from *United Nations Environment Programme, 2002, 2003, 2004, 2005, 2006* (New York: United Nations Development Programme), at http://maps.grida.no/go/graphic/human_impact_year_1700_approximately and http://maps.grida.no/go/graphic/human_impact_year_2002.]

Humans and the Environment

The same forces that created globalization have resulted in enormous impacts on physical environments. Mass consumption of resources has altered the environment most profoundly, but even less wasteful human ways of life have environmental effects. Figure 1.14 shows the impact of humans on the planet's land surface. The intensity and nature of the impact varies greatly, but human impact can be found virtually everywhere.

Mounting awareness of environmental impacts has prompted numerous proposals to limit the damage, from relying on technological advances to reducing resource consumption. Halting and reversing environmental damage may be the greatest challenge our species has yet faced, in part because our societies have become so transformed by how we use the earth's resources. The issue at the forefront now is global warming.

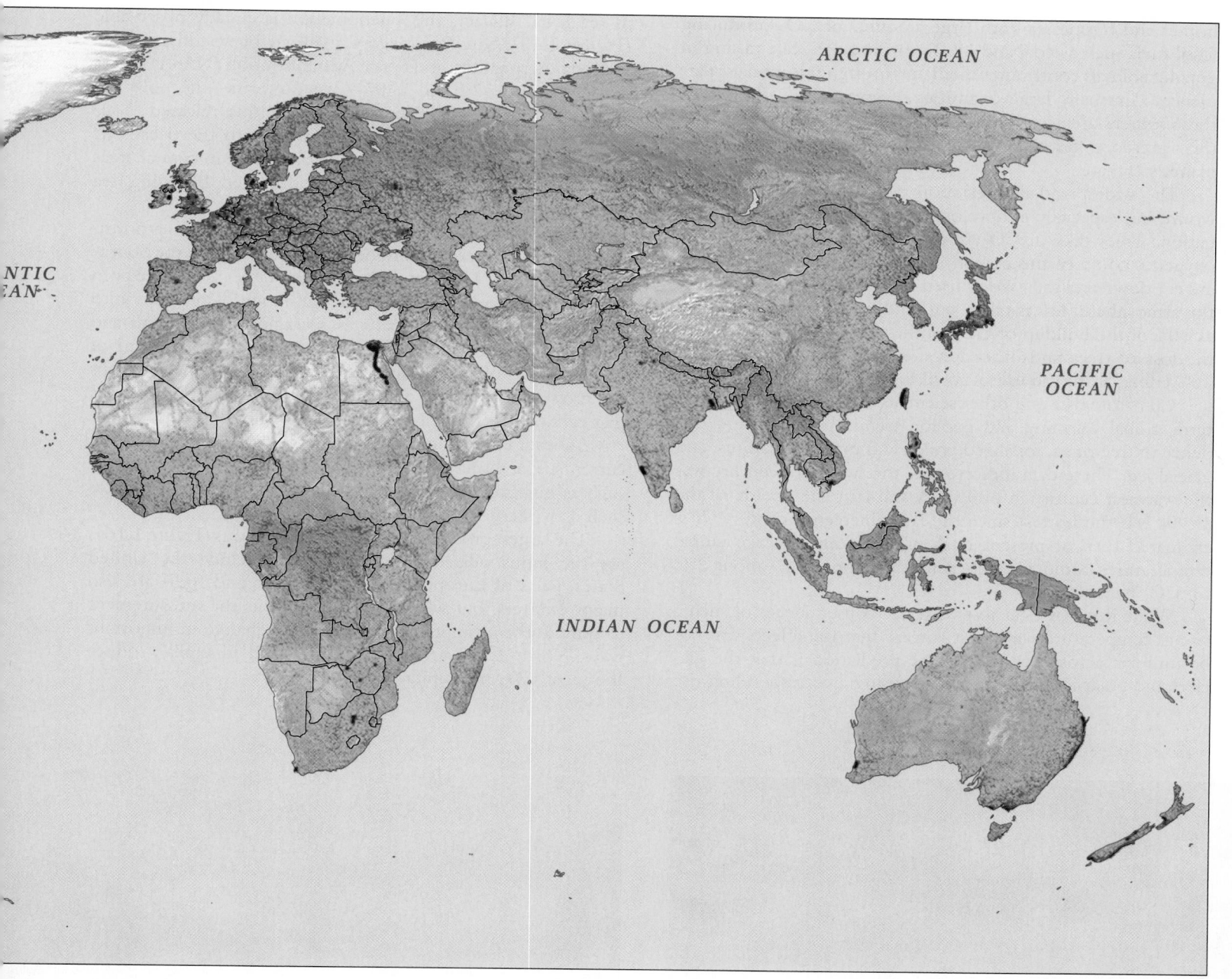

GLOBAL WARMING

The term **global warming** refers to the observed warming of the earth's surface and climate in recent decades. Human activities have increased atmospheric levels of carbon dioxide (CO_2), methane, water vapor, and other gases above natural levels. These gases are collectively known as "greenhouse gases" because their presence allows large amounts of heat from sunlight to be trapped in the earth's atmosphere in much the same way that heat is trapped in a greenhouse or a car parked in the sun.

Greenhouse gases exist naturally in the atmosphere. In fact, it is their heat-trapping ability that makes the earth warm enough for life to exist. Increase their levels, as humans are doing now, and the earth becomes warmer still.

Over the last several hundred years, humans have greatly intensified the release of greenhouse gases. Electricity generation, vehicles, industrial processes, and the heating of homes and businesses burn large amounts of CO_2-producing fossil fuels such as coal and oil. Even the large-scale raising of grazing animals contributes methane through the animals' flatulence. Unusually large quantities of greenhouse gases from these sources are accumulating in the earth's atmosphere, and their presence has already led to significant warming of the planet's climate.

The widespread deforestation occurring throughout the world, but especially in developing countries, worsens the situation. Trees take in CO_2 from the atmosphere, release the oxygen, and store the carbon in their bodies. As more trees are cut down and their wood used for fuel, more carbon enters the atmosphere, less is taken out, and less is stored. As much as 30% of the buildup of CO_2 in the atmosphere results from the loss of trees and other forest organisms. The remaining 70% comes from the use of fossil fuels.

Climatologists and other scientists are documenting long-term global warming and cooling trends by examining evidence in tree rings, fossilized pollen and marine creatures, and glacial ice. These data indicate that the twentieth century was the warmest century in 600 years and that the decade of the 1990s was the hottest since the late nineteenth century. It is estimated that, at present rates of emissions, average global temperatures could rise between 2.5°F and 10°F (about 2°C to 5°C) by 2100.

While it is not clear just what the consequences of such a rise in temperature will be, it is clear that the effects will not be uniform across the globe. One prediction is that the glaciers and polar ice caps will melt, causing a corresponding rise in sea level. In fact, this phenomenon is already observable (Figure 1.15). Satellite imagery analyzed by scientists at the National Aeronautics and Space Administration (NASA) shows that between 1979 and 2005—just 26 years—the polar ice caps shrank by about 23 percent The melting released thousands of trillions of gallons of meltwater into the oceans. If this trend continues, at least 60 million people in coastal areas and on low-lying islands could be displaced by rising sea levels. ◼

Scientists also forecast a shift of warmer climate zones northward in the Northern Hemisphere and southward in the Southern Hemisphere. This pattern is also observable: there have been sightings of robins in Alaska, and the range of the mosquito that carries the West Nile virus is spreading to the north and south. Such climate shifts might lead to the displacement of huge numbers of people, because the zones where specific crops can grow would change dramatically. Animal and plant species that cannot adapt rapidly to the change will disappear.

Another effect of global warming could be a shift in ocean currents. The result would be more chaotic and severe weather, such as hurricanes, and possible changes in climate for places such as western Europe. ◼

The largest producers of total CO_2 emissions (Figure 1.16a) are the industrialized countries, which include the United States, parts of Europe, and Russia. China and India are also major emitters, but when tons *per capita* is the measurement (Figure 1.16b), the geographic patterns change in important ways. North America, Europe, Russia, and Australia produce the most CO_2 per capita.

Figure 1.15 Effects of global warming: The Muir Glacier (Alaska) in 1941 (left) and 2004 (right). Both photos were taken from the same vantage point. Geologist Bruce Molnia, with the U.S. Geological Survey, reports that in 63 years, the glacier retreated 7 miles (12 km) and thinned more than 875 yards (800 m). [National Snow and Ice Data Center: (a) W. O. Field; (b) B. F. Molnia.]

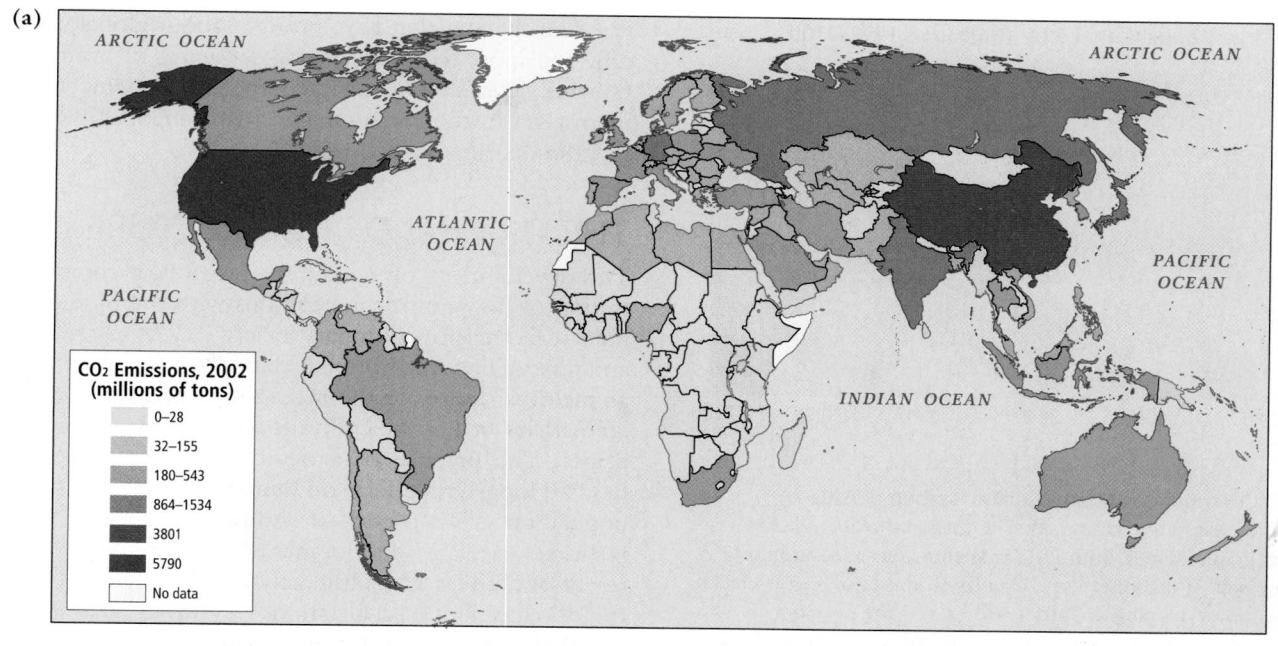

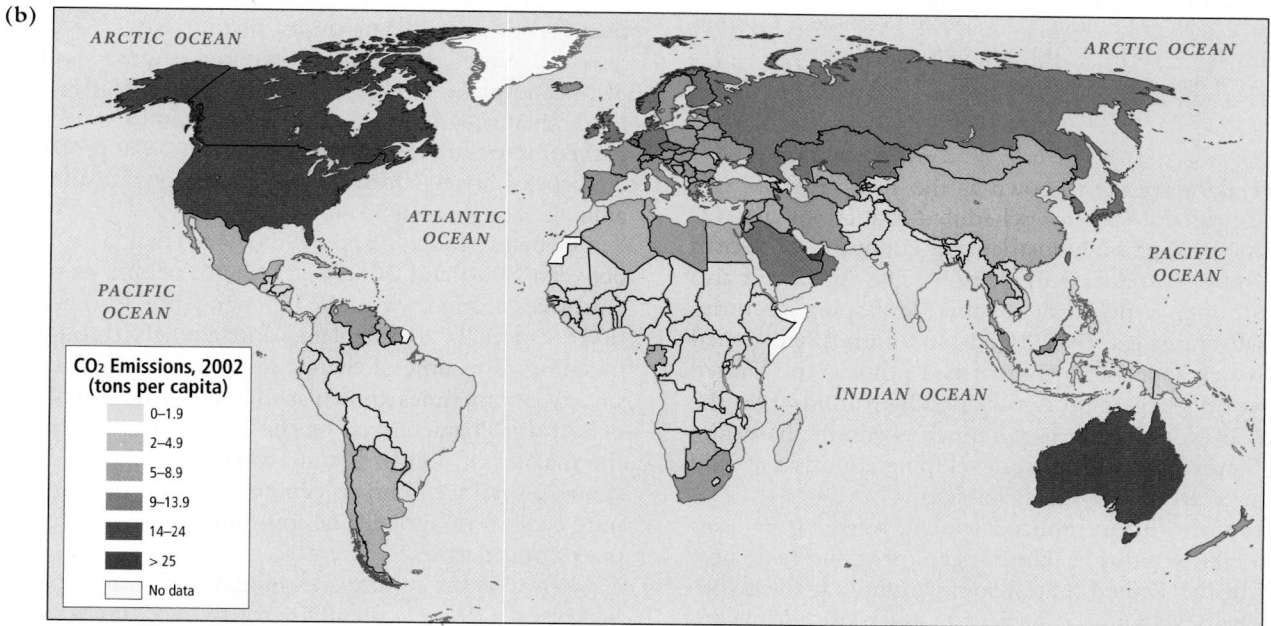

Figure 1.16 Carbon dioxide emissions around the world. **(a)** Total emissions by tons, 2002. The United States leads the entire world (23.32 percent of the world total), with China second (15.28 percent). **(b)** Tons of emissions per capita. Some high-emitting countries are Qatar (41 tons per capita), the United States (20 tons per capita), Australia (17.2 tons per capita), and Canada (16.5 tons per capita). Note that by this measurement, China (3 tons per capita) is not a leading emitter. [Adapted from World Resources Institute at http://cait. wri.org/cait.php?page=yearly&mode=view&sort=val-desc&pHints= shut&url=form&year=2002&sector=natl&co2=1.]

For the period 1859–1995, developed countries produced roughly 80 percent of the greenhouse gases from industrial sources, and developing countries produced 20 percent. But by 2000, the developing countries were catching up, accounting for nearly 30 percent of total CO_2 emissions. As developing nations industrialize over the next century and continue to cut down their forests, they will release more and more greenhouse gases every year (Figure 1.17). If present patterns hold, greenhouse gas contributions by the developing countries will exceed those of the developed world by 2040.

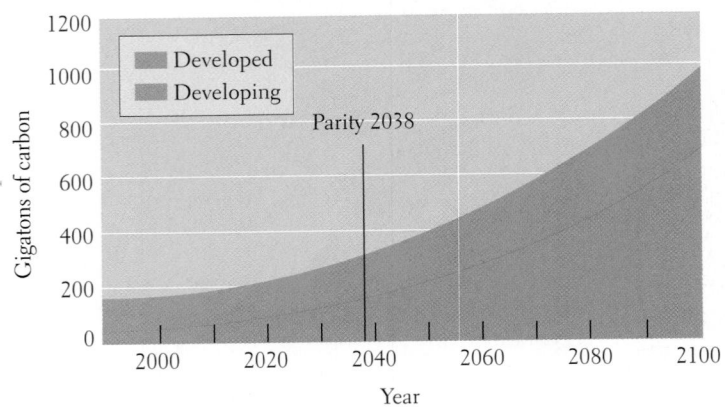

Production of CO₂ Emissions, 1990–2100

Figure 1.17 Contributions to atmospheric carbon dioxide by developed and developing countries. The carbon dioxide releases shown in this graph include both industrial emissions and amounts released as a result of deforestation. When both these sources are taken into account, the developing countries will exceed the developed countries in CO₂ production after 2038. [Adapted from Duncan Austin, José Goldemberg, and Gwen Parker, "Contributions to climate change: Are conventional metrics misleading the debate?" *Climate Notes* (World Resource Institute Climate Protection Initiative) (October 1998).]

In 1992, an agreement known as the *Kyoto Protocol* was drafted. The protocol calls for scheduled reductions in CO₂ emissions by the highly industrialized countries of North America, Europe, East Asia, and Oceania. The agreement also encourages, though it does not require, developing countries to curtail their emissions. By 2008, 174 countries had signed on. The only developed country that had not was the United States, the world's largest producer of CO₂. Arguing that the Kyoto Protocol would cause it too much economic hardship, the United States demanded that developing countries should first be required to reduce their emissions.

Evidence for human-induced global warming is now accepted virtually worldwide. The disagreement hinges on how to respond. In the United States, some argue that the Kyoto Protocol allows developing countries too much latitude in controlling emissions, even though their emissions per capita are far lower than those in the United States. Those in the scientific community see the currently agreed-upon reductions as far too low. They call for stepped-up energy conservation and for more research into such alternatives as solar, wind, and geothermal energy. They also note that even if comprehensive measures are taken, it will take perhaps a century or more to reverse the present trends. Hence, this book will repeatedly visit the factors shaping the vulnerability of regions to global warming. This calls for a deeper understanding of how places are exposed to the hazards created by global warming, such as flooding or drought, and how resilient they are to those hazards. ▰◗

The changes that have brought about global warming are only the most recent in a long sequence of events in which humans have altered, adapted to, and become transformed themselves by the Earth's many environments. One such event was the development of agriculture.

THE ORIGINS OF AGRICULTURE

The development of agriculture provides a compelling illustration of how human interactions with the physical environment can transform human society and ways of life. Agriculture includes animal husbandry, or the raising of animals, as well as the cultivation of plants. The practice of agriculture has had long-term effects on human population growth, rates of natural resource use, the development of towns and cities, and ultimately on the development of civilization.

> Agriculture has had long-term effects on human population growth, rates of natural resource use, the development of towns and cities, and ultimately on the development of civilization.

Where and when did plant cultivation and animal husbandry first develop? Very early humans hunted animals and gathered plants and plant products (seeds, fruits, roots, fibers) for their food, shelter, and clothing. The transition from hunting and gathering to tending pastures and gardens was probably a gradual process arising from a long familiarity with the plants and animals that humans liked to use.

Genetic studies support the view that at varying times between 8000 and 20,000 years ago, people in many different places around the globe independently learned to domesticate especially useful plants and animals through selective breeding. This time of change in the economic base of human society is sometimes known as the **Neolithic Revolution.** The period was characterized by the expansion of agriculture and the making of polished stone tools. The map in Figure 1.18 shows several well-known centers of domestication. A continuing process of agricultural innovation also occurred in many places outside these centers.

Why did agriculture and animal husbandry develop in the first place? Certainly the desire for more secure food resources played a role, but the opportunity to trade may have been just as important. Many of the known locations of agricultural innovation lie near early trade centers. There, people would have had access to new information and new plants and animals brought by traders, and would have needed products to trade. Perhaps, then, agriculture was at first a profitable hobby for hunters and gatherers that eventually, because of the desire for food security *and* market demand, grew into a "day job" for some—their primary source of sustenance.

This link between agriculture and trade also provides a glimpse of how cities may have emerged at trading crossroads. People were attracted to these centers, and some developed occupations that served the needs of others who gathered to trade.

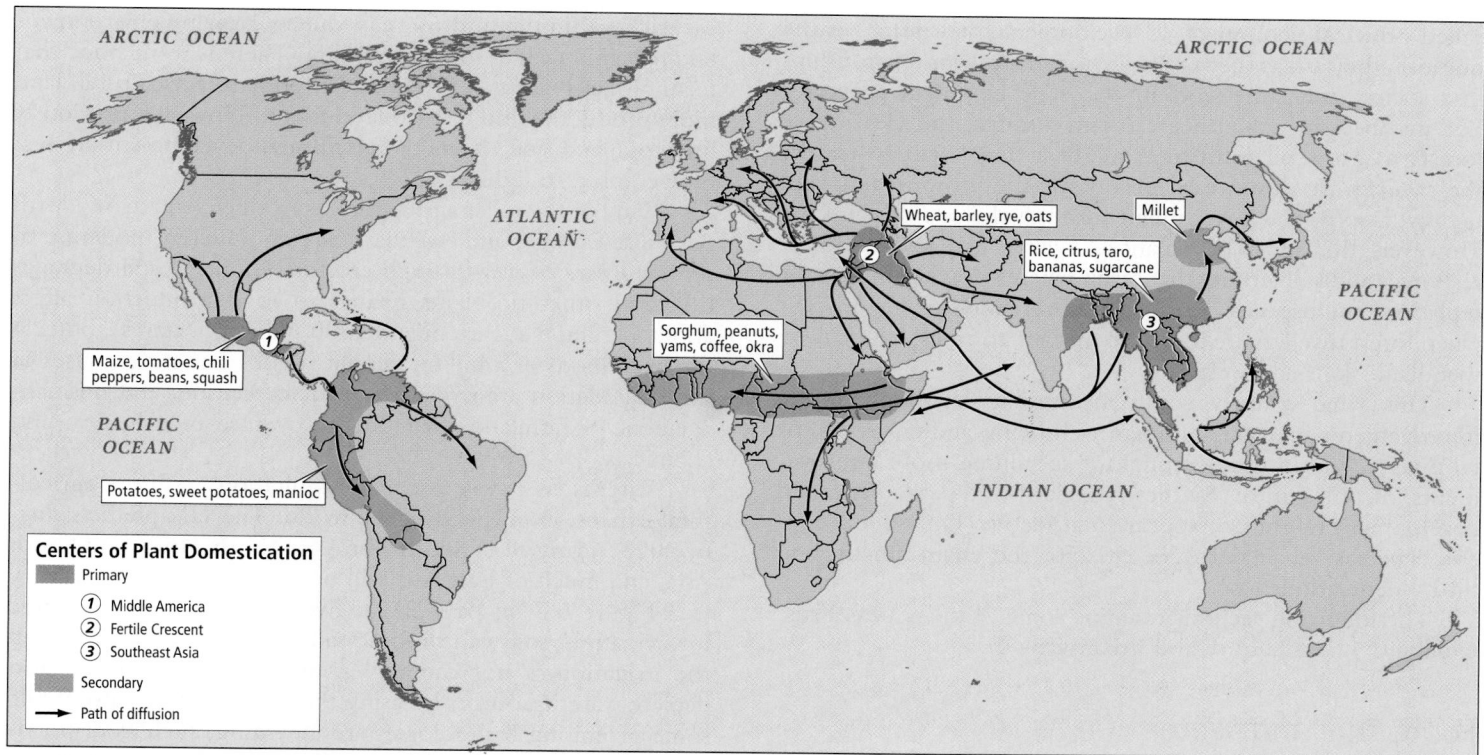

Figure 1.18 The origins of agriculture. Scientists have identified six main areas of the world where agriculture emerged. For lengthy periods, people in these different places tended plants and animals and selected for the genetic characteristics they valued. This knowledge eventually spread around the world. Domesticated plants and animals were then further adapted to new locations. This selection and adaptation process continues in the present. [Adapted from Terry G. Jordan-Bychkov, Mona Domosh, Roderick P. Neumann, and Patricia L. Price, *The Human Mosaic*, 10th ed. (New York: W. H. Freeman, 2006), pp. 274–275.]

AGRICULTURE AND LANDSCAPE TRANSFORMATION

Agriculture made it possible to amass surplus stores of food for lean times and allowed some people to specialize in activities other than food procurement. It also led to several developments now regarded as problems: rapid population growth, extreme social inequalities, environmental degradation, and famine.

As groups turned to raising animals and plants for their own use or for trade, more labor was needed. As the population expanded to meet this need and more resources were used to produce food, natural habitats were destroyed, and hunting and gathering were gradually abandoned. One possible consequence was that the quality of human diets may have declined as people stopped eating diverse wild plants and began to eat primarily cultivated corn, wheat, or rice. Moreover, land clearing increased vulnerability to drought and other natural disasters that could wipe out an entire harvest. Thus, as ever-larger populations depended solely on agriculture, famine became more common.

The potential for human impact on the world's environments has increased markedly as fields of cultivated plants and pastures for cattle, sheep, and goats have replaced forests and grasslands. In addition, we apply ever-larger amounts of chemicals and irrigation water to keep our fields and pastures productive. The trend of increasingly intense human impact has become even more pronounced over the past few centuries as the human population has doubled and redoubled. Today, housing tracts are replacing cultivated fields in many parts of the world, raising concerns about the sustainability of our species.

SUSTAINABLE DEVELOPMENT AND POLITICAL ECOLOGY

The United Nations defines **sustainable development** as the effort to improve present living standards in ways that will not jeopardize those of future generations. By destroying resources (as in deforestation) or poisoning them (as in pollution of water and air), we may be depriving future generations of resources they will need. The goal of sustainability is particularly important for the vast majority of earth's people who do not yet enjoy an acceptable level of well-being.

Geographers who study the interactions among development, politics, human well-being, and the environment are called **political ecologists.** "Development for whom?" is the question they ask as they examine how the power relationships in a society affect the ways in which development proceeds. For instance, in a Southeast Asian country, the clearing of forests to grow oil palm trees might at first seem to benefit the country. It would earn profits for the growers and raise tax revenue for the government through the sale of palm oil. However, this must be balanced against the loss of forest resources and soil fertility that result when a single plant species replaces a multispecies forest. Moreover, ways of life are lost when forest dwellers are forced to migrate to cities, where their woodland skills are useless.

This same country could measure its development by improvements in average human well-being and environmental quality, and in the potential for sustaining those improvements into the future. By these standards, oil palm plantation development might not appear so attractive. Only a few benefit, whereas the majority of citizens, the environment, and future generations lose.

The following sections examine some of the issues of sustainability in agriculture and urban growth.

Food, Soil, and Water

Farming that meets human needs without poisoning the environment or using up water and soil resources is called **sustainable agriculture.** This term is related to **carrying capacity,** which refers to the maximum number of people a given place can support sustainably.

> Farming that meets human needs without poisoning the environment or using up water and soil resources is called sustainable agriculture.

Technology has increased food production on earth remarkably, especially over the last several decades. In the 25 years between 1965 and 1990, total food production rose between 70 and 135 percent, depending on the region. But population also rose quickly during this period, so the gains were much less per capita. By 2007, growth in global agricultural production was slowing, but overall the global system is still capable of producing more than enough food for all. However, according to the United Nations Food and Agriculture Organization, one-fifth of humanity subsists on a diet too low in total calories and vital nutrients to sustain adequate health and normal physical and mental development. While undernutrition has fallen in Asia and Latin America in recent years, it rose in Southwest Asia and North Africa and in sub-Saharan Africa (Figure 1.19). Much research has shown that the problem of hunger is really a problem of inadequate or politically manipulated food distribution systems. Nevertheless, growing populations combined with skyrocketing food prices in recent years call into question many current food production systems. ▟

Just how sustainable are the world's present food production systems? The answer is unclear. Scientists from many disciplines think the world will reach the limit of its carrying capacity within the next 50 years due to growing environmental problems such as global warming. There is some hope that technological advances will make present agricultural land more productive and unused land useful. However, previously unrecognized side effects of agricultural development are just now coming to light.

Many of the most agriculturally productive parts of North America, Europe, and Asia have already suffered moderate to serious losses of soil through erosion. Globally, soil degradation and other problems related to food production affect about 7 million square miles (2000 million hectares), putting the livelihoods of a billion people at risk. The main causes of soil degradation are overgrazing, deforestation, and mismanagement of farmland through the overuse of irrigation and chemicals.

Water is emerging as a major limitation on future agricultural expansion in much of the world. The UN predicts that, by 2025, nearly all of Southwest Asia. North Africa, and South Asia, and much of East Asia will be in a state of water scarcity (see Figure 6.7 on page 212), Even when water is available for irrigation, soils can become salty and infertile over time if the irrigation is not carefully managed. Irrigation can also deplete water resources by using them at too fast a rate. While some promising technologies are emerging, such as drip irrigation (see page 213 in Chapter 6), many of those experiencing water shortage are too poor to afford them.

Modern agricultural techniques pioneered in the United States—such as the use of fertilizers, pesticides, and herbicides—are now spreading throughout the world. Though they may increase crop yield, they have produced some catastrophic side-effects, such as massive die-offs of birds, insect pollinators, fish, and other life-forms.

Few truly sustainable agricultural solutions have been developed so far. On one hand, technologically sophisticated mass production almost always results in environmental problems and favors wealthier farmers who can afford new technologies over poorer farmers who can't. Some small-scale production techniques used by millions of poor farmers may actually be more environmentally sound and able to meet local needs. However, it is not clear that these techniques can meet the huge food demands of growing urban populations. The agricultural successes and failures of various countries and agencies will be examined further throughout this book. ▟

Urbanization and Water Use

Our world is being rapidly transformed by **urbanization,** the process whereby cities, towns, and suburbs grow as populations shift from rural to urban livelihoods. In 1700, fewer than 7 million people, or 10 percent of the world's total population, lived in cities. Only five cities had populations as high as several hundred thousand people. By 2005, 47 percent of the world's population lived in cities. There were more than 400 cities of over 1 million and about 25 cities of over 10 million. ▟

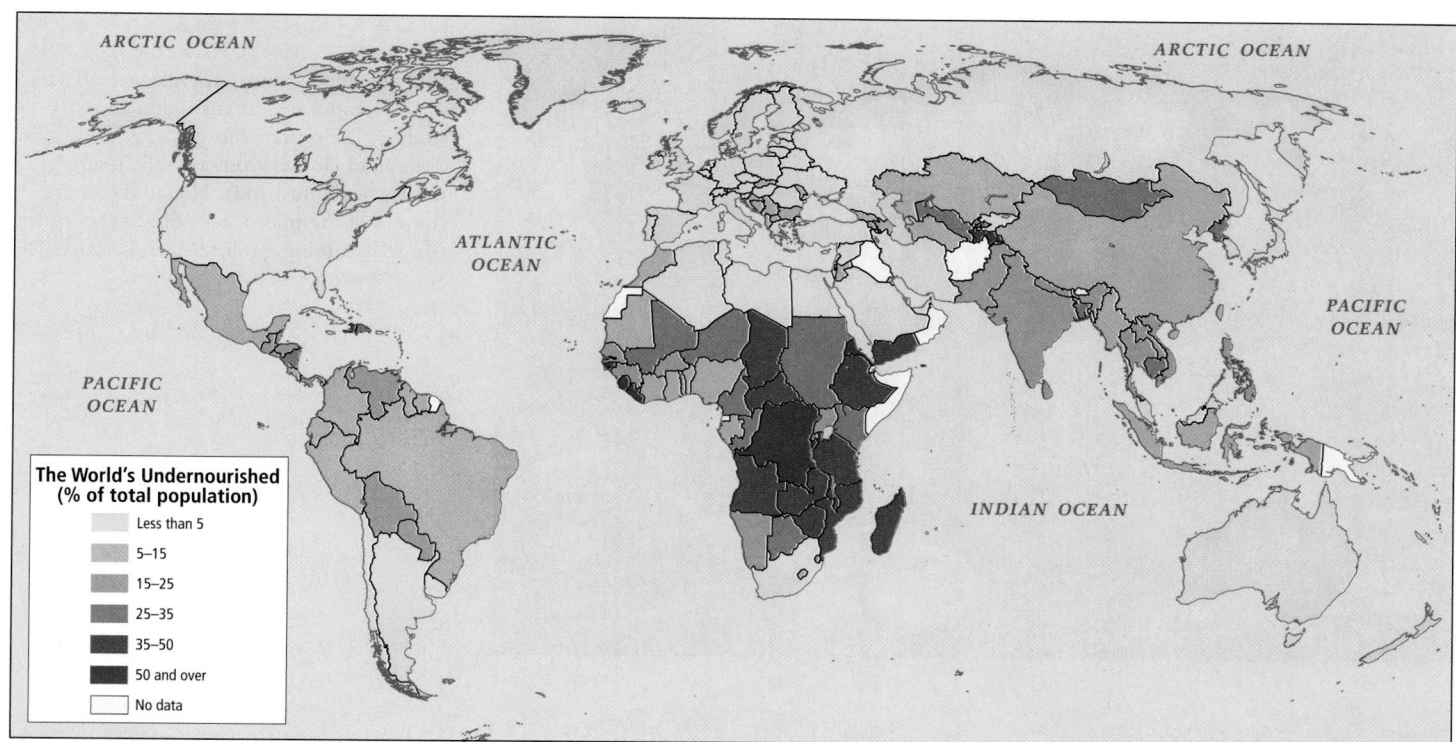

Figure 1.19 Undernourishment in the world. The proportion of people suffering from undernourishment—the lack of adequate nutrition to meet their daily needs—has declined in the developing world over the past several years. However, an estimated 824 million people were still affected by chronic hunger in 2003. As you can see from the map, people in much of Africa, parts of Central Asia, Mongolia, Bangladesh, North Korea, Cambodia, Nicaragua, and parts of the Caribbean suffer the most. To see a country-by-country animated map of undernourished populations between 1970 and 2003, go to http://www.fao.org/es/ess/faostat/foodsecurity/FSMap/flash_map.htm. [Adapted from the Food and Agriculture Organization of the United Nations at http://www.fao.org/es/ess/faostat/foodsecurity/.]

The World's Undernourished (% of total population)
- Less than 5
- 5–15
- 15–25
- 25–35
- 35–50
- 50 and over
- No data

> Our world is being rapidly transformed by urbanization, the process whereby cities, towns, and suburbs grow as populations shift from rural to urban livelihoods.

Most urban dwellers today cope with some difficult realities. Water and sanitation are cases in point. In rapidly growing cities in Asia, Africa, and Middle and South America, widespread poverty means that many people acquire their water with a pail from a communal faucet (Figure 1.20). Usually this water should be boiled before use, even for bathing, and truly safe drinking water must be purchased. But most cannot afford to buy clean water or the fuel to boil it. As a result, adults are often chronically ill, and waterborne diseases cause many children to die before the age of 5.

Urban housing is often self-built of scavenged materials, and sanitation systems are absent. Even the sewage and other wastewater from modern high-rise apartments and hotels are often pumped untreated into a nearby river, swamp, or ocean. This method of waste disposal causes serious health hazards and widespread ecological damage. Building adequate waste-water collection and water purification systems in cities already housing several million inhabitants is so costly that this option is rarely considered.

As we will see in later chapters, clean water is only one of several problems faced by the world's urban dwellers. Technological and other advances will no doubt help alleviate some problems, but we are now far short of the sustainable ideal for the world's cities.

Changing Patterns of Resource Consumption

As people move from rural agricultural work to industry or service sector jobs in cities, they begin to use more resources per capita, and they draw them from a wider and wider area. Water once fetched from nearby village wells may now be piped hundreds of miles into urban apartments or shantytowns. Clothing once made laboriously by hand at home is now purchased from manufacturers half a world away. Consumers in wealthy countries may have access to a variety of imported products at lower prices than they would pay for locally produced items.

Resource use has become so skewed toward the affluent that the relatively rich minority (about 20 percent) of the world's population consumes more than 80 percent of the available world resources. The poorest 80 percent of the population are left with less than 20 percent of the resources.

Figure 1.20 Lack of safe water. In sections of Kolkata (Calcutta), India, the only available public water supply comes from small pipes located at curbsides. Here a woman (left center) fills buckets for washing dishes and clothes from a public spigot (under her right hand). Notice the entire context of the photo, including what other people are doing. [Dilip Mehta, Contact/The Stock Market.]

Human consumption of natural resources is increasingly being looked at through the concept of the **ecological footprint.** This is a method of estimating the amount of biologically productive land and sea area needed to sustain a human population at its current standard of living. It is particularly useful for drawing comparisons. For example, the worldwide average biologically productive area per person—in other words, one individual's ecological footprint—is about 4.5 acres. However, in the United States ecological footprints average about 24 acres, and in China about 4 acres. You can calculate your own footprint at http://www.earthday.net/Footprint/index.asp. A similar concept more related to global warming is the "carbon footprint" which measures the greenhouse gas emissions that a person's activities produce. ◼▶

Physical Geography: Perspectives on the Earth

To better appreciate the environmental contexts in which humans operate, it is helpful to recognize two components of the physical environment that are of particular interest to physical geographers: landforms and climate.

LANDFORMS: THE SCULPTING OF THE EARTH

The processes that create the world's varied **landforms**—mountain ranges, continents, and the deep ocean floor—are some of the most powerful and slow-moving forces on earth. Originating deep beneath the earth's surface, these *internal processes* can move entire continents, often taking hundreds of millions of years to do their work. Many of the earth's features, however, such as a beautiful waterfall or a dramatic rock formation, are formed by more rapid and delicate processes that take place on the surface of the earth *(external processes).* All of these forces are studied by geomorphologists, the geographers who focus on the processes that constantly shape and reshape the earth's surface.

Plate Tectonics

Two key ideas related to internal processes in physical geography are the Pangaea hypothesis and plate tectonics. The **Pangaea hypothesis** was first suggested by geophysicist Alfred Wegener in 1912. It proposes that all the continents were once joined in a single vast continent called *Pangaea* (meaning "all lands"), which fragmented over time into the continents we know today (Figure 1.21). As one piece of evidence for his theory, Wegener pointed to the neat fit between the west coast of Africa and the east coast of South America.

For decades, most scientists rejected Wegner's hypothesis. We now know, however, that the earth's continents have been assembled into supercontinents at least three different times, only to break apart again. All of this activity is made possible by plate tectonics, a process of continental motion discovered in the 1960s, long after Wegener's time.

The premise of **plate tectonics** is that the earth's surface is composed of large plates that float on top of an underlying layer of molten rock. The plates are of two types. *Oceanic plates* are dense and relatively thin, and they form the floor beneath

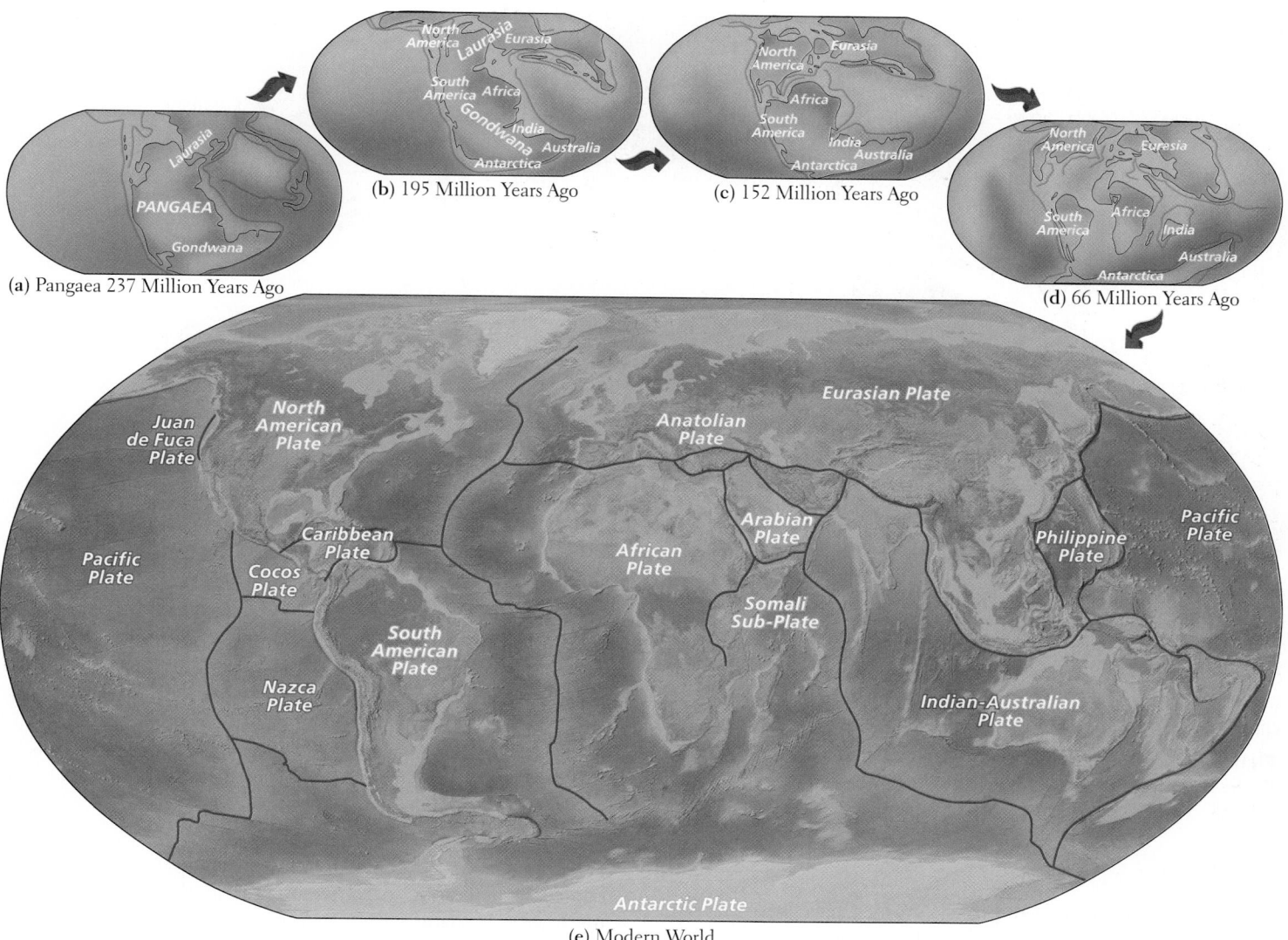

Figure 1.21 The breakup of Pangaea. The Modern World map (e) depicts the current boundaries of the major tectonic plates. Pangaea is only the latest of several global configurations that have coalesced and then fragmented over the last billion years. [Adapted from Frank Press, Raymond Siever, John Grotzinger, and Thomas H. Jordan, *Understanding Earth,* 4th ed. (New York: W. H. Freeman, 2004), pp. 42–43.]

the oceans. *Continental plates* are thicker and less dense. Much of their surface rises above the oceans, forming continents. These massive plates drift slowly, driven by the circulation of the underlying molten rock flowing from hot regions deep inside the earth to cooler surface regions and back. The creeping movement of tectonic plates created the continents we know today by fragmenting and separating Pangaea (see Figure 1.21).

Plate movements influence the shapes of major landforms, such as continental shorelines and mountain ranges. The continents have piled up huge mountains on their leading edges as the plates carrying them collided with other plates, folding and warping in the process. Hence, the theory of plate tectonics accounts for the long, linear mountain ranges extending from Alaska to Chile in the Western Hemisphere and from Southeast Asia to the European Alps in the Eastern Hemisphere. The highest mountain range in the world, the

Himalayas of South Asia, was created when what is now India, at the northern end of the Indian-Australian Plate, ground into Eurasia. The only continent that lacks these long, linear mountain ranges is Africa. Often called the "plateau continent," Africa is believed to have been at the center of Pangaea and to have moved relatively little since the breakup.

Humans encounter tectonic forces most directly as earthquakes and volcanoes. Plates slipping past each other create the catastrophic shaking of the landscape we know as an earthquake. When plates collide and one slips under the other, this is known as **subduction.** Volcanoes arise at zones of subduction or sometimes in the middle of a plate, where gases and molten rock (called magma) can rise to the earth's surface through fissures and holes in the plate. Volcanoes and earthquakes are particularly common around the edges of the Pacific Ocean, an area known as the **Ring of Fire** (Figure 1.22).

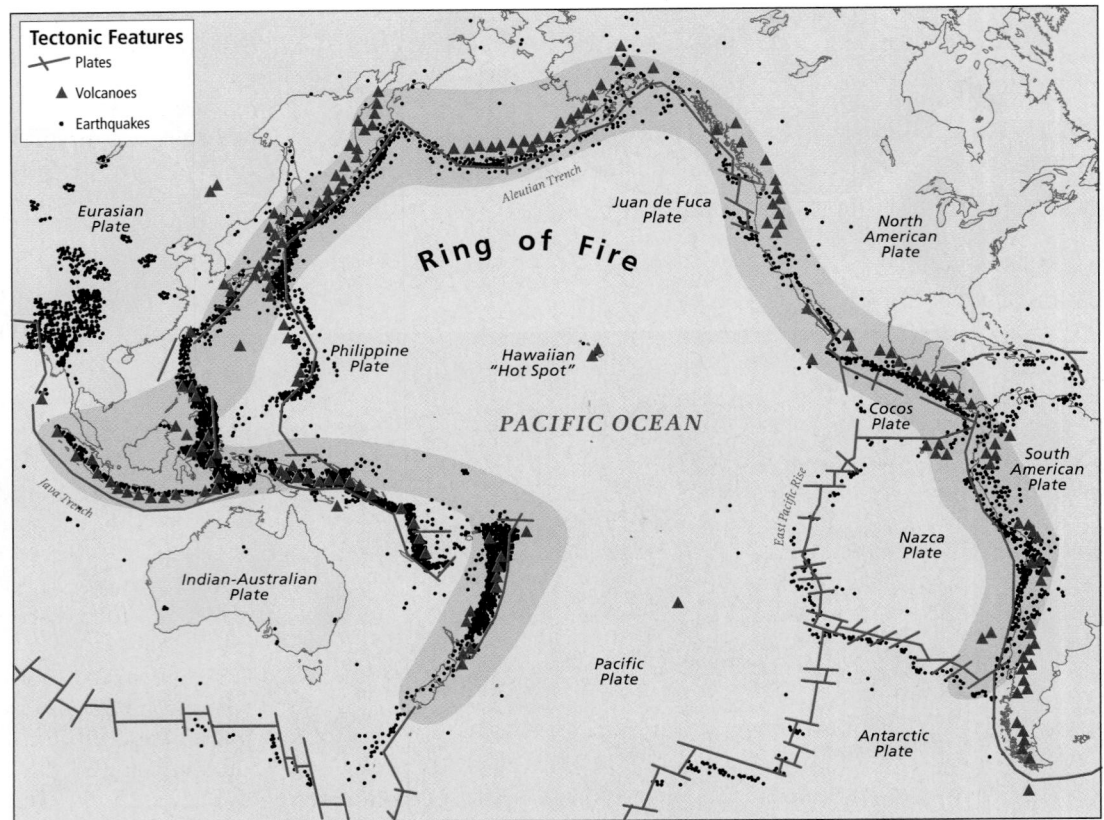

Figure 1.22 The Ring of Fire. Volcanic formations encircling the Pacific Basin form the Ring of Fire, a zone of frequent earthquakes and volcanic eruptions. [Adapted from http://vulcan.wr.usgs.gov/Glossary/ PlateTectonics/Maps/map_plate_tectonics_world.html; and Frank Press, Raymond Siever, John Grotzinger, and Thomas H. Jordan, *Understanding Earth,* 4th ed. (New York: W. H. Freeman, 2004), p. 27.]

Landscape Processes

The landforms created by plate tectonics have been further shaped by external processes, which are more familiar to us because we can observe them daily. One such process is **weathering.** Rock, exposed to the onslaught of sun, wind, rain, snow, ice, and the effects of life-forms, fractures and decomposes into tiny pieces. These particles then become subject to another external process, **erosion.** During erosion, wind and water carry rock particles away and deposit them in new locations. The deposition of eroded material can raise and flatten the land around a river, where periodic flooding spreads huge quantities of silt. As small valleys between hills are filled in by silt, a **floodplain** is created. Where rivers meet the sea, floodplains often fan out roughly in the shape of a triangle, creating a **delta.** External processes tend to smooth out the dramatic mountains and valleys created by internal processes.

Human activity often contributes to external landscape processes. By altering the vegetative cover, agriculture and forestry expose the earth's surface to sunlight, wind, and rain. These agents in turn increase weathering and erosion. Flooding becomes more common because the removal of vegetation limits the ability of the earth's surface to absorb rainwater. As erosion increases, rivers may fill with silt, and deltas may extend into the oceans.

Urban development also brings many changes. Buildings and roads made of concrete, asphalt, and steel often cover formerly wooded land with impervious surfaces. Again, flooding is the result because rainwater runs over the surface to the lowest point instead of being absorbed into the ground. The physical effects of human activities vary in degree from one culture to another depending in part on the tools used. Mechanized earthmovers that were used to build roads change the earth's surface more rapidly and profoundly than machetes used to clear a path.

CLIMATE

The processes associated with climate are generally more rapid than those that shape landforms. **Weather,** the short-term expression of climate, can change in a matter of minutes. **Climate** is the long-term balance of temperature and precipitation that keeps weather patterns fairly consistent from year to year. By this definition, the last major global climate change took place 15,000 years ago, when the glaciers of the last ice age began to melt. As we have seen, human activity is producing a new global climate change in our own time.

Energy from the sun gives the earth a temperature range hospitable to life. The earth's atmosphere, oceans, and land surfaces absorb huge amounts of solar energy. The atmosphere traps much of that energy at the earth's surface, insulating the earth from the deep cold of space. Solar energy is also the engine of climate. The most intense, direct sunlight falls in a broad band stretching about 30° north and south of the equator. The highest average temperatures on earth occur within this band. Moving away from the equator, sunlight becomes less intense, and average temperatures drop.

Temperature and Air Pressure

The wind and weather patterns we experience daily are largely a product of complex patterns of air temperature and air pressure. To understand **air pressure,** you can think of air as existing in a particular unit of space—for example, a column of air above a square foot of the earth's surface. Air pressure is the amount of force exerted by that column on that square foot of surface. Air pressure and temperature are related: the gas molecules in warm air are relatively far apart and are associated with low air pressure. In cool air, the gas molecules are relatively close together (dense) and are associated with high air pressure.

As a unit of cool air is warmed by the sun, the molecules move farther apart. The air becomes less dense and exerts less pressure. Air tends to move from areas of higher pressure to areas of lower pressure, creating wind. If you have been to the beach on a hot day, you may have noticed a cool breeze blowing in off the water. Land heats up (and cools down) faster than water, so on a hot day, the air over the land warms, rises,

and becomes less dense than the air over the water. This causes the cooler, denser air to flow inland. At night the breeze often reverses direction, blowing from the now cooling land onto the now relatively warmer water.

These air movements have a continuous and important influence on global weather patterns. Over the course of a year, continents heat up and cool off much more rapidly than the oceans that surround them. Hence, the wind tends to blow from the ocean to the land during summer and from the land to the ocean during winter. It is almost as if the continents were breathing once a year, inhaling in summer and exhaling in winter.

Precipitation

Perhaps the most tangible way we experience changes in air temperature and density is through rain or snow. Precipitation occurs primarily because warm air holds more moisture than cool air. When this moist air rises to a higher altitude, its temperature drops, which reduces its ability to hold moisture. The moisture condenses into drops to form clouds and may eventually fall as rain or snow.

Several conditions that encourage moisture-laden air to rise influence the pattern of precipitation observed around the globe (Figure 1.23). When moisture-bearing air is forced to rise as it passes over mountain ranges, the air cools and the moisture condenses to produce rainfall. This process, known

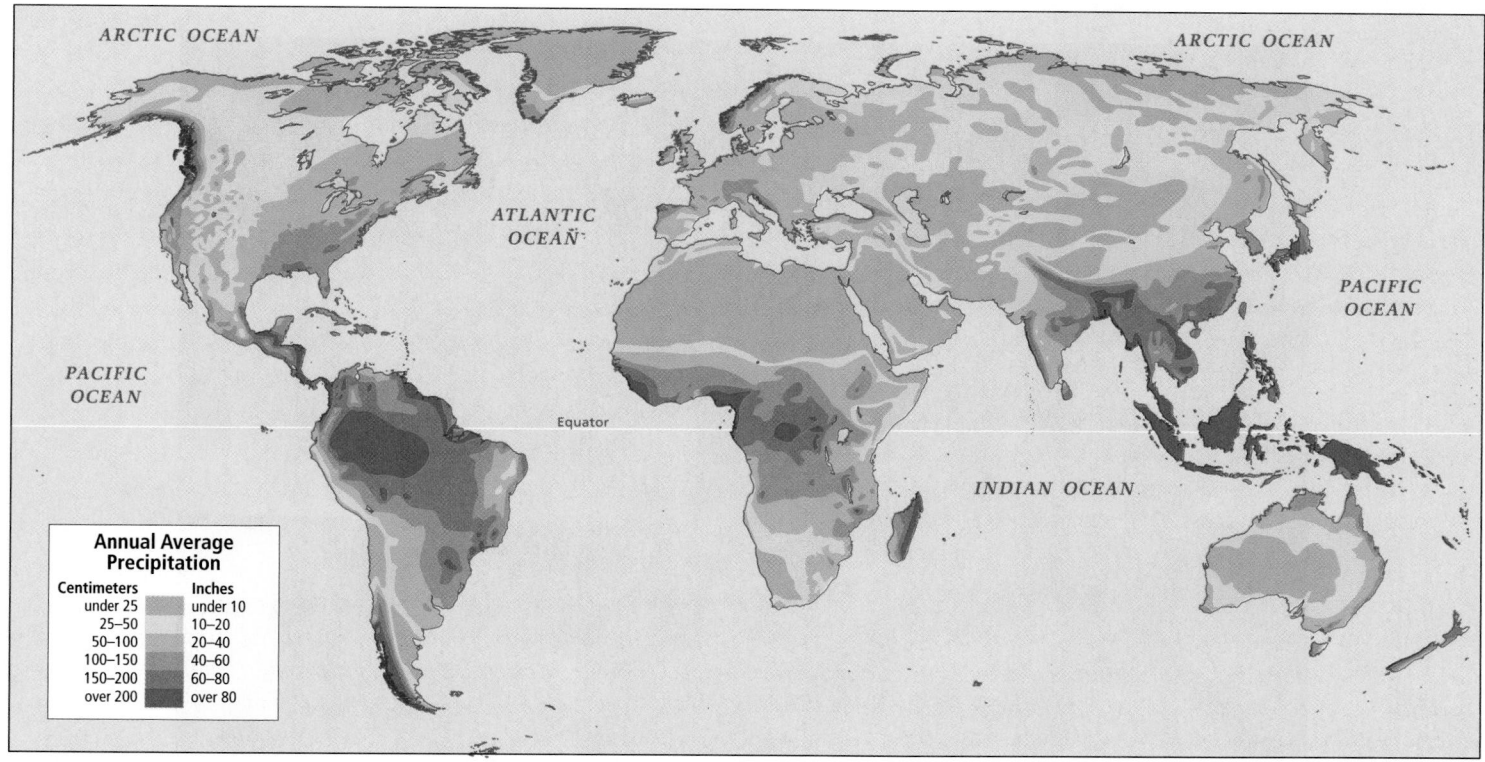

Figure 1.23 World average annual precipitation. Notice the concentration of the heaviest precipitation primarily in a wide irregular band on both sides of the equator. [Adapted from *Goode's World Atlas,* 21st ed. (Chicago: Rand McNally, 2005), pp. 20–21.]

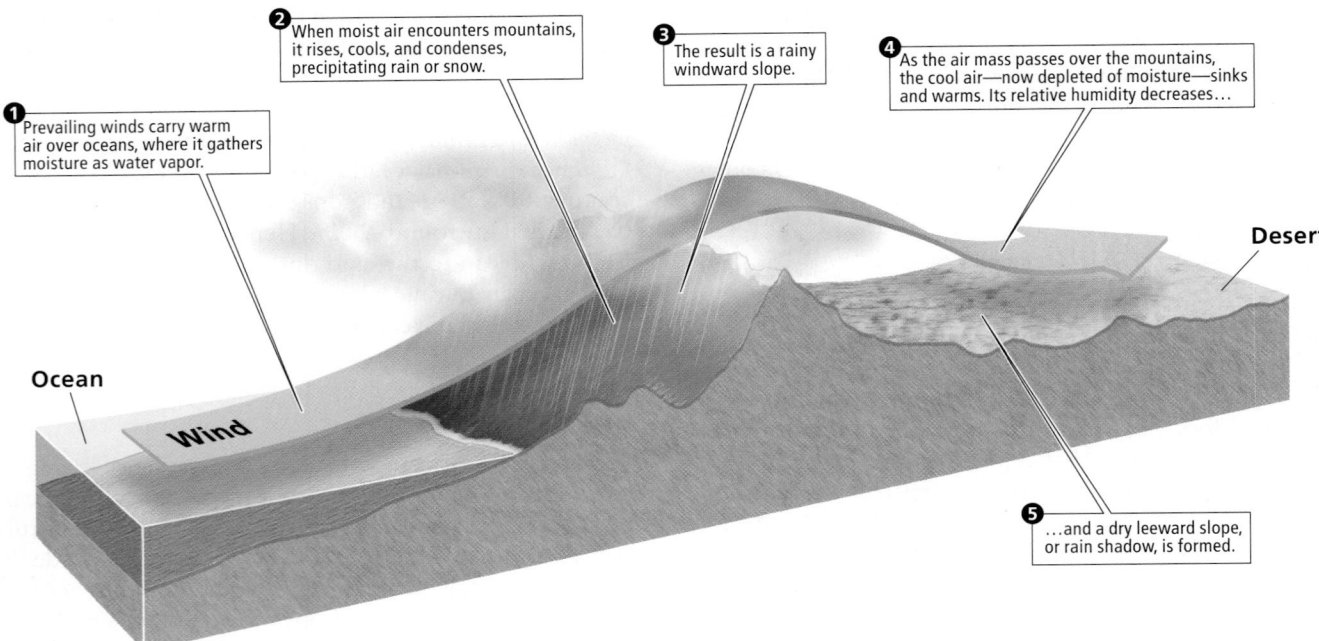

❶ Prevailing winds carry warm air over oceans, where it gathers moisture as water vapor.

❷ When moist air encounters mountains, it rises, cools, and condenses, precipitating rain or snow.

❸ The result is a rainy windward slope.

❹ As the air mass passes over the mountains, the cool air—now depleted of moisture—sinks and warms. Its relative humidity decreases...

❺ ...and a dry leeward slope, or rain shadow, is formed.

Ocean

Wind

Desert

Figure 1.24 Orographic rainfall (rain shadow). [Adapted from Frank Press, Raymond Siever, John Grotzinger, and Thomas H. Jordan, *Understanding Earth,* 4th ed. (New York: W. H. Freeman, 2004), p. 281.]

Figure 1.25 An Asian monsoon. This photo shows a summer monsoon rainstorm approaching Sri Lanka from the southeast coast of India. [Mark Henley/Panos.]

as **orographic rainfall,** is most common in coastal areas where wind blows moist air from above the ocean onto the land and up the side of a coastal mountain range (Figure 1.24). Most of the moisture falls as rain as the air is rising along the coastal side of the range. On the inland side, the descending air warms and ceases to drop its moisture. The drier side of a mountain range is said to be in the **rain shadow.** Rain shadows may extend for hundreds of miles across the interiors of continents, as they do on the Mexican Plateau, east of California's Pacific coast, or north of the Himalayas of Eurasia.

Near the equator, moisture-laden tropical air is heated by the strong sunlight and rises to the point where it releases its moisture as rain. This produces the "rain belt" in equatorial areas in Africa, Southeast Asia, and South America, which you can see in Figure 1.24. Neighboring nonequatorial areas also receive some of this moisture when seasonally shifting winds blow the rain belt north and south of the equator. The huge downpours of the Asian summer **monsoon** are an example (Figure 1.25).

In a monsoon, the Eurasian continental landmass heats up during the summer, causing the overlying air to expand, become less dense, and rise. The somewhat cooler, yet moist, air of the Indian Ocean is drawn inland. The effect is so powerful that the equatorial rain belt is sucked onto the land (see Figure 8.5 on page 289). The result is tremendous, sometimes catastrophic, rains throughout virtually all of South and Southeast Asia and much of coastal and interior East Asia. Similar forces pull the equatorial rain belt south during the Southern Hemisphere's summer.

Much of the moisture that falls on North America and Eurasia is **frontal precipitation** caused by the interaction of large air masses of different temperatures and densities. These masses develop when air stays over a particular area long enough to take on the temperature of the land or sea beneath it. Often when we listen to a weather forecast, we hear about warm fronts or cold fronts. A front is the zone where warm and cold air masses come into contact, and it is always named after the air mass whose leading edge is moving into an area. At a front, the warm air tends to rise over the cold air, carrying warm clouds to a higher altitude. Rain or snow may follow. Much of the rain that falls along the outer edges of a hurricane is the result of frontal precipitation.

Climate Regions

Geographers have several systems for classifying the world's climates that are based on the patterns of temperature and precipitation just described. This book uses a modification of the widely known Köppen classification system, which divides the world into several types of climate regions, labeled A, B, C, D, and E on the climate map in Figure 1.26 (page 30). As you look at the regions on this map and read the accompanying climate descriptions, the importance of climate to vegetation becomes evident. Each regional chapter will contain a climate map; when reading these maps, you can refer to the verbal descriptions in Figure 1.26 if necessary. Keep in mind that the sharp boundaries shown on climate maps are in reality much more gradual transitions.

Cultural Geography

Culture is an important distinguishing characteristic of human societies. It comprises everything we use to live on earth that is not directly part of our biological inheritance. Culture is represented by the ideas, materials, and social arrangements that people have invented and passed on to subsequent generations. Among other things, culture includes language, music, gender roles, belief systems, and moral codes (for example, those prescribed in Confucianism, Islam, and Christianity).

Material culture comprises all the things that people use: clothing, houses and office buildings, axes, guns, computers, earthmoving equipment (from hoes to work animals to bulldozers), books, musical instruments, domesticated plants and animals, agricultural and food-processing equipment—the list is virtually endless.

In this section, we explore the concept of culture groups and some of the cultural attributes, such as value systems and languages, that help to define them. We also examine gender roles and perceptions about race as cultural phenomena.

ETHNICITY AND CULTURE: SLIPPERY CONCEPTS

A group of people who share a location, a set of beliefs, a way of life, a technology, and usually a common ancestry form an **ethnic group.** The term **culture group** is often used interchangeably with ethnic group. The concepts of culture and ethnicity are imprecise, however, especially as they are popularly used. For instance, as part of the modern globalization process, migrating people often move well beyond their customary cultural or ethnic boundaries to cities or even distant countries. In these new places they take on new ways of life

Figure 1.26 Climate regions of the world.

Tropical Humid Climates (A). These climates occupy a wide band reaching 15° to 20° north and south of the equator and extending to higher latitudes when moderated by marine influences. Here we have simplified the variations to just two distinct climates: tropical wet and tropical wet/dry.

In **tropical wet climates,** rain falls predictably every afternoon and usually just before dawn. The natural vegetation is the tropical rain forest, consisting of hundreds of species of broad-leafed evergreen trees that form a several-layered canopy above the forest floor.

The **tropical wet/dry climate,** also called a **tropical savanna,** experiences a wider range of temperatures than the tropical wet climate. It may also receive more total rainfall, but the rain comes in great downpours during the heat of the summer. The vegetation is mixed grassland and tropical forest. All species have to survive long dry periods, during which they may drop their leaves to conserve moisture.

Arid and Semiarid Climates (B). Arid and semiarid climates are either deserts or steppes.

Deserts generally receive very little rainfall (two inches or less per year). Most of that rainfall comes in downpours that are extremely rare and unpredictable, but that are capable of bringing a brief, beautiful flourishing of desert life. Usually, deserts have sparse vegetation and almost no cloud cover, which leads to wide swings in temperature between day and night. Life is a struggle for both plants and animals because they must be able to survive heat stress during the day and freezing temperatures at night.

Steppes, such as the pampas of Argentina or the Great Basin of the American West, have climates similar to those of deserts, but more moderate. They usually receive about 10 inches more rain per year than deserts and are covered with grass or scrub.

Arid and semiarid climates are found primarily in two locations: the subtropics (slightly poleward of the tropics) and the midlatitudes. Subtropical deserts and steppes are found between 20° and 30° north and south latitudes, where high-pressure air descends in a belt around the planet. They are generally much warmer than midlatitude deserts and steppes. These midlatitude regions are found farther toward the poles in the interiors of continents, often in the rain shadows of high mountains. The soil is generally thin and unproductive in most deserts and steppes. However, some midlatitude steppes, such as the Great Plains of North America, have some of the thickest and richest soil in the world. The slightly colder temperatures and pronounced seasonality of these steppes keep down rates of decay and hence encourage the accumulation of organic matter in the soil over time.

Temperate Climates (C). In this book, we distinguish three temperate climates.

Midlatitude climates, such as those in southeastern North America and China, are moist all year and have short, mild winters and long, hot summers. A variant of the midlatitude climate is the **marine west coast climate,** such as that of western Europe, which is noted for fine drizzling rains.

Subtropical climates differ from midlatitude climates in that winters are dry.

Mediterranean climates have moderate temperatures but are dry in summer and wet in winter. Plants do not get much moisture when temperature and evaporation rates are highest, so the plant species that live in this climate tend to have adapted to dry conditions, with scrubby, waxy leaves capable of storing moisture. California, Portugal, northwestern Africa, southern Italy, Greece, and Turkey are examples of places with this climate type.

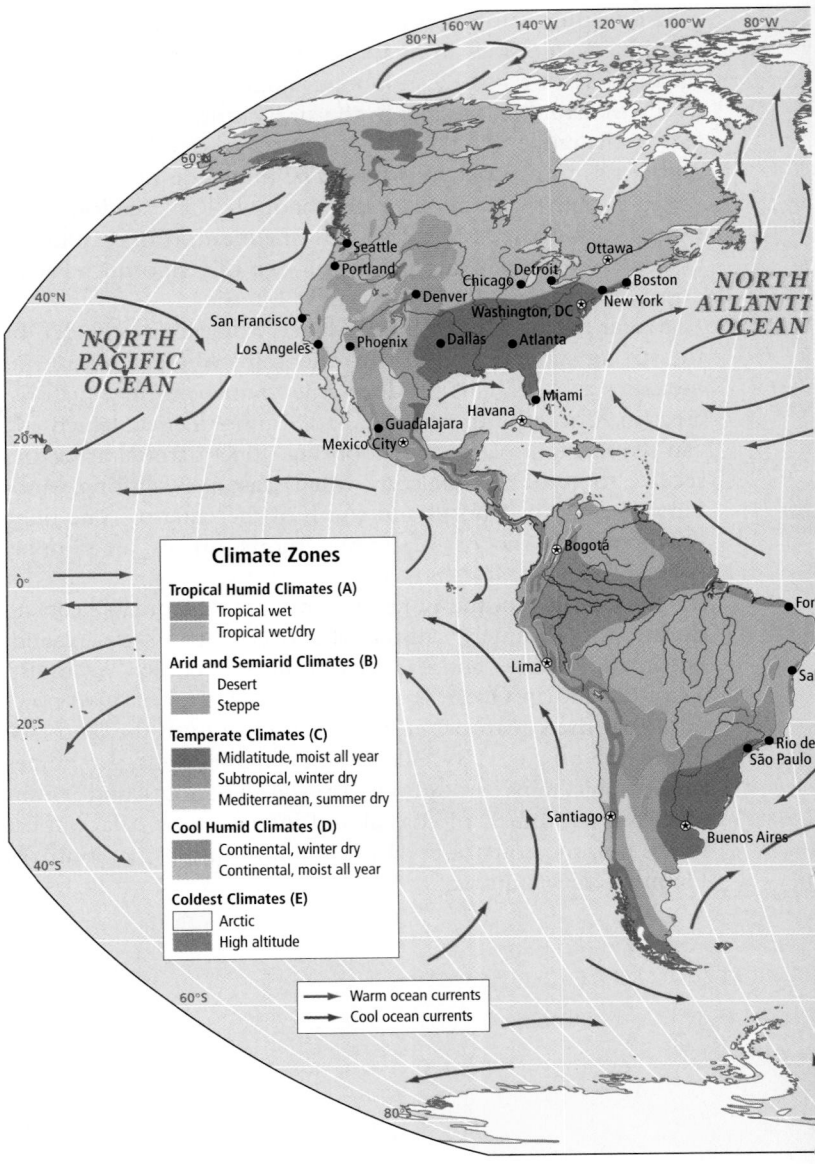

Climate Zones

Tropical Humid Climates (A)
Tropical wet
Tropical wet/dry

Arid and Semiarid Climates (B)
Desert
Steppe

Temperate Climates (C)
Midlatitude, moist all year
Subtropical, winter dry
Mediterranean, summer dry

Cool Humid Climates (D)
Continental, winter dry
Continental, moist all year

Coldest Climates (E)
Arctic
High altitude

→ Warm ocean currents
→ Cool ocean currents

Cool Humid Climates (D). Stretching across the broad interiors of Eurasia and North America are continental climates, either with **dry winters** (northeastern Eurasia) or **moist all year** (North America and north-central Eurasia). Summers in cool humid climates are short but can have very warm days. The natural vegetation of southern cool humid climates is broad-leafed deciduous and evergreen forest. Here the soil is deep and rich as a result of seasonally low temperatures that inhibit decay, as in the midlatitude steppes. In the more northerly areas, winters are long and cold. Vast needle-leafed evergreen forests called *taiga* stretch across the cold interior. In the taiga, the soil can be

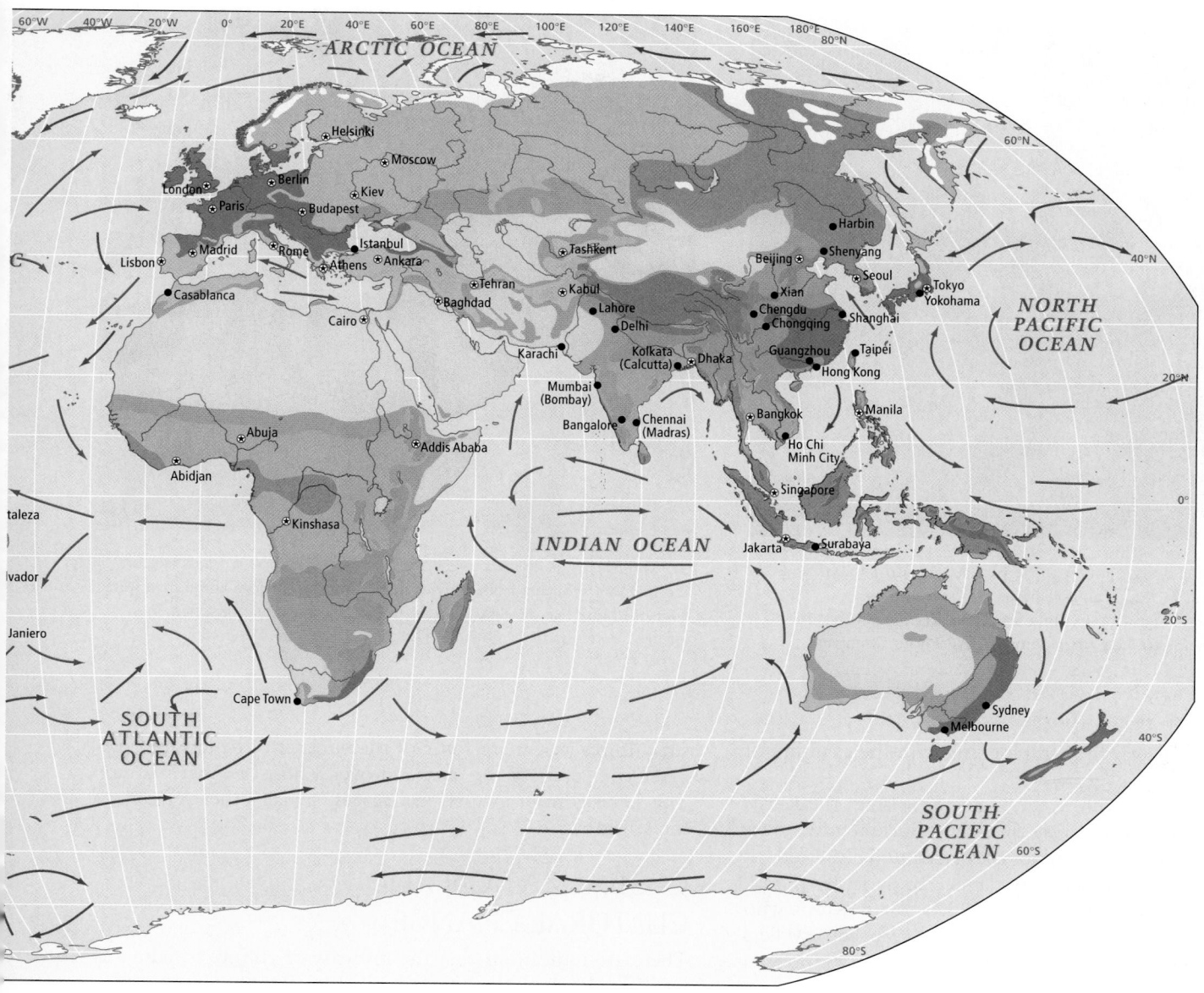

deep, but it is not as rich as the soil farther south. Growing seasons are short, so cultivation is minimal.

Coldest Climates (E). Arctic and high-altitude climates are by far the coldest and are also among the driest. Although moisture is present, there is little evaporation because of the low temperatures.

The **Arctic** climate is often called *tundra,* after the low-lying vegetation that covers the ground. This dwarfed vegetation is a response to the 7 to 11 months of below-freezing temperatures. What little precipitation there is usually comes during the warmer months, and even this may fall as snow.

The **high-altitude** version of this climate, which may occur far from the Arctic, is more widespread and subject to greater daily fluctuations in temperature. High-altitude microclimates, such as those in the Andes and the Himalayas, can vary tremendously depending on factors such as available moisture, orientation to the sun, and vegetation cover. As one ascends in altitude, the climate changes loosely mimic those found as one moves from lower to higher latitudes. These changes are known as temperature-altitude zones (see Figure 3.7 on page 103).

Figure 1.27 What does it mean to be Kurdish? **(a)**In rural areas away from the war zone in Iraq, Kurdish life is still peaceful and agriculturally based. Here, two young women carry sacks of grain home from the fields to feed their animals. [AP Photo/Brennan Linsley.]

(b) The Mazi supermarket in the Kurdish city of Duhok is Iraq's largest, and it draws shoppers from all over the country. The store and a mall were built on the site of a military facility where Kurds were imprisoned and tortured under Saddam Hussein. [Ed Kashi/National Geographic.]

or even new beliefs, yet they still identify with their culture of origin.

For example, long before the U.S. war in Iraq, the Kurds in Southwest Asia were asserting their right to create their own country in the territory where they traditionally lived as nomadic herders. (This area is now claimed by Syria, Iraq, Iran, and Turkey; see Figure 6.31 on page 237). Many Kurds who actively support the cause of the herders are now urban dwellers living and working in modern settings in Turkey, Iraq, or even London. Although these people think of themselves as ethnic Kurds and are so regarded in the larger society, they do not follow the traditional Kurdish way of life (Figure 1.27). Hence, we could argue that these urban Kurds have a new identity within the Kurdish culture or ethnic group.

Another problem with the concept of culture is that it is often applied to a very large group that shares only the most general of characteristics. For example, one often hears the terms *American culture, African-American culture,* or *Asian culture.* In each case, the group referred to is far too large to share more than a few broad characteristics.

It might fairly be said, for example, that U.S. culture is characterized by beliefs that promote individual rights, autonomy, and individual responsibility. But, when we look at specifics, contradictions emerge. For example, in most U.S. states the terminally ill do not have the right to use medications to achieve a "managed death," and almost all states restrict who can marry. In fact, U.S. culture encompasses many

subcultures that share some of the core set of beliefs, but disagree over parts of the core and over a host of other matters. The same is true, in varying degrees, for all other regions of the world.

GLOBALIZATION AND CULTURAL CHANGE

There are indications that the diversity of culture is fading as trends and fads circle the globe via the instant communication now available. American fast food, popular music, and clothing styles can now be found from Mongolia to Mozambique. At the same time, a wide variety of ethnic music, textiles, cuisines, and dress from distant places now graces the lives of consumers in the United States and across the world. As globalization proceeds, people migrate and ideas spread. Inevitably, some measure of **cultural homogeneity**—more overall similarity between culture groups—will occur.

Are we all drifting toward a common material culture, and perhaps even similar ways of thinking? Possibly, but there are also countervailing trends. It is now possible for people to reinforce their feeling of cultural identity with a particular group through the same channels that are encouraging homogenization. Consider the example of the people of Aceh in Southeast Asia.

Though Aceh is a province within the country of Indonesia, it has a distinctive cultural identity. The people of Aceh desire greater *autonomy*—the right to control their own affairs and especially to retain control of their own resources. The central Indonesian government, however, views all resources as national, not local. The Indonesian government has sent military troops to enforce its interests, with fatal consequences to the Acehese.

To build awareness of their plight and a sense of identity among Acehese migrants worldwide, the Acehese established several Web sites and Internet chat groups. After a tsunami devastated Aceh in December 2004, the cultural solidarity already built through the Internet helped raise funds efficiently and distribute assistance to victims. Hence the Internet has helped maintain and strengthen the ties that bind the Acehese together, especially during times of adversity.

The ability to communicate easily over the Internet and to travel quickly can reaffirm cultural identity, but they also enhance the conditions necessary for **multiculturalism,** the state of relating to, reflecting, or being adapted to several cultures. For example, the young webmaster from Aceh who helped us to understand the perspective of the Acehese in Indonesia is now an artist in the United States. He earns a living painting not only scenes from his tropical homeland, but also the mansions of U.S. corporate executives, as well as portraits of these executives.

CULTURAL MARKERS

Members of a particular culture group share features, such as language and common values, that help to define the group. These shared features are called *cultural markers*. The following sections examine the roles of values, religion, language, and material culture as cultural markers. Notice how colonization and modern communications are causing some cultural markers to disappear and others to become more dominant.

Values

Occasionally you will hear someone say, "After all is said and done, people are all alike," or "People ultimately all want the same thing." It is a heartwarming sentiment, but an oversimplification. Culturally, people are not all alike, and that is in large part what makes the study of geography interesting. We would be wise not to expect or even to want other people to be like us. It is often more fruitful to look for the reasons behind differences among people than to search hungrily for similarities. Cultural diversity has helped humans to be successful and adaptable animals. The various cultures serve as a bank of possible strategies for responding to the social and physical challenges faced by the human species. The reasons for differences in behavior from one culture to the next are usually complex, but they are often related to differences in values.

Let us look at an example that contrasts the values and norms (accepted patterns of behavior based on values) held by modern urban individualistic culture with those held by rural community-oriented culture. One recent rainy afternoon, a beautiful forty-something Asian woman walked alone down a fashionable street in Honolulu, Hawaii. She wore high-heeled sandals, a flared skirt that showed off her long legs, and a cropped blouse that allowed a glimpse of her slim waistline. She carried a laptop case and a large fashionable handbag. Her long, shiny black hair was tied back. Everyone noticed and admired her because she exemplified an ideal Honolulu businesswoman: beautiful, self-assured, and rich enough to keep herself well-dressed.

In the village of this woman's grandmother—whether it be in Japan, Korea, Taiwan, or rural Hawaii—her clothes would breach a widespread traditional value that no individual should stand out from the group. Furthermore, the dress that exposed her body to open assessment and admiration by strangers of both sexes would signal that she lacked modesty. The fact that she walked alone down a public street—unaccompanied by her father, husband, or female relatives—might even indicate that she was not a respectable woman. Thus a particular behavior may be admired when judged by one set of values and norms, yet considered questionable or even despicable when judged by another.

If culture groups have different sets of values and standards, does that mean that there are no overarching human values or standards? This question increasingly worries geographers, who try to be sensitive both to the particularities of place and to larger issues of human rights. Those who lean too far toward appreciating difference could be led to the tacit acceptance of inhumane behavior, such as the oppression of minorities and women, or even torture and genocide. Acceptance of difference does not mean that we cannot make judgments about the value of certain extreme customs or points of view. Nonetheless, although it is important to take a stand against cruelty, deciding when and where to take that stand is rarely easy.

Religion and Belief Systems

The **religions** of the world are *formal* and *informal institutions* that embody value systems. Most have roots deep in history, and many include a spiritual belief in a higher power (God, Yahweh, Allah) as the underpinning for their value systems. These days, religions often focus on reinterpreting age-old values for the modern world. Some formal religious institutions—such as Islam, Buddhism, and Christianity—*proselytize;* that is, they try to extend their influence by seeking converts. Others, such as Judaism and Hinduism, accept converts only reluctantly. Informal religions, often called *belief systems,* have no formal central doctrine and no firm policy on who may or may not be a practitioner.

Religious beliefs are often reflected in the landscape. For example, settlement patterns often demonstrate the central role of religion in community life: village buildings may be grouped around a mosque (Figure 1.28) or synagogue, or an urban neighborhood may be organized around a Catholic church. In some places, religious rivalry is a major feature of the landscape.

Figure 1.28 Religion at the center of community life. This mosque, in Banja Luka, Serbia, is one of twenty-one mosques that were destroyed in the town during the 1992–1995 war. Rebuilding the mosque became a central goal of the Muslim community, as a symbol of their survival. [AFP/Getty Images.]

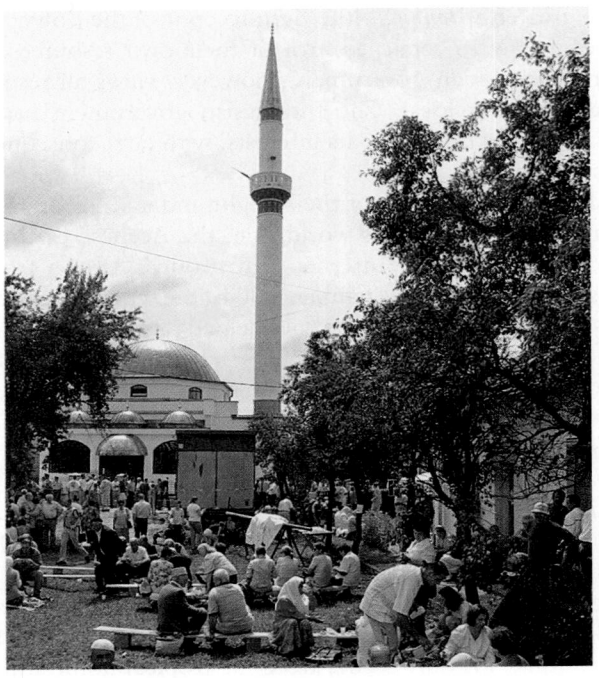

Certain spaces may be clearly delineated for the use of one group or another, as in Northern Ireland's Protestant and Catholic neighborhoods.

Religion has also been used to wield power. For example, during the era of European colonization, religion was a way to impose a change of attitude on conquered people. And the influence lingers. Figure 1.29, which shows the distribution of the major religious traditions on earth today, demonstrates some of the religious consequences of colonization. Note, for instance, the distribution of Roman Catholicism in the parts of the Americas, Africa, and Southeast Asia colonized by European Catholic countries.

Religion can also spread through trade contacts. In the seventh and eighth centuries, Islamic people used a combination of trade and political power (and less often, actual con-

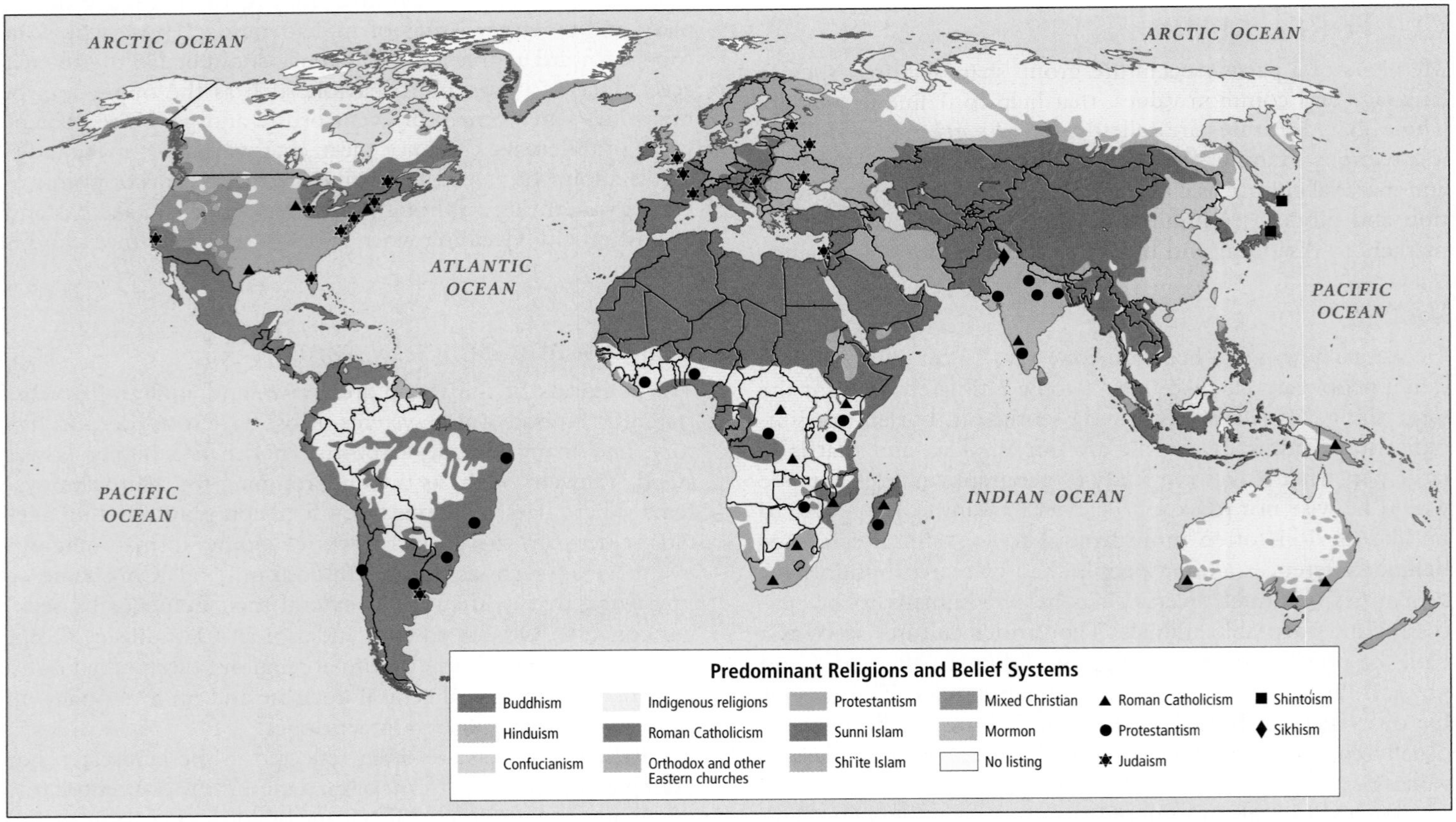
ARCTIC OCEAN
ARCTIC OCEAN
ATLANTIC OCEAN
PACIFIC OCEAN
PACIFIC OCEAN
INDIAN OCEAN
Predominant Religions and Belief Systems
Buddhism
Hinduism
Confucianism
Indigenous religions
Roman Catholicism
Orthodox and other Eastern churches
Protestantism
Sunni Islam
Shi'ite Islam
Mixed Christian
Mormon
No listing
▲ Roman Catholicism
● Protestantism
✱ Judaism
■ Shintoism
◆ Sikhism

Figure 1.29 Major religions around the world. The small symbols indicate a localized concentration of a particular religion within an area where another religion is predominant. [Adapted from *Oxford Atlas of the World* (New York: Oxford University Press, 1996), p. 27.]

quest) to extend their influence across North Africa, throughout Central Asia, and eventually into South and Southeast Asia.

The history and distribution of belief patterns throughout the world is complex. The distribution of major religions has changed many times over the course of history. Moreover, a world map is too small in scale to convey detailed religious spatial patterns, such as where two or more religious traditions intersect at the local level. And as the world's cultural traditions become increasingly mixed and urban life spreads, **secularism,** a way of life informed by values that do not derive from any one religious tradition, is spreading.

Language

Language is one of the most important criteria in delineating cultural regions. The modern global pattern of languages (Figure 1.30) reflects the complexities of human interaction

and isolation over several hundred thousand years. But the map does not begin to depict the actual details of language distribution. Between 2500 and 3500 languages are spoken on earth today, some by only a few dozen people in isolated places. Many languages have several dialects—regional variations in grammar, pronunciation, and vocabulary.

The geographic pattern of languages has continually shifted over time as people have interacted through trade and migration. The pattern changed most dramatically around 1500 when the age of European exploration and colonization began. From that point on, the languages of European colonists often replaced the languages of the colonized people. This is why we find large patches of English, Spanish, Portuguese, and French in the Americas, Africa, Asia, and Oceania. In North America, European languages largely replaced Native American languages. In Middle and South America, Africa, and Asia, by contrast, European and native tongues coexisted.

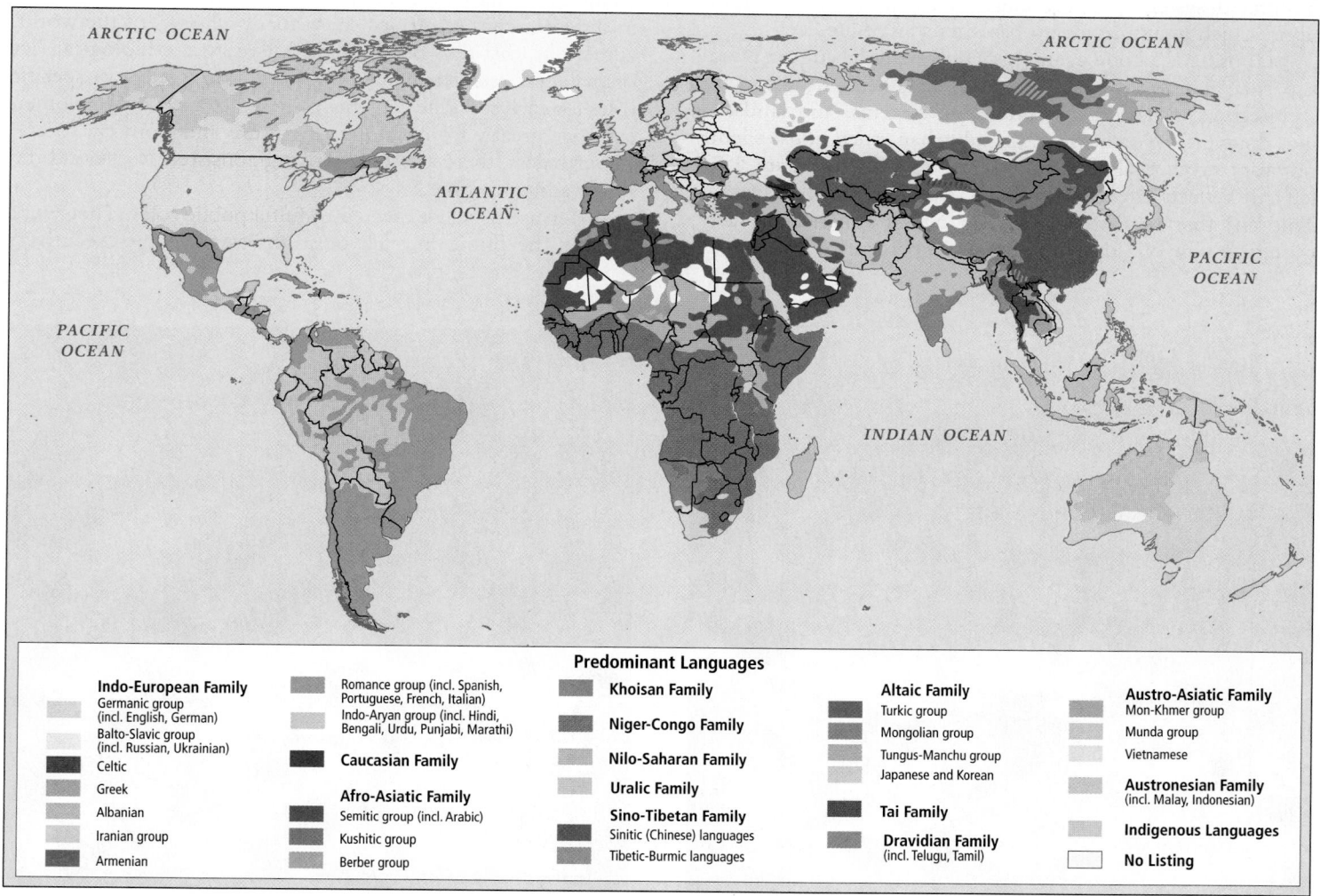

Figure 1.30 The world's major language families. Distinct languages (Spanish and Portuguese, for example) are part of a larger language group (Romance languages), which in turn is part of a language family (Indo-European). [Adapted from *Oxford Atlas of the World* (New York: Oxford University Press, 1996), p. 27.]

Many people became bilingual or trilingual. Today, with increasing trade and instantaneous global communication, a few languages have become dominant—Chinese, Hindi, Spanish, and English are the four main native languages spoken in the world today. At the same time, other languages are becoming extinct because children no longer learn them. 📹

Arabic is an important *lingua franca*, or language of international trade, as are English, Spanish, and Chinese. Among these, English dominates, largely because the British colonial empire introduced English as a second language to many places around the globe. U.S. economic influence is another factor that led to the dominance of English, and it is currently the second language of an estimated 1.5 billion people. The need for a common world language in the computerized information age is reinforcing English as the primary lingua franca.

Material Culture and Technology

A group's material culture reflects its *technology*, which is the integrated system of knowledge, skills, tools, and methods upon which a culture group bases its way of life.

Housing is a good example of how material culture reveals a particular culture group's way of life. With its distinctive architecture, electrical and plumbing systems, and landscaping, the typical North American suburban ranch house silently reveals a great deal about the culture's values (Figure 1.31, left). It reflects the nuclear family structure (mother, father, children) that remains an ideal in North American society, though it now constitutes less than 25 percent of the region's

families. The ranch house also reflects values about privacy (multiple bedrooms) and gender roles (Mom's special spaces may be the kitchen and laundry room; Dad's, the TV room, the garage, and the toolshed). This house embodies a certain level of affluence and leisure, equipped as it is with labor-saving devices and conveniences and set apart by a green lawn requiring constant maintenance. It also symbolizes ideas about private property, polite neighborliness, and mobility.

An astute foreign observer could learn a great deal about North Americans simply by "reading" the material culture of their homes and surrounding landscapes. Similarly, you could learn about another culture, such as Mongolia, by closely observing its homes. What does Figure 1.31, right, teach you about that culture's notions of proper family structure, gender roles, intimacy rules, aesthetic values, property rights, use of resources, and trading patterns?

Gender Issues

Geographers have begun to pay more attention to gender roles in different culture groups. In virtually all parts of the world, and for at least tens of thousands of years, the biological fact of maleness and femaleness has been translated into specific roles for each sex. The activities assigned to men and to women can vary greatly from culture to culture and from era to era. Nevertheless, there are some striking consistencies around the globe and over time.

Men are usually expected to fulfill public roles. They work outside the home in such positions as traveling executives,

Figure 1.31 Material culture. This home in Canyon, California (left), belongs to the Cavin family. They are pictured (in the foreground) with their possessions. [Peter Menzel/Material World.] The six members of the Batsuur family live in a *ger* (right), the traditional tentlike Mongolian house that can easily be dismantled and moved to another place. The house has electricity (for the hot plate and television) and a coal-burning stove. [Leong Ka Tai and Peter Menzel/Material World.]

animal herders, hunters, or government workers. Women are usually expected to fulfill private roles. They keep house, bear and rear children, care for the elderly, and prepare the meals, among many other tasks. In nearly all cultures, women are defined as dependent on men—their fathers, husbands, brothers, or adult sons—even when the women may produce most of the family sustenance.

Because their activities are focused on the home, women typically have less access to education and paid employment, and hence less access to wealth and political power. When they do work outside the home (as is the case increasingly in every world region), women tend to fill lower-paid positions, whether as laborers, service workers, or professionals. Despite working, women retain their household duties.

Gender—the sexual category of a person—is both a biological and a cultural phenomenon. Men have larger muscles,

> Gender—the sexual category of a person—is both a biological and a cultural phenomenon.

can lift heavier weights, and can run faster than women (but not necessarily for longer periods). In some physical exercises, the average woman has more endurance and are capable of more precise movements than the average man. In populations that enjoy overall good health, women tend to outlive men by an average of 3 to 5 years.

Women's physical capabilities are somewhat limited during pregnancy and nursing—and, for some, during menstruation—but from the age of about 45, women are no longer subject to these limits. Most contribute in some significant way to the well-being of their adult children and grandchildren. A growing number of evolutionary biologists postulate that the evolutionary advantage of menopause in midlife is that it gives women the time and energy and freedom to help succeeding generations thrive—an idea sometimes labeled the grandmother hypothesis.

Although average physical gender differences exist, in most cultures they carry greater social significance than the biological facts would warrant. Customary ideas about masculinity and femininity, proper gender roles, and sexual orientation are handed down from generation to generation and have enormous effects on the everyday lives of men, women, and children. Perhaps more than for any other culturally defined human characteristic, significant agreement exists across places and over time that gender is important.

The historical and modern global gender picture is a puzzlingly negative one for women. In nearly every culture, in every region of the world, and for a great deal of recorded history, women have had (and still have) an inferior status. It is hard to find exceptions, although the intensity of this second-class designation varies considerably. People of both sexes routinely accept the idea that males are more productive and intelligent than females. In nearly all cultures, families prefer boys over girls because, as adults, boys have greater earning capacity (Table 1.1), have more power in society, and will perpetuate the family name (because of patrilineal naming customs). Around the globe, females have less access to food,

Table 1.1 Comparisons of male and female income in countries where average education levels are higher for females than for males

Country	Female income (PPP[a] U.S.$, 2003)	Male income (PPP U.S.$, 2003)	Female income as percent of male income
Austria	15,878	45,174	35
Barbados	11,976	19,687	61
Canada	23,922	37,572	64
Japan	17,795	38,612	46
Jordan	2,004	6,491	31
Kuwait	8,448	24,204	35
Poland	8,769	14,147	62
Russia	7,302	11,429	62
Saudi Arabia	4,440	20,717	21
Sweden	21,842	31,722	69
United Kingdom	20,790	33,713	62
United States	29,017	46,456	62

[a]PPP, purchasing power parity, is the amount that the local currency equivalent of U.S.$1 will purchase in a given country.

Source: United Nations Human Development Report 2005 (New York: United Nations Development Programme), Table 25, "Gender-related development index," pp. 299–302, and Table 27, "Gender inequality in education," pp. 307–310.

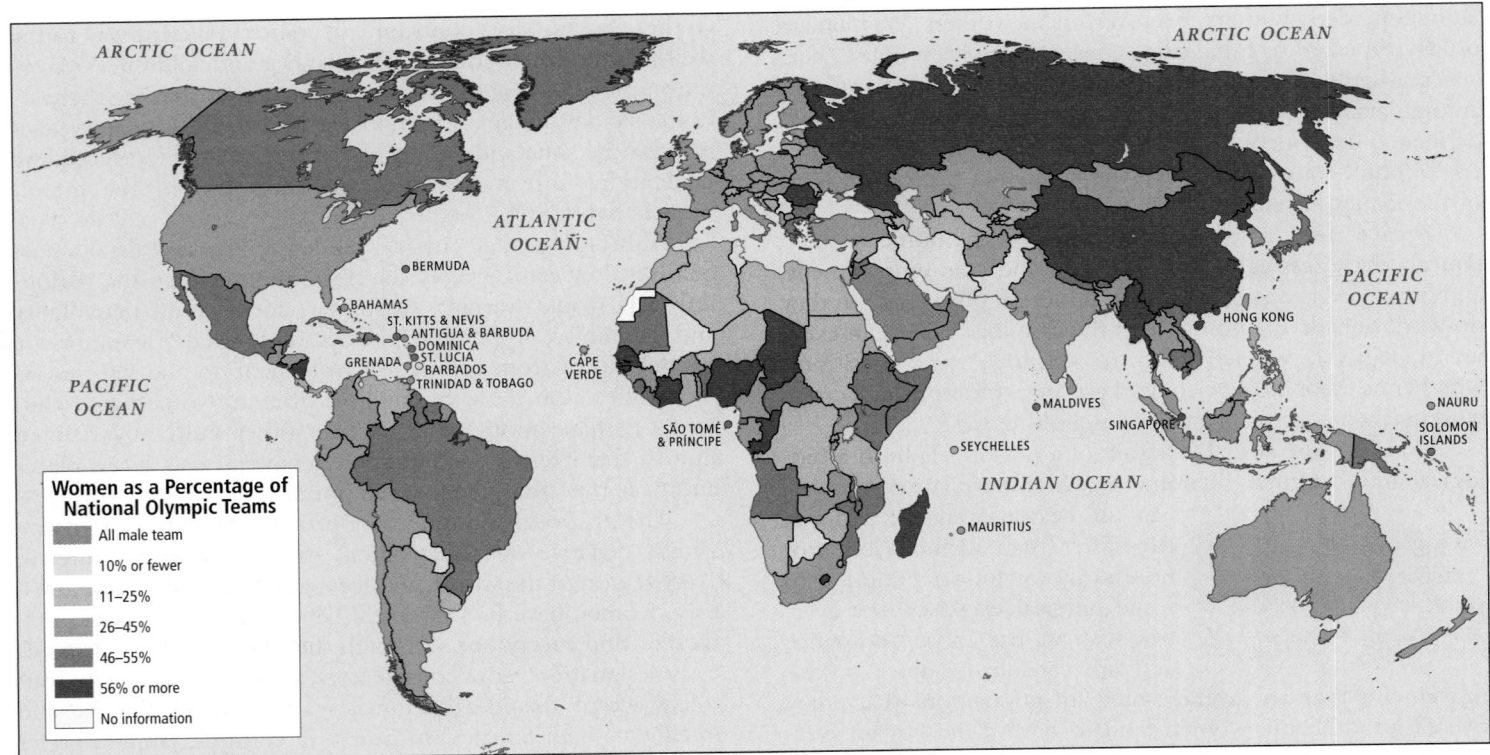

Figure 1.32 Women's participation in the 2004 Olympics. Traditional notions of femininity, female roles, and female strength, speed, and endurance are being strongly challenged by Olympic athletes. In the first modern Olympic Games in 1896, women were barred from participating. In 2004, women made up 44 percent of the Olympic competitors. In some countries (China, for instance), more than 50 percent of the athletes were women. [Adapted from Joni Seager, *The State of Women in the World Atlas* (New York: Penguin, 2003), pp. 50–51; updated from http://multimedia.olympic.org/pdf/en_report_1000.pdf.]

medical care, and education. They start work at a younger age and work longer hours than males. The question of how and why women became subordinate to men has not yet been well explored because, oddly enough, few thought the question significant until recently.

There are growing challenges to traditional notions of femininity. In many countries around the world, females are now acquiring education at higher rates than males. This fact should eventually make women competitive with men for jobs. Unless discrimination persists, women should also begin to earn pay equal to that of men. Between 2000 and 2004, female athletes increased their participation in the summer Olympic games from 30 to 40 percent of the competitors. In some countries (China, for example), more than half the competitors were female (Figure 1.32).

Yet looking only at women's woes misses half the story. Men are also disadvantaged by strict gender expectations. For most of human history, young men have borne the lion's share of burdensome physical tasks and dangerous undertakings. Until recently, usually only young men left home to migrate to distant, low-paying jobs. Overwhelmingly, it is young men who die in wars or suffer physical and psychological injuries from combat.

This book will return repeatedly to the question of gender disparities in an effort to investigate this most perplexing cross-cultural phenomenon. Our examination will also reveal that many societies are addressing gender inequalities and that, in every country on earth, what it means to be a man or a woman is being renegotiated, even if only in small ways.

Race

Like ideas about gender roles, ideas about race affect human relationships everywhere on earth. However, according to the science of biology, all people now alive on earth are members of one species, *Homo sapiens sapiens*. The popular markers of what we call **race** (skin color, hair texture, face, and body shapes) have no biological significance. For any supposed racial trait, such as skin color, there are wide variations within human groups. In addition, many invisible biological characteristics, such as blood type and DNA patterns, cut across skin color distributions and are shared by what are commonly viewed as different races. In fact, over the last several thousand years there has been such massive gene flow among human populations that no modern group presents a discrete set of biological characteristics. Although we may look quite different, biologically speaking we are all closely related.

It is likely that some of the easily visible features of particular human groups evolved to help them adapt them to environmental conditions. For example, biologists think that darker skin evolved in regions close to the equator, where sunlight is most intense. All humans need the nutrient vitamin D, and sunlight striking the skin helps the body to produce vitamin D. Too much of the vitamin, however, can result in improper kidney functioning. Dark skin produces less vitamin D than light skin and thus would be a protective adaptation in equatorial zones. In higher latitudes, where the sun's rays are more dispersed, sufficient vitamin D can be produced by light skin without danger.

Many physical characteristics, such as big ears, deep-set eyes, or high cheekbones, do not serve any apparent adaptive purpose. They are probably the result of random chance and ancient inbreeding within isolated groups. Similarly, there is no evidence that any "race" has particularly high math ability or athletic ability. Such characteristics are present in individuals in all human populations and may become enhanced by cultural practices.

While race may be biologically meaningless, it has aquired enormous social and political significance. Over the last sev-

eral thousand years, humans from different parts of the world have increasingly encountered each other in situations of unequal power. Some researchers have suggested that European colonizers adopted *racism*—the negative assessment of unfamiliar, often darker-skinned, people—to justify taking land and resources away from those supposedly inferior beings. But racism is hardly peculiar to Europe. Humans have long committed atrocities against their own kind, often in the name of race, ethnicity, or even gender. Race and its implications in North America will be a focus in Chapter 2, and the topic will be discussed in several other world regions as well.

While recognizing all the ills that have emerged from racism and similar prejudices, we should not infer that human history has been marked primarily by conflict and exploitation. Actually, humans have probably been so successful because of a strong inclination toward *altruism*, the willingness to sacrifice one's own well-being for the sake of others. On a small scale, altruism can be found in the sacrifices individuals make to help family, neighbors, and community. On a larger scale, it includes charitable giving to help anonymous people in need. It is probably our capacity for altruism that causes us such deep distress over the relatively infrequent occurrences of inhumane behavior.

Measures of Development

Until recently, the term *development* was used to describe only economic changes that lead to better standards of living. These changes often accompany the greater productivity in agriculture and industry that comes from such technological advances as mechanization and computerization. However, merely raising national productivity may benefit primarily those who are already economically well off. The majority may be left in circumstances that are little improved, or even worsened. Furthermore, development as measured by economic gains often results in environmental side effects that reduce the quality of life for everyone.

Some development experts (for example, the Nobel Prize–winning economist from India, Amartya Sen) advocate a broader definition of development that includes measures of **human well-being** (a healthy and socially rewarding standard of living), as well as measures of environmental quality. This perspective is elaborated on below.

GROSS DOMESTIC PRODUCT PER CAPITA

The most popular economic measure of development is **gross domestic product (GDP) per capita,** which is the total value of all goods and services produced in a country in a given year, divided by the number of people in the country. This figure

(Table 1.2, column II) is often used as a crude indicator of how well people are living in a given country.

There are, however, several problems with GDP per capita as a measure of overall well-being. First is the matter of wealth distribution. Because GDP per capita is an *average*, it can hide the fact that a country has a few fabulously rich people and a mass of abjectly poor people. For example, a GDP per capita of U.S.$20,000 would be meaningless if a few lived on millions per year and most lived on $5000 per year.

Second, the purchasing power of currency varies widely around the globe. A GDP of U.S.$15,000 per capita in Barbados might represent a middle-class standard of living, whereas that same amount in New York City could not buy even basic food and shelter. Because of these purchasing power variations, in this book we use GDP per capita figures that have been adjusted for **purchasing power parity (PPP).** PPP is the amount that the local currency equivalent of U.S.$1 will purchase in a given country. For example, according to *The Economist,* on January 12, 2006, a Big Mac at McDonald's in the United States cost U.S.$3.15. In Australia it cost the equivalent of U.S.$2.44, and in China it cost just U.S.$1.30. Of course, for the consumer in China, where annual per capita income averages about $4000, this would be a very expensive meal.

Table 1.2 Sample human well-being table[a]

Ranking among Selected countries (I)	GDP per capita, adjusted for PPP[b] 177 countries, 2005 (II)	Human Development Index (HDI) 177 countries, 2005[c] (III)	Gender Development Index (GDI) 140 countries, 2005 (IV)	Gender Empowerment Measure (GEM) 80 countries, 2005 (V)
Barbados	15,720 (36)	30 (high)	29	25
Japan	27,967 (13)	11 (high)	14	43
Kenya	1,037 (155)	154 (low)	117	ND
Kuwait	18,047 (30)	44 (high)	39	ND
Malaysia	9,512 (54)	61 (medium)	61	51
United States	37,562 (4)	10 (high)	8	12

[a]Rankings are in descending order; i.e., low numbers indicate high rank.

[b]PPP = purchasing power parity, figured in 2003 U.S. dollars.

[c]The high, medium, and low designations indicate where the country ranks among the 177 countries classified into three categories by the United Nations.

ND = No data available.

Sources: United Nations Human Development Report 2005 (New York: United Nations Development Programme), Table 1, "Human development index," Table 25, "Gender-related development index," and Table 26, "Gender empowerment measure," at http:/hdr.undp.org/reports/global/2005/pdf/HDR05_complete.pd

A third problem with GDP per capita is that it measures only what goes on in the formal economy. In some places, however, the informal economy actually accounts for more activity. Researchers who examined all types of societies and cultures have shown that, on average, women perform about 60 percent of all the work done, much of it unpaid. Nonetheless, only their paid work performed in the formal economy appears in the statistics. Statistics also neglect the contributions of millions of men and children who work in the informal economy as subsistence farmers, traders, service people, or seasonal laborers.

The most important failing of GDP per capita as a measure for comparing countries is that it ignores all aspects of development other than economic ones. For example, there is no way to tell from GDP per capita figures how fast a country is consuming its natural resources, or how well it is educating its young or maintaining its environment. Therefore, along with the traditional GDP per capita figure, geographers are increasingly using several other measures of development (see Table 1.2). These include the United Nations Human Development Index (HDI), the United Nations Gender Development Index (GDI), and the United Nations Gender Empowerment Measure (GEM). Together, these measures reveal some of the subtleties and nuances of well-being and make comparisons between countries somewhat more valid. Because these more sensitive indexes are also more complex than the purely economic GDP per capita, they are all still being refined by the United Nations.

Population Geography

The changing levels of well-being associated with economic development have had dramatic effects on population patterns, such as growth, decline, and distribution. Because geographers are concerned with the interaction between people and their environments, **demography,** the study of population patterns and changes, is an important part of geographic analysis.

GLOBAL PATTERNS OF POPULATION GROWTH

It took between 1 million and 2 million years, or at least 40,000 generations, for humans to evolve and to reach a population of 2 billion. This happened in about 1945. Then, in just 55

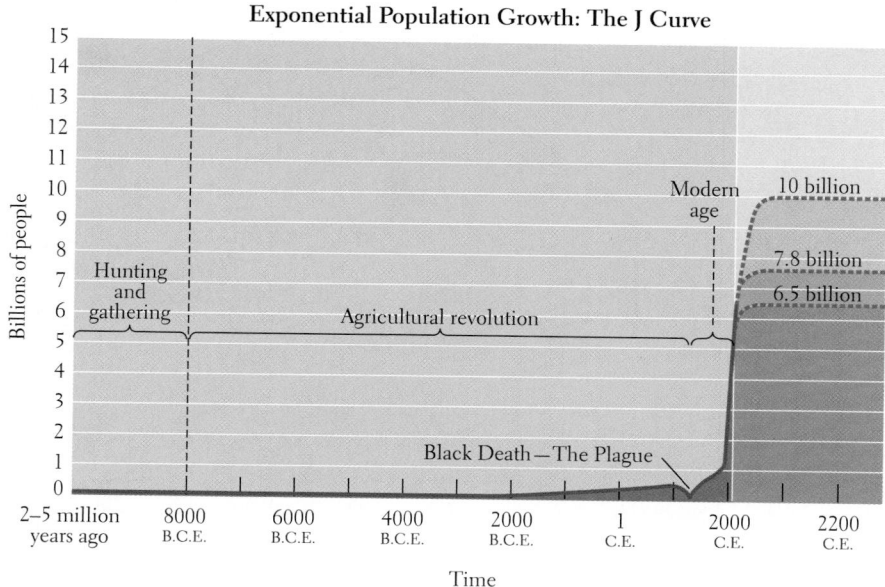

Exponential Population Growth: The J Curve

Figure 1.33 Exponential growth of the human population. The curve's J shape is a result of successive doublings of the population: it starts out nearly flat, but as doubling time shortens, the curve bends ever more sharply upward. Note that B.C.E. (before the common era) is equivalent to B.C. (before Christ); C.E. *(common era)* is equivalent to A.D. *(anno domini)*. [Adapted from G. Tyler Miller, Jr., *Living in the Environment*, 8th ed. (Belmont, Calif.: Wadsworth, 1994), p. 4.]

years, by the year 2000, the world's population more than tripled to 6.1 billion (Figure 1.33). What happened to make the population grow so quickly in such a short time?

The explanation lies in changing relationships between humans and the environment. For most of human history, fluctuating food availability, natural hazards, and disease kept human death rates high, especially for infants. Periodically, there would even be crashes in human population—as happened in the 1300s throughout Europe and Asia during the pandemic known as the Black Death.

An astonishing upsurge in human population began about 1500, at a time when the technological, industrial, and scientific revolutions were beginning in some parts of the world. Human life expectancy increased dramatically, and more and more people lived long enough to reproduce successfully, often many times over. The result was an exponential pattern of growth (see Figure 1.33). This pattern is often called a J curve because the ever-shorter periods between doubling and redoubling of the population cause an abrupt upward swing in the growth line when depicted on a graph.

Today, the human population is growing in virtually all regions of the world, but more rapidly in some places than in others. Even if all couples agreed to have only one child, the world population would probably grow to at least 7.8 billion before zero growth would set in. The reason is simply that there are currently many people who are about to reach the age of reproduction.

Nevertheless, there are indications that the *rate* of growth is slowing globally. In 1993, the world's population was 5.5 billion, and the world growth rate was 1.7 percent per year. By 2005, the world's population was 6.5 billion, but the growth rate had decreased to 1.2 percent. If present slower growth trends continue, the world population may level off at about 7.8 billion before 2050. However, this projection is con-

tingent on couples in less-developed countries having access to birth control technology and information on why smaller families might be advisable for them.

In a few countries, especially in Central Europe, the population is actually declining and aging, due to low birth and death rates. This situation could prove problematic as younger workers become scarce and those who are elderly and dependent become more numerous and pose a financial burden to working-age people. HIV–AIDS is affecting population patterns to varying extents in all world regions. In Africa, the epidemic is severe; several countries are beginning to show sharply lowered life expectancies, and some countries are beginning to show declines in population. The effects of HIV–AIDS are discussed in several world region chapters.

LOCAL VARIATIONS IN POPULATION DENSITY AND GROWTH

If the more than 6.5 billion people on earth today were evenly distributed across the land surface, they would produce an **average population density** of about 113 people per square mile (43.6 per square kilometer). People are not evenly distributed across the face of the earth, however (Figure 1.34). Nearly 90 percent of all people live north of the equator, and most of them live between 20°N and 60°N. Even within that limited territory, people are concentrated on about 20 percent of the available land. They live mainly in zones that have climates warm and wet enough to support agriculture, along rivers, in lowland regions, or fairly close to the sea. In general, people are located where resources are available.

Usually, the variable that is most important for understanding population growth in a region is the **rate of natural increase** (often called *growth rate*). The rate of natural increase

Figure 1.34 World population density. [Data courtesy of Deborah Balk, Gregory Yetman, et al., Center for International Earth Science Information Network, Columbia University; at http://www.ciesin.columbia.edu.]

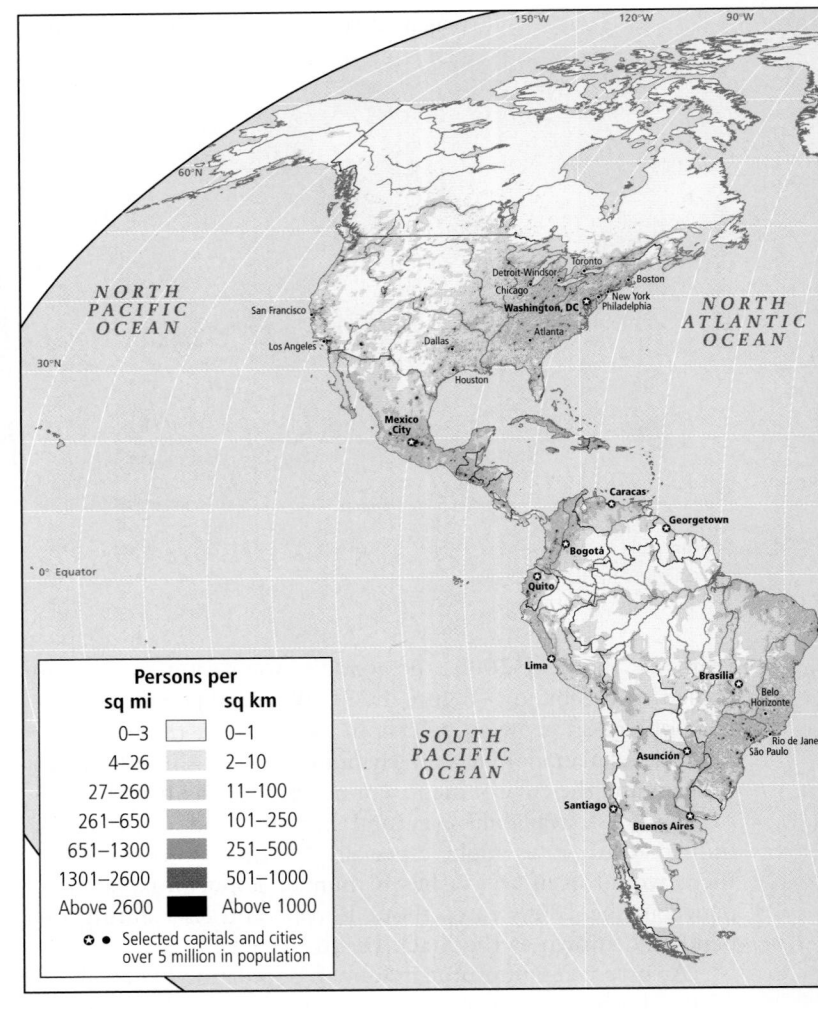

is the relationship between the number of people being born (**birth rate**) and the number dying (**death rate**) in a given population, without regard to the effects of migration.

The rate of natural increase is usually expressed as a percentage. For example, in Austria (in Europe), which has 8,200,000 people, the annual birth rate in 2005 was 10 per 1000 people, and the death rate was 9 per 1000 people. Therefore, the annual rate of natural increase is 1 per 1000 (10 − 9 = 1), or 0.1 percent. Only 16 percent of Austrians are under 15 years of age, and 15 percent are over 65 years of age.

For comparison, consider Jordan (in Southwest Asia), which has 5,800,000 people. The birth rate is 29 per 1000 and the death rate is 5 per 1000. The annual rate of natural increase is thus 24 per 1000 (29 − 5 = 24), or 2.4 percent per year. At this rate, Jordan will double its population in just 29 years. As you might expect, the population of Jordan is very young: 37 percent are under 15 years of age, and only 3 percent are over 65.

Total fertility rate (TFR) is another term used to indicate trends in population. TFR is the average number of children

women in a country are likely to have at the present rate of natural increase.

Migration can also be a powerful contributor to population growth. Of all the world's regions, this is most true in Europe. Though the rate of natural increase is generally quite low, the region's economic growth attracts immigrants from throughout the world. In 2005, international migration accounted for 85 percent of the European Union's population growth.

AGE AND GENDER STRUCTURES

The age distribution, or **age structure**, of a population is the proportion of the total population in each age-group. The **gender structure** is the proportion of males and females in each age-group. Age and gender structures reflect past and present social conditions, and knowing these structures helps us to predict future population trends.

The **population pyramid** is a graph that depicts age and gender structures. Careful study of the population pyramids

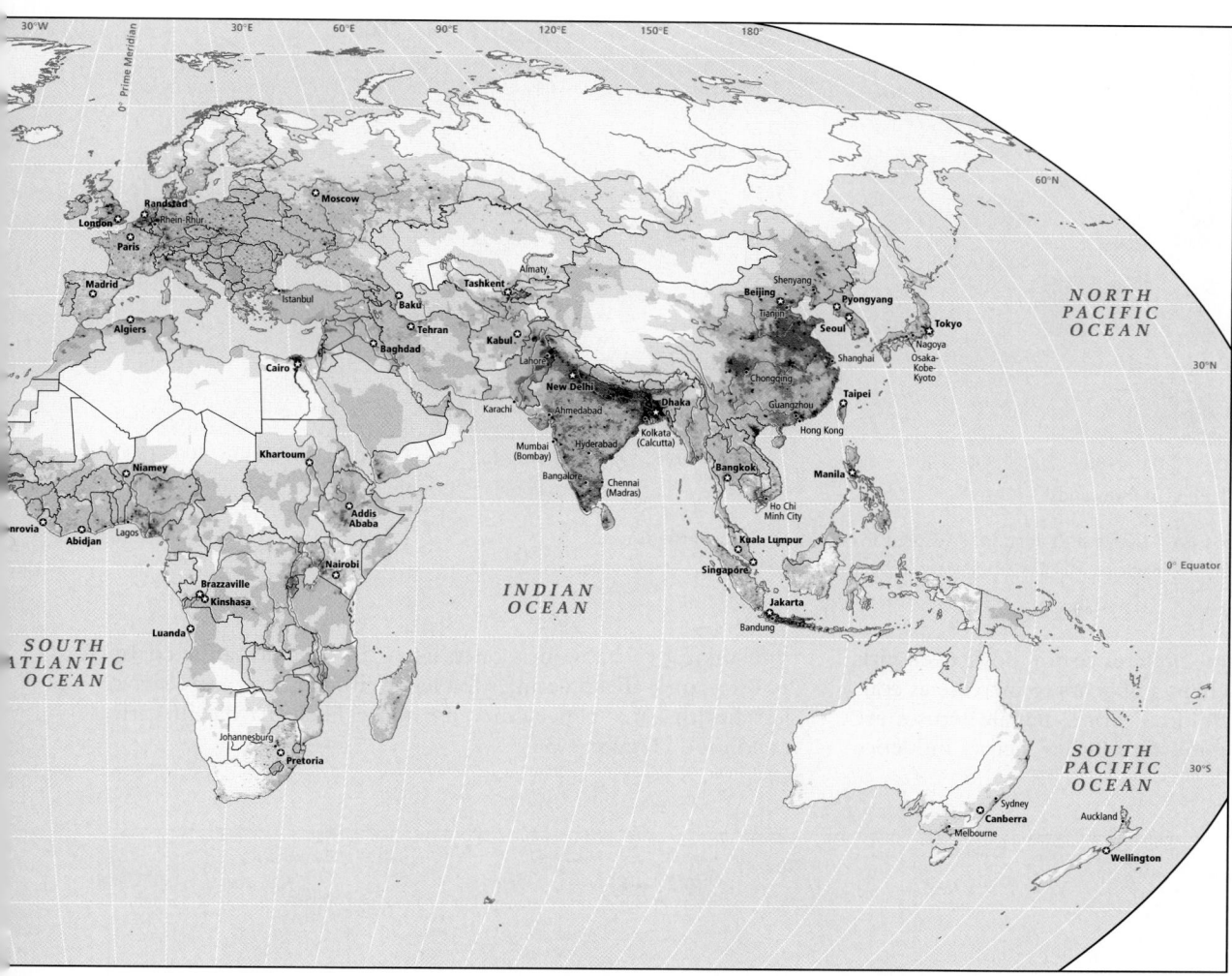

of places such as Austria and Jordan (Figure 1.35) reveals the age and gender distribution of the populations of those countries. As we have noted, nearly 40 percent in Jordan are very young. You can see that they are clustered toward the bottom of the pyramid, with the largest groups in the age categories 0 through 9. The pyramid tapers off quickly as it rises above age 34, showing that in Jordan most people die before they reach old age.

In contrast, Austria's pyramid has an irregular shape, indicating alternately increased or decreased population growth. The narrow base indicates that there are now fewer people in the younger age categories than in young adulthood or middle age, and those over 70 greatly outnumber the youngest (ages 0 to 4). This age distribution tells us that many Austrians now live to an old age. It also illustrates that in the last several decades Austrian couples have been choosing to have only one child, or none. Austrians, like many Europeans, now worry that their population will be weighted with elderly people who will need care and financial support from an ever-declining group of working-age people.

Population pyramids also reveal subtle gender differences within populations. Look closely at the right (female) and left (male) halves of the pyramids in Figure 1.35. In several age categories the sexes are not evenly balanced on either side of the line. In the Austrian pyramid, there are more women than men near the top (age 70 and older). This reflects both the deaths among male soldiers in World War II and the as-yet poorly understood trend in wealthy countries for women to live about 5 years longer than men. In Jordan, the gender imbalances occur at younger ages. For instance, there are about 230,000 women and about 290,000 men age 25 to 29.

Because gender-based research is so new, explanations for the imbalances are only now being proposed. The normal ratio worldwide is for about 95 females to be born for every 100 males. Because baby boys are somewhat weaker than girls, the ratio evens out naturally within the first 5 years. Over the last 100 years, the ratio of females to males at birth has declined further to 92.6 females to 100 males globally. The ratio is as low as 80 females to 100 males in parts of South and East

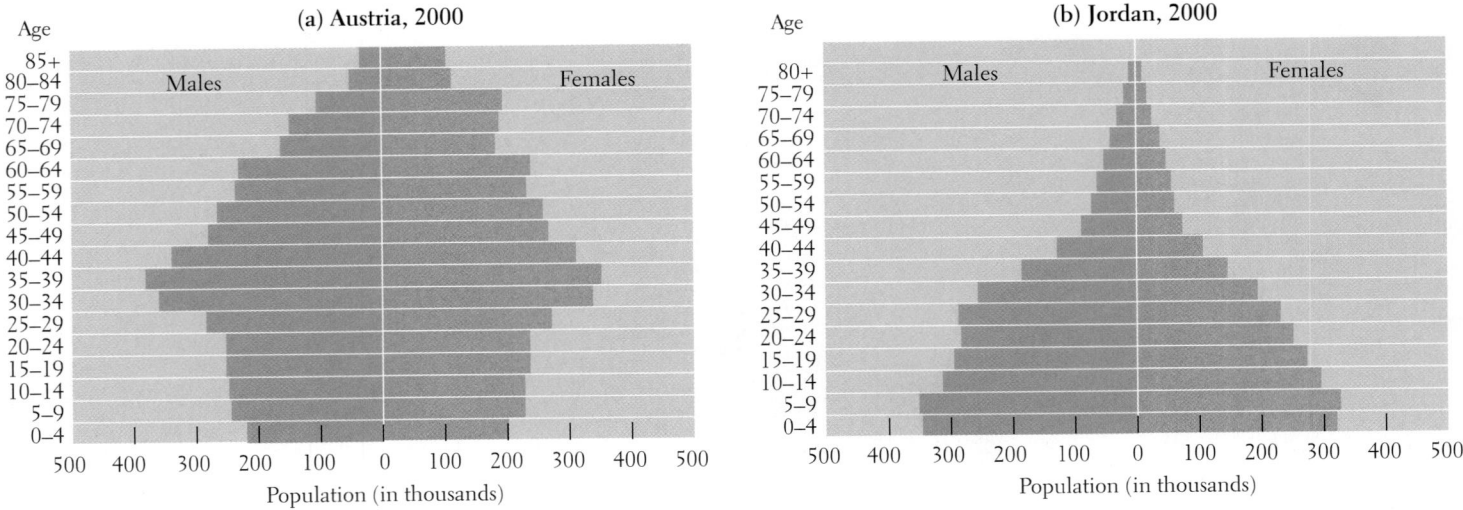

Figure 1.35 Population pyramids for Austria and Jordan. [Adapted from U.S. Bureau of the Census, International Data Base (IDB), at http://www.census.gov/ipc/www/idbpyr.html.]

Asia. The widespread cultural preference for boys over girls that we noted earlier in this chapter appears to increase as couples choose to have fewer children. Some female fetuses are never born because they are aborted. Other factors influence the survival of baby girls once born. In societies afflicted by poverty, girls are sometimes fed less well than boys and receive less health care; hence, they are more likely to die in early childhood (Figure 1.36).

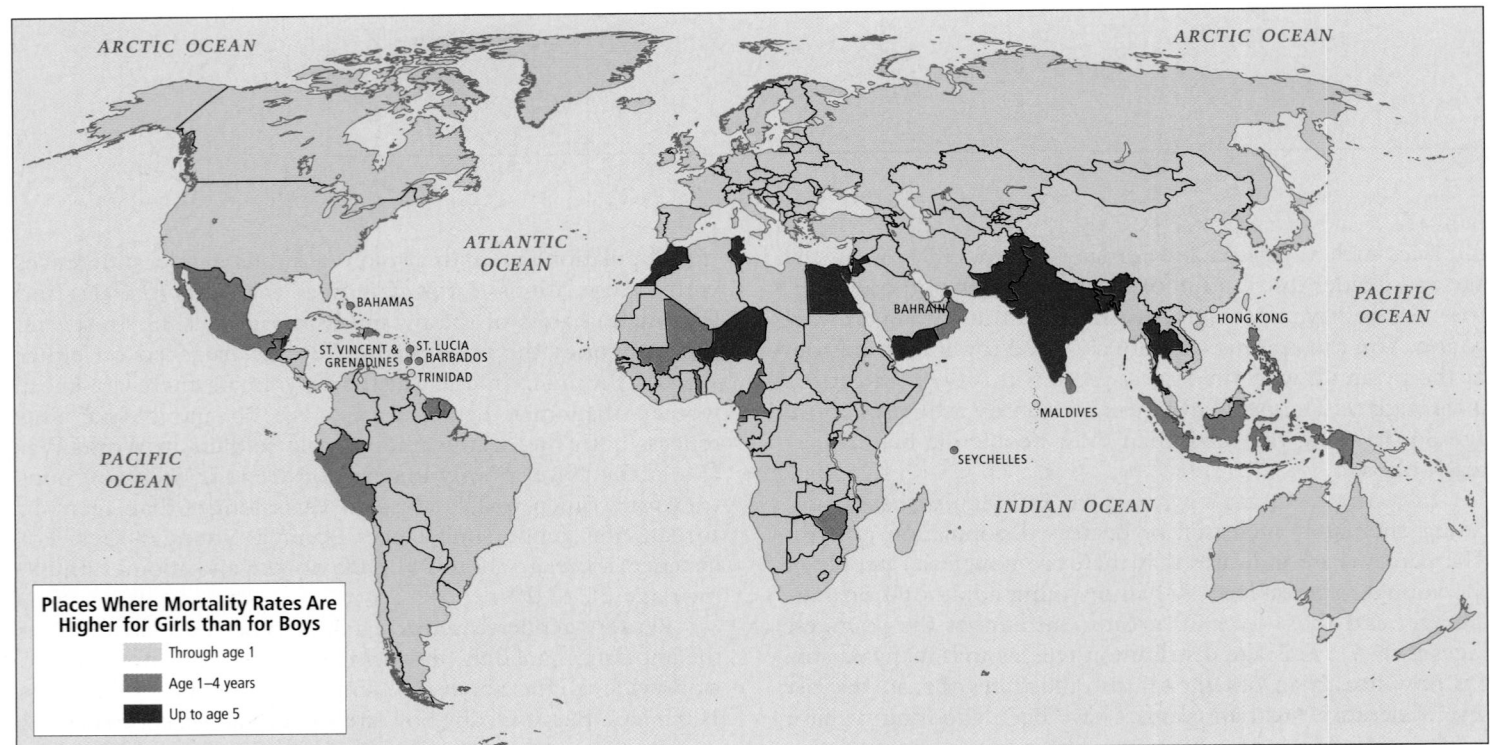

Figure 1.36 Young female mortality rates. In the countries shown in shades of purple, the mortality rate for girls is abnormally high. The darker the color, the longer the risk to girls lasts. [Adapted from Joni Seager, *The State of Women in the World Atlas* (New York: Penguin, 1997), p. 35.]

POPULATION GROWTH RATES AND WEALTH

Although there is a wide range of variation, regions with slow population growth rates usually tend to be affluent, and regions with fast growth rates tend to have widespread poverty. The reasons are complicated; again, Austria and Jordan serve as examples.

Austria has a GDP per capita (PPP) of $30,094 and a low infant mortality rate of 5 per 1000. Its highly educated population is 100 percent literate and is employed largely in high-tech industry and services. Large amounts of time, effort, and money are required to educate a child to compete in this economy. Moreover, it is highly likely that a child will survive to adulthood and hence be available to care for aging parents. As a result, many couples to choose to have only one or two children.

By contrast, Jordan has a GDP per capita (PPP) of $4320 and a high infant mortality rate of 22 per 1000. Much of the everyday work is still done in the informal economy by hand, making each new child a potential contributor to the family income at a young age. There is also a much greater risk of children not surviving into adulthood, so having more children ensures that someone will be there to provide care for aging parents.

However, the situation is changing in Jordan. Just 25 years ago, the GDP per capita was $993, infant mortality was 77 per 1000, and the average woman had 8 children. Now she has on average only 3.7 children.

Geographers would say that Jordan is going through the **demographic transition,** meaning that a period of high population growth rate is giving way to a period of much lower (or no) growth rate (Figure 1.37). This transition occurs gradually as people make the shift from subsistence to cash economies.

In a **subsistence economy,** a family produces most of its own food, clothing, and shelter, so many children to share the work are an asset. Subsistence economies are disappearing today because cash is needed to buy such goods as television sets and canned goods. In **cash economies,** there is less need for many people to do agricultural or other subsistence labor. Instead, there is a demand for fewer, well-trained specialists. Each child needs years of education to qualify for a good cash-paying job and does not contribute to the family budget while in school. Having many children, therefore, is a drain on the family's resources. Perhaps most important, cash economies are more likely to have better health care, increasing the likelihood that each child will survive to adulthood.

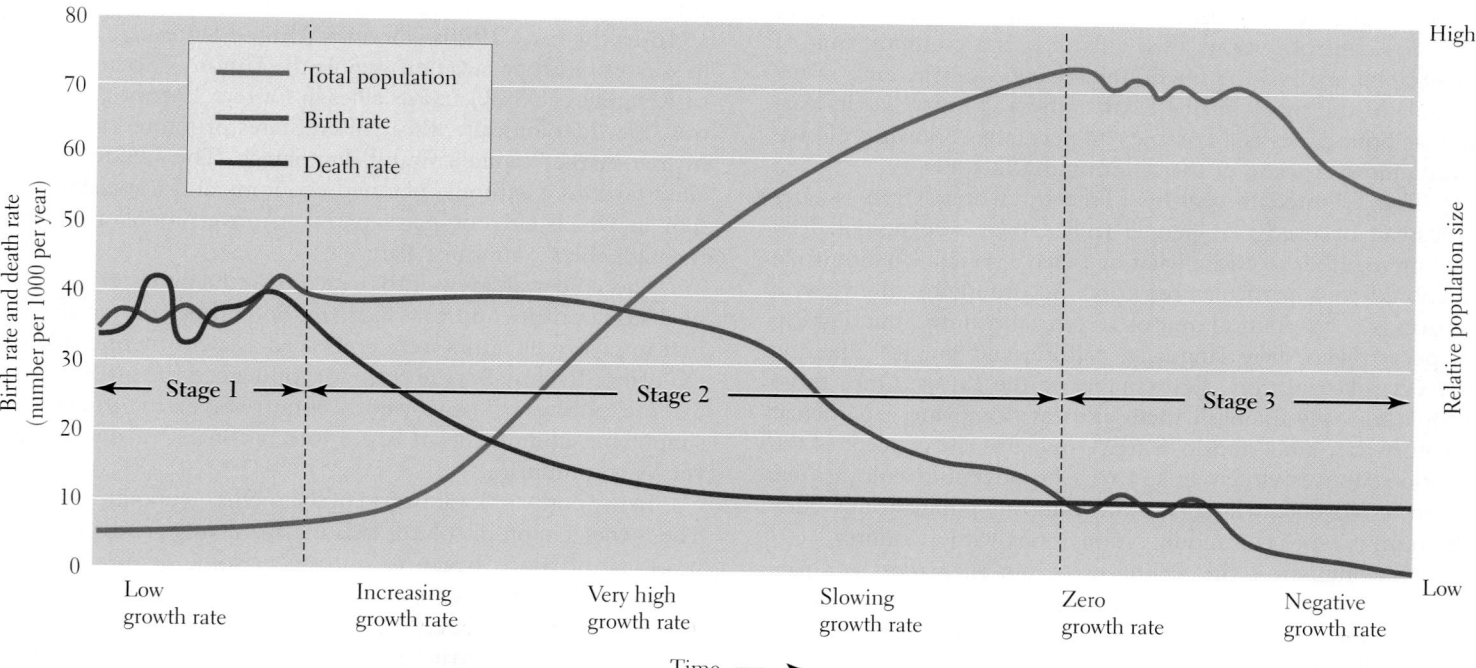

Figure 1.37 The demographic transition. In traditional societies (Stage 1), both birth rates and death rates are usually high (left vertical axis), and population numbers (right vertical axis) remain low and stable. With advances in food production, education, and health care (Stage 2), death rates usually drop rapidly. However, strong cultural values regarding reproduction remain, often for generations, so birth rates drop much more slowly. The result is that for decades or longer, the population continues to grow significantly.

Eventually, changed social and economic circumstances enable most children to survive to adulthood and it is no longer necessary to produce a cadre of family labor (Stage 3). Population growth rates slow and may eventually drift into negative growth. At this point, demographers say that the society has gone through the demographic transition. [Adapted from G. Tyler Miller, Jr., *Living in the Environment,* 8th ed. (Belmont, Calif.: Wadsworth, 1994), p. 218.]

Political Geography

Power—its exercise, its allocation to different segments of society, and its spatial distribution within and among world regions—is the interest of political geography. An issue commonly discussed in geography is **geopolitics**—the jockeying among countries for territory, resources, or influence. **Democratization**, or the transition toward political systems guided by competitive elections, is an area of increasing interest for political geographers. Political geographers also look at how these issues are addressed at smaller scales, such as local and state governments, and how international organizations and social movements play a role in the allocation of power.

> Democratization, or the transition toward political systems guided by competitive elections, is an area of increasing interest for political geographers.

COUNTRIES, NATIONS, AND BORDERS

In everyday speech, we use many terms to describe areas of the globe, often imprecisely. It is helpful to clarify the meaning of these terms from a geographer's perspective.

In this book, you will see regions defined primarily as a set of countries. A **country** is a political division of territory that has control over its own affairs. It is a common unit of geographic analysis and the boundaries of countries are a major feature of the maps in this book. Why a country has its particular boundaries is a complex subject; the following discussion addresses some of the determining factors.

The eighteenth-century French writer Jean-Jacques Rousseau believed that people realize their true potential by joining together to create what he called a cohesive nation-state anchored to a particular territory. According to this view, a **nation** is not a political unit or an official country, but a group of people who share language, culture, and political identity: the Cherokee Nation, for example, or the Kurds. Once those people formally establish themselves as occupying a particular territory they are a **nation-state.** Originally, the concept of the nation-state was closely linked to a homogenous culture (like Japan or Slovenia, which have tiny minority populations). Now the term is more commonly synonymous with a country.

Nationalism is the feeling of strong allegiance to a particular country. Nationalism developed in Europe, spreading to the Americas and beyond. Cultural diversity usually prevented the establishment of culturally homogenous nation-states in the former colonies of Europe, but nearly all became independent countries eventually. Some of the new countries that emerged from colonies, such as Indonesia, were **pluralistic states** in which power was shared among several distinct culture groups. Often one group managed to monopolize power, leaving others at a disadvantage.

As countries were created where none had existed before, territory traditionally occupied by indigenous peoples was often claimed by the new national governments. The indigenous groups were left without a homeland, relocated to a tiny section of their original territory, or forced to migrate to entirely new territory. Such was the case with almost all the indigenous peoples of North, South, and Middle America; Australia; and New Zealand. In many cases, indigenous groups remain in conflict with national governments over rights to territory.

The emergence of the idea of the nation-state was linked with the concept of **sovereignty.** A sovereign country is self-governing (though not necessarily according to democratic principles) and can conduct its internal affairs as it sees fit without interference from outside. Sovereignty increased the importance of precise legal boundaries as demarcations of power and control. A country's borders mark the extent of its territory, and hence its sovereignty. When we look at a map, we often are viewing a spatial representation of the outcomes of struggles over control of a territory; this is why political maps can change over time.

GEOPOLITICS

Geopolitics encompasses the strategies that countries use to ensure that their own interests are served in relations with other countries. Geopolitics typified the *cold war era,* the period from 1946 to the early 1990s when the United States and its allies in Western Europe faced off against the Union of Soviet Socialist Republics (USSR) and its allies in Eastern Europe and Central Asia. Ideologically, the United States promoted a version of free market capitalism and democracy. The USSR and its allies favored a centrally planned economy and a socialist system in which citizens participated in government indirectly through the Communist Party.

The cold war grew into a race to attract the loyalties of unallied countries and to arm them. Sometimes the result was that unsavory dictators were embraced as allies by one side or the other. Eventually, the cold war influenced the internal and external policies of virtually every country on earth, often oversimplifying complex local issues into a contest of democracy versus communism.

In the post–cold war period of the 1990s, geopolitics shifted. The Soviet Union dissolved, creating many independent states, nearly all of which began to implement some democratic and free market reforms. Globally, countries jockeyed for a better position in what appeared to some to be a new era of trade and amicable prosperity, rather than war. But throughout the 1990s, while the developed countries enjoyed unprecedented prosperity, there were many unresolved political conflicts in Asia, Africa, Southwest Asia, and Europe. Too often these disputes erupted into bloodshed and **genocide,** the systematic attempt to kill all members of a particular ethnic or religious group.

The terrorist attacks on the United States on September 11, 2001, ushered in a new geopolitical era that is still evolving. Because of the size and the geopolitical power of the United States, the attacks and the U.S. reactions to them affected virtually all international relationships, public and private. The ensuing adjustments are directly or indirectly affecting the daily lives of billions of people around the world. We will discuss the ramifications of these various geopolitical changes in the regional chapters, particularly Chapter 2 and Chapter 6.

DEMOCRATIZATION

The last century saw a steady expansion of democracy throughout the world, though some notable reversals in this trend also occurred. Recent decades have shown an acceleration of democratization in Middle and South America, Sub-Saharan Africa, Europe, and the Newly Independent States of the former Soviet Union. While there is considerable debate about what forces encourage the general public to assert a more active role in their own governance, the most widely agreed on factors include:

- **Broad prosperity:** As countries become wealthier, and as prosperity is shared by a broad segment of the population (usually the "middle class"), there is generally a shift toward giving more power to citizens through free elections of leaders. Whether greater prosperity must occur before truly stable democracy can be established is still widely debated.

- **Education:** Better-educated people tend to want more input on how they are governed. While democracy has frequently spread to countries with relatively undereducated populations, leaders there sometimes become more authoritarian once elected.

- **Civil society:** Institutions that encourage a sense of unity and informed common purpose among the general population are widely seen as supportive of democracy. Such institutions include academia, unions, the media, non-governmental organizations (discussed below), and human rights organizations.

As we will see in the chapters ahead, these factors are just a few of the forces guiding the global shift toward democratization.

INTERNATIONAL COOPERATION

Increasingly, people, goods, and capital are moving freely across national borders, a trend that favors international cooperation over national sovereignty. It also creates a need for some way to enforce laws governing business, trade, and human rights at the international level.

The prime example today of international cooperation is the **United Nations (UN),** an assembly of 185 member states. The member states sponsor programs and agencies focusing on scientific research, humanitarian aid, economic development, general health and well-being, and peacekeeping assistance in hot spots around the world. Thus far countries have been unwilling to relinquish sovereignty. Consequently, the United Nations has limited legal power and can often enforce its rulings only through persuasion. Even in its peacekeeping mission, there are no true UN forces. Rather, troops from member states wear UN designations on their uniforms and take orders from temporary UN commanders.

The World Bank, the International Monetary Fund, and the World Trade Organization, discussed earlier (see page 15), are important international organizations that affect economies and trade practices throughout the world.

Nongovernmental organizations (NGOs) are an increasingly important embodiment of globalization. In such associations, individuals, often from widely differing backgrounds and locations, agree on a political, economic, social, or environmental goal. For example, some NGOs work to protect the environment (for instance, the World Wildlife Fund). Others provide medical care to those who need it most (for instance, Doctors Without Borders [Figure 1.38]). The

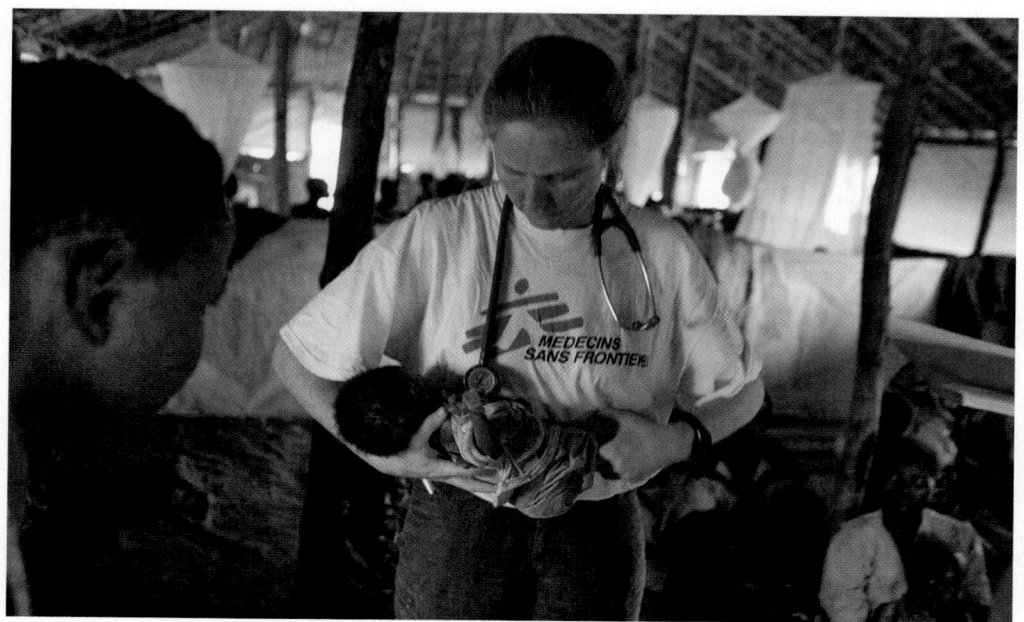

Figure 1.38 Help from a nongovernmental organization (NGO). *Médecins Sans Frontières* (Doctors Without Borders), an international humanitarian aid organization, provides emergency medical assistance to populations in danger in more than 70 countries. Here, a doctor holds a malnourished and dehydrated child in eastern Democratic Republic of Congo in 2005. [AP Photo/Ron Haviv.]

educational efforts of an NGO such as Rotary International can raise awareness of important issues among the global public.

There is some concern that huge international NGOs wield such power that they undermine democratic processes, especially in small countries. Some critics feel that NGO officials are a powerful do-gooder elite that does not interact sufficiently well with local people. On the other hand, NGOs such as OXFAM International (actually a group of NGOs) have provided essential relief and rebuilding services in times of great need. OXFAM was a major provider of relief after the Indian Ocean tsunami (2004) and during the conflicts in Lebanon, Israel, and Gaza in 2006. At the local level, the best NGOs solicit input from a wide range of individuals—a feature that political scientists consider essential to participatory democracy. 🎥

Reflections on the Geography of the World

The concepts discussed in this chapter give you a foundation for understanding the rest of the book. Here are a few questions to keep in mind as you study the geography of the world.

1. Some people argue that it is acceptable for people in the United States to consume at high levels because their consumerism keeps the world economy going. What are the weaknesses in this idea?

2. What are the causes of the huge increases in migration, legal and illegal, that have taken place over the last 25 years?

3. What would happen in the global marketplace if all people had a living wage?

4. As people live longer and decide not to raise large families, how can they beneficially spend the last 20 to 30 years of their lives?

5. Given the threats posed by global warming, what are the most important steps to take now?

6. As you read about differing ways of life, values, and perspectives on the world, reflect on the appropriateness of force as a way of resolving conflicts.

7. What are some possible careers that would address one or more of the issues raised in particular chapters?

8. Reflect on the reasons why some people have much and others have little.

9. What would be some of the disadvantages and advantages of abolishing gender roles in any given culture?

10. Consider the ways that access to the Internet enhances prospects for world peace and the ways that it contributes to discord.

Chapter Key Terms

age structure 42
air pressure 27
average population density 41
birth rate 42
capital 14
carrying capacity 22
cash economy 45
climate 26
country 46
cultural homogeneity 32
culture 29
culture group 29
death rate 42
delta 26
democratization 46
demographic transition 45
demography 40
digital divide 10
ecological footprint 24
erosion 26
ethnic group 29
European colonialism 9
exchange or service sector 13

extractive resource 13
fair trade 15
floodplain 26
formal economy 13
free trade 14
frontal precipitation 29
gender 37
gender structure 42
genocide 46
geopolitics 46
global economy 11
global warming 17
globalization 10
gross domestic product (GDP) 13
gross domestic product (GDP) per capita 39
human geography 5
human well-being 39
import quota 15
Industrial Revolution 13
informal economy 13
International Monetary Fund (IMF) 15

interregional linkage 9
landform 24
latitude 5
living wage 15
longitude 5
material culture 29
monsoon 29
multiculturalism 33
multinational corporation 14
nation 46
nationalism 46
nation-state 46
Neolithic Revolution 20
nongovernmental organization (NGO) 47
orographic rainfall 29
Pangaea hypothesis 24
physical geography 5
plate tectonics 24
pluralistic state 46
political ecologist 22
population pyramid 42
purchasing power parity (PPP) 39

CHAPTER 2

North America

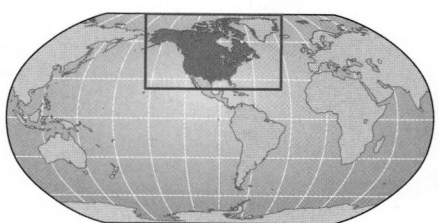

Global Patterns, Local Lives Fifteen women friends from Vera Cruz, Mexico, are settling down in front of the television in the living room of their group quarters in Terrebonne Parish, Louisiana. They all have H1B temporary work visas arranged for them by Motivatit Seafood, one of the state's largest harvesters and processors of oysters. In two hours the women will join a few local workers to start their shift in the chilly workroom where they shuck, wash, and package oysters for between $5.50 and $6.00 an hour.

In the months after hurricanes Katrina and Rita hit the U.S. Gulf Coast (Figure 2.1) in August and September of 2005, businesses struggled to find enough workers to reopen. With local, state, and national governments unable to meet their immediate needs following the storm's destruction, hundreds of thousands of coastal residents who had fled the Gulf Coast decided to stay away permanently. Soon some of the demand for workers along the Gulf Coast was being filled by workers from Mexico, documented and undocumented. The influx of immigrants was so great that the mayor of New Orleans worried that the culture of his city would be changed forever. ▭

Forty percent of the nation's oysters come from Louisiana, and the oyster beds suffered serious storm damage, but by February of 2006, nature was repairing itself. With the help of the Mexican women, business for Motivatit Seafood was returning to normal. Still, the company needed at least thirty more workers. Owner Mike Voisin says that even before the storms it was nearly impossible to keep local employees because they spurned the hard labor of oyster work.

Mike's grandson, Kevin Voisin, says that he was originally against the idea of Mexican workers, but he changed his mind when he met them. He visits them often with his wife, and praises their dedication to hard work and their frugality. "These are great people," Voisin says. "This is what America is all about. [Many of]

Figure 2.1 Regional map of North America.

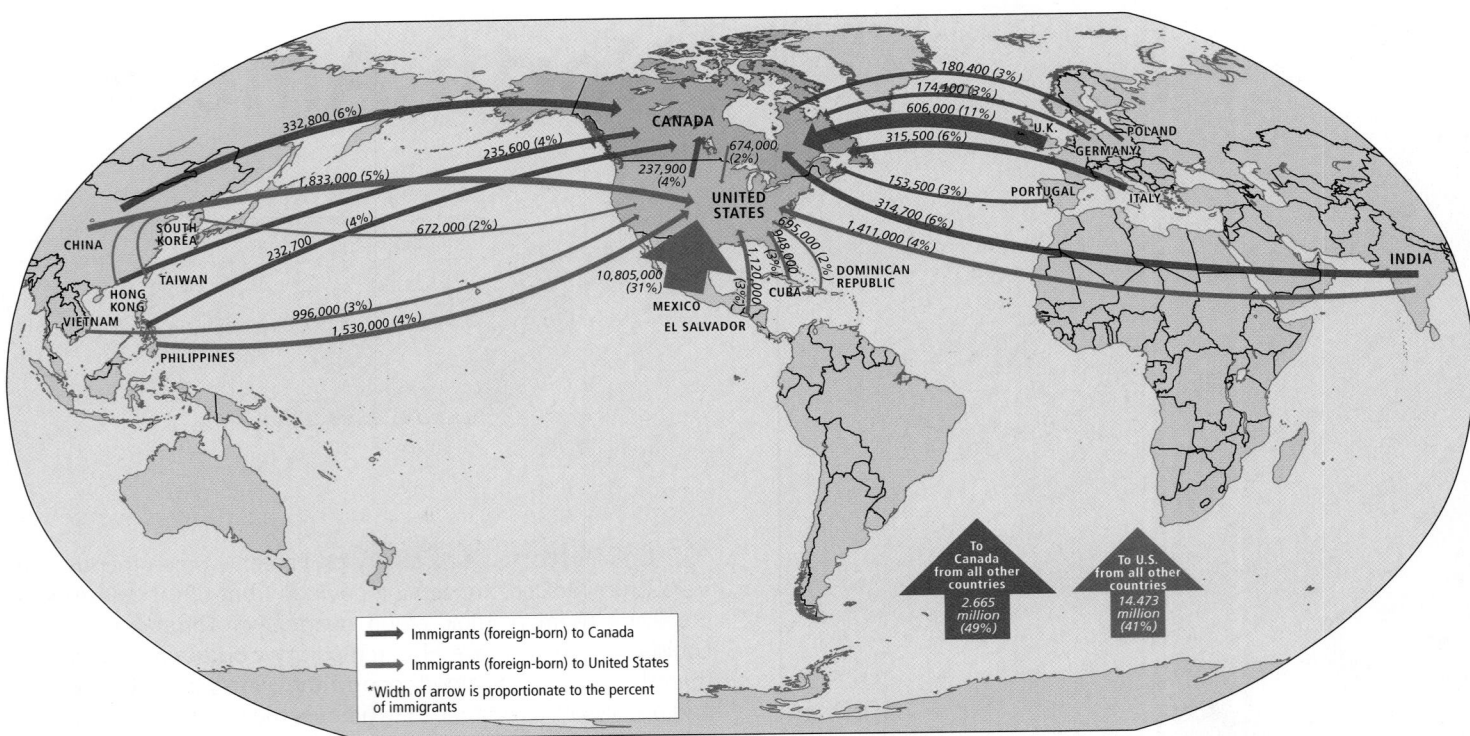

Figure 2.2 Total foreign-born residents of Canada and the United States by country of birth. Foreign-born residents include legal and illegal immigrants, temporary workers, students, and refugees. The numbers for Canada are as of the 2001 census (the latest available); those for the United States are as of 2005. Percentages refer to proportion of total immigrants to a particular country.

my people came from Ireland. People worried then that the Irish would take over America, but this [hasn't become] the Irish Republic of America."

Adapted from the National Public Radio's "Sunday Morning Edition" series on Motivatit Seafood, by Lianne Hanson et al., September 11 and 25, and October 2, 2005, and February 12 and 19, 2006. ■

The influx of Mexican and other Hispanic workers into Louisiana and Mississippi in the aftermath of hurricanes Katrina and Rita is one of many cases in which North American employers have turned to immigrant workers. Some immigrants are highly skilled technical specialists such as computer engineers and medical doctors; some are less skilled laborers.

Figure 2.2 shows recent patterns of immigration to North America. Immigrants from Asia and Middle and South America are bringing their skills and culture to even the most remote communities of North America. Rural towns in the mountains of North Carolina have Mexican grocery stores catering to farm laborers from Michoacan in central Mexico. Physicians from India live in eastern Kentucky hill-country towns and treat predominantly Appalachian patients (Figure 2.3).

Americans are ambivalent about new immigrants and guest workers and the major social and economic changes that they represent. In 2006, powerful anti-immigration sentiment inspired the U.S. congress to authorize the creation of a 700-mile-long barrier between the United States and Mexico. The barrier, which is a wall in some places and a fence in others, has been criticized as a danger not only to Mexican immigrants but also to wildlife, which can no longer move between border area habitats. While only a few miles of the barrier have been built so far, opinion polls in 2007 showed that 56 percent of U.S. citizens favored completing it entirely, while 31 percent opposed it.

North America's new immigrants and guest workers are part of larger shifts in the region. Manufacturing is no longer the mainstay of North America's economy. Sophisticated technology and the high-speed global exchange of information, goods, and money are bringing rapid changes in the types and locations of jobs across the continent.

If you are a resident of North America, you can bring your own experience to understanding the evolving character of North America. If you grew up in North America, you might want to look back at the childhood landscape map that you drew in Chapter 1 and place it in its broader regional context. Think of this chapter as a guide to enhance your understanding of what you already know and to help you construct an ever more complex understanding of the geography of North America.

Figure 2.3 Indian immigrants bringing medical care to the rural United States. Doctors Rakesh Sachdeva and Seema Sachdeva, both from India but now U.S. citizens, examine Connor Mayhorn at their medical office in Pikeville, Kentucky. Pikeville is a small eastern Kentucky city in a county that for years was underserved by doctors. [Dena Potter/*Appalachian News-Express.*]

Questions to consider as you study North America (also see Views of North America, pages 54–55)

1. How is globalization contributing to changing patterns of immigration to North America? How has globalization changed patterns of employment in the region? North America is a strong advocate of the economic aspects of globalization, especially through the free trade agreement known as NAFTA (see page 80). What kinds of jobs have become more scarce in this region as a result of NAFTA?

What kinds of jobs have become more common? What forces are bringing immigrants to this region?

2. How have North American patterns of urbanization increased this region's contribution to global warming? North America creates more greenhouse gases than any other region in the world, thanks in part to its long love affair with the automobile. How is the extensive suburban development that characterizes this region connected to the automobile? Will efforts to reduce greenhouse gas emissions require changes in North American cities?

3. How have North American food production systems contributed to water pollution? This region's farms are some of the most productive in the world, thanks in part to the extensive use of fertilizers and pesticides. What environmental problems have these chemicals created? How extensive is the damage?

4. What social and economic problems are likely to occur as the North American population ages? With people living longer and families having fewer children, more retirees must now be supported by fewer working-age people. Will health care systems and living arrangements have to change? How are different parts of this region experiencing unique age-related problems?

5. How have women been empowered by North America's democratic institutions in recent years? Why are women still relatively less politically powerful than their numbers might warrant? Women are a rising political force in North American politics; yet female political leaders at the national level are vastly outnumbered by males. What other aspects of North American gender relations might help explain this situation?

The Geographic Setting

Terms to Be Aware Of

The world region discussed in this chapter consists of Canada and the United States. The term *North America* is used to refer to both countries. Other terms relate to the growing cultural diversity in this region. The text uses the term **Hispanic** to refer to all Spanish-speaking people from Middle and South America, although their ancestors may have been black, white, Asian, or Native American. In Canada, the **Québecois,** or those French Canadians living in Québec, are an ethnic group that is distinct from the rest of Canada. They are the largest of an increasingly complex mix of minorities in that country, most of whom are still content simply to be called Canadians.

PHYSICAL PATTERNS

The continent of North America is a huge expanse of mountain peaks, ridges and valleys, expansive plains, long winding rivers, myriad lakes, and extraordinarily long coastlines. We cannot cover the entire physical geography of this large territory here, but we will focus on a few of the most significant landforms.

Landforms

A wide mass of mountains and basins, known as the Rocky Mountain zone, dominates western North America (see Figure 2.1). It stretches down from the Bering Strait in the far north,

Views of North America

For a tour of these places, visit www.whfreeman.com/pulsipher

[Background image: Lori K. Perkins NASA/GSFC/Scientific Visualization Studio]

1. Food. Most farms, such as this one outside Spokane WA, have become highly mechanized operations that need few workers but require huge investments in land, machinery, fertilizers and pesticides to be profitable. See page 75 for more. Lat 47°N, Lon 117°W. [USDA]

2. Globalization and employment. Technology and service-based industries are the backbone of the North American economy. They now dwarf manufacturing industries which have scaled back or moved abroad. This man is servicing computers aboard a US Navy ship in Bremerton, WA. See pages 78–79 for more. Lat 47°33'N, Lon 122°39'W. [MCSA Matthew A. Lawson; USN]

Bremerton **2**

Spokane **1**

7 San Luis

7. Immigration and globalization. The US-Mexico Barrier is a controversial effort to prevent illegal movement across the US-Mexico border. These US military personnel are building the barrier in San Luis, AZ. See pages 80–81 and 86–88 for more. Lat 32°28'03.06"N, Lon 114°42'58.71"W. [U.S. Air Force photo by Staff Sgt. Dan Heaton]

3. Urbanization and global warming. Urban Sprawl increases North America's contribution to global warming by making people more dependent on their automobiles which are a major source of greenhouse gases. Shown here is new suburban development outside Des Moines, IA. See pages 60 and 84–85 for more. Lat 41°36'38"N, Lon 93°52'41"W. [Lynn Betts/ USDA Natural Resources Conservation Service : JohnnyAlbert10]

4. Gender and democracy. Women's participation in the economy and politics has transformed North America, though women are still under-represented in state and national legislatures. Hillary Clinton's (D-NY) run for the U.S. presidency gave many gender issues a higher profile. See pages 83–84 for more. Lat 40°44'N, Lon 73°59'W. [TSGT Lisa M. Zunzanyika; USAF]

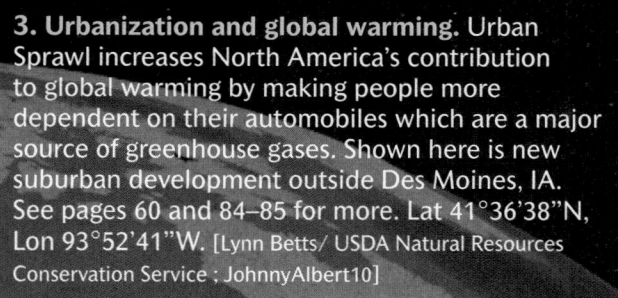

4

New York City

3

Des Moines

5. Population and aging. Growing numbers of the elderly will need more care from a shrinking population of younger people. This elderly man is receiving health care in Punta Gorda, one of Florida's many retirement communities where the average age is over 60. See pages 69–72 for more. Lat 26°54'N, Lon 82°02'W. [FEMA Photo/ Andrea Booher]

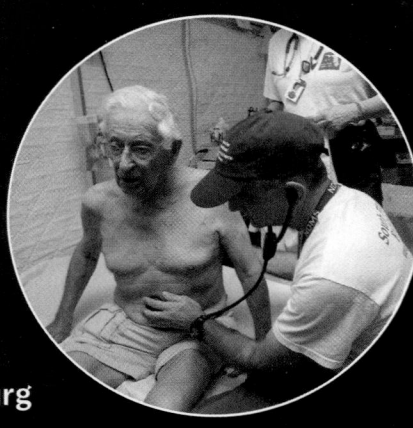

6

Mouth of Mississippi River

5 **St. Petersburg**

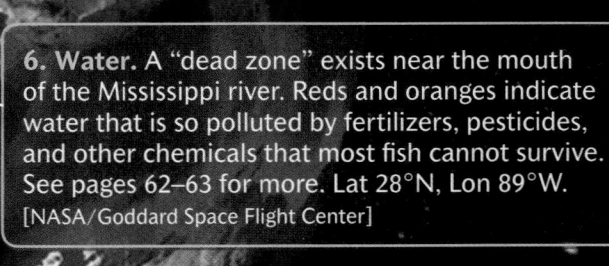

6. Water. A "dead zone" exists near the mouth of the Mississippi river. Reds and oranges indicate water that is so polluted by fertilizers, pesticides, and other chemicals that most fish cannot survive. See pages 62–63 for more. Lat 28°N, Lon 89°W. [NASA/Goddard Space Flight Center]

through Alaska, and into Mexico. This zone formed about 200 million years ago, when, as part of the breakup of the supercontinent Pangaea (see Figure 1.21 on page 25), the Pacific Plate pushed against the North American Plate, thrusting up mountains. These plates still rub against each other, causing earthquakes along the Pacific coast of North America.

The much older, and hence more eroded, Appalachian Mountains stretch along the eastern edge of North America from New Brunswick and Maine to Georgia. This range resulted from earlier collisions of the North American Plate with North Africa.

Between these two mountain ranges lies a huge central lowland of undulating plains stretching from the Arctic to the Gulf of Mexico. This landform was created by the deposition of material eroded from the mountains and carried to the North American region by wind and rain and by the rivers flowing east and west into what is now the Mississippi drainage basin.

Glaciers have covered the northern portion of the region during periodic ice ages over the last 2 million years. During the most recent ice age (between 25,000 and 10,000 years ago) the glaciers, sometimes as much as 2 miles (about 3 kilometers) thick, moved south from the Arctic, picking up soil and scouring depressions in the rock. When the glaciers later melted, these depressions filled with water, forming the Great Lakes. Thousands of smaller lakes, ponds, and wetlands that stretch from Minnesota to Massachusetts were formed in the same way. Melting glaciers also dumped huge quantities of soil throughout the central United States. This soil, often many meters deep, provided the basis for large-scale agriculture.

East of the Appalachians, a coastal lowland stretches from New Brunswick to Florida. It then sweeps west to the southern reaches of the central lowland along the Gulf of Mexico. In Louisiana and Mississippi, much of this lowland is filled in by the Mississippi delta, a low, flat, swampy transition zone between land and sea. The delta was formed by huge loads of silt deposited by North America's largest river system during floods over the past 150 million years. It originally began at what is now the junction of the Mississippi and Ohio rivers at Cairo, Illinois, and then slowly advanced 1000 miles (1600 kilometers) to the Gulf of Mexico.

Over the centuries, human activities such as deforestation, deep plowing, and heavy grazing have led to erosion and added to the silt load of the rivers (Figure 2.4). At the same time, the construction of levees along riverbanks has drastically reduced flooding. Because this flood control reduced the huge loads of silt that used to spread widely across the delta region, much of the southern part of the Mississippi delta is sinking into the Gulf of Mexico. Meanwhile, the silt is now contained by levees, which force it into the deep waters of the Gulf of Mexico.

Climate

The landforms over this continental expanse contribute to its enormous climate variety by influencing the movement and interaction of air masses (Figure 2.5). The West Coast of North America, especially north of San Francisco, receives moderate to heavy rainfall; climates are much drier to the east of the coastal mountain ranges. (See Figures 1.23, 1.24, and the related discussion on pages 27–29 for an explanation of this

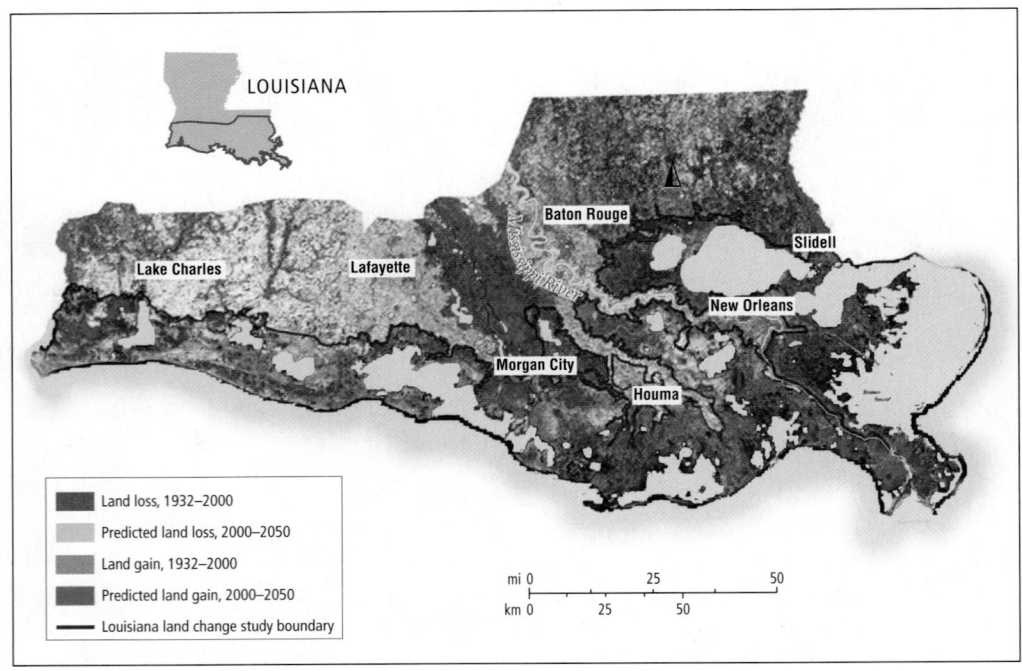

Figure 2.4 100+ years of land change for coastal Louisiana. The Louisiana coastline and the lower Mississippi River basin are vital to the nation's interests. They provide coastal wildlife habitat, recreational opportunities, transport lanes that connect the vast interior of the country with the world ocean, and access to offshore oil and gas. Most important, the wetlands provide a buffer against damage from hurricanes. However, Louisiana has lost one-quarter of its total wetlands over the last century, largely due to human impacts on natural systems. The remaining 3.67 million acres constitute 14 percent of the total wetland area in the lower 48 states. [USGS/National Wetlands Research Center.]

LOUISIANA

Baton Rouge
Slidell
Lake Charles   Lafayette
New Orleans
Morgan City
Houma

Land loss, 1932–2000
Predicted land loss, 2000–2050
Land gain, 1932–2000
Predicted land gain, 2000–2050
Louisiana land change study boundary

mi 0 · 25 · 50
km 0 · 25 · 50

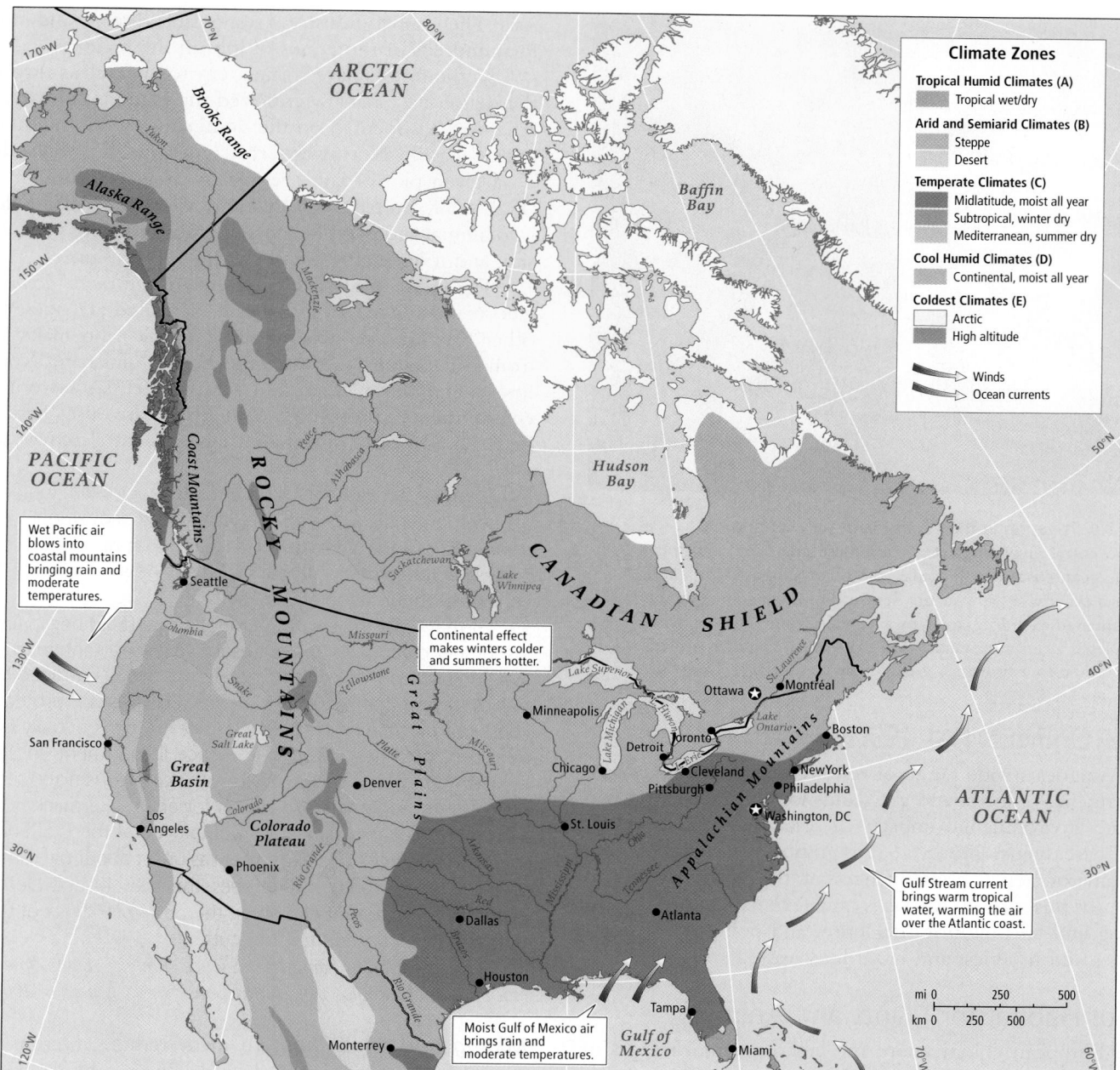

Figure 2.5 Climates of North America.

phenomenon.) In the arid region, many expensive dams and reservoirs for irrigation projects have been built to make agriculture and large-scale human habitation possible (Figure 2.6).

On the eastern side of the Rocky Mountains, the main source of moisture is the Gulf of Mexico. When the continent is warming in the spring and summer, the air masses above it rise, sucking in moist, buoyant air from the Gulf. This air interacts with cooler, drier, heavier air masses moving into the central lowland from the north and west, often creating violent thunderstorms and tornadoes. Generally, central North America is wettest in the eastern and southern parts and driest in the north and west. Along the Atlantic coast, moisture is supplied by warm, wet air above the Gulf Stream, a warm ocean current that flows up the eastern seaboard from the tropics.

Because land heats up and cools off more rapidly than water, temperatures in the continental interior are hotter in the summer and colder in the winter than areas along the coast, where temperatures are moderated by the oceans.

Figure 2.6 Irrigation in the Great Plains and continental interior. Central pivot irrigators create these great circles in the landscape near Tuscarora, Nevada. The water supports crops that would not otherwise grow in this semiarid environment. Often a well at the center of each irrigator taps an underground aquifer, which is then depleted at an unsustainable rate. When the aquifer is depleted, the circles fade to a brownish green. [Alex S. MacLean/Landslides.]

ENVIRONMENTAL ISSUES

North America's wide range of resources and seemingly limitless stretches of forest and grasslands long diverted attention from the environmental impacts of settlement and development. Increasingly, however, it is impossible to ignore the many environmental consequences of the North American lifestyle. In this section, we focus on a few: habitat loss, global warming and air pollution, depletion and pollution of water resources and fisheries, and hazardous waste.

Loss of Habitat for Plants and Animals

Before European colonization, the environmental impact of humans was significantly lower in North America (Figure 2.7a). Though North America was by no means a pristine paradise when Europeans arrived, millions of acres of forests and grasslands that had served as habitats for native plants and animals were subsequently cleared to make way for European-style farms, cities, and industries. This was particularly true in what became the United States. Though widespread clearing for agriculture is now rare, other human activities are having significant effects on habitat across the region (Figure 2.7b).

Logging. Logging occurs in many of North America's forested areas, including the northern Pacific coast. Logging in the Pacific Northwest provides most of the construction lumber and an increasing amount of the paper used in North America. Lumber and wood products are also important exports to

Asia. The lumber industry is responsible directly and indirectly for hundreds of thousands of jobs in the region.

As the forests have shrunk, environmentalists throughout the region have harshly criticized the logging industry. Much attention has focused on the dominant logging method used in the Pacific Northwest, **clear-cutting,** in which all trees on a given plot of land, regardless of age, health, or species (Figure 2.8, on page 60) are cut down. Clear-cutting destroys wild animal and plant habitats, thereby reducing species diversity, and it leaves the soil susceptible to erosion.

The battle lines have been drawn. On one side are those who make a living from logging or related activities. On the other side are two groups: those people who make a living from occupations that tout the beauty of Pacific Northwest forests and those urban and rural residents who want strict environmental protection. But there are also battle lines between advocates of different environmental activist tactics.

Vignette Jim Hermack lives in Eugene, Oregon, long a center of environmental activism. He joined the movement when he returned from the Vietnam War to find that his favorite camping places had been clear-cut, but in the mid-1990s, Hermack became conflicted about his role as an activist.

At that time, the Earth Liberation Front (ELF) began firebombing lumber companies and ski resorts to draw attention to the devastation caused by clear-cutting. Hermack was outraged by these tactics, saying that by using violence, ELF was not protecting the liberty and justice he held dear. However, he is equally outraged by recent activities of the U.S. Justice Department. In 2005, the Department charged ELF activists with ecoterrorism and sentenced one of them to 22 years in prison for arson. Hermack notes that a far more flagrant arsonist—an employee of the U.S. Forest Service who set 35 fires because she was upset about overtime pay—received only probation. He sees the U.S. Justice Department action as a witch hunt exploiting the post–9/11 fears of terrorism to discredit opposition to clear-cutting.

Source: "Harsh Sentences Silence Radical Environmentalists," National Public Radio, "All Things Considered," March 8, 2006. ∎

Mining and oil drilling. In many remote interior areas of North America, mining and oil drilling are by far the largest and most environmentally damaging industries. Huge piles of mining waste called "tailings" can pollute waterways and threaten communities that depend on well water. Oil-drilling operations have also had a dramatic effect on humans and the environment.

Along the northern coast of Alaska, the Trans-Alaska Pipeline runs southward for 800 miles to the port of Valdez. Often running above ground to avoid shifting as the earth freezes and thaws, the pipeline constitutes a major ecological disruption. It interferes with caribou migrations and always poses the threat of oil spills. Oil spills are a danger offshore as well. In 1989, a giant spill from the oil tanker *Exxon Valdez* devastated 1100 miles of Alaskan shoreline, killing much wildlife and destroying native livelihoods and commercial fishing.

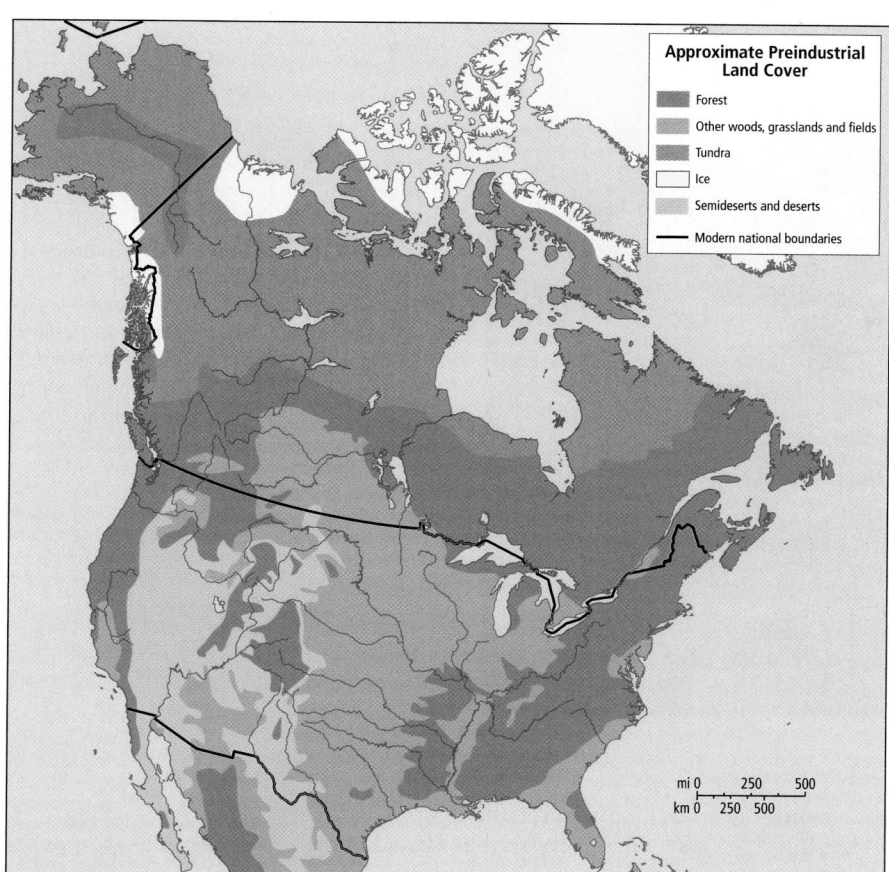

(a) Preindustrial

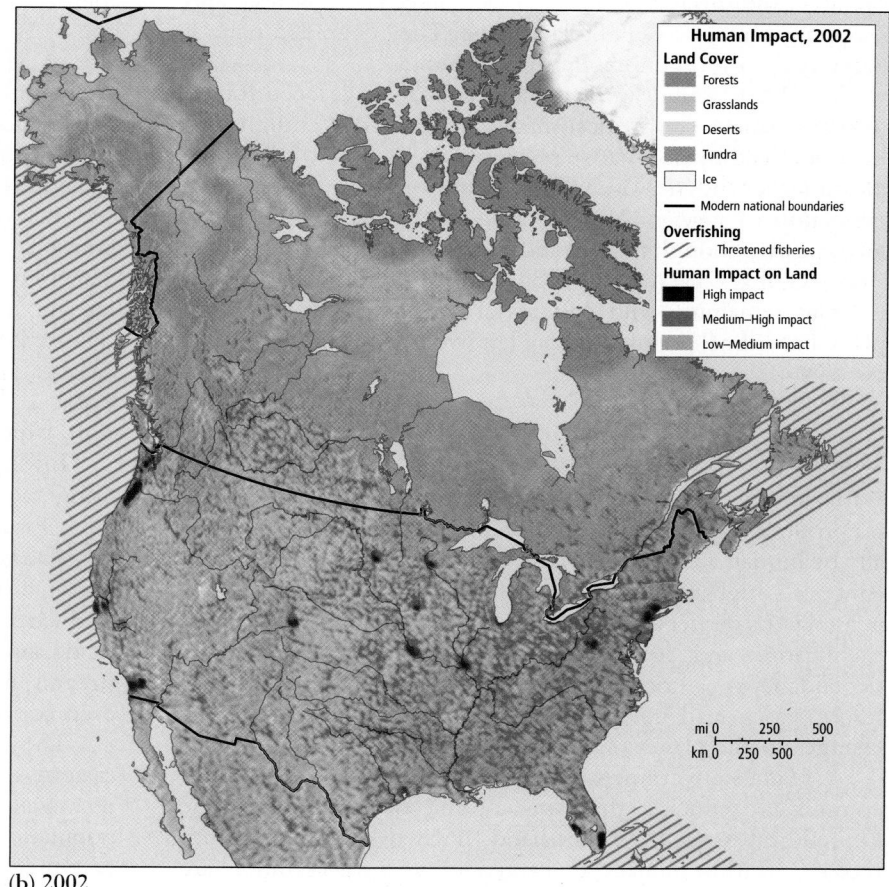

(b) 2002

Figure 2.7 Land cover (preindustrial) and human impact (2002). In the preindustrial era (**a**), humans had cleared forests, extended grasslands (by burning), and planted agricultural fields. About 350 years later (**b**), some impact of human agency was observable virtually everywhere in North America. [Adapted from *United Nations Environment Programme, 2002, 2003, 2004, 2005, 2006* (New York: United Nations Development Programme), at http://maps.grida.no/go/graphic/human_impact_year_1700_approximately and http://maps.grida.no/go/graphic/human_impact_year_2002.]

Figure 2.8 Clear-cutting in Washington. This landscape near Cathlamet, in central Washington State, shows various stages of clear-cutting. The dark green is uncut forest, the bright green is regrowing forest, and the gray is newly clear-cut forest. Logging roads are also visible. In the Pacific Northwest, logging companies have been known to leave wide strips of forest along public highways so that the citizenry remains unaware of the extent of the forest removal. [Jim Wark/Airphoto.]

Other threats to habitat. In many areas, farms are giving way to expansive low-density urban and suburban development commonly called **urban sprawl.** Here, residential developments, highways, golf courses, office complexes, and shopping centers cover the landscape. They degraded natural habitats even more intensely than farming.

As North American plants and animals have been forced into ever smaller territories, many have died out entirely and been replaced by non-native species. ▰ Because of the complex interdependence of the biological world, the depletion or extinction of species will have long-term negative effects for life on earth, though the specific effects are not yet well understood. Global warming will undoubtedly lead to further extinctions as remaining wild habitats are transformed and many plant and animal populations are left without a place to migrate to. ▰

Global Warming and Air Pollution

With only 5 percent of the world's population, North America produces 26 percent of the greenhouse gases released globally by human activity. This large share can be traced to North America's high consumption of fossil fuels, which is in turn related to its oil-dependent industrial and agricultural processes, the heating and cooling of its homes and offices, and its dependency on automobiles. As discussed in Chapter 1, greenhouse gases lead to global warming and bring profound changes to the planet (see pages 17–20).

Canada's government was one of the first in the world to commit to reducing the consumption of fossil fuels. The United States, on the other hand, resisted such moves until recently, fearing damage to its economy. Both countries are now exploring alternative sources of energy, such as solar, wind, nuclear, and geothermal. ▰ However, neither Canada nor the United States has so far been able to reduce its levels of greenhouse gas emissions, or even the rate at which these emissions are growing.

In addition to global warming, most greenhouse gases contribute to various forms of air pollution, two of which we describe here.

Though Canada and the United States are exploring alternative sources of energy, neither has yet been able to reduce either its levels of greenhouse gas emissions or even the rate at which these emissions are growing.

Smog is a combination of industrial emissions and car exhaust that frequently hovers over many North American cities, causing a variety of health problems for their inhabitants (Figure 2.9). In Los Angeles, the intensity of the smog is due in large part to the city's warm land temperatures and West Coast seaside location. This often results in a **thermal inversion,** which occurs when a mass of warm air settles over cooler air blowing in from the ocean and prevents the cool air from rising and dissipating pollution. The inversion is held in place, often for days, by the mountains that surround the city.

The burning of fossil fuels also releases sulfur dioxide and nitrogen oxides into the air. **Acid rain** (and acid snow) is created when these gases dissolve in falling rain and make it acidic. Acid rain can kill trees and, when concentrated in lakes, streams, and snow cover, can destroy fish and wildlife. It can even corrode buildings and other structures.

The eastern half of the continent, which includes the entire eastern seaboard from the Gulf Coast to Newfoundland, is greatly affected by acid rain. The United States, with its larger population and more extensive industry, is responsible for the vast majority of this acid rain. Due to continental weather patterns, however, the area most affected by acid rain

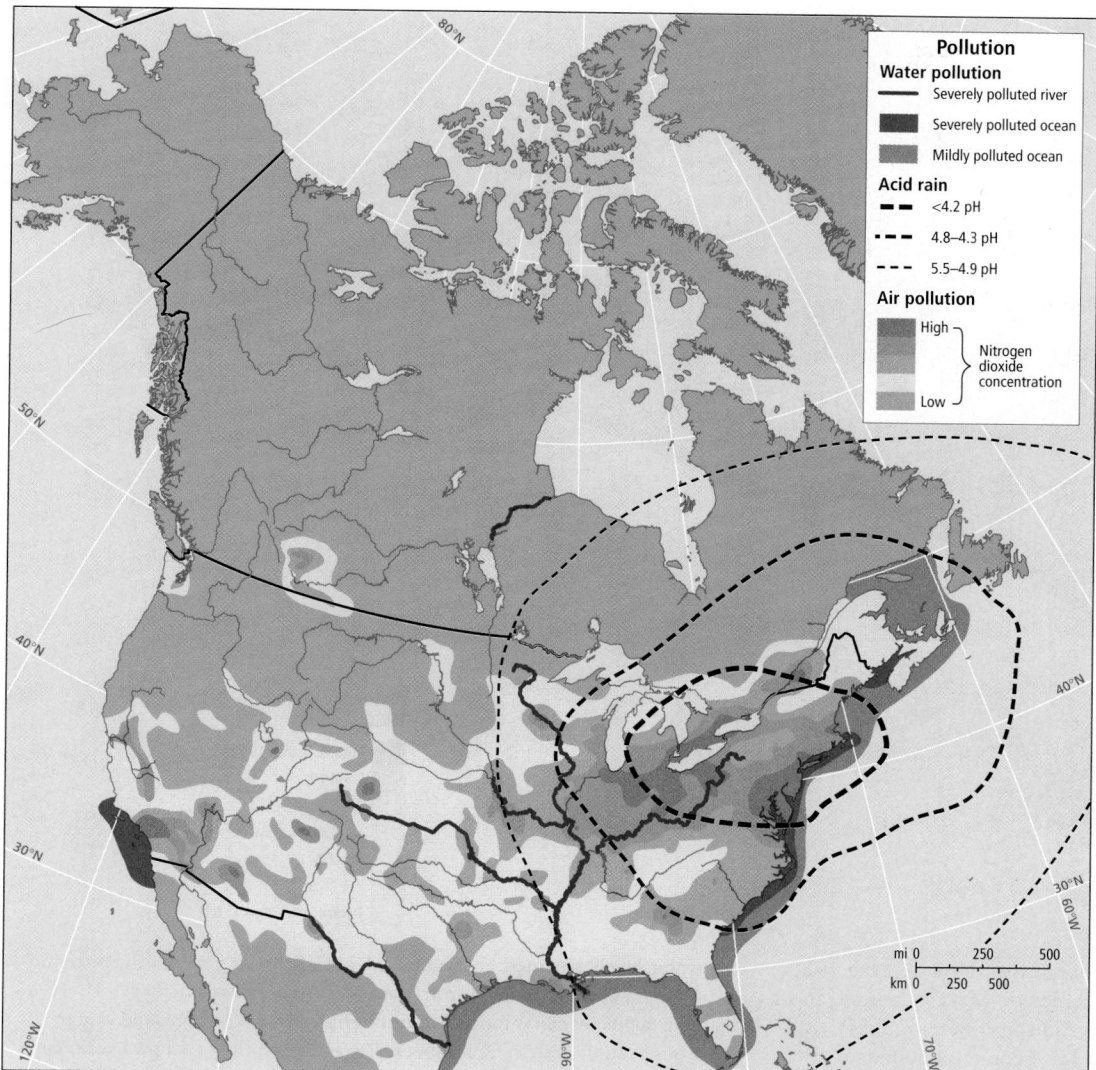

Figure 2.9 Air and water pollution in North America. Nitrogen dioxide (NO_2), a toxic gas that comes primarily from the combustion of fossil fuels by motor vehicles and power plants, transforms in air to gaseous nitric acid and toxic organic nitrates.

encompasses a wide area on both sides of the eastern U.S.–Canada border (see Figure 2.9).

Water Resource Depletion and Pollution

People who live in the humid eastern part of North America find it difficult to believe that water is becoming scarce even there. But as populations grow and per capita water usage increases, it has become necessary to look farther and farther afield for sufficient water resources. New York City, for example, obtains most of its water from the distant Catskill Mountains in upstate New York. Atlanta, Georgia, with over 5 million residents, absorbs water so insatiably that downstream users in Alabama and Florida have sued Atlanta for depriving them of their rights to water. Disputes over water rights exist around the world and are difficult to resolve, as we will see in our discussion of many world regions.

In North America, water becomes increasingly precious the farther west one goes. On the Great Plains, rainfall in any given year may be too sparse to support healthy crops and animals. To make farming more secure and predictable, taxpayers across the continent have subsidized the building of massive aqueducts, pumps, stock tanks, and reservoirs.

Irrigation is increasingly based on "fossil water" that has been stored over the millennia in natural underground reservoirs called **aquifers.** The **Ogallala aquifer** (Figure 2.10) underlying the Great Plains is the largest in North America. In parts of the Ogallala, water is being pumped out at rates that exceed natural replenishment by 10 to 40 times.

Irrigation is also a major issue in California, which supplies much of the fresh fruit and vegetables consumed in the United States. Billions of dollars of federal and state funds have paid for massive engineering projects that bring enormous quantities of water from Washington, Oregon, Colorado,

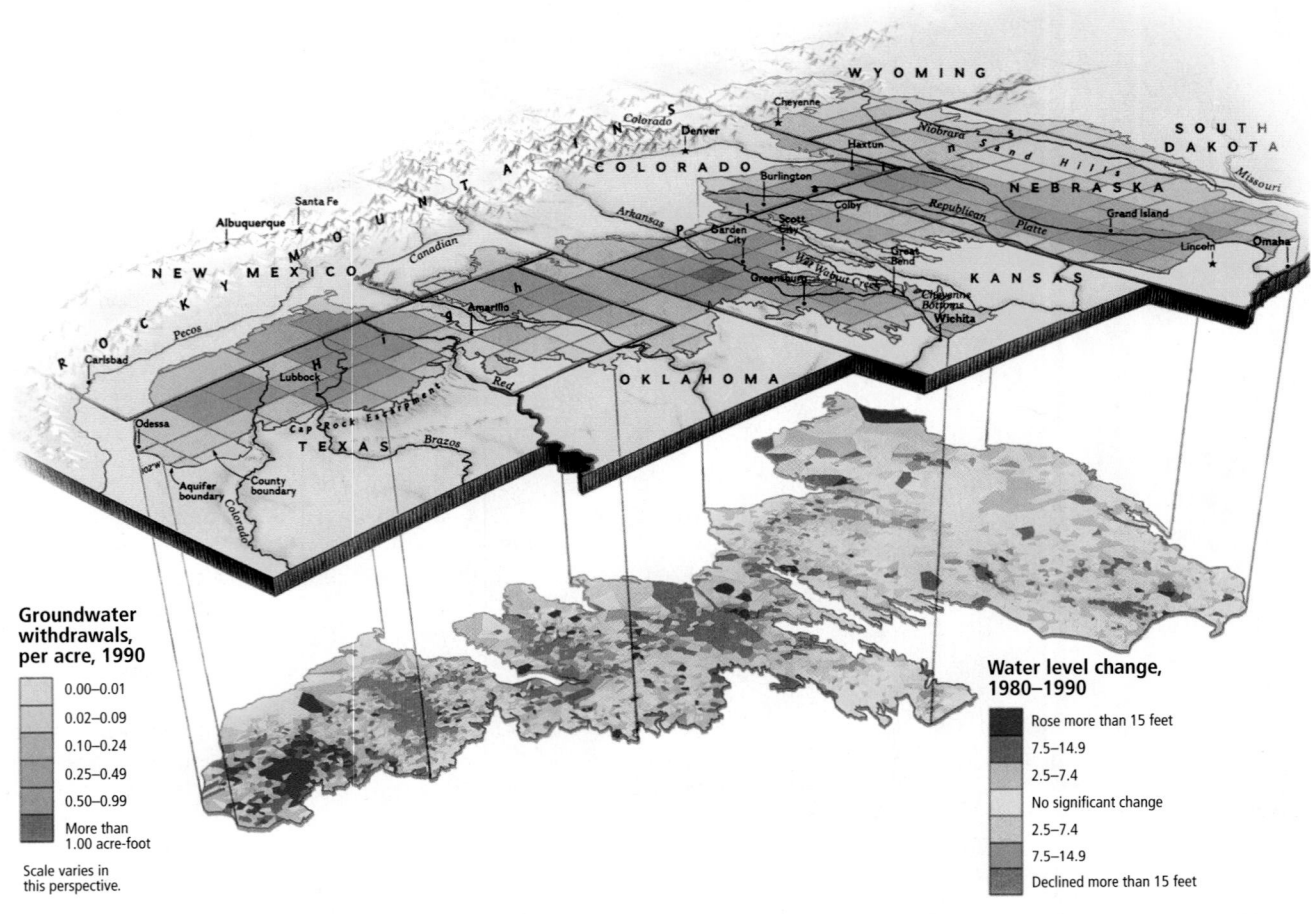

Groundwater withdrawals, per acre, 1990

- 0.00–0.01
- 0.02–0.09
- 0.10–0.24
- 0.25–0.49
- 0.50–0.99
- More than 1.00 acre-foot

Scale varies in this perspective.

Water level change, 1980–1990

- Rose more than 15 feet
- 7.5–14.9
- 2.5–7.4
- No significant change
- 2.5–7.4
- 7.5–14.9
- Declined more than 15 feet

Figure 2.10 The Ogallala aquifer. Between the 1940s and the 1980s, the aquifer lost an average of 10 feet (3 meters) of water overall, and more than 100 feet (30 meters) of water in some parts of Texas. During the 1980s, the aquifer declined less because of abundant rain and snow. However, this is an area where the climate fluctuates from moderately moist to very dry. When a drought began in mid-1992 (continuing until late 1996), large agribusiness firms pumped water from the aquifer for irrigation to supplement precipitation. As a result, water levels declined an average of 1.35 feet per year during the mid-1990s. [Graphic adapted from *National Geographic* (March 1993): 84–85, with supplemental information from High Plains Underground Water Conservation District 1, Lubbock, Texas, at http://www.hpwd.com; and Erin O'Brian, Biological and Agricultural Engineering, National Science Foundation Research Experience for Undergraduates, Kansas State University, 2001.]

Arizona, and northern California to farms in southern and central California. This water also supplies the cities of southern California, which are built on land that was once desert. Water is pumped from hundreds of miles away and over entire mountain ranges. California uses more energy to move water than some states use for all purposes. Irrigation in Southern California also deprives Mexico of this much-needed resource. The mouth of the Colorado River, which used to be navigable, is now dry and sandy. Only a mere trickle gets to Mexico.

Increasingly, citizens in western North America are recognizing that the use of scarce water for irrigated agriculture is unsustainable and uneconomical. Conflicts over transporting water from wet regions to dry ones, or from sparsely inhabited to urban areas, have halted some new water projects.

Nevertheless, because of government subsidies that keep water artificially cheap and past successes in harnessing new water supplies, there is a disincentive to change.

Water pollution. In the United States, 40 percent of rivers are too polluted for fishing and swimming, and over 90 percent of riparian areas (the interface between land and flowing surface water) have been lost or degraded. Pollution in U.S. rivers comes mainly from urban and suburban storm-water runoff, agriculture, and industry. ◼

In the 1970s, scientists studying coastal areas began noticing "dead zones" where water is so polluted that it supports almost no life. Dead zones occur near the mouths of major river systems that have been polluted by fertilizers washed off

Dead zones occur near the mouths of major river systems that have been polluted by fertilizers and other chemicals washed off farms and lawns when it rains.

farms and lawns when it rains. The largest dead zone is in the Gulf of Mexico near the mouth of the Mississippi, but similar zones have been found in all U.S. coastal areas. Even Canada, where much lower population density means that rivers are generally cleaner, has dead zones on its western coast.

Depleted Fisheries

Abundant fish was the major attraction for the first wave of Europeans to North America. In the 1500s, hundreds of fishermen from Europe's Atlantic coast came to the Grand Banks, offshore of Newfoundland and Maine, to take huge catches of cod and other fish. The fishing lasted for over 400 years. Now the fish stocks of the Grand Banks are badly depleted by modern fishing vessels (some from outside North America) that use equipment which destroys the marine ecosystem (see Figure 2.7b). Fisheries in the Pacific Northwest and the Gulf of Mexico are also threatened by pollution and overfishing. ▣◄

The fishing industry has also been hurt by competition from fish farming, a rapidly growing industry in many parts of North America. There is a global connection here: farmed fish are fed with millions of tons per year of wild ocean fish from around the world. The harvesting of these wild fish poses a threat to oceanic resources in countless places around the globe where the poor are dependent on fish for their basic nutrition.

Hazardous Waste

Hazardous wastes are those that pose a risk to public health or the environment. They are produced by a wide range of industries in North America, including nuclear power generation, weapons manufacturing, mineral mining and drilling, and waste incineration. While industry produces 80 percent of all liquid hazardous waste, small businesses and private homes also contribute through the use of cleaning and paint products, weed killers, and gasoline.

Each year the United States, with a population of roughly 300 million, generates five times the amount of hazardous waste generated by the entire European Union (with a population of about 460 million). Canada generates much less hazardous waste per capita each year than the United States, although its citizens generate several times the global average of about 2.2 pounds (1 kilogram) per person, per day.

The disposal of hazardous waste within the United States has a geographic pattern. Sociologist Robert D. Bullard, who studies environmental justice issues, has shown that a disproportionate amount of hazardous waste is disposed of in the South and in locations inhabited by poor Native American, African-American, and Hispanic people. Bullard writes that "nationally, 60 percent of African-Americans and 50 percent of Hispanics live in communities with at least one uncontrolled toxic-waste site."

Ten Key Aspects of North America

- North America produces 25 percent of the world's greenhouse gases, even though it has only 5 percent of the world's population.
- Water pollution is a major problem, especially in the United States, where 40 percent of the rivers are too polluted for fishing or swimming.
- As more of North America's population reaches retirement age (65 and older), there are fewer young people to take care of them.
- While Canada and the United States have similar democratic governments, they take very different approaches to issues such as unemployment and health care.
- Most North American farms have become highly mechanized operations that need few workers but require huge investments in land, machinery, fertilizers, and pesticides to be profitable.
- Since World War II, North America's urban populations have increased by about 150 percent, but the amount of land they occupy has increased almost 300 percent.
- North America is a major promoter of globalization and free trade throughout the world.
- North America's job market has become more oriented toward knowledge-intensive jobs requiring more education and specialized professional training in technology and management.
- For every dollar that a male worker in the United States earns, a female worker earns on average only about 62 cents (64 cents in Canada).
- A steady increase in migration from Middle and South America and parts of Asia promises to make North America a region where most people are of non-European descent.

HUMAN PATTERNS OVER TIME

The human history of North America is a series of arrivals and dispersals of people across the vast continent. In prehistoric times, humans came from Eurasia via Alaska. From the 1600s on, waves of European immigrants, enslaved Africans, and their descendants, spread over the continent, primarily from east to west. Today, immigrants from Asia and Middle and South America are arriving, mainly in the West. Internal migration is still a defining characteristic of life for North Americans, who are among the most mobile people in the world.

The Peopling of North America

Recent evidence suggests that humans first came to North America from northeastern Asia between 25,000 and 14,000 years ago, during an ice age. At that time, the global climate was cooler, polar ice caps were thicker, and sea levels were lower. The Bering land bridge, a huge, low landmass more than 1000 miles (1600 kilometers) wide, connected Siberia to

Alaska. Bands of hunters crossed by foot or small boats into Alaska and traveled down the west coast of North America.

By 15,000 years ago, humans had reached nearly to the tip of South America and had spread deep into that continent. By 10,000 years ago, global temperatures began to rise, and as the ice caps melted, the Bering land bridge sank beneath the sea. In North America a mid-continent corridor opened through the glaciers, allowing more people to pass to the south. Soon virtually every climate region throughout the Americas was occupied.

Over thousands of years, the people settling in the Americas domesticated plants, created paths and roads, cleared forests, built permanent shelters, and sometimes created elaborate social systems. The domestication of corn is thought to be closely linked to settled life and population growth in North America. About 3000 years ago, corn was introduced from Mexico into what is now the U.S. southwestern desert along with other Mexican domesticates, particularly squash and beans.

These foods provided surpluses that allowed some community members to engage in activities other than agriculture, hunting, and gathering, making possible large, city-like regional settlements. For example, by 1000 years ago, the urban settlement of Cahokia (in what is now central Illinois) covered 5 square miles (12 square kilometers) and was home to an estimated 30,000 people (Figure 2.11). Here people could specialize in crafts, trade, or other activities beyond the production of basic necessities.

The arrival of the Europeans. North America was completely transformed by the sweeping occupation of the continent by Europeans. In the early seventeenth century, the British established colonies along the Atlantic coast in what is now Virginia (1607) and Massachusetts (1620). Over the next two centuries, colonists from northern Europe built villages, towns, port cities, and plantations along the east coast. By the mid-1800s, they had occupied most Native American lands throughout the Appalachian Mountains and into the central part of the continent.

Disease, technology, and the Native Americans. The rapid expansion of European settlement was assisted by the vulnerability of Native American populations to European diseases. Having long been isolated from the rest of the world, Native Americans had no immunity to diseases such as measles and smallpox. Transmitted by Europeans and Africans who had built up immunity to them, these diseases killed around 90 percent of Native Americans within the first 100 years of contact.

Technologically advanced European weapons, trained dogs, and horses also took a large toll. Often the Native Americans had only bows and poisoned arrows. While some Native Americans in the Southwest

> In 1492, roughly 18 million Native Americans lived in North America. By 1542, after only a few Spanish expeditions, there were half that many. By 1907, slightly more than 400,000, or just 0.02 percent, remained.

Figure 2.11 Cahokia in 1150, an artist's interpretation. Long before European settlements in North America, 30,000 people lived in this urban area in what became central Illinois. [Painting by William R. Iseminger.]

acquired horses from the Spanish and learned to use them in warfare against the Europeans, their other technologies could not compete. Numbers reveal the devastating effect of European settlement on Native American populations. Roughly 18 million Native Americans lived in North America in 1492. By 1542, after only a few Spanish expeditions, only half that many survived. By 1907, slightly more than 400,000, or just 0.02 percent, remained.

The European Transformation

European settlement erased many of the landscapes familiar to Native Americans and imposed new ones that fit the varied physical and cultural needs of the new occupants. As a result, different subregions developed as settlement proceeded, and they still exist today. One way that geographers view the subregions of North America is shown in Figure 2.12. Refer to this figure as you read about the subregions below.

The southern settlements. European settlement of eastern North America began with the Spanish in Florida in 1565 and establishment of the British colony of Jamestown in Virginia in 1607. By the late 1600s, the colonies of Virginia, the Carolinas, and Georgia were cultivating cash crops such as tobacco and rice on large plantations.

To secure a large, stable labor force, Europeans brought enslaved African workers into North America beginning in 1619. Within 50 years, enslaved Africans were the dominant labor force on some of the larger southern plantations. By the start of the Civil War in 1861, slaves made up about one-third of the population in the southern states and were often a majority where plantations were located. North America's largest concentrations of African-Americans are still in the southeastern states (Figure 2.13).

The plantation system concentrated wealth in the hands of a small class of landowners, who made up just 12 percent of southerners in 1860. Planter elites kept taxes low and

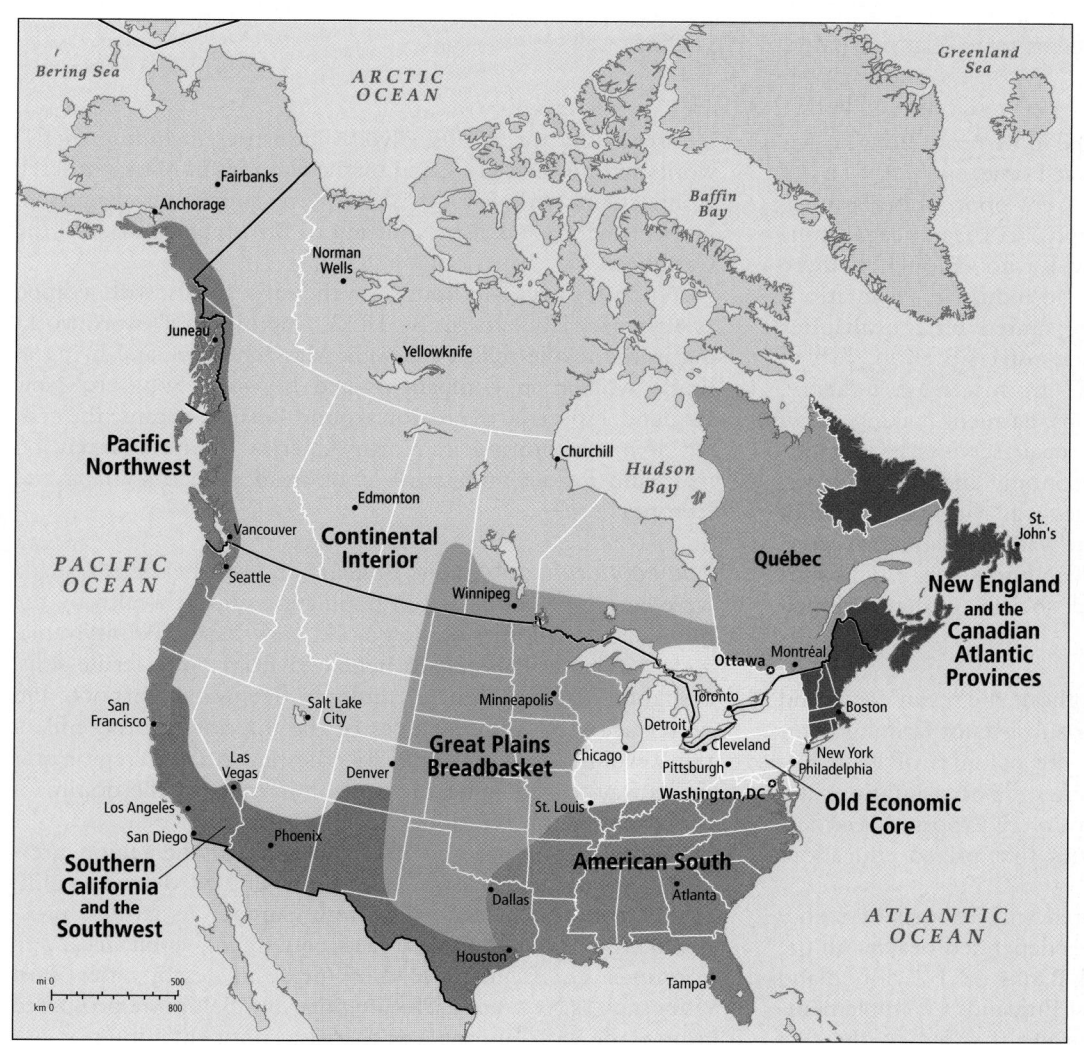

Figure 2.12 Subregions of North America. There are many schemes for dividing North America into subregions. The scheme we use here is partly indebted to Joel Garreau and his book *The Nine Nations of North America* (1981). Garreau proposed a set of regions that cut not only across state boundaries but also across national boundaries. Though Garreau includes parts of Mexico and the Caribbean, our examination is limited to the United States and Canada. [Adapted from Joel Garreau, *The Nine Nations of North America* (Boston: Houghton Mifflin, 1981).]

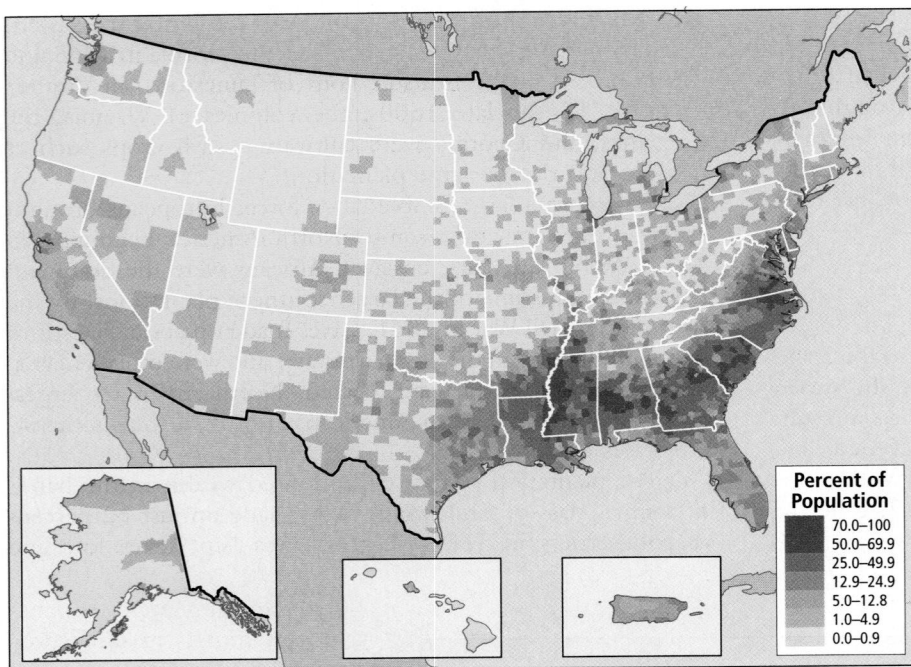

Figure 2.13 African-American population, 2000. While many African-Americans live in cities across the United States, the majority continue to live in the southeastern parts of the country. As of 2000, African-Americans were returning to the Southeast in greater numbers than they were leaving, many retiring on pensions earned elsewhere. The three inset maps here and in other U.S. maps show, left to right, Alaska, Hawaii, and Puerto Rico (a U.S. territory). [U.S. Census Bureau, Mapping Census 2000: The Geography of Diversity.]

Percent of Population
- 70.0–100
- 50.0–69.9
- 25.0–49.9
- 12.9–24.9
- 5.0–12.8
- 1.0–4.9
- 0.0–0.9

invested their money in Europe or the more prosperous northern colonies, instead of in industry at home.

More than half of southerners were poor white farmers. Both they and the slaves lived simply, so their meager consumption did not provide a demand for goods. Hence, there were few market towns and almost no industries. Plantations tended to be self-sufficient and generated little **multiplier effect**—economic development and diversification, which results when enterprises "spin off" from a main industry. (Examples would have included shops, garment making, small manufacturing, and transport and repair services.) Competition between the weak southern economy and the stronger, more diversified northern economies was the main cause of the Civil War (1861–1865). After the war, the plantation economy declined, and the South sank deep into poverty. This subregion remained economically and socially underdeveloped well into the twentieth century.

The northern settlements. Throughout the seventeenth and eighteenth centuries, relatively poor subsistence farming communities dominated agriculture in the colonies of New England and southeastern Canada. There were no plantations and few slaves, and not many cash crops were exported. Farmers lived in interdependent communities that prized education, ingenuity, and thrift.

At first, incomes were augmented with exports of timber and animal pelts. Some communities depended heavily on the rich fishing grounds of the Grand Banks off Newfoundland and Maine. By the late 1600s, New England was implementing ideas and technology from Europe that led to the first

industries. By the 1700s, diverse industries including metalworks and pottery, glass, and textile factories in Massachusetts, Connecticut, and Rhode Island were supplying markets in North America and also exporting to British plantations in the Caribbean.

Industry began to flourish in the early 1800s, with women as a primary labor force. By 1822, "factory girls" were working in the textile mills of Lowell, Massachusetts, and living as single women in company-owned housing. Southern New England, especially the region around Boston, became the center of manufacturing in North America. It drew largely on male and female immigrant labor from French Canada and Europe.

The economic core. New England and southeastern Canada were eventually surpassed in population and in wealth by the mid-Atlantic colonies of New York, New Jersey, Pennsylvania, and Maryland. This region benefited from more fertile soils, a slightly warmer climate, multiple deepwater harbors, and better access to the resources of the interior. By the end of the Revolutionary War in 1783, the mid-Atlantic region was on its way to becoming the **economic core,** or the dominant economic region, of North America.

Both agriculture and manufacturing in the region grew and diversified in the early nineteenth century, drawing immigrants from much of northwestern Europe. As farmers prospered, they bought mechanized equipment, appliances, and consumer goods, many of them made in nearby cities. Port cities such as New York, Philadelphia, and Baltimore prospered from trade with Europe and the vast continental interior.

By the mid-nineteenth century, the economy of the core was increasingly based on the steel industry, which diffused westward to the Great Lakes industrial cities of Cleveland, Detroit, and Chicago, and Pittsburgh. The steel industry further stimulated the mining of local deposits of coal and iron ore. Steel became the basis for mechanization, and the region was soon producing heavy farm and railroad equipment, including steam engines.

By the late nineteenth century, the economic core stretched from the Atlantic to Chicago and from Ottawa to Washington, D.C., and it dominated North America economically and politically well into the twentieth century. Most other areas produced food and raw materials for the core's markets and depended on the core's factories for manufactured goods.

Expansion West of the Mississippi

The east-to-west trend of settlement continued as land in the densely settled eastern parts of the continent became too expensive for new immigrants. By the 1840s, immigrant farmers from central and northern Europe, as well as some older settlers, were pushing their way west across the Mississippi River.

The Great Plains. Much of the land west of the Mississippi River was dry grassland, or prairie. The soil usually proved very fertile in wet years, and the area became known as the region's breadbasket. But the naturally arid character of this land eventually created an ecological disaster for Great Plains farmers. In the 1930s, after ten especially dry years, a series of devastating dust storms blew away topsoil by the ton; animals died, and entire crops were lost (Figure 2.14). This hardship was made worse by the widespread economic depression of the 1930s. Many Great Plains farm families packed up what they could and left what became known as the Dust Bowl, heading west to California and other states on the Pacific Coast.

The mountain West and Pacific coast. As the Great Plains filled up, other settlers were alerted to the possibilities farther west. By the 1840s, settlers were coming to the valleys of the Rocky Mountains (especially the Great Basin) and to the well-watered and fertile coastal zones of what was then known as the Oregon Territory and California. News of the discovery of gold in California in 1849 created the **Gold Rush,** drawing thousands with the prospect of getting rich quick. The vast majority of gold seekers were unsuccessful, however, and by 1852, they had to look for employment elsewhere. Further north, logging eventually became a major industry.

The extension of railroads across the continent in the nineteenth century facilitated the transportation of manufactured goods to the West and raw materials to the East (Figure 2.15). Today, the coastal areas of this region, often called the Pacific Northwest, have booming, diverse high-tech economies and growing populations. Perhaps in response to their history, residents of the Pacific Northwest are on the forefront of so many efforts to reduce human impacts on the environment that the region has been nicknamed "eco-topia."

The Southwest. The Southwest was first colonized by people from the Spanish colony of Mexico at the end of the sixteenth century. Their settlements were sparse, and as

Figure 2.14 Dust storm approaching Stratford, Texas, April 18, 1935. [NOAA/George E. Marsh Album.]

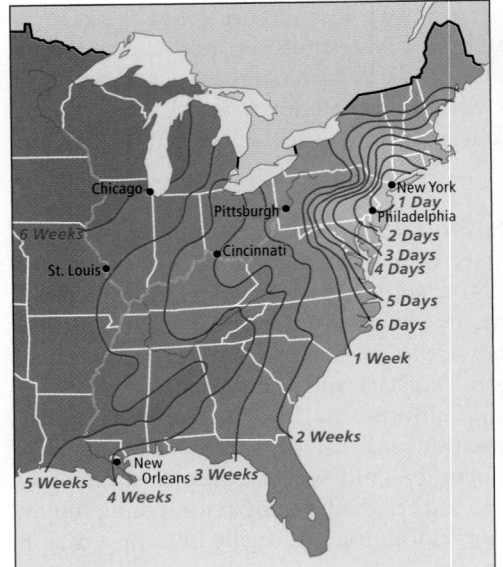

(a) Travel Times from New York City, 1800.
It took a day to travel by wagon from New York City to Philadelphia and a week to go to Pittsburgh.

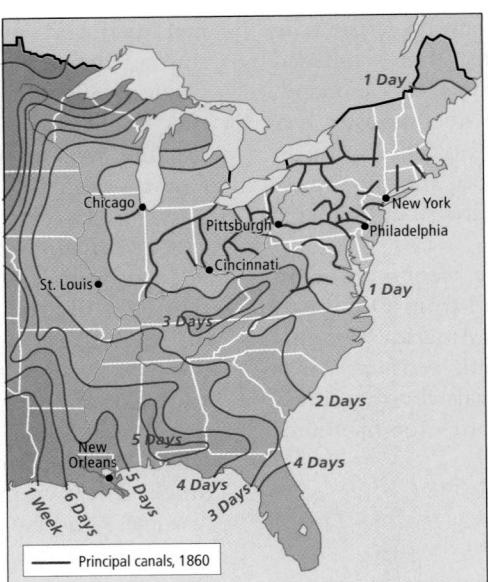

(b) Travel Times from New York City, 1857.
The travel time from New York to Philadelphia was now only 2 or 3 hours and to Pittsburgh less than a day because people could go part of the way via canals (dark blue). On the canals they could easily reach principal cities along the Mississippi River, and travel was less expensive and onerous.

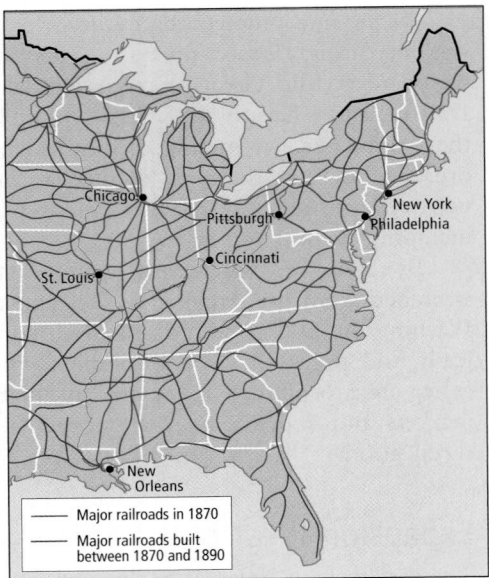

(c) Railroad Expansion by 1890.
With the building of railroads, which began in the decade before the Civil War, the mobility of people and goods increased dramatically. By 1890, railroads crossed the continent, though the network was most dense in the eastern half.

Figure 2.15 Nineteenth-century transportation. [Adapted from James A. Henretta, W. Elliot Brownlee, David Brody, and Susan Ware, *America's History,* 2nd ed. (New York: Worth, 1993), pp. 400–401; and James L. Roark, Michael P. Johnson, Patricia Cline Cohen, Sarah Stage, Alan Lawson, and Suan M. Hartmann, *The American Promise: A History of the United States,* 3rd ed. (Boston: Bedford/St. Martin's, 2005), p. 601.]

immigrants from the United States expanded into the region, Mexico found it increasingly difficult to maintain control. By 1850, nearly all of the Southwest was under U.S. control.

By the twentieth century, a vibrant agricultural economy was developing in central and southern California, made possible by massive government-sponsored irrigation schemes. There the mild Mediterranean climate made it possible to grow vegetables almost year-round. With the advent of refrigerated railroad cars, fresh California vegetables could be sent to the major population centers of the East. Southern California's economy rapidly diversified to include oil, entertainment, and a variety of engineering- and technology-based industries.

European Settlement and Native Americans

Native Americans living in the eastern part of the continent, who had survived early encounters with Europeans, occupied land that Europeans wished to use. Almost all the surviving Native Americans were forcibly relocated to relatively small reservations in the west with few resources. The largest relocation, in the late 1830s, involved the Creek and Cherokee of the southeastern states. These people had already adopted many European methods of farming, building, education, government, and religion. Nevertheless, they were rounded up by the U.S. Army and marched to Oklahoma along a route that became known as the Trail of Tears because of the many who died along the way.

As European settlers occupied the Great Plains and prairies, many of the reservations were further shrunk or relocated on ever less desirable land. Today, reservations cover just over 2 percent of the land area of the United States.

Reservations now cover 20 percent of Canada, mostly due to the creation of the Nunavut territory in 1999 and the Dogrib territory in 2003, both in the far north of Canada (see Figure 2.1). The Nunavut and Dogrib stand out as having won the right to legal control of these lands. In contrast to the United States, it is unusual for native groups in Canada to have legal control of their territory.

Most reservations in North America have insufficient resources to support their populations at the standard of living enjoyed by other citizens. After centuries of mistreatment, many Native Americans are more familiar with poverty than with true native ways. Alcoholism, suicide, and homicide rates are high, especially in the United States.

Beginning in the 1990s, several tribes found avenues to greater affluence on the reservations by establishing manufacturing industries; developing fossil fuel, uranium, and other mineral deposits underneath reservations; or opening gambling casinos. Although these activities have produced enormous income, they have also brought corruption. Environmental pollution has also increased, especially in mining areas.

Native American populations overall are now expanding from a low of 400,000 in 1900 to almost 4 million in 2006, most living in the United States.

The Changing Regional Composition of North America

The subregions of European-led settlement still remain in North America, but they are now less distinctive. The economic core no longer dominates as industry has spread to other parts of the continent. Some regions that were once dependent on agriculture, logging, or mineral extraction now have high-tech industries as well. The west coast in particular has boomed with a high-tech economy and a rapidly growing population. The west coast also benefits from trade with Asia, which now surpasses trade with Europe in volume and value.

POPULATION PATTERNS

The population map of North America (Figure 2.16) shows the uneven distribution of the more than 334 million people who live on the continent. Canadians account for just under one-tenth (33 million) of North America's population. They live primarily in southeastern Canada, close to the border with the United States. The population of the United States is about 301 million, with the greatest concentration of people still in the old economic core. ◼ However, other regions of the country are now growing much faster (Figure 2.17, on page 71). The 2000 U.S. census shows that the Northeast grew by just 5.5 percent in the 1990s, while the West grew by 20 percent and the South, east of the Mississippi River by 17 percent.

Many farm towns in the Middle West (the large central farming region of North America) have shrinking populations. As farms consolidate under corporate ownership, labor needs are decreasing and young people are choosing better-paying careers in cities. (Notice the changes in the area labeled *Middle West* in Figure 2.17.) Middle Western cities are growing modestly and becoming more ethnically diverse, with rising populations of Hispanics and Asians in such places as Indianapolis, St. Louis, and Chicago.

In the continental interior, settlement remains light. The mountainous topography, lack of rain, and in northern or high-altitude zones, a growing season that is too short to sustain agriculture are the main reasons for this low population density. Some population clusters exist in irrigated agricultural areas, such as in the Utah Valley, and near rich mineral deposits and resort areas. Las Vegas in Nevada, at the southern end of the region, is one of the fastest growing cities in the United States.

Along the Pacific coast, a band of population centers stretches north from San Diego to Vancouver and includes Los Angeles, San Francisco, Portland, and Seattle. These are all port cities engaged in trade around the **Pacific Rim**—all the countries that border the Pacific Ocean on the west and east. Over the past several decades, these cities also have become centers of innovation in computer technology. ◼

The rate of natural increase in North America (0.6 percent per year) is low, less than one-half the rate of the rest of the Americas (1.6 percent). Still, North Americans are adding to their numbers fast enough through births and immigration that they will reach 420 million by 2050. Many of the important social issues now being debated in North America, such as legal and illegal immigration, the language of instruction in public schools, and the social effects of the aging of the population, reflect changing population patterns.

Mobility in the United States

Every year, almost one-fifth of the U.S. population relocates. Some are changing jobs, others are attending school or retiring to a warmer climate, others are merely moving across town or to the suburbs or countryside. Still other people are arriving from outside the region; immigrants enter North America at the rate of about 5000 per day.

Dynamic economies and warmer climates are drawing both employers and retirees to cities throughout the South, Southwest, and Pacific Northwest. Cities in these areas have sprouted satellite or "edge" cities around their peripheries, often based on businesses dealing with technology and international trade. Here and elsewhere, urbanization has taken the form of urban sprawl, discussed further on pages 84–85.

Aging

During the twentieth century, the number of older North Americans grew rapidly. In 1900, 1 in 25 individuals was over the age of 65; by 1994, 1 in 8 was. By 2050, when most of the current readers of this book will be over 65, 1 in 5 North Americans will be elderly. The number of elderly people will shoot up especially fast between the years 2010 and 2030. This spike results from a marked jump in birth rate that took place after World War II, from 1947 to 1964. The so-called **baby boomers** born in those years constitute the largest age-group in North America, as can be seen in Figure 2.18 (page 71). As this group reaches age 65 and retires, the outflow of money from Social Security (the taxpayer-funded program that provides money to the elderly) and private pensions will be high, and medical costs will leap upward.

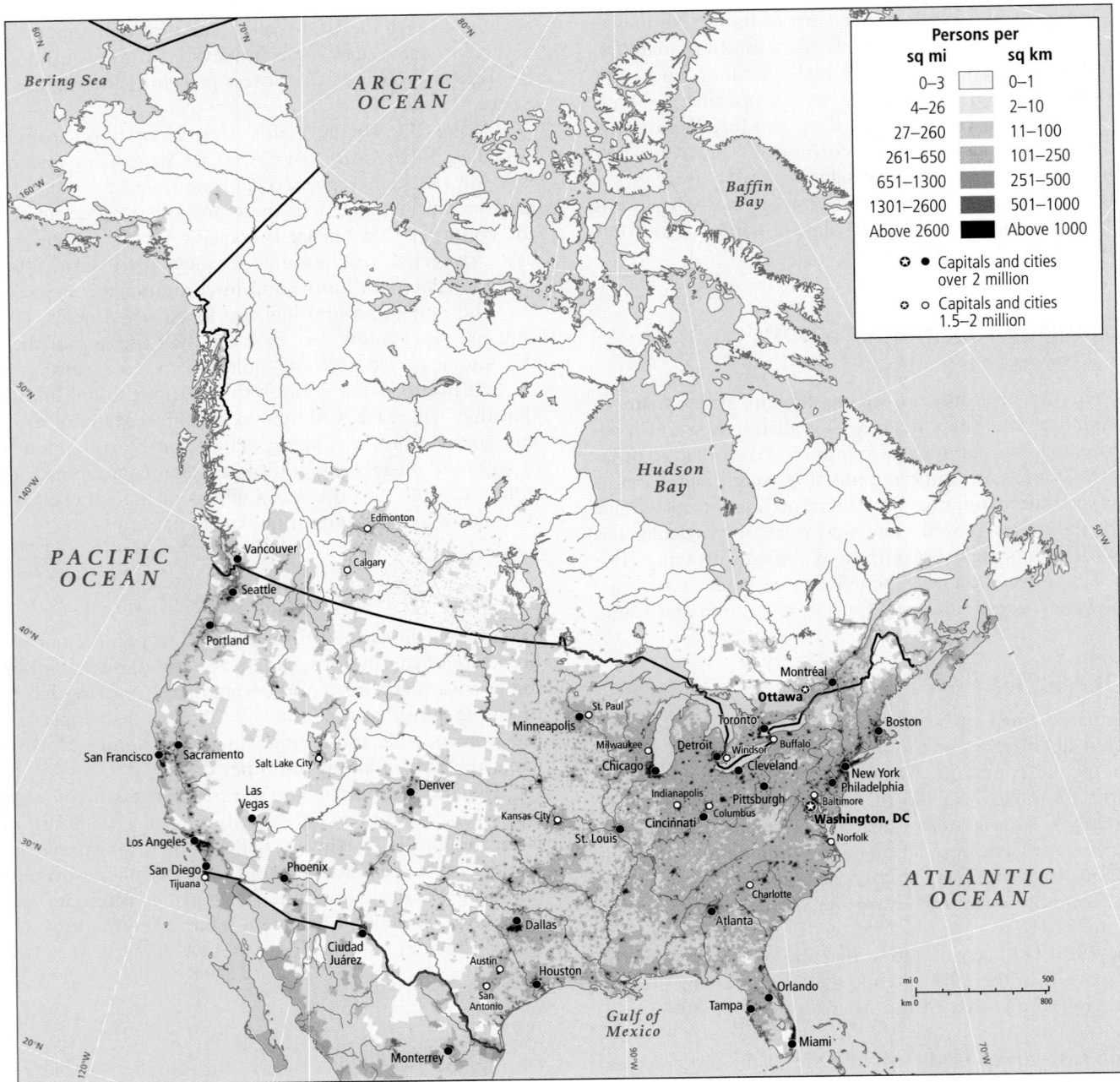

Figure 2.16 Population density in North America. [Data courtesy of Deborah Balk, Gregory Yetman, et al., Center for International Earth Science Information Network, Columbia University, at http://www. ciesin.columbia.edu.]

Because the boomers had fewer children than their parents had, there will be fewer people of working age to pay taxes and make pension fund contributions. Moreover, fewer people will be available to care for and provide companionship to elderly kin. Most families will not be able to afford assisted living and residential care, which in 2006 cost between $30,000 and $70,000 per year for one person. We might expect that once the boomers begin to retire in large num-

bers, living arrangements will change to reflect people's efforts to find humane and economical means to care for elderly family members.

The maps in Figure 2.19 show where the region's elderly were concentrated in the 1990s and how these patterns are expected to look by 2020. Many young people have moved from rural areas through the Middle West and Northeast to other parts of the country, leaving aging parents behind. In

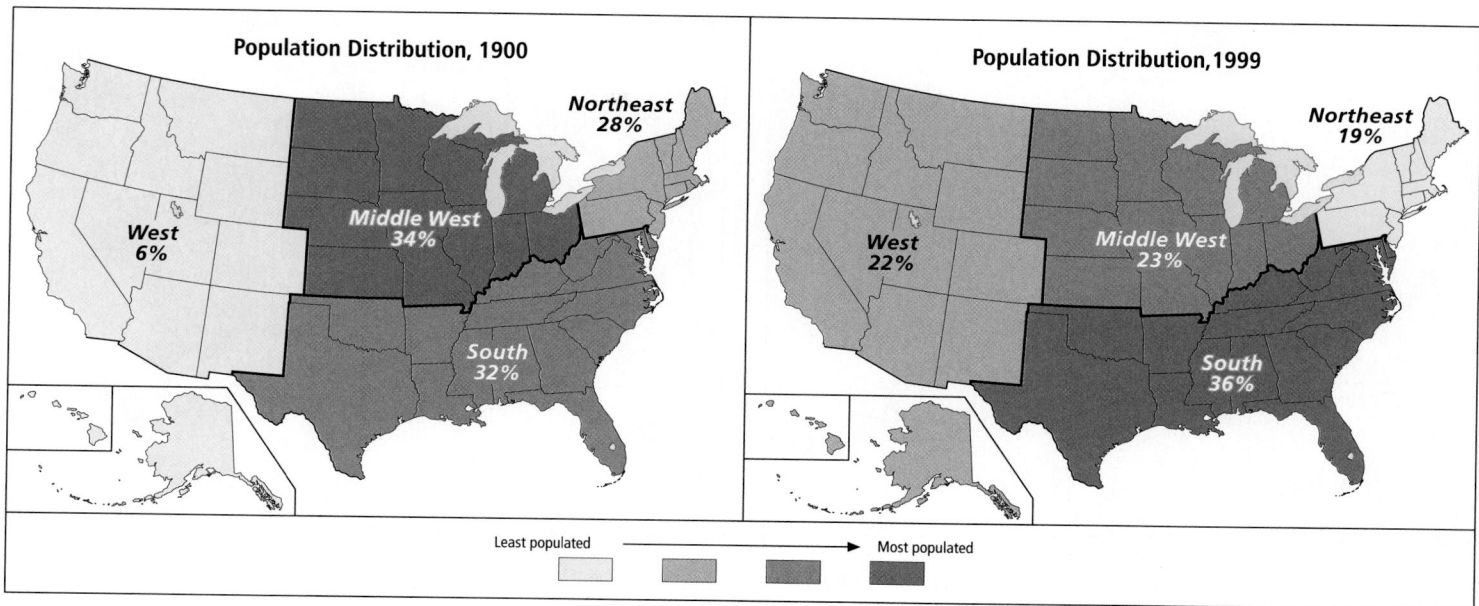

Figure 2.17 Percentage of population by region in the United States, 1900 and 1999. Notice how much the percentage has increased in the West and decreased in the Northeast and Middle West over the past 100 years. [Adapted from "America's diversity and growth," *Population Bulletin* (June 2000): 12. See http://www.statcan.ca/english/research/11F0019MIE/11F0019MIE2005254.pdf.]

other areas—especially the southern United States—the elderly are the new residents, attracted by the warm, pleasant climate of the Sunbelt.

The problems presented by aging populations reveal a dilemma. On one hand, it is widely agreed that population growth should be reduced to lessen the environmental impact of human life on earth, especially that of the societies that consume the most. On the other hand, slower population growth means that there will be fewer working-age people to keep the economy going and to provide the financial and physical help the increasing number of elderly people will require.

> Slower population growth means that as more of the world's citizens age, there will be fewer young, working-age people to keep the economy going and to provide the financial and physical help that the elderly require.

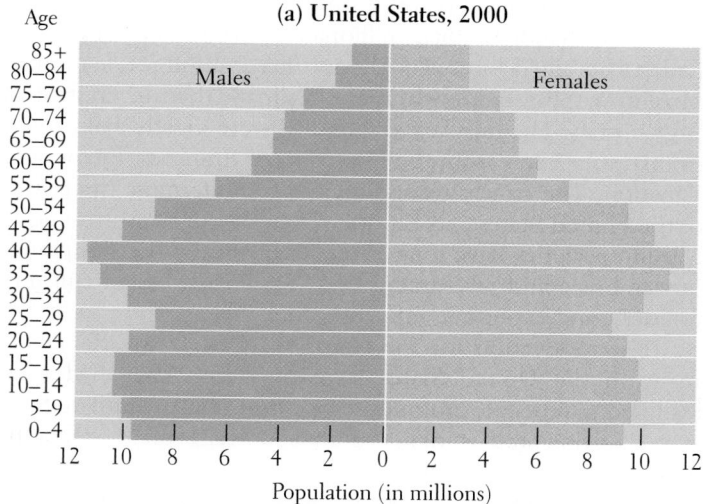

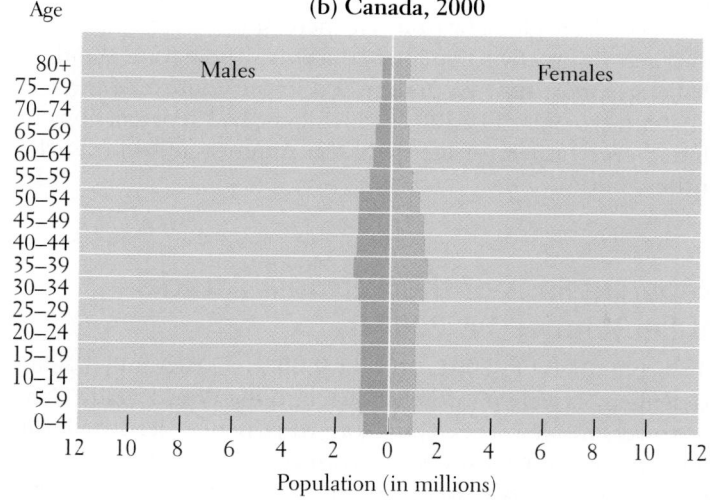

Figure 2.18 Population pyramids for the United States and Canada, 2000. The "baby boomers," born between 1947 and 1964, constitute the largest age group in North America, as indicated by the wider middle portion of these population pyramids. [Adapted from "Population pyramids of the United States, 2000" and "Population pyramids of Canada, 2000" (Washington, D.C.: U.S. Bureau of the Census), International Data Base, at http://www.census.gov/ipc/www/idb/.]

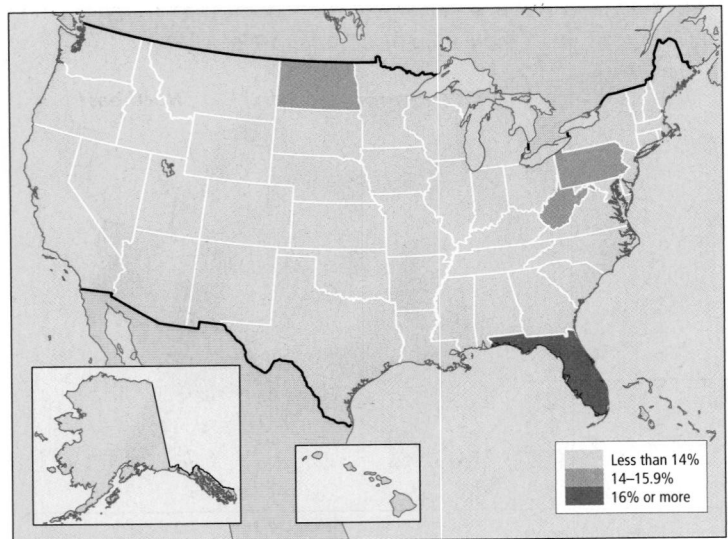

(a) Percent of total state population 65 years and over, 2004

Figure 2.19 Changing distribution of the elderly in the United States. The proportion of elderly people in the population is expected to increase across the country in the coming decades. The large numbers of elderly in California are masked on these maps by the fact that this state also has many young people. [Data from U.S. Census

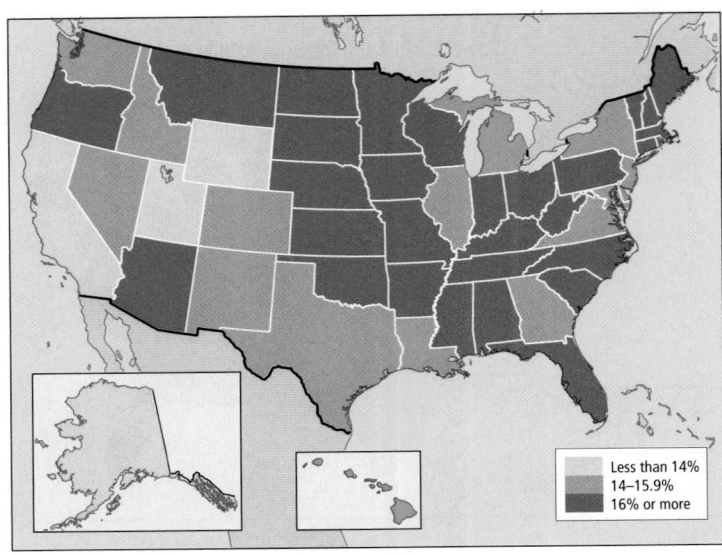

(b) Percent of total state population 65 years and over, 2020 (projected)

Bureau, 2000 census and 2004, at http://factfinder.census.gov/servlet/ ThematicMapFramesetServlet?_bm=y&-geo_id=01000US&-tm_name= ACS_2004_EST_G00_M00623&-ds_name=ACS_2004_EST_G00_&-_ dBy=040.]

Current Geographic Issues

Although North America is privileged by its huge economy and still-plentiful resources, the region faces complex problems posed by globalization, an increasingly diverse population, and rising environmental concerns. These long-standing challenges must now be understood in the context of the terrorist attacks of 9/11/2001, which have had repercussions in almost every aspect of life in North America and indeed the entire world.

Repercussions of the Terrorist Attacks of September 11, 2001

On September 11, 2001, around 9:00 A.M., two fully loaded passenger jets were purposely flown into the World Trade Center towers in New York City. A third passenger jet struck the Pentagon, and a fourth plane crashed in rural Pennsylvania, forced down by its passengers before it could reach its target in Washington, D.C. All four planes had been commandeered by suicide hijackers. Within the hour, the World Trade Center towers collapsed, killing 2792 people. Altogether, 3016 people died as a direct consequence of the attacks on 9/11.

Soon it was learned that the attacks were masterminded by a radical Muslim organization known as Al Qaeda, led by a dissident Saudi Arabian millionaire, Osama bin Laden. Bin Laden had allied himself with, and lived among, a faction of Muslim extremists known as the Taliban, then in control of Afghanistan. The hijackers themselves were mostly from Saudi Arabia, with a few from the United Arab Emirates, Egypt, and Lebanon. The attacks were designed to undermine the power of the United States and Europe. Especially in Southwest Asia, Muslim societies have long been conflicted by the simultaneous embrace and resentment of U.S. and European cultural, political, and economic influence.

The motivation for the terrorist attacks was difficult for many in the United States to understand. They had long assumed that people around the world sought to emulate their way of living and thinking. Public outrage fueled an anti-Muslim (and antiforeign) backlash that at times led to violent vigilantism. The U.S. government hastily passed the Patriot Act and used it to incarcerate hundreds of Muslim citizens and legal residents as well as illegal aliens for years without due process. Several years after the attack, it remains difficult for Americans to agree on which security strategies are appropriate, given the strong value placed on personal freedom in North American society.

Even more divisive has been the issue of what strategies the United States should implement on the international level. Should the United States focus primarily on its own interests and take aggressive stands against perceived enemies and possible terrorists? Or should it cooperate with the rest of the world to alleviate the festering resentment and outrage (justified or not) that led to the 9/11 events?

The war on terror. In the fall of 2001, the administration of George W. Bush, with the advice and consent of Congress, launched what was called the War on Terror. The first target was Afghanistan, thought to be host to the elusive Al Qaeda network. Though the aim was to capture bin Laden, his whereabouts are still unknown. Then, in the spring of 2003, President Bush brought the War on Terror to Iraq (Figure 2.20). The U.S. justified its invasion of that country with since-disproven claims that Iraq's President Saddam Hussein was involved in the 9/11 events, had chemical weapons, and was seeking to acquire materials for nuclear weapons.

The world is outraged. The international community extended warm sympathy to the United States after 9/11 and generally supported the war in Afghanistan. However, the manner in which the war in Iraq was conceived and carried out led to the most severe diplomatic isolation in U.S. history. The United Nations and all major U.S. allies, with the exception of the United Kingdom, advised against attacking Iraq and did not cooperate in the attack. World opinion was overwhelmingly against U.S. policy in Iraq. The failure of the United States to substantiate any of its prewar justifications for the invasion brought about high levels of opposition throughout the world and within the United States as well. Throughout 2008, over 60 percent of U.S. survey respondents said that the Iraq War was not worth fighting.

After its initial success in removing Saddam Hussein from power, the United States was unable to establish order and democracy in Iraq. Iraqi resistance to the U.S. occupation grew, and tension among Iraq's different political, ethnic, and religious groups brought the country to the brink of full-scale civil war. The most thorough surveys of civilian deaths due to the war suggest that between 600,000 and 1.2 million Iraqi civilians had died by early 2008. More than 4000 U.S. armed service people were killed and 128,000 wounded during this same period.

Equally damaging has been the U.S. decision not to treat captured prisoners in the War on Terror according to the **Geneva Conventions** (treaties that protect the rights of prisoners of war). Repeated discoveries of physical and psychological torture of prisoners by the U.S. military have sparked outrage both at home and abroad. Indeed, many critics of the U.S. responses to 9/11 point out that the various actions may have made the country more vulnerable to terror, not less, because of the loss of international support that resulted. By

Figure 2.20 A U.S. Army soldier on patrol in Baghdad, Iraq, in February 2007. [U.S. Arrmy photo by Sgt. Tierney Nowland.]

2007, opinion polls around the world revealed little support for the U.S. War on Terror.

The economy is shaken. The 9/11 attacks had immediate negative effects on the economy. The recession that was already looming in the spring of 2001 deepened as investors lost confidence, and hundreds of thousands of people lost jobs. Because the attacks made many people afraid to fly, cities and businesses dependent on tourism or business travel suffered huge losses. It took until the end of 2005 for U.S. air traffic to reach the level it had been at in the summer of 2001.

The economic impacts of the 9/11 attacks are now felt mainly through the cost of the wars in Afghanistan ($1 billion a month as of March 2006) and Iraq ($12 billion a month, or $200 billion per year by mid-2008). These costs are 14 times higher than the Bush administration had estimated in 2003. Because the war is now being fought with money the United States is borrowing from abroad, its economic impacts will linger for decades.

Dependence on oil is questioned. The 9/11 attacks led to alarms about U.S. dependence on oil produced in predominantly Muslim countries. In 2003, roughly 30 percent of the oil used in the United States came from countries with large Muslim populations. By 2005, it was up to 34 percent (Figure 2.21).

It is doubtful that all Muslim countries would participate in disrupting flows of oil to the United States because they are too dependent on the income. Most oil price analysts argue that over the long term, global oil prices will be most affected not by the aftermath of 9/11, but by spiraling demand for oil in growing Asian economies and worries over the effects of oil-based emissions on global warming. Nevertheless, the ever-increasing price of oil has sparked greater interest in alternatives to oil, especially renewable energy sources.

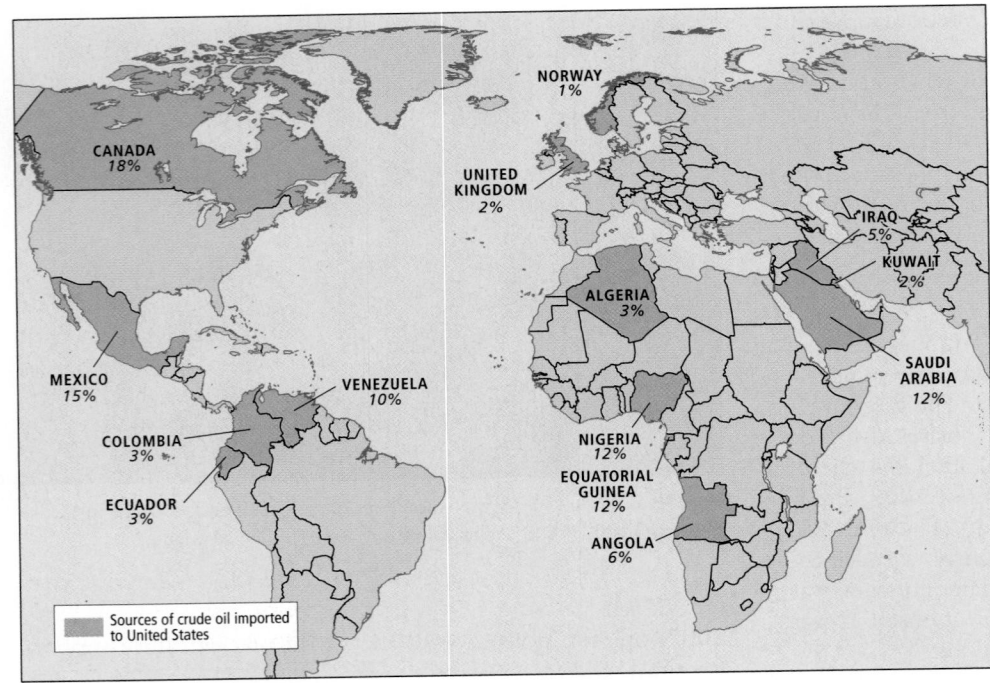

Figure 2.21 Sources of crude oil imported to the United States. The United States is dependent on crude oil from many locations around the world. This map shows the percentage of oil imported to the U.S. from specific suppliers as of November 2005. [Adapted from http://www.gravmag.com/oil.html.]

Relationship between Canada and the United States

Citizens of Canada and the United States share many characteristics and concerns. Indeed, in the minds of many people—especially those in the United States—the two countries are one. Yet that is hardly the case. Three key words characterize the interaction between Canada and the United States: *asymmetries, similarities,* and interdependencies.

Asymmetries

Asymmetry means "lack of balance." Although the United States and Canada occupy about the same amount of space (Figure 2.22), much of Canada's territory is sparsely inhabited, cold country. The U.S. population is nearly ten times the Canadian population. And although Canada's economy is one of the largest and most productive in the world (producing U.S.$1.18 trillion purchasing power parity (PPP) in goods and services in 2006), it is dwarfed by the U.S. economy, which is more than ten times larger ($13.13 trillion PPP in 2006).

> Although Canada's economy is one of the largest and most productive in the world, it is dwarfed by the U.S. economy, which is more than ten times larger.

There is also asymmetry in international affairs. The United States is an economic, military, and political superpower preoccupied with maintaining a world leadership role. Canada is only an afterthought in U.S. foreign policy, in part because the country is so secure an ally. But for Canada, managing its relationship with the United States is the top foreign policy priority. As former Canadian Prime Minister Pierre Trudeau once told the U.S. Congress, "Living next to you is in some ways like sleeping with an elephant: no matter how friendly and even-tempered the beast, one is affected by every twitch and grunt."

Figure 2.22 Political map of North America. Note that the U.S. state of Hawaii is not shown here because it is not part of North America; it is discussed in Chapter 11.

Similarities

The United States and Canada do have much in common. Both are former British colonies, and from this experience they have developed comparable democratic political traditions. Both are federations (of states or provinces), and both are representative democracies. Their legal systems are also alike.

Not the least of the features they share is a 4200-mile (6720-kilometer) border, which contains the longest sections of unfortified border in the world. There are now 1000 U.S. border guards along the Canadian border, far fewer than the nearly 10,000 agents along the border with Mexico, which is half as long. Although recent years have seen greater attention to securing the U.S.–Canada border, there are still thousands of miles without even a fence. Some rural residents pass out of one country and into the other many times in the course of a day simply by going about their usual business. ▣

Well beyond the border country, Canada and the United States share many other landscape similarities. Their cities and suburbs look much the same. The billboards that line their highways and freeways advertise the same brand names. Shopping malls have followed suburbia into the countryside, encouraging similar patterns of mass consumption. The two countries also share similar patterns of ethnic diversity that developed in nearly identical stages of immigration from abroad.

Interdependencies

Canada and the United States are perhaps most intimately connected by their long-standing economic relationship. The two countries engage in mutual tourism, direct investment, migration, and most of all, trade. By 2005, that trade relationship had evolved into a two-way flow of U.S.$1 trillion annually (Figure 2.23). Each country is the other's largest trading partner. Canada sells 85 percent of its exports to the United States and buys 59 percent of its imports from the United States. The United States, in turn, sells 23 percent of its exports to Canada and buys 17 percent of its imports from Canada.

Notice, however, that asymmetry exists even in the realm of interdependencies: Canada's smaller economy is much more dependent on the United States than the reverse. Nonetheless, if Canada were to disappear tomorrow, as many as one million American jobs would be threatened.

ECONOMIC AND POLITICAL ISSUES

The economic and political systems of Canada and the United States have much in common. Both countries evolved from societies based primarily on family farms. After an era of industrialization, both now have primarily service-based economies with important technology sectors and economic influence that reaches worldwide. While Canada and the United States have similar democratic governments and face similar problems, they often take very different approaches to issues such as unemployment, health care, and international relations.

North America's Changing Food Production Systems

North America benefits from an abundant supply of food, and it is an important producer of food for foreign as well as domestic consumers (Figure 2.24, page 77). However, though exports of agricultural products were once the backbone of the North American economy, agriculture now accounts for just 1.7 percent of the region's gross domestic product.

The shift to mechanized agriculture in North America has brought about sweeping changes in employment and farm ownership. In 1790, agriculture employed 90 percent of the American workforce; in 1890 it employed 50 percent. Thousands of highly productive family-owned farms, spread over much of the United States and southern Canada, provided for most domestic consumption and the majority of all exports until 1910. Today, agriculture employs less than 2 percent of the North American workforce.

Farms have become highly mechanized operations that need few workers but require huge investments in land, machinery, fertilizers, and pesticides to be profitable. Large **agribusiness** corporations, with better access to loans and cash, have an advantage over individual farmers, who now produce only 27 percent of the region's agricultural output.

Corporate agriculture provides a wide variety of food at low prices for North Americans. However, in many rural areas, corporate farms have depressed local economies and created social problems. Communities in such places as rural Iowa, Nebraska, and Kansas were once made up of farming families with similar incomes, similar social standing, and a commitment to the region. Today, farm communities are increasingly composed of a few wealthy farmer-managers amid a majority of poor, often migrant Hispanic or Asian laborers working for low wages on corporate farms and in food-processing plants.

Even families that find stable work in one place, such as a meatpacking plant, face considerable disadvantages. Many workers are paid only minimum wage and have difficulty affording housing. Locals are often reluctant to rent to immigrant workers, so many live in makeshift housing. For example, Laotian families in Storm Lake, Iowa, occupy a series of old railroad cars and shanties. Down the road, Hispanic workers live in two dilapidated trailer parks.

The children of migrant workers may attend school only occasionally (Figure 2.25). The Reverend Tom Lo Van, a Laotian Lutheran pastor in Storm Lake, sees little chance that Laotian youth will prosper from their parents' toil. "This new generation is worse off," he says. "Our kids have no self-identity, no sense of belonging . . . no role models. Eighty percent of [them] drop out of high school."

Under the assumption that the family farm has inherent value, the U.S. federal government and some states are increasing efforts to protect family farms and the rural communities of which they are a part. The states of Iowa, Kansas, Minnesota, Missouri, Nebraska, North and South Dakota, Oklahoma, and Wisconsin all have laws that restrict corporate involvement in agriculture.

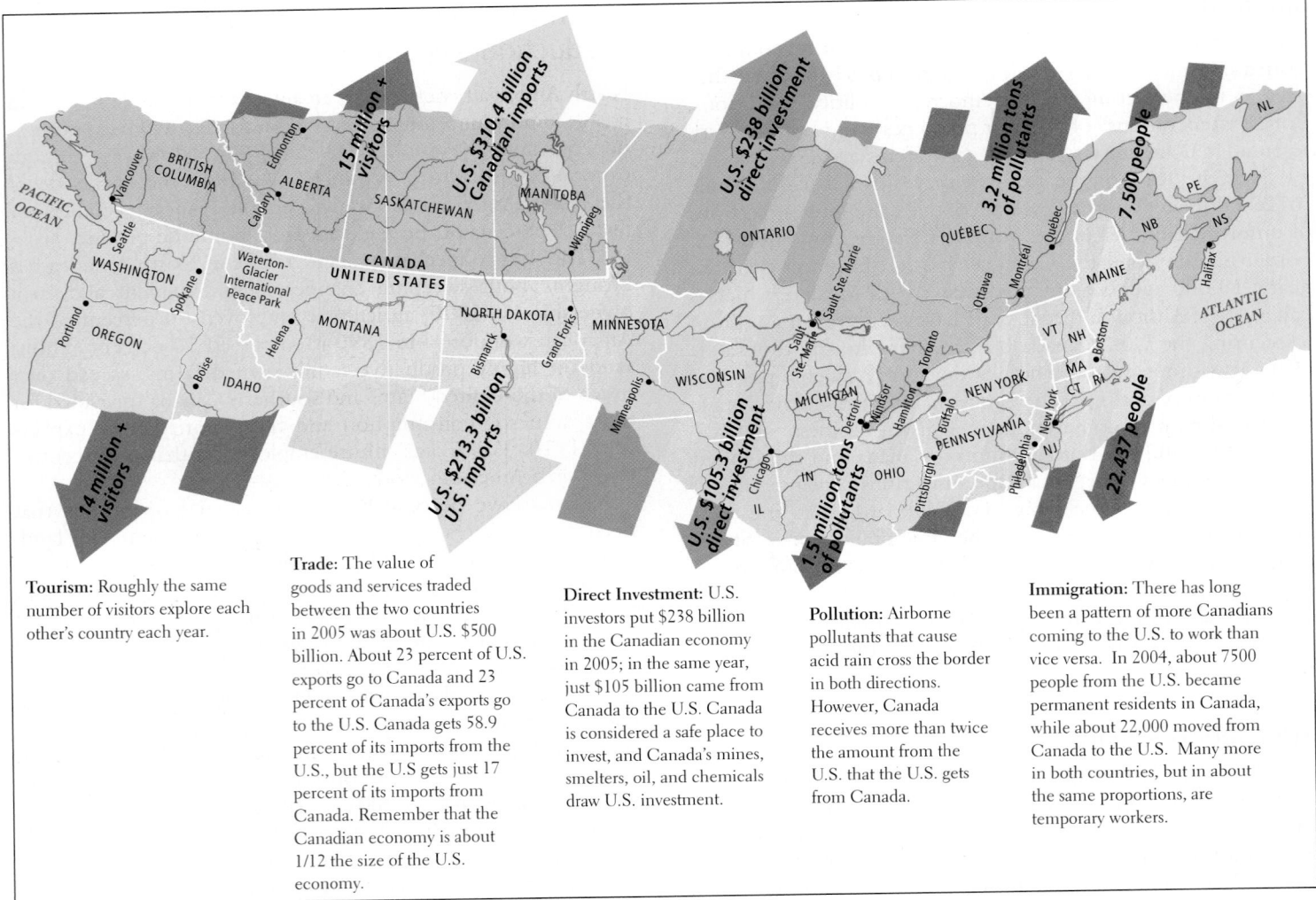

Figure 2.23 Transfers of tourists, goods, investment, pollution, and immigrants between the United States and Canada. Canada and the United States have the world's largest trading relationship. The flows of goods, money, and people across the long Canada–U.S. border are essential to both countries. However, because of its relatively small population and economy, Canada is more reliant on the United States than the reverse. All amounts shown are in U.S. dollars. [Adapted from *National Geographic* (February 1990): 106–107, and augmented with data from the Office of Travel and Tourism Industries, at http://tinet.ita.doc. gov/view/f-2000-04-001/index.html?ti_cart_cookie=20030901. 194057.17337; International Travel Forecasts, Fourth Quarter Update, 2003, at http://www.statcan.ca/Daily/English/080520/d080520c.htm; http://www.international.gc.ca/eet/pdf/FDI-stock-outward-by-Country-data-en.pdf; and http://www.international.gc.ca/eet/pdf/PFACT_Ann_Merch_Trade_Country_2007_apr_2008-en.pdf; and CIA Factbook, 2002, 2003; The Green Lane, Acid Rain, at http://www.ec.gc.ca/acidrain/acidhealth.html; and Population Today 30(2) (February–March 2002).]

Within the figure, the following annotations appear:

Tourism: Roughly the same number of visitors explore each other's country each year.

Trade: The value of goods and services traded between the two countries in 2005 was about U.S. $500 billion. About 23 percent of U.S. exports go to Canada and 23 percent of Canada's exports go to the U.S. Canada gets 58.9 percent of its imports from the U.S., but the U.S gets just 17 percent of its imports from Canada. Remember that the Canadian economy is about 1/12 the size of the U.S. economy.

Direct Investment: U.S. investors put $238 billion in the Canadian economy in 2005; in the same year, just $105 billion came from Canada to the U.S. Canada is considered a safe place to invest, and Canada's mines, smelters, oil, and chemicals draw U.S. investment.

Pollution: Airborne pollutants that cause acid rain cross the border in both directions. However, Canada receives more than twice the amount from the U.S. that the U.S. gets from Canada.

Immigration: There has long been a pattern of more Canadians coming to the U.S. to work than vice versa. In 2004, about 7500 people from the U.S. became permanent residents in Canada, while about 22,000 moved from Canada to the U.S. Many more in both countries, but in about the same proportions, are temporary workers.

Food Production and Sustainability

North American agriculture relies heavily on the use of highly modified seeds and chemical fertilizers, pesticides, and herbicides to increase crop yields. These farming methods can contaminate food, pollute nearby rivers, and even affect distant coastal areas (see pages 62–63). In addition, many North American farming areas have lost as much as one-third of their topsoil due to farming methods that create soil erosion.

Part of the problem may be the shift to corporate agriculture. Compared with owner-operators who have a personal stake in the long-term sustainability of a farm (Figure 2.26), corporate farms are more oriented toward short-term profits and hence are less likely to take environmental precautions. However, consumers are also to blame. Although an increasing number of North Americans are buying food grown "organically," without the use of pesticides or chemical fertilizers, most consumers still prefer food that is cheaper to food grown more sustainably.

Perhaps an even more controversial aspect of corporate agriculture is **genetically modified organisms (GMOs),** in which the DNA of animals and crop plants is modified. GMOs

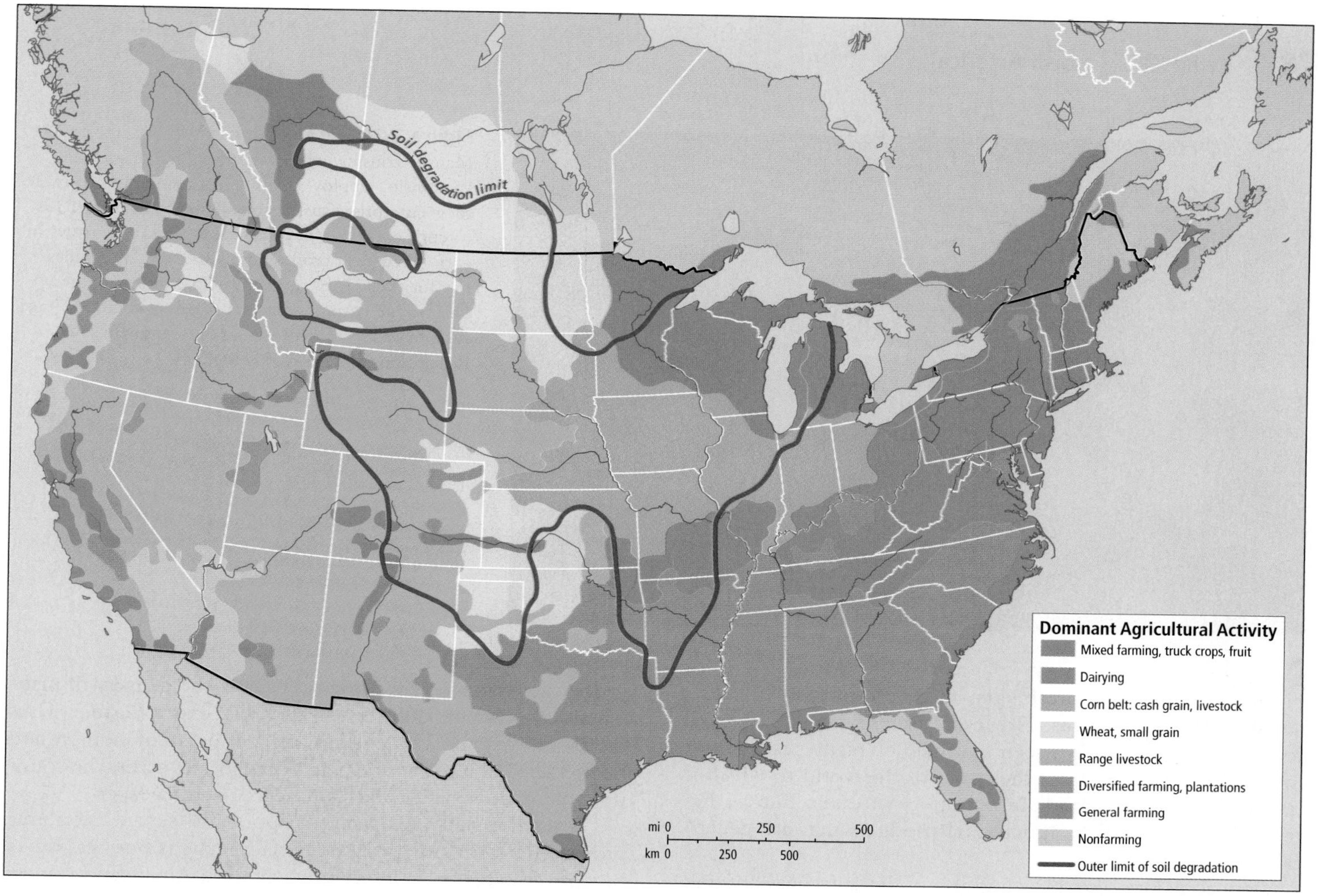
Soil degradation limit
Dominant Agricultural Activity
Mixed farming, truck crops, fruit
Dairying
Corn belt: cash grain, livestock
Wheat, small grain
Range livestock
Diversified farming, plantations
General farming
Nonfarming
Outer limit of soil degradation
mi 0
250
500
km 0
250
500

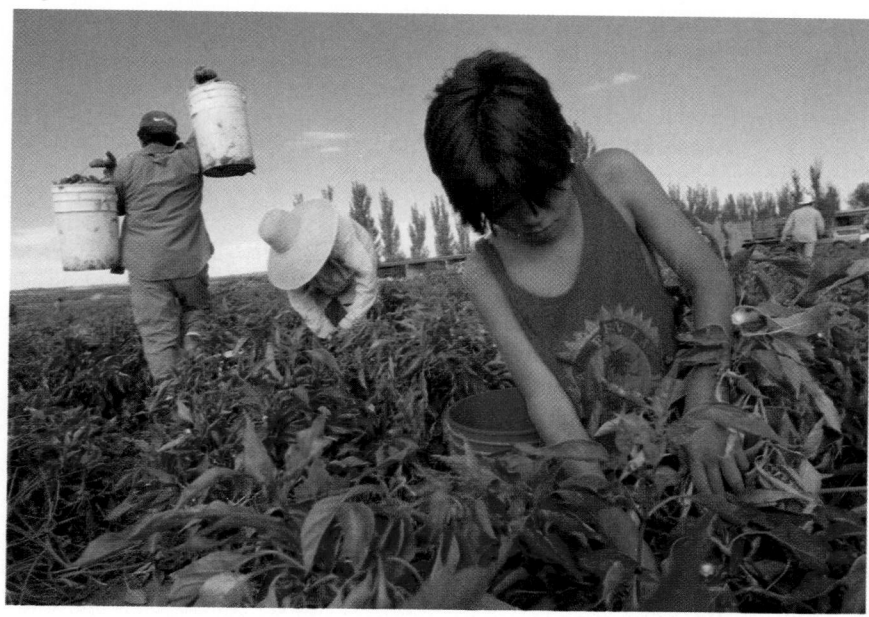

Figure 2.24 Agriculture in North America. North America is remarkable in that some type of agriculture is possible throughout much of the continent. The major exceptions are the northern reaches of Canada and Alaska and the dry mountain and basin region (the continental interior) lying between the Great Plains and the Pacific coastal zone. However, even some marginal areas, such as southern California, southern Arizona, and the Utah Valley, are cultivated with the help of irrigation. (Hawaii is not included here because it is covered in Oceania, Chapter 11.) [Adapted from Arthur Getis and Judith Getis, eds., *The United States and Canada: The Land and the People* (Dubuque, Iowa: William C. Brown, 1995), p. 165.]

Figure 2.25 Migrant farmworkers in the Southwest. José Madrid, 11, picks green chilis in New Mexico. "I'm not good at math, but I'm good at money," he says. Like many child migrant workers in the United States, he goes to school only intermittently. [Eric Draper/AP/Wide World Photos.]

Figure 2.26 The Manske farms in Wisconsin. The Manske cousins, who own adjoining farms in Wisconsin, employ a number of techniques to make their enterprises sustainable. They contour furrows to curb erosion. They also grow alternating rows of corn (their cash crop) and alfalfa (which provides hay for their dairy cows) to help control runoff. Large-scale corporate farms are rarely willing to take such measures to protect the environment. [Jim Richardson/Richardson Photography.]

produce varieties that are bigger, more productive, or resistant to pests and diseases. However, genetic manipulation of food crops and animals is seen as unsafe by many people both in North America and in other parts of the world to which it has traditionally exported. In a number of cases, Europe has refused to import U.S. agricultural products that are genetically modified.

Concerns are also growing about the sustainability and safety of food imported into North America. Currently, 80 percent of seafood, 45 percent of fresh fruit, and 17 percent of fresh vegetables consumed in the U.S. are imported. Recent years have seen a number of food safety scares related to food imported from China and elsewhere. Importing so much food from around the globe also contributes to global warming, as large amounts of fossil fuels are burned for transport.

Changing Transport Networks and the North American Economy

Many of North America's current transformations are related to changes in technology and trade. Manufacturing is fading in economic importance, just as agriculture did earlier, and a vibrant new service economy is emerging. Based on technology and information exchange, new economic centers are springing up across the continent. Essential to their productivity is an extensive network of road and air transportation that enables high-speed delivery of people and goods.

The development of the inexpensive mass-produced automobile in the 1920s changed North American transport radically. Soon trucks were delivering cargo more quickly and conveniently than the railroads could, and the number of miles of railroad track decreased between 1930 and 1980. The Interstate Highway System, a 45,000-mile (72,000-kilometer) network of high-speed, multilane roads, was begun in the 1950s and completed in 1990. Because this network was connected to the vast system of local roads, it made the delivery of manufactured products quicker, cheaper, and more flexible. Thus the highway system made possible the dispersal of industry and related services into suburban and rural locales across the country, where labor, land, and living costs were cheaper.

After World War II, air transport also served economic growth in North America. Its primary niche is business travel because face-to-face contact is essential to American business culture despite the growth of telecommunications and the Internet. Because many industries are widely dispersed in numerous medium-size cities, air service is organized as a **hub-and-spoke network**. Hubs are strategically located airports, such as the ones in Atlanta, Chicago, and Dallas, that are used as collection and transfer points for ongoing passengers and cargo. Most airports are also located near major highways, which provide an essential link for high-speed travel and cargo shipping.

Flying is now a way of life in North America, used for personal as well as business travel. In 2007, airlines in North America carried nearly 769 million passengers, more than double the continental population.

The New Service and Technology Economy

North America's job market has become more oriented toward knowledge-intensive jobs requiring education and specialized professional training in technology and management. Meanwhile, low-skill, mass-production industrial jobs are increasingly being moved abroad.

Decline in manufacturing employment. By the 1960s, the geography of manufacturing was changing. In the old economic core, higher pay and benefits and better working conditions won by labor unions led to increased production costs. Many companies began moving their factories to the southeastern United States, where wages were lower due to the absence of labor unions, and where the warmer climate meant lower energy costs.

In 1994, the **North American Free Trade Agreement (NAFTA)** was passed. In response, many manufacturing industries, such as clothing, electronic assembly, and auto parts manufacturing, began moving farther south to Mexico or overseas. Here labor was vastly cheaper. Further, employers could save because laws mandating environmental protection and safe and healthy workplaces were absent or less strictly enforced.

An equally significant factor in the decline of manufacturing employment is automation. The steel industry provides an illustration. In 1980, huge steel plants, most of them in the economic core, employed more than 500,000 workers. At that time, it took about 10 person-hours and cost about $1000 to produce 1 ton of steel. Spurred by more efficient foreign competitors in the 1980s and 1990s, the North American steel industry applied new technology to lower production costs, improve efficiency, and increase production. By 2006, steel was being produced at the rate of just 0.44 person-hour per ton and at a cost of about $165 per ton. As a result, the steel industry in the United States now employs fewer than half the workers it did in 1980. Throughout North America, far fewer people are now producing more of a given product at a far lower cost than was the case 20 years ago. Therefore, although the share of the gross domestic product (GDP) produced by manufacturing has declined, the actual level of industrial production has not.

Growth of the service sector. The economic base of North America is now a broad and varied **service sector.** Here people are engaged in the sale of services such as transportation, utilities, wholesale and retail trade, health, leisure, maintenance, government, and education.

As of 2000, in both Canada and the United States, about three-fourths of the GDP and a majority of the jobs were in the service sector. In Canada, 70 percent of workers, and in the United States 80 percent, now work in services. High-paying jobs exist in all the service categories, but low-paying jobs are more common. The largest employers in the United States are Wal-Mart and Manpower Inc. (an agency that arranges temporary contracts between workers and employers, often with no health care or retirement benefits).

Service jobs are often connected in some way to international trade. They involve the processing, transport, and trading of agricultural and manufactured products and information that are either imported to or exported from North America. Hence, international events can shrink or expand the numbers of these jobs.

An important subcategory of the service sector involves the creation, processing, and communication of information—what is often labeled the **knowledge economy.** The knowledge economy includes workers who manage information, such as those employed in finance, journalism, higher education, research and development, and many aspects of health care.

Industries that rely on the use of computers and the Internet to process and transport information—banks, software companies, and medical technology companies, for instance—are increasingly called **information technology (IT)** industries. They are freer to locate where they wish than were the manufacturing industries of the old economic core, which depended on locally available steel and energy, especially coal. Because IT industries depend on highly skilled thinkers and technicians, they often locate near major universities and research institutions. ◼▶

The Internet is emerging as an economic force more rapidly in North America than in any other world region. It was here that the Internet was first widely available. Though North America has only 5 percent of the world's population, it accounts for 22 percent of the world's Internet users. As of 2006, roughly 68 percent of the population of Canada and the United States used the Internet, compared to 50 percent in the European Union and 16 percent for the world as a whole.

However, a **digital divide** has also developed as important portions of the population are not able to afford computers and Internet connections. For the poor, the elderly, and many women and minorities, the public library or county courthouse may be the only access they have to the Internet.

Globalization and the Economy

North America is wealthy, technologically advanced, and hugely influential in the global economy. The United States is the world's second largest economy, after the European Union, and both it and Canada and have some of the highest per capita incomes in the world. North America's advantageous position in the global economy is reflected in the pro-globalization policies of free trade promoted by major corporations and the governments of Canada and the United States.

This wasn't always the case. Before its rise to prosperity and global dominance, trade barriers were important aids to North American development. For example, when it achieved independence from Britain in 1776, the new U.S. government imposed tariffs and quotas on imports and gave subsidies to domestic producers. This protected fledgling domestic industries and commercial agriculture, allowing its economic core region to flourish.

Having become wealthy, technologically advanced, and globally competitive exporters, both Canada and the United States are now powerful advocates for the reduction of trade barriers worldwide. Business leaders in both countries believe that tariffs and quotas on imports are no longer needed to protect their economic interests at home, and they see tariffs and quotas in other countries as obstacles to North America's economic expansion abroad.

Critics of free trade policies point out that many poorer countries still need tariffs and quotas, just as North America once did. They also argue that even in North America, the benefits of free trade go mostly to large businesses and their managers; many workers end up losing jobs or seeing their incomes stagnate.

Free trade is promoted regionally through NAFTA and globally through the United Nations, the World Bank, the International Monetary Fund (IMF), and the World Trade Organization (WTO)—(see the discussion in Chapter 1, pages 14–16).

NAFTA. Trade between the United States and Canada has long been relatively unrestricted by tariffs or quotas. The process of trade barrier reduction formally began with the Canada–U.S. Free Trade Agreement of 1989. The creation of NAFTA in 1994 brought in Mexico as well. The major long-term goal of NAFTA is to increase the amount of trade among Canada, the United States, and Mexico. Today, it is the world's largest trading bloc in terms of the GDP of its member states.

The effects of NAFTA are hard to assess because it is difficult to tell whether many changes are due to the agreement or to other changes in regional and global economies. However, a few things are clear. NAFTA has increased trade, and many U.S., Canadian, and Mexican companies are making higher profits because they have larger markets to sell in. Wal-Mart, the world's largest retailer, for example, expanded aggressively into Mexico after NAFTA was passed. Mexico now has more Wal-Marts than any country except the United States (Figure 2.27). On the other hand, NAFTA seems to have worsened the long-standing tendency for the United States to spend more money on imports than it earns from exports. This imbalance is called a **trade deficit.** Between 1993 and 2004, the value of U.S. exports to Canada and Mexico increased 77 percent, while the value of imports increased 137 percent.

NAFTA has also resulted in a net loss of around 1 million jobs in the United States. Increased imports from Canada and Mexico have displaced about 2 million U.S. jobs, while increased exports to these countries has created only about 1 million. Some new NAFTA-related jobs do pay up to 18 percent more than the average North American wage. However, those jobs are usually in different locations from the ones that were lost, and the people who take them tend to be younger and more highly skilled than those who lost jobs. Former factory workers often end up with short-term contract jobs or low-skill jobs that pay minimum wage and carry no benefits. 🎦

NAFTA and immigration from Mexico. In Mexico, NAFTA appears to have increased exports and levels of foreign investment. However, these gains have been concentrated in only a few firms along the U.S. border and have not increased growth in the Mexican economy. Moreover, stiff competition from

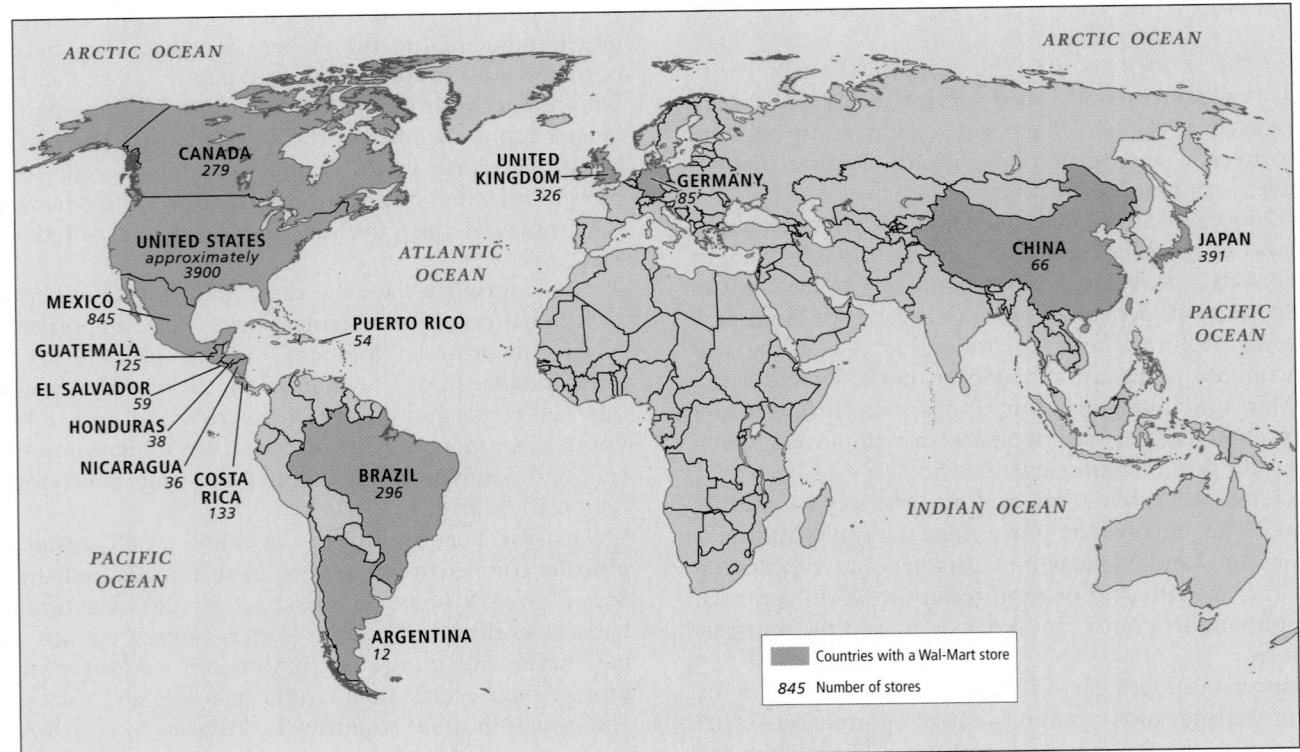

Figure 2.27 Wal-Mart on the global scale. In late 2006, Wal-Mart had 3944 store operations in the United States and 2745 in 14 other countries. [Adapted from http://www.wal-martchina.com/english/walmart/wm_world.htm.]

U.S. agribusiness has resulted in about 1.3 million job losses in the Mexican agricultural sector. These losses, in turn, have fueled legal and illegal immigration from Mexico to the United States, which increased from 350,000 per year in 1992 to 500,000 to 600,000 per year in 2005.

Even as the benefits and drawbacks of NAFTA are being assessed, there is talk of extending it to the entire Western Hemisphere. Such an agreement, which would be called the Free Trade Area of the Americas (FTAA), would have its own drawbacks and benefits. A number of countries, such as Brazil, Bolivia, Ecuador, and Venezuela, are wary of being overwhelmed by the U.S. economy. This emerging trade agreement is discussed in Chapter 3.

The Asian link to globalization.

NAFTA is only one way in which the North American economy is becoming globalized. The lowering of trade barriers has encouraged the growth of trade between North America and Asia, which surpassed growth in trade with Europe during the early 1990s.

One huge category of trade with Asia is the seemingly endless variety of goods imported from China—everything from underwear to chemicals used to make prescription drugs. China's lower wages make its goods cheaper than similar products imported from Mexico, despite Mexico's proximity and membership in NAFTA. Indeed, many factories that first relo-

> Stiff competition from U.S. agribusiness has resulted in about 1.3 million job losses in the Mexican agricultural sector, fueling huge increases in legal and illegal immigration from Mexico to the United States.

cated to Mexico from the southern United States are now moving to China to take advantage of its enormous supply of cheap labor. While recent concerns about lead and other toxins in Chinese imports have made many North Americans wary of these goods, trade with China promises to remain quite robust for some time.

Asian investment in North America is also growing. Japanese and Korean automotive companies, for instance, are locating plants in North America to be near their most important pool of car buyers—commuting North Americans. In rural locations close to the Interstate Highway System, the companies have found a ready, inexpensive labor force among rural North Americans. These workers are willing to commute 20 miles (32 kilometers) or more for secure jobs that pay reasonably well and include health and retirement benefit packages.

Toyota North America, for example, employs 38,000 North Americans. The Toyota Camry plant, constructed during the 1980s in Georgetown, Kentucky, is Toyota's largest plant outside Japan, employing 7500 people who turn out more than 500,000 vehicles and engines per year (Figure 2.28). Toyota recently announced a new Camry installation in Indiana with 1000 new jobs.

How is it that Japanese and Korean car makers can prosper in manufacturing in North America, even as U.S. auto manufacturers such as General Motors are downsizing their plants? One factor is more advanced Japanese automated productions systems that require fewer but better educated workers. These production systems are also capable of producing higher quality cars, which sell better both in North America and globally.

Figure 2.28 Toyota plant in Georgetown, Kentucky. Donna Jones puts control panels into Camrys as they move along the assembly line. Toyota leads foreign automakers in U.S. sales, and in 2006 had about 13 percent of the U.S. auto market. Eleven Toyota plants are located in the United States, and one is located in Canada. [AP Photo/Al Behrman.]

New competition from developing economies for IT jobs. By the early 2000s, globalization was resulting in the **offshore outsourcing** of information technology (IT) jobs. A range of jobs, from software programming to telephone-based customer support services, were shifted to lower-cost areas outside North America. By mid-2003, 500,000 IT jobs had been outsourced, and forecasts are that 3.3 million more will follow by 2020. New IT centers are developing in India, China, Southeast Asia, Central Europe, and Russia. Here large pools of highly trained, English-speaking young people work for wages that are less than 15 percent of their American counterparts'. Some argue that rather than depleting jobs, outsourcing will actually help job creation in North America by saving corporations money, which will then be reinvested in new ventures. The truth of this argument remains to be seen.

The Social Safety Net: Canadian and U.S. Approaches

The Canadian and U.S. governments have responded differently to the displacement of workers by economic change. For many decades, Canada has spent more than the United States on social programs that lessen the financial burdens of working people in times of economic crisis. These policies have made the financial lives of working people more secure. However, until recently, the higher taxes and greater economic regulation they entail have also made Canada slightly less attractive than the United States to new businesses. This has contributed to Canada's slightly higher unemployment rate and its slightly lower rate of economic growth relative to the United States.

By contrast, the U.S. approach provides little job protection or government unemployment assistance. Many poor rural and urban areas in the United States that have lost jobs have experienced increases in ill health, violent crime, drug abuse, and family disintegration. All of these problems are associated with declining incomes and an inadequate **social safety net,** the services provided by the government—such as welfare, unemployment benefits, and healthcare—that prevent people

from falling into extreme poverty. Nonetheless, for years the prevailing political position has been that the U.S. system offers lower taxes for businesses, which in turn attract new investment and new jobs, the benefits of which will trickle down to those most in need.

Health care. Some U.S. firms are beginning to eye Canada as a desirable place to relocate precisely because of the country's social safety net. Particularly attractive is Canada's government-sponsored health-care system. At present, among those U.S. firms that provide benefits, about 53 percent of the cost of every new employee is health insurance—a cost that places U.S. firms at a disadvantage in the global economy.

The contrasts between the two systems are striking. The Canadian health-care system is heavily subsidized and covers 100 percent of the population. In the United States, health care is largely private and relatively expensive. The government subsidizes care for the elderly, disabled, children, veterans, and the poor, though for these groups the quality of care is often much lower than in Canada. About 40 percent of the U.S. population is left with no health-care coverage. At the same time, the United States spends 16 percent of its gross domestic product on health care, while Canada spends only 10.6 percent. Nevertheless, Canada exceeds the United States on most indicators of overall health (Table 2.1). The taxes to cover Canada's system are substantial, but because employers don't have to pay for private care plans, they are offset by the lower cost of creating new jobs. ▄

Democratic Systems of Government: Shared Ideals, Different Trajectories

Canada and the United States have similar democratic systems of government, but there are differences in the way power is divided between the federal government and provincial or state governments. There are also differences in the way the division of power has changed since each country achieved independence.

Table 2.1 Health-related indexes for Canada and the United States

Country	Health-care costs as a percentage of GDP[a]	Percentage of population fully insured	Deaths per 1000	Infant mortality per 1000 live births	Maternal mortality per 100,000 live births	Life expectancy at birth (years)	Health expenditures per capita (PPP U.S.$)[b]
Canada	9.6	100	7.5	5	6	79	2931
United States	14.6	66	8.3	7.3	17	77	5274

[a]Data from 2002.   [b]PPP = purchasing power parity.

Sources: United Nations Human Development Report 2005 (New York: United Nations Development Programme), Table 6 and Table 10; World Resources Institute, "Population, health, and human well-being," at http://earthtrends.wri.org/pdf_library/data_tables/pop2_2005.pdf, in *Health Care Spending in 23 Countries,* at http://www.thirdworld traveler.com and http://www.thirdworldtraveler.com/Health/O_Canada.html.

Both countries have a federal government, in which a union of states (provinces in Canada) recognizes the sovereignty of a central authority while many governing powers are retained by state/provincial or local governments. In both Canada and the United States, the federal government has an elected executive branch, elected legislatures, and an appointed judiciary. In Canada, the executive branch is more closely bound to follow the will of the legislature. At the same time, the Canadian federal government has more and stronger powers (at least constitutionally) than does the U.S. federal government.

Over the years, both the Canadian and U.S. federal governments have moved away from the original intentions of their constitutions. Canada's originally strong federal government has become somewhat weaker. This is largely in response to demands by provinces, such as the French-speaking province of Québec, for greater autonomy over local affairs.

Meanwhile, the initially more limited federal government in the United States has expanded its powers. The U.S. federal government's original source of power was its mandate to regulate trade between states. Over time, this mandate has been interpreted ever more broadly. Now the U.S. federal government powerfully affects life even at the local level. This power is exercised primarily through its ability to dispense federal tax monies in programs such as grants for school systems, federally assisted housing, military bases, and the building of interstate highways. Money for these programs is withheld if state and local governments do not conform to federal standards. This practice has made some poorer states dependent on the federal government. However, it also has encouraged some state and local governments to enact more enlightened laws than they might have done otherwise. For example, in the 1960s, the federal government promoted civil rights for African-American citizens by requiring states to end racial segregation in schools.

Gender in National Politics and Economics

There are some powerful political contradictions in North America with regard to gender. Women cast the deciding votes in the U.S. presidential election of 1996, voting overwhelmingly for Bill Clinton. In the 2006 interim elections, 55 percent of women voted for the Democratic Party candidates, registering strong anti-war sentiment. Such women's issues as family leave, health care, equal pay, day care, and reproductive rights were forced into the national debate. 🎥

> Women cast the deciding votes in the U.S. presidential election of 1996, voting overwhelmingly for Bill Clinton.

However, of the 538 people in the U.S. Congress as of 2006, only 85, or 16 percent, were women. At the state level, 23 percent of legislators were women. In the Canadian Houses of Parliament, 21 percent of the members are women and 20 percent of provincial elected officials are women. Both the United States and Canada are well behind several other countries in their percentages of women in national legislatures: Cuba (36), Norway (38.2), and Sweden (45.3). Canada's governor-general (a largely ceremonial post appointed by the sovereign of the United Kingdom) is a woman (Figure 2.29).

On the economic side, in the United States, men still hold 70 percent of the top executive positions in business and government. On average, U.S. female workers earn only about

Figure 2.29 Canada's Governor-general, Michaëlle Jean. Canada is both a parliamentary democracy and a constitutional monarchy that recognizes the sovereign of the United Kingdom as head of state, though not as head of government. Canada's governor-general is appointed by the Queen to "carry out Her Majesty's duties." Here, Governor-general Michaëlle Jean (left, in beige suit) is being greeted by the lieutenant governor of British Columbia in March 2006. [Sgt. Eric Jolin, Rideau Hall/Government General of Canada.]

62 cents for every dollar that male workers earn (64 cents in Canada). In both countries, however, women tend to work more hours than men, so their hourly earnings are actually even lower. It is estimated that if women and men earned equal wages, the poverty rate would be reduced by half.

Women are now about half the North American labor force, with most working for male managers. In both countries, women entrepreneurs are increasingly active, starting close to half of all new businesses. Women-owned businesses tend to be small, however, and less financially secure than those in which men have dominant control. This is true in part because it is harder for women business owners to obtain loans and to land large contracts.

In education, North American women have attained virtually the same level as men in most categories. In the United States, 25 percent of women age 25 and over hold an undergraduate degree, while 29 percent of men do. This imbalance is likely to disappear because by 2000, U.S. women were receiving slightly more undergraduate degrees than men.

SOCIOCULTURAL ISSUES

North American attitudes about urbanization, immigration, race, ethnicity, and religion are changing rapidly. Changing gender roles are also redefining the North American family, and an aging population is raising new concerns about the future.

Urbanization and Sprawl

A dramatic change in urban form has transformed the way most North Americans live. Since World War II, North America's urban populations have increased by about 150 percent, but the amount of land they occupy has increased almost 300 percent.

Today close to 80 percent of North Americans now live in **metropolitan areas,** cities of 50,000 or more plus their surrounding suburbs and towns. Most of these people live in car-dependent suburbs built since WWII that bear little resemblance to the central cities of the past.

In the nineteenth and early twentieth centuries, cities in Canada and the United States consisted of dense inner cores and less-dense urban peripheries. Starting in the early 1900s, central cities began losing population and investment while urban peripheries, the suburbs, began growing. Workers were drawn by the opportunity to raise their families in single-family homes with large lots in secure and pleasant surroundings. Some continued to travel to the city to work, usually on streetcars. After World War II, suburban growth accelerated dramatically as cars became affordable to more workers.

As North American suburbs grew and spread out, nearby cities eventually coalesced into a single urban mass. The term **megalopolis** was originally coined to describe the 500-mile (800-kilometer) band of urbanization stretching from Boston through New York City, Philadelphia, and Baltimore, to the south of Washington, D.C. (Figure 2.30). Other megalopolis formations in North America include the San Francisco Bay area, Los Angeles and its environs, the region around Chicago, and the stretch of urban development from Eugene, Oregon, to Vancouver, British Columbia.

This pattern of urban sprawl requires residents to drive automobiles to complete most daily activities such as grocery shopping or commuting to work. Increasing air pollution and emission of greenhouse gases that contribute to global warming are two major side-effects of the dependence on vehicles

Figure 2.30 Megalopolis: Must development destroy natural landscapes? In a megalopolis, there is a continuing battle between urban developers and those who want to preserve landscapes such as wetlands. In 1969, the Hackensack (New Jersey) Meadowlands Development Commission was charged with overseeing development while protecting nearly 20,000 acres of wetlands. Today railroad yards, toxic waste sites, highways, power plants, and other industrial developments have eaten up more than 11,000 of those wetland acres. [Melissa Farlow/ National Geographic Image Collection.]

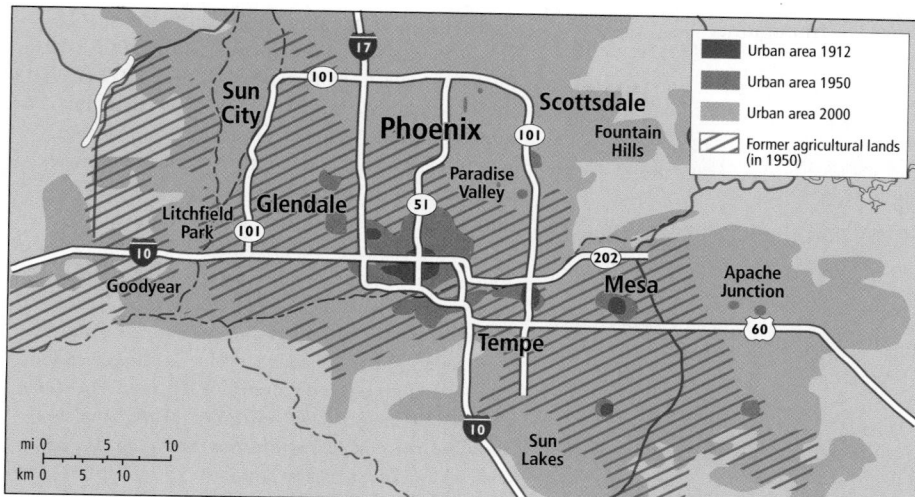

Figure 2.31 Urban sprawl in Phoenix, Arizona, 2004. To see an informative animation of the rate and extent of urban sprawl in Phoenix, go to http://sciencebulletins.amnh.org/bio/v/sprawl.20050218/. [American Museum of Natural History 2004.]

> Increasing air pollution and emission of greenhouse gases that contribute to global warming are two major side-effects of the dependence on vehicles that comes with urban sprawl.

that comes with urban sprawl. Another important environmental consequence is habitat loss brought about by the invasion of farmland, forest, grassland, and desert by sub-urban development.

Farmland and urban sprawl.

Urban sprawl drives farmers off land that is located close to urban areas. Farms on the urban fringe are very attractive to real estate developers. The land is cheap compared to urban land, and it is easy to build roads and houses on since farms are generally flat and already cleared. Each year, 2 million acres of agricultural and forest lands make way for urban sprawl. A striking example is Phoenix, Arizona, one of the fastest-growing areas in the United States (Figure 2.31).

Advocates of farmland preservation argue that beyond food and fiber, farms also provide economic diversity, soul-soothing scenery, and even habitat for some wildlife. The town of Pittsford, New York, a suburb of Rochester, decided in 1997 that farms were a positive influence on the community. Mark Greene's 400-acre, 200-year-old farm lay at the edge of town. As the population grew, the chances of the farm remaining in business for another generation looked dim. As land prices rose, so did property taxes, and the Greene family could not meet its tax payments. Residents in the new suburban homes sprouting up on what had been neighboring farms pushed local officials to halt normal farm practices, such as noisy nighttime harvesting or planting, spreading smelly manure, or importing bees to pollinate fruit trees. Pittsford decided to stand by the farmers by issuing $410 million in bonds so that it could pay Greene and six other farmers for promises that they would not sell their 1200 acres to developers, but would continue to farm them.

Pittsford is an exception to a trend throughout North America. Urban sprawl is now an issue virtually everywhere,

and open flat and rolling spaces are most vulnerable. An urban sprawl index that ranks cities on the severity of the problem can be found at http://smartgrowthamerica.org.

Inner city abandonment and reoccupation.

The same forces that drive urban sprawl leave behind huge tracts of abandoned inner-city land. Old industrial sites that once held factories or rail yards are called **brownfields.** Because they are often contaminated with chemicals and covered with obsolete structures, they can be very expensive to redevelop for other uses.

Also left behind in the inner cities are the least-skilled and least-educated citizens, many of whom are members of racial and ethnic minorities. In the early 1990s, 70 of the 100 largest U.S. cities had white majorities, but by 2000 almost half of the largest U.S. cities had nonwhite majorities. The majority population in these cities is a mixture of African-Americans, Asians, Hispanics, and other groups who identify themselves as nonwhite. Often these populations are in great need of the very services—health care, social support (including churches, synagogues, and mosques), and schools—that have moved with more affluent people to the suburbs.

Some inner cities are becoming densely populated again. In what has been labeled the **New Urbanism** movement, people are moving back to the city from the suburbs. Urban life offers less dependence on automobiles, as housing and jobs are often within walking distance, and greater diversity of residents and cultural activities.

Gentrification of old, urban residential districts is increasingly common as more affluent people invest substantial sums of money in renovating old houses and apartments, often displacing poor residents in the process. The effect of gentrification on the displaced poor appears to be somewhat less harsh in Canada than in the United States, primarily because of Canada's stronger social safety net that better ensures housing and social services.

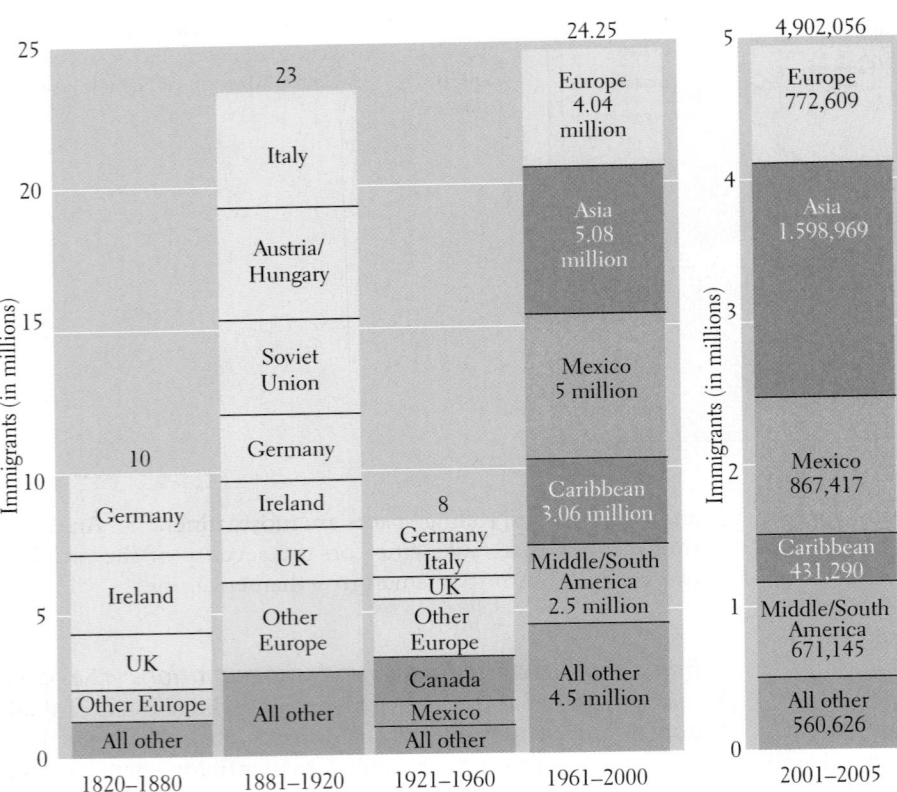

Regions and Countries of Origin for U.S. Immigrants

Figure 2.32 Changing national origins of legal U.S. immigrants. The graph for 2001–2005 is at a different scale from those for 1820–2000 because the time span is only five years. Note that the number of legal immigrants from Asia, Mexico, and other Middle and South American countries continues to outnumber European and other immigrants. [Adapted from Philip Martin and Elizabeth Midgley, "Immigration to the United States: Journey to an uncertain destination," *Population Bulletin* 49 (September 1994): 25; Martha Farnsworth Riche, "America's diversity and growth," *Population Bulletin* 55 (June 2000): 11; Table 2, "Immigration by region and selected country of last residence: fiscal years 1820–1998," *1998 Statistical Yearbook of the Immigration and Naturalization Service*, p. 10; and *Yearbook of Immigration Statistics: 2005*, Table 2, at http://www.uscis.gov/graphics/shared/statistics/yearbook/LPR05.htm.]

Immigration and Diversity

Immigration has played a central role in populating both the United States and Canada, and most of the region's people have roots in some other part of the globe, primarily Europe. However, new waves of migration from Middle and South America and parts of Asia promise to make this a region where most people are of non-European descent (Figure 2.32). Most major North American cities are already characterized by wide ethnic diversity. In Canada, with its relatively small population, the recent surge in immigration has led to near majorities of foreign-born residents in a number of leading cities.

In the United States, the spatial pattern of immigration is also changing. For decades immigrants settled mainly in coastal or border states such as New York, Texas, or California (Figure 2.33a). However, since about 1990, immigrants are increasingly settling in interior states such as Tennessee, Iowa, and Utah (Figure 2.33b).

Immigration and cultural diversity are topics of increasing public debate in the United States. By 2006, polls showed that a majority of U.S. residents felt future immigration should be controlled. At the same time, a clear majority also felt that immigration is a strength of the country, and 76 percent supported the idea that illegal immigrants should have a chance to become citizens. Below we examine some of the main issues in the immigration debate in the United States.

Do new immigrants cost U.S. taxpayers too much money? Repeated studies have shown that over the long run, immigrants contribute more to the U.S. economy than they cost. Most immigrants start to work and pay taxes within a week or two of their arrival in the country. Immigrants who draw on taxpayer-funded services such as welfare tend to do so only in the first few years after they arrive. Even illegal immigrants play important roles as payers of payroll taxes, sales taxes, and indirect property taxes through rent. Perhaps most noteworthy is the role of immigrants in support of the elderly. As the U.S. population ages and the base of native-born young workers shrinks, Social Security contributions by young new immigrant workers may provide essential support for the elderly.

A 2004 study by the National Institutes of Health (NIH) reports that immigrants are healthier and live longer than native U.S. residents. Hence, they are less of a drain on the healthcare and social service systems. The NIH attributes this difference to a stronger work ethic, a healthier lifestyle, and more nutritious eating patterns of new residents compared to U.S. society at large.

Do immigrants take jobs away from U.S. citizens? The least-educated, least-skilled American workers are the most likely to find themselves competing with immigrants for jobs. In a local area, a large pool of immigrant labor can drive down

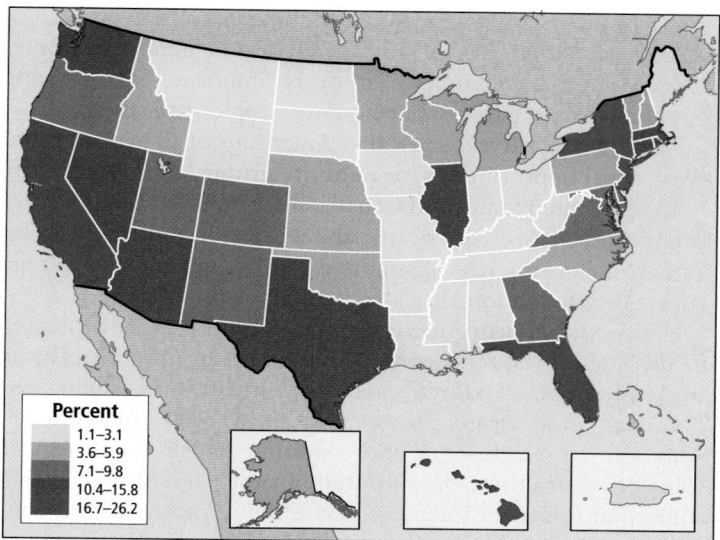

(a) Percent of total foreign-born within each state as of 2000. Includes all living foreign-born as of that year.

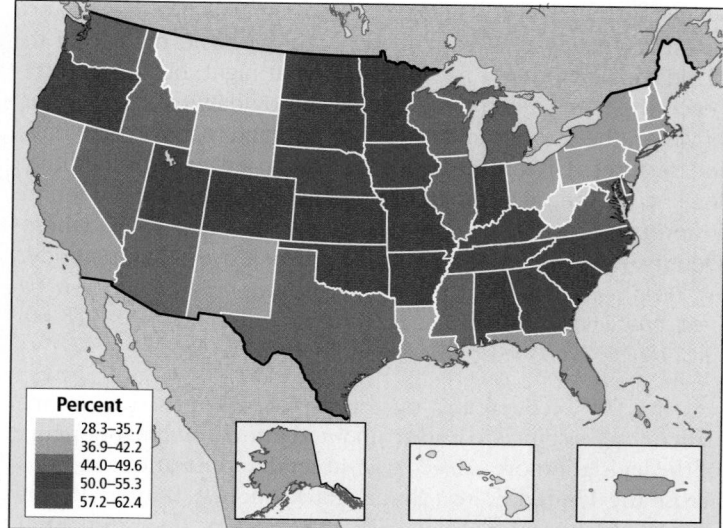

(b) Percent of foreign-born who entered each state during the 1990–2000 decade.

Figure 2.33 Changing patterns of foreign-born settlements in the United States. [U.S. Census Bureau.]

wages in fields like roofing, landscaping, and general construction. Immigrants with little education now fill many of the very lowest paid service and agricultural jobs.

It is often argued that U.S. citizens have rejected these jobs because of their low pay, and hence immigrants are needed to fill them. Others argue that these jobs might pay more, and hence be more attractive to U.S. citizens, if there weren't a large pool of immigrants ready to do the work for less pay. Research has not added much clarity. Some studies show that immigrants have driven down wages for U.S. natives without a high-school diploma by 7.4 percent. However, other studies show no drop in wages at all.

Professionals in the United States are also occasionally in competition with highly trained immigrants for jobs. Such competition usually occurs in occupations where there are not enough native-born people who are trained to fill these positions. The computer technology industry, for example, regularly recruits abroad in such places as India, where there is a surplus of high-tech workers. In this case also it is unclear whether immigrants are driving down wages.

Are too many immigrants being admitted to the United States?
Several circumstances contribute to increasing calls for curtailing immigration into the United States. First, after a mid-century lull, the immigration rate has picked up speed again (see Figure 2.32), particularly since the 1990s. Immigrants and their children accounted for 78 percent of U.S. population growth in the 1990s. At present rates, by the year 2050, the U.S. population will be over 420 million, nearly 25 percent larger than if all immigration stopped now. ▱

Second, undocumented (illegal) immigration has reached unprecedented levels over the past thirty years. The Census

Bureau estimates that the undocumented immigrant population in the United States is about 11 million, with 700,000 more entering each year. Undocumented immigrants tend to lack skills, and they cannot be screened for criminal backgrounds. The vast majority do not engage in criminal activity, however. Illegal immigrants take tremendous risks coming to North America. Around 6000 immigrants have died since 1994, mostly due to dehydration and exhaustion from long journeys through the deserts of the U.S.–Mexico border region.

Vignette Sheriff's deputy Michael Walsh works along the Arizona–Mexico border. Walsh describes a recent encounter with two bereaved young Mexican men holding the body of their relative: "Mostly you just find skeletons in the desert. This time there was a lot more emotion, family emotion. . . . Obviously they were close, they were crying. You have a name, you know that he had a brother, a cousin, a family in Mexico. . . ."

Matias Garcia, age 29, died after walking 32 miles through the desert. He was a Zapotec Indian who lived near Oaxaca (southern Mexico) with his wife and three children, as well as his parents, younger brothers, and several cousins. Matias's cash crop of chili peppers was ruined by a spring frost just as they were ripening, leaving him in debt. He reluctantly decided to risk a trip to the United States to work in some vineyards where he had worked on and off since he was a teenager. From there he could send money home to his family to keep the house in repair and to send his children to school. But it takes money just to cross the border, and it took months for Matias, his younger brother, and a cousin to save the necessary amount.

Since the NAFTA agreement of 1994, the border has been more carefully patrolled, so the men would attempt to cross the less-patrolled but more perilous Arizona desert on a route known as the "Devil's Highway." May is one of the hottest and driest months in this part of the Americas. The men tried to avoid the worst of the heat by walking at night, but they didn't reach the highway on the Arizona side by dawn. They ran out of water, the sun became especially hot, and Matias began having seizures. His brother and cousin carried him, desperately looking for the highway, but he died shortly before they found it and could flag down someone to call for help. That's where Deputy Walsh found the two men grieving over Matias's body.

Find out more about Matias Garcia, why he migrated, how he died, and what his death has meant back in his village. Watch the FRONTLINE/World video "Mexico: A Death in the Desert." ■

In 2006, federal and state laws were proposed to criminalize any assistance to undocumented immigrants. Opponents of the legislation organized large street demonstrations in cities across the United States. Some states, such as Illinois, Washington, Idaho, and New Mexico, support undocumented immigrants with programs that help them and their children. Virginia, Kentucky, South Carolina, and Arizona, on the other hand, have passed laws that crack down on illegal immigrants. Along the U.S.–Mexico border, the work of some 10,000 border guards was judged insufficient by conservative private militia groups, such as the Minutemen, who placed themselves along the border as vigilante enforcers. Similar sentiment has led to the creation of the U.S.–Mexico border barrier shown on page 54. Immigration is now a major political issue throughout the United States. ▰

Race and Ethnicity in North America

Throughout the world, people perceive skin color and other visible anatomical features to be significant, even though the science of biology tells us otherwise (see Chapter 1, pages 38–39). The same may be said for **ethnicity,** which is the cultural counterpart to race in that people may ascribe overwhelming (and often unwarranted) significance to cultural characteristics, such as religion or family structure or gender customs. Hence, race and ethnicity are very important factors not because they *have* to be, but because people *make* them so.

Numerous surveys show that Americans of all backgrounds favor equal opportunities for minority groups. Nonetheless, in both the United States and Canada, many middle-class African-Americans, Native Americans, and Hispanics (and to a lesser extent, Asian-Americans) report experiencing both overt and covert discrimination that affects them economically.

Although North America is ethnically and racially diverse, in this region the term *race* usually has been used in relation to deeply embedded discrimination against African-Americans. Prejudice has clearly hampered the ability of African-Americans to reach social and economic equality with Americans of other ethnic backgrounds. In the United States, and somewhat less so in Canada, despite the removal of legal barriers to equality, African-Americans as a group still experience lower life expectancies, higher infant mortality rates, lower levels of academic achievement, higher poverty rates, and greater unemployment than other groups.

In 2001, African-Americans were overtaken by Hispanics as the largest minority group in the United States. Because of a higher birth rate and a high immigration rate, the Hispanic population increased by 58 percent in the 1990s. Though Asian-Americans made up only 3.6 percent of the U.S. population in 2000, their population had increased by 48 percent. Figure 2.34 shows the changes in the ethnic composition of the U.S. population in 1950, 2004, and projected in 2050. ▰

> In 2001, African-Americans were overtaken by Hispanics as the largest minority group in the United States.

There are increasing discrepancies in income among Asian-Americans, Euro-Americans, Native Americans, Hispanics, and

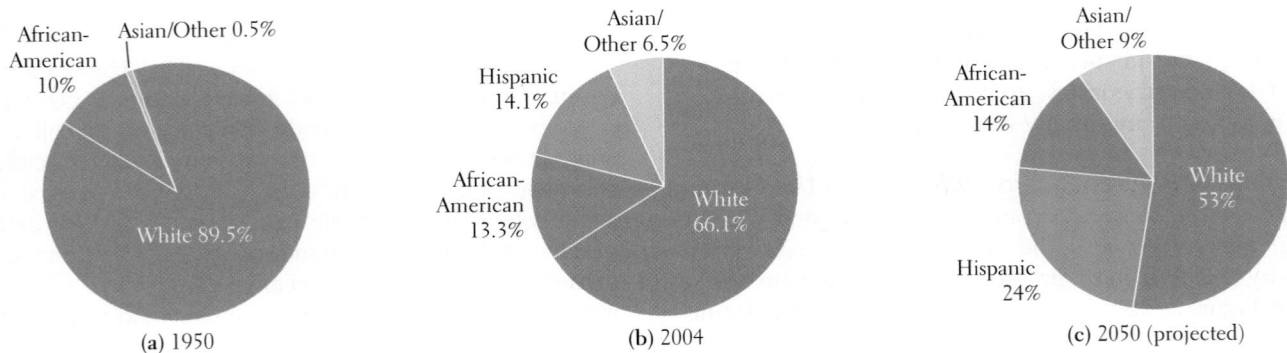

U.S. Population by Race and Ethnicity, 1950, 2004, and 2050 (projected)

(a) 1950 — African-American 10%, Asian/Other 0.5%, White 89.5%

(b) 2004 — Asian/Other 6.5%, Hispanic 14.1%, African-American 13.3%, White 66.1%

(c) 2050 (projected) — Asian/Other 9%, African-American 14%, Hispanic 24%, White 53%

Figure 2.34 The changing U.S. ethnic composition. The U.S. Census did not begin counting Hispanics as a separate category until 1970, so they do not appear in panel (a). [Adapted from Jorge del Pinal and Audrey Singer, "Generations of diversity: Latinos in the United States," *Population Bulletin* 52 (October 1997): 14; "U.S. Census Bureau, 2004 Updated Report," at http://www.census.gov/Press-Release/www/releases/archives/population/005164.html; and "Census of Population," volume 2, part 1, "United States Summary," Table 36, at http://www2.census.gov/prod2/decennial/documents/21983999v2p1ch3.pdf.]

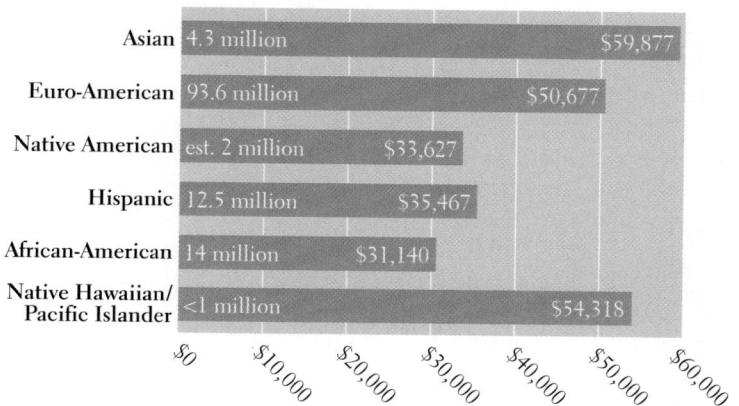

Figure 2.35 U.S. median household income by race and ethnicity, average for 2003–2005. [Adapted from "Income of households by race and Hispanic origin using 3-year-average medians: 2003 to 2005," *Income, Poverty, and Health Insurance Coverage in the United States: 2005,* Table 2, U.S. Census Bureau, August 2006, at http://www.census.gov/prod/2006pubs/p60-231.pdf.]

African-Americans in the United States (Figure 2.35). Over the past few decades, many non-Euro-Americans have joined the middle class and achieved success in the highest ranks of government and business. And yet, African-Americans, Hispanics, and Native Americans remain the country's poorest people.

Is anything other than prejudice holding back African-Americans, Hispanics, and Native Americans? Some argue that there is. They point to the experience of first- and second-generation African-Caribbean immigrants (such as Wyclef Jean or Colin Powell), who have been more successful than other African-Americans even though their experience of discrimination and past enslavement is very similar. A similar observation can be made about North Americans of Chinese and South Asian background. Though they often started out poor and suffered severe discrimination, these groups are now among the most prosperous in North America (see Figure 2.35).

Observations such as these have led to the argument that African Americans, Hispanics, Native Americans, and other persistently disadvantaged groups suffer from a "culture of poverty." The argument is that poverty and low social status breed the perception among their victims that there is no hope for them and hence no point in trying to succeed. This perception is supported by the fact that, relative to wealthier North Americans, the poorest, no matter what their background, have fewer opportunities to get a decent education, a well-paying job, or a nice home to live in.

A major aspect of the "culture of poverty" is the single-parent family. In 2005, while only 23 percent of Euro-American children lived in single parent families, 65 percent of African Americans, 49 percent of American Indians, and 36 percent of Hispanics did. Only 17 percent of Asian Americans lived in single parent families. Usually children stay with their mothers, and fathers often are not active in their support and upbringing. The enormous responsibilities of both child rearing and breadwinning are often left in the hands of undereducated young mothers who are unable to help their children advance.

However, another explanation of persistent poverty among some of North America's minorities is that it is part of a larger problem of economic and spatial segregation based on class. In both the United States and Canada, the increasingly prosperous middle class, of whatever race or ethnicity, has moved to the suburbs. Hence, the very poor rarely have the chance to associate with models of success. Evidence for this class-based explanation is that when privileged Americans of all racial backgrounds share middle- and upper-class neighborhoods, workplaces, places of worship, and marriages, attention paid to skin color decreases markedly. Meanwhile, the poor of all races see the material evidence of the success of others all around them but have little access to the life choices that made that success possible for others.

Religion

Because so many early immigrants to North America were Christian in their home countries, Christianity is currently the predominant religious affiliation claimed by North Americans. Nonetheless, virtually every medium-size city has at least one synagogue, mosque, and Buddhist temple. In some localities, adherents of Judaism, Islam, or Buddhism are numerous enough to constitute a prominent cultural influence (Figure 2.36).

There are many versions of Christianity in North America, and their geographic distributions are closely linked to the settlement patterns of the immigrants who brought them (see Figure 2.36). Roman Catholicism dominates in regions where Hispanic, French, Irish, and Italian people have settled—in southern Louisiana, the Southwest, and the far Northeast in the United States, and in Québec and other parts of Canada. Lutheranism is dominant where Scandinavian people have settled, primarily in Minnesota and the eastern Dakotas. Mormons dominate in Utah.

Baptists, particularly Southern Baptists and other evangelical Christians, are prominent in the religious landscapes of the South, which has come to be called the Bible Belt. Christianity—especially the Baptist version—is such an important part of community life in the South that newcomers to the region are frequently asked what church they attend. (This question is rarely asked in most other parts of the continent.)

The proper relationship of religion and politics has long been a controversial issue in the United States. This is true in large part because the framers of its Constitution, in an effort to ensure religious freedom, supported the idea that church and state should remain separate. In the past three decades, however, many conservative Christians have successfully pushed for a closer integration of religion and public life. Their political

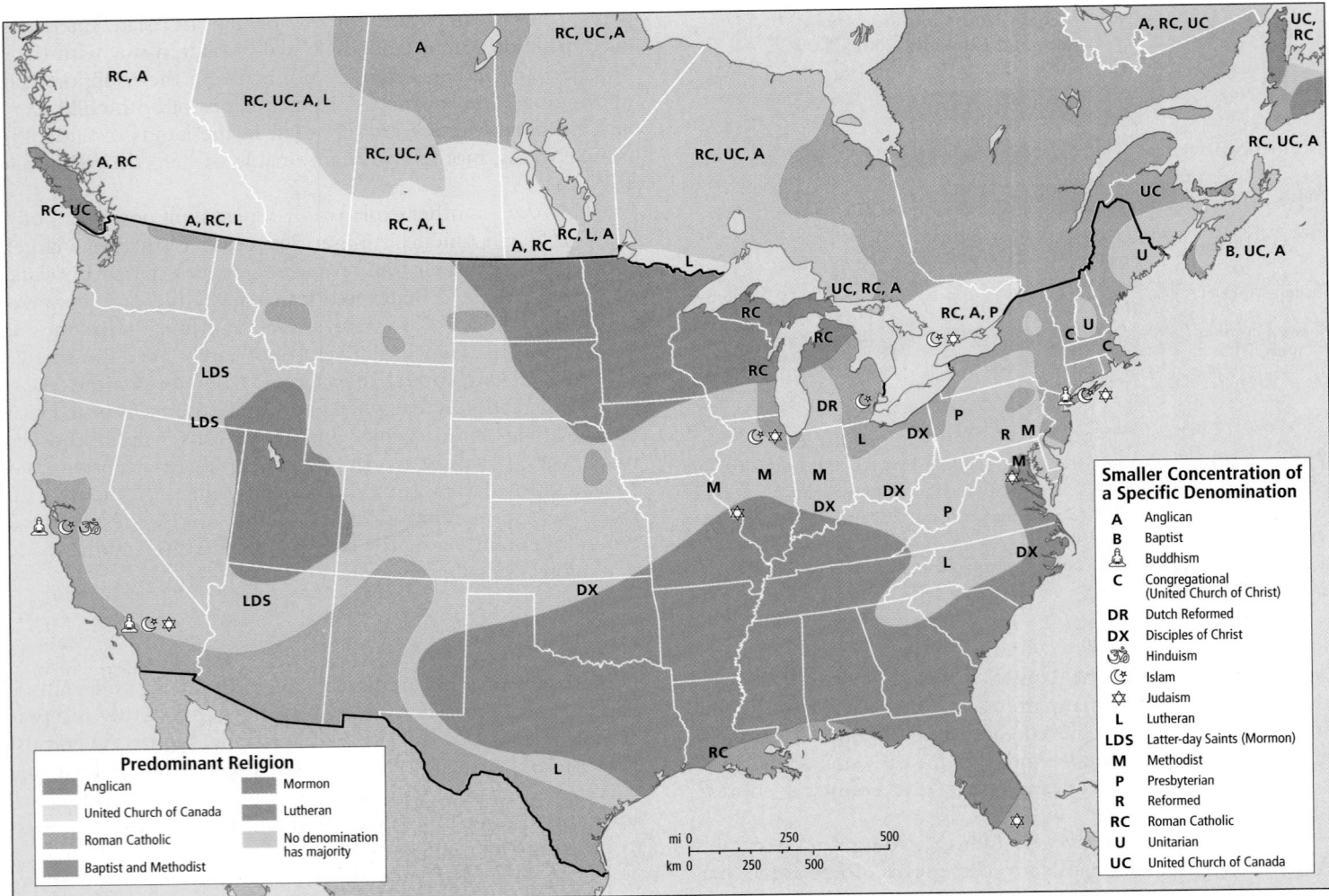

Figure 2.36 Religious affiliations across North America. [Adapted from Jerome Fellmann, Arthur Getis, and Judith Getis, *Human Geography* (Dubuque, Iowa: Brown & Benchmark, 1997), p. 164.]

goals include promoting prayer in the public schools, teaching the biblical version of creation instead of evolution, and banning abortion. The policies of conservative Christians have met with the most success in the southern United States, but their goals are shared by a minority scattered across the country.

New immigrants have brought their own faiths and belief systems, and they are contributing to the debate about religion and public life. The long-term outcome of these conflicts in American religious and political life is not yet apparent. Recent national surveys indicate that a substantial majority of Americans favor the continued separation of church and state and personal choice in belief and behavior.

Gender and the American Family

The family has repeatedly been identified as the institution most in need of support in today's fast-changing and ever more impersonal North America. A century ago, most North Amer-

icans lived in extended families of several generations. Families pooled their incomes and shared chores. Aunts, uncles, cousins, siblings, and grandparents were almost as likely to provide daily care for a child as its mother and father were. The **nuclear family,** consisting of a married father and mother and their children, is a rather recent invention of the industrial age.

Beginning after World War I, and especially after World War II, many young people left their large kin groups on the farm and migrated to distant cities, where they established new nuclear families. Soon suburbia, with its many similar single-family homes, seemed to provide the perfect domestic space for the emerging nuclear family.

This small, compact family-suited industry and business, too, because it had no firm ties to other relatives and hence was portable. Many North Americans born since 1950 have moved as many as ten times before reaching adulthood. The grandparents, aunts, and uncles who were left behind missed helping to raise the younger generation, and they had no one to look after them in old age. Nursing homes for the elderly proliferated.

In the 1970s, the whole system began to come apart. It was a hardship to move so often. Suburban sprawl meant onerous commutes to jobs for men and long, lonely days at home for women. Women began to want their own careers, and rising consumption patterns made their incomes increasingly useful to family economies. By the 1980s, 70 percent of the females born between 1947 and 1964 were in the workforce, compared with 30 percent of their mothers' generation.

Once employed, however, women could not easily move to a new location with an upwardly mobile husband. Nor could working women manage all of the family's housework and child care as well as a job. Some married men began to handle part of the household management and child care, but the demand for commercial child care grew sharply. With husband and wife both spending long hours at work, many people of both sexes found that their social life increasingly revolved around work, while family life receded in importance. With kinfolk no longer around to strengthen the marital bond, and with the new possibility of self-support available to women in unhappy marriages, divorce rates rose drastically. In 2006, about 50 percent of marriages ended in divorce in the United States and 40 percent in Canada.

By 2003, the nuclear family, which had been dominant for only a few decades, accounted for less than 24 percent of

> There is no longer a typical American household, only an increasing diversity of forms.

U.S. households (Figure 2.37). The most common household type— 28.2 percent—was a married couple *without* children. The fastest-growing household type, a single person living alone, had reached 26.4 per-

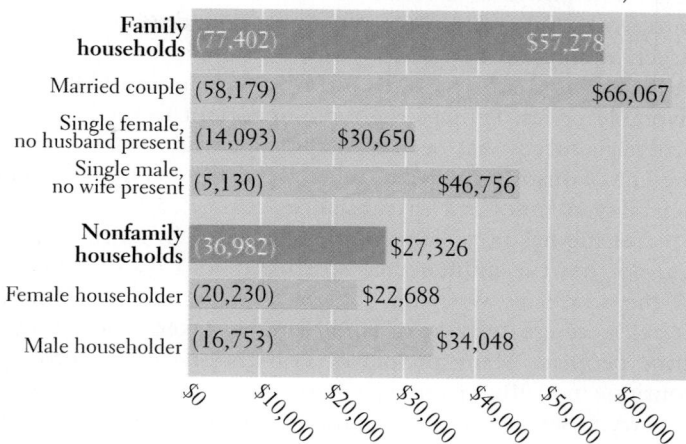

Median U.S. Income by Type of Household, 2005 (number of households in thousands)

Household type	Number	Median income
Family households	(77,402)	$57,278
Married couple	(58,179)	$66,067
Single female, no husband present	(14,093)	$30,650
Single male, no wife present	(5,130)	$46,756
Nonfamily households	(36,982)	$27,326
Female householder	(20,230)	$22,688
Male householder	(16,753)	$34,048

Figure 2.38 U.S. median income by household type, 2005. Nonfamily households can mean a single person living alone, an unmarried same-sex couple, or unrelated acquaintances living together. Notice the gender differences in income. [Adapted from "Income of households by race and Hispanic origin using 3-year-average medians: 2003 to 2005," *Income, Poverty, and Health Insurance Coverage in the United States: 2005,* Table 1, U.S. Census Bureau, August 2006, at http://www.census.gov/prod/2006pubs/p60-231.pdf.]

cent. More Americans than ever before are living alone; the majority are over the age of 45 and were once part of a nuclear family that dissolved due to divorce or death. There is no longer a typical American household, only an increasing diversity of forms.

Some of these new forms do not necessarily represent an improvement over the nuclear family, especially in providing for the welfare of children. In 2006, more than 29 percent of U.S. children lived in single-parent households. Although most single parents are committed to rearing their children well, the responsibilities can be overwhelming.

Single-parent families tend to be hampered by economic hardship and lack of education. The vast majority are headed by young women, whose incomes are, on average, 35 percent lower than those of single male heads of household (Figure 2.38). One result is that children in the United States are disproportionately poor. In 2005, 18 percent of U.S. children lived in poverty, whereas only 11.4 percent of adults did. In Canada, 14.7 percent of children were poor. In Sweden, by comparison, 2.4 percent of children lived in poverty; in Ireland, 12.4 percent; in Spain, 12.4 percent; and in Poland, 12.7 percent. 📹

Reflections on North America

It is not hard to rhapsodize about North America: the sheer size of the continent, its great wealth in natural and human resources, its superior productive capacities, and its powerful

Household Composition, 1970–2003 (percent)

	1970	1980	1990	1995	2000	2003	
	40.3	30.9	26.3	25.5	24.1	23.3	**Family households** Married couples with children (nuclear family)
	30.3	29.9	29.8	28.9	28.7	28.2	Married couples without children
	10.6	12.9	14.8	15.6	16.0	16.4	Other family households
	5.6	8.6	9.7	10.2	10.7	11.2	**Nonfamily households** Men living alone
	11.5	14.0	14.9	14.7	14.8	15.2	Women living alone
	1.7	3.6	4.6	5.0	5.7	5.6	Other nonfamily households

Figure 2.37 U.S. households by type, 1970–2003. [Adapted from Jason Fields, "America's family and living arrangements: 2003," *Current Population Reports,* November 2004, U.S. Census Bureau, at http://www.census.gov/prod/2004pubs/p20-553.pdf.]

political position in the world are attributes possessed by no other world region. North America enjoys this prosperity, privilege, and power as a result of fortunate circumstances, the hard work of its inhabitants, the diversity and creativity of its largely immigrant population, astute planning on the part of early founders, and access to worldwide resources and labor at favorable prices. Perhaps the most important factor in North America's success has been its democratic institutions, such as the U.S. Constitution, which allow for individual freedom and flexibility as times and circumstances change.

Yet life has not been good to everyone in North America, nor has the influence of North America on other parts of the world always been benign. Canada and the United States were created out of lands forcibly taken from indigenous peoples. Many people of non-European background continue to suffer from racial prejudice and a "culture of poverty." Settlers of all origins and their descendants had, and continue to have, a significant negative impact on the continent's environments. The standard of living now expected by all North Americans promises to increase the strain on the environment.

There is no guarantee that North America will continue its leadership role into the future. In fact, in the post-9/11 world, challenges to that leadership occur in every corner of the globe. Some critics say that the recent U.S. tendency to use military force before adequate consultation with allies has been attempted is a major cause for concern. Others cite insensitive trading policies or mistreatment of employees in the overseas workplaces of U.S.-based corporations such as Wal-Mart or Nike. North American models for development—based on democracy and on assumptions of rich and inexhaustible resources—are being challenged as inappropriate for much of the rest of the world. As we shall see in subsequent chapters, societies elsewhere are beginning to prosper without following North American examples, and sometimes without first installing democratic institutions.

Chapter Key Terms

acid rain 60
agribusiness 75
aquifer 61
baby boomer 69
brownfield 85
clear-cutting 58
digital divide 79
economic core 66
ethnicity 88
genetically modified organism
  (GEM) 76
Geneva Conventions 73
gentrification 85

Gold Rush 67
hazardous waste 63
Hispanic 53
hub-and-spoke network 78
information technology
  (IT) 79
knowledge economy 79
megalopolis 84
metropolitan area 84
multiplier effect 66
New Urbanism 85
North American Free Trade
  Agreement (NAFTA) 79

nuclear family 90
Ogallala aquifer 61
offshore outsourcing 82
Pacific Rim (Basin) 69
Québecois 53
service sector 79
social safety net 82
smog 60
thermal inversion 60
trade deficit 80
urban sprawl 60

Critical Thinking Questions

1. Discuss the ways in which North American culture has adapted to its residents' high rate of spatial mobility.

2. The North American population is changing in many ways. Explain why the aging of North America is of concern generally, and how you personally are likely to be affected.

3. North American family types are changing. Explain the general patterns and discuss whether or not the nuclear family is or should be the sought-after norm.

4. The influence of globalization is now felt even in small, isolated places in North America. Choose an example from the text or from your own experience and explain four ways in which this place is now connected to the global economy or global political patterns.

5. North Americans profit from having undocumented workers produce goods and services. Why is this the case? Discuss if and how the situation should be changed.

6. People like to say that the attacks on 9/11 changed everything for North Americans (especially for those living in the United States). What do they mean by this? How were U.S. attitudes toward the rest of the world modified? Do you think these modifications will be useful in the long run?

7. When comparing the economies of the United States and Canada and the ways they are related, what are the most important factors to mention?

8. Why do producers and farmers in poor countries say that the U.S. government's practice of giving subsidies to U.S. farmers hurts them?

9. Examine the connections between urban sprawl, rising consumption of gasoline, and rising air pollution in North American urban areas. Then propose a solution that you could abide by yourself.

10. The United States and Canada have quite different approaches to treatment of citizens who experience difficulties in life. Explain those different approaches and the philosophies behind them.

ATLANTIC OCEAN

Canary
Islands

Western
Sahara

MAURITANIA
Nouakchott

CAPE
VERDE

SENEGAL
Praia
Dakar

THE GAMBIA Banjul

GUINEA-BISSAU

Belém

Fortaleza

I L

Recife

São Francisco

Brazilian Highlands

Brasília

Salvador

Belo
Horizonte

Rio de Janeiro
São Paulo

ATLANTIC OCEAN

Meridian

CHAPTER 3

Middle and South America

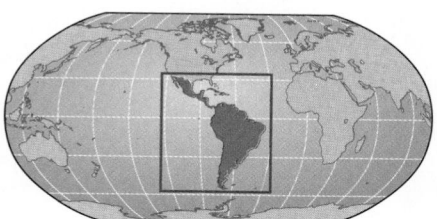

The oil industry has a long history of polluting environments and transforming cultures in some of the planet's most remote and ecologically sensitive places. Alex Pulsipher encountered one example of this tendency on a trip through the Ecuadorian Amazon.

Global Patterns, Local Lives The boat trip down the Aguarico River in Ecuador took me into a world of magnificent trees, river canoes, and houses built high up on stilts to avoid floods. I was there to visit the Secoya, a group of 350 indigenous people locked in negotiations with the U.S. oil company Occidental Petroleum over its plans to drill for oil on Secoya lands. Oil revenues supply 40 percent of the Ecuadorian government's budget and are essential to paying off its national debt. The government had threatened to use military force to compel the Secoya to allow drilling.

The Secoya wanted to protect themselves from pollution and cultural disruption. As Colon Piaguaje, chief of the Secoya, put it to me, "A slow death will occur. Water will be poorer. Trees will be cut. We will lose our culture and our language, alcoholism will increase, as will marriages to outsiders, and eventually we will disperse to other areas. There is no other way. We will negotiate, but these things will happen." Given all the impending changes, Chief Piaguaje asked Occidental to use the highest environmental standards in the industry. He also asked the company to establish a fund to pay for the educational and health needs of the Secoya people.

Like the Secoya, indigenous peoples around the world are facing environmental and cultural disruption arising from economic development efforts (Figure 3.2). Chief Piaguaje based his predictions for the future on what has happened in parts of the Ecuadorian Amazon that have already experienced several decades of oil development.

The U.S. company Texaco was the first major oil developer to establish operations in Ecuador. From 1964 to 1992, its pipelines and waste ponds leaked almost 17 million gallons of oil into the

Figure 3.1 Regional map of Middle and South America.

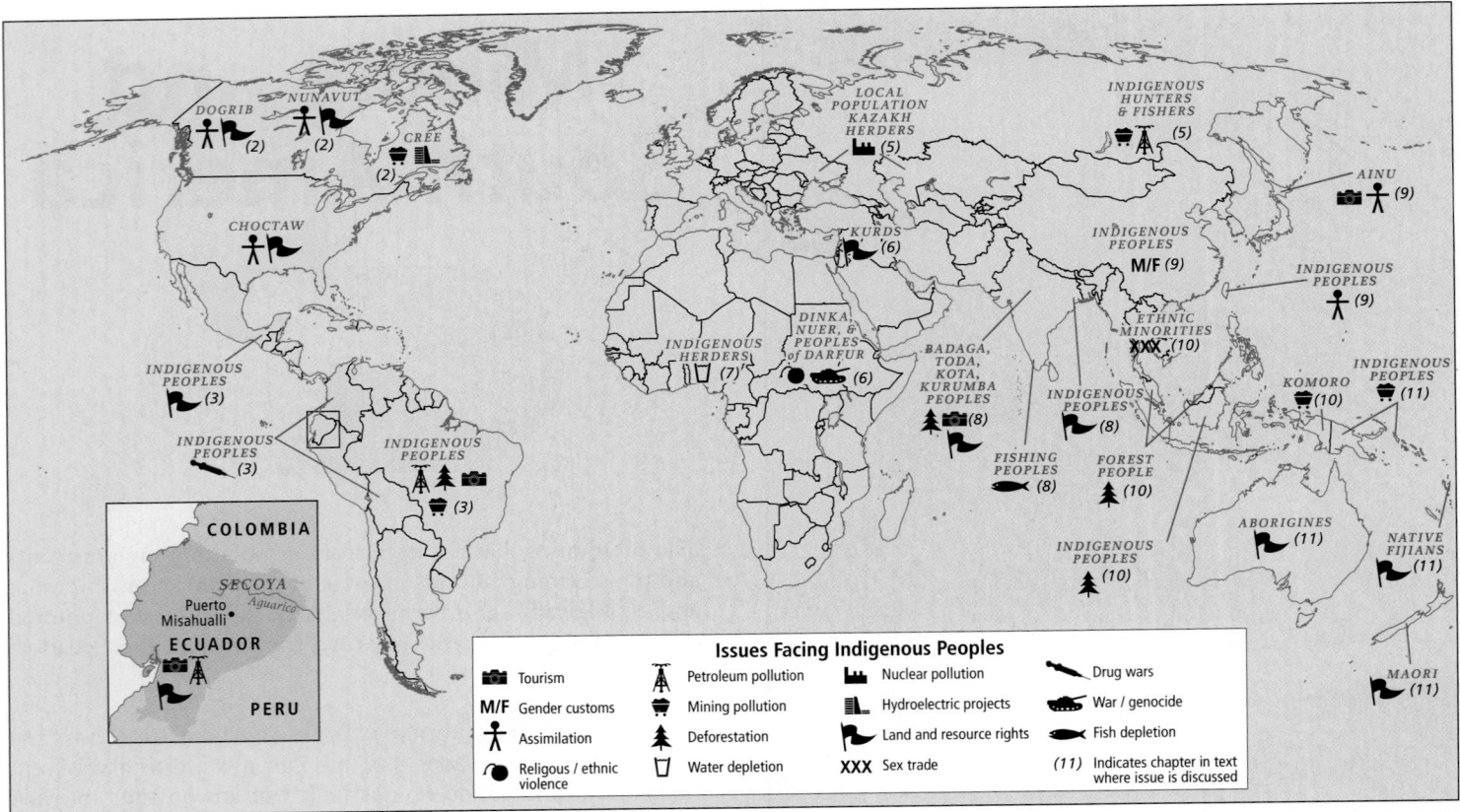

Figure 3.2 Issues facing indigenous peoples. Issues relating to indigenous peoples are mentioned in many places in this book. In all cases, the issues are in some way related to interactions with the outside world.

Amazon Basin, enough to fill about 1900 fully loaded oil tanker trucks, or 35 Olympic size swimming pools. Although Texaco sold its operations to the government and left Ecuador in 1992, its oil wastes continue to leak into the environment from hundreds of open pits (Figure 3.3).

In 1993, some 30,000 people sued Texaco in New York State, where the company (now called ChevronTexaco) is headquartered, for damages from the pollution. Those suing were both indigenous people and settlers who had established farms along Texaco's service roads. Several epidemiological studies concluded that oil contamination has contributed to higher rates of childhood leukemia, cancer, and spontaneous abortions among people who live near pollution created by Texaco.

Six years after my visit to the Secoya, many of Chief Piaguaje's worries have been borne out. Occidental did establish a fund to help the Secoya deal with the disruption of oil development. However, part of this fund was distributed to individual households, with varying results. Some people invested their money in ecotourism and other commercial enterprises and have prospered. Others simply spent their money and are now working for those who invested. Such employer-employee relationships are new to the Secoya, changing what was once an egalitarian culture.

Oil development has had many negative effects on the envi-ronment. Air and water pollution have increased rates of illness. The animals that the Secoya used to depend on, such as tapirs, have disappeared almost entirely due to overhunting by new settlers from the highlands.

In 2002, the Ecuadorian suit against ChevronTexaco was dismissed by the U.S. Court of Appeals. It was refiled in Ecuador in 2003, and by 2008 it looked as if ChevronTexaco might lose the case and have to pay between 7 and 16 billion dollars in damages.

Sources: Alex Pulsipher field notes; Amazon Watch 2006; Oxfam America 2005. ■

The rich resources of Middle and South America have attracted outsiders since the first voyage of Christopher Columbus in 1492. Europe's encounter with this region marked a major expansion of the global economy. However, for several hundred years Middle and South America occupied a disadvantaged position in global trade, supplying cheap raw materials that aided the Industrial Revolution in Europe but left this region poor. In recent years, however, investment from Europe, East Asia, and North America is now supporting new and more profitable manufacturing and service industries. Meanwhile trade blocs within the region are making countries better able to prosper from trade with each other.

Figure 3.3 Pollution from oil development in Ecuador. A worker samples one of the several hundred open waste pits that Texaco left behind in the Ecuadorian Amazon. Wildlife and livestock trying to drink from these pits are often poisoned or drowned. After heavy rains, the pits overflow, polluting nearby streams and wells. [Alex Pulsipher.]

Middle and South America differ from North America in several ways. Physically, this world region is larger and has more diverse environments. Culturally, there are larger **indigenous** (native) populations and more highly stratified social systems based on class, race, and gender. Politically, the region includes more than three dozen independent countries with models of self-government ranging from socialist Cuba to ultra-capitalist Chile. Economically, the gap between rich and poor is much wider than in North America. Yet there are many commonalities within the region, most arising from shared experience as colonies of Spain or Portugal.

Questions to consider as you study Middle and South America (also see Views of Middle and South America, pages 98–99)

1. How are the processes of global warming and deforestation linked? Forests absorb carbon dioxide from the atmosphere and enormous amounts of carbon are stored in the bodies of trees. What strategies are being proposed to limit deforestation and curb greenhouse gas emissions?

2. In this region of abundant water resources, what factors have caused a water crisis? Industrial and urban pollution are degrading many rivers. Meanwhile, unplanned urbanization has left many people without access to clean water. What efforts have been made to address the problems of the water crisis?

3. What drove the population explosion that occurred in this region during the early twentieth century? What led to declining birthrates by the 1980s? Improvements in health care lowered death rates while cultural and economic factors maintained high birth rates, especially in rural areas. How has urbanization and changing gender roles resulted in lower population growth rates?

4. How is the shift toward large-scale, mechanized agriculture related to urbanization? How has this transition been resisted throughout the region? Many governments see large-scale, mechanized agriculture as the best way both to provide food for booming urban populations and to increase exports. What forces have made small farmers leave their land? How have some resisted, creating world-famous popular movements?

5. Why has there been so much dissatisfaction with SAPs and the patterns of globalization they bring? How has this been expressed through the democratic process? While SAPs (see page 119) have met some of their goals, they have failed to achieve others. Leaders who oppose SAPs have been elected throughout much of the region since 1999. How have the outcomes of SAPs created this voter backlash? How have the new leaders impacted this region's relationships with the United States?

6. What factors in the region's development are challenging traditional gender roles? For decades, the Catholic Church has promoted conservative gender roles throughout this region. Now its influence is challenged by new religious movements. At the same time, urbanization is giving women new employment opportunities outside the home. What other factors are changing gender roles?

Views of Middle and South America

For a tour of these places, visit www.whfreeman.com/pulsipher
[Background image: Lori K. Perkins NASA/GSFC Scientific Visualization Studio]

 **1 Mexico City**

 **2 Belize**

1. Urbanization. Fueled by massive migration from rural areas, has given this region some of the largest cities on the planet. Mexico City, shown here, has a population 20 Million. See page 113–116 for more. Lat 19°23'N, Lon 99°06'W.
[Alejandro Covarrubias 2007]

2. Food. Agriculture has shifted towards increasingly large scale operations such as this banana plantation in Belize. See page 123–125 for more. Lat 16°58'59"N, Lon 88°17'24"W. [Paul Stokstad]

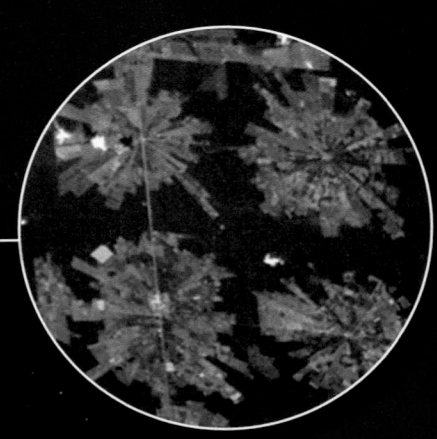

6. Global warming and deforestation. Deforestation in eastern Bolivia, as seen from a satellite, takes on a radial pattern as swaths of vegetation are cleared from a central point. Forest clearing often releases huge amounts of carbon into the atmosphere through burning. See page 106–107 for more. Lat 16°40'S, Lon 62°47'W. [Jami Dwyer; NASA]

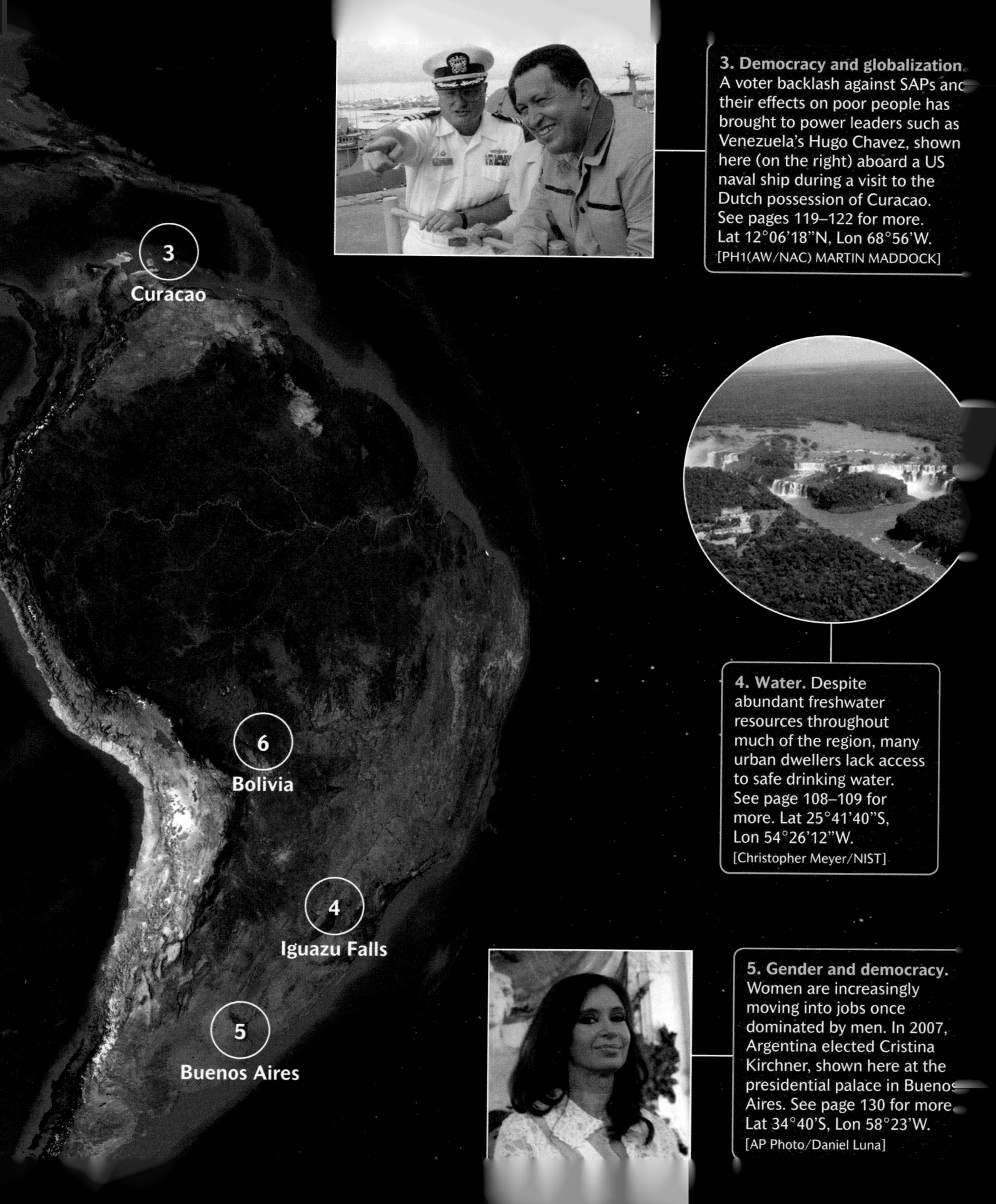

3

Curacao

3. Democracy and globalization. A voter backlash against SAPs and their effects on poor people has brought to power leaders such as Venezuela's Hugo Chavez, shown here (on the right) aboard a US naval ship during a visit to the Dutch possession of Curacao. See pages 119–122 for more. Lat 12°06'18"N, Lon 68°56'W.
[PH1(AW/NAC) MARTIN MADDOCK]

6

Bolivia

4

Iguazu Falls

4. Water. Despite abundant freshwater resources throughout much of the region, many urban dwellers lack access to safe drinking water. See page 108–109 for more. Lat 25°41'40"S, Lon 54°26'12"W.
[Christopher Meyer/NIST]

5

Buenos Aires

5. Gender and democracy. Women are increasingly moving into jobs once dominated by men. In 2007, Argentina elected Cristina Kirchner, shown here at the presidential palace in Buenos Aires. See page 130 for more. Lat 34°40'S, Lon 58°23'W.
[AP Photo/Daniel Luna]

The Geographic Setting

Terms to Be Aware Of

In this book, **Middle America** (the narrow ribbon of land south of Mexico that makes up Central America) refers to Mexico and the islands of the Caribbean (Figure 3.4). **South America** refers to the vast continent south of Central America. The term *Latin America* is not used in this book because it describes the region only in terms of the Roman (Latin-speaking) origins of the former colonial powers of Spain and Portugal. It ignores the region's large indigenous, African, Asian, and Northern European populations, as well as the many mixed cultures, often called mestizo cultures, that have emerged here. In this chapter, we use the term *indigenous groups* or *peoples* rather than *Native Americans* to refer to the native inhabitants of the region.

PHYSICAL PATTERNS

Middle and South America extend south from the midlatitudes of the Northern Hemisphere across the equator through the Southern Hemisphere, nearly to Antarctica (see Figure 3.1).

Figure 3.4 Political map of Middle and South America.

This vast north-south expanse combines with variation in altitude to create the wide range of climates in the region. Tectonic forces have shaped the primary landforms of this huge territory to form an overall pattern of highlands to the west and lowlands to the east.

Landforms

Highlands. A nearly continuous chain of mountains stretches along the western edge of the American continents for more than 10,000 miles (16,000 kilometers) from Alaska to Tierra del Fuego at the southern tip of South America. The southern part of this long mountain chain is known as the Sierra Madre in Mexico and the Andes in South America (Figure 3.1). It was formed by a lengthy **subduction zone** (a zone where one tectonic plate slides under another) that runs thousands of miles along the western coast of the continents (see Figure 1.21 on page 25). Here, two oceanic plates—the Cocos Plate and the Nazca Plate—plunge beneath three continental plates—the North American Plate, the Caribbean Plate, and the South American Plate.

In a process that continues today, the overriding plates crumple to create mountain chains. Often they develop fissures that allow molten rock from beneath the earth's crust to ascend to the surface and form volcanoes (see Figure 1.22 on page 26). These volcanic highlands constitute a major barrier to transportation and communication. In northern and central South America, where the population is dense, they pose a greater challenge.

The chain of high and low mountainous islands in the eastern Caribbean is similarly volcanic in origin, created as the Atlantic Plate thrusts under the eastern edge of the Caribbean Plate. It is not unusual for volcanoes to erupt in this active tectonic zone. On the island of Montserrat, for example, people have been living with an active, and sometimes deadly, volcano for more than a decade. Eruptions have taken the form of violent blasts of superheated rock, ash, and gas that move down the volcano's slopes with great speed and force.

Lowlands. Vast lowlands extend over most of the land to the east of the western mountains. In Mexico, east of the Sierra Madre, a coastal plain borders the Gulf of Mexico. Farther south, in Central America, wide aprons of sloping land descend to the Caribbean coast. In South America, a huge wedge of lowlands, widest in the north, stretches from the Andes east to the Atlantic Ocean. These South American lowlands are interrupted in the northeast and the southeast by two modest highland zones: the Guiana Highlands and the Brazilian Highlands. Elsewhere in the lowlands, grasslands cover huge, flat expanses, including the llanos of Venezuela and Colombia and the pampas of Argentina.

The largest feature of the South American lowlands is the Amazon Basin, drained by the Amazon River and its tributaries (Figure 3.5). This basin lies within Brazil and the neighboring countries to its west. Here, the earth's largest remaining expanse of tropical rain forest gives the Amazon Basin global significance as a reservoir of **biodiversity,** or variety of life forms. Hundreds of thousands of plant and animal species live here.

Figure 3.5 The Amazon lowlands. The Tigre River, a tributary of the Amazon, meanders through the Peruvian lowland rain forest. This scene is typical of those parts of the Amazon Basin that have not been disturbed by roads or forest removal. [Layne Kennedy/CORBIS.]

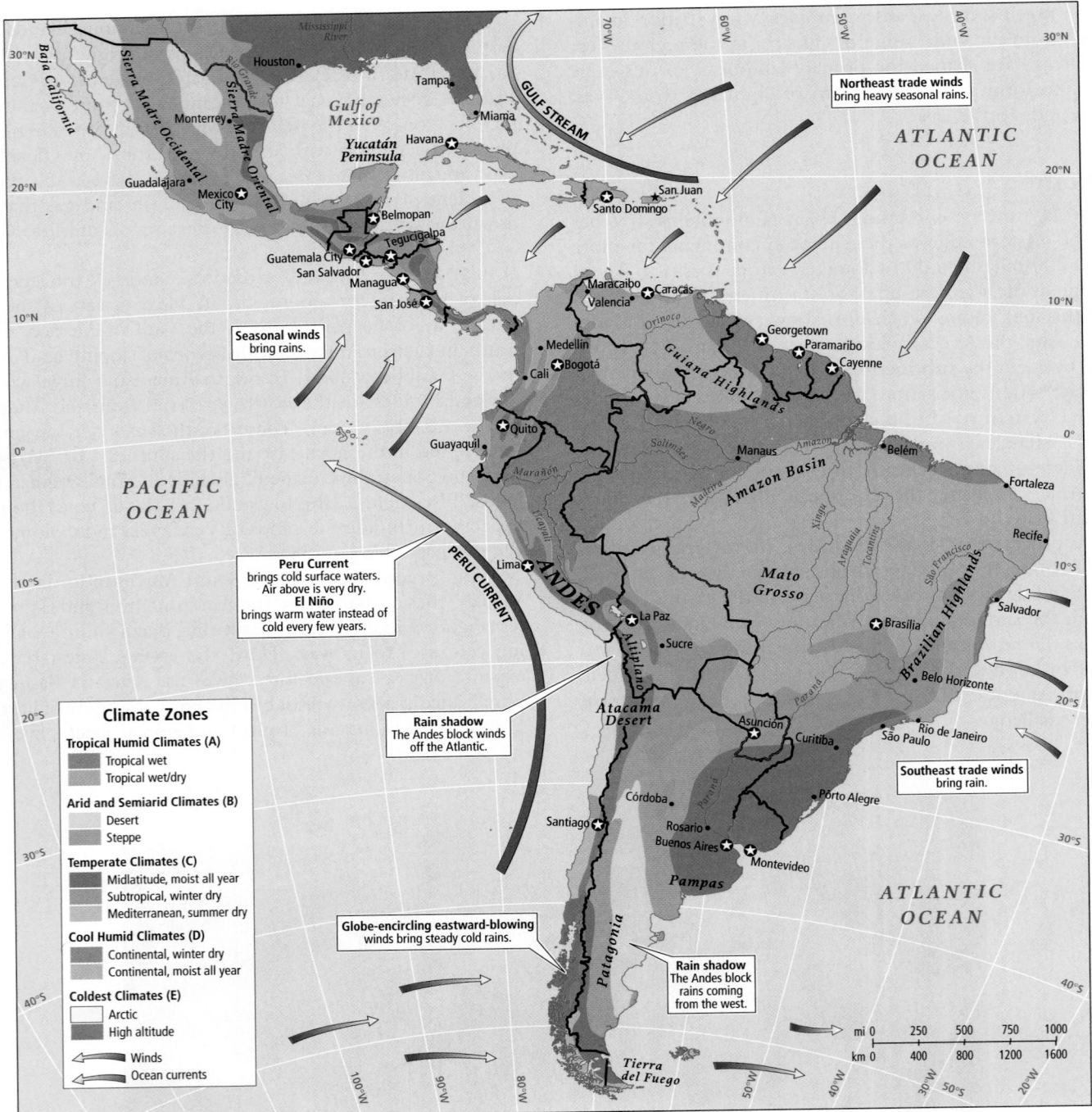

Figure 3.6 Climates of Middle and South America.

The basin's water resources are also astounding. Twenty percent of the earth's flowing surface waters exist here, running in rivers so deep that ocean liners can steam 2300 miles (3700 kilometers) upriver from the Atlantic Ocean all the way to Iquitos, jokingly referred to as Peru's "Atlantic seaport." The vast Amazon River system starts as streams high in the Andes. These streams eventually unite as rivers that flow eastward toward the Atlantic. Once they reach the flat land of the Amazon Plain, their velocity slows abruptly, and fine soil particles, or **silt,** sink to the riverbed. When the rivers flood, silt and organic material transported by the floodwaters renew the soil of the surrounding areas, nourishing millions of acres of

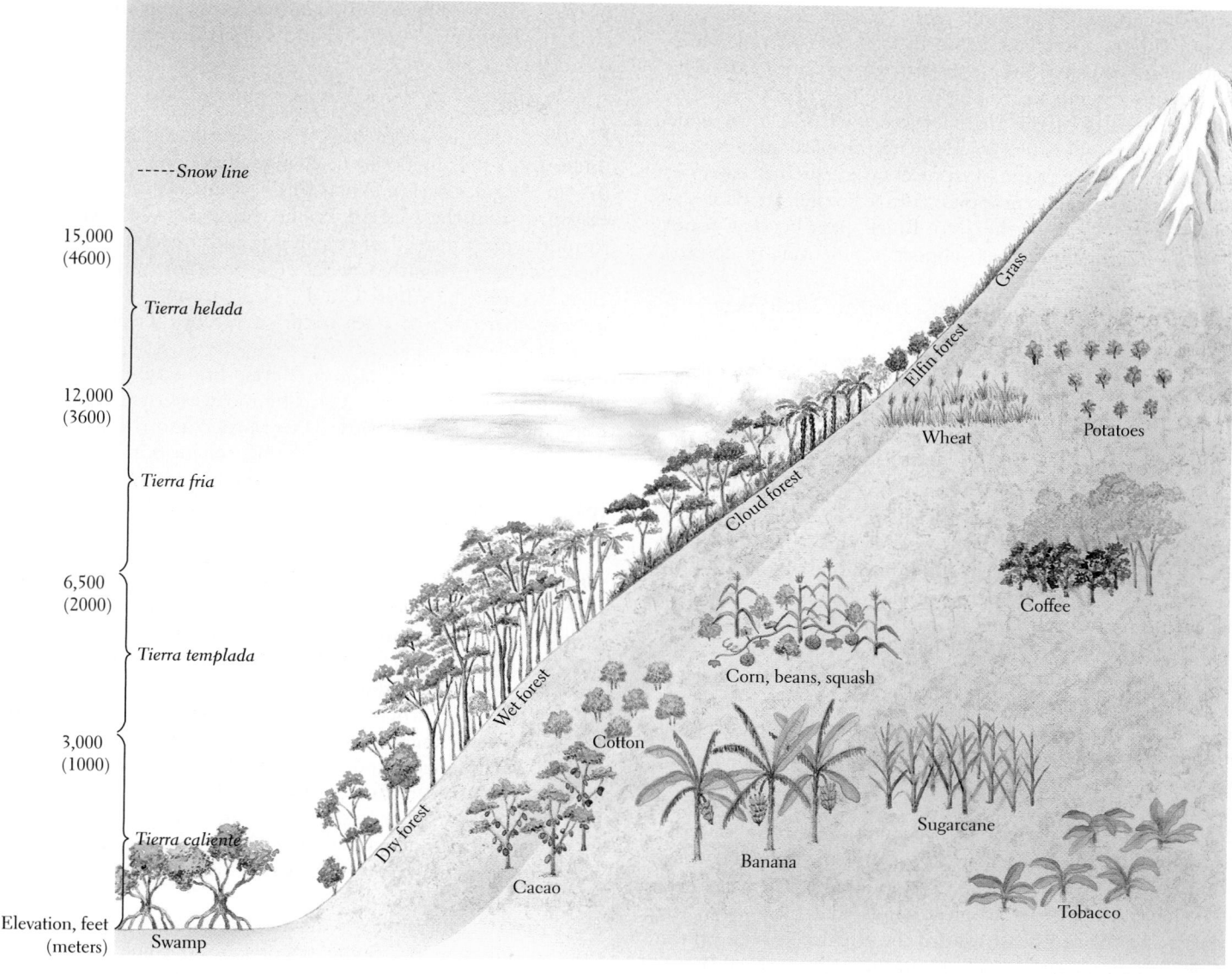

Figure 3.7 Temperature-altitude zones of Middle and South America. As the temperature changes with altitude, the natural vegetation on mountain slopes also changes, as shown along the edge of the mountain here. The same is true for crops, some of which are suited to lower, warmer elevations and some to higher, cooler elevations. [Illustration by Tomo Narashima, based on fieldwork and a drawing by Lydia Pulsipher.]

tropical forest. Not all of the Amazon Basin is rain forest, however. Variations in weather and soil types, as well as human activity, have created grasslands and seasonally dry deciduous tropical forests in some areas.

Climate

From the jungles of the Amazon and the Caribbean to the high, glacier-capped peaks of the Andes to the parched moonscape of the Atacama Desert, the climate variety of Middle and South America is astounding (Figure 3.6). This variety results from several factors. The wide range of temperatures and tremendous changes in altitude reflect the great distance the landmass spans on either side of the equator. In addition, global patterns of wind and ocean currents result in a distinct pattern of precipitation.

Temperature-altitude zones. Four main **temperature-altitude zones** are commonly recognized in the region (Figure 3.7). As altitude increases, the temperature of the air decreases by about 1°F per 300 feet (1°C per 165 meters) of elevation. Thus temperatures are warmest in the lowlands, which are known in Spanish as the *tierra caliente,* or "hot land." The *tierra caliente* extends up to about 3000 feet (1000 meters), and in some parts of the region these lowlands cover wide expanses. Where moisture is adequate, tropical rain forests thrive, as does a wide range of tropical crops, such as bananas, sugarcane, cacao, and pineapples. Many coastal areas of the *tierra caliente,* such as northeastern Brazil, have become zones of plantation agriculture that support populations of considerable size.

Between 3000 and 6500 feet (1000 to 2000 meters) is the cooler *tierra templada* ("temperate land"). The year-round, spring-like climate of this zone drew large numbers of indigenous people in the distant past and later, Europeans. Here, such crops as corn, beans, squash, various green vegetables, wheat, and coffee are grown.

Between 6500 and 12,000 feet (2000 to 3600 meters) is the *tierra fria* ("cool land"). Many crops such as wheat, fruit trees, root vegetables, and cool-weather vegetables—such as cabbage and broccoli—do very well at this altitude. Many animals—such as llamas, sheep, and guinea pigs—are raised for food and fiber. Several modern population centers are in this zone, including Mexico City, Mexico, and Quito, Ecuador.

Above 12,000 feet (3600 meters) is the *tierra helada* ("frozen land"). In the highest reaches of this zone, vegetation is almost absent, and mountaintops emerge from under snow and glaciers. The remarkable feature of such tropical mountain zones is that in a single day of strenuous hiking, one can encounter most of the climate types known on earth.

Precipitation. The pattern of precipitation throughout the region (see Figure 1.23 on page 27) is influenced by the interaction of global wind patterns with mountains and ocean currents. The **trade winds** (tropical winds that blow from the northeast and the southeast toward the equator) sweep off the Atlantic, bringing heavy seasonal rains (see Figure 3.6). Winds blowing in from the Pacific bring seasonal rain to the west coast of Central America, but mountains block the rain from reaching the Caribbean side.

The Andes are a major influence on precipitation in South America. They block the rains borne by the northeast and southeast trade winds off the Atlantic, creating a rain shadow on the western side of the Andes in northern Chile and western Peru. Southern Chile is in the path of eastward-trending winds that sweep north from Antarctica, bringing steady, cold rains that support forests similar to those of the Pacific Northwest of North America. The Andes block this flow of wet, cool air and divert it to the north. They thereby create an extensive rain shadow on the eastern side of the mountains along the southeastern coast of Argentina (Patagonia).

The pattern of precipitation is also influenced by the adjacent oceans and their currents. Along the west coasts of Peru and Chile, the cold surface waters of the Peru Current bring cold air that cannot carry much moisture. The combined effects of the Peru Current and the central Andes rain shadow have created what is possibly the world's driest desert, the Atacama of northern Chile.

El Niño. One aspect of the Peru Current that is only partly understood is its tendency to change direction every few years (on an irregular cycle). When this happens, warm water flows eastward from the western Pacific, bringing warm water and torrential rains instead of cold water and drought, to parts of the west coast of South America. The phenomenon was named **El Niño,** or "the Christ Child," by Peruvian fishermen, who noticed that when it does occur, it reaches its peak around Christmas time.

El Niño also has global effects, bringing cold air and drought to normally warm and humid western Oceania, and unpredictable weather patterns to Mexico and the southwestern United States. The El Niño phenomenon in the western Pacific is discussed further in Chapter 11, where Figure 11.8 (page 409) illustrates its transpacific effects.

Hurricanes. In this region, many coastal areas are threatened by powerful storms that can create extensive damage and loss of life. These form annually, primarily in the Atlantic Ocean north of the equator and close to Africa. A tropical storm begins as a group of thunderstorms. When enough warming wet air comes together, the individual storms organize themselves into a swirling spiral of wind that moves across the earth's surface. The highest wind speeds are found at the edge of the eye, or center, of the storm. Once wind speeds reach 75 miles (120.7 kilometers) per hour, such a storm is officially called a hurricane. Hurricanes usually last about one week; they slow down and eventually dissipate as they move over cooler water or land.

Human vulnerability to hurricanes is increasing as coastal populations increase. Some scientists also think that global warming is leading to an increase in the number and intensity of hurricanes.

ENVIRONMENTAL ISSUES

Environments in Middle and South America were among the first to inspire concern about the use and misuse of the earth's resources. Beginning in the 1970s, construction of the Trans-Amazon Highway provided migrant farmers with access to the Amazon Basin. They followed the new road into the rain forest and began clearing it to grow crops. Scholars warned of an impending crisis in the tropical rain forests of this region. We now know that even in prehistory, every human settlement has had consequences for environments across Middle and South America (Figure 3.8a). Today's impacts, however, are particularly severe because both local resource consumption and exports to the global economy have increased so dramatically (Figure 3.8b).

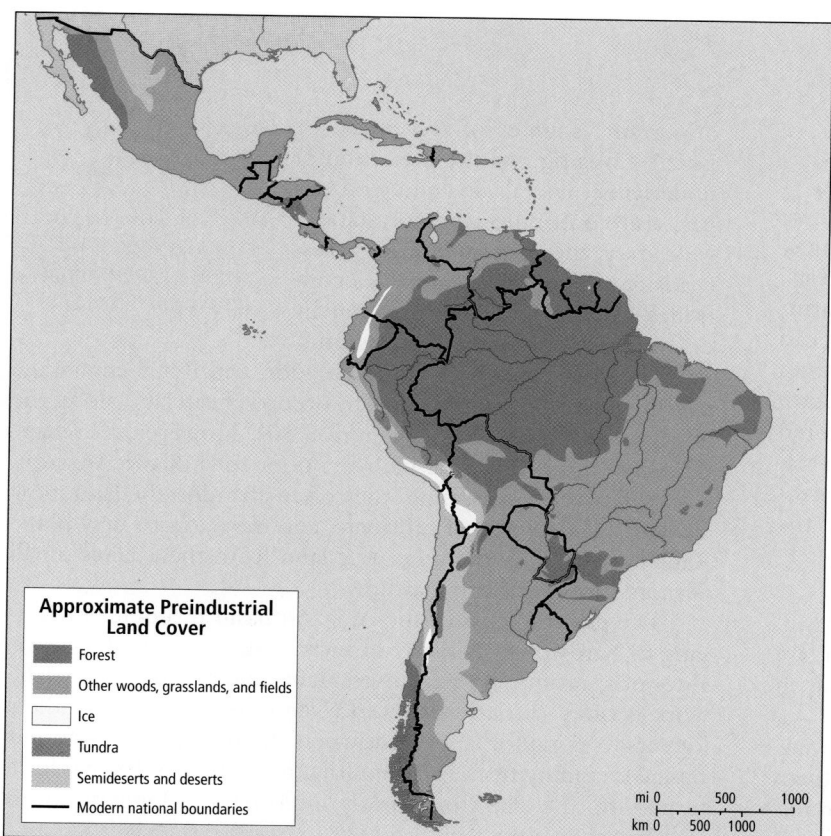

(a) Preindustrial

Approximate Preindustrial Land Cover

- Forest
- Other woods, grasslands, and fields
- Ice
- Tundra
- Semideserts and deserts
- Modern national boundaries

mi 0 500 1000
km 0 500 1000

Figure 3.8 Land cover (preindustrial) and human impact (2002). In the preindustrial era **(a)**, humans had cleared some forest, planted plantation crops, and built cities, but the overall human impact is not visible at this scale. By 2002 **(b)**, human impact was observable in most of the region, with the possible exception of some parts of the Amazon. [Adapted from *United Nations Environment Programme, 2002, 2003, 2004, 2005, 2006* (New York: United Nations Development Programme), at http://maps.grida.no/go/graphic/human_impact_year_1700_approximately and http://maps.grida.no/go/graphic/human_impact_year_2002.]

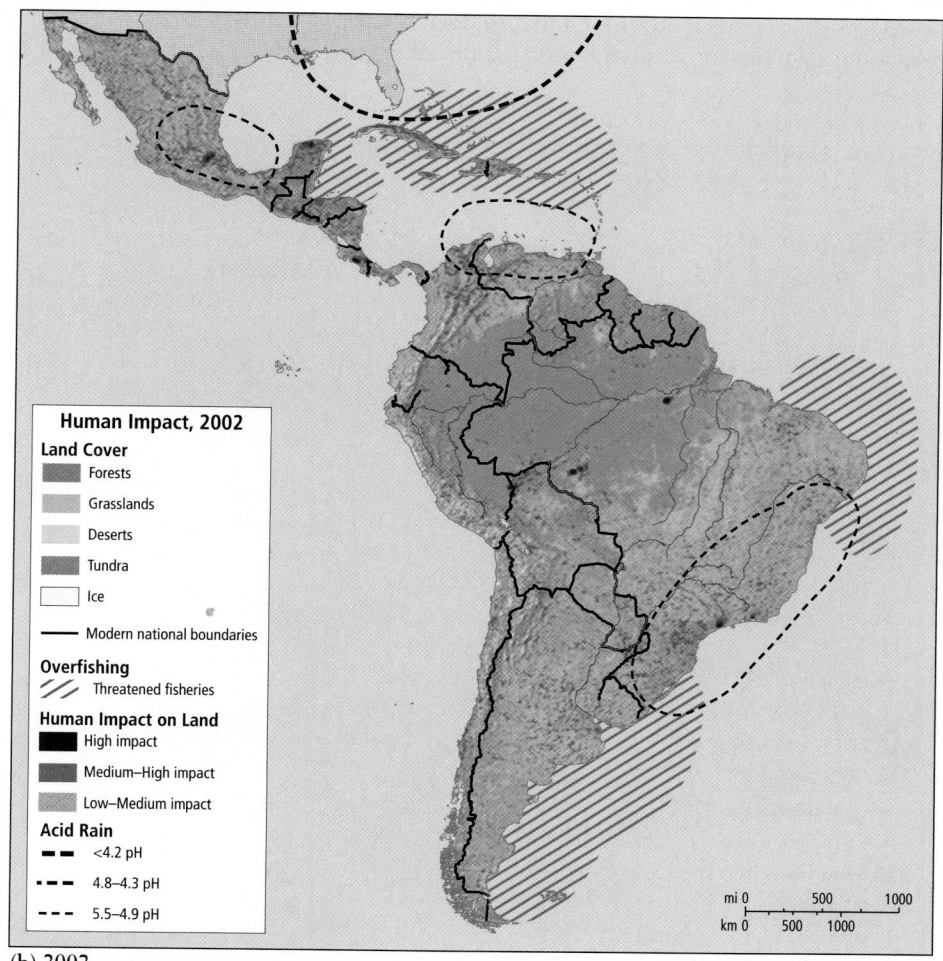

(b) 2002

Human Impact, 2002

Land Cover

- Forests
- Grasslands
- Deserts
- Tundra
- Ice
- Modern national boundaries

Overfishing

- Threatened fisheries

Human Impact on Land

- High impact
- Medium–High impact
- Low–Medium impact

Acid Rain

- <4.2 pH
- 4.8–4.3 pH
- 5.5–4.9 pH

mi 0 500 1000
km 0 500 1000

Tropical Forestlands, Global Warming, and Globalization

As we explained in Chapter 1 (see pages 17–18), forests release oxygen and absorb carbon dioxide, the greenhouse gas most responsible for global warming. Hence, the loss of large forests, such as those in the Amazon Basin, contributes to global warming because it reduces the amount of carbon dioxide that can be taken out of the atmosphere. And since many of the trees are burned rather than used as building material, deforestation also often releases huge amounts of carbon that were stored in the bodies of the trees. Because of the massive deforestation occurring in the Amazon, Brazil now ranks as the fourth-largest emitter of greenhouse gasses in the world after the United States, China, and Indonesia.

The Amazon Basin. The rain forests of the Amazon Basin, some of the oldest and largest remaining forests on earth, are planetary treasures of biodiversity. They also play a key role in regulating the earth's climate.

Though still vast, Amazon rain forests are being diminished by multiple threats (Figure 3.9). Primary among them is the clearing of land to grow crops such as soybeans for animal feed. Brazil is now the world's largest exporter of soybeans, most of which go to China to feed that country's increasing appetite for meat. These agricultural exports are an important source of revenue for Brazil. Logging and the extraction of underlying minerals, including oil and gas, also contribute to deforestation. Moreover, the construction of access roads to support these activities continues to accelerate deforestation by opening new forest areas to migrants.

The governments of Peru, Ecuador, and Brazil encourage overflowing urban populations to occupy cheap land along the newly built logging roads (Figure 3.10). However, these settlers have a difficult time with the poor soils of the Amazon. After a few years of farming, they often abandon the land, now eroded and depleted of nutrients, and move on to new plots. Ranchers may buy the worn-out land from these failed small farmers to use as cattle pastures.

The governments of the Amazon basin promote the logging of hardwoods as a way of increasing exports, which provide jobs, profits, foreign investment, and tax revenues. The work is often carried out by local entrepreneurs and landless loggers working for large landowners. Increasingly, investment capital is coming from Asian multinational companies that have turned to the Amazon forests after having logged up to 50 percent of the tropical forests in Southeast Asia.

Much of the wood from these forests ends up in Asia, Europe, and North America, where its value is vastly increased

> Brazil is now the world's largest exporter of soybeans, most of which go to China to feed that country's increasing appetite for meat.

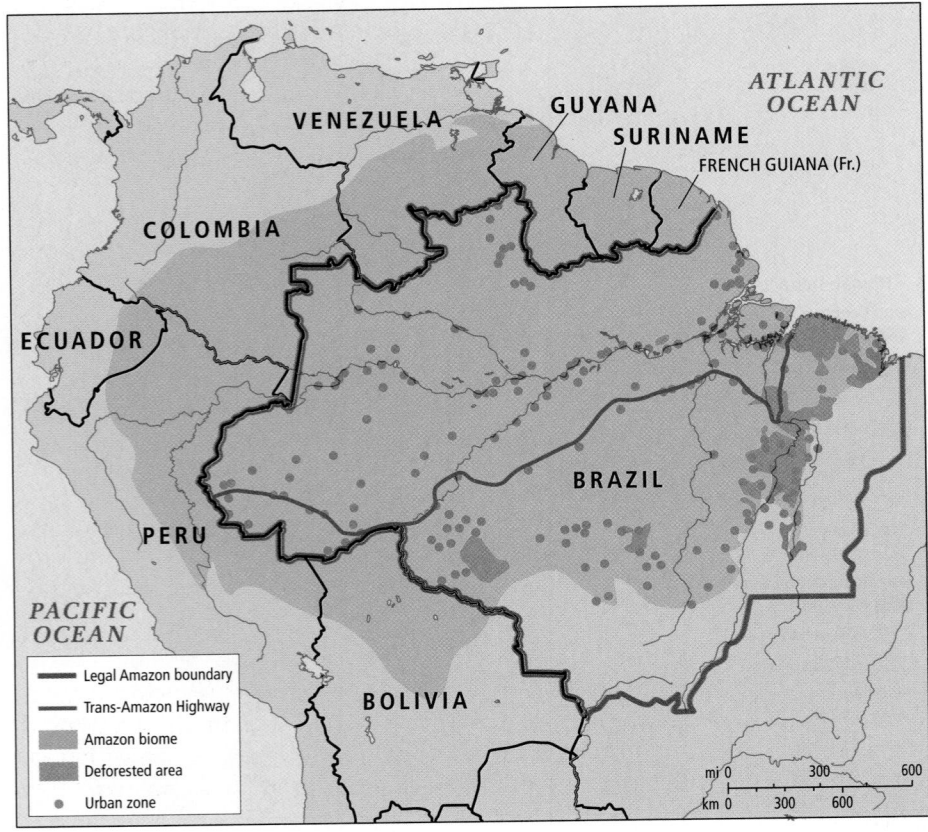

Figure 3.9 Effects of development in the Amazon Basin. The Amazon biome is huge and extends into many countries, as shown in green. Brazil legally defines its part of the Amazon (which it calls Amazonia) somewhat differently, as shown by the dark boundary. [Adapted from "Human pressure in the Brazilian Amazon, all indicators," World Resources Institute, at http://www.wri.org/publication/human-pressure-brazilian-amazon-forests.]

Figure 3.10 Logging in the Amazon Basin. In the Brazilian state of Rondônia, in the Amazon Basin, colonists from other areas clear the rain forest for settlement and large-scale agriculture. New roads, like the one at the upper left, provide the settlers with access to the forest and become the conduit for many other changes. Some of the wood is harvested for local use or sold, but much is wasted. [Randall Hyman.]

by processing. A single tree may be worth $6000 to the barefoot men who fell the tree, cut it up, and raft it to a transshipment point. The same tree can be worth $300,000 once it is turned into flooring, paneling, or furniture in the United States or Europe. The World Wildlife Fund estimates that 50,000 such trees ended up in the U.S. market in 2002. According to an industry spokesperson, U.S. consumers typically do not inquire about the origins of the wood in the items they buy.

Central America and the Caribbean. The forests in other parts of Middle and South America are also at risk. About 65 percent of the forest clearing in Central America is intended to create pastures for beef cattle grown primarily for the U.S. fast-food industry. The forests of Central America and the Caribbean have also been cleared for export crops, such as sugar, cotton, and tobacco in the past and bananas today. If deforestation continues in these areas at the present rates, the natural forest cover will be entirely gone in 20 years. ▪️

Environmental Protection and Economic Development

In the past, governments in the region argued that economic development was so desperately needed that environmental regulations were an unaffordable luxury. Now, there are increasing attempts to embrace economic development as necessary to raise standards of living while trying to minimize its negative effects on the environment. Environmental regulations are being developed in all the countries of the Amazon.

Some state and local governments in Brazil are encouraging the planting of fast-growing trees in forest plantations. This could reduce deforestation's impact on climate change and save some natural forests from cutting in the future. However, forest plantations fall far short of replicating the complex biodiversity of the rain forest. Tropical rain forest may have as many as 300 different species of trees on a single hectare (2.5 acres) of land, compared with the one or two tree species that would cover a plantation of hundreds of hectares. It is estimated that tropical rain forests contain 60 percent of all species found on earth, so losing these forests at the present rates means losing an enormous part of the planet's genetic inheritance.

Ecotourism. Other economic development efforts focus on earning money from the beauty of undegraded natural environments. Many countries are now promoting **ecotourism,** which encourages visitors from relatively rich countries to appreciate ecosystems and wildlife that don't exist in the developed world. Travel experiences in unfamiliar natural and cultural environments can sensitize both travelers and hosts to the complexity of environmental issues. Sustainable use and conservation of resources can be achieved while providing a livelihood to local people and the broader host community. Ecotourism is now the most rapidly growing segment of the global tourism and travel industry, which itself is the world's largest industry, with $3.5 trillion spent annually. Many Middle and South American nations also have spectacular national parks that can provide a basis for ecotourism. ▪️

Ecotourism has its downsides, however. It can be little different from other kinds of tourism that damage the environment and give little back to the surrounding community. While there is the potential to use the profits of ecotourism to benefit local communities and environments, the margins of profit may be small.

Vignette Puerto Misahualli is a small river boomtown that is currently enjoying significant economic growth. Its prosperity is due to the many European, North American, and other foreign travelers who come for experiences that will bring them closer to the now legendary rain forests of the Amazon.

Figure 3.11 A typical ecotourism lodge in Ecuador. [Gary Irvine/Alamy.]

The array of ecotourism offerings can be perplexing. One indigenous man offers to be a visitor's guide for as long as desired, traveling by boat and on foot, camping out in "untouched forest teeming with wildlife." His guarantee that they will eat monkeys and birds does not seem to promise the nonintrusive, sustainable experience the visitor might be seeking. At a well-known "eco-lodge," visitors are offered a plush room with a river view, a chlorinated swimming pool, and a fancy restaurant serving "international cuisine." All this is on a private 740-acre nature reserve separated from the surrounding community by a wall topped with broken glass. It seems more like a fortified resort than an eco-lodge.

By contrast, the solar-powered Yachana Lodge (Figure 3.11) has simple rooms and local cuisine. Its knowledgeable resident naturalist is a veteran of many campaigns to preserve Ecuador's wilderness. Profits from the lodge fund a local clinic and various programs that teach sustainable agricultural methods that protect the fragile Amazon soils while increasing farmers' earnings from surplus produce. The "local" cultivators are actually poor migrants from Ecuador's densely populated cool highlands who have been given free land in the Amazon by the government but no training in how to farm in the lowland rain forest. Many of their attempts at cultivation have resulted in extensive soil erosion and deforestation. The nonprofit group running the Yachana Lodge—the Foundation for Integrated Education and Development—is earning just barely enough to sustain the clinic and agricultural programs.

[Source: Alex Pulsipher's field notes in Ecuador.] ■

The Water Crisis

Although this world region receives more rainfall than any other and has three of the world's six largest rivers (in terms of volume), it is experiencing a water crisis. Rapid urbanization combined with a wide array of misguided policies and corruption have left many people without access to clean water or sanitation.

Some of the problem relates to environmental policies that allow industries to pollute with few restraints. Waterways along the Mexican border with the United States are polluted by numerous factories set up to take advantage of trade with the United States and Canada (see discussion of maquiladoras on page 121). Only 10 percent of their often highly toxic water discharges are properly treated and disposed of. In Brazil, which has more fresh water than any country in the world, a largely unregulated gold mining industry dumps huge amounts of mercury into the rivers. Brazil now has some of the most contaminated waterways on the planet.

Rapid and often unplanned urbanization has left many poor people without access to clean drinking water and sanitation. Also in need of an urgent solution is inadequate infrastructure. In the largest cities, as much as 50 percent of freshwater is lost due to leaky pipes. Moreover, 80 percent of the population has no access to decent sanitation, which means that local water resources are polluted and cities must bring in water from distant areas.

Behind these immediate problems lie systemic failures. Political corruption diverts many public funds that should be going to water infrastructure improvements. In many countries, poorly managed changes in the economy and agriculture are creating massive migration to urban areas. Some efforts have been made to increase access to clean water in cities by selling once government-run water utilities to private companies. In some cases, however, profit-driven private companies have increased water prices to levels that few urban residents can afford, while doing little to improve water supply or delivery systems. In Bolivia, for example, a subsidiary of the global engineering firm Bechtel even charged urban residents for water taken from their own wells, and rainwater harvested off their roofs! Popular protests have recently forced many countries to take back control of water services from private companies.

In Bolivia, a subsidiary of the global engineering firm Bechtel charged urban residents for water taken from their own wells and rainwater harvested off their own roofs.

Some efforts by local grassroots environmental movements, often assisted by international nongovernmental organizations (NGOs), are paying off. For example, in Bolivia, Agua para Todos (Water for All), a UN-funded agency, has helped local people to create their own sustainable water distribution systems. The effort has reduced both the cost of water and its wasteful use. Nevertheless, a host of problems related to water supply and sanitation remain throughout the region.

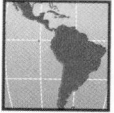

Ten Key Aspects of Middle and South America

- The rain forests of the Amazon Basin are planetary treasures of biodiversity that also play a key role in regulating the earth's climate.
- This region receives more rainfall than any other and has three of the world's six largest rivers (in terms of volume).
- As of 2007, about 570 million people were living in Middle and South America—more than ten times the population of the region in 1492.
- With the exception of a few small countries in Middle America, the gap between rich and poor in this region is one of the largest on earth.
- Vast amounts of wealth have been removed from the region over the past 500 years, prolonging its dependence on exporting cheap raw materials.
- In the last 25 years, there have been repeated peaceful and democratic transfers of power in countries once dominated by rulers who had seized power by force.
- Food production is increasingly shifting away from small-scale, labor-intensive farming and toward large-scale, mechanized agriculture.
- Since the early 1970s, Middle and South America have led the world in migration from rural to urban communities.
- Gender roles in the region have been strongly influenced by the Catholic Church through the ideals of *marianismo* and *machismo.*
- Evangelical Protestantism came in from North America and is now the fastest-growing religious movement in the region.

HUMAN PATTERNS OVER TIME

The conquest of Middle and South America by Europeans set in motion a series of changes that helped create the ways of life found in this region today. The conquest wiped out much of indigenous civilization and set up colonial regimes in its place. It introduced many new cultural influences and led to the disproportionate distribution of power and wealth that continues to this day.

The Peopling of Middle and South America

Recent evidence suggests that between 25,000 and 14,000 years ago, groups of hunters and gatherers from northeastern Asia spread throughout North America after crossing the Bering land bridge on foot, moving along shorelines in small boats, or both. Some of these groups remained in North America, while others ventured south across the Central American land bridge, reaching the tip of South America by about 13,000 years ago.

By 1492, there were 50 to 100 million indigenous people in Middle and South America. In some places, population densities were high enough to threaten sustainability. People altered the landscape in many ways. They modified drainage to irrigate crops, constructed raised fields in lowlands, terraced hillsides, and built paved walkways across swamps and mountains. They constructed cities with sewer systems and freshwater aqueducts, and raised huge earthen and stone ceremonial structures that rivaled the pyramids of Egypt.

The indigenous people also perfected the system of **shifting cultivation** that is still common in wet, hot regions in Central America and the Amazon Basin. Small plots are cleared in forestlands, the brush is dried and burned to release nutrients in the soil, and the clearings are planted with multiple crop species. Each plot is used for only two or three years and then abandoned for many years of regrowth. This system is highly productive per unit of land and labor if there is sufficient land for long periods of regrowth. If population pressure increases to the point that a plot must be used before it has fully regrown, yields will decrease drastically.

Although they lacked the wheel and gunpowder, the **Aztecs** of the high central valley of Mexico had some technologies and social systems that rivaled or surpassed those of Asian and European civilizations of the time. Particularly well developed were urban water supplies, sewage systems, and elaborate marketing systems. Historians have concluded that on the whole, Aztecs probably lived more comfortably than their contemporaries in Europe.

In 1492, the largest state in the region was that of the **Incas.** It stretched from southern Colombia to northern Chile and Argentina, with the main population clusters in the Andes highlands. The cooler temperatures at these high altitudes eliminated the diseases of the tropical lowlands, yet proximity to the equator guaranteed mild winters and long growing seasons. For several hundred years, the Inca empire was one of the most efficiently managed empires in the history of the world. Highly organized systems of labor were used to construct paved road systems and great stone cities in the Andean highlands. Incan agriculture was advanced, particularly in the development of crops such as numerous varieties of potatoes and grains (Figure 3.12).

Figure 3.12 Ancient Incan terraces rediscovered and put to use. These circular sunken terraces at 12,000 feet in Moray, Peru, were built long before the Spanish arrived. The terraces had a system of canals that delivered rainwater for irrigation. Because they were below grade level, the terraces protected the plants from the wind and cold temperatures. Each descending level had a different microclimate so that elaborate experimental agriculture—including crops of wheat, quinoa, and other plants—was possible. [Tom Dempsey, http://www.photoseek.com.]

The Conquest

The European conquest of Middle and South America was one of the most significant events in human history. It rapidly altered landscapes and cultures and ended the lives of millions of indigenous people through disease and slavery.

Columbus established the first Spanish colony in 1492 on the Caribbean island of Hispaniola, now occupied by Haiti and the Dominican Republic. This initial seat of the Spanish empire expanded to include the rest of the Greater Antilles—Cuba, Puerto Rico, and Jamaica. After learning of Columbus's exploits, other Europeans, mainly from Spain and Portugal on Europe's Iberian Peninsula, conquered Middle and South America.

The first part of the mainland to be conquered was Mexico, home to several advanced indigenous civilizations, most notably the Aztecs. The Spanish were unsuccessful in their first attempt to capture the Aztec capital of Tenochtitlán, but they succeeded a few months later after a smallpox epidemic decimated the native population. The Spanish demolished the grand Aztec capital in 1521 and built Mexico City on its ruins.

The conquest of the Incas in South America was also achieved by a tiny band of Spaniards, again aided by a smallpox epidemic. Out of the ruins of the Inca empire, the Spanish created the Viceroyalty of Peru, which originally encompassed all of South America except Portuguese Brazil. The newly constructed capital of Lima flourished, in large part as a trans-shipment point for enormous quantities of silver extracted from mines in the highlands of Bolivia.

Diplomacy by the Roman Catholic Church prevented conflict between Spain and Portugal over the lands of the region. The Treaty of Tordesillas of 1494 divided Middle and South America at approximately 46°W longitude (Figure 3.13). Portugal took all lands to the east and eventually acquired much of what is today Brazil; Spain took all lands to the west.

The conquest of Brazil by the Portuguese was unique in some key respects. Brazil was only sparsely populated by indigenous people and there were no highly organized urban cultures. Most Atlantic coastal cultures were annihilated early on, and the populations of the huge Amazon Basin declined sharply as contagious diseases spread through trading. Because it was difficult to penetrate the lowland tropical forests, the Portuguese focused on extracting gold and precious gems from the Brazilian Highlands and on establishing plantations along the Atlantic coast.

The superior military technology of the Iberians speeded the conquest of Middle and South America. A larger factor, however, was the vulnerability of the indigenous people to diseases such as smallpox and measles, which were carried by the Europeans. In the 150 years following 1492, the total population of Middle and South America was reduced by more than 90 percent to just 5.6 million. To obtain a new supply of labor to replace the indigenous people, the Spanish initiated the first shipments of enslaved Africans to the region in the early 1500s.

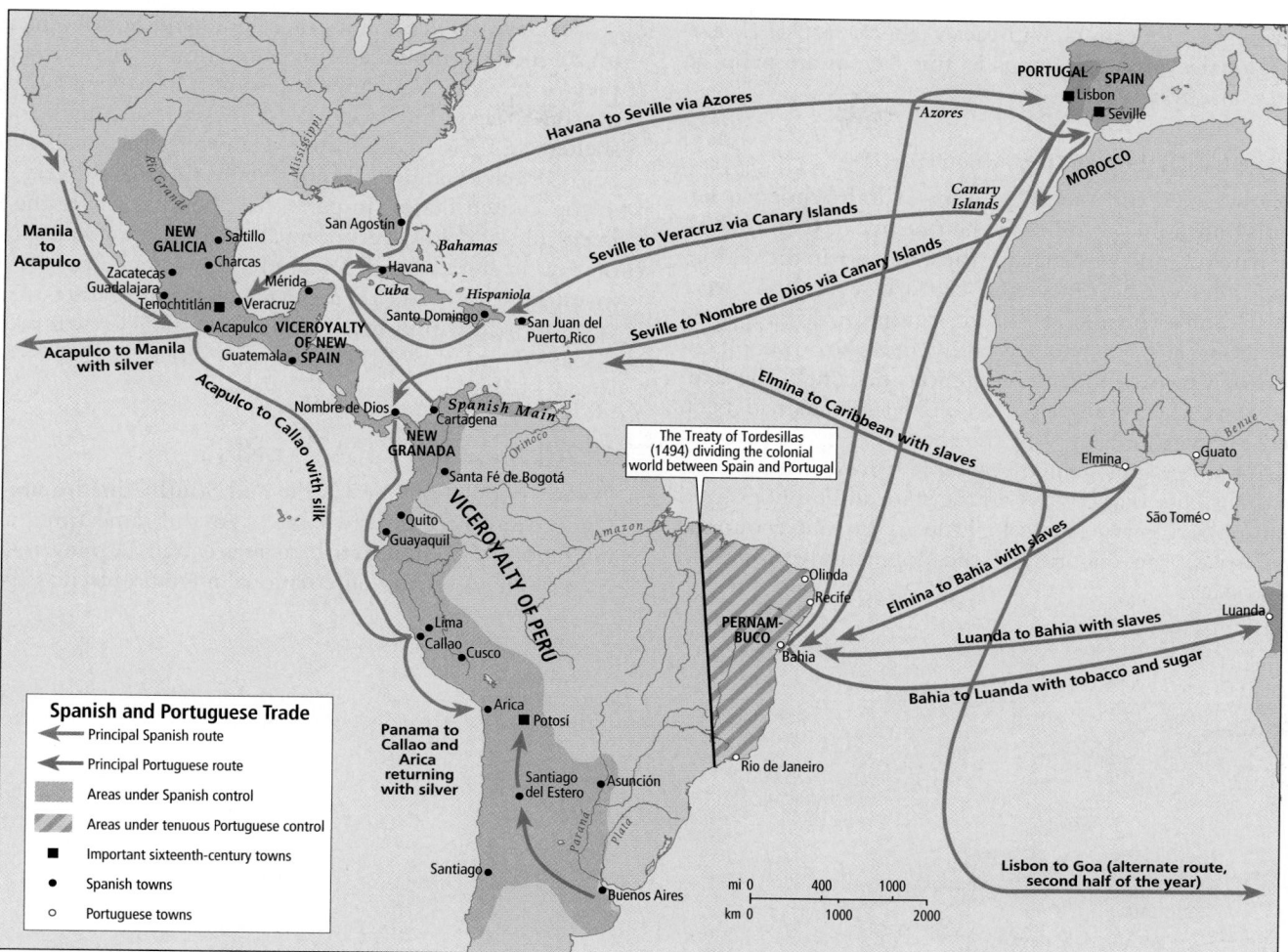

Figure 3.13 Spanish and Portuguese trade routes and territories in the Americas, circa 1600. The major trade routes from Spain to its colonies led to the two major centers of its empire, Mexico and Peru. The Spanish colonies could trade only with Spain, not directly with one another. By contrast, there were direct trade routes from Portuguese colonies in Brazil to Portuguese outposts in Africa. Many millions of Africans were enslaved and traded to Brazilian plantation and mine owners (as well as to Spanish, British, French, and Dutch colonies in the Caribbean and Middle and South America). [Adapted from *Hammond Times Concise Atlas of World History* (Maplewood, N.J.: Hammond, 1994), pp. 66–67.]

By the 1530s, a mere 40 years after Columbus's arrival, all major population centers of Middle and South America had been conquered and were rapidly being transformed by Iberian colonial policies. The colonies soon became part of extensive trade networks within the region and with Europe, Africa, and Asia (see Figure 3.13). By 1570, Spain was trading from its Mexican colony across the Pacific with its colony in the Philippine Islands.

A Global Exchange of Crops and Animals

From the earliest days of the conquest, plants and animals were exchanged between Middle and South America, Europe, Africa, and Asia via the trade routes illustrated in Figure 3.13.

Many plants essential to agriculture in Middle and South America today—rice, sugarcane, bananas, citrus, melons, onions, apples, wheat, barley, and oats, for example—were all originally imports from Europe, Africa, or Asia. When disease decimated the native populations of the region, the colonists turned the abandoned land into pasture for herd animals imported from Europe, including sheep, goats, oxen, cattle, donkeys, horses, and mules.

Plants first domesticated by indigenous people of Middle and South America have changed diets everywhere and have become essential components of agricultural economies around the globe. The potato, for example, had so improved the diet of the European poor by 1750 that it fueled a population explosion. Manioc (cassava) played a similar role in

West Africa. Corn, peanuts, and cacao (the source of chocolate) are globally important crops to this day, as are peppers, pineapples, and tomatoes.

The Legacy of Underdevelopment

In the early nineteenth century, wars of independence left Spain with only a few colonies in the Caribbean. Figure 3.14 gives the dates of independence for the countries in the region. The supporters of the nineteenth-century revolutions were primarily **Creoles** (people mostly of European descent born in the region) and relatively wealthy **mestizos** (people of mixed European, African, and indigenous descent). The Creoles' access to the profits of the colonial system had been restricted by mercantilism, and the mestizos' by racist colonial policies. Once they gained power, however, these groups became a new elite that controlled the state and monopolized economic opportunity. They did little to expand economic development or the majority of the population's access to political power.

Today, the economies of Middle and South America are much more complex and technologically sophisticated than they once were. Nevertheless, around 30 percent of the population is poor and lacks access to land, adequate food and shelter, and basic education. Meanwhile, the small elite class enjoys levels of affluence equivalent to those of the very wealthy in the United States. In part, these conditions are the lingering results of colonial economic policies that favored the export of raw materials and fostered the dominance of outside investors who spend their profits elsewhere rather than reinvesting them within the region. These policies will be further discussed in the Economic and Political Issues section (page 117).

POPULATION PATTERNS

Today, populations in Middle and South America are climbing because of high birth rates. At the same time, a major migration from rural to urban areas is transforming traditional ways of life. A second, international migration trend is growing

Figure 3.14 The colonial heritage of Middle and South America. Most of Middle and South America was colonized by Spain and Portugal, but important and influential small colonies were held by Britain, France, and the Netherlands. Nearly all the colonies had achieved independence by the late twentieth century. Those for which no date appears on the map are still linked in some way to the colonizing country. [Adapted from *Hammond Times Concise Atlas of World History* (Maplewood, N.J.: Hammond, 1994), p. 69.]

as well: many people are leaving their home countries, temporarily or permanently, to seek opportunities in the United States and Europe.

Population Numbers and Distribution

As of 2007, about 570 million people were living in Middle and South America, close to ten times the population of the region in 1492. This number is nearly two times the number of people that presently live in North America.

Population distribution. The population density map of this region reveals a very unequal distribution of people. If you compare Figure 3.15 on page 114 with the regional map (Figure 3.1), you will find areas of high population density in a variety of environments. Some of the places with the highest densities, such as those around Mexico City and in Colombia and Ecuador, are in highland areas. But high concentrations are also found in lowland zones along the Pacific coast of Central America and especially along the Atlantic coast of South America. The cool uplands (*tierra templada*) are particularly pleasant and were densely occupied even before the European conquest. Most coastal lowland concentrations, in the *tierra caliente,* are near seaports that attract people with their vibrant economies and social life.

Population growth. Rates of natural population increase were high in Middle and South America in the twentieth century. Although they are now declining, as shown in Figure 3.16 (page 115), the region's population could still double in just 43 years. (For comparison, doubling in North America is expected to take 116 years.) This is a disturbing prospect for a region struggling to increase living standards. Any gains might be checked by the costs of supplying more and more people with food, water, homes, schools, and hospitals.

During the twentieth century, cultural and economic factors combined with improvements in health care to create a population explosion. The Catholic Church discouraged systematic family planning, and cultural mores encouraged men and women to reproduce prolifically. Furthermore, most people until recently lived in agricultural areas, where children were seen as sources of wealth. They could do useful farm and household work at a young age and eventually would care for their aging elders. Low living standards and poor healthcare meant high infant death rates, which encouraged parents to have many children to be sure of raising at least a few to adulthood. Access to medical care began improving in the 1930s as incomes rose and as more people moved to the cities. By 1975, death rates were one-third what they had been in 1900.

By the 1980s, people began to limit their family sizes, and population growth rates started to fall. Now the region is beginning to undergo a demographic transition (see Figure 1.37 on page 45). Between 1975 and 2005, the annual rate of natural increase for the entire region fell from about 1.9 percent to 1.3 percent—still a higher rate of growth than the world average of 1.2 percent.

> When medical care began improving in the 1930s, death rates declined rapidly, and by 1975, they were one-third what they had been in 1900.

HIV-AIDS

The global epidemic of HIV-AIDS is now taking a significant toll on populations throughout this region, accounting for 9.5 percent of all deaths in 2005. In 2005, more than 1.8 million people were HIV-positive. Of these, at least 60,000 were children. The impoverished island of Haiti has one of the highest HIV rates outside of Africa. In the rest of the Caribbean, aggressive education programs about HIV, high literacy rates, and the relatively high status of women are limiting the number of infections.

The Venezuelan journalist Silvana Paternostro, in her book *In the Land of God and Man: Confronting Our Sexual Culture* (1998), argues that cultural practices contribute to HIV infection in this region. Cultural mores discourage the open discussion of sex. Men are rarely expected to be monogamous, and many visit prostitutes, making their wives and any children they might bear vulnerable to infection. Condom use is rare, and a wife would be loath to ask her husband to use one. Drug use is also a significant factor in the spread of HIV.

Migration and Urbanization

Since the early 1970s, Middle and South America have led the world in migration from rural to urban communities. Cities have grown remarkably quickly, and more than 75 percent of the people in the region now live in towns with populations of at least 2000. Increasingly, one city is vastly larger than all the others, accounting for a large percentage of the country's total population. Such cities are called **primate cities.** Examples in the region are Mexico City, Mexico (with 21 million people and about 20 percent of Mexico's population); Managua, Nicaragua (20 percent of that country's population); Lima, Peru (30 percent); Santiago, Chile (34 percent); and Buenos Aires, Argentina (38 percent).

The concentration of people into just one or two primate cities in a country leads to uneven spatial development, and government policies and social values that favor urban areas. ◢ Wealth and power are concentrated in one place, while distant rural areas, and even other towns and cities, have difficulty competing for talent, investment, industries, and government services. Many provincial cities languish as their most educated youth leave for the primate city.

Gender and urbanization. Interestingly, rural women are just as likely to migrate to the city as rural men. This is especially true when employment is available in foreign-owned

Figure 3.15 Population density in Middle and South America. [Data courtesy of Deborah Balk, Gregory Yetman, et al., Center for International Earth Science Information Network, Columbia University; at http://www.ciesin.columbia.edu.]

Trends in Average Rates of Natural Population Increase, 1975–2003, and Projected for 2003–2015

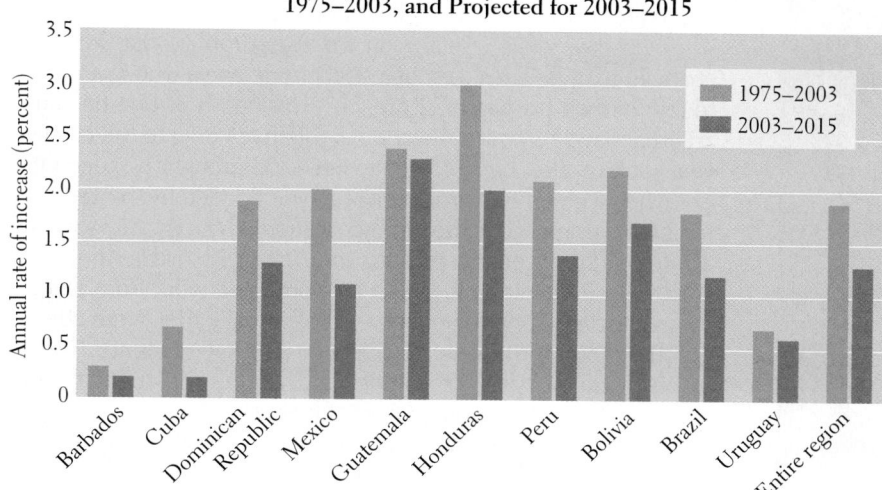

Figure 3.16 Trends in natural population increases, 1975–2015. Comparison of the orange and blue columns indicates that rates of natural increase have declined steadily throughout the region and are projected to continue to do so into the future. Nevertheless, in many countries natural population increase remains high enough to outstrip efforts at improving standards of living. Note that the rates of natural increase given here are for ranges of time. [Adapted from *United Nations Human Development Report 2005* (New York: United Nations Development Programme), Table 5.]

factories that produce goods for export. Companies prefer women for such jobs because they are a low-cost and usually passive labor force. Ironically, rural development projects may actually force female migration to urban areas. Due to the mechanization of formerly labor-intensive agricultural systems, these projects often end up decreasing available jobs. Women are rarely considered for training as farm equipment drivers and mechanics. In urban areas, unskilled women migrants usually find work as domestic servants. Low wages force them to live in the households where they work, where they are often subject to the sexual advances of their employers, yet they are themselves blamed if they fail to adhere to rigid standards of behavior.

> Ironically, rural development projects may actually force female migration.

Male urban migrants tend to depend on short-term, low-skill day work in construction, maintenance, small-scale manufacturing, and petty commerce. Many work in the informal economy as street vendors, errand runners, car washers, and trash recyclers, and some turn to crime. The loss of family ties and village life is sorely felt by both men and women, and the chances for recreating the family life they once knew in the urban context are extremely low.

Brain drain. It always takes some resourcefulness to move from one place to another. Usually those people who already have some advantages—some years of education and strong ambition—are the ones who migrate to cities. When young adults leave rural communities that have invested years in nurturing and educating them, the loss is often referred to as **brain drain**. Brain drain happens at several scales in Middle and South America: through rural-to-urban migration from villages to regional towns, and from towns and small cities to primate

cities. There is also international migration from the many countries in the region to North America and Europe. Parents often encourage their children to migrate so that they can benefit from the remittances, goods, and services that migrants provide to their home communities.

Favelas: unplanned neighborhoods. A lack of planning to accommodate the massive rush to the cities has created urban landscapes that are very different from the common U.S. pattern. In the United States, a poor inner city is usually surrounded by affluent suburbs, with clear and planned spatial separation by income. In Middle and South America, by contrast, rich and working class areas have become unwilling neighbors to unplanned slums filled with poor migrants. Too destitute even to rent housing, these "squatters" occupy parks and small patches of vacant urban land wherever they can be found.

The best known of these unplanned communities are Brazil's **favelas** (Figure 3.17). In other countries they are known as *colonias, barrios,* or *barriadas.* The settlements often spring up overnight, without water or electricity and with housing made out of whatever is available. Once they are established, efforts to relocate squatters usually fail. The impoverished are such a huge portion of the urban population that even those in positions of power will not challenge them directly. Hence, nearby wealthy neighborhoods simply barricade themselves with walls and security guards.

The squatters are frequently enterprising people who work hard to improve their communities. They often organize to press governments for social services. Some cities, such as Fortaleza in northeastern Brazil, even contribute building materials so that favela residents can build more permanent structures with basic indoor plumbing. Over time, as shacks

Figure 3.17 A favela in Rio de Janeiro. These people are crossing a bridge in the Rocinha favela to cast their votes in Brazil's national election of October 29, 2006. [AP Photo/Silvia Izquierdo.]

and lean-tos are transformed through self-help into crude but livable suburbs, the economy of slums can become quite vibrant. Housing may become intermingled with shops, factories, warehouses, and other commercial enterprises (Figure 3.18). Slums can become centers of pride and support for their residents, where community work, music, folk belief systems, and crafts flourish. For example, many of the best steel bands of Port of Spain, Trinidad, have their homes in such shantytowns.

Vignette Favelas are everywhere in Fortaleza, Brazil. The city grew from 30,000 to 300,000 residents during the 1980s. By 2006, there were more than 3 million residents, most of whom had fled drought and rural poverty in the interior. City parks of just a square block or two in middle-class residential areas were suddenly invaded by squatters. Within a year, 10,000 or more people were occupying crude stacked concrete dwellings in a single park, completely changing the ambience of the neighborhood. In the early

days of the migration, lack of water and sanitation often forced the migrants to relieve themselves on the street.

One day, while strolling on the waterfront, Lydia Pulsipher chanced to meet a resident of a beachfront favela who invited her to join him on his porch. There, he explained how one becomes an urban favela dweller. He and his wife had come to the city five years before, after being forced to leave the drought-plagued interior when a newly built irrigation reservoir flooded the rented land their families had cultivated for generations. With no way to make a living, they set out on foot for the city. In Fortaleza, they constructed the building they used for home and work from objects they collected along the beach, and eventually they were able to purchase roofing tiles, which gave it an air of permanency. Today, he maintains a small refreshment stand, and his wife runs a beauty parlor that caters to women from the area.

Source: Lydia Pulsipher's field notes in Brazil, updated with the help of John Mueller, Fortaleza, 2006. ■

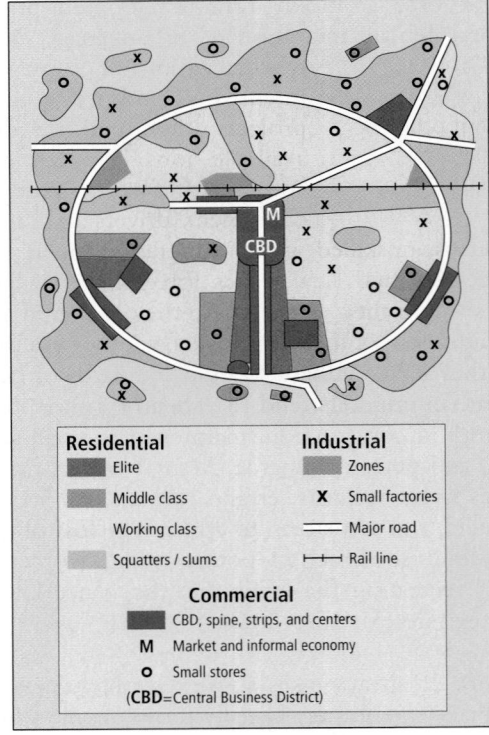

Figure 3.18 A model of urban land use in mainland Middle and South America. William Crowley, an urban geographer specializing in Middle and South America, developed this model. It depicts how residential, industrial, and commercial uses are mixed together, with people of widely varying incomes living in close proximity to one another and to industries. Squatters and slum dwellers ring the city in an irregular pattern. [From *Yearbook of the Association of Pacific Coast Geographers* 57 (1995): 28; printed with permission.]

Current Geographic Issues

Historically, power and wealth in the countries of Middle and South America have been concentrated in the hands of a few. Globalization, economic modernization, and the transformation from rural to urban societies have not changed that reality, though recent shifts toward democratization might. Culturally, this region retains diverse influences from Europe, Africa, and indigenous peoples. Meanwhile, shifting gender roles and new religious movements are powerful agents of change.

ECONOMIC AND POLITICAL ISSUES

Although Middle and South America are not as poor on average as sub-Saharan Africa, South Asia, or Southeast Asia, they face serious economic challenges. Many of these challenges revolve around **income disparity,** the gap in wealth between rich and poor. With the exception of a few small countries in Middle America, the gap between rich and poor is one of the largest on earth. In 2005, the richest 20 percent of the population was between 12 and 26 times richer than the poorest 20 percent, depending on which country was analyzed (Table 3.1). In 2005, though disparities were shrinking slightly in Brazil and Venezuela due to new government policies, nearly 40 percent of the regional population was living in poverty. In Haiti, Nicaragua, Bolivia, Paraguay, Guatemala, and Peru, more than half the population lives in poverty. ▭◄

Phases of Economic Development: Early Extractive and ISI Phases

The current economic and political situation in Middle and South America derives from the region's history, which can be divided into three major phases: the early extractive phase, the import substitution industrialization phase, and the structural adjustment phase of the present. All three phases have helped entrench wide income disparities.

The early extractive phase. The **early extractive phase** began with the European conquest and continued until the early twentieth century. Economic development was guided by a policy of **mercantilism**, in which European rulers sought to increase the power and wealth of their realms by managing all aspects of production, transport, and commerce in their colonies. Nearly all commercial activity in Middle and South America was focused solely on enriching the colonizing, or "mother," country. Little effort was made to build stable local economies.

A small flow of foreign investment and manufactured goods entered the region, but a vast flow of raw materials left for Europe and beyond. The foreign money to fund the farms, plantations, mines, and transport systems that enabled the extraction of resources for export came first from Europeans

and later from North Americans and other international sources. One example is the still highly lucrative Panama Canal, which was attempted first by the French in 1880, later completed by the United States, and turned over to Panamanian control in 1999. ▭◄ The profits from these ventures were usually banked abroad, depriving the region of investment that could have made it more economically independent. Industries did not develop, and hence even essential items, such as farm tools and household utensils, had to be purchased from Europe and North America at relatively high prices. Many people simply did without.

Compare this with the economic development of North America at the time. After the American Revolution, the

Table 3.1 Income disparity in selected countries[a]

Country	Ratio of wealth of richest 20% to poorest 20%[b] of the population		
	1987	1998–2000[c]	2005
Middle and South America			
Bolivia	9:1	12:1	12:1
Brazil	26:1	29:1	26:1
Chile	17:1	19:1	19:1
Colombia	20:1	20:1	23:1
Guatemala	30:1	16:1	24:1
Mexico	16:1	17:1	19:1
Peru	12:1	12:1	18:1
Other selected countries			
China	8:1	8:1	11:1
France	6:1	6:1	6:1
Jordan	6:1	6:1	6:1
Philippines	10:1	10:1	10:1
South Africa	33:1	18:1	18:1
Thailand	8:1	8:1	8:1
Turkey	8:1	8:1	8:1
United States	9:1	9:1	8:1

[a]The UN used data from 1987, 1998 to 2000, and 2005 on either income or consumption to calculate an approximate representation of how much richer the wealthiest 20 percent of the population is than the poorest 20 percent. The lower the ratio, the more equitable the distribution of wealth in the country.

[b]Decimals rounded up or down.

[c]Survey years fall within this range.

Source: United Nations Human Development Report 2000, Table 13; 2003, Table 13; 2005, Table 15.

profits from U.S. extractive industries went to owners who tended to live in North America. They invested their money locally in industries that processed the region's raw materials into more valuable finished products. These manufactured goods were then bought by American workers or exported, providing the foundation for North America's history of economic stability and high living standards.

A number of economic institutions arose in Middle and South America to supply food and raw materials to Europe and North America. Large rural estates called **haciendas** were granted to colonists as a reward for conquering territory and people for Spain. These estates were then passed down through the families of those colonists for generations. Over time, the owners, who often lived in a distant city or in Europe, lost interest in the day-to-day operations of the haciendas. Hence, productivity was generally low and hacienda laborers remained extremely poor. Nevertheless, haciendas did produce a diverse array of crops for local consumption and export.

Plantations, large factory farms growing a single crop such as sugar, coffee, cotton, or (more recently) bananas, were more profitable and brought in more investment. Plantation owners made larger investments in equipment than did hacienda owners, but they, too, contributed little to local economies. Instead of employing local populations, they imported slave labor from Africa.

First developed by the European colonizers of the Caribbean and northeastern Brazil in the 1600s, plantations became more common throughout South America by the late nineteenth century. Unlike haciendas, which were often established in the continental interior in a variety of climates, plantations were for the most part situated in tropical coastal areas with year-round growing seasons. Their coastal location also gave them easier access to global markets via ocean transport. Because most profits were either reinvested in the plantations themselves or taken abroad, few connections with local economies developed.

As markets for meat, hides, and wool grew in Europe and North America, the livestock ranch emerged, specializing in raising cattle and sheep. Today, commercial ranches are found in the drier grasslands and savannas of South America, Central America, northern Mexico, and even in the wet tropics on cleared rain-forest lands.

Mining was another early extractive industry. Important mines (at first primarily gold and silver) were located on the island of Hispaniola, in north-central Mexico, in the Andes, in the Brazilian Highlands, and in many other locations. Today, oil and gas have been added to the mineral extraction industry, but rich mines throughout the region continue to produce gold, silver, copper, tin, precious gems, titanium, bauxite, and tungsten (Figure 3.19).

Profits from the region's mines, ranches, plantations, and haciendas continued to leave the region even after the countries gained independence in the nineteenth century. One of the main reasons for this was that wealthy European and North American private investors had purchased many of the extractive enterprises.

The import substitution industrialization phase. In the early and mid-twentieth century, there were waves of protest against the domination of the economy and society by local

Figure 3.19 The Yanacocha gold mine in Cajamarca, Peru. This mine, the largest in the Americas, is run by a subsidiary of the U.S. mining group Newmont. Mining is one of the most polluting of all extractive industries, and this mine has been the focus of widespread protests by local people who say it is polluting their water supply. The protests caused the suspension of mining in August 2006. [STR/AFP/Getty Images.]

elites and foreign businesses. Many governments—Mexico and Argentina most prominent among them—proclaimed themselves socialist democracies. To keep money and resources within the region in order to foster economic self-sufficiency, they enacted a set of policies that became known as **import substitution industrialization (ISI)** policies. Import substitution policies encouraged local production of machinery and other items that had long been imported at great expense from abroad. These worked in a few cases but generally failed to lift the region out of poverty.

First, national governments seized the most profitable extractive industries from foreign owners (usually with some payment). The intention was to use any profits to create local manufacturing industries that could supply the goods once purchased from Europe and North America. These goods could also be exported to foreign markets to earn more money. To encourage local people to buy manufactured goods from local suppliers, governments placed high tariffs on imported manufactured goods. The money and resources kept within each country were expected to provide the basis for further industrial development. This would create well-paying jobs, raise living standards for the majority of people, and ultimately replace the extractive industries as the backbone of the economy.

The goal of lifting large numbers of people out of poverty through jobs and general economic growth was never realized. The state-owned manufacturing sectors on which the success of ISI depended were never able to produce high-quality goods that could compete with those produced in Asia, Europe, and North America. This was largely because they couldn't afford the technologies and didn't have the managerial skills to run globally competitive factories. Further, local populations weren't large or prosperous enough to support these industries. And because exports never took off, employment stagnated, tax revenues remained low, and social programs were not adequately funded.

Not all state-owned corporations were losing propositions, however. ISI still survives in some countries and is being considered for reimplementation in a few others, such as Bolivia and Venezuela. Brazil—with its aircraft, armament, and auto industries—and Mexico—with its oil and gas industries—both experienced successes with some ISI programs. 🎥 Interest in state-supported manufacturing industries continues, but the general trend is now toward more market-oriented management and global competitiveness. Dependence on the export of raw materials persists.

Other policies enacted during the ISI phase also had some success. Income disparities were reduced as governments increased spending on public health, education, and infrastructure. Meanwhile, **land reform** broke up some large landholdings and distributed them among poor landless farmers.

The debt crisis. Ultimately, a global economic crisis diminished the ISI phase. In the 1970s, increases in oil prices and decreases in global prices of raw materials ended a period of global prosperity that had begun in the early 1950s. While earnings from exports were falling, governments and private interests continued to pursue ambitious plans to modernize and industrialize their national economies. They paid for these projects by borrowing millions of dollars from major international banks, most of which were in North America or Europe.

The beginning of a global **recession** (a slowing of economic activity) in 1980 put a halt to the development plans of many governments in the region. Dragged down by the recession, the Middle and South American economies could not meet their targets for growth, and thus the governments were unable to repay their loans. The damage to the region was worsened by the fact that the biggest borrowers—such as Mexico, Brazil, and Argentina—also had the largest economies in the region. Hence, huge **external debts** (debts owed to foreign banks or governments and repayable only in foreign currency) now burdened the very countries that had been the most likely to grow (Figure 3.20).

Phases of Economic Development: Structural Adjustment and Globalization

Structural adjustment policies. In the 1980s, foreign banks that had made loans to governments in the region took action. The International Monetary Fund developed and enforced structural adjustment policies (SAPs) that mandated profound changes in the organization of economies, all of which supported globalization (see Chapter 1, pages 14–16). **Privatization,** the selling of formerly government-owned industries to private investors, was considered one of the soundest ways to achieve economic expansion—and hence repayment of debts—during the SAP period. Often these investors were multinational corporations located in Asia, North America, and Europe. The other main policy was free trade. In order to obtain further loans, governments were required to remove tariffs on imported goods of all types.

SAPs also reversed the ISI-era trend toward the expansion of government social programs. To free up funds for debt repayment, governments were required to fire many civil servants and drastically reduce spending on public health, education, and infrastructure. 🎥 SAPs encouraged the expansion of industries that were already earning profits by lowering taxes on their activities. In the countries of Middle and South America, as in most developing countries, the most profitable industries were still based on the extraction of raw materials for export.

Outcomes of SAPs. SAPs have increased some kinds of economic activity in Middle and South America, resulting in modest rates of national GDP growth (averaging about 5 percent per year). Lowered trade barriers have expanded opportunities for local entrepreneurs. They have also encouraged investors from other countries, especially multinational corporations, to invest in the region, an activity called **foreign direct investment (FDI)**. U.S.-, European-, and Asian-owned companies have invested in the mining, agriculture, forestry, and fishing industries. 🎥 Partnerships between foreign companies

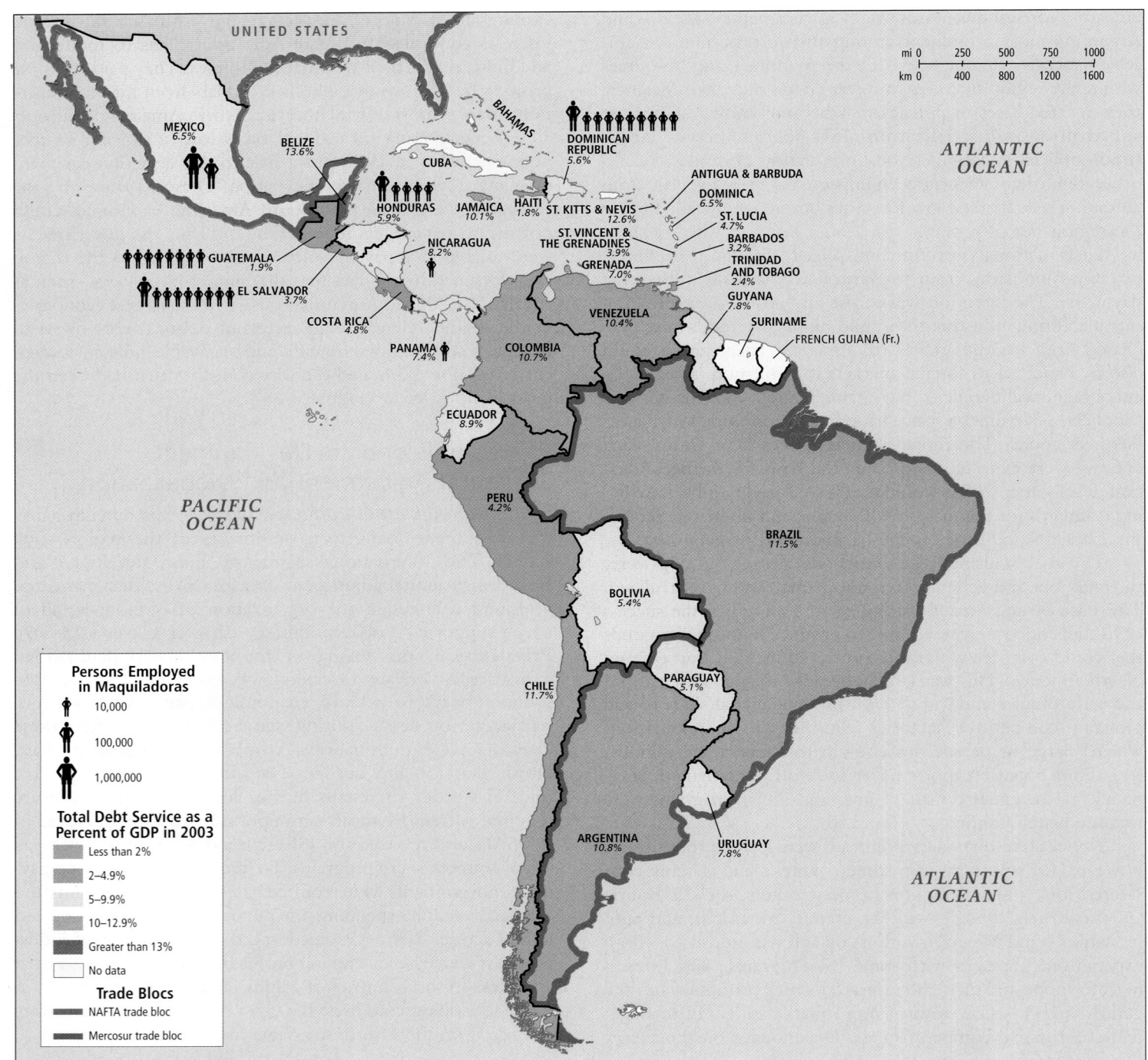

Figure 3.20 Debt, maquiladoras, and trade blocs in Middle and South America. The high rate of debt in the region stems from countries borrowing money to finance industrial and agricultural development projects intended to replace imports and reduce poverty by providing jobs. Debt is presented as total debt service (repayment of public and private loans) as a percentage of a country's exports of goods and services. Maquiladoras have played a major role in efforts to reduce government debts. The creation of trade blocs such as NAFTA and Mercosur is another means countries in the region are using to reduce debt by lowering tariffs and reducing dependence on higher-cost imports from outside the region. [Adapted from *United Nations Human Development Report 2005* (New York: United Nations Development Programme), Table 20.]

and local businesspeople operate the largely unregulated gold mining, forestry, and other extractive industries of the Amazon Basin. However, here, as elsewhere in the region, SAPs encouraged industries to expand in ways that raised concerns about environmental impacts, worker safety, and sustainability.

A major aspect of SAPs was the expansion of manufacturing industries in **Export Processing Zones (EPZs)**, also known as *free trade zones*. EPZs are specially created areas within a country where taxes on imports and exports are not charged in order to attract foreign-owned factories. The main benefit to the host country is the employment of local labor, which eases unemployment and brings money into the economy. Products are often assembled strictly for export to foreign markets.

EPZs exist in nearly all countries on the Middle and South American mainland and on some Caribbean islands. However, the largest of the EPZs is the conglomeration of assembly factories, called **maquiladoras,** located along the Mexican side of the U.S.–Mexico border (Figures 3.20 and 3.21). While these factories do provide employment, they might not be as beneficial to the Mexican economy as once thought, as the following vignette illustrates.

Vignette Orbalin Hernandez has just returned to his self-built shelter in the town of Mexicali on the border between Mexico and the United States. He smiles as he enters the yard, and his wife breathes a sigh of relief. Recently fired for taking off his safety goggles while loading TV screens onto trucks at Thompson Electronics, he went to the personnel office to ask for his job back. Because there were no previous problems with him, he was rehired at his old salary of $300 per month ($1.88 per hour). Thompson is a

French-owned electronics firm that took advantage of NAFTA (see page 120) when it moved to Mexicali from Scranton, Pennsylvania in 2001. There, its 1100 workers had been paid an average of $20.00 per hour. Now 20 percent of the Scranton workers are unemployed, and many feel bitter toward the Mexicali workers.

Though Orbalin's salary at Thompson-Mexicali is barely enough for him to support his wife, Mariestelle, and four kids, they are grateful for the job. They are from a farming community in Tabasco State, where, for people with no high school education, wages averaged $60 per month. Thompson is known for treating its Mexican workers well. It pays into a housing fund, supports local educational facilities, and provides bus transport for workers to and from home. But over the last 20 years of the maquiladora boom, Mexican wages have lost 81 percent of their buying power. A typical maquiladora worker has to work more than 60 minutes to buy one kilo (2.2 pounds) of rice. By contrast, a dockworker in Los Angeles can buy that rice after just 3 minutes of work, and an undocumented minimum-wage worker in L.A. after 12 minutes of work.

Nonetheless, Orbalin and his fellow Mexicali workers are now being told that their wages are too high. Maquiladora firms claim that to compete globally they must cut costs even further. Fourteen Mexicali plants have closed and moved to Asia, where in 2005, workers with the same skills as Orbalin earned just $0.35 an hour. Those firms remaining in Mexicali have cut wages and reduced benefits in response. But the tide may be turning: because of a growing labor shortage in China, wages there are increasing by 10 to 40 percent per year, and soon it will make little sense to move a Mexican factory to China.

Source: NPR reports by John Idste, August 14, 2001, August 25, 2003, August 16, 2003, and by Gary Hadden, August 27, 2003; David Bacon, "Anti-China campaign hides maquiladora wage cuts,"

Figure 3.21 Maquiladora workers. In March 2006, these young Mexican women were assembling car radios for Delphi Delco Electronics de Mexico in Matamoros, Mexico. The plant, a maquiladora across the border from Brownsville, Texas, is one of seven Delphi plants in the vicinity. Together they employ a total of 11,000 Mexican workers. [Bob Daemmrich/CORBIS.]

February 2, 2003, http://www.globalpolicy.org/socecon/labor/2003/ 0203maq.htm; April, 2006; http://www.maquilaportal.com/cgi-bin/ public/index.pl; http://www.businessweek.com/magazine/content/ 06_13/b3977049.htm. ■

The economic growth that was expected to relieve the debt crisis and achieve broader prosperity never occurred. Basic economic indicators, such as per capita growth in income, showed a marked decline during the SAP phase as compared with the ISI phase. During the last two decades of the ISI phase (1960 to 1980), per capita GDP grew by 82 percent for the region as a whole. When SAPs were implemented across the region (1980 to 2000), growth in per capita GDP slowed to just 12.6 percent, and between 2000 and 2005 it slowed to only 1 percent.

A major reason that SAPs failed to stimulate economic growth was that they encouraged greater dependence on exports of raw materials just when prices for these items were falling in the global market. Ironically, prices fell largely because of SAPs! SAPs forced so many indebted countries around the globe to boost raw materials exports at the same time that a global glut was created that prices were driven down.

Voter backlash against SAPs. Millions of voters within this region have registered their opposition to SAPs in recent years. Since 1999, presidents that explicitly oppose SAPs have been elected in eight countries in the region: Argentina, Bolivia, Brazil, Chile, Ecuador, Nicaragua, Uruguay, and Venezuela. Both Brazil and Argentina have made considerable sacrifices in order to pay off their debts and hence be liberated from the restrictions of SAPs. Under President Hugo Chavez, Venezuela has emerged as a leader of the SAP backlash, His administration has **nationalized** (taken over by the government, with compensation) foreign oil companies operating in Venezuela. ▶ It has also used its own considerable oil wealth to help other countries, such as Bolivia, pay off their debt. Since then, Bolivia, led by Evo Morales (a major Chavez ally), has reversed the SAP policy of privatization by nationalizing the country's national gas industry. ▶

> Since 1999, presidents who explicitly oppose SAPs have been elected in eight countries in the region.

The Informal Economy

For centuries, low-profile businesspeople throughout Middle and South America have operated outside the law in the informal economy. As described in Chapter 1 (see pages 11–13), they support their families through inventive entrepreneurship. Most are small-scale operators involved with street vending or recycling used items such as clothing, glass, or waste materials. In some instances, the informal economy can serve as an incubator for businesses that might eventually provide legitimate jobs for family and friends.

Figure 3.22 A vegetable vendor on a street corner in Lima, Peru. [© Lin MacDonald.]

The informal economy is huge throughout this region, with most middle- and low-income families depending on it in some way. Critics argue that informal workers are just treading water, making too little to ever expand their businesses significantly. Moreover, the bribes they have to pay to avoid arrest or fines are much less beneficial to the economy as a whole than the taxes paid by legitimate businesses. Work in the informal economy is also risky because there is no protection of workers' health and safety and no retirement or disability benefits.

Nevertheless, the informal economy can be a lifesaver during times of economic recession. For example, after recession hit Peru in 2000, 68 percent of urban workers were in the informal sector. They generated 42 percent of the country's total gross domestic product. Most of them worked as street vendors (Figure 3.22).

Geographer Maureen Hays-Mitchell found street vending in Peruvian cities to be a vibrant, organized, and highly competitive sector of the informal economy. Street vendors were found to be better than the formal retailing system at making goods available in convenient places and at affordable prices. For instance, they sell food and small gifts in front of hospitals during visiting hours, and candy, games, and toys in front of schools during recess. Outside jails, visitors and guards can buy food, souvenirs, and handicrafts produced by inmates. Few opportunities are left unexploited. A single city block may contain between 15 and 30 street vendors on average.

Regional Trade and Trade Agreements

The growth in international trade and foreign investment in the region, encouraged in part by SAPs, has been joined by growth in regional free trade agreements. Such agreements reduce tariffs and other barriers to trade among a group of neighboring countries. The two largest free trade agreements within the region are NAFTA and Mercosur (see Figure 3.20).

The **North American Free Trade Agreement (NAFTA),** created in 1994, is by far the larger of the two. It links the economies of Mexico, the United States, and Canada in a free trade bloc containing more than 435 million people. Its total annual economy is worth at least $14 trillion. In the 1990s, the United States began pushing for the creation of a NAFTA-like trade bloc for all of North, Middle, and South America. One purpose of this Free Trade Agreement of the Americas (FTAA) would be to check, at least partially, European trading power in the region. However, rising opposition to SAPs, which many in the region see as working hand in hand with U.S.-dominated free trade agreements such as NAFTA, has stalled the FTAA in recent years.

Increasingly, **Mercosur** is seen as a more equitable alternative to NAFTA and the FTAA. Mercosur is older than NAFTA, having been created in 1991, and smaller. It links the economies of Argentina, Brazil, Paraguay, Uruguay, and Venezuela to create a common market with 250 million potential consumers and an economy worth nearly $2 trillion per year. (Bolivia, Chile, Colombia, Ecuador, and Peru are associate Mercosur members and are not included in these figures.)

The record of regional free trade agreements so far is mixed. While they have increased trade, the benefits of that trade usually are not spread evenly among regions or among richer and poorer sectors of society. In Mexico, for instance, the benefits of NAFTA have been concentrated in the northern states that border the United States. There, the agreement facilitated the growth of maquiladoras by easing cross-border finances and transit. Meanwhile, as many as one-third of small-scale Mexican farmers have lost their jobs as a result of increased competition from U.S. corporate agriculture, which now has unrestricted access to Mexican markets. These unemployed agricultural workers make up a significant number of the undocumented workers coming into the United States.

Vignette Reporter Sam Quinones speaks to his audience: "In the current crisis, peasant coffee growers have to learn the Starbucks lesson and focus on quality. Consumers, meanwhile, have to be willing to pay extra for the best coffee, searching out regional coffees, the way they do with wine."

The Guatemalan coffee farmer listens to the speaker and frowns down at a slick package of Green Mountain organic coffee. It has his picture on the front. It's the first time he has ever seen how the coffee he grows is marketed to upscale consumers in North America. He has never tasted coffee himself, preferring sweet soda drinks. Only a harsh, low-quality coffee is sold in his village, and virtually no one drinks it.

Coffee used to be a primary commercial crop in the Pacific coastal uplands of Guatemala. Now, many of the coffee plantations (*fincas*) are abandoned. The workers have migrated to Mexico and the United States seeking some other way to support their families. Too many poor tropical farmers on 10 or 12 acres trying to make a living in coffee produced a glut on the global coffee market. As a consequence, most are paid only a tiny amount per pound by the itinerant coffee trader (called a coyote) who is their only access to the market. When you buy a $1.00 cup of coffee, the grower gets 10 to 12 cents, the trader 3 cents, and the shipper 4 cents. The roaster—usually a large multinational company—gets 65 to 70 cents. The retailer typically gets 10 to 15 cents.

Green Mountain and other *fair trade* coffee marketers are trying to change that lopsided allocation of profits by buying directly from the growers. The direct-to-market approach of fair trade means that growers can get as much as $1.26 per pound for their coffee. But it also means that they have to produce high-quality coffee without chemical pesticides and fertilizers—coffee that passes the high standards of professional coffee tasters. Therefore, the farmers themselves must learn to judge coffee by its taste. The goal of the fair trade coffee movement is to teach them to be coffee connoisseurs (as vintners are connoisseurs of wine) rather than just producers of beans.

For more on how fair trade affects the coffee market and the lives of growers, see the FRONTLINE/World video "Coffee Country." ■

Food Production and Contested Space

The agricultural lands of Middle and South America (Figure 3.23) are an example of what geographers call **contested space:** various groups are in conflict over the right to use a specific territory as they see fit.

For centuries, while people labored for very low wages on haciendas, the owners would at least allow them a bit of land to grow a small garden or some cash crops. In recent years, however, SAPs have encouraged a shift to large-scale, mechanized, export-oriented farms and plantations. The rationale is that these operations can earn larger profits and hence can help countries pay off debts faster. SAPs have made it easier for foreign multinational corporations, such as Del Monte, with their highly efficient, state-of-the-art operations, to dominate the most profitable agribusinesses. Meanwhile, smaller farmers are squeezed out because they cannot afford the newest equipment.

> In recent years, SAPs have encouraged a shift to large-scale, mechanized, export-oriented farms and plantations.

At the personal scale: plantation labor. Throughout Middle and South America, increasing numbers of rural people find themselves displaced from lands they once cultivated and forced to work as migrant laborers for low wages, as the following vignette illustrates.

Vignette Aguilar Busto Rosalino used to work on a Costa Rican hacienda. He had a plot on which to grow his own food and in return worked three days a week for the hacienda. Since a banana plantation took over the hacienda, he rises well before dawn and

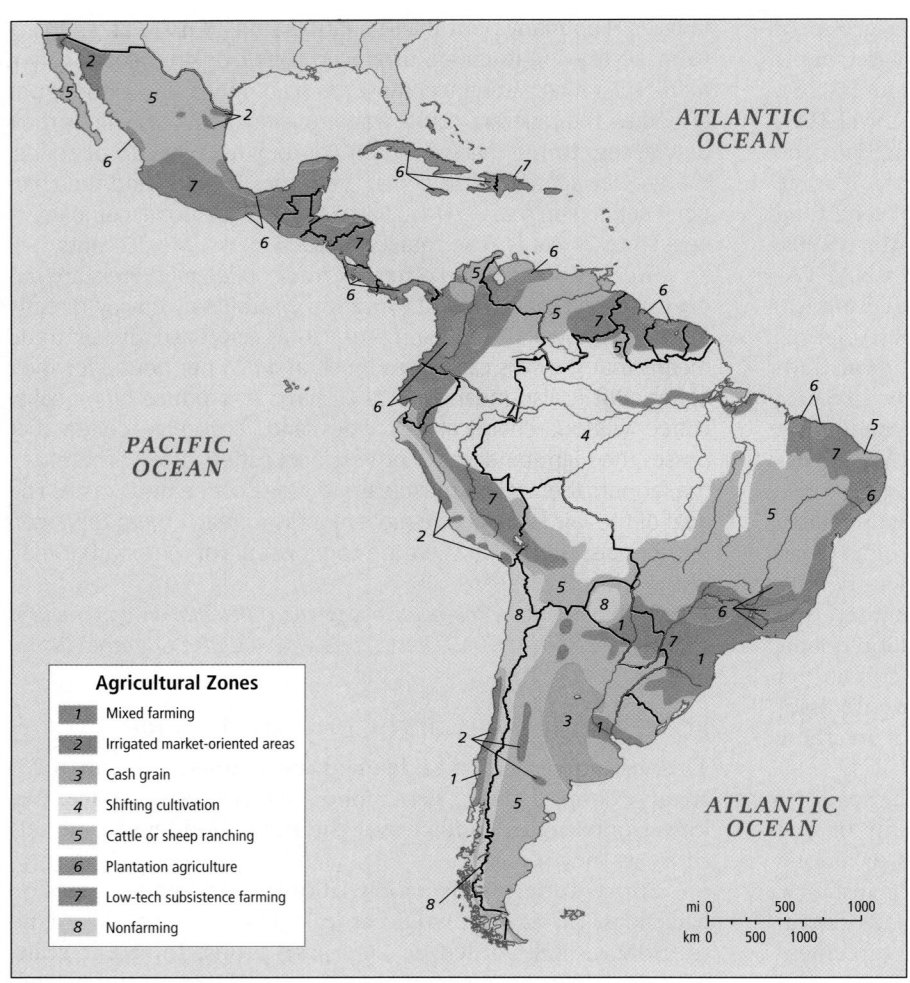

Figure 3.23 Agricultural zones in Middle and South America. Subsistence farming (7) remains widespread, but this traditional mode of agriculture is losing ground to modern, mechanized, export-oriented agriculture (2, 3, 5, 6). The displaced farmers often join the stream of migration to urban areas and North America. [Adapted from Edward F. Bergman, *Human Geography: Cultures, Connections, and Landscapes* (Englewood Cliffs, N.J.: Prentice Hall, 1995), p. 194.]

Agricultural Zones

1	Mixed farming
2	Irrigated market-oriented areas
3	Cash grain
4	Shifting cultivation
5	Cattle or sheep ranching
6	Plantation agriculture
7	Low-tech subsistence farming
8	Nonfarming

works five days a week from 5:00 A.M. to 6:00 P.M., stopping only for a half-hour lunch break. Because he doesn't have time to farm, he now has to buy most of his food.

Aguilar places plastic bags containing pesticide around bunches of young bananas. He prefers this work to his last assignment of spraying a more powerful pesticide, which left him and 10,000 other plantation workers sterile. He works very hard because he is paid according to how many bananas he treats. Usually he earns between $5.00 and $14.50 a day.

Right now Aguilar is working for a plantation that supplies bananas to the Del Monte corporation, but he thinks that in a few months he will be working for another plantation nearby. It is common practice for these banana operations to fire their workers every three months so that they can avoid paying the employee benefits that Costa Rican law mandates. Although Aguilar makes barely enough to live on, he has no plans to press for higher wages because he knows that he would be put on a blacklist of people that the plantations agree not to hire.

Source: Adapted from Andrew Wheat, "Toxic bananas," Multinational Monitor 17 (9) (September 1996): 6–7. ∎

At the provincial scale: the Zapatista rebellion. In the southern Mexican state of Chiapas, agricultural activists and indigenous leaders have mobilized armed opposition to the economic and political systems that have left them poor and powerless. The Mexican government redistributed some hacienda lands to poor farmers early in the twentieth century, but most land in Chiapas is still held by a wealthy few who grow cash crops for export. The poor majority farm tiny plots on infertile hillsides. In 2000, about three-fourths of the rural population was malnourished, and one-third of the children did not attend school.

The Zapatista rebellion (named for the Mexican revolutionary hero Emiliano Zapata) began on the day the North American Free Trade Agreement took effect in 1994. The Zapatistas view NAFTA as a threat because it diverts the support of the Mexican federal government from land reform to large-scale mechanized export agriculture. The Mexican government used the army to suppress the rebellion.

In 2003, after nine years of armed resistance, the Zapatista movement redirected part of its energies. It began a nonviolent political campaign to democratize local communities by setting up people's governing bodies parallel to local official governments. Before the national elections in 2006, the Zapatistas toured the country in an effort to turn the political climate against globalization and toward greater support for indigenous people and the poor.

At the national scale: Brazil's landless movement.

The trends in agriculture that we have been describing have sparked rural resistance in Brazil as well. Sixty-five percent of Brazil's arable and pasture land is owned by wealthy farmers who make up just 2 percent of the population. Since 1985, more than 2 million small-scale farmers have been pressed to sell their land to large-scale, mechanized farms. Because the larger farms specialize in major export items, such as cattle and soybeans, they have been favored by governments under pressure from SAPs to increase exports. As a result, many poor farmers have been forced to migrate.

To help these farmers, organizations such as the Movement of Landless Rural Workers (MST) began taking over unused portions of some large farms. They argue that it is wrong for agricultural land to lie unused while poor people go hungry. Since the mid-1980s, the MST has coordinated the occupation of more than 51 million acres (21 million hectares) of Brazilian land (an area about the size of Kansas). Some 250,000 families have gained land titles, while the elite owners have been paid off by the Brazilian government and have moved elsewhere. MST played a major role in the political success of Brazilian President Lula da Silva in 2002 and 2006. Public opinion polls show that 77 percent of Brazilians support the landless movement. Movements with goals similar to those of MST now exist in Ecuador, Venezuela, Colombia, Peru, Paraguay, Mexico, and Bolivia.

The continuing growth of large-scale mechanized agriculture.

The instances of contested use of agricultural land we have been describing are exceptions to overall trends that favor the growth of large-scale mechanized agriculture. Because of the money-making potential and political power of big agriculture, conflicts are only rarely resolved in favor of small farmers and agricultural workers. Moreover, in rapidly urbanizing countries, large-scale mechanized agriculture is seen as the only way to supply sufficient food for the millions of city dwellers. Ironically, many of these new urbanites were once farmers capable of feeding themselves, but they were unable to compete with mechanized agriculture.

Zones of modern mixed farming, including the production of meat, vegetables, and specialty foods for sale in urban areas, are located on large and small farms around most major urban centers. Many areas have had large farms for centuries. For example, a wide belt of large-scale mechanized grain production, similar to that found in the midwestern United States, stretches through the Argentine pampas (see Figure 3.23). And as we saw in the earlier discussion of environmental issues (see

pages 102–3), soybeans, which are often produced on huge mechanized farms in and around the Amazon Basin, are a major export for Brazil. Hence, despite the recent political tensions, large-scale agriculture appears to have a solid place in the future of this region.

Democratization and Political Reform

After decades of elite and military rule, almost all countries in the region now have multiparty political systems and democratically elected governments. In the last 25 years, there have been repeated peaceful and democratic transfers of power in countries once dominated by rulers who seized power by force. These **dictators** often claimed absolute authority, governing with little respect for the law or the rights of their citizens. Their authority was based on alliances between the military, wealthy rural landowners, wealthy urban entrepreneurs, foreign corporations, and even foreign governments such as the United States.

Although democracy has transformed the politics of the region, problems still remain. Elections are sometimes poorly or unfairly run, and their results are frequently contested. Elected governments are sometimes threatened with a **coup d'état,** in which the military takes control of the government by force or by similar efforts by civilians. Such coups are usually a response to policies that are unpopular with large segments of the population, powerful elites, the military, or the United States. Venezuela, Ecuador, Colombia, and Bolivia have been threatened by coups in the last decade.

Change has been especially dramatic in Mexico. The entrenched political machine of the Institutional Revolutionary Party (PRI) ruled the country for almost three-quarters of a century. But in 2000, it was unseated by a coalition led by Vicente Fox, a wealthy Coca-Cola executive. Fox managed to pass a number of reform-oriented laws in his 6-year term. In the 2006 presidential election, the conservative Felipe Calderón (from Fox's political party) narrowly defeated the populist mayor of Mexico City, Andres Lopez Obrador. Lopez Obrador, with significant public support, created a constitutional crisis by refusing to accept the election results. He organized a so-called parallel government, declaring himself president and appointing ministers. However, he failed to force President Calderón to resign. ◼

In Venezuela, the landslide election of Hugo Chavez as president in 1998 appeared to advance the cause of democracy. Elites had long dominated Venezuelan politics, and profits from the country's rich oil deposits had not trickled down to the poor. One-third of the population lived on $2.00 (PPP) a day, and per capita GDP (PPP) was lower in 1997 than it was in 1977. Chavez campaigned as a champion of the poor, calling for the government to provide jobs, community health care, and subsidized food for the large underclass. These policies, all of which were enacted soon after his election, eroded Chavez's original support from the middle class and alarmed the government of the United States. Chavez also became a major critic of U.S. and other foreign dominance in the region, leading the backlash against SAPs and condemning the long history of U.S. intervention in the region's national political scenes.

Chavez was reelected in 2000, and then briefly deposed in a coup d'état in 2001. He was quickly reinstated, however, by a groundswell of popular support among the poor, many of whom had benefited from his policies. His position was sustained in a referendum in 2004 and again in the landslide election of 2006. Nevertheless, in 2007, voters rejected a constitutional amendment that would have removed presidential term limits, allowing Chavez to run for president indefinitely. 🎥

Political corruption. Despite recent progress, democracy is still fragile and not particularly *transparent*—meaning that official decisions are not open to public input and review. Much of the region suffers from *political corruption*, in which elected officials use their power for personal gain through bribery, cronyism, nepotism, embezzlement, or other illicit means. Virtually every country in the region has had a serious scandal in recent years involving high-level officials performing million-

dollar favors for friends or family. 🎥 Some have simply stolen taxpayers' money from national treasuries. International banks in Miami, Florida, are well-known as depositories for such stolen funds. Meanwhile, Middle and South American countries have been robbed of much-needed public funds and private investment capital.

The Drug Trade

The international illegal drug trade is a major factor contributing to corruption, violence, and subversion of democracy throughout the region. Cocaine and marijuana are the primary drugs of trade, with most coming from Mexico, Central America, and northwestern South America. 🎥 Figure 3.24 graphically illustrates the geographic distribution of cocaine seizures between 2002 and 2003. (Such seizures may not reveal activity in areas where law enforcement is lax or coopted by the drug traders.)

Most coca growers are small-scale farmers of indigenous or mestizo origin in remote locations who can make a better income for their families from these plants than from other cash crops. Production of addictive drugs is illegal in all of

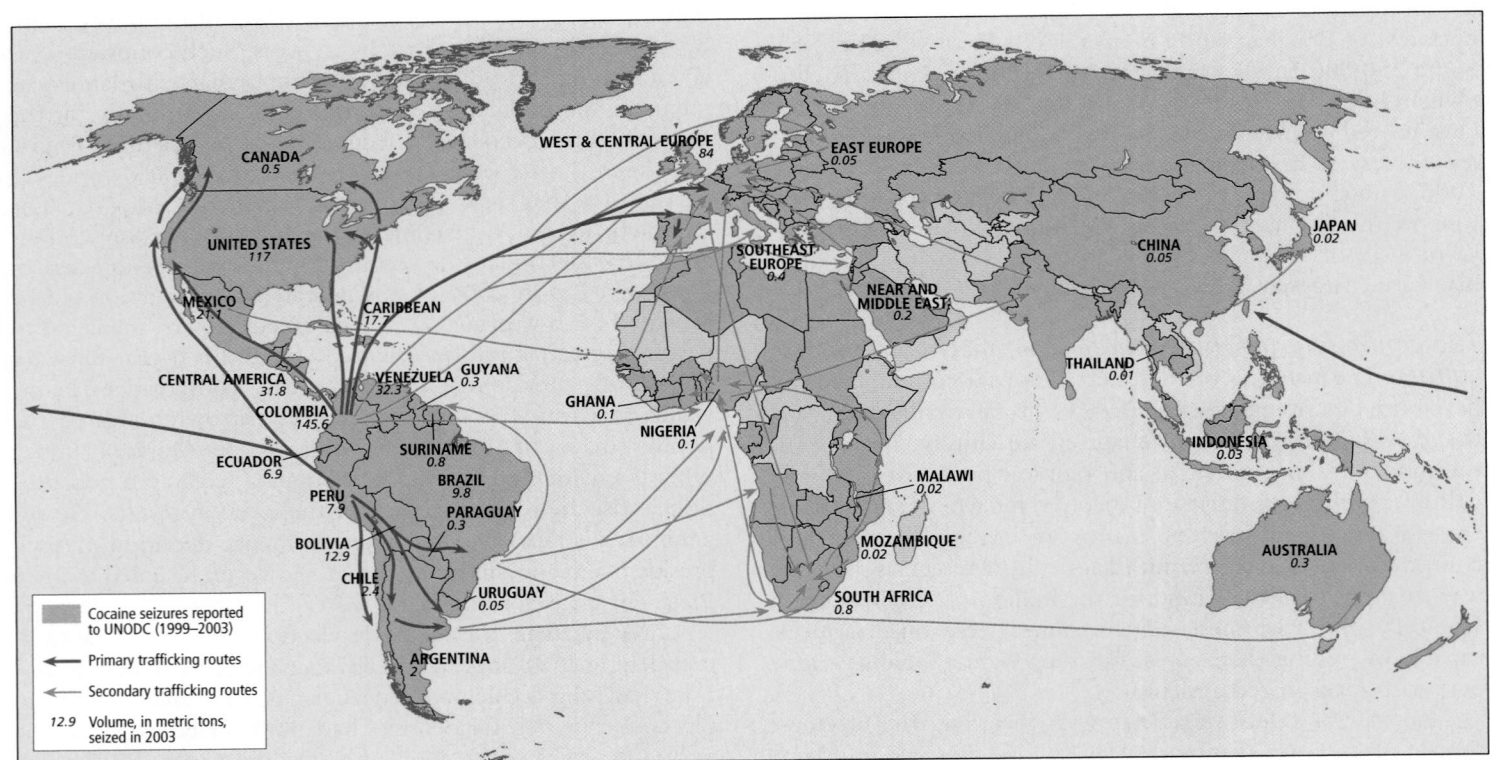

Figure 3.24 Interregional linkages: Cocaine sources, trafficking routes, and seizures worldwide. Colombia, Peru, and Bolivia are the largest producers of cocaine in the world. The big cocaine markets are in North America (primarily the United States) and in Europe (primarily in Spain and the United Kingdom). The drug is also widely used in western South America. The red lines show main trafficking routes. The green color shows that cocaine was seized in almost every country in the world between 1999 and 2003. [Adapted from *2005 World Drug Report*, Volume 1: *Analysis* (Vienna: United Nations Office on Drugs and Crime, 2005), pp. 65, 75, and 78; available at http://www. unodc.org/pdf/WDR_2005/volume_1_web.pdf.]

Middle and South America. However, public figures, from the local police on up to high officials, are paid to turn a blind eye to the industry.

In Colombia, the illegal drug trade has financed all sides of an ongoing civil war that has threatened the country's democratic traditions and displaced more than 1.5 million people over the past several decades. Meanwhile, competing drug-smuggling rings and paramilitaries bribe, intimidate, kidnap, and murder Colombian citizens and government officials who seek to reduce or eradicate the drug trade. 📹

The war on drugs. The effort to stop the production of illegal drugs in Middle and South America has led to the largest U.S. military presence in the region's history. The U.S. "war on drugs" focuses on supplying intelligence, eradication chemicals, military equipment, and training to military forces in the region. U.S. military aid to Middle and South America is now about equal to U.S. aid for education and other social programs in the region. Meanwhile, drug production in Middle and South America is now so great that it exceeds demand in the U.S.

> The effort to stop the production of illegal drugs in Middle and South America has led to the largest U.S. military presence in the region's history.

market, where street prices for many drugs have fallen in recent years. Much of the production is now also traded within Middle and South America, where drug use is increasing.

The U.S. war on drugs is controversial in the region because it focuses on halting production and destroying enough crops to stop the flow of drugs into the United States. A more direct approach would be to reduce demand by acting on U.S.-sponsored research showing that social programs providing emotional and physical support for youth are most likely to reduce drug usage.

Foreign Involvement in the Region's Politics

Interventions in the region's politics by outside powers have frequently compromised democracy and human rights. Although the former Soviet Union, Britain, France, and other European countries have wielded much influence, by far the most active foreign power has been the United States.

In 1823, the United States proclaimed the Monroe Doctrine to warn Europeans that the United States would allow no further colonization in the Americas. Subsequent U.S. administrations interpreted this policy more broadly to mean that the United States itself had the right to intervene in the affairs of the countries of Middle and South America, and it has done so many times. The official goal for such interventions was usually to make countries safe for democracy, but in most cases the driving motive was to protect U.S. political and economic interests.

During the past 150 years, U.S.-backed, unelected political leaders, many of them military dictators, have been installed at some point in many countries in the region. In recent decades, the United States funded armed interventions in Cuba (1961), the Dominican Republic (1965), Nicaragua (1980s), and Panama (1989). Perhaps the most infamous intervention occurred in Chile in 1973. With U.S. aid, the elected socialist-oriented government of Salvador Allende was overthrown and a military dictator, General Augusto Pinochet, installed in his place. Over the next 17 years, the Pinochet regime imprisoned and killed thousands of Chileans who protested the loss of democracy. 📹

Cuba and the cold war. Since 1959, the Caribbean island of Cuba has been governed by a radically socialist government led first by Fidel Castro and then, starting in 2006, by his brother Raúl. The revolution that brought Fidel Castro to power transformed a plantation and tourist economy once known for its extreme income disparities into one of the most egalitarian in the region. However, because Castro adopted socialism and allied Cuba with the Soviet Union, relations with the United States became extremely hostile. The United States has funded many efforts to destabilize Cuba's government, and it actively discourages other countries from trading with Cuba.

With help from the Soviet Union, Castro managed to dramatically improve the country's life expectancy, literacy, and infant mortality rates. Unfortunately, he also imprisoned or executed thousands of Cubans who disagreed with his policies. Much of the Cuban upper class fled to southern Florida, which remains a popular destination for exiled Cubans to this day. Following the demise of the Soviet Union, Castro opened the country to foreign investment. Many countries have responded, and Cuba is now a major European tourist destination. However, political repression in Cuba persists, and relations with the United States remain cold. 📹

The Post–Cold War era. Since the end of the cold war between the United States and the Soviet Union in 1991 (see Chapter 1, page 46), the United States has curtailed direct military interventions in Middle and South America. Instead, it funds local opposition groups that support U.S. interests. These groups sometimes attempt to seize power with implicit U.S. support, as was the case in Venezuela in 2002. Although resentment of the United States has increased dramatically throughout the region in recent years, the country still wields considerable influence.

The Political Impacts of Information Technology

The Internet already plays a role in the politics of Middle and South America by helping activists, such as the Zapatistas of Chiapas or the MST of Brazil, get their message out to the rest of the world. However, few poor or even middle class people in the region have Internet access. Many countries are trying to increase this number, and they have had some success

Figure 3.25 Internet use in Middle and South America. The numbers on the map indicate the number of Internet users in each country; percents indicate the percentage of the country's population that uses the Internet. By March 2006, 14.4 percent of the people (nearly 80 million) in Middle and South America were using the Internet, a 342 percent increase since 2000. This region has the highest percentage of Internet users in the developing world. Chile has the highest percentage of users (35.7) in South America, Barbados (56.2 percent) the highest in the Caribbean, and Costa Rica (22.7) the highest in Middle America. [Data from "Internet usage statistics—the big picture" and "Internet usage statistics for the Americas," from Internet World Stats at http://www.internetworldstats.com/stats.htm and http://www.internetworldstats.com/stats2.htm.]

(Figure 3.25). Internet users made up just 3.2 percent of the region's population in 2000 but rose to 14.4 percent by 2006. The highest rate of use was in the Caribbean island of Barbados, where 56.2 percent of the population used the Internet in that year.

Overall, Middle and South America is more advanced in technological achievement than other developing regions of the world. A 2000 study by *Wired* magazine considered Mexico, Argentina, Costa Rica, Chile, and Brazil to have excellent potential for leadership in global technological innovation. In 2006, Brazil ranked tenth in the world for total number of Internet users. It has two world-class technology hubs in the environs of São Paulo, and is attempting to cover the whole country with broadband fiber-optic cable networks for telecommunications and Internet service. Mexico (which ranks fifteenth in number of Internet users worldwide) has recently launched a program to give its citizens access to training and higher education via the Internet. This is part of an effort to stem the flow of migration to the United States.

SOCIOCULTURAL ISSUES

Under colonialism, a series of social structures evolved that guided daily life—standard ways of organizing the family, community, and economy. They included rules for gender roles, race relations, and religious observance. These social structures are still widely accepted in the region, but they are changing in response to urbanization, economic development, and globalization (in this context, the diffusion of ideas from outside the region). The results are varied. In the best cases, change is leading to a new sense of initiative on the part of women, men, and the poor. In the worst cases, the result is the breakdown of family life.

Cultural Diversity

The region of Middle and South America is one of the most culturally complex on earth. Many distinct indigenous groups were present when the Europeans arrived and many cultures were introduced during and after the colonial period. In the Caribbean, the Guianas (Guyana, Suriname, and French Guiana), and Brazil, the arrival of a relatively small number of European newcomers resulted in the almost total annihilation of indigenous cultures. These areas then became populated by various people from outside the region.

From 1500 to the early 1800s, some 10 million Africans were brought to plantations on the islands and in the coastal zones of Middle and South America. After the emancipation of African slaves in the British-controlled Caribbean islands in the 1830s, more than half a million Asians were brought there from India, Pakistan, and China as indentured agricultural workers. Their cultural impact remains most visible in Trinidad and Tobago, Jamaica, and the Guianas (Figure 3.26). In some

Figure 3.26 Javanese Muslim women in Suriname. These women, descendants of indentured servants brought by the Dutch from their colony in Indonesia in the 1850s, are celebrating the end of Ramadan, a period of religious observance. [Robert Caputo/Aurora.]

parts of Mexico, Central America, the Amazon Basin, and the Andean Highlands, indigenous people have remained numerous. To the unpracticed eye, they may appear little affected by colonization.

Mestizos are now the majority in Mexico, Central America, and much of South America. In some areas, such as Argentina, Chile, and southern Brazil, people of European descent are also numerous. The Japanese, though a tiny minority everywhere in the region, increasingly influence agriculture and industry, especially in Brazil, the Caribbean, and Peru.

In some ways, diversity is increasing as the media and trade introduce new influences. However, the processes of **acculturation** (cultural borrowing) and **assimilation** (the loss of old ways and adoption of new ones) are also accelerating. This is especially true in the biggest cities, where people of widely different backgrounds live in close proximity.

Race and the Social Significance of Skin Color

People from Middle and South America, especially those from Brazil, often proudly claim that race and color are of less consequence in this region than in North America. They are right in certain ways. Skin color is less associated with status than in the colonial past. A person of any skin color, by acquiring an education, a good job, a substantial income, the right accent, and a high-status mate, may become recognized as upper class.

Nevertheless, the ability to erase the significance of skin color through one's actions is not quite the same as race having no significance at all. Overall, those who are poor, less well-educated, and of lower social standing tend to have darker skin than those who are educated and wealthy. And while there are poor light-skinned people throughout the region (often the descendants of migrants from Central Europe over the last century), most light-skinned people are middle and upper class. Indeed, race and skin color have not disappeared as social factors in the region. In some countries—Cuba, for example, where overt racist comments are socially unacceptable—it is common for a speaker to use a gesture (tapping his forearm with two fingers) to indicate that the person referred to in the conversation is African-Caribbean.

The Family and Gender Roles

The **extended family**—which includes cousins, aunts, uncles, grandparents, and more distant relatives—is the basic social institution in the region. It is generally accepted that the individual should sacrifice many of his or her personal interests to those of the extended family and community, and that individual well-being is best secured by doing so.

The arrangement of domestic spaces and patterns of socializing illustrate these strong family ties. Families of adult siblings, their mates and children, and their elderly parents frequently live together in domestic compounds of several houses surrounded by walls. Social groups in public are most likely to be family members of several generations rather than unrelated groups of single young adults or married couples, as would be the case in Europe or the United States. A woman's best friends are likely to be her female relatives. A man's social or business circles will include male family members or long-standing family friends.

Gender roles in the region have been strongly influenced by the Catholic Church. The Virgin Mary, the mother of Jesus, is held up as the model for women to follow through a set of values known as *marianismo. Marianismo* puts emphasis on chastity, motherhood, and service to the family. The ideal woman is the day-to-day manager of the house and of the family's well-being. She trains her sons to enter the wider world and her daughters to serve within the home. Over the course of her life, a woman's power increases as her skills and sacrifices for the good of all are recognized and enshrined in family lore.

Her husband, the official head of the family, is expected to work and to give most of his income to his family. Men have much more autonomy and freedom to shape their lives than women because they are expected to move about the community and establish relationships, both economic and personal. A man's social network is deemed just as essential to the family's prosperity and status in the community as his work.

In addition, there is an overt double sexual standard for males and females. While expecting strict fidelity in mind and body from his wife, a man is freer to associate with the opposite sex. Males measure themselves by the model of *machismo,*

which considers manliness to consist of honor, respectability, fatherhood, leadership of the household, attractiveness to women, and the ability to be a charming storyteller. Traditionally the ability to acquire money was secondary to other symbols of maleness. Increasingly, however, a new market-oriented culture prizes visible affluence as a desirable male attribute.

Many factors are transforming these family and gender roles. With infant mortality declining steeply, couples are now having only two or three children instead of five or more. Moreover, because most people still marry young, parents are generally free of child-raising responsibilities by the time they are 40. Left with 30 or more years of active life to fill in other ways, middle-aged women are increasingly working in urban factory or office jobs that put to use the organizational and problem-solving skills they perfected while managing a family. Employment outside the home is a way to gain a measure of independence as a woman and also contribute to the needs of the extended family. Some women are even moving into high-level jobs, such as the presidencies of Chile and Argentina, that have traditionally gone to men. ▶ Nevertheless, for many families, relationships are strained when women work outside the home, especially when a job requires moving to a distant city.

> Left with 30 or more years of active life to fill, middle-aged women are increasingly working in factory or office jobs, where they put to use the organizational and problem-solving skills they perfected while supervising a family.

Religion in Contemporary Life

While the Roman Catholic Church remains highly influential, it has had to contend both with popular efforts to reform it and with increasing competition from other religious movements. ▶ From the beginning of the colonial era, the church was the major partner of the Spanish and Portuguese colonial governments. It received extensive lands and resources from those governments, and it sent thousands of missionary priests to convert indigenous people (Figure 3.27).

For centuries, the Catholic Church encouraged working people to accept their low status, be obedient to authority, and postpone their rewards until heaven. Furthermore, the church ignored those teachings of Christ that admonish the privileged to share their wealth and attend to the needs of the poor. Nonetheless, poor people throughout Spanish and Portuguese America embraced the faith and still make up the majority of the church members. Many are indigenous people who put their own spin on Catholicism, creating multiple folk versions with music, worship services, and interpretations of Scripture that vary greatly from European ones.

The Catholic Church began to see its power erode in the nineteenth century in places such as Mexico. **Populist movements**—popularly-based efforts seeking relief for the poor—seized and redistributed church lands. They also

Figure 3.27 For centuries the Catholic church was the largest landowner in Middle and South America, and it is still a major landowner throughout the region today. This cathedral is located in San Cristóbal de las Casas, in Chiapas, Mexico. (Odysseus79/ WikimediaCommons)

canceled the high fees the clergy had been charging for simple rites of passage such as baptisms, weddings, and funerals. Over the years, the church became less obviously connected to the elite and more attentive to the needs of poor and non-European people. By the mid-twentieth century, the church was abandoning many of its racist policies and ordaining indigenous and African-American clergy. Women were also given a greater role in religious ceremonies.

In the 1970s, a radical Catholic movement known as **liberation theology** was begun by a small group of priests and activists. They sought to reform the church into an institution that could combat the extreme inequalities in wealth and power common in the region. The movement portrayed Jesus Christ as a social revolutionary who symbolically spoke out for the redistribution of wealth when he divided the loaves and fish among the multitude. The movement viewed the perpetuation of gross economic inequality and political repression as sinful, and it promoted social reform as liberation from evil.

At its height in the 1970s and early 1980s, liberation theology was the most articulate movement for region-wide social change. It had more than 3 million adherents in Brazil alone. Today, its influence has declined and its positions have been severely criticized by the Vatican. In countries such as Guatemala and El Salvador, liberation theology became the target of state-sponsored attacks. Its participants were vilified as communist collaborators with the Soviet Union aiming to overthrow governments and establish a classless, stateless society based on the common ownership of all means of production. Liberation theology has also had to compete with newly emerging evangelical Protestant movements.

Evangelical Protestantism has diffused into Middle and South America from North America and is now the region's fastest-growing religious movement. About 10 percent of the population, or about 50 million people, are adherents. Like liberation theology, it appeals to the rural and urban poor. It does not, however, share liberation theology's emphasis on combating the region's extreme inequalities in wealth and power. Some evangelical Protestants teach a "gospel of success," stressing that those who are true believers and give themselves to Christ and to a new life of hard work and clean living will experience prosperity of the body (wealth) as well as of the soul.

The movement is charismatic, meaning that it focuses on personal salvation and empowerment of the individual through miraculous healing and transformation. The movement has no central authority but consists of a host of small, independent congregations whose leadership often consists of both men and women.

Evangelical Protestantism is growing rapidly in Brazil and Chile, and is also strong in the Caribbean, Mexico, and Middle America among both the poor and middle classes. The revival tents of North American evangelical pastors are not an uncommon sight on the fringes of urban areas throughout the region.

Other religions found across the region include Judaism, Islam, Hinduism, and indigenous beliefs. A range of African-based belief systems (Condomble, Umbanda, Santeria, Obeah, and Voodoo) combined with Christian beliefs are found in Brazil, the Caribbean, and wherever the descendants of Africans have settled. These African-based religions have attracted adherents of European or indigenous backgrounds as well, especially in urban areas.

on Middle and South America

...nialism in Middle and South America launched the modern global economic system. It was in this region that large-scale extractive industries were inaugurated. Raw materials were shipped at low prices to distant locales, where they were turned into high-priced products, the profits of which went to Europe.

In most of the region, this pattern has persisted for 500 years. After the massive outflows of raw materials during the colonial era, this region tried a number of economic development strategies, with mixed results. Import substitution industrialization projects worked in a few cases but generally failed to lift the region out of poverty. The huge debts that many governments accumulated pursuing these projects later led to the imposition of structural adjustment programs. SAPs also failed to produce economic growth as they relied heavily on raw materials exports just when prices for these exports were falling on global markets. Now deforestation, much of it driven by the need for export earnings, is contributing to both global warming and the loss of the region's great biodiversity.

Massive rural-to-urban migrations resulted from the shift from small-scale agriculture for local consumption to large-scale, mechanized agriculture for export. Most cities were unprepared for the influx of migrants, and a water crisis developed as existing supply and delivery systems were overwhelmed. While the informal economy has been a lifesaver for many poor urbanites, its ability to lift people out of poverty may be limited. Traditional gender roles have also changed with urbanization, giving new opportunities to some, but also straining family ties.

However, there are many positive signs in the region. Environmental regulations and alternatives to deforestation hold the potential to reduce pressure on the region's forests. Increasingly, regional trade organizations are emerging that may keep more wealth within the region. Finally, almost all countries have taken major steps toward democracy and away from corrupt dictatorships.

As you read about other regions, it may be useful to reflect on the issues related to globalization discussed in this chapter. For example, notice how Europe's situation today—its wealth, its position as a world leader, and its emerging commitment to help its former colonies—is related in part to its colonizing experiences, which began in the Americas. You will see that Africa and South and Southeast Asia experienced some conditions under colonial rule that were similar to those in Middle and South America. Recently, they too have felt the sting of SAPs.

Chapter Key Terms

acculturation 129
assimilation 129
Aztecs 109
biodiversity 101
brain drain 115
contested space 123
Creole 112
coup d'état 125
dictator 125
early extractive phase 117
ecotourism 107
El Niño 104
evangelical Protestantism 131
Export Processing Zones (EPZs) 121
extended family 130
external debts 119

favelas 115
foreign direct investment (FDI) 119
hacienda 118
import substitution industrialization (ISI) 119
Incas 109
income disparity 117
indigenous 97
land reform 119
liberation theology 131
machismo 130
maquiladoras 121
marianismo 130
mercantilism 117
Mercosur 123
mestizos 112

Middle America 100
nationalize 122
North American Free Trade Agreement (NAFTA) 123
plantation 118
populist movements 130
primate city 113
privatization 119
recession 119
shifting cultivation 109
silt 102
South America 100
subduction zone 101
temperature-altitude zones 103
trade winds 104

Critical Thinking Questions

1. Imagine that the European colonists had come to Middle and South America in a different frame of mind—say, simply looking for a new place to settle and live quietly. How do you think the human and physical geography of the region would be different today?

2. How are the processes of globalization linked to global warming in this region?

3. How did a water crisis develop in a region of such abundant water resources?

4. Discuss the ways in which the problems created by colonization are similar to those created by SAPs.

5. Describe the main patterns of migration in this region and discuss the effects of migration on gender roles and families within both the societies sending and the societies receiving immigrants.

6. How might migration from rural to urban areas help slow overall population growth in this region?

7. How have the drug trade, corruption, and U.S. intervention influenced democracy in this region?

8. Explain how the Amazon Basin and its resources constitute an example of contested space.

9. Argue for or against the proposition that free trade blocs, such as NAFTA and Mercosur, help farmers earn more money.

10. How would you respond to someone who suggested that European colonization brought development and modernization to Middle and South America?

The map labels (Figure 4.1) include: RUSSIA, Ufa, Kazan, Nizhniy Novgorod, Kuybyshev, Moscow, Rybinsk Reservoir, Gorkiy Reservoir, Kuybyshev Reservoir, Volga, Kama, Sukhona, N. Dvina, Pechora, Ob, KAZAKHSTAN, Volgograd Reservoir, Volgograd, Don, Depression, Tsimlyansk Reservoir, Kharkiv, Kremenchug Reservoir, Dnieper, Dnipropetrovsk, Donetsk, Rostov-na-Donu, Kakhovka Reservoir, Sea of Azov, Caucasus Mountains, Caspian Sea, Caspian Depression, Odessa, Constanta, GEORGIA, Tbilisi, AZERB., ARMENIA, Yerevan, IRAN, Black Sea, Bosporus Straits, Istanbul, Ankara, TURKEY, Lake Van, Lake Tuz, Izmir, Tigris, Euphrates, Aleppo, SYRIA, Damascus, Beirut, LEBANON, Nicosia, CYPRUS, IRAQ, ISRAEL, Amman, SAUDI, vargas

Land Elevations

meters	feet
4877	16,000
3353	11,000
2134	7000
914	3000
305	1000
152	500
0	0

mi 0 — 100 — 200 — 300
km 0 — 100 — 200 — 300 — 400 — 500
1:14,800,000
Albers Equal Area Conic Projection

CHAPTER 4

Europe

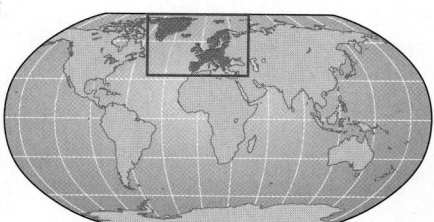

Global Patterns, Local Lives It is a chilly, overcast late October day in Riga, Latvia. Janis Neulans sits on a creaky bed in a smelly, cheap hotel. He is about to take the midnight flight from Riga to Dublin, Ireland, where he hopes to find work. His last job was sandblasting the hulls of freighters in the Riga shipyard. Well-built, with powder blue eyes and a thatch of blond hair (Figure 4.2), Neulans comes from a distant village of eight houses near the Russian border. Although he is 39, he is not yet married. His schooling is minimal, and his entire net worth is $420, including cash from the recent sale of his 1984 Volvo.

Neulans and Vladimir, whom he meets on the plane, arrive in Dublin in the middle of the night. Vladimir's friend, also a Latvian, meets them at the airport. He takes them to two bare mattresses in spare quarters in a vast new housing development in a west Dublin suburb. By mid-morning, Neulans has paid the rent on the room he shares with Vladimir and begun his job search with the help of other Latvians who arrived some months ago. They help him write job ads and a resume in English.

For the first two weeks, Neulans subsists on bread, cheddar cheese, and diet cola while visiting dozens of newly built warehouses and factories seeking work. On the eleventh day, down to his last $8, he agrees to work on a farm 40 miles (64.4 kilometers) from Dublin, milking cows and doing other chores for 14 hours a day. But soon Neulans receives a call from a door-frame factory, where he applied before taking the farm job. The farmer pays him just half of what he is owed.

Just two weeks after arriving in Ireland, Neulans has his dream job. He works for 10 hours a day at a drill press, making prefabricated door frames. He earns the Irish minimum wage of $9.20 an hour (three times the wage in Latvia), plus overtime pay. In addition, he serves as the factory's night watchman, for which he gets the use of an on-site three-bedroom mobile home. There, he outlines his plans: he will work here for several years while saving to return to his Latvian village, where he will buy some calves to raise

Figure 4.1 Regional map of Europe.

Figure 4.2 Janis Neulans in Ireland. Neulans, seen here on the subway in Dublin, polished his job-search skills in that city. Within two weeks after arriving from Latvia, he had landed a job at a door-frame factory in southeastern Ireland. [Kevin Sullivan.]

and enjoy the company of his family. "It's where my heart is," he says.

Source: Kevin Sullivan, "East-to-west migration remaking Europe," Washington Post Foreign Service, November 28, 2005. A video of Kevin Sullivan accompanying Janis Neulans on his journey is available at http://www.washingtonpost.com/video. ■

Janis Neulans was able to move easily between Latvia and Ireland because both countries are members of the **European Union (EU).** The European Union is a supranational organization that unites most of the countries of West, South, North, and Central Europe. In principle, throughout the European Union, people, goods, and money can move freely. In fact, only Ireland, Britain, and Sweden actually have truly open borders. Ireland, with a newly booming economy based on high-tech and light manufacturing, is a favorite destination for these migrants. 📹

Altogether, 27 countries were members of the European Union as of January 2007, and more are scheduled to join in the next several years. Most new members are formerly communist countries in Central Europe, with lower standards of living and higher unemployment rates than Western Europe. Hundreds of thousands of workers from these countries have taken advantage of the EU principle that citizens of all member states have the right to move to any other EU state.

Many current EU residents fear that expansion of the European Union into Central Europe and beyond will bring new political tensions and a throng of immigrants requiring social and educational services. Others fear that if the European Union succeeds, **cultural homogenization** will result as distinctive local ways of life disappear. Meanwhile, employers across Europe are saying that without the immigrant workers, economic growth and competitiveness in the global market is impossible. 📹

Questions to consider as you study Europe (also see Views of Europe, pages 138–139)

1. How did Europe's colonization of much of the world over the past 400 years accelerate globalization? As European powers like Spain, the United Kingdom, and the Netherlands conquered vast overseas territories, they created trade relationships that laid the foundation for the modern global economy. What strategies guided this trade? How was Europe transformed as a result?

2. How has urbanization influenced the development of democracy in Europe? As Europe's industrial cities grew, poverty and squalid living conditions for the vast majority of people created pressure for change in Europe's political order. How did the needs of urban people shape European democracy? What responsibilities did governments assume as a result?

3. In what ways is Europe responding to global warming? Europe has emerged as the world's leader on combatting global warming. What commitments have been made to reduce Europe's production of greenhouse gasses? What countries are doing the most within the EU? What technologies are being pursued?

4. How is Europe's aging population linked to changing gender roles? Europeans are choosing to have fewer children, and as a result the population as a whole is aging. Small families are in part a result of women pursuing careers. What aspects of European society create difficulties for working mothers?

5. How is food production in Europe changing in response to EU expansion? Throughout the EU, smaller family-run farms are giving way to larger farms run by corporations. This move is strongest in the recently admitted EU states of Central Europe. How has this shift impacted rural areas in Central Europe?

6. What factors complicate efforts to reduce water pollution in Europe's seas? Europe's many seas are increasingly threatened by water pollution from agriculture, industry, and cities. Why is pollution in nearly landlocked seas hard to deal with? How does the diverse array of countries, both European and non-European, make pollution in the Mediterranean particularly difficult to address?

The Geographic Setting

Terms to Be Aware Of

In this book, Europe is divided into four subregions (Figure 4.3)—*North, West, South,* and *Central Europe.*

For convenience, we occasionally use the term *western Europe* to refer to all the countries that were not part of the experiment with communism in the Soviet sphere and Yugoslavia. That is, *western Europe* comprises the combined subregions of North Europe (except Estonia, Latvia, and Lithuania), West Europe (except the former East Germany), and South Europe. When we refer to the countries that were part of the Soviet sphere up to 1989, we use the pre-1989 label *eastern Europe.* When it is necessary to refer to the group of countries that includes Albania, Bosnia and Herzegovina, Bulgaria, Croatia, Macedonia, Montenegro, Romania, Serbia, and Slovenia (once known collectively as the *Balkans*) we use the term *southeastern Europe.*

PHYSICAL PATTERNS

Europe is a region of peninsulas upon peninsulas (see Figure 4.1). The entire European region is one giant peninsula extending off the Eurasian continent. The whole of its very long coastline itself has many peninsular appendages, large and

Figure 4.3 Political map of Europe showing the four subregions.

Views of Europe

For a tour of these places, visit www.whfreeman.com/pulsipher
[Background image: Lori K. Perkins NASA/GSFC Code 610.3/587 Scientific Visualization Studio]

1.Globalization. European colonization of much of the world accelerated globalization, often in ways that benefited Europe more than its colonies. Shown here is a mural commissioned by the former British East India company in London in 1778. The painting's title is "The East offering its riches to Britannia." See pages 150–151 for more. Lat 51°30'09"N, Lon 0°07'38"W. [Roma Spiridone/British Library]

2. Gender and democracy. Gender roles are changing significantly in Europe, where women are moving into jobs that once went to men. For example, in Germany's capitol of Berlin, the government is led by Angela Merkel, shown below. See pages 166–168 for more. Lat 52°31'12"N, Lon 13°22'E. [NATO Photo]

3
Baltic

1
London

6
Paris

North Atlantic Drift

7. Global warming, North Atlantic Drift. A cooler climate in Europe could result if the North Atlantic Drift is weakened by global warming. This current, which is the eastern end of North America's Gulf Stream, brings huge amounts of warm water (shown in this satellite image as yellow, green, and blue-green) to Europe. See page 140 for more. [MODIS Oceans Group, NASA Goddard Space Flight Center]

3. Water. Pollution is a growing problem in Europe's many seas. Shown here is an algal bloom in the Baltic Sea. Such ecologically devastating events occur when water drains into the sea off of agricultural and urban land where fertilizers and other chemicals are in use. See pages 147–148 for more. Lat 58°N, Lon 10°E. [Jeff Schmaltz, MODIS Rapid Response Team, NASA/GSFC]

4. Immigration. Immigration is an increasingly controversial subject in Europe. Some right wing political parties, such as Sweden's nationalist party (shown here at a demonstration in Stockholm in 2007) advocate strict limits on immigration. See pages 164–166 for more. Lat 59°19'30"N, Lon 18°04'20"E. [Peter Isotalo]

4

Stockholm

2

Berlin

5. Population and aging. An aging population may pose challenges to Europe's continued growth and prosperity. This woman in Serbia is having her face checked for signs of skin cancer. See pages 155–156 for more. Lat 43°16'19"N, Lon 20°47'30"E. [DoD photo by Staff Sgt. Milton H. Robinson, U.S. Army]

5 **Serbia**

6. Urbanization. Cities have been central to Europe's industrialization and the expansion of democracy to the working class. Cities like Paris (shown here) are now the focus of Europe's modern economy. See pages 149–150, 152, and 155 for more. Lat 48°51'30"N, Lon 2°17'40"E. [SvG. Blick vom Tour Montparnasse Ÿber das abendliche Paris. http://commons.wikimedia.org/wiki/Image:Paris_SvG.jpg]

small. Norway and Sweden share one of the larger appendages. The Iberian Peninsula (shared by Portugal and Spain), Italy, and Greece are other large peninsulas. One result of these many fingers jutting into oceans and seas is that much of Europe feels the climate-moderating effect of the large bodies of water that surround it.

Landforms

Although European landforms are fairly complex, the basic pattern is mountains, uplands, and lowlands, all stretching roughly east to west in wide bands. As you can see in Figure 4.1, Europe's largest mountain chain stretches west to east through the middle of the continent, from southern France through Switzerland and Austria. It extends into the Czech Republic and Slovakia, and curves southeast into Romania. The *Alps* are the highest and most central part of this formation. This network of mountains is mainly the result of pressure from the collision of the northward-moving African Plate with the southeasterly moving Eurasian Plate (see Figure 1.21 on page 25). Europe lies on the westernmost extension of the Eurasian Plate.

South of the main Alps formation, mountains extend into the peninsulas of Iberia and Italy to the west, and along the Adriatic Sea through Greece to the east. The northernmost mountainous formation is shared by Scotland, Norway, and Sweden. These northern mountains are old (about the age of the Appalachians in North America) and have been worn down by glaciers.

Extending northward from the central mountain zone is a band of low-lying hills and plateaus curving from Dijon (France) through Frankfurt (Germany) to Krakow (Poland). These uplands form a transitional zone between the high mountains and lowlands of the *North European Plain,* the most extensive landform in Europe. The plain begins along the Atlantic coast in western France and stretches in a wide band around the northern flank of the main European peninsula, reaching across the English Channel and the North Sea to take in southern England, southern Sweden, and most of Finland. The plain continues east through Poland, then broadens to the south and north to include all the land east to the Ural Mountains (in Russia).

The coastal zones of the North European Plain are densely populated all the way east through Poland. Crossed by many rivers and holding considerable mineral deposits, this coastal lowland is an area of large industrial cities and densely occupied rural areas. Over the past thousand years, people have transformed the natural seaside marshes and vast river deltas into farmland, pastures, and urban areas by building dikes and draining the land with wind-powered pumps. This is especially true in the Netherlands.

The rivers of Europe link its interior to the surrounding seas. Several of these rivers are navigable well into the upland zone, and Europeans have built large industrial cities on their banks. The Rhine carries more traffic than any other European river, and the course it has cut through the Alps and uplands to the North Sea also serves as a route for railways and motorways. The area where the Rhine flows into the North Sea is considered the economic core of Europe. It is here that Rotterdam, Europe's largest port, is located. The larger and much longer Danube River flows southeast from Germany, connecting the center of Europe with the Black Sea. As the European Union expands to the east, the economic and environmental roles of the Danube River basin will be getting increased attention.

Vegetation

Europe was once covered by forests and wild grasslands, most of which were cleared at one time or another for farmland, pasture, towns, and cities. Today, forests with very large and old trees exist only in scattered areas, especially on the more rugged mountain slopes and in the northernmost parts of **Scandinavia** (the area occupied by Norway, Sweden and Finland). In parts of central and southeastern Europe, forests have been sustainably managed for generations, and more are regenerating where small farms have been abandoned. Today, forests cover about one-third of Europe, but elsewhere, the dominant vegetation is now crops and pasture grass. Vast areas are covered with industrial sites, railways, roadways, parking lots, canals, cities, suburbs, and parks.

Climate

Europe has three main climate types: temperate midlatitude, Mediterranean, and humid continental (Figure 4.4). The **temperate midlatitude climate** dominates in northwestern Europe, where the influence of the Atlantic Ocean is very strong. A broad warm-water current called the North Atlantic Drift brings large amounts of warm water to Europe. The North Atlantic Drift is really just the easternmost end of the Gulf Stream, which carries water from the Gulf of Mexico along the eastern coast of North America and across the North Atlantic (see Figure 2.5 on page 57).

The air above the North Atlantic Drift is warm and wet. Eastward-blowing winds push it over northwestern Europe and the North European Plain, bringing moderate temperatures and rain deep into the Eurasian continent. These factors create a climate that, though still fairly cool, is much warmer than elsewhere in the world at similar latitudes. To minimize the effects of precipitation runoff, people in these areas have developed elaborate drainage systems for their houses and communities—steep roofs, rain gutters, storm sewers, drain fields, and canals. They also grow crops, such as potatoes, beets, turnips, and cabbages, that thrive in cool, wet conditions.

There is some concern that global warming (see pages 17–20 in Chapter 1) could weaken the North Atlantic Drift, leading to a significantly cooler Europe. However, recent years have actually experienced abnormally warm temperatures in Europe.

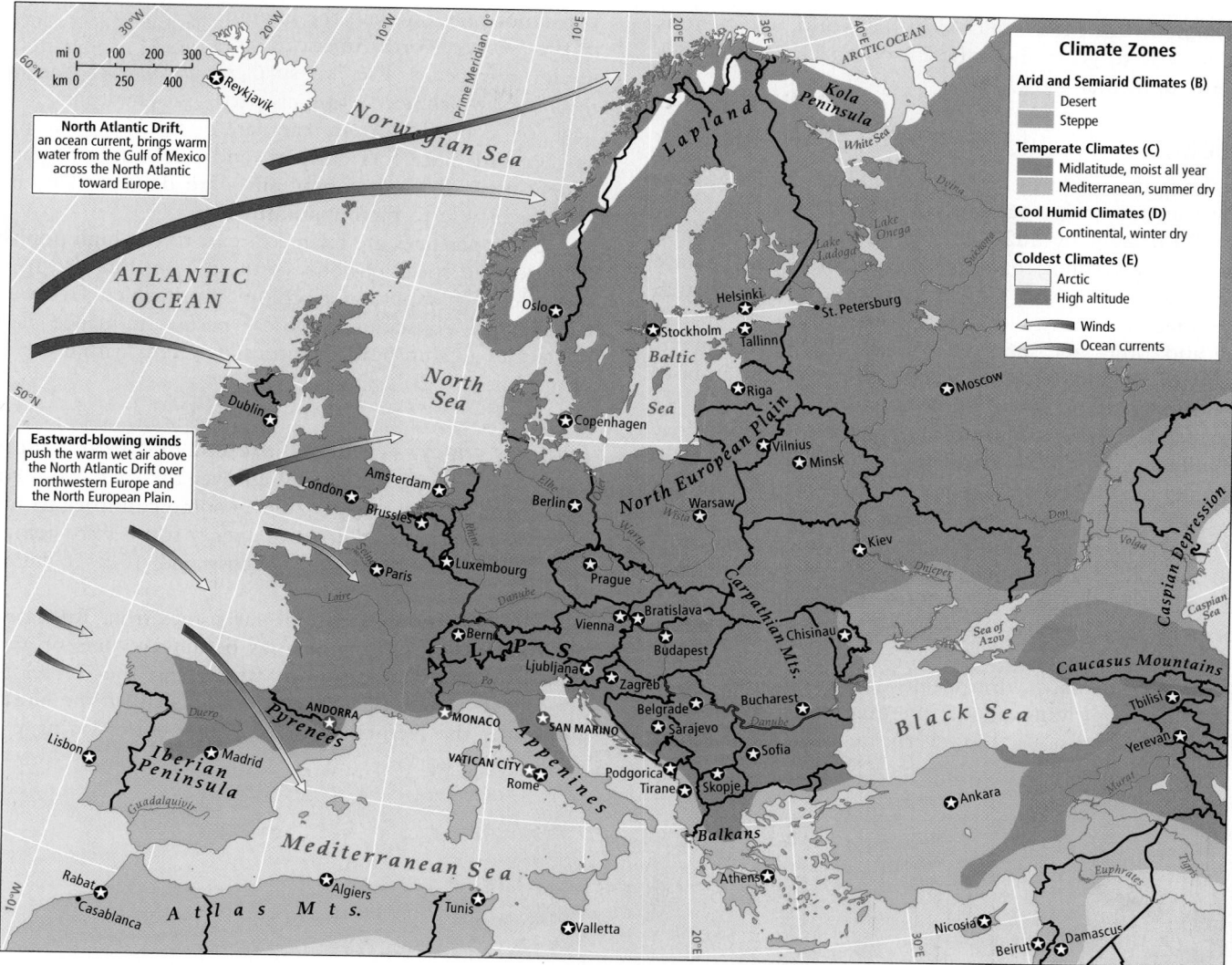

Figure 4.4 Climates of Europe.

Farther to the south, the **Mediterranean climate** of warm, dry summers and mild, rainy winters prevails. In the summer, warm, dry air from North Africa shifts north over the Mediterranean Sea as far north as the Alps, bringing high temperatures and clear skies. Crops grown in this climate, such as olives, citrus fruits, and wheat, must be drought-resistant or irrigated. In the fall, this warm, dry air shifts to the south and is replaced by cooler temperatures and thunderstorms sweeping in off the Atlantic. Overall, the climate here is mild, and houses along the Mediterranean coast are often open and airy to afford comfort in the hot, sunny summers.

In eastern Europe, without the moderating influences of the Atlantic Ocean and the Mediterranean Sea, the climate is more extreme. In this region of **humid continental climate,** summers are fairly hot, and winters are longer and colder the farther north or deeper into the interior of the continent one

goes. Here, houses tend to be well insulated, with small windows and low ceilings. Crops must be adapted to much shorter growing seasons.

ENVIRONMENTAL ISSUES

Having dramatically transformed their environments over the past 10,000 years, Europeans are now increasingly taking action on issues of local and global environmental concern.

Opinion polls suggest that EU citizens see changes in energy production, transport, and most efforts at environmental protection, as an incentive for innovation, not a hindrance to economic performance. Perhaps reflecting this perspective, EU environmental reports over the past decade have shown slow improvement in most environmental categories in nearly all 27 countries. Nevertheless, Europe's air, seas, and rivers remain some of the most polluted in the world, and there is still a long way to go to meet the European Union's various stated environmental goals.

European Leadership on Global Warming

Europe leads the world in response to global warming, with EU governments having agreed to cut greenhouse gas emissions by 20 percent by 2020. Europe has been more willing to act on global warming than any other region largely because it recognizes the economic sense in doing so. Recent research suggests that investments in energy conservation, alternative energy, and other measures would cost EU governments 1 percent of their GDP. By contrast, doing nothing about global warming would *shrink* GDP by 20 percent.

Europe's increasing concern about global warming may also be influenced by public alarm at recent abnormal weather. The summer of 2003 broke all temperature records for Europe. Crops failed, river levels sank, forests burned, and deaths soared. In France alone, 3000 people died. However, in 2002 and 2006, rainfall and snowfall in central Europe reached record levels. In the spring of 2006, the rivers of central Europe—the Elbe, the Danube, and the Morava—flooded for weeks (Figure 4.5).

By global standards, Europeans use large amounts of resources and contribute about one-quarter of the world's greenhouse gas emissions. However, one European resident consumes only one-half the energy of the average North American resident. Europeans live in smaller dwellings, which need less energy to heat or cool. European cars are smaller and more fuel efficient, and public transportation is widely used. Because communities are denser, many people walk or bicycle to their appointments.

> Europeans produce about one-quarter of the world's greenhouse gas emissions.

These practices are related in part to the high population densities and social customs of the region, but also to widespread explicit support for ecological principles. **Green** (environmentally conscious) political parties influence national policies in all European countries as well as within the European Union.

Europe's energy resources: present and future. Europe's main energy sources have shifted over the years from coal to petroleum and natural gas, and in some countries to nuclear power. Increasingly, alternative energy sources are being pursued in response to rising energy costs and efforts to cut greenhouse gas emissions.

Most of Europe's natural gas comes from Russia, which makes many Europeans nervous because of present and historical political tensions with Russia and the former Soviet Union. 📹 Large oil and gas deposits exist in the North Sea and under the Netherlands (see Figure 4.19b on page 160),

Figure 4.5 Flooding on the Elbe River, 2006. Heavy snowmelt in the mountains of the Czech Republic and a lot of rain across Central Europe in April 2006 brought floodwaters to the Elbe River in eastern Germany. The village of Gohlis, near Dresden, shown here, remained flooded for several days. Most experts believe global warming lies behind the periods of flood and drought that have taken hundreds of lives and caused billions of dollars in damage in Europe in the past few years. [AP Photo/Fabian Bimmer.]

but much of the European Union's oil comes from the Middle East, another political hot spot.

The use of nuclear power to generate electricity has been more common in Europe than in North America. In France, it accounts for 78 percent of the electricity generated (compared with only 20 percent in the United States). However, support for nuclear power has declined, partly in response to the 1986 explosion of a nuclear power plant in Chernobyl, Ukraine, which sent a cloud of deadly radiation across Europe.

The European Union has set a goal of obtaining 12 percent of its energy from renewable sources by 2010. Germany is the region's leader, with 14 percent of its energy already coming from renewables. A wide array of alternative energy projects are now attracting massive investment. Throughout Europe, wind power is generally the favored technology, but a diverse array of technologies are being pursued. Solar power plants are being built in Spain, geothermal power generation abounds in Iceland, and energy from biomass and biofuels are increasingly popular throughout the region.

Changes in transportation. Europeans have long favored fast rail networks for both passengers and cargo rather than multilane highways for cars and trucks. Yet despite high and rising gasoline prices over the last two decades, there was a noticeable trend toward less energy-efficient but more flexible motorized road transport. Now, however, fuel costs are increasingly being considered in the design of *multimodal transport* that links high-speed rail to road, air, and water transport (Figure 4.6). A 2005 EU transport report notes that 1 kilogram of gasoline can move 50 tons of cargo a distance of 1 kilometer by truck, but the same amount of gasoline can move 90 tons by rail and 127 tons by waterway.

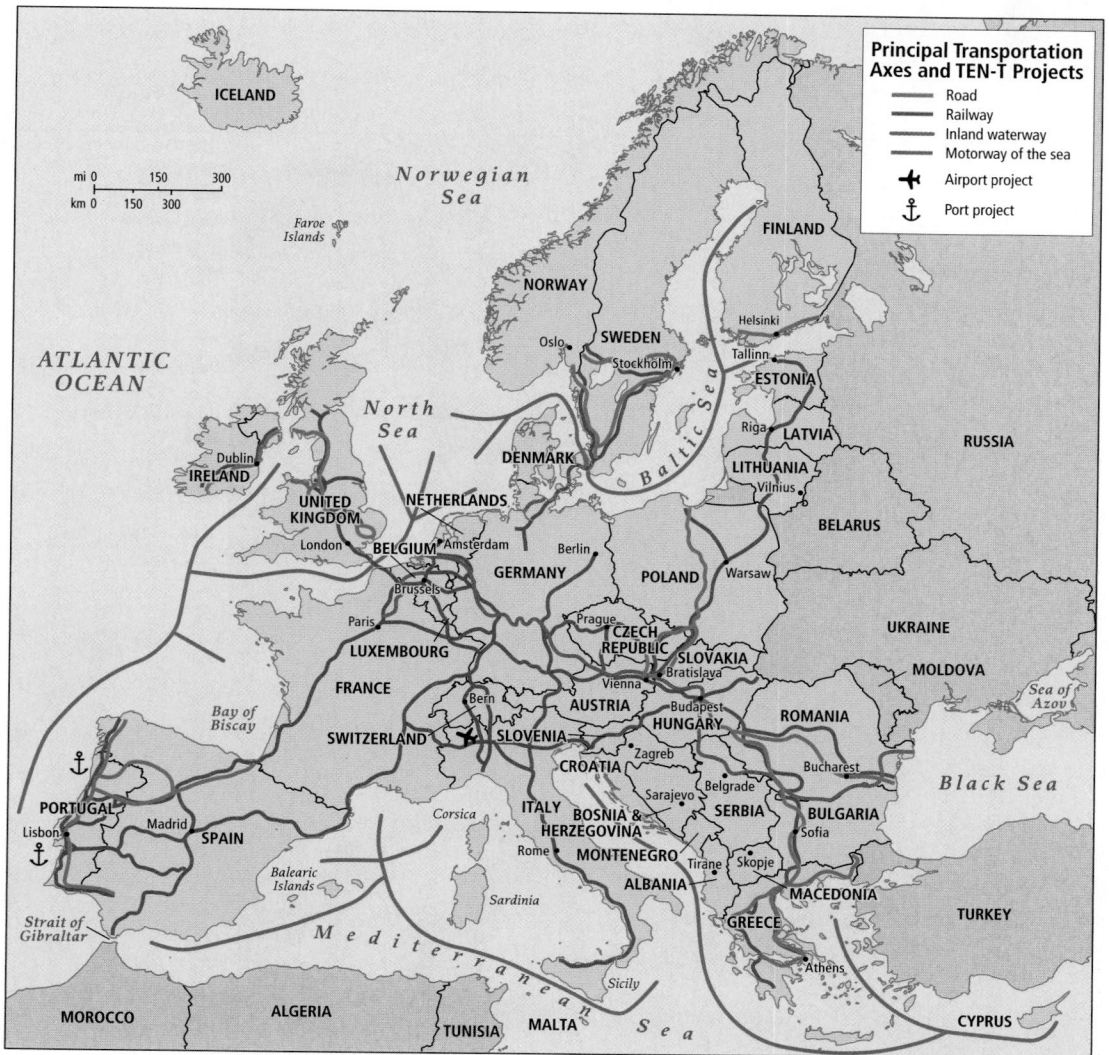

Figure 4.6 The Trans-European transport network (TEN-T): Priority axes and projects. As of 2007, Europeans were reevaluating transport policies in light of rising fuel costs. Cheaper rail and water transport were getting renewed attention over cars and trucks, which, for a time, were favored for flexibility. [Adapted from "Transport, a driving force for regional development," *Inforegio panorama*, No. 18 (December 2005), p. 11, at http://ec.europa.eu/comm/regional_policy/sources/docgener/panora_en.htm.]

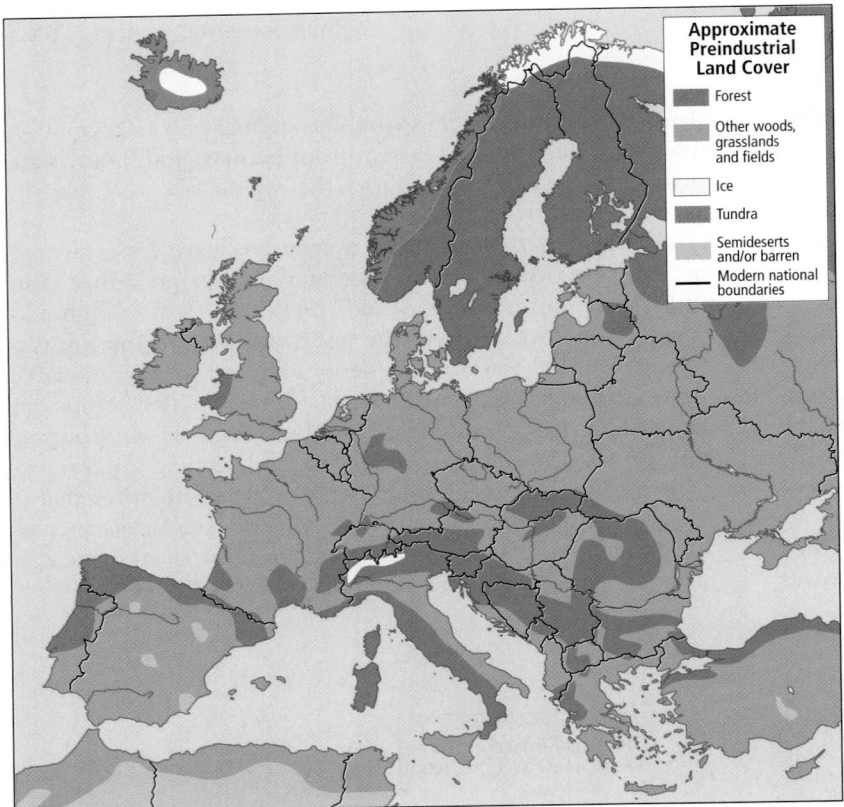

(a) Preindustrial

Figure 4.7 Land cover (preindustrial) and human impact (2002) in Europe. In the preindustrial era **(a)**, humans had converted much forest to grassland and farmland, but there was as yet little impact from industrialization. By 300 years later **(b)**, human impact on the land in Europe was much greater, with virtually no place devoid of human influence. [Adapted from *United Nations Environment Programme, 2002, 2003, 2004, 2005, 2006* (New York: United Nations Development Programme), at http://maps.grida.no/ go/graphic/human_impact_year_1700_approximately and http://maps.grida.no/go/graphic/human_impact_year_2002.]

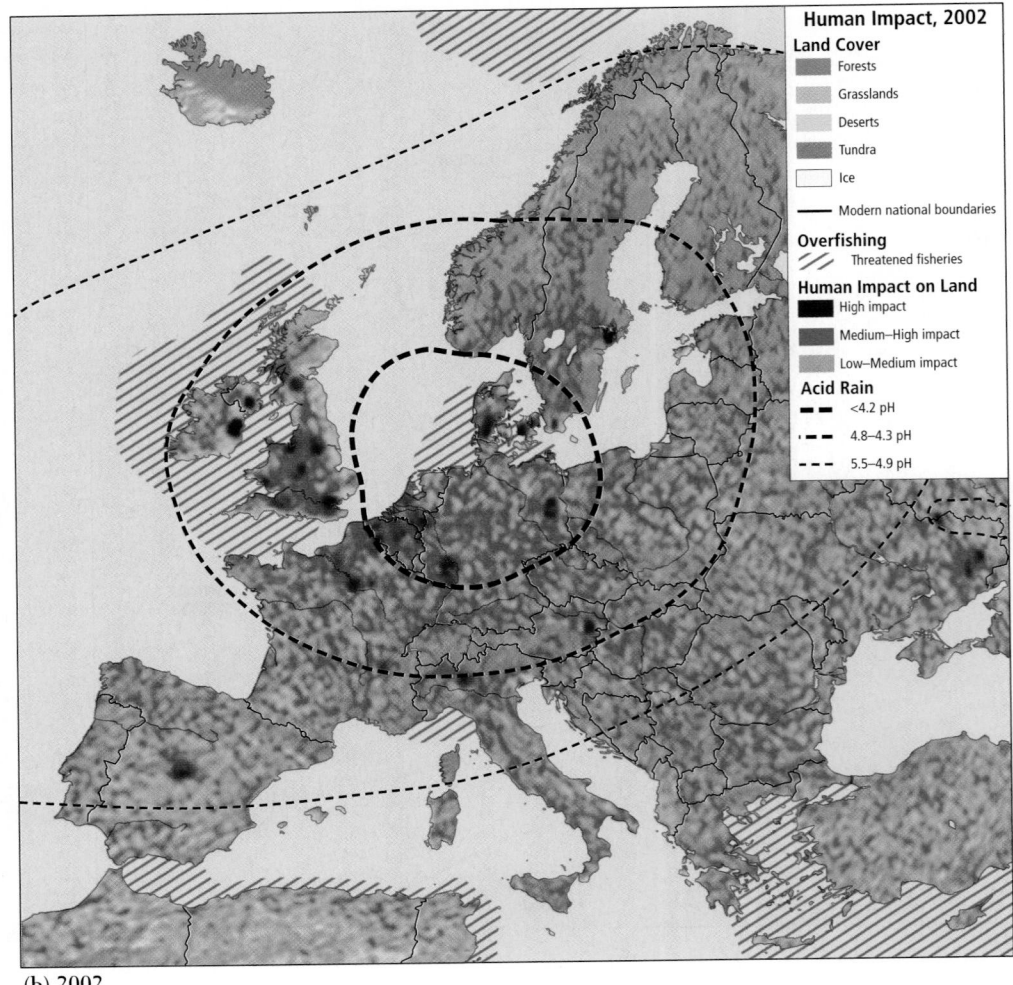

(b) 2002

Europe's Transformed Landscapes

Nearly all of Europe's original forests are gone, and some have been gone for more than a thousand years. Many of the region's seemingly natural landscapes are largely the creation of people, who have changed the landforms, drainage systems, and vegetation cover repeatedly over time (compare Figures 4.7a and 4.7b). The Netherlands is an extreme example, where almost no natural landscapes are left.

The Netherlands: a human-made country. For almost a thousand years, the people of the Netherlands have been reclaiming land that was previously under the sea (Figure 4.8).

As its population grew during and after the **medieval period** (450–1500 C.E.), people created more living space by draining and filling in a large natural coastal wetland. To protect themselves from devastating North Sea surges, they built dikes, dug drainage canals, pumped water with windmills, and constructed artificial dunes along the ocean. Today, a train trip through the Netherlands between Amsterdam and Rotterdam takes one past raised rectangular fields crisply edged with narrow drainage ditches and wide transport canals. The fields are filled with commercial flower beds of tulips and daffodils, all of which supply two-thirds of the world's fresh cut plant, flower, and bulb exports. There are also grazing cattle and vegetable gardens, which feed the primarily urban population.

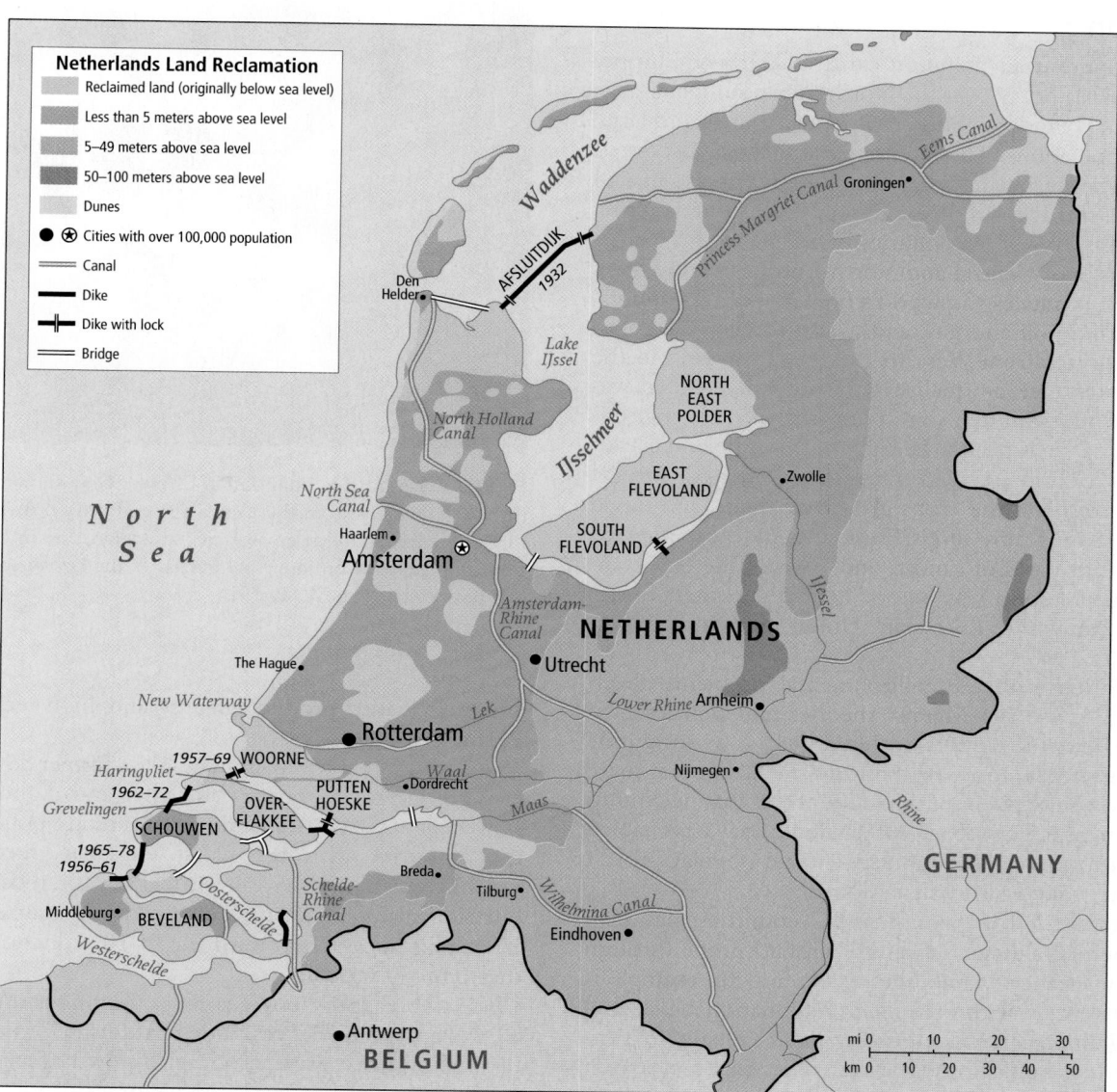

Figure 4.8 Land reclamation areas in the Netherlands. The Netherlands has a landscape that is almost entirely human-made. As its populations grew, people created more living space by filling in a large natural coastal wetland. Filling in wetlands is now prohibited in many parts of the world because of the negative environmental impacts. [Adapted from William H. Berentsen, *Contemporary Europe: A Geographic Analysis* (New York: Wiley, 1997), p. 317.]

Today, the Netherlands has 16.3 million people, who enjoy a high standard of living. It is the most densely settled country in Europe (with the exception of tiny Malta). There is practically no land left for urban expansion without intruding on agricultural space, and not nearly enough space for recreation. As picturesque as they might appear, transport canals and carefully controlled raised fields are not a venue for picnics or soccer games. Even for weekend getaways, people usually leave the Netherlands for a neighboring country that has more natural areas. The choice to maintain agricultural space is largely psychological and environmental, as agriculture accounts for only 2 percent of both GDP and employment.

Air Pollution

At present, significant air pollution exists in much of Europe, but it is particularly heavy over the North European Plain. This is a region of heavy industry, dense transport routes, and large and affluent populations. The intense fossil fuel use associated with such lifestyles results not only in pollution but also in acid rain, which is a threat for most of Europe (see Figure 4.7b).

The highest level of air pollution is found in the former Communist states of Central and North Europe (Figure 4.9). Central Europe produces the world's highest per capita emissions from burning oil and gas, and it also receives air pollution blown eastward from Western Europe. In Poland in the late 1980s, for instance, air quality in the industrial centers was so poor that young children were relocated to rural environments to preserve their health. In Upper Silesia, Poland's leading coal-producing area, acid rain has destroyed forests, contaminated soils and the crops grown on them, and raised water pollution to deadly levels. Residents have experienced birth defects, high rates of cancer, and lowered life expectancies. Industrial pollution was one of Poland's greatest obstacles to entry into the European Union, which it barely achieved.

Central Europe's severe environmental problems developed in part because the Marxist theories and policies promoted by the Soviet Union portrayed nature as existing only to serve human needs. Moreover, during the Soviet era, there was little opportunity for public outrage to be channeled into constructive change. The recent shift to democracy has enabled greater action in places like Hungary, where popular protest has resulted in reductions in air pollution.

New investment from wealthier EU countries may help reduce environmental degradation in Central Europe through the use of new cleaner technologies. One hopeful example is Poland's joint venture with the French multinational glass and high-tech building materials company Saint-Gobain to construct modern glass-making factories. The first factory, in Silesia, is one of the most environmentally responsible plants of its kind in Europe. Its emissions are well below the standards set by Germany, the industry leader. All glass waste is completely recycled. By 2006, Saint-Gobain had 19 factories

Figure 4.9 Industrial pollution in Estonia. Gray snow floats down through emissions from the Kunda cement factory. As in most other former Soviet republics, environmental degradation in Estonia resulted from the emphasis on industrial output as the key to economic growth and prosperity. [Larry C. Price.]

and subsidiaries producing glass and high-tech materials in Poland.

The new market economies in the former Soviet bloc countries are improving energy efficiency and reducing emissions. Power plants, factories, and agriculture are polluting less, and the countries with the worst emissions records, such as Estonia and Latvia, have been making the most progress. All but Slovenia, which had relatively low emissions to start with, are making progress toward the 2010 reduction levels set by the Kyoto Protocol.

However, progress is slow for a number of reasons. Since the collapse of Soviet era trade relationships, people have been more concerned with economic growth than improved air quality. Many governments in Central Europe subsidize fossil fuels, which encourages wasteful use by consumers. As a result, emissions from motorized transport have increased. Meanwhile, environmental activists are disorganized and poorly

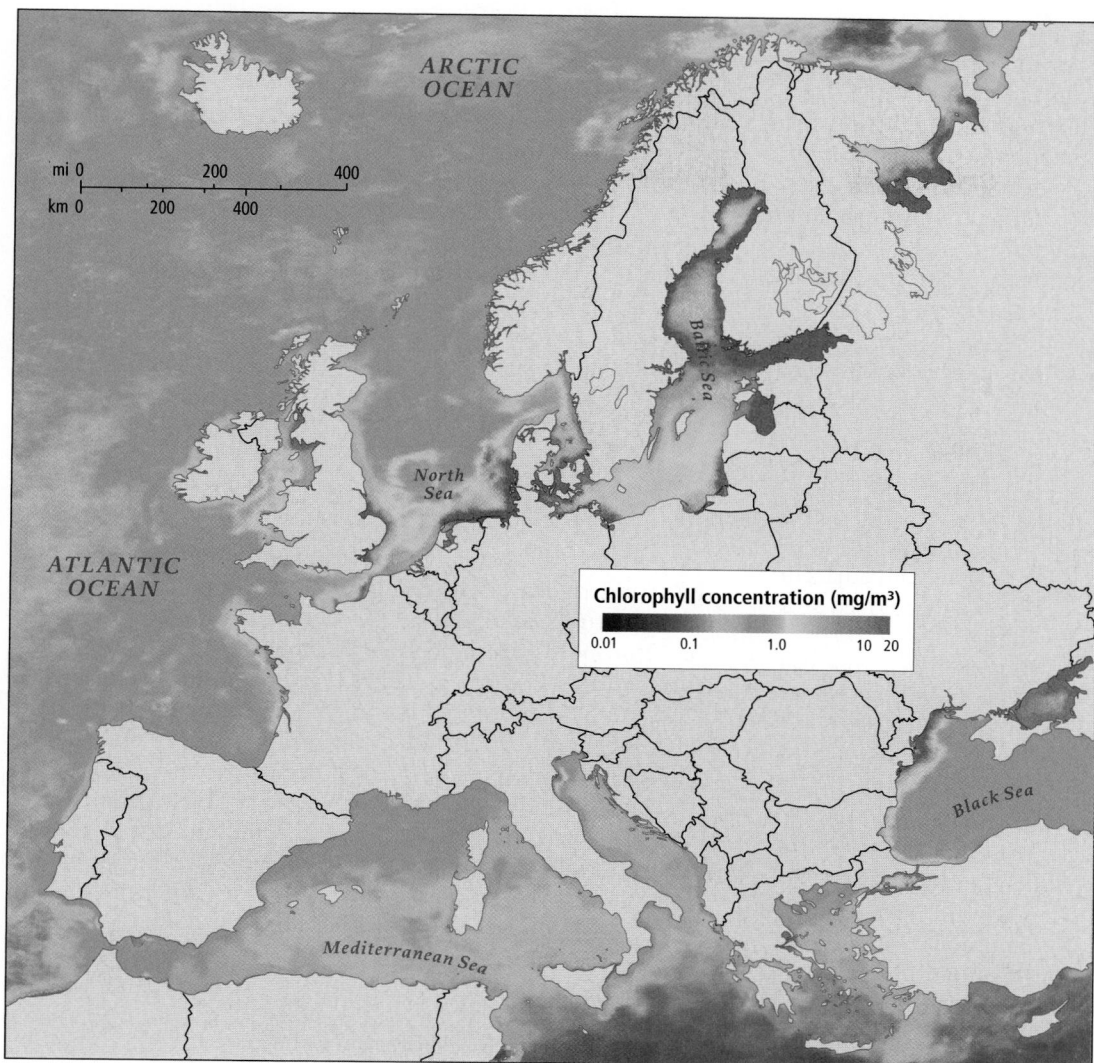

Figure 4.10 Pollution of the seas. Europe's exceptionally long and convoluted coast affords easy access to the world's oceans. However, pollution of the nearly landlocked Baltic, Black, and Mediterranean seas is causing increasing concern. Chlorophyll concentration, one measure of pollution levels, is significant in the Atlantic and Arctic oceans as well. [Adapted from a NASA image created by Jesse Allen, Earth Observatory, using data provided courtesy of the SeaWiFS Project, NASA/Goddard Space Flight Center and ORBIMAGE, available at http://earthobservatory.nasa.gov/Newsroom/NewImages/images.php3?img_id=17332.]

funded. They often find it hard to mobilize public support for regulations that seem as if they might slow economic growth.

Water Pollution

Europe is surrounded by seas, and any pollutants that enter Europe's rivers, streams, and canals eventually reach the surrounding seas (Figure 4.10). The Atlantic Ocean, the Arctic Ocean, and the North Sea are able to disperse most pollutants dumped into them because they are part of, or closely connected to, the circulating flow of the world ocean. In contrast, the Baltic, Mediterranean, and Black seas are nearly landlocked bodies of water that do not have the capacity to flush themselves out quickly; thus, all three are prone to accumulating pollution.

There are 34 countries with coastlines on Europe's many seas, all with different economies, politics, and cultural traditions. Such diversity often makes it difficult for them to coop-

erate to minimize pollution or even reduce the risk of severe pollution.

Vignette "All of us who have been cleaning these beaches," says Belen Piniero, "know that even if we get everything clean, our land will be dead for many years. Maybe our children will be able to enjoy the beaches again. I don't know. . . ." She stops short as tears well in her eyes. "Sorry, but I don't want to talk any more."

It's December 2002. Belen is one of hundreds of volunteers who wade through the thick black oil that covers over 350 miles (563 kilometers) of Atlantic coastline in northwestern Spain. Millions of fish and birds are smothered in the gunk. Dead and dying animals lie all around. Help is slow to come, so the volunteers are inventing their own methods for cleaning up the poisonous oil by hand. It is a hopeless and depressing task, and nerves are frayed.

A month earlier, the oil tanker *Prestige* had broken apart in a heavy storm and sunk. It spilled 20 million gallons of highly toxic crude oil, twice the amount dumped by the *Exxon Valdez* off Alaska in 1989. The cause of the spill was complex, as was finding out who was responsible. The tanker was so old and decrepit that its previous captain had deemed the ship unsafe and resigned in protest. It was owned by a Greek family who registered it in the Bahamas but under a Liberian corporation. It was certified as "in compliance" with all applicable laws by a U.S. corporation. Lawsuits seeking compensation for the spill are currently stuck in a maze of international laws governing oil tankers and pollution of the seas. So far no damages have been paid by the responsible parties in what has become the worst environmental disaster in Spain's history.

Watch the FRONTLINE/World video "The Lawless Sea" to learn the facts. ∎

Pollution in the Mediterranean Sea. The lands that drain into the Mediterranean are the sources of multiple pollutants. Municipal and rural sewage (over half of it untreated), agricultural chemicals, and industrial wastes pour in from adjacent lands. As a result, fish catches have declined and beloved seaside resorts have become unsafe for swimmers.

Pollution in the Mediterranean is exacerbated by the fact that it has just one tiny opening to the world ocean (see Figure 4.1). At the surface, seawater flows in from the Atlantic through the narrow Strait of Gibraltar and moves eastward. At the bottom of the sea, water exits through the same narrow opening, but only after it has been in the Mediterranean for 80 years! The natural ecology of the Mediterranean is attuned to this lengthy cycle, but the balance has been upset by the 320 million people now living in the countries surrounding the sea. Their pollution stays in the Mediterranean for decades.

Although most of the Mediterranean's pollution is generated by Europe, rapidly growing populations to the south and east of the sea also pose an environmental threat. Since 1995, the European Union has been working with the much poorer countries of Southwest Asia and North Africa to lessen pollution in the Mediterranean Basin. However, the latter regions still lack adequate urban sewage treatment and environmental regulations to control agricultural and industrial wastes. And although EU countries have the necessary technologies, funding, and legal framework to enforce regulations, they often lack the political will to reduce environmental impacts.

Ten Key Aspects of Europe

- The European Union leads the world in response to global warming with a goal of cutting greenhouse gas emissions by 20 percent by 2020.
- The ecology of the Mediterranean is threatened by water pollution from the 320 million people now living in the countries surrounding this sea.
- Most of the countries in the world have been ruled by a European colonial power at one point in their history.
- In West, North, and South Europe, around 80 percent of the population lives in urban areas.

- Europe's population is aging, as families are choosing to have fewer children.
- A new role for the European Union as a global peacemaker and peacekeeper is developing through the North Atlantic Treaty Organization (NATO).
- Agriculture is heavily subsidized throughout much of Europe.
- In Central Europe, democratic institutions were not adopted until the 1990s, after the fall of the Soviet Union.
- Throughout the European Union, women are paid on average 15 percent less than men for equal work.
- In Germany, Austria, Italy, and France, right-wing political parties now openly advocate forcing non-European immigrants to leave.

HUMAN PATTERNS OVER TIME

Over the last 500 years, Europe has profoundly influenced how the world trades, fights, thinks, and governs itself. Attempts to explain this influence are wide ranging. One argument is that Europeans are somehow a superior breed of humans. Another is that Europe's many bays, peninsulas, and navigable rivers have promoted commerce to a greater extent there than elsewhere. In fact, much of Europe's success is based on technologies and ideas it borrowed from elsewhere. For example, the concept of the peace treaty, so vital to current European and global stability, was first documented not in Europe but in ancient Egypt. In an effort to explain how Europe gained the leading global role it continues to play, it is worth taking a look at the broad history of this area.

Sources of European Culture

Starting about 10,000 years ago, the practice of agriculture and animal husbandry gradually spread into Europe from the Tigris and Euphrates river valleys in Southwest Asia and from farther east in Central Asia and beyond (see Figure 1.18 on page 21). Mining, metalworking, and mathematics also came to Europe from these places and from North Africa. All these innovations opened the way for a wider range of economic activity in Europe, most notably trade.

The first European civilizations were ancient Greece (800–86 B.C.E.) and Rome (100 B.C.E.–450 C.E.). Located in southern Europe, Greece and Rome initially interacted more with the Mediterranean rim, Southwest Asia, and North Africa than with the rest of Europe, which then had a small and relatively poor population. Later European traditions of science, art, and literature were heavily based on Greek ideas, which were themselves derived from yet earlier Egyptian and Southwest Asian sources.

The Romans, perhaps the greatest borrowers of Greek culture, also left important legacies in Europe. Many Europeans today speak Romance languages, such as Spanish, Portuguese, Italian, and French, which are largely derived from Latin, the language of the Roman Empire. Rome was the origin of

European laws that determine how individuals own, buy, and sell land. These laws have been spread throughout the world by Europeans.

Roman practices used in colonizing new lands also shaped much of Europe. After a military conquest, the Romans secured control in rural areas by establishing large plantation-like farms. Politics and trade were centered on new Roman towns built on a grid pattern that facilitated both commercial activity and military repression of rebellions. These same systems were later used when Europeans colonized the Americas, Asia, and Africa.

The influence of Islamic civilization on Europe is often overlooked. After the fall of Rome, while Europe slumbered during the *Dark Ages* (roughly 450–1000), pre-Muslim (Arab and Persian) and Muslim scholars preserved learning from Rome and Greece. Muslims originally from North Africa ruled Spain from 711 to 1492, and from the 1400s through the early 1900s, the Islamic Ottoman Empire dominated much of southeastern Europe and Greece. The Muslims brought new technologies, food crops, architectural principles, and textiles to Europe from Arabia, China, India, and Africa. Muslims also brought Europe its numbering system, mathematics, and significant advances in medicine, engineering, and architecture, building on ideas they themselves picked up in South Asia.

The Inequalities of Feudalism

As the Roman Empire declined, a social system known as **feudalism** evolved during the *medieval period* (450–1500 C.E.). This system originated from the need to defend rural areas against local bandits and raiders from Scandinavia and the Eurasian interior. The objective of feudalism was to have a sufficient number of heavily armed, professional fighting men, or knights, to defend the **serfs,** who were legally bound to live on and cultivate plots of land for them. Over time, some of these knights became a wealthy class of warrior-aristocrats,

called the **nobility,** who controlled certain territories. Some nobles gained so much power that they became centralized rulers—called "kings" or "monarchs"—of a certain area.

The often lavish lifestyles and elaborate castles (Figure 4.11) of the wealthier nobility were supported by the labors of the serfs. Most serfs lived in poverty outside castle walls, and much like slaves, were legally barred from leaving the lands they cultivated for their protectors. During later colonial times, aspects of feudalism were brought to the Americas, where the Spanish crown expropriated land from Native Americans and granted its use to colonists. The native inhabitants were treated as serfs.

Urbanization and the Transformation of Europe

While rural life followed established feudal patterns, new political and economic institutions were developing in Europe's towns and cities. Here, thick walls provided defense against raiders, and commerce and crafts supplied livelihoods. Thus the people could be more independent from feudal knights and kings.

Located along trade routes, Europe's urban areas were exposed to new ideas, technologies, and institutions from Southwest Asia, India, and China. Some institutions, such as banks, insurance companies, and corporations, provided the foundations for Europe's modern economy. Over time, Europe's urban areas established a pace of social and technological change that left the feudal rural areas far behind.

Urban Europe flourished in part because of laws that granted basic rights to urban residents. With adequate knowledge of these laws, set forth in legal documents called *town charters,* people with few resources could protect their rights even if challenged by those more wealthy and powerful than themselves.

Figure 4.11 Podsreda Castle, Slovenia. Podsreda Castle in eastern Slovenia—begun in the twelfth century, modified over the centuries since, and now renovated and preserved as a national landmark in Kozjanski Park—exemplifies the feudalism out of which modern Europe eventually emerged. [Mac Goodwin.]

Urban Europe flourished in part because of laws that granted basic rights to urban residents.

Town charters provided a basis for European notions of *civil rights,* which have proved hugely influential throughout the world. With strong protections for their civil rights, some of Europe's townsfolk grew into a middle class whose prosperity moderated the feudal system's extreme divisions of status and wealth. A related outgrowth of urban Europe was a philosophy known as **humanism,** which emphasized the dignity and worth of the individual regardless of wealth or social status.

The liberating influences of European urban life transformed the practice of religion. Since late Roman times, the Catholic Church had dominated not just religion but also politics and daily life throughout much of Europe. In the 1500s, however, a movement known as the **Protestant Reformation** arose in the urban centers of the North European Plain. Reformers such as Martin Luther challenged Catholic practices that stifled public participation in religious discussions, such as holding church services in Latin, which only a tiny educated minority understood. Protestants also promoted individual responsibility and more open public debate of social issues. These ideas spread faster with the invention of the European version of the printing press, which enabled widespread literacy.

European Colonialism: An Acceleration of Globalization

A direct outgrowth of the greater openness of urban Europe was the exploration and subsequent colonization of much of the world by Europeans. The increased commerce and cultural exchange began a period of accelerated *globalization* (see the discussion in Chapter 1 on pages 13–14) that persists today.

In the fifteenth and sixteenth centuries, Portugal took advantage of advances in navigation, shipbuilding, and commerce to set up a trading empire in Asia and a colony in Brazil. ▶ Spain soon followed, founding a vast and profitable empire in the Americas and the Philippines. By the seventeenth century,

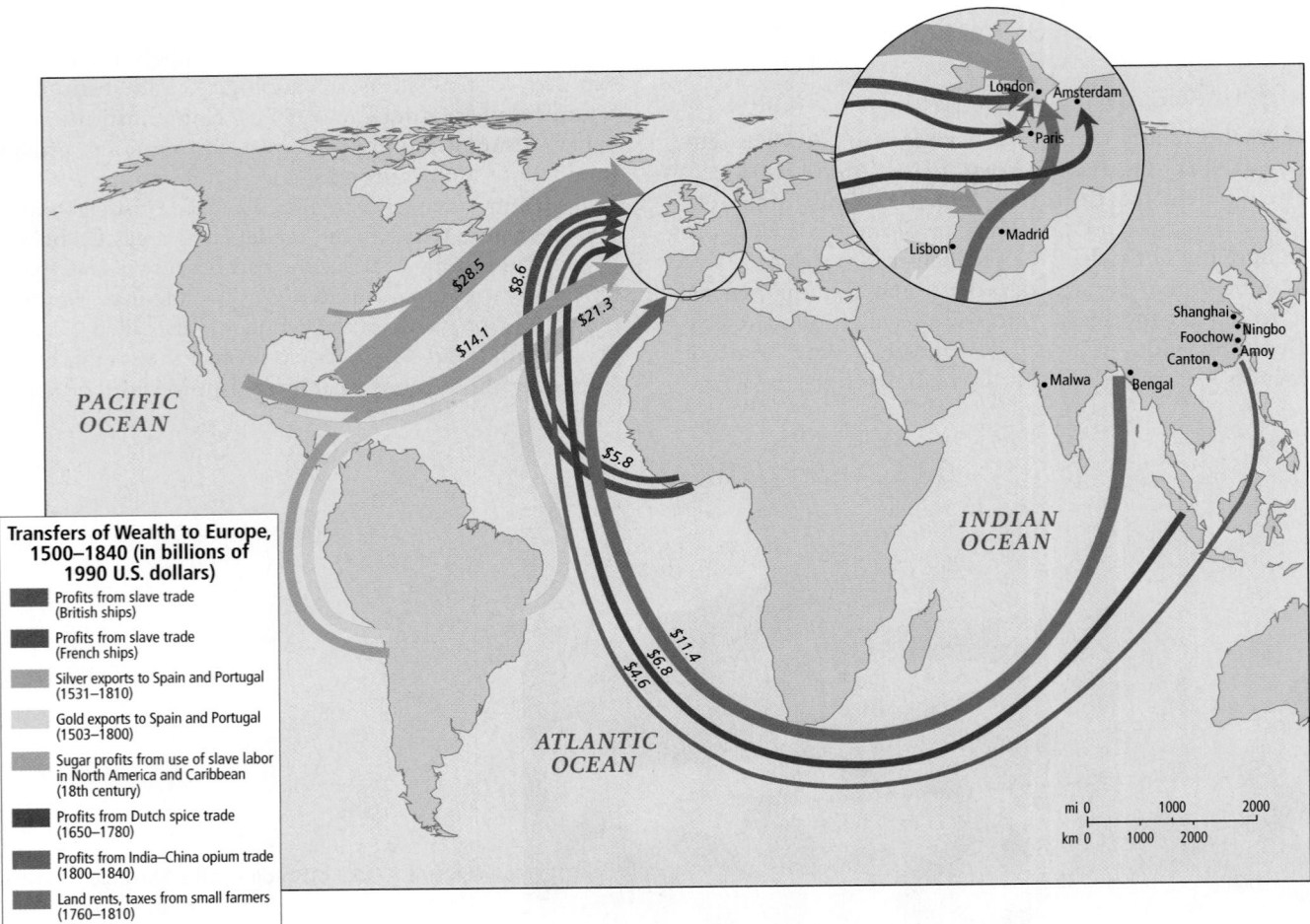

Figure 4.12 Transfers of wealth from the colonies to Europe, 1500–1840. Europe received billions of dollars of income from its overseas colonies during this period of mercantilism. (For European colonial holdings in various regions of the world, see Figures 3.14, 6.15, 6.16, and 7.12.) [Adapted from Alan Thomas, *Third World Atlas* (Washington, D.C.: Taylor & Francis, 1994), p. 29.]

however, England and the Netherlands had seized the initiative from Spain and Portugal. They perfected **mercantilism,** a strategy for increasing a country's power and wealth by acquiring colonies and managing all aspects of their production, transport, and trade for the colonizer's benefit. Mercantilism supported the Industrial Revolution (page 152) in Europe by supplying cheap resources from around the globe for Europe's new factories. The colonies also supplied markets for European manufactured goods (Figure 4.12).

By the nineteenth century, the Spanish and Portuguese colonial empires were overshadowed by those of the English, French, and Dutch (the people who live in the Netherlands), who extended their influence into Asia and Africa. By the twentieth century, European colonial systems had strongly influenced nearly every part of the world.

Globalization and urbanization. The overseas empires of England, the Netherlands, and eventually France shifted wealth and investment away from Southern Europe and the Mediterranean and toward Western Europe (Figures 4.13a and 4.13b).

The regional trading networks developed by some Italian cities during the medieval period (450–1500 C.E.) were overshadowed by the global trading networks of cities such as Amsterdam, London, and Paris.

By the mid-1700s, wealthy merchants in these and other cities were investing in new industries. Workers from rural areas poured into urban centers in England, the Netherlands, Belgium, France, and Germany to work in new mining and manufacturing industries (Figure 4.13c). Wealth and raw materials flowed in from ports in the Americas, Asia, and Africa. Some cities, such as Paris and London, were elaborately rebuilt in the 1800s to reflect their roles as centers of global empires.

By 1800, London and Paris, each of which had a million people, were Europe's largest cities, a status that eventually brought them to their present standing as **world cities** (cities of worldwide economic or cultural influence). London is a global center of finance, and Paris is a cultural center that has influence over global consumption patterns from food to fashion to tourism.

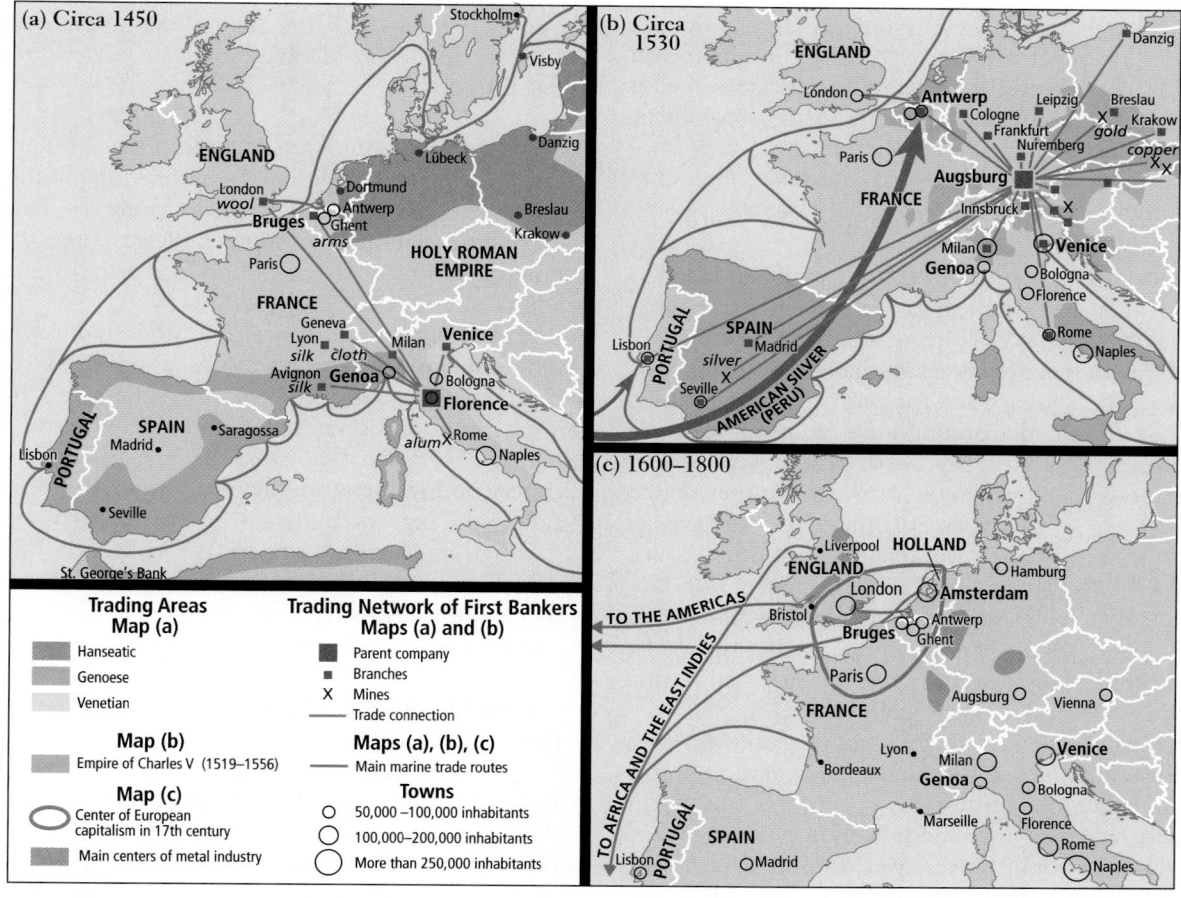

Figure 4.13 Shifts of power among urban centers, 1450–1800. [This map was prepared for this text with the assistance of geographer John Agnew.]

Urban Revolutions in Industry and Democracy

The wealth derived from Europe's colonialism helped fund two of the most dramatic transformations in a region already characterized by rebirth and innovation: the industrial and democratic revolutions. Both took place in urban Europe.

The Industrial Revolution. Europe's Industrial Revolution—particularly Britain's ascendancy as the leading industrial power of the nineteenth century—was intimately connected with colonial expansion. In the seventeenth century, Britain developed a small but growing trading empire in the Caribbean, North America, and South Asia, which provided it with access to a wide range of raw materials and to markets for British goods. Sugar, produced by British colonies in the Caribbean, was an especially important trade crop (see Figure 4.12).

Sugar production was a complex process requiring major investments in equipment and slaves forcibly brought in from Africa. The skilled management and large-scale organization needed later provided a production model for the Industrial Revolution. The mass production of sugar also generated enormous wealth that helped fund industrialization.

By the late eighteenth century, Britain was introducing mechanization into its industries, first in textile weaving and then in the production of coal and steel. By the nineteenth century, Britain was the world's greatest economic power, with a huge and growing empire, expanding industrial capabilities, and the world's most powerful navy. Its industrial technologies spread throughout Europe, North America, and elsewhere, transforming millions of lives in the process.

Urbanization and democratization. Industrialization led to massive growth in urban areas in the eighteenth and nineteenth centuries. Extremely low living standards in Europe's cities created tremendous pressures for change in the political order. Most industrial jobs were dangerous and unhealthy, demanding long hours and offering little pay. Most people were packed into tiny apartments that they shared with many others. Children were often weakened and ill due to poor nutrition and inadequate health care. Water was often contaminated, and most sewage ended up in the streets. Opportunities for advancement through education were restricted to the tiny few who could afford it.

> The road to democracy in Europe was rocky and violent, just as it is now in many democratizing parts of the world.

However, through their own efforts, more people did learn to read. The ideas and information they gained gave them the incentive to organize and protest for change. After lengthy struggles, democracy was expanded to Europe's huge and growing working class, and power, wealth, and opportunity were distributed more evenly throughout society. However, the road to democracy in Europe was rocky and violent, just as it is now in many parts of the world.

In 1789, the French Revolution led to the first major inclusion of common people in the political process in Europe. Angered by the extreme disparities of wealth in French society, and inspired by news of the popular revolution in North America, the poor rebelled against the monarchy and the elite-dominated power structure that still controlled Europe. As a result, the general populace, especially in urban areas, became involved in governing through democratically elected representatives. The democratic expansion created by the French Revolution ultimately proved short lived as elite-dominated governments soon regained control in France. Nevertheless, the French Revolution provided crucial inspiration to later urban democratic political movements.

Communism. During the struggles that resulted in the expansion of democracy, popular discontent erupted periodically in the form of new revolutionary political movements that threatened the established civic order. The political philosopher and social revolutionary Karl Marx framed the mounting social unrest in Europe's cities as a struggle between socioeconomic classes. His treatise *The Communist Manifesto* (1848) helped social reformers across Europe articulate ideas on how wealth could be more equitably distributed. East of Europe in Russia, Marx's ideas inspired the creation of a revolutionary Communist state in 1917, the Soviet Union. Eventually the Soviet Union extended its ideology and state power throughout Central Europe.

Popular democracy and nationalism. In the cities of western Europe, while communism gained many followers, political movements among workers were more successful at expanding democracy. Innumerable struggles throughout the nineteenth and early twentieth centuries eventually gave all adults the right to vote.

Throughout the nineteenth and twentieth century, the development of democracy was also linked to the idea of **nationalism,** or allegiance to the state (see Chapter 1, page 46). The notion spread that all the people who lived in a certain area formed a nation and that loyalty to that nation should supersede loyalties to individual kings or queens. Eventually, the whole map of Europe was reconfigured, and the mosaic of kingdoms gave way to today's collection of nation-states. All of these new nations were slowly but relentlessly transformed into democracies by the political movements arising in Europe's industrial cities.

Democracy and the welfare state. Channeled through the democratic process, continued efforts for improved living standards forced most governments to become **welfare states** by the mid-twentieth century. Such governments accept responsibility for the well-being of their people, guaranteeing basic necessities such as education, employment, and health care for all citizens. In time, government regulations on wages, hours, safety, and vacations established more harmonious relations between workers and employers. The gap between rich and poor declined somewhat, and overall civic peace increased. And although welfare states were funded primarily through taxes on industry, industrial productivity did not decline.

Modern Europe's welfare states have yielded generally high levels of well-being for rich and poor. Just how much support the welfare state should provide is still a subject of debate, however, and one that has been resolved in different ways across Europe (see pages 168–170).

When assessing progress toward democracy in other world regions, it is helpful to remember Europe's long and difficult path to democracy. Women gained the right to vote only after considerable delay and political agitation. In Switzerland, women did not obtain the vote until 1971! In Central Europe, democracy was not adopted until the 1990s, after the fall of the Soviet Union.

Two World Wars and Their Aftermath

Despite Europe's many advances in industry and politics, the region still lacked a system of collective security that could prevent war between its rival nations. Between 1914 and 1945, two horribly destructive world wars left Europe in ruins, no longer the dominant region of the world. At least 20 million people died in World War I and 70 million in World War II. During World War II, 15 million civilians were killed by Germany's Nazi government during its failed attempt to conquer the Soviet Union. Eleven million civilians died at the hands of the Nazis during the **Holocaust,** a massive execution of 6 million Jews and 5 million **Roma** (Gypsies), disabled and men-

tally ill people, homosexuals, political dissidents, and ethnic Poles and other Slavs (Figure 4.14). ▣◀

The defeat of Germany, seen as the instigator of both world wars, resulted in a number of enduring changes in Europe. After the end of World War II, in 1945, Germany was divided into two parts. West Germany became an independent democracy allied with the rest of western Europe, especially Britain and France, and the United States. The Soviet Union controlled East Germany and the rest of what was then called eastern Europe (Latvia, Lithuania, Estonia, Poland, Czechoslovakia, Hungary, Romania, Bulgaria, Ukraine, Moldova, and Belarus). The line between East and West Germany was part of what was called the **iron curtain,** a long, fortified border zone that separated western Europe from (then) eastern Europe.

The cold war. The division of Europe created a period of conflict, tension, and competition between the United States and the Soviet Union. This **cold war** lasted from 1945 to 1991. During this time, once-dominant Europe, and indeed the entire world, became a stage on which the United States and the Soviet Union competed for dominance.

In most of what we now call Central Europe, the Soviet Union forcibly implemented its model of socialism, in which the state owned all farms, industry, land, and buildings. In contrast with the market economies of western Europe, these economies were **centrally planned:** a central bureaucracy dictated prices and output with the stated aim of allocating goods

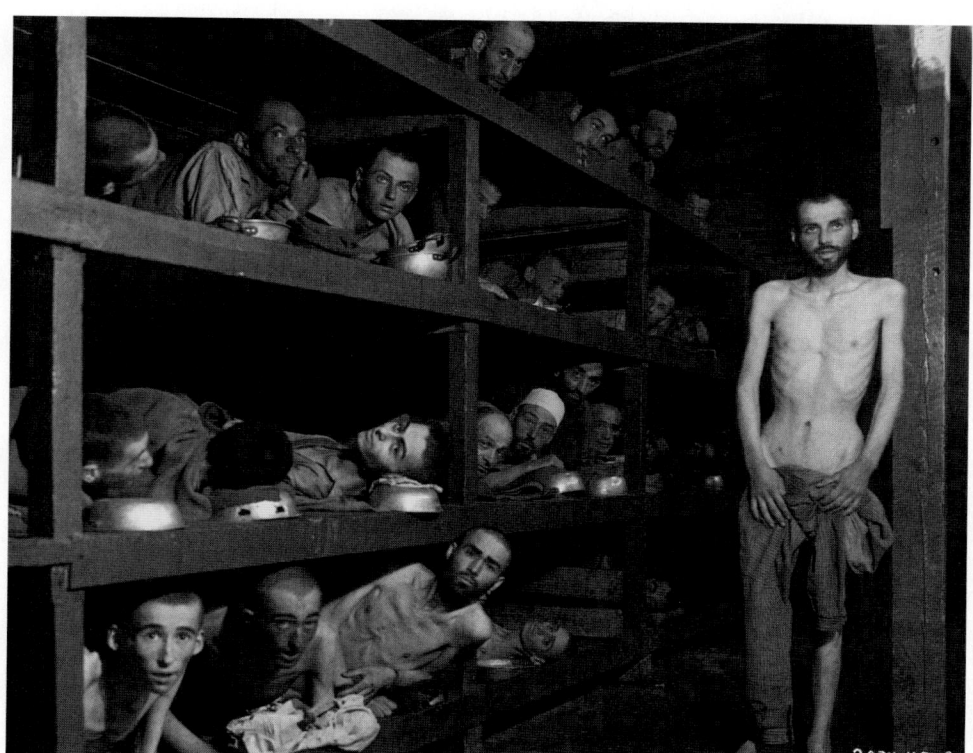

Figure 4.14 The Holocaust. On April 16, 1945, at the end of World War II, the Allies liberated the Buchenwald (Germany) concentration camp and found many scenes like this. Many of the camp inmates were on the edge of starvation; the soldiers found the bodies of many others who had died shortly before the liberation. Prisoners in Buchenwald were mostly Jews, but there were also resisters of Nazi domination and ethnic minorities. [Pvt. H. Miller/CORBIS.]

equitably across society according to need. This system worked for a time, but it ultimately collapsed in the 1990s due to inefficiency, high levels of environmental pollution, and public demands for democracy. ▰

In the rest of Europe, especially Western Europe, the United States provided financial assistance for rebuilding basic facilities, such as roads, housing, and schools, through the *Marshall Plan*. Governments continued their free market economic systems, characterized by privately owned businesses that adjusted prices and output to match the demands of the market. Economic reconstruction proceeded rapidly in the following decades. ▰

Decolonization. Europe's decline also led to the loss of its colonial empires. Many European colonies had participated in the two world wars, with some suffering extensive casualties, and almost all emerged economically devastated. After the war, Europe could not afford to help these areas rebuild, and demands for independence grew. By the 1960s, most former European colonies had gained independence, often after bloody wars fought against European powers and their local allies.

POPULATION PATTERNS

There are currently about 531 million Europeans. Their highest population densities stretch in a discontinuous band from the United Kingdom south and east through the Netherlands and central Germany (Figure 4.15). Northern Italy is another zone of density, along with pockets in many countries along the Mediterranean coast. Overall, Europe is one of the more

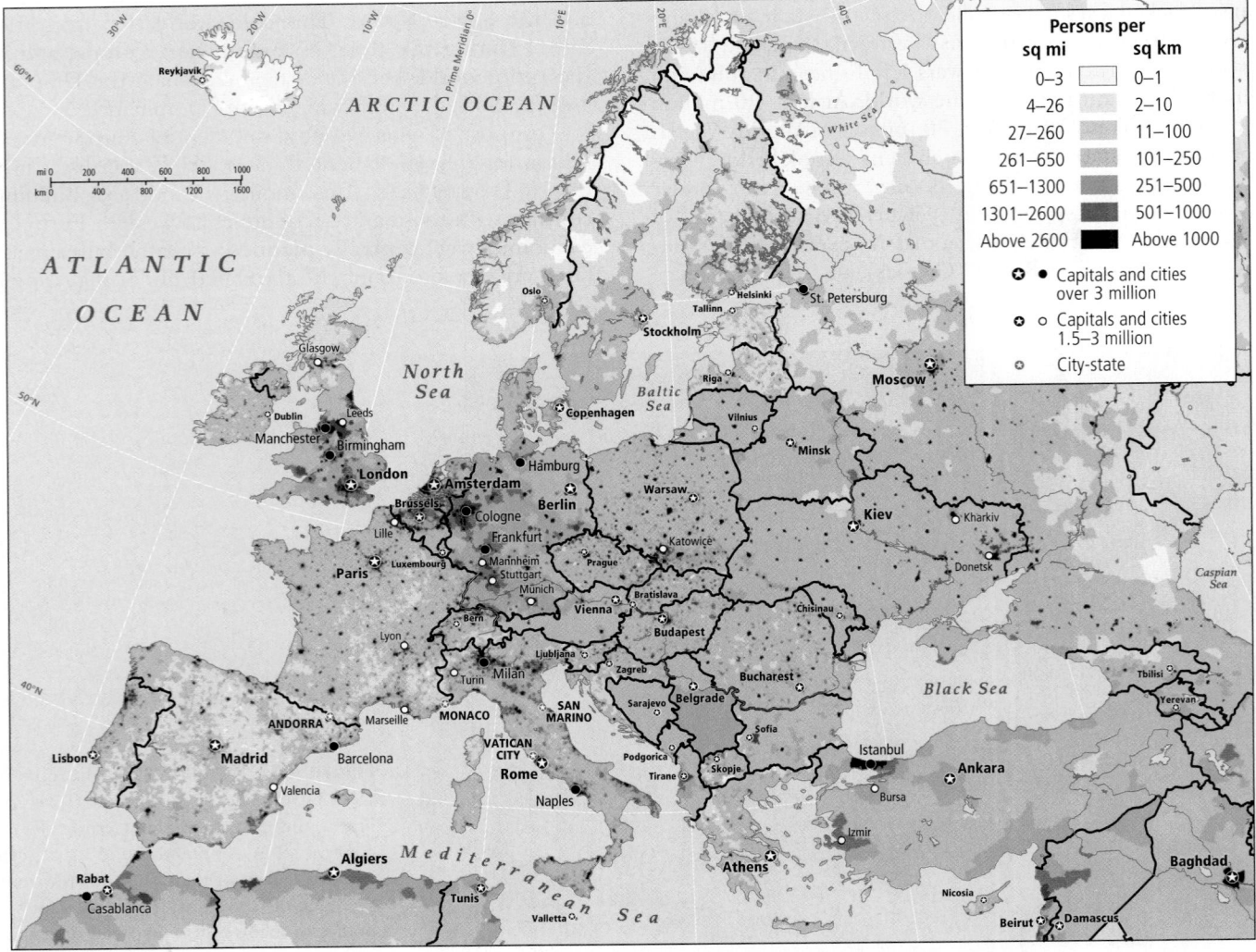

Figure 4.15 Population density in Europe. [Data courtesy of Deborah Balk, Gregory Yetman, et al., Center for International Earth Science Information Network, Columbia University; at http://www.ciesin. columbia.edu.]

densely occupied regions on earth (see Figure 1.34 on page 42). Most of this population now lives in cities.

Modern Urbanization in Europe

Today, Europe is a region of cities surrounded by well-developed rural hinterlands. These cities are the focus of the modern European economy, which, though long grounded in agriculture, trade, and manufacturing, is now primarily service-based. In West, North, and South Europe, 80 percent of the population lives in urban areas. Even in Central Europe, the least urbanized part of the region, around 75 percent of people live in cities. As noted earlier, many European cities began as trading centers more than a thousand years ago and still bear the architectural marks of medieval life in their historic centers. These old cities are located either on navigable rivers in the interior or along the coasts because water transport figured prominently (and still does) in Europe's trading patterns (see Figures 4.13a and 4.13b).

Since World War II, nearly all the cities in Europe have expanded around their perimeters in concentric circles of apartment blocks. Well-developed rail and bus lines link the blocks to one another and to the old central city. Land is scarce and expensive in Europe, so only a small percentage of Europeans live in single-family homes, although the number is growing. Even single-family homes tend to be attached or densely arranged on small lots. Rarely does one see the sweeping lawns that many North Americans are accustomed to. Publicly funded transportation is widely available, so many people live without cars in apartments near city centers. However, many others commute by car daily from suburbs or ancestral villages to work in nearby cities.

Although deteriorating housing and slums do exist, substantial public spending (on sanitation, water, utilities, education, housing, and public transportation) help people maintain a generally high standard of urban living. Tourists from around the world are drawn to Europe's large cities by their architectural heritage, outdoor spaces, art, music, museums, and generally pleasant, walkable ambience.

Europe's Aging Population

Europe's population is aging as families are choosing to have fewer children. As a region, Europe has a negative rate of natural increase (−0.1), the lowest on Earth. Birth rates are lowest in the countries that were part of the former Soviet Union or Yugoslavia. In addition, death rates are high in this area because the quality of health care has deteriorated. The one-child family is increasingly common throughout Europe. In western Europe, immigrants are a major source of population growth. However, once they have assimilated to European life, immigrants, too, choose to have small families.

The declining birth rate is illustrated in the population pyramids of European countries, which look more like lumpy towers than pyramids. The population pyramid of Germany (East and West, reunited in 1990, are combined here) is an example (Figure 4.16a). The pyramid's narrowing base indicates that for the last 35 years, far fewer babies have been born than in the 1950s and 1960s, when there was a baby boom across Europe. By 2000, 35 to 40 percent of Germans were choosing to have no children at all.

The reasons for these trends are complex. For one thing, more and more women desire professional careers. This alone could account for late marriages (25 percent of Germans were

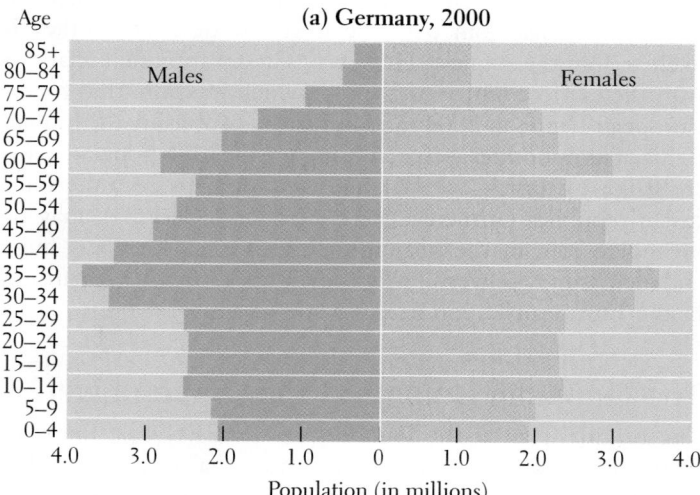

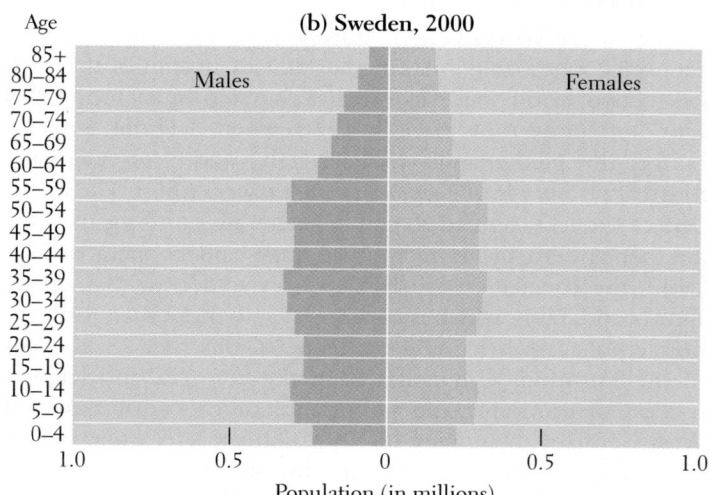

Figure 4.16 Population pyramids for Germany and Sweden, 2000. These two population pyramids have quite different shapes, but both exhibit a narrow base, indicating low birth rates. Observe that the scales along the bottom of the pyramids are different for the two countries, indicating that Sweden's total population is much smaller than Germany's. [Adapted from "Population pyramids of Germany" and "Population pyramids of Sweden" (Washington, D.C.: U.S. Census Bureau, International Data Base, 2000).]

choosing to remain unmarried well into their thirties) and lower birth rates. Governments also make few provisions for working mothers beyond paid maternity leave. In Germany, for example, there is little day care available for children under three and school days are short, ending at noon. Hence, many women choose not to become mothers because they would have to settle for part-time jobs. To encourage higher birth rates, there is a move within the European Union to lengthen school days and to give one parent (mother or father) a full year off with reduced pay after a child is born or adopted.

As a result of the low birth rate, the vast majority of Germans are older than 30. As the older generations die, the population may eventually settle into a stable age structure, as Sweden's population has already done (Figure 4.16b). In this structure, each age category has roughly similar numbers of people, tapering only at the top after age 60.

A stable population with a low birth rate has several consequences. Because consumers are dying faster than they are born, economies may contract over time. Demand for new workers, especially highly skilled ones, may go unmet. Further, the number of younger people available to provide expensive and time-consuming health care for the elderly, either personally or through tax payments, will decline. Currently, for example, there are just two German workers for every retiree. Immigration provides one solution to the dwindling number of young people. However, Europeans are reluctant to absorb large numbers of immigrants, especially from distant parts of the world which have very different cultures from their own. 📹

> Because consumers are dying faster than they are produced, economies may contract over time.

Current Geographic Issues

Europe today is in a state of transition as a result of two major changes that occurred during the 1990s: the demise of the Soviet Union (discussed above and in Chapter 5) and the rise of the European Union. These developments could ultimately bring greater peace, prosperity, and world leadership to Europe. However, many problems and tensions remain to be resolved.

ECONOMIC AND POLITICAL ISSUES

At the end of World War II, European leaders felt that closer economic ties would prevent the kind of hostilities that had led to two world wars. The first major step in achieving this economic unity took place in 1958, when Belgium, Luxembourg, the Netherlands, France, Italy, and West Germany formed the *European Economic Community (EEC)*. The members of the EEC agreed to eliminate certain tariffs against one another and to promote mutual trade and cooperation. In 1992, the concept of the EEC was expanded to that of the European Union, which is concerned with more than just economic policy. An ongoing challenge to this broader concept of the European Union has been the complex and lengthy process of economic and political reunification of East and West Germany, beginning in 1990, and the later expansion of the European Union into Central Europe. Figure 4.17 shows the current members of the European Union and the dates they joined, as well as new candidates for membership. 📹

There is one non-EU member in the center of Europe (Switzerland) and two on its northern periphery (Norway and Iceland). These three countries have long treasured their neu-

tral role in world politics. Moreover, as wealthy countries, they are concerned about losing control over their domestic economic affairs.

Two countries in southeastern Europe, Romania and Bulgaria, joined the European Union in January 2007 and are undergoing many changes in order to align themselves with EU standards. The former states of Yugoslavia are also being prepared to become part of the European Union.

Future EU expansion. Several countries on the perimeter of Europe may also join over the next few decades. Turkey is likely to join the soonest. However, Turkey's strained relations with the island country of Cyprus (which was admitted to the European Union in 2004) and its history of abusing the human rights of its minorities—especially its large Kurdish populations—are strikes against it. 📹 Ukraine, and perhaps even the Caucasian republics (Armenia, Azerbaijan, and Georgia), may be invited to join in the near future. However, Europe's huge and increasingly powerful neighbor, Russia, opposes expansion of the European Union in these areas.

The European Union: A Rising Superpower?

The original plan at the founding of the EEC was simply to work toward a level of economic and social integration that would make possible the free flow of goods and people across national borders. While this goal remains central, some Europeans feel that the European Union, already a global economic power, should become a global counterforce to the United States in political and military affairs.

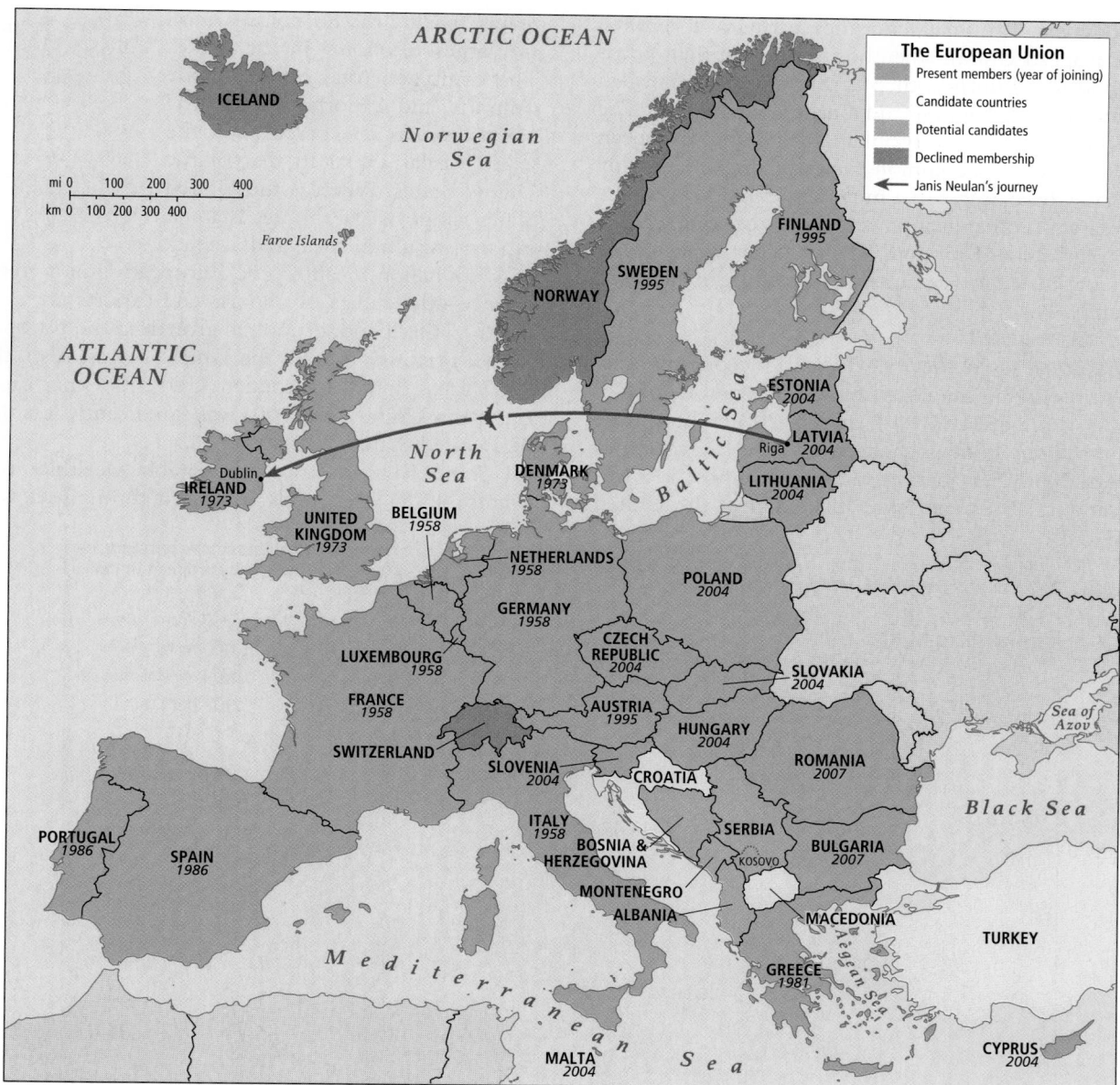

Figure 4.17 The European Union (EU). The EU, formed with the initial goal of economic integration, has led the global movement toward greater regional cooperation. It is older and more deeply integrated than its closest competitor, the North American Free Trade Agreement (NAFTA), which was largely a response to the EU.

EU governing institutions. Somewhat similar to the United States, the European Union has one executive branch and two legislative bodies. The *European Commission* acts like an executive branch of government, proposing new laws, implementing decisions, and generally running the European Union on a day-to-day basis. The highest-ranking members are appointed by EU member states, but the commission also includes about 25,000 civil servants who work in Brussels.

The *European Parliament* is directly elected by EU citizens, with each country electing a proportion of seats based on its population, much like the U.S. House of Representatives. The *Council of the European Union* is similar to the U.S. senate in that it is the more powerful of the two legislative bodies. However, its members are not elected but consist of ministers from the national governments of EU member states. A formal constitution for the European Union has been developed but not yet adopted (ratification by all members is required).

Economic integration. Individual European countries have far smaller populations than their competitors in North America and Asia. Because they have smaller markets for their products, companies in small countries earn lower profits. Before the European Union, when businesses sold their products to neighboring countries, their earnings were diminished by tariffs, currency exchanges, and border regulations. The European Union solved this problem by joining European national economies into a common market. Companies in any EU country now have access to a much larger market and the potential for larger profits through **economies of scale** (reductions in the unit costs of production that occur when goods or services are produced in large amounts, resulting in increased profits per unit).

The world's largest economy. The EU economy now encompasses close to 500 million people (out of a total of 525 million in the whole of Europe)—roughly 200 million more than live in the United States. Collectively, the EU countries are wealthy. In 2008 their joint economy was almost $15 trillion (PPP), about 10 percent larger than that of the United States, making the European Union the largest economy in the world. The combined total external trade (imports and exports) of the EU countries was 19 percent of the world's total, equal to that of the United States. Whereas the United States imports far more than it exports, resulting in a trade deficit of $711 billion in 2007, the European Union enjoys a trade balance—the values of imports and exports are nearly equal. On the other hand, economic growth in the European Union is usually slower than in the United States. Also, the average 2007 gross domestic product (GDP, PPP) per capita for the European Union ($32,900) was significantly less than that of the United States ($46,000).

Some EU countries are notably wealthier than others (Figure 4.18). EU funds are raised through an annual 1.27

> In 2008, the EU's economy was almost $15 trillion (PPP), about 10 percent larger than that of the United States, making the EU the largest economy in the world.

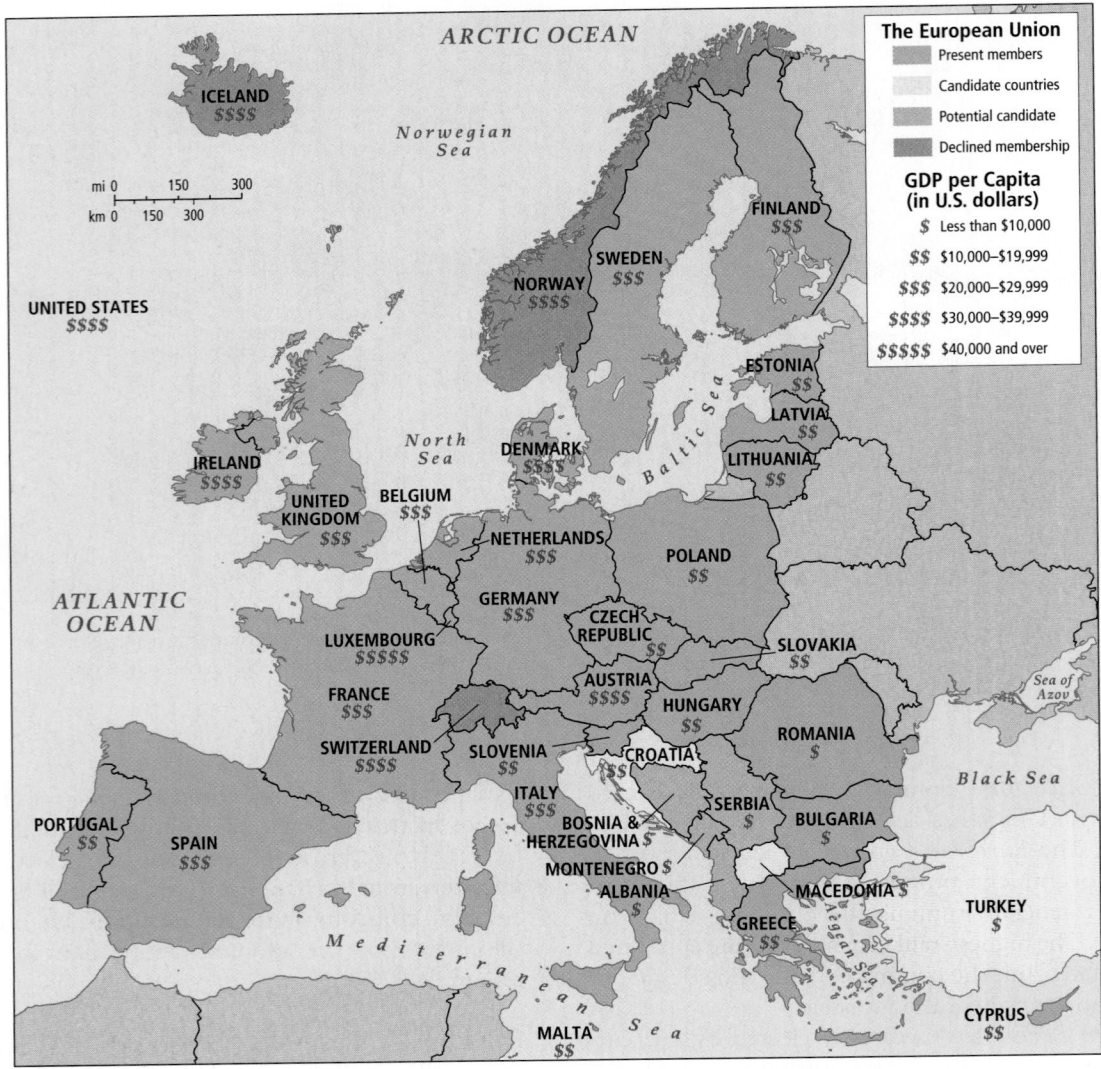

Figure 4.18 The EU economy: GDP per capita.

percent tax on the gross national product (GNP) of all members. Although financial allotments change from year to year, most member countries in North and West Europe contribute more than they receive back in grants. Most countries in South and Central Europe, on the other hand, tend to receive more than they contribute.

Since 1993, the EU agenda has expanded to include the creation of a common European currency, the defense of Europe's interests in international forums, and negotiation of EU-wide agreements on human rights and social justice. The European Union is also experimenting with various forms of political unification, including the creation of a common European military.

A common European currency. The official currency of the European Union is the **euro** (€). About half the EU countries use it as their currency, and all but Sweden and the United Kingdom have currencies whose value is determined by that of the euro. Countries that use the euro have a greater voice in the creation of EU economic policies. The euro now rivals the U.S. dollar as the preferred currency of international trade and finance.

The European Union and globalization. The European Union has a number of strategies designed to ensure that it continues to be economically competitive with the United States, Japan, and the developing economies of Asia, Africa, and South America. One strategy it is pursuing is the relocation of factories from the wealthiest EU countries to the relatively poorer, lower-wage member states (Figure 4.18). It is hoped that these efforts will help poorer European countries prosper and will keep the costs of doing business low enough to restrain European companies from moving to developing countries where costs are lower still. However, the resulting reduction of industrial capacity (**deindustrialization**) in Europe's wealthiest countries has led to higher unemployment there (Figure 4.19).

Like the United States, the European Union exerts a powerful influence in the global trading system. The European Union often negotiates privileged access to world markets for European firms and farmers and for former European colonies (Figure 4.20, page 161). It also employs protectionist measures that favor European producers over cheaper imports. The higher-priced goods create added expense for European consumers, and at the same time, non-European producers lose access to EU markets. Poor countries throughout the world have united to protest the European Union's failure to open its economies to foreign competition. So far such protests have met with little success.

NATO and the rise of the European Union as a global peacemaker. A new role for the European Union as a global peacemaker and peacekeeper is developing through the

North Atlantic Treaty Organization (NATO), which is based in Europe. During the cold war, European and North American countries cooperated militarily through NATO to counter the influence of the Soviet Union. NATO originally included the United States, Canada, the countries of western Europe, and Turkey; it now includes almost all the EU countries as well.

Since the breakup of the Soviet Union, NATO has focused mainly on providing the international security and cooperation needed to expand the European Union. However, with the United States preoccupied with Iraq, NATO is also becoming more of a global peacekeeper. It already provides a majority of the troops in Afghanistan. Worldwide opposition to the U.S. invasion of Iraq has elevated the global status of the European Union, most of whose members opposed the war. NATO, or a successor alliance with the European Union playing a leading role, may one day join the United States in the very difficult task of addressing global security issues. 🎦

Ethnic cleansing in the former Yugoslavia—a failure of EU leadership. The European Union's slow response to a bloody war resulting from the breakup of Yugoslavia in 1991 left many skeptical of its ability to lead on the global stage. Until the European Union can rapidly and forcefully respond to such situations, its ability to join the United States as a global leader will remain limited.

After World War II, five ethnically distinct territories—Serbia, Croatia, Slovenia, Macedonia, Montenegro, and Bosnia-Herzegovina—were brought together to form the country of Yugoslavia. In 1991, the first free elections ever held in Yugoslavia resulted in declarations of independence by several territories. Slovenia and Macedonia managed to separate relatively peacefully. Serbia and Montenegro, however, fought protracted wars to retain Croatia and Bosnia-Herzegovina.

Both Croatia and Bosnia-Herzegovina had Serb-populated provinces, which Serbia argued should be part of one Serbian-led country. To rid the coveted provinces of non-Serbs, genocidal **ethnic cleansing** campaigns were enacted against civilians in Bosnia-Herzegovina and in Croatia. A similar genocidal campaign took place later in Kosovo, a province of Serbia. The goal was to create ethnically "pure" nation-states, or independent countries consisting of just one nationality out of a very complex ethnic mosaic (Figure 4.21, page 161).

Only after considerable delay did the EU countries and the United States both send military peacekeepers. Peace accords were finally brokered by the United States in 1995, granting independence to Croatia and Bosnia-Herzegovina. By this time, however, hundreds of thousands had died. In Bosnia-Herzegovina alone, 200,000 people—5 percent of the population—died. Serbia gained no new territory and its economy was ruined. Kosovo declared its independence from Serbia in 2008. 🎦

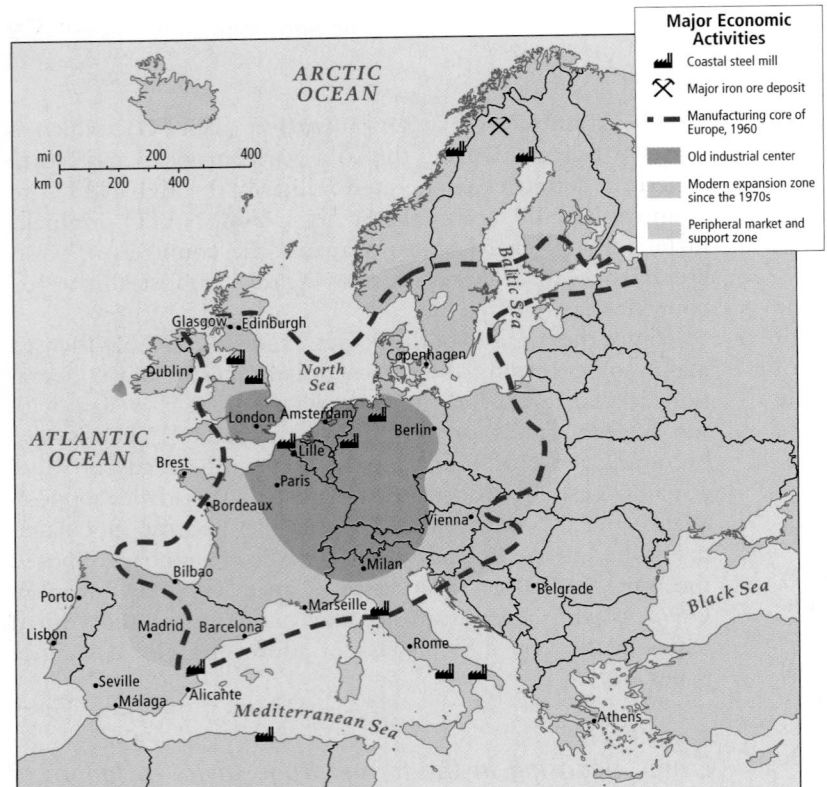

(a) 1960–1970s

Major Economic Activities

- Coastal steel mill
- Major iron ore deposit
- Manufacturing core of Europe, 1960
- Old industrial center
- Modern expansion zone since the 1970s
- Peripheral market and support zone

Figure 4.19 Europe's principal industrial and manufacturing centers: Shifts from 1960 (a) to 2000 (b). [Adapted from Terry G. Jordan-Bychkov and Bella Bychova Jordan, *The European Culture Area: A Systematic Geography,* 4th ed. (Lanham, Md.: Rowman & Littlefield, 2002), pp. 288 and 300.]

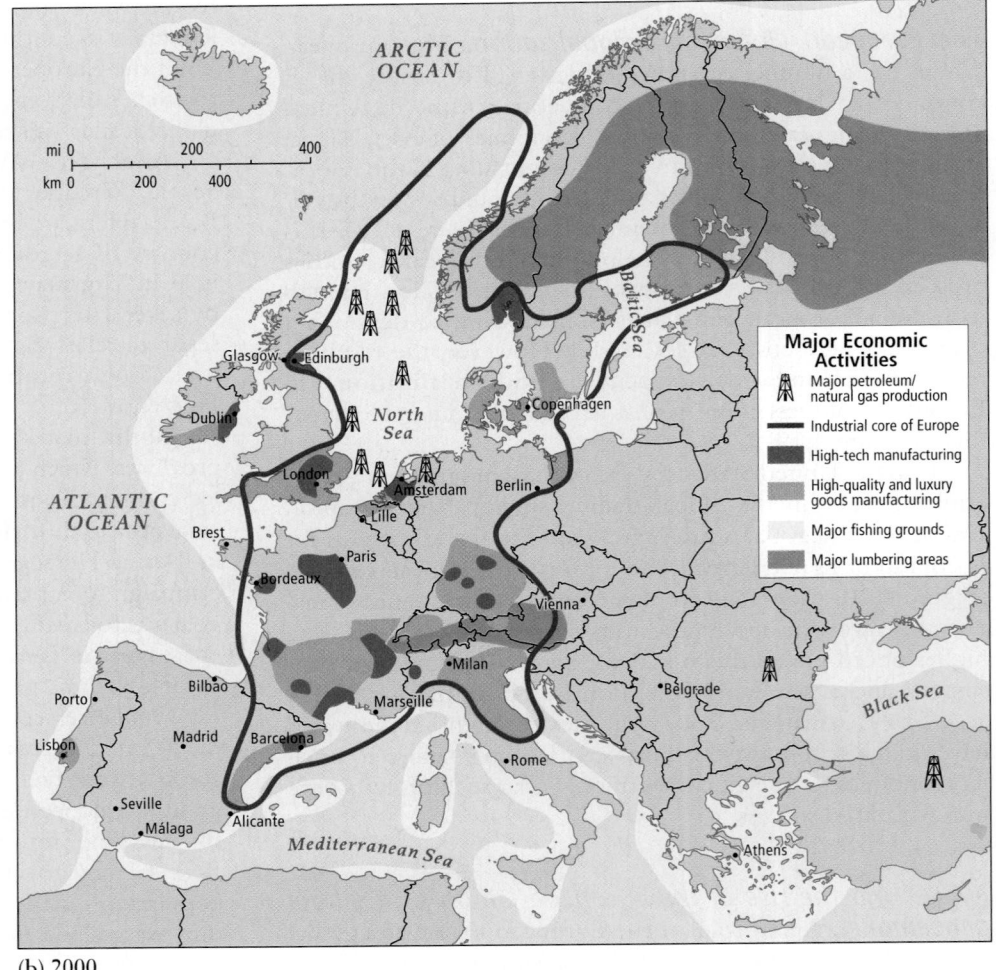

(b) 2000

Major Economic Activities

- Major petroleum/natural gas production
- Industrial core of Europe
- High-tech manufacturing
- High-quality and luxury goods manufacturing
- Major fishing grounds
- Major lumbering areas

EU's Share of World Trade, 2003

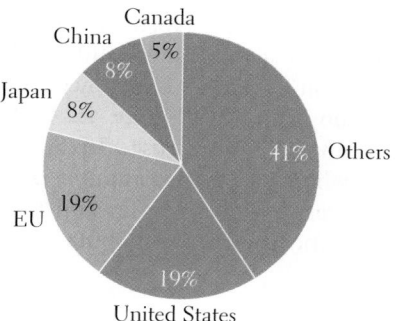

Figure 4.20 The European Union's share of world trade, 2003. At 19 percent, the EU's share of world trade (imports and exports combined) was equal to that of the United States and more than those of China and Japan combined. [Adapted from *Europe in Figures: Eurostat Yearbook 2005* (Luxembourg: European Commission, 2005), p. 187, at http://bookshop.europa.eu/eubookshop/FileCache/PUBPDF/ KSCD05001ENC/KSCD05001ENC_002.pdf.]

Central Europe and EU Expansion

Membership in the European Union became especially attractive to countries in Central Europe after the demise of the Soviet Union, when their economies and socialist safety nets began to deteriorate. The global shift away from socialist, centrally planned economies and toward more open and competitive market systems has placed many of these countries in jeopardy. Many workers have lost their jobs and in some of the poorest countries, such as Romania, Bulgaria, and Serbia, social turmoil and organized crime threaten stability. Investments by neighboring EU member countries in West and North Europe have reduced many of these problems in the Central European EU economies.

Standards for EU membership, however, are exacting. A country must achieve political stability and a democratically elected government. Each country has to adjust its constitution to EU standards that guarantee the rule of law, human rights, and respect for minorities. Each must also have a functioning market economy open to investment by foreign-owned

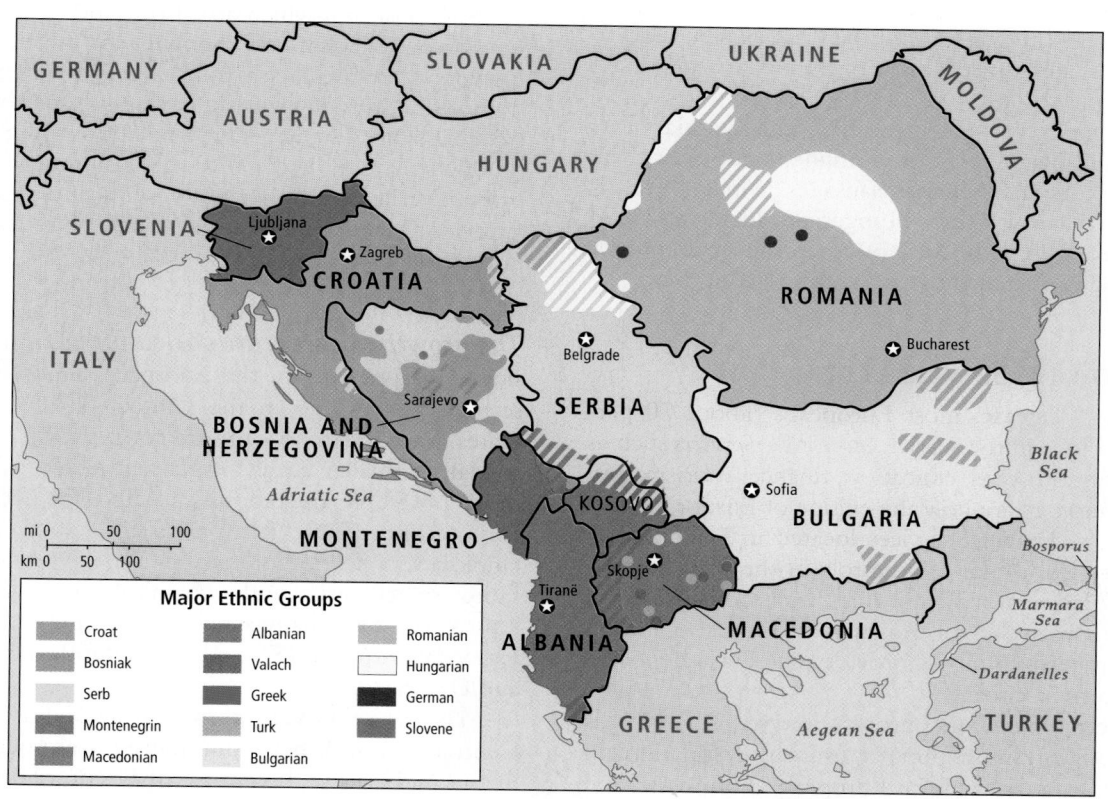

Figure 4.21 Ethnic groups in southeastern Central Europe. Southeastern Central Europe is a patchwork of culture and religious groups that entered the region at various times over the last several thousand years. Normally, relations were amicable and intermarriage was common, but occasionally hostilities arose, often as a result of outside pressures. [Adapted from Philippe Rekacewicz, UNEP/GRID- Arendal, *The Military Balance, 2002–2003* (London: The World Bank, 2003); and Catherine Samary and Jean-Arnault Dérens, *Les conflits yougoslaves de A á Z*, (Paris: Editions de l'Atelier, 2000), and available at http://maps.grida.no/go/graphic/ethnic_groups_in_the_ south_eastern_europe.]

companies. Finally, farms and industries must comply with strict regulations governing the finest details of their products. Meeting these standards can be both annoying and painful. For example, in 2004, Slovenia was abuzz with the news that the country's cucumbers would henceforth have to be a precise size and shape to be sold in EU markets. And sure enough, in short order, the "EU Protocol for Distinctness, Uniformity and Stability for *Cucumis sativus* L. (cucumber)" was adopted and published. Regulated cucumbers may be only the beginning. Throughout Europe, many fear that some EU policies will erase distinguishing cultural features, encourage boring homogenization, and even diminish well-being.

Case study: the Hungarian sausage experience and the European Union. The Hungarian writer and humorist Dork Zygotian (probably a pseudonym) assesses the disadvantages of joining the European Union from the perspective of a sausage lover. Zygotian writes that when you bite into a Hungarian sausage (*kolbasz*), "great torrents of paprika colored grease and juice should explode into the atmosphere around you. If you eat more than two, you should expect to bite on some piece of bone or possibly find a tooth or hair sometime during your meal. There will be a large yellow gelatinous bit somewhere in your sausage that you should not be able to identify." This is all part of the tasty Hungarian kolbasz experience.

But now Hungary has joined the EU, and Zygotian fears that overzealous EU standards of cleanliness and purity will kill the special flavors of Hungarian sausages. Worse, "Eurofication" (his term) may well leave Hungarians unable to afford their own beloved kolbasz because prices in the small, poorer countries will rise to match those in wealthier countries.

Europe's Growing Service Economies

As industrial jobs decrease, most Europeans (about 70 percent) have found jobs in the service economy. *Services* such as the provision of health care, education, finance, tourism, and information technology are now the engine of Europe's economy. For example, financial services located in London and serving the entire world play a huge role in the British economy, and many transnational companies are headquartered in London. Service jobs in the government sector are also numerous because European countries provide many tax-supported social services to their citizens.

A major component of Europe's service economy is *tourism*. Europe is the most popular tourism destination on earth, and one job in eight in the European Union is related to tourism. Tourism generates 13.5 percent of the European Union's gross domestic product and 15 percent of its taxes. Europeans are themselves enthusiastic travelers, visiting one another's countries as well as many exotic world locations frequently. This travel is made possible by the long paid vacations—usually 4 to 6 weeks—that Europeans are granted by employers.

Europe has lagged behind North America in the development and use of personal computers and the Internet. However, the information economy is advanced in Europe, with West Europe leading the way (Figure 4.22). In fact, Europe leads the world in cell phone use. In South Europe and Central Europe, where personal computer ownership is lowest, public computer facilities in cafés and libraries are common, and surfing the Internet is popular, especially among schoolchildren.

Food Production and the European Union

Only about 4 percent of Europeans are now engaged in full-time farming. However, Europeans like the idea of being self-sufficient in food. Toward this end, the European Union aids farmers with tariffs on imported agricultural goods and with **subsidies** (payments to farmers to lower their costs of production). Such measures are expensive—payments to farmers are the largest expense category in the EU budget. While these policies do provide a decent living standard for farmers, they also raise food costs for millions of consumers.

Such policies, which are also found in the United States, Canada, and Japan, are unpopular in the developing world. Farmers there are locked out of major potential markets by the tariffs. Subsidies also encourage farmers to overproduce (to collect more payments), thereby causing a glut of farm products that are then sold cheaply on the world market. This practice, called **dumping,** lowers global prices and thus hurts farmers in developing countries more significantly than those in developed countries.

The growth of corporate agriculture. Small family farms are slowly disappearing in the European Union, just as they did several decades ago in the United States. Now the trend is toward consolidating smaller farms into larger, more profitable operations run by corporations. These farms tend to employ very few laborers and use more machinery and chemical inputs. Even so, many small farms still exist, and the average European farm (at 45 acres) is one-tenth the size of the average U.S. farm.

> Small family farms are slowly disappearing in the EU, just as they did several decades ago in the United States.

The move toward corporate agriculture is strongest in Central Europe. When Communist governments gained power in the mid-twentieth century, they consolidated many small, privately owned farms into large collectives. After the breakup of the Soviet Union, these farms were rented to large corporations, which in turn further mechanized the farms and laid off all but a few laborers. Rural poverty rose and small towns shrank as farm workers and young people left for the cities.

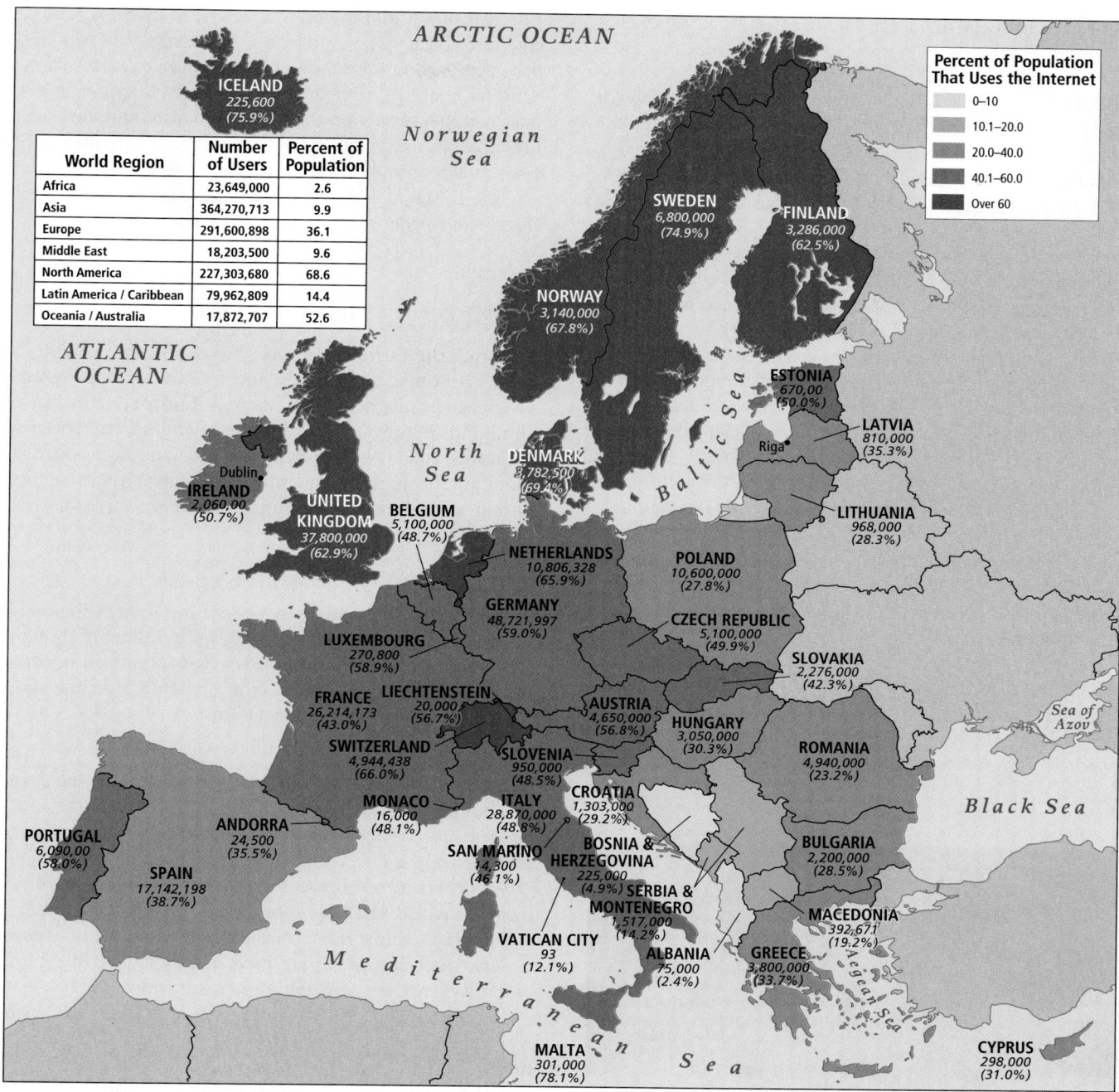

Figure 4.22 Internet use in Europe. The numbers on the map indicate the number of Internet users in the country as of 2006. The percents indicate the percentage of the country's population that uses the Internet.

Many fear that a similar pattern would sweep across the European Union without its generous subsidies to farmers.

Case study: organic agriculture in Slovenia. In Slovenia, unlike most of the rest of Central Europe, farms were not collectivized in the Communist era. Therefore, it has not been necessary to redistribute land. The problem, instead, is that farms are too small for efficient production: the average farm size is just 8.75 acres (3.5 hectares). Although it has plenty of rich farmland, Slovenia is a net importer of food, mostly from EU countries such as Italy, Spain, and Austria. Nonetheless, Slovenia's new emphasis on private entrepreneurship, combined with a growing demand throughout Europe for organic foods, has encouraged some Slovene farmers to carve out a niche for themselves in local markets. The case of Vera Kuzmic is illustrative.

Vignette Vera Kuzmic (a pseudonym) lives two hours by car south of Ljubljana, Slovenia's capital. Her family has farmed 12.5 acres (5 hectares) of fruit trees near the Croatian border for generations. After Slovenia became independent in 1991, first her husband and then Vera lost their government jobs due to economic restructuring. The Kuzmic family decided to try earning its living in vegetable market gardening because vegetable farming could be more responsive to market changes than fruit tree cultivation. By 2000, the adult children and Mr. Kuzmic were working on the land, and Vera was in charge of marketing their produce and that of neighbors she had also convinced to grow vegetables.

Vera secured market space in a suburban shopping center in Ljubljana, where she and one employee maintained a small vegetable and fruit stall (Figure 4.23). Her produce had to compete with much less expensive Italian-grown produce sold elsewhere in the same shopping center—all of it produced on large corporate farms in northern Italy and trucked in daily. But Vera gained market share by bringing her customers special orders and by guaranteeing that only animal manure, no pesticides or herbicides, was used on the fields. For a while, her special customer services and her organically grown produce kept her in business. But when Slovenia joined the European Union in 2004, more had to be done to compete with produce growers and marketers across Europe.

Anticipating the challenges to come, the Kuzmics' daughter Lili completed a marketing degree at the University of Ljubljana. The family incorporated their business, and Lili is now its Ljubljana-based director, while Vera manages the farm. Lili's market

research shows that it would be wisest to diversify. The Kuzmics continue to focus on Ljubljana's expanding professional population, whose food habits are changing and who are willing to pay extra for fine organic vegetables and fruits. But now, in a recently built banquet facility on the farm, Vera also prepares special dinners for bus-excursion groups interested in traditional Slovene dishes made from homegrown organic crops.

Source: Lydia Pulsipher's conversations with Vera Kuzmic and Dusan Kramberger, 1993 through 2006. ■

SOCIOCULTURAL ISSUES

Although the European Union was conceived primarily to promote economic cooperation and free trade, its programs have social implications as well. Across Europe, attitudes toward immigration and gender roles are changing and social welfare programs are evolving. Religion and language, once divisive issues in the region, are now fading as a focus of disputes. Immigration however, continues to be a source of tension.

Immigration: Needs and Fears

In the 1990s, the European Union and many of its neighbors approved the **Schengen Accord,** an agreement allowing free movement of people and goods across common borders. The accord has facilitated trade, employment, tourism, and, most controversially, immigration (Figure 4.24).

Attitudes toward immigrants. Immigration, legal and illegal, has had a huge impact on the EU. Many Turks and North Africans come legally as **guest workers,** who are expected to stay for only a few years, fulfilling Europe's need for temporary workers in certain sectors. Other immigrants are refugees from the world's trouble spots, such as Afghanistan, Iraq, Haiti, and Sudan. Many also come illegally from all of these areas. Today, non-Europeans are increasingly visible in schools, the workplace, sports, and religious institutions. ▄▶

Some Europeans see immigrants as important contributors to their economies, providing needed skills and making up for the low birth rates of many countries. Others, however, are alarmed by recent increases in immigration. Studies of public attitudes in Europe show that immigration is least tolerated in areas where incomes are low, suggesting that many people fear an influx of poor migrants who may drive down wages. Central and Southern Europe have the lowest incomes and are the least tolerant of new immigration. North and West Europe, which have higher incomes and generally more stable economies, are the most tolerant.

Cultural issues also influence attitudes toward immigrants. Indeed, the presence of so many new culture groups in Europe raises some thorny questions. Is Germany no longer a German place? Should Islam, the religion of many immigrants,

Figure 4.23 Vera Kuzmic in her market stall in Ljubljana, Slovenia. [Lydia Pulsipher.]

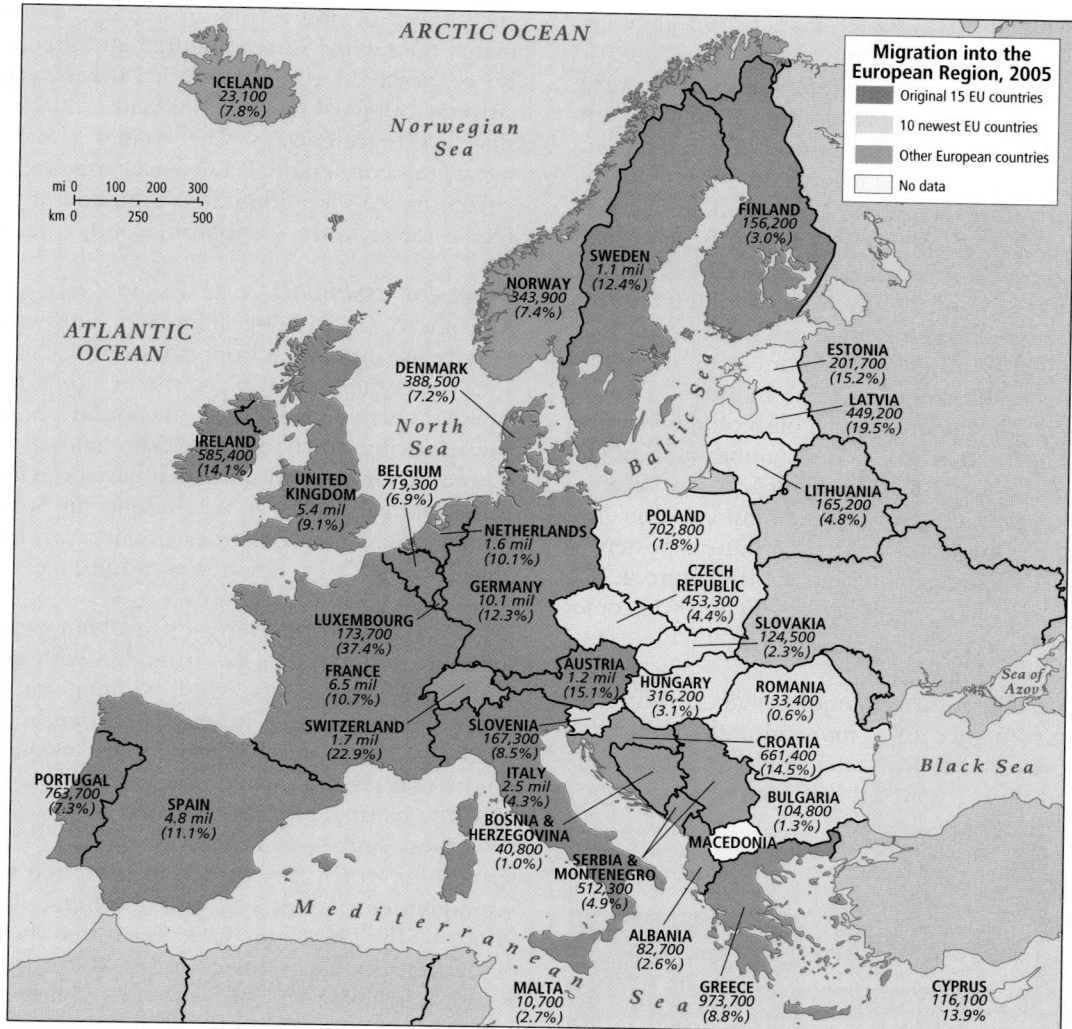

(a) Immigrants in Europe, 2005

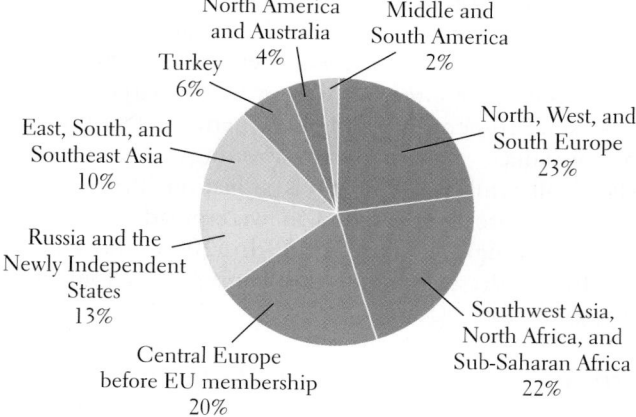

(b) Immigrant population in the 15 original EU countries, by country of origin, 2004

Figure 4.24 Migration to western Europe. Migration to western Europe increased in the 1990s and continued to increase into the 2000s, becoming a crucial issue in EU debates. The numbers in part **(a)** indicate the numbers of immigrants living in western European countries in 2005 and their percentage of the total population of each country. [Data from (a) "World migrant stock: The 2005 revision population database," United Nations Department of Economic and Social Affairs, Population Division, at http://esa.un.org/migration/p2k0data.asp; (b) Christina Boswell, "Migration in Europe," a paper prepared for the Policy Analysis and Research Programme of the Global Commission on International Migration, September 2005, p. 4, at http://www.gcim.org/attachments/RS4.pdf.]

have equal footing with Christianity? How will Europeans cope with unfamiliar value systems and family types?

Across Europe, anti-immigration views, especially toward non-Europeans, are becoming more common. Mainstream politicians increasingly support stricter control of immigration. In Germany, Austria, Italy, and France, right-wing political parties now openly advocate forcing non-European immigrants to leave. ◀ In response, the European Union is increasing its efforts to curb illegal immigration from outside Europe. EU countries are also limiting the number of migrants from other EU countries, especially those from Central Europe.

An incident in France shows how humor can diffuse some of the tensions surrounding immigration. In 2005, French nationalists complained in the press that Polish plumbers were taking French jobs. (In fact, there was a significant shortage of plumbers in France.) The "Polish Plumber" quickly became shorthand across Europe for issues related to immigration and jobs. Poland responded by featuring on its tourism posters a "hunky" male model wielding plumbing tools (Figure 4.25) and saying seductively, "I'm staying in Poland, won't you come over?"

Citizenship. Recent relaxations in requirements for citizenship suggest that acceptance of a multicultural, multiracial Europe is growing. Until recently, achieving legal citizenship in a European country was very difficult for outsiders, espe-

cially those of non-European heritage. For example, in Germany, the German-born children and grandchildren of Turkish or North African immigrant workers were not considered citizens. But as of January 2000, all children born in Germany since 1975 are citizens. The United Kingdom, probably the most multicultural of all European countries, recently granted citizenship to several hundred thousand immigrants from former colonies in the Caribbean, South Asia, and Africa.

Rules for assimilation. In Europe, race and skin color play less of a role in defining differences between people than does culture. An immigrant from Asia or Africa may be fully accepted into the community if he or she has gone through a comprehensive change of lifestyle. **Assimilation** in Europe usually means giving up the home culture and adopting the ways of the new country. Minorities that have been in Europe for thousands of years, such as the Basques in Spain and the Roma who reside in many countries, find it nearly impossible to blend into society if they retain their traditional ways. ◀

Muslims in Europe. Europe's small but growing Muslim population (Figure 4.26) has struggled with assimilation. Deepening alienation among Muslim immigrants and their children has boiled over in recent years, resulting in protests, riots, and terrorism. Some tensions relate to exclusion from employment and social services, while others revolve around the Iraq war and the cultural aspects of Islam.

Riots broke out in Paris in the fall of 2005 when young people of North African decent protested their lack of access to higher education, jobs, and housing. Subsequent investigations by the French media revealed that their complaints were indeed legitimate. However, many Europeans viewed these events as linked to deadly terrorist bombings in Madrid in 2004 and London in 2005. Both of those events were carried out by young Muslims, some of them born in Europe, protesting British and Spanish involvement in the war in Iraq.

Conflicts have also arisen over the cultural aspects of Islam. For example, in France in 2004, wearing of the *hijab* (traditional covering for women) by observant Muslim schoolgirls became the center of a national debate about civil liberties, religious freedom, and national identity. French authorities wanted to ban the *hijab* but were wary of charges of discrimination since students had long been permitted to wear crosses and yarmulkes (Jewish head covering). Eventually they declared all symbols of religious affiliation illegal in French schools, a move that resulted in widespread protest among Muslims throughout the world. In 2006, movements to ban Muslim dress arose in the United Kingdom and the Netherlands. ◀

European Ideas About Gender

Gender roles in Europe have changed significantly from the days when most women married young and worked in the home or on the family farm. Increasing numbers of European

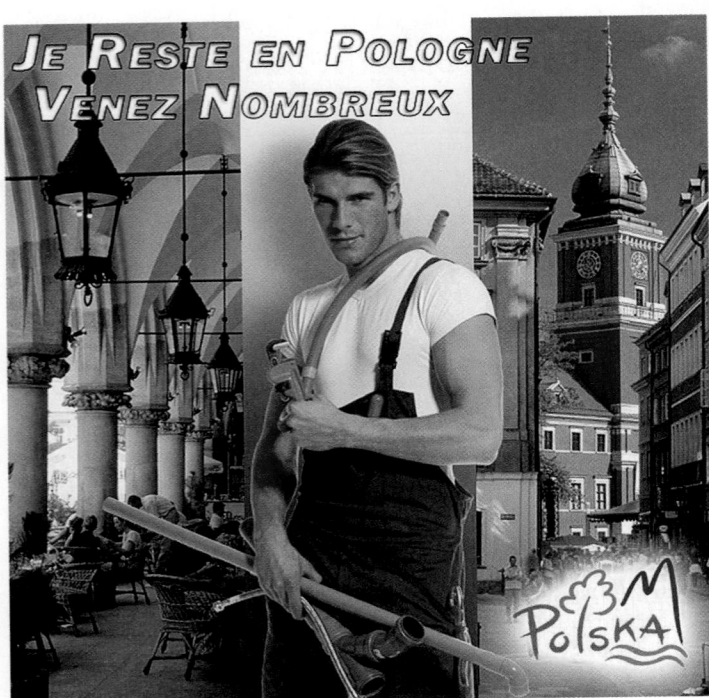

Figure 4.25 The Polish plumber. This poster, inviting the French to meet the plumber in Poland, became an emblematic reaction to anti-immigrant sentiment. [SIPA.]

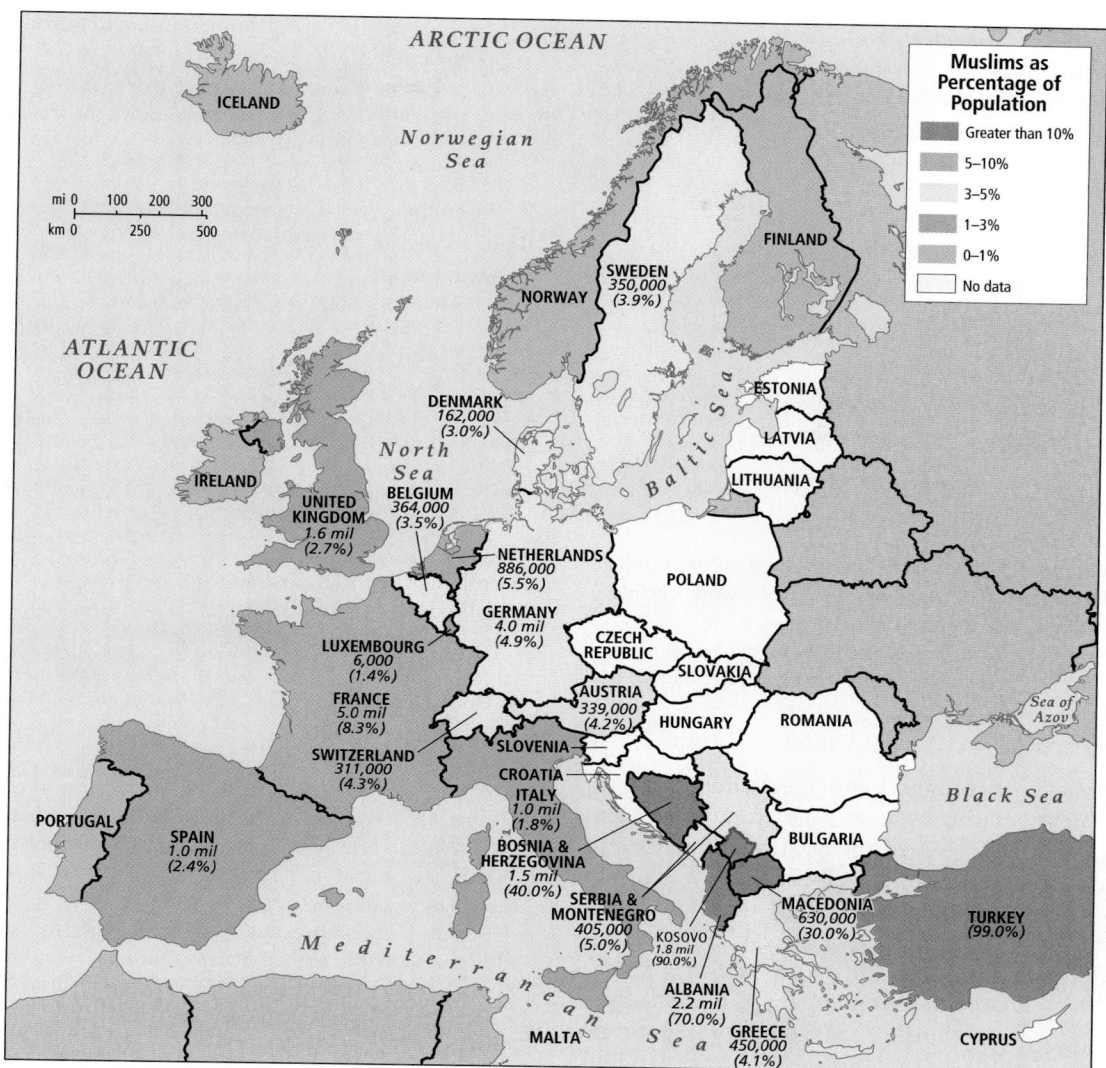

Figure 4.26 Muslims in Europe. Muslims are a tiny minority in Europe, but many intend to make the region their permanent home. Just how or if assimilation of Muslims into largely secular Europe will occur is a major topic of public debate. [Data from "Muslims in Europe," *The Times,* at http://images.thetimes.co.uk/TGD/picture/ 0,,216394,00.jpg; "Muslims in Europe: Country guide," *BBC News,* at http://news.bbc.co.uk/2/hi/europe/4385768.stm#france; and "Muslims in Europe," *Financial Times,* at http://www.ft.com/cms/s/89b5eccc-f48c-11d9-9dd1-00000e2511c8,ft_acl5s0151.html.]

women are working outside the home, though the percentages vary considerably among different parts of the region (Figure 4.27). Nevertheless, European public opinion among both women and men largely holds that women are less able than men to perform the types of work typically done by men and that men are less skilled at domestic duties. In most places, men have greater social status, hold more managerial positions, earn higher pay, and have greater autonomy in daily life (more freedom of movement, for example) than women. These male advantages have a stronger hold in Central and South Europe today than they do in West and North Europe.

Usually, women who work outside the home face what is called a **double day** in that they are expected to do most of the domestic work in the evening in addition to their job outside the home during the day. United Nations research shows that in most of Europe, women's workdays, including time spent in housework and child care, are three to five hours longer than men's. (Iceland and Sweden reported that women and men there share housework equally.) Women burdened by the double day generally operate with somewhat less efficiency in a paying job than do men. They also tend to choose employment that is closer to home and that offers more flexibility in the hours and skills required. These more flexible jobs (often erroneously classed as part-time) almost always offer lower pay and less opportunity for advancement, though not necessarily fewer working hours, than typical male jobs.

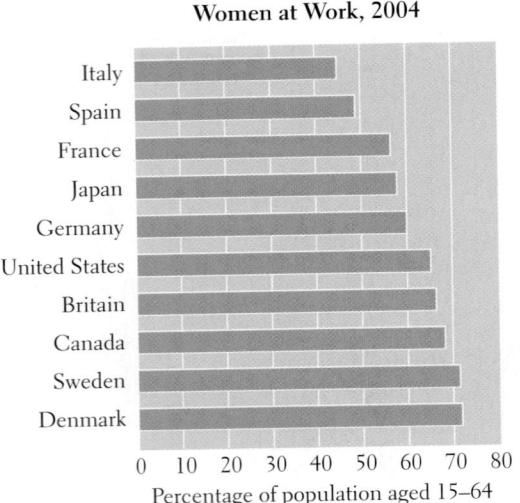

Women at Work, 2004

Percentage of population aged 15–64

Figure 4.27 Women at work in Europe (and selected other countries for comparison), 2004. A majority of women in Europe work outside the home, and their numbers are increasing. A related trend is declining birth rates in these countries. [Data from "Women and the world economy—A guide to womenomics," *The Economist*, April 12, 2006, p. 74.]

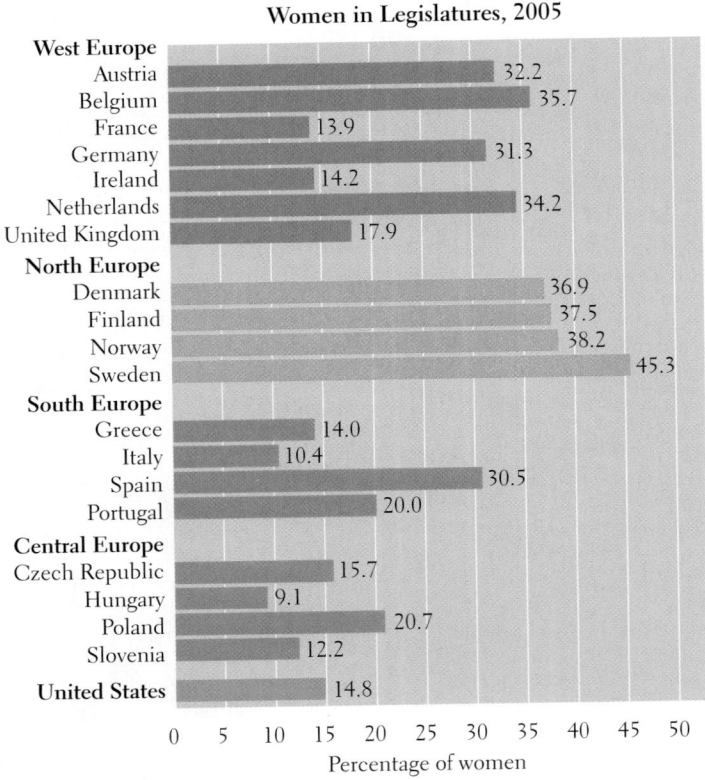

Women in Legislatures, 2005

Percentage of women

Figure 4.28 Women in the legislatures of selected countries, 2005. Women are about half the adult population of the EU, but they do not have anywhere near their fair share of representation in European legislatures; hence, their influence on legislation is seriously restricted. Notice that some regions are closer to equity than others. Also note how the United States compares. [Data from *United Nations Human Development Report 2005* (New York: United Nations Development Programme), Table 26.]

Many EU policies encourage gender equality, and well over half the university graduates in Europe are now women. However, the political influence and economic well-being of European women lag behind those of European men. As Figure 4.28 illustrates, in most European national parliaments, women make up less than a third of elected representatives. Only in North Europe do women come anywhere close to filling 50 percent of the seats in the legislature. Although Germany elected Angele Merkel in 2005 as its first woman chancellor (prime minister), women generally serve in the lower ranks of government bureaucracies. Such positions give them little voice in the formation of national policies.

> The political influence and economic well-being of European women lag behind those of European men.

Because women are largely absent from policy-making positions, their progress has been slow on many fronts. For example, in 2006, female unemployment was higher in every EU country than male unemployment, and 32 percent of women's jobs were part time, as opposed to only 7 percent of men's jobs. Throughout the European Union, women are paid on average 15 percent less than men for equal work, and young women, who tend to be more highly educated than young men, have higher unemployment rates.

Social Welfare Systems

In many European countries, elaborate, tax-supported systems of **social welfare** provide all citizens with health care, free or low-cost higher education, affordable housing, disability payments, and generous unemployment and pension benefits.

Europeans generally pay much higher taxes than North Americans, and in return they expect more.

Europeans do not agree on the goals of these welfare systems, or on just how generous they should be. Some argue that Europe can no longer afford high taxes if it is to remain competitive in the global market. Others maintain that Europe's economic success and high standards of living are the direct result of the social contract to take care of basic human needs for all. The debate has been resolved differently in different parts of Europe, and the resulting regional differences have become a source of worry in the European Union. With open borders, unequal benefits might encourage those in need to flock to a country with a generous welfare system and overburden the taxpayers there. European welfare systems can be classified into four basic categories (Figure 4.29):

1. Social democratic welfare systems, common in Scandinavia, are the most generous systems. They attempt to achieve equality across gender and class lines by providing extensive health care, education, housing, and child and elder care benefits to all citizens from cradle to grave. Child

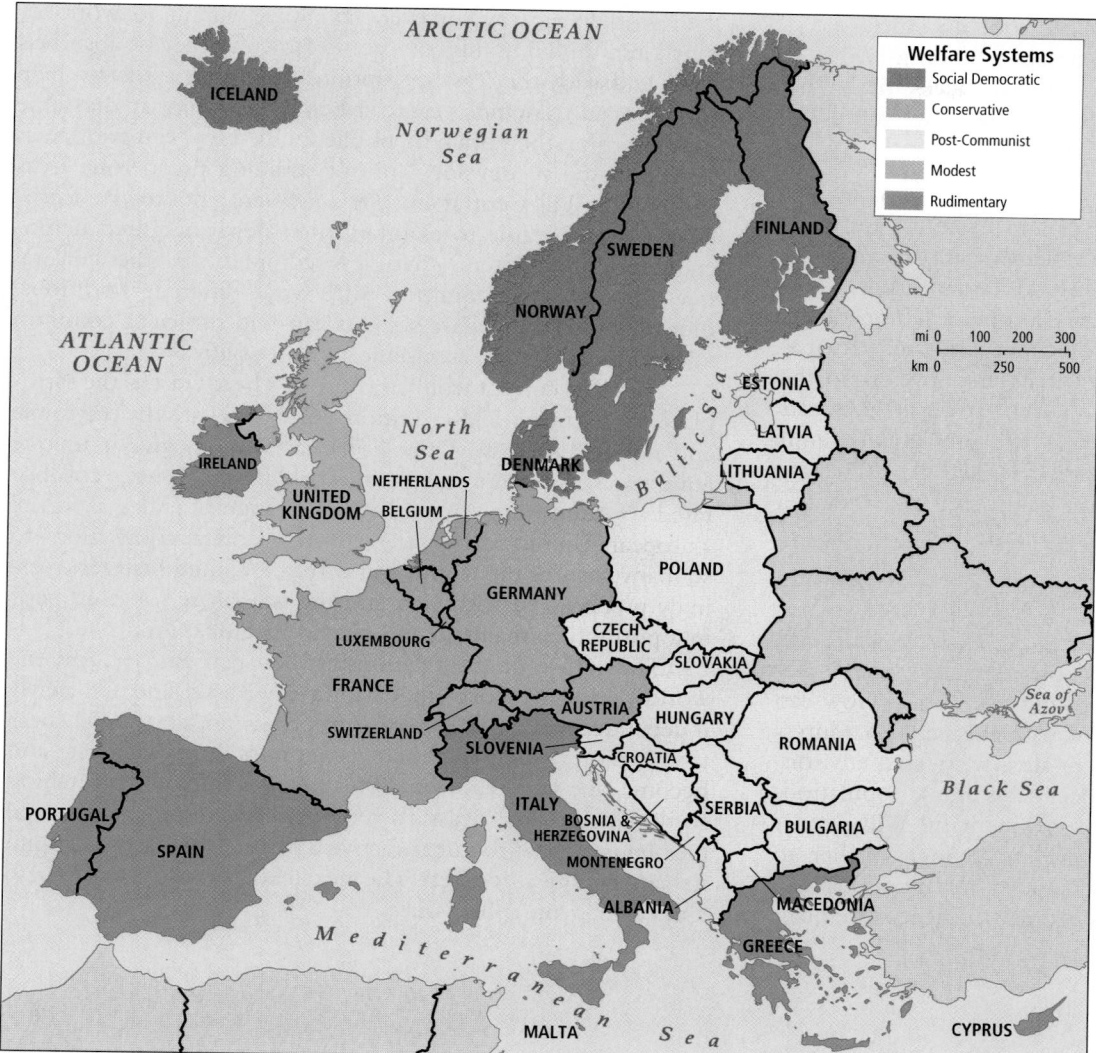

Figure 4.29 European social welfare systems. The basic categories of social welfare systems shown here and described in the text should be taken as only an informed approximation of the existing patterns.

care is widely available, in part to help women enter the labor market. But early childhood training, a key feature of this system, is also meant to ensure that in adulthood every citizen will be able to contribute to his or her highest capabilities and that criminal behavior or drug abuse will not develop. Although gender equality is a stated goal, traditional gender roles are officially emphasized throughout this lifelong support system.

2. Conservative and modest welfare systems seek to provide a minimum standard of living for all citizens and are common in the countries of West Europe. The state assists those in need, but does not see assisting upward mobility as its mission. For example, college education is free or heavily subsidized for all, but strict entrance requirements in some disciplines can be hard for the poor to meet. State-supported health care and retirement pensions are available to all, but these systems reinforce the traditional "housewife contract" by assuming that women will stay home and take care of children, the sick, and the elderly. The UK system is

considered slightly less generous than those found elsewhere in West Europe, so it is described in Figure 4.29 as "modest."

3. Rudimentary welfare systems do not accept the idea that citizens have inherent rights to government-sponsored support. They are found primarily in South Europe and in Ireland. Here, local governments provide some services or income for those in need, but the availability of such services varies widely, even within a country. The state assumes that when people are in need, their relatives and friends will provide the necessary support. The state also assumes that women work only part time, and hence are available to provide child care and other social services for free.

4. Post-Communist welfare systems prevail in the countries of eastern and southeastern Europe. During the Communist era, these systems were comprehensive, resembling the cradle-to-grave social democratic system in Scandinavia, except that women were pressured to work outside the home. Benefits often extended to nearly free apartments, health care, and subsidized food and fuel. However, in the post-Communist era state funding has collapsed, forcing many to do without basic necessities. Hence, those with skills often try to migrate to western Europe, where many basics are still provided.

As the European Union evolves and expands, Europe's social welfare systems will probably become more similar to one another, but it is still unclear just which models will prevail. Increasing awareness across the European Union of the link between welfare provisions, women's economic participation, and general economic growth may lend support to the more comprehensive Scandinavian model. However, the costs of welfare policies will be a factor, as will pressures to limit benefits to immigrants. Nevertheless, it is likely that the state provision of health care and a social safety net will remain a feature of life in Europe for decades to come.

Reflections on Europe

Many of the issues confronting people in Europe are similar to those encountered elsewhere around the world. How can economic development be balanced with the need to address global warming and the problems of already stressed environments? Will Europe's economies decline as aging populations strain the ability to maintain high productivity? Will gender roles and support for childcare have to change further to encourage higher population growth? How can a country create and retain satisfying jobs for its people when there are qualified workers nearby, or even far away, willing to work for much less? And if immigrants come, how can they be absorbed fairly and with the least disruption? How can a society help the poor and unemployed to a better life while at the same time lessening their drain on public funds? How can poor areas be stimulated to develop without straining prosperous areas too much? Will Central Europe's new entrants to the European Union be able to maintain their newly acquired democratic political systems? No less complex are the cultural questions of how countries with very different traditions, mores, and social welfare systems can find sufficient common ground to cooperate economically and politically.

For all its current troubles, Europe's head start as the birthplace of the Industrial Revolution, combined with the economic and political success of the European Union, give it unique advantages. Nowhere else in the world have countries collaborated so extensively on such a broad range of issues as in the European Union. Moreover, with its historical connections to so many parts of the world, and with the United States increasingly preoccupied with Iraq and the war on terror, Europe is in a position to maintain and extend its global influence.

The future of the relationship between Europe and the world region we cover in Chapter 5—Russia and the newly independent states of eastern Europe—is not clear. Will these two regions integrate their economies and societies and become one large Europe, 800 million strong and stretching to the Pacific? Will Russia's very serious economic and social problems make it too unattractive a partner for Europe despite its vast energy resources? The next chapter will discuss these and other geographic issues.

Chapter Key Terms

assimilation 166
centrally planned economies 153
cold war 153
cultural homogenization 136
deindustrialization 159
double day 167
dumping 162
economies of scale 158
ethnic cleansing 159
euro 159
European Union (EU) 136
feudalism 149

Green 142
guest workers 164
Holocaust 153
humanism 150
humid continental climate 141
iron curtain 153
medieval period 145
Mediterranean climate 141
mercantilism 151
nationalism 152
nobility 149

North Atlantic Treaty Organization (NATO) 159
Protestant Reformation 150
Roma 153
Scandinavia 140
Schengen Accord 164
serfs 149
social welfare 168
subsidies 162
temperate midlatitude climate 140
welfare states 152
world cities 151

Critical Thinking Questions

1. How do the issues of immigration in Europe compare with those in the United States? If you were a poor, undocumented immigrant searching for a way to support your family, would you choose the United States or Europe as a possible destination? Why?

2. How was Europe transformed by its colonial empires? How is Europe still linked to its former colonies?

3. How have urban areas been central to changes in Europe before and after the Industrial Revolution?

4. What are some of the potential consequences of negative population growth rates in Europe? Why might an understanding of a region's population age structure affect immigration policy?

5. How might the European Union's status as the world's largest economy, and the rising importance of NATO, lead to changes in global power relationships?

6. What does the evolution of democracy in Europe suggest about current efforts to establish democracy in Iraq?

7. Why is the European Union's generous support for European farmers criticized by developing countries?

8. What state-supported welfare systems have offered women the greatest opportunities for employment outside the home?

9. What is the evidence that European ways of life contribute less to global warming than North American ways?

10. What could the European Union do to reduce its contribution to pollution in Europe's seas? What other efforts might be effective at addressing this problem?

Greenland

North Pole 90°N

ARCTIC OCEAN

Svalbard

Franz Josef Land

Barents Sea

North Land

Norwegian Sea

Arctic Circle

70°N

80°N

60°N

Manchester
Birmingham
London
UNITED KINGDOM

NORTH SEA

FRANCE
BELGIUM
Brussels
Amsterdam
NETHERLANDS
LUX
Luxembourg
Hamburg
GERMANY
Berlin
Rhine
Prague
CZECH REP.
Vienna
POLAND
Bratislava
SLOVAKIA
Budapest
HUNGARY
Danube
ROMANIA
MOLDOVA
Chisinau

NORWAY
Oslo
DENMARK
Copenhagen
SWEDEN
Stockholm
Gulf of Bothnia
Baltic Sea
Kaliningrad
Warsaw
LITHUANIA
Vilnius
LATVIA
Riga
ESTONIA
Tallinn
FINLAND
Helsinki
Gulf of Finland
Gulf of Riga

Vistula
Carpathian Mts.
Lviv
BELARUS
Minsk
Brest
UKRAINE
Kiev
Chernobyl
Homyel
Smolensk

Murmansk
Kola Peninsula
Kirovsk
White Sea
Arkhangelsk
Lake Ladoga
Lake Onega
St. Petersburg
North European Plain
Tver
Moscow
Yaroslavl
Divina
Pechora

Novaya Zemlya

Kara Sea

Kara Strait

Dikson

Taymyr Peninsula

Lena R. De

Laptev Sea

North Siberian Lowland

Yamal Peninsula

Vorkuta
Salekhard
Pechora Basin
Pechora
Norilsk
Olenek

Central

S i b e

Siberian

Plateau

Vilyuy Res.

Odesa
Mykolayiv
Kherson
Dnipropetrovsk
Donetsk
Sevastopol
Sea of Azov
Black Sea
Rostov-na-Donu
Volgograd

Bryansk
Tula
Kursk
Lipetsk
Voronezh
Penza
Saratov
Volga
Ivanovo
Nizhniy Novgorod
Kazan
Kama
Perm
Samara
Tolyatti
Ufa
Orenburg
Orsk

Kirov

Nizhniy Tagil
Yekaterinburg
Magnitogorsk
Chelyabinsk
Tyumen

Ural Mountains

West Siberian Plain

Urengoy

Surgut
Ob

RUSSIAN FEDERATION
(RUSSIA)

Tunguska Basin

Yenisey
Seversk
Tomsk
Omsk
Irtysh
Novosibirsk
Novokuznetsk
Krasnoyarsk
Gladkaya
Tayshet
Bratsk
Bratsk Res.
Angara
E. Sayan Mts.
Irkutsk Basin
Angarsk
Irkutsk
Lake Baikal
Yabl
Chita
Ulan-Ude
Kyzyl
W. Sayan Mts.

Caspian Depression
Astrakhan
Caspian Sea
Caucasus Mts.
GEORGIA
Baturi
Tbilisi
Groznyy
TURKEY
ARMENIA
Yerevan
AZER
AZERBAIJAN
Baku
Lake Orma
IRAQ
IRAN
Tehran
Elburz Mts.
Zagros Mts.
Esfahan
Mashhad

Steppes

Aral Sea

Aktogay
Leninsk
Astana

KAZAKHSTAN
Qaraghandy

Lake Balkhash

Syr Darya
Amu Darya
TURKMENISTAN
Ashkhabad
UZBEKISTAN
Samarkand
Tashkent
Dushanbe
TAJIKISTAN
Hindu Kush
Pamirs
Shymkent
Bishkek
KYRGYZSTAN
Almaty
Tien Shan
Tarim
Junggar Basin
Urumqi

MONGOLIA
Ulan Bator
Altai Mts.

Gol

Baot

Persian Gulf
QATAR
Ad Dawhah (Doha)
Dubayy
Abu Zaby
UAE
Muscat
OMAN

AFGHANISTAN
Kabul
Islamabad
PAKISTAN
Lahore
Faisalabad
INDIA

Tarim Basin
Taklimakan Desert

C H I N A

Lanzhou

Land Elevations

meters	feet
4877	16,000
3353	11,000
2134	7000
914	3000
305	1000
152	500
0	0

1:26,000,000
Azimuthal Equidistant Projection

[Map of northeastern Russia, Japan, Korea, and northeastern China with labeled features including East Siberian Sea, Chukchi Sea, Bering Sea, Sea of Okhotsk, Kamchatka Peninsula, Sakhalin, Kolyma Range, Cherskiy Range, Verkhoyansk Range, Stanovoy Range, Sikhote Alin, Sea of Japan, and various cities.]

CHAPTER 5

Russia and the Newly Independent States

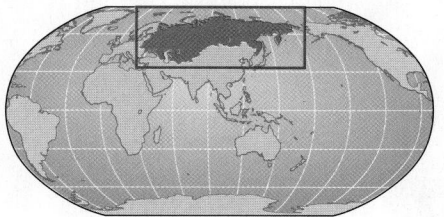

Global Patterns, Local Lives When a coffin arrived at the head offices of the Moscow catering firm Na Ilyinke bearing the name of the still quite lively Alexei Likhachev, director of the company, he sensed it was not a harmless prank. Soon the phones began ringing with condolence messages from shareholders who had received invitations to his memorial service. Likhachev and his partners knew immediately what the real message was: "Sell or else!"

Low-profile, medium-sized companies like Na Ilyinke have become the focus of mafia-style corporate raiders. Intimidation, corrupt judges, bureaucrats, and police are used to frighten business owners into selling their successful firms at prices that are less than half their true value. Often it is the real estate that the raiders want. Na Ilyinke occupies prime property in the heart of Moscow. Shortly after the coffin arrived, the partners were notified that their 58 percent share of the company had been sold, via a forged power of attorney, first to a woman in Ukraine and then to a man in New York. The Na Ilyinke partners persevered in getting the government to investigate the fraud and received a favorable ruling. The raiders' assaults continued, however, in the form of three judgments returned against the company in court proceedings. The partners were not notified of the hearing and so had not been there to defend themselves.

Unsure that official law enforcement agencies will help them, the Na Ilyinke partners now hope to fend off a physical takeover of their property with armed guards, an alarm system, razor wire, and steel doors. As one partner puts it, "If you lose physical possession of your property, you are in serious trouble. So far we've

Figure 5.1 Regional map of Russia and the Newly Independent States.

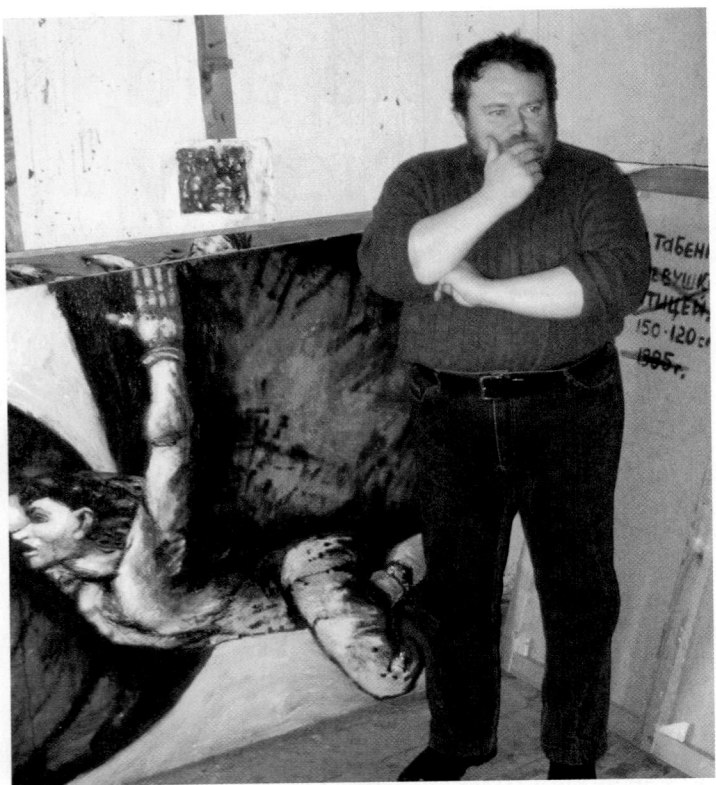

Figure 5.2 Forced takeovers in Russia. Lev Tabenkin, a Russian painter, stands in front of some of his canvases at a temporary studio. He lost possession of his old studio after the ownership was transferred without his knowledge. Tabenkin and the Union of Artists are battling in the courts to reclaim a number of studios at a Moscow artists' cooperative. [Peter Finn/*The Washington Post*.]

kept them out." Other businesspeople, and even artists, have not been so lucky (Figure 5.2).

Adapted from Peter Finn, "Hostile takeovers: Russian financial predators use legal tactics to seize prized real estate," The Washington Post National Weekly Edition (April 24–30, 2006): 18–19. ∎

Na Ilyinke operates in a region that has changed its political and economic systems entirely in just a few short years. Barely two decades ago, the *Union of Soviet Socialist Republics* (USSR), more commonly known as the **Soviet Union,** was the largest political unit on earth, stretching from eastern Europe to the Pacific Ocean. It covered one-sixth of Earth's land surface. In 1991, the Soviet Union broke apart, ending a 70-year experiment with nearly complete government control of the economy, society and politics. (This history will be discussed in more detail later in the chapter.) Over the course of a few years, an economy controlled by government bureaucrats was replaced by a capitalist system similar to that of the United States, based on competition among private businesses.

The Soviet Union has been replaced by a loose alliance of Russia and 11 newly independent states—Ukraine, Belarus,

Moldova, the Caucasian republics of Georgia, Armenia, Azerbaijan, and the Central Asian republics of Kazakhstan, Kyrgyzstan, Tajikistan, Turkmenistan, and Uzbekistan (Figure 5.3). Russia, which was always the core of the Soviet Union, remains predominant in the region and the world because of its size, population, military, and huge oil and gas reserves.

Today, this region faces many challenges as it struggles to adjust to market-based economies, globalization, a shift toward democracy, global warming, and some of the worst pollution on the planet. The economic transition has caused great hardship and anxiety. Many have suffered dramatic drops in income and health as formerly state-run industries have closed, resulting in lost jobs and social services. Alcoholism and violent crime are widespread. On the other hand, many people are exhilarated by their new opportunities. The changes offer them greater freedom of expression and movement, and they also offer consumer products in far greater variety and quality than ever before.

Geopolitically, this region is going through major changes as well. The cold war between the Soviet Union and the United States and its allies is over. Some former Soviet allies in Central Europe have already joined the European Union, and some western parts of this region may eventually do the same. Meanwhile, the Central Asian republics currently allied with Russia may align themselves with neighbors in Southwest or South Asia. The far eastern parts of Russia are already finding common trading ground with East Asia and Oceania. The map in Figure 5.4 shows these potential regional realignments.

Questions to consider as you study Russia and the Newly Independent States (Also see Views of Russia and the Newly Independent States, pages 176–177)

1. Why does this region have such severe environmental problems? Water and air pollution, especially in urban and

Figure 5.3 Political map of Russia and the Newly Independent States.

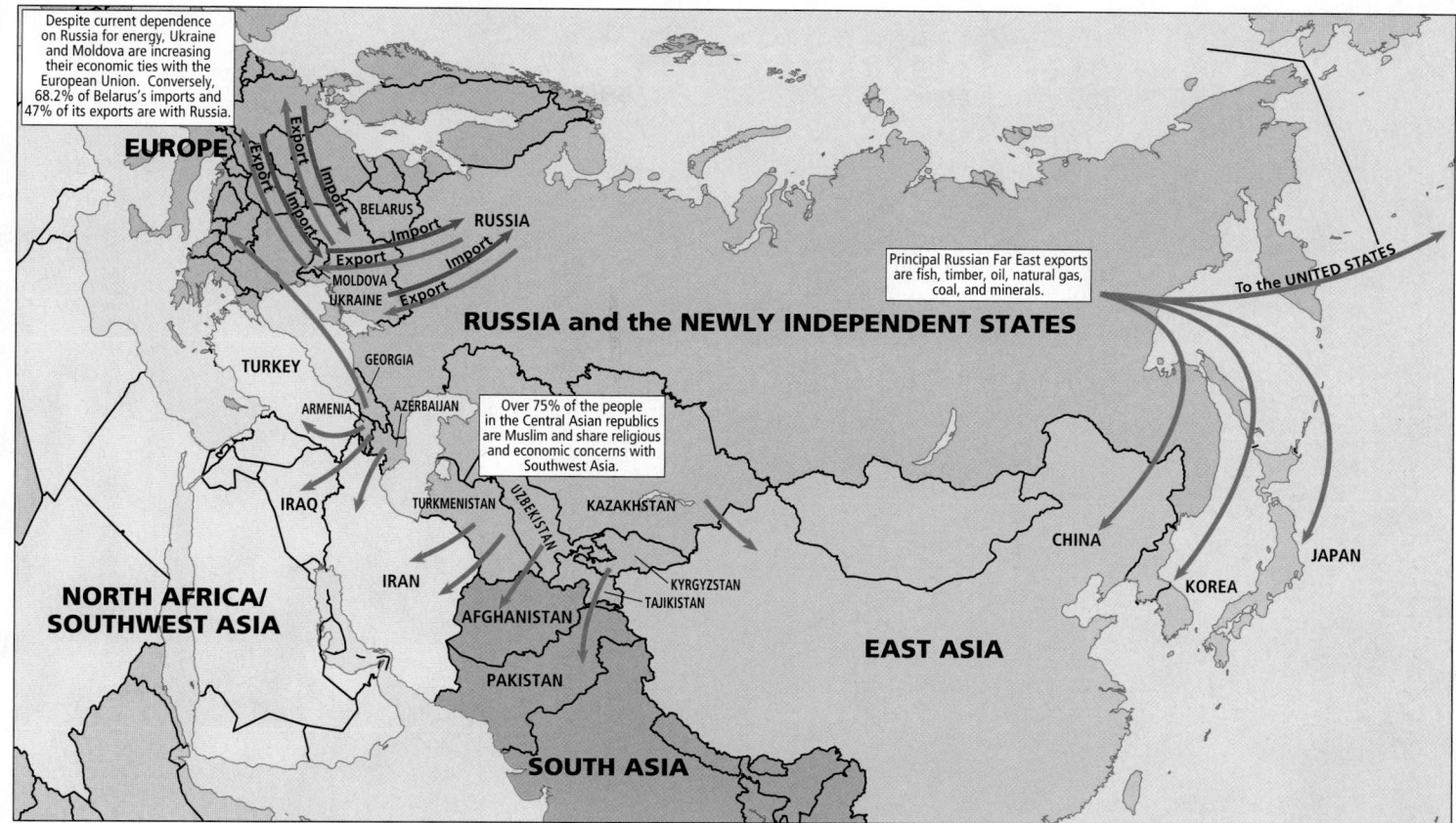

Despite current dependence on Russia for energy, Ukraine and Moldova are increasing their economic ties with the European Union. Conversely, 68.2% of Belarus's imports and 47% of its exports are with Russia.

EUROPE

BELARUS
RUSSIA
Export
Import
Export
Import
MOLDOVA
UKRAINE
Export

TURKEY
GEORGIA
ARMENIA
AZERBAIJAN

RUSSIA and the NEWLY INDEPENDENT STATES

Principal Russian Far East exports are fish, timber, oil, natural gas, coal, and minerals.

To the UNITED STATES

Over 75% of the people in the Central Asian republics are Muslim and share religious and economic concerns with Southwest Asia.

IRAQ
TURKMENISTAN
UZBEKISTAN
KAZAKHSTAN

CHINA

JAPAN

IRAN
KYRGYZSTAN
TAJIKISTAN

KOREA

NORTH AFRICA/
SOUTHWEST ASIA

AFGHANISTAN

EAST ASIA

PAKISTAN

SOUTH ASIA

Figure 5.4 Russia and the Newly Independent States: Contacts with other world regions. The region of Russia and the Newly Independent States is in flux as all of the component countries rethink their geopolitical positions relative to one another and to adjacent regions. The arrows indicate various types of contact, ranging from economic to religious and cultural.

industrial areas, are some of the worst in the world. Global warming is also a rising concern, with Russia a major contributor of greenhouse gases. Parts of this region are highly vulnerable to the changes that global warming may bring. How did these problems develop over time? How might solutions be found?

2. What changes have economic reform and globalization brought to this region since the fall of the Soviet Union? New global investment and exports of fossil fuels have made some people rich, but millions have lost jobs and access to health care. Will new global connections lead to lasting economic stability? Will the differences between rich and poor continue to grow?

3. Is democracy emerging? Regular and orderly elections are increasingly the norm, but economic and political instability have led many to support strong leaders who tolerate little criticism. Meanwhile, crime and corruption have exploded. Are these new democracies strong enough to endure the challenges they face?

4. Why are populations shrinking in some places and growing in others? Russia and countries bordering Europe have the most rapidly declining populations on the planet. Meanwhile, the Central Asian and Caucasian countries are growing. How might different levels of pollution, recent societal transformations, culture, and other factors explain these differences?

5. How have the massive economic and political changes in this region affected men and women differently? The post-Soviet era has seen increasing gender gaps in employment and political empowerment. Why have these differences emerged? Who has been affected the most?

6. How are identities changing in the Post-Soviet era? Regional identities, religious affiliations, and the political empowerment of women and ethnic minorities have shifted, sometimes dramatically. What are the forces driving these changes in the ways people see themselves?

Views of Russia and the Newly Independent States

For a tour of these places, visit www.whfreeman.com/pulsipher

[Background image: Lori K. Perkins NASA/GSFC Code 610.3/587 Scientific Visualization Studio]

2. Globalization. Natural gas exports to Europe are financing Russia's economic recovery and accelerating globalization throughout the region. Moscow is now sprouting glittering new skyscrapers, shown below, thanks to a boom in foreign investment. See page 194 for more. Lat 55°45'N, Lon 37°32'19"E. [BOLSCHOI]

1. Food. Agriculture is recovering after a severe decline during the 1990s brought on by the collapse of the Soviet Union. This grape grower in Ukraine's Crimean peninsula benefits from a moist temperate climate that is rare in this region of long harsh winters. See pages 194–197 for more. Lat 44°34'N, Lon 33°42'E. [USAID]

2&3
Moscow

1
Crimea

7
Georgia

5
Azerbaij

7. Democratization. Unfair elections and authoritarian leaders have slowed the pace of democratization in this region. However, in Georgia a youth movement known as "Enough!" (written above in Georgian on the woman's bandana) fiercely protested unfair elections and brought a new, reform minded regime to power. See pages 197–198 for more. Lat 41°41'36"N, Lon 44°48'05"E. [USAID: USAID]

3. Urbanization. Cities grew rapidly during the Soviet era and without much concern for the environment. Today this region has some of the worst urban and industrial pollution in the world. Shown below are the cooling towers of a large gas-fired powerplant in Moscow located near residential apartment blocks. See pages 180–182 for more. Lat 55°53'50"N, Lon 37°30'54"E. [Sergei Rubliov]

4. Gender. Women were strongly encouraged to work outside the home during the soviet era. Shown below is a postage stamp commemorating the 1963 voyage, beginning in Baikonur Kazakhstan, of Valentina Tereshkova, the first woman to fly in space. The post-Soviet era has brought higher unemployment and reduced political empowerment, but also new opportunities for women. See pages 201–202 for more. Lat 45°55'13"N, Lon 63°20'32"E.

ПЕРВАЯ В МИРЕ
ЖЕНЩИНА-КОСМОНАВТ
ВАЛЕНТИНА
ТЕРЕШКОВА
10к ПОЧТА СССР

5. Population. Populations are declining in some countries and rising in others. The post-Soviet era brought higher death rates to Russia, Belarus, and Ukraine. However, death rates remained low and birthrates stayed high in Caucasia. These children live in Azerbaijan. See pages 188–190 for more. Lat 40°21'58"N, Lon 49°50'E. [Thatcher Cook/Mercy Corps/USAID]

6. Global warming and water. Central Asia's irrigated agriculture systems are dependent on water melting off of glaciers high in the mountains of Tajikistan (shown here) and Kyrgyzstan. Higher temperatures are now shrinking these glaciers, threatening the agricultural areas that depend on them. See page 184 for more. Lat 38°46'37"N, Lon 72°15'24"E. [Simon Garbutt; NASA]

4
Baikonur

6 Tajikistan

The Geographic Setting

Terms to Be Aware Of

There is no entirely satisfactory new name for the former Soviet Union. The formal name used in this chapter is *Russia and the Newly Independent States*. Russia is still closely associated with all of these states economically, but they are independent countries, and their governments are legally separate from Russia's. Russia itself is formally known as the **Russian Federation** because it includes more than 30 (mostly) ethnic *internal republics*—such places as Chechnya, Tatarstan, and Tuva—which constitute about one-tenth of its territory and one-sixth of its population.

PHYSICAL PATTERNS

The physical features of Russia and the Newly Independent States vary greatly over the huge territory they encompass. The region bears some resemblance to North America in size, topography, climate, and vegetation. In fact, Russia is the largest country in the world, nearly twice the size of the second largest country, Canada.

Landforms

Because this is such a physically complex region, a brief summary of its landforms is useful (see Figure 5.1). Moving west to east, there is first the eastern extension of the North European Plain, then the Ural Mountains, then the West Siberian Plain, followed by an upland zone called the Central Siberian Plateau, and finally, in the far east, a mountainous zone bordering the Pacific. To the south of these territories from west to east is an irregular border of mountains (the Caucasus), semiarid grasslands, or **steppes** (in western Central Asia), and barren uplands and high mountains (in eastern Central Asia).

The eastern extension of the North European Plain rolls low and flat from the Carpathian Mountains in Ukraine and Romania 1,200 miles east (about 2,000 kilometers) to the Ural Mountains. This part of the region is often called *European Russia* because the Ural Mountains are traditionally considered part of the unclear border between Europe and Asia. European Russia is the most densely settled part of the entire region and is its agricultural and industrial core. Its most important river is the Volga, which flows into the Caspian Sea. The Volga River is a major transport route connecting many parts of the North European Plain to the Baltic and White seas in the north and to the Black Sea in the southwest.

The Ural Mountains extend in a fairly straight line south from the Arctic Ocean into Kazakhstan. The Urals are not much of a barrier to humans or to nature. There are several easy passes across the mountains, and winds carry moisture all the way from the Atlantic and Baltic across the Urals and into Siberia. Much of the Urals' once-dense forest has been felled to build and fuel new industrial cities.

The West Siberian Plain, lying east of the Urals, is the largest plain in the world. A vast, mostly marshy lowland about the size of the eastern United States, it is drained by the Ob River and its tributaries, which flow north into the Arctic Ocean. Long, bitter winters mean that in the northern half of this area a layer of permanently frozen soil (**permafrost**) lies just a few feet beneath the surface. In the far north, the permafrost comes to within a few inches of the surface. Because water doesn't sink through this layer, swamps and wetlands form in the summer months, providing habitats for many migratory birds. In the far north lies the **tundra,** where only mosses and lichens can grow due to extreme cold and shallow permafrost. The West Siberian Plain has some of the world's largest reserves of oil and natural gas, although their extraction is made difficult by the harsh climate and the permafrost.

The Central Siberian Plateau and the Pacific mountain zone farther to the east together equal the size of the United States. Permafrost prevails except along the Pacific coast. There, the ocean moderates temperatures, and additional heat is supplied by the many active volcanoes created as the Pacific Plate sinks under the Eurasian Plate. Places warmed by these forces are the Kamchatka Peninsula (Figure 5.5), Sakhalin Island, and Sikhote-Alin on Russia's southeastern coast. Only lightly populated, these places are havens for wildlife.

To the south of the West Siberian Plain is an irregular band of steppes and deserts stretching from the Caspian Sea to the mountains bordering China. Farther to the south and east of these grasslands is a wide, curving band of mountains, including the Caucasus, Elburz, Hindu Kush, Tien Shan, and Altai. Their rugged terrain has not deterred people from crossing these mountains and exchanging plants, animals, technologies, and religious belief systems such as Islam and Buddhism.

Climate and Vegetation

No inhabited place on earth has as harsh a climate as the northern part of the Eurasian landmass occupied by Russia (Figure 5.6). The winters are long and cold, with only brief hours of daylight. Summers are short and cool to hot, with long days. The short growing season generally lowers agricultural production.

Most rainfall in the region comes from storms that blow in from the Atlantic Ocean far to the west. By the time these initially rain-bearing air masses arrive, most of their moisture has been squeezed out over Europe. A fair amount of rain does reach Ukraine, European Russia, and the Caucasian republics.

East of the Urals, the lands of Siberia receive moderate precipitation (primarily from the east) but experience long, cold winters. Huge expanses of Siberia are covered with **taiga** (northern coniferous forest), which stretches to the Pacific.

Figure 5.5 Kamchatka. In autumn, the stone birch, alder, and elfin wood in Kamchatka's Nalychevo Nature Park erupt in a spectacular display. The Nalychevo River valley is surrounded by snow-capped volcanoes, and there are hot and cold mineral springs in the upper reaches of the river. In 1996, the park was placed on UNESCO's World Heritage List. [© Christian Gluckman/http://www.edouardas.com.]

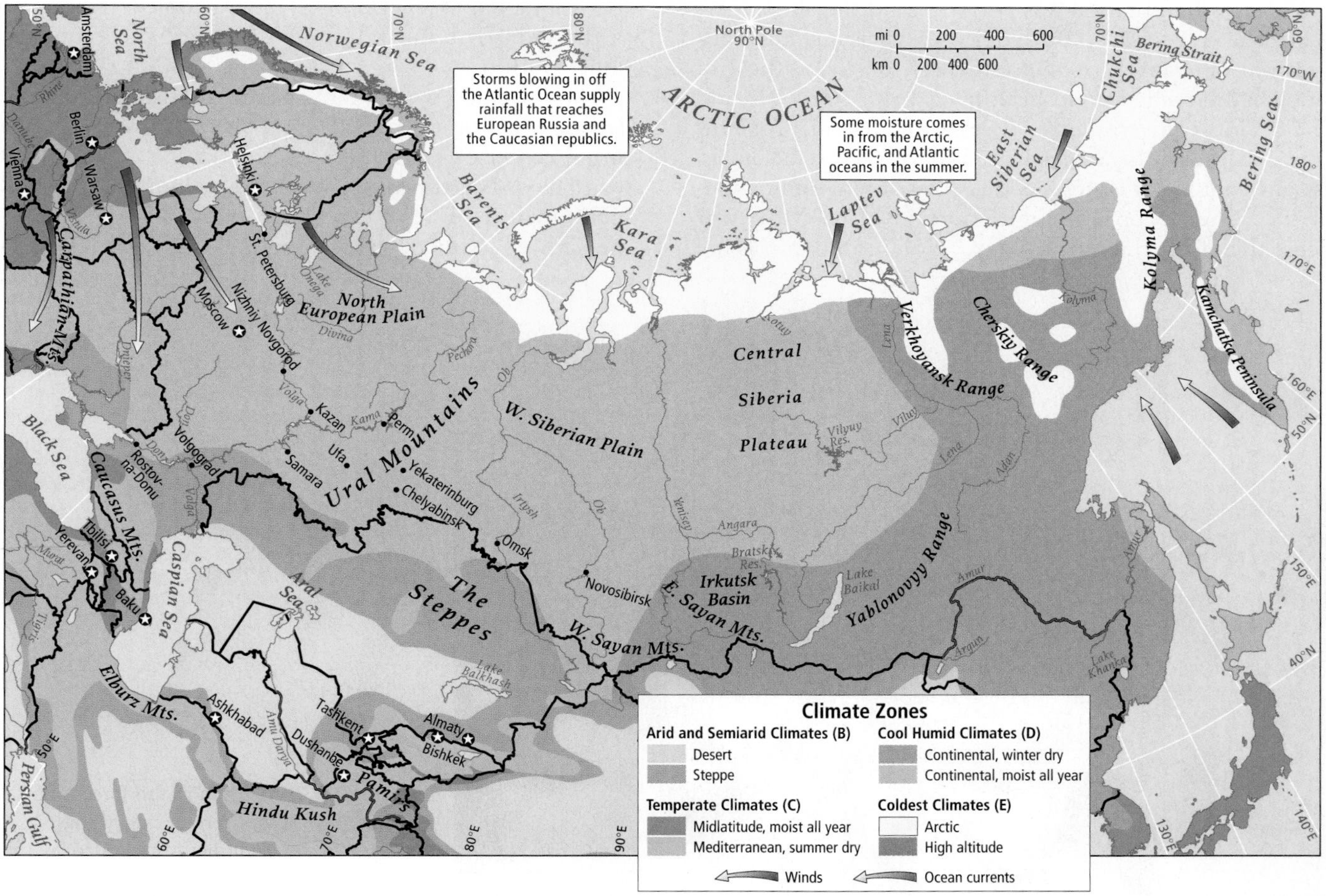

Figure 5.6 Climates of Russia and the Newly Independent States.

Agriculture is generally not possible, though reindeer are tended in the tundra of the far north.

East of the Caucasus Mountains, the lands of Central Asia have semiarid to arid climates influenced by their location in the middle of a very large continent. The summers are scorching and short, the winters intense. The more southern areas support grasslands, which are used for herding and agriculture where irrigation is possible.

ENVIRONMENTAL ISSUES

"We cannot expect charity from Nature. We must tear it from her."
—Joseph Stalin, General Secretary of the Communist Party of the Soviet Union 1922–1953

Soviet ideology held that nature was the servant of industrial and agricultural progress, and that humans should dominate nature on a grand scale. Hence, huge dams, factories, and other industrial facilities were built without regard to their effect on the environment or public health. Now Russia and the Newly Independent States have some of the worst environmental problems in the world. By 2000, more than 35 million people in the region (15 percent of the population) were living in areas where the soil was poisoned and the air dangerous to breathe (Figure 5.7). Birth defects such as missing limbs or hands are rampant. By some estimates, only one-third of all schoolchildren enjoy good health.

The region's governments, beset with myriad problems since the collapse of the Soviet Union, have been reluctant to address environmental issues. As one Russian environmentalist put it, "When people become more involved with their stomachs, they forget about ecology." Pollution controls are complicated by a lack of funds to correct even a few of the past environmental abuses. Figure 5.8 shows changes in human impacts on the region's environment over time. Notice that by 2002, environmental degradation had crossed international borders, making the resolution of pollution issues especially difficult.

Urban and Industrial Pollution

Urban and industrial pollution was ignored during Soviet times as cities expanded quickly to accommodate new industries. As workers flooded into the cities from the countryside, lethal levels of pollutants were generated. Today, this region has some of the most polluted cities on the planet.

It is often difficult to link urban pollution directly to health problems because the sources of contamination are diffuse. Such **nonpoint sources of pollution** include untreated automobile exhaust, raw sewage, and agricultural chemicals that drain from fields into urban water supplies. Moscow, for example, is located at the center of a large industrial area, where infant mortality and birth defects are particularly high. Researchers are convinced that these effects result from a complex mixture of pollutants that are difficult to trace. In all urban areas of the region, air pollution resulting from the burning of fossil fuels is skyrocketing as more people purchase cars and as the industrial and transport sectors of the economy continue to grow.

Some cities were built around industries that produce harmful by-products. The former chemical weapons manufacturing center of Dzerzhinsk is listed in the Guinness book of world records as the most chemically polluted city in the world. Here, men have a life expectancy of 42 and women, 47. In the city of Norilsk, not a single living tree exists within

Figure 5.7 Pollution in the former Soviet Union. "Everything rots; everything dies." These are the sentiments expressed by this Azerbaijani woman, referring to her garden near the Baku oil fields. [Reza Deghati/National Geographic Image Collection.]

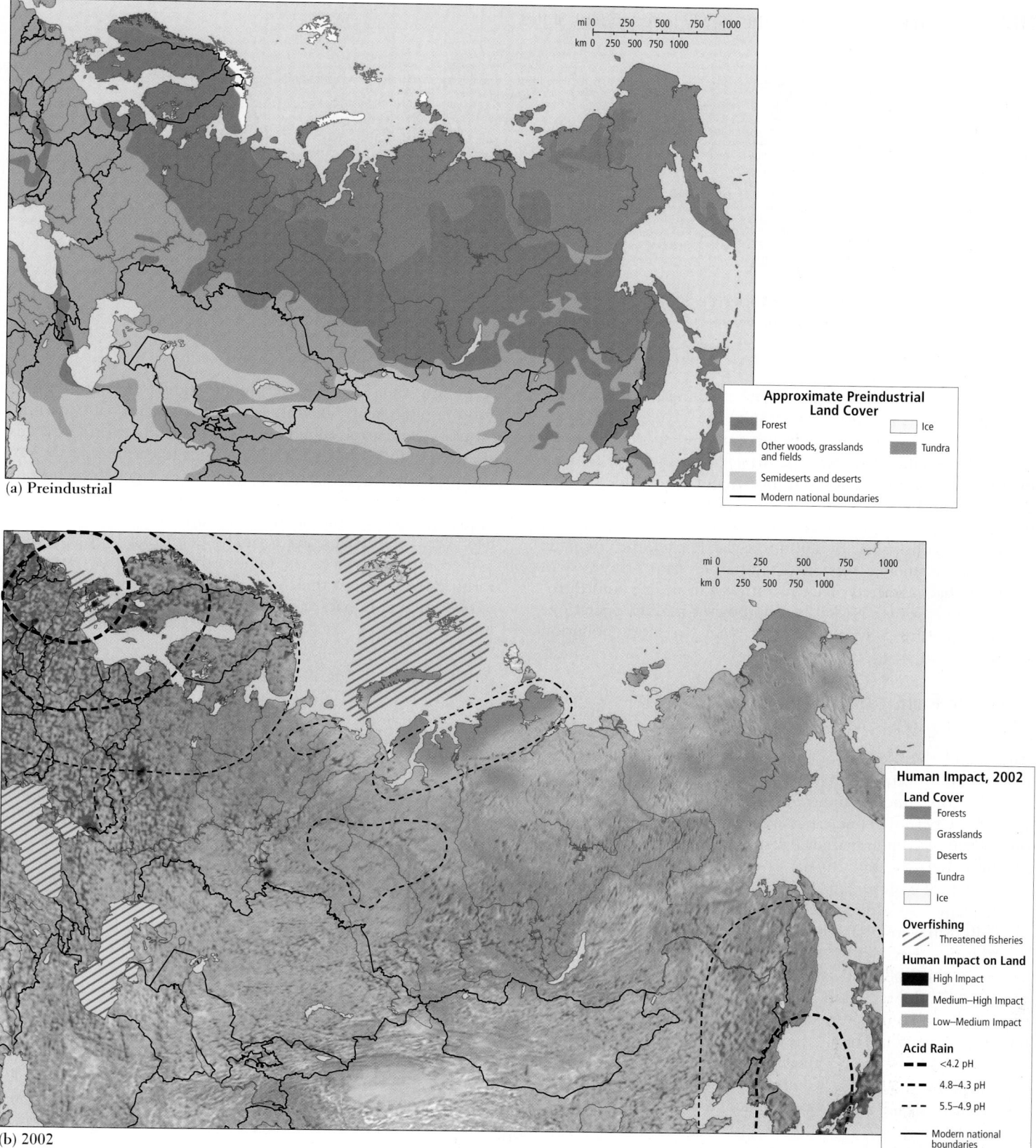

(a) Preindustrial

**Approximate Preindustrial
Land Cover**

Forest   Ice

Other woods, grasslands
and fields   Tundra

Semideserts and deserts

—— Modern national boundaries

(b) 2002

Human Impact, 2002

Land Cover

Forests

Grasslands

Deserts

Tundra

Ice

Overfishing

Threatened fisheries

Human Impact on Land

High Impact

Medium–High Impact

Low–Medium Impact

Acid Rain

– –   <4.2 pH

– · –   4.8–4.3 pH

– – –   5.5–4.9 pH

—— Modern national
boundaries

Figure 5.8 Land cover (preindustrial) and human impact (2002). In the preindustrial era **(a),** human impact on the lands and waters of Russia and its neighbors was significant but generally not visible at the scale of this map. By 2002 **(b),** human impact on land, water, and air was considerable, and in some areas had reached high intensity.

[Adapted from *United Nations Environment Programme, 2002, 2003, 2004, 2005, 2006* (New York: United Nations Development Programme), at http://maps.grida.no/go/graphic/human_impact_year_1700_ approximately and http://maps.grida.no/go/graphic/human_impact_ year_2002.]

181

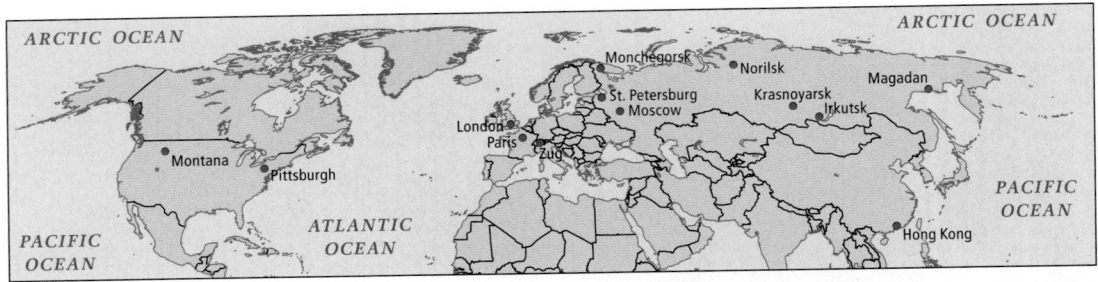

Figure 5.9 Norilsk Nickel's worldwide operations. [Adapted from "About Norilsk Nickel," at http://www.nornik.ru/en/about.]

30 miles of the world's largest metal smelting complex. Male life expectancy is just 50 years, despite the city's general prosperity and availability of free health care and sports clubs.

And yet cities like Norilsk, which sits on huge mineral deposits, are attracting foreign investment crucial to the new Russian economy. Thirty-five percent of the world's nickel supply, 10 percent of its copper, and 40 percent of its platinum are in the Norilsk area. By 2006, Norilsk Nickel, the company that runs the smelter, was producing 2 percent of the Russian GDP and had attracted major investment from U.S. and European banks. Norilsk Nickel is now buying mining operations overseas (Figure 5.9), which may result in more pollution abroad.

Nuclear Pollution: Was Chernobyl the Tip of the Iceberg?

Russia and the Newly Independent States are also home to nuclear pollution, and its effects have spread globally. The world's worst nuclear disaster occurred in Ukraine in 1986, when the Chernobyl nuclear power plant exploded. The explosion severely contaminated a vast area in northern Ukraine, southern Belarus, and Russia. It spread a cloud of radiation over much of eastern Europe, Scandinavia, and eventually the entire planet. As a direct result of this incident, 5000 people died, 30,000 were disabled, and 100,000 were evacuated from their homes. ▄▀

Even the pollution released at Chernobyl pales in comparison with the radiation leaking from former Soviet military sites such as Tomsk-7 (a closed city east of the Urals, now renamed Seversk). There, the soil alone holds 20 times the amount of radiation released at Chernobyl. These facilities have been linked to radiation pollution recorded thousands of miles away in the Arctic, carried by rivers and in the bodies of migrating ducks.

The Arctic Ocean and the Sea of Okhotsk in the northwestern Pacific (Figure 5.1) are also polluted with nuclear waste dumped at sea.

> Although the Soviet government signed an international antidumping treaty, it sank 14 nuclear reactors and dumped thousands of barrels of radioactive waste into the world's oceans.

Although the Soviet government signed an international antidumping treaty, it sank 14 nuclear reactors and dumped thousands of barrels of radioactive waste in the world's oceans.

Russia and Kazakhstan have sought to earn money by taking in the nuclear waste of other countries eager to be rid of it (France is a major client). No reliably safe system has yet been found for storing nuclear waste until it is no longer radioactive. Yet the Russians and Kazakhs claim that they will safely and economically store the imported nuclear waste, using the earnings to clean up their own nuclear waste dumps. Reliable environmental impact studies, however, have not been done.

Resource Extraction and Environmental Degradation

Russia and the Newly Independent States have considerable natural resources (see Figures 5.17 and 5.19, pages 192 and 195). Russia alone has the world's largest natural gas reserves, major oil deposits, and forests, which stretch across the northern reaches of the continent. Russia also has major deposits of coal and industrial minerals such as iron ore and nickel. The Central Asian republics share substantial deposits of oil and gas, which are centered on the Caspian Sea and extend east toward China.

For a while after the demise of the Soviet Union, general pollution levels fell simply because the economy slowed. But now, as economies rebound, demand for resources is growing. Air and water pollution levels are on the rise, and the government is issuing contracts to foreign timber companies for rapid and unsustainable clear-cutting.

Russia's booming oil and gas industries are also creating some of the world's worst oil spills. In Siberia, inland oil spills have contaminated lakes, rivers, and wildlife. Recently Lake Baikal, one of the world's largest and least damaged freshwater lakes, was threatened by the construction of a 2500-mile (4000-kilometer) pipeline to carry Russian oil to Asian Pacific markets. In April 2006, after local and international protests, President Vladimir Putin unexpectedly agreed to divert the pipeline a safe distance around the lake. Still, the future success of public environmental protests is uncertain.

Water, Irrigation, and the Aral Sea

Once the fourth-largest lake in the world, the Aral Sea is disappearing as a result of large-scale irrigation projects in Central Asia (Figure 5.10). For millions of years, this landlocked inland sea was fed by the Syr Darya and Amu Darya rivers, which brought snowmelt from nearby mountains. In 1918, the Soviet leadership decreed that water would be diverted from the two rivers to irrigate millions of acres of cotton in Kazakhstan and Uzbekistan. So much water was consumed by these projects that within four years, the Aral Sea had shrunk measurably. By the early 1980s, no water at all was reaching the Aral Sea, and by early 2001, the sea had lost 75 percent of its volume and had shrunk into three smaller lakes.

The native fish died out due to increasing water salinity, and port cities were marooned far from the water.

The shrinkage of the Aral Sea may also have caused changes in climate and human health. The country around the sea has become drier, summers are hotter, and winters cooler and longer. Winds sweep across the newly exposed seabed, picking up salt and chemical residues and creating poisonous dust storms. At the southern end of the Aral Sea in Uzbekistan, 69 of every 100 people report chronic illness. In some villages, life expectancy is just 38 years (the national average is 66 years).

Efforts to increase water flows to the sea have confronted continuing high demands for irrigation. Uzbekistan is the world's fifth-largest cotton grower, and cotton accounts for one-third of its total exports and employs 40 percent of its

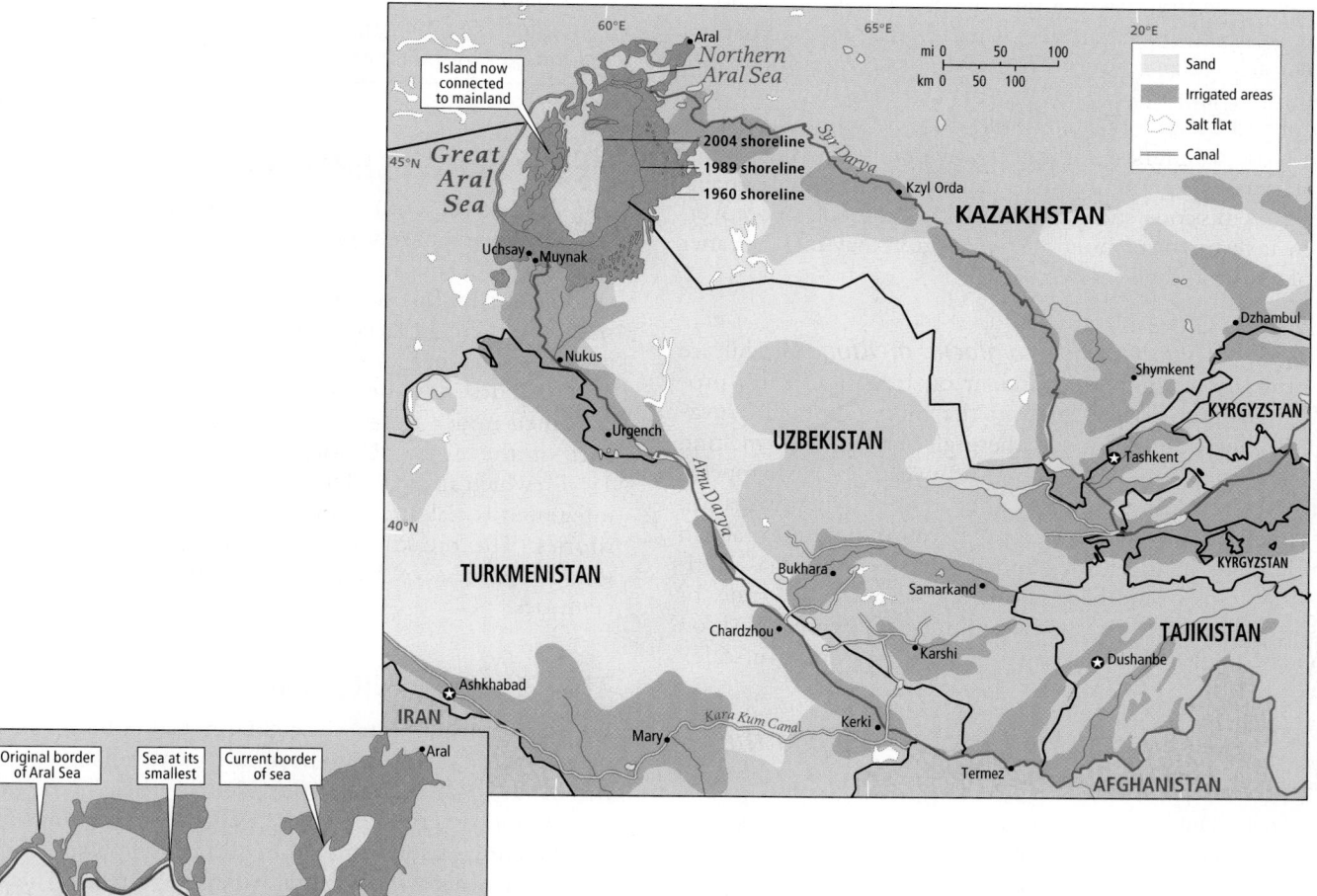

Figure 5.10 The decline and disappearance of the Aral Sea. Once the fourth-largest lake in the world, the Aral Sea is disappearing as a result of large-scale irrigation projects in Central Asia. The inset map shows the northern Aral Sea. [Adapted from *National Geographic* (February 1990): 72, 80–81; satellite images of the Aral Sea, 2003.]

labor force. Nevertheless, some actions have been taken to restore the Aral Sea. Kazakhstan, which relies on oil more than cotton, has built a dam to keep water in the northern Aral Sea (see inset map, Figure 5.10). By 2006, the water had risen 10 feet, and the fish catch was improving. Kazakh fishers note that there is now more open public debate about what to do next regarding the Aral Sea. This is a significant change from Soviet times, when there was no questioning of the wisdom of grand-scale irrigation and no public awareness of its likely environmental effects. ▆◀

Global Warming

Although Russia has recently made moves to reduce its greenhouse gas emissions, it is still a major contributor. Many wasteful practices date from the Soviet Era, such as burning off, or "flaring," natural gas that comes to the surface when oil wells are drilled. Many oil wells are too remote to justify the expense of capturing natural gas in a pipeline and selling it for home heating or cooking. As a result, Russia flares off more natural gas than any other country in the world, contributing more carbon to the atmosphere this way than all the vehicles in New England and New York State. Still, in some ways, Russia has shown more willingness to limit its own emissions than other big polluters by signing international treaties to reduce greenhouse gas emissions such as the Kyoto Protocol. Moreover, the new investment coming into the region may make cleaner technologies more affordable.

Central Asia's vulnerability to glacial melting. Agriculture in Central Asia is highly dependent on irrigation water provided by melting glaciers high in the mountains of Kyrgyzstan and Tajikistan. However, these glaciers are now melting so fast that they could be gone in fifty years. If this happens, the rivers would run dry during the summer, when irrigation is most needed for growing crops. Because most rain falls in the mountains in the winter and spring, Central Asia's agricultural systems would have to adapt to grow at this time, or else store water for use in the summer. Either proposition demands complex and expensive changes on a massive scale.

Ten Key Aspects of Russia and the Newly Independent States

- The Soviet Union did more than any other country to defeat Hitler's armies in World War II, bearing the brunt of Nazi Germany's war machine and ultimately exhausting it.
- Since the breakup of the Soviet Union, Russia's population has shrunk by about 5 percent to 142 million.
- The benefits of economic reforms in the post-Soviet era have been dampened by widespread corruption, lower incomes for many, and the loss of social services.
- The struggle to control this region's huge crude oil and natural gas resources is increasing international tensions, while the need to profit from them is creating new incentives for international cooperation.
- Almost all countries in the region have held democratic elections, but many forms of authoritarian control remain, such as limits on media freedom.
- Since the fall of the Soviet Union the political empowerment of women has advanced the least where democracy has developed the most.
- Religion has revived since the fall of the Soviet Union and its atheistic ideology, with greater public roles developing for Orthodox Christianity, Islam, and evangelical Christianity.
- Food production has started to recover in recent years after falling by 20–30 percent in the 1990s compared to Soviet times.
- Urban and industrial pollution is intense and widespread. One example is the former chemical weapons manufacturing center of Dzerzhinsk, where pollution is so bad that men have a life expectancy of only 42 and women, 47.
- Global warming could result in the disappearance of glaciers that feed Central Asia's two major rivers, resulting in water shortages for many irrigation projects.

HUMAN PATTERNS OVER TIME

The core of the entire region has long been European Russia, the most densely populated area and the homeland of the ethnic Russians. Expanding gradually from this center, the Russians conquered a large area inhabited by a variety of other ethnic groups. These conquered territories remained under Russian control as part of the Soviet Union (1917–1991), which attempted to create an integrated social and economic unit out of the disparate territories. The breakup of the Soviet Union has reversed this gradual process of Russian expansion for the first time in centuries.

> The breakup of the Soviet Union has reversed the gradual process of Russian expansion for the first time in centuries.

The Rise of the Russian Empire

For thousands of years, the militarily and politically dominant people in the region were **nomadic pastoralists** who lived on the meat, milk, and fiber provided by their herds of sheep, horses, and other grazing animals. Their movements followed the changing seasons across the wide grasslands stretching from the Black Sea to the Central Siberian Plateau. The nomads would often take advantage of their superior horsemanship and hunting skills to plunder settled communities. To defend themselves, permanently settled peoples gathered in fortified towns.

Towns arose in two main areas: the dry lands of Central Asia and the moister forests of Ukraine and Russia. As early as 5000 years ago, Central Asia had settled communities supported by irrigated croplands and enriched by trade along the

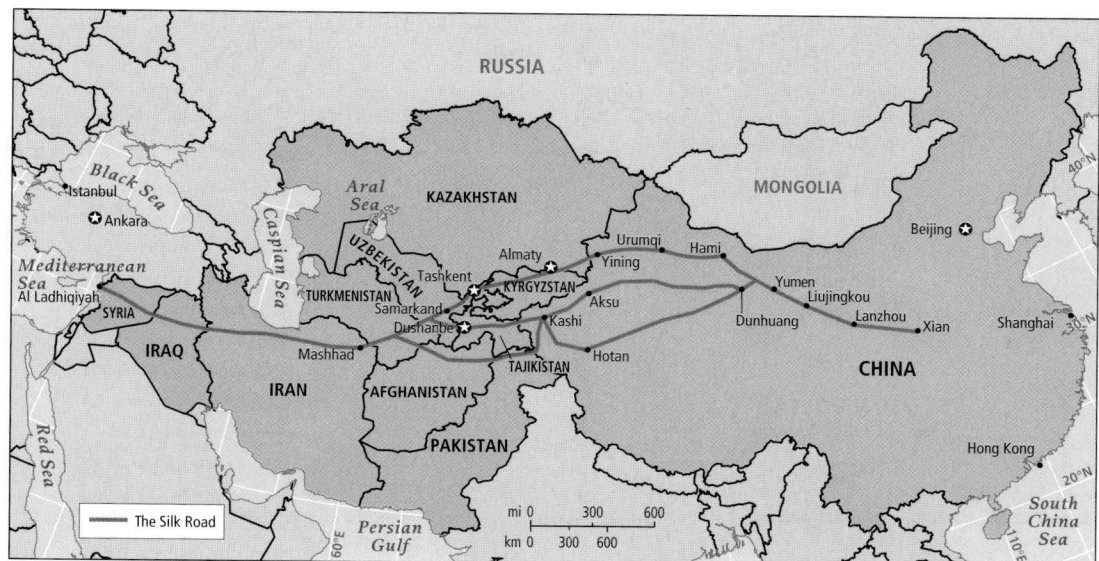

Figure 5.11 The ancient Silk Road. Merchants who worked the Silk Road rarely traversed the entire distance. Instead, they moved back and forth along part of the road, trading with merchants to the east or west. [Adapted from *National Geographic* (March 1966): 14–15.]

famed Silk Road, the ancient trading route between China and the Mediterranean (Figure 5.11).

About 1500 years ago, the **Slavs,** a group of farmers including those known as the Rus (of possible Scandinavian origin), emerged in what is now Poland, Ukraine, and Belarus. They moved east, founding numerous settlements including the towns of Kiev in about 480 and Moscow in 1100. By 600, Slavic trading towns were located along all the rivers west of the Ural Mountains. The Slavs prospered from a lucrative trade route along the Volga River that connected Scandinavia (North Europe) and Southwest Asia (via Constantinople, modern-day Istanbul). Powerful kingdoms developed in Ukraine and European Russia. Greek missionaries introduced both Christianity and the Cyrillic alphabet, still used in most of the region's countries.

In the twelfth century, the Mongol armies of Genghis Khan conquered the forested lands of Ukraine and Russia. The **Mongols** were a loose confederation of nomadic pastoral people centered in East and Central Asia. Moscow's rulers became tax gatherers for the Mongols, dominating neighboring kingdoms and eventually growing powerful enough to challenge local Mongol rule. The Slavic ruler Ivan the Terrible conquered the Mongols in 1552, marking the beginning of the Russian empire. St. Basil's Cathedral, a major landmark in Moscow, commemorates the victory (Figure 5.12).

By 1600, Russians centered in Moscow had conquered many former Mongol territories, integrating them into their empire (Figure 5.13). The first major non-Russian area to be annexed was western Siberia (1598–1689). Russian expansion into Siberia resembled the spread of European colonial powers throughout Asia and the Americas. Russian colonists forcibly took Siberian resources from the local populations, whose cultures were treated as inferior. Moreover, massive

Figure 5.12 St. Basil's Cathedral, Moscow. The cathedral was erected between 1555 and 1561 to commemorate Ivan the Terrible's defeat of the Mongols. The central church, capped by a pyramidal tower, stands amid eight smaller churches with colorful onion domes. [Stock Connection Distribution/Alamy.]

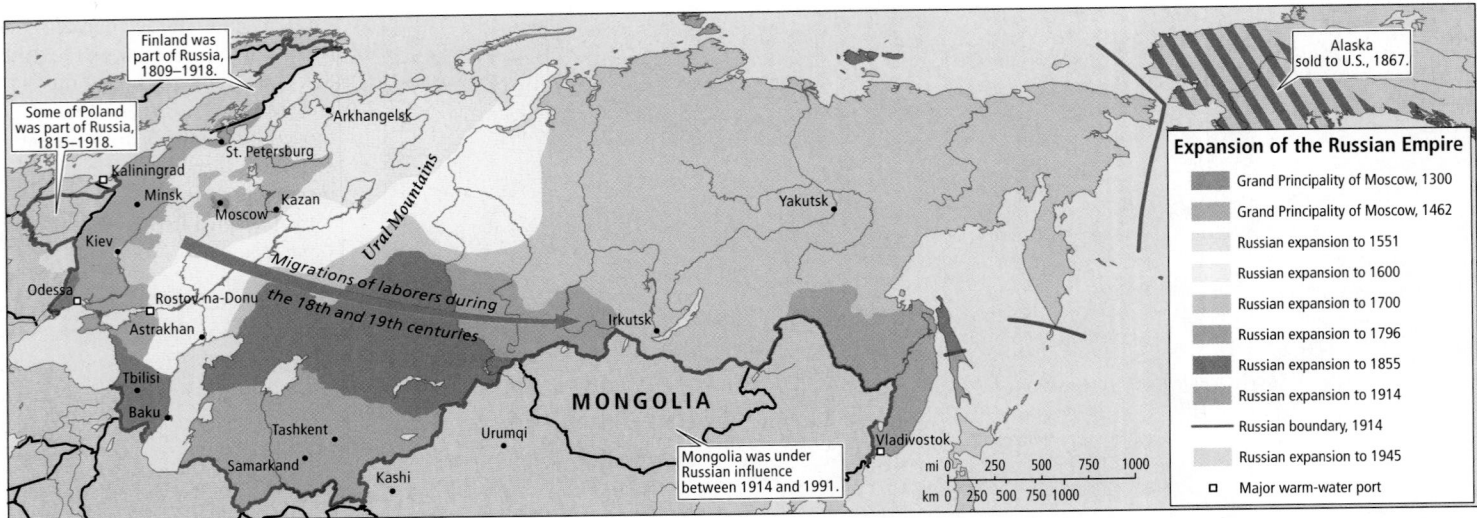

Figure 5.13 Russian imperial expansion, 1300–1945. A long series of powerful entities expanded Russia's holdings across Eurasia to the west and east. Expansion was particularly vigorous after 1700 when Siberia, the largest area, was acquired. [Adapted from Robin Milner-Gulland with Nikolai Dejevsky, *Cultural Atlas of Russia and the Former Soviet Union,* rev. ed. (New York: Checkmark Books, 1998), pp. 56, 74, 128–129, 177.]

migrations of laborers from Russia meant that indigenous Siberians were vastly outnumbered by the eighteenth century. By the mid-nineteenth century, Russia had also conquered Central Asia to gain control of cotton, its major export crop.

The Russian empire was ruled by a powerful leader, the **czar,** who lived in splendor (along with a tiny aristocracy) while the vast majority of the people lived in poverty. Many Russians were *serfs,* who were legally bound to live on and farm land owned by an aristocrat. If the land was sold, the serfs were transferred with it. Serfdom was ended legally in the mid-nineteenth century. However, the brutal inequities of Russian society persisted into the twentieth century, fueling opposition to the czar. By the early twentieth century, a number of violent uprisings were under way.

The Communist Revolution and Its Aftermath

In 1917, at the height of Russian suffering during World War I, Czar Nicholas II was overthrown in a revolution led by a group called the **Bolsheviks.** The Bolsheviks were inspired by the principles of **communism** as explained by the German revolutionary philosopher Karl Marx. Marx criticized the societies of Europe as inherently flawed because of domination by **capitalists**—a wealthy minority who owned the factories, farms, businesses, and other means of production. The impoverished, propertyless majority worked for low wages that undervalued their labor. Communism called on workers to unite to overthrow the capitalists and establish a completely egalitarian society without government or currency. The philosophy held that people would work out of a commitment to the common good, sharing whatever they produced.

The Bolshevik leader Vladimir Lenin declared that the people of the former Russian empire needed a transition period in which to realize the ideals of communism. Accordingly, Lenin's Bolsheviks formed the **Communist Party,** which set up a powerful government centered in Moscow. The government would run the economy, taking land and resources from the wealthy and using to benefit the poor majority. This strategy did not work well, however, and soon production was not meeting demand.

Stalin's rule: success and failure on a grand scale. After Lenin's death, Joseph Stalin began his 31-year rule of the Soviet Union as party chairman and premier (1922–1953). Stalin brought to the country a mixture of brutality and revolutionary change that set the course for the rest of the Soviet Union's history. He sought to cure production shortfalls through rapid industrialization made possible by a **centrally planned,** or **socialist** economy. The state would own all real estate and means of production, while government bureaucrats in Moscow directed all economic activity. This included the locating of all factories, residences, and transport infrastructure, and the management of all production, distribution, and pricing of products. The idea was that under a socialist system the economy would grow more quickly and hasten the transition to the idealized communist state in which everyone shared equally. This notion was reflected in the new name chosen for the country: the Union of Soviet Socialist Republics.

> In the Soviet Union's centrally planned economy, the state owned all real estate and means of production, while government bureaucrats in Moscow directed all economic activity.

Stalin used the powers of the command economy with fervor and cruelty. To increase agricultural production, he forced farmers to join large government-run collectives. Those who resisted were relocated or executed. For Stalin, the true key to achieving economic growth was increasing industrial production. Accordingly, he ordered massive government investments in gigantic development projects, such as factories, dams, and chemical plants, some of which are still the largest of their kind in the world. The labor was supplied largely by former farmers who had lost their farms to collectivization. Eventually, government-controlled companies monopolized every sector of the economy, from agriculture to mining to clothing design and production.

This strategy of government control resulted in significant successes. For those millions of farmers who peacefully went to work in urban industries, their wages brought higher standards of living. The schools provided for their children contributed to major technological and social advances. During the Great Depression of the 1930s, the Soviet Union's industrial productivity grew steadily even while the economies of other countries stagnated.

However, Stalin's model had some serious flaws. One problem was that production was geared largely toward heavy industry (the manufacture of machines and transport equipment) and supplying the military with armaments. Less attention was paid to the demand for consumer goods and services that could have dramatically improved daily life for the Russian people. These activities could also have served as a major source of economic growth, as they did in Europe and North America. The most destructive aspect of Stalin's rule, however, was his ruthless use of the secret police, starvation, and mass executions to silence anyone who dared to oppose him. The lucky ones were those merely sentenced to labor camps in remote Siberia. Twenty million others were killed. Stalin's atrocities created a climate of fear that squelched the possibility of political empowerment for ordinary people.

World War II and the Cold War

The Soviet Union did more than any other country to defeat Hitler's armies in World War II. Nazi Germany's war machine was exhausted in a failed attempt to conquer the Soviet Union. In the process, 23 million Soviets were killed, more than all the other European combatants combined. After the war, the Stalin was determined to erect a buffer of allied Communist states in Central Europe that would be the battleground of any future war with Europe. The United States and its allies were afraid that the Soviets would seek to extend their power even further into Europe. The result was the **cold war,** a fifty-year-long global geopolitical rivalry that pitted the Soviet Union and its allies against the United States and its allies (Figure 5.14).

In an attempt to match the global military power of the United States, the Soviets diverted resources to their military and away from much-needed economic and social development. Internationally, the Soviets promoted Communism far and wide, with major efforts in China, Mongolia, North Korea, Afghanistan, Cuba, Vietnam, Nicaragua, and various African nations. Closer to home, in Central Europe, the Soviet Union maintained a pervasive influence through political, economic, and military coercion. The economies of individual countries

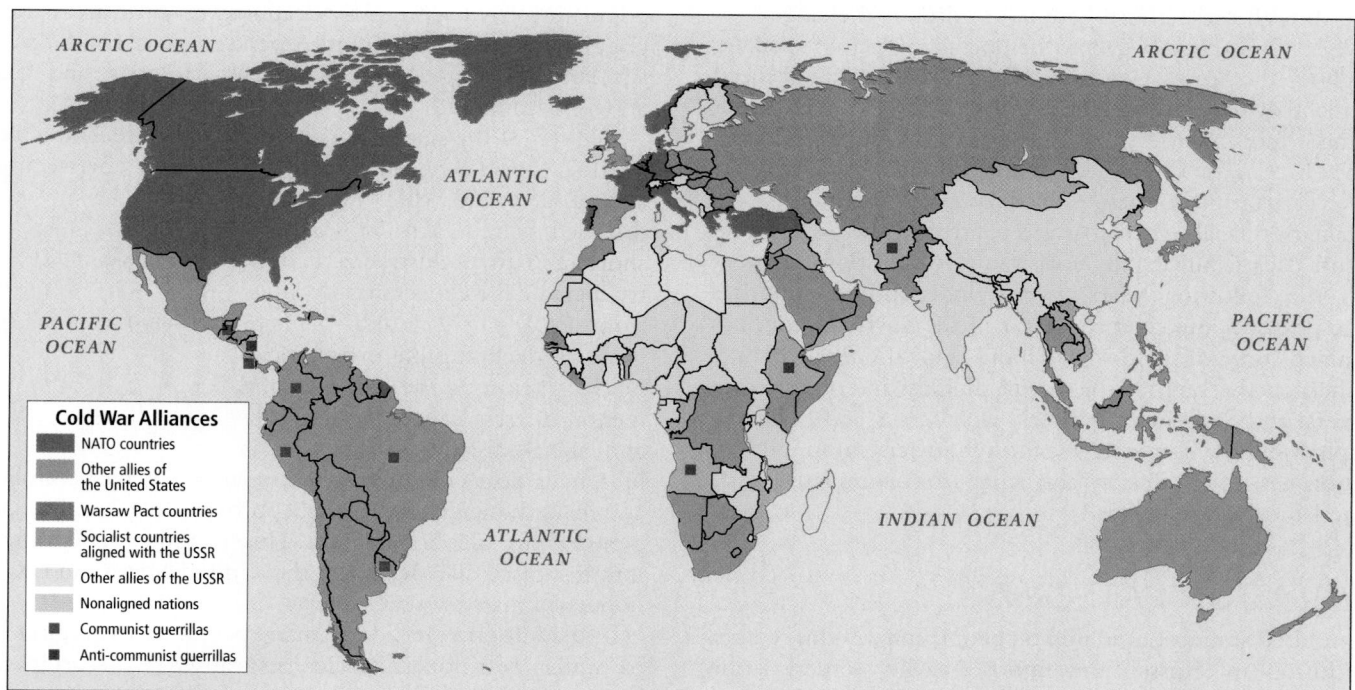

Figure 5.14 The cold war in 1980. In the post–World War II contest between the Soviet Union and the United States, both sides enticed allies through economic and military aid. The group of countries militarily allied with the Soviet Union were known as the Warsaw Pact. Some countries remained unaligned. [Adapted from Clevelander, at http://en.wikipedia.org/wiki/Image:New_Cold_War_Map_1980.png.]

were manipulated to meet the needs of the Soviet Union as a whole. Political systems were tightly controlled from Moscow, and Soviet troops crushed any popular resistance.

The arms race and Afghanistan: drains on finances and morale. By the late 1960s, the Soviet Union was locked in a race against the United States and its massive military to build ever more sophisticated weaponry, including thousands of nuclear missiles. Severe financial strains emerged by the 1980s as the inefficiency of a centrally planned economy proved no match for the dynamic free market economy of the United States. Soon many of the Soviet Union's allies were drifting toward the U.S. mode of economic development. Soviet finances and morale were further drained by a war launched in 1979 to prop up a Soviet-allied regime in Afghanistan. The next decade saw the Soviets badly beaten by highly motivated Afghan freedom fighters, the mujahedeen, who were financed, armed and trained by the United States and Pakistan. (The Soviet–Afghanistan conflict is discussed further in Chapter 8.)

The Post-Soviet Years: Democratization and Economic Reform

In 1985, the Soviet Union's reform-minded president, Mikhail Gorbachev, responded to the various pressures for change. He opened up public discussions of social and economic problems, an innovation known as **glasnost.** He also attempted to revitalize the Soviet economy through **perestroika,** or restructuring, though these efforts resulted in little real change.

When Gorbachev began to democratize decision making throughout the Soviet Union in the late 1980s, long-silenced resentment of the government in Moscow boiled over. Independence movements soon surfaced throughout Central Europe. In August of 1991, a counter reaction against Gorbachev's liberalizing policies, led by a group of Communist Party and military officials, tried to seize control of the Soviet government. Their failure led to the dissolution of the Soviet Union.

In the following years, twelve independent countries emerged. Russia, the chief successor of the Soviet Union, has maintained some of the Soviet Union's global influence, having inherited the bulk of the Soviet military and a vast store of mineral and fossil fuel resources. Democracy and free market capitalism now dominate the former Soviet Union, but the transition has not been easy and is far from complete, as we will see later in this chapter.

POPULATION PATTERNS

Two hundred seventy-seven million people inhabit this region, with European Russia the most heavily settled zone (Figure 5.15). Nevertheless, with an average density of 22 people per square mile (8.5 per square kilometer), the region is still much less dense than the United States (80 per square mile; 30 per square kilometer). A broad area of moderately dense population forms a wedge stretching from Ukraine on

the Black Sea north to St. Petersburg on the Baltic Sea and east to Novosibirsk, the largest city in Siberia.

Beyond Novosibirsk on the Western Siberian Plain, settlement follows an irregular pattern of industrial and mining development across Siberia. These activities in turn are linked primarily to the course of the Trans-Siberian Railroad (see Figure 5.17 on page 192). Although Siberia is often seen as a desolate, lonely place, nearly 90 percent of its people are concentrated in a few large urban areas.

In the west, a secondary spur of dense settlement extends south from European Russia into **Caucasia,** the mountainous region between the Black Sea and the Caspian Sea. Another patch of relatively dense settlement is centered on Tashkent and Almaty and along major rivers in the Central Asian republics. Here the development of irrigated cotton farming and mineral extraction has resulted in patches of high rural density, fueled partly by ethnic Russian immigration.

Shrinking Populations: High Death Rates and Low Birth Rates

This region is experiencing a unique variant of the demographic transition (see Chapter 1, page 45) as many countries' populations are shrinking faster than those in any other world region. During the Soviet era, population growth was moderated by increased opportunities for women to work outside the home. Free health care and adequate retirement pensions also helped lower incentives for large families. But since the economic crisis brought on by the breakup of the Soviet Union, population has been rapidly declining in some areas. Russia's population has shrunk about 5 percent to 142 million in the last 16 years. Populations in Belarus, Moldova, and Ukraine are also shrinking.

Much of the reason for the population decline is declining life expectancy. In Russia, for example, between 1990 and 2005, male life expectancy dropped from 63.9 to 58 years, the shortest in any industrialized country. Female life expectancy dropped from 74.4 to 72 years. A major cause of declining life expectancies in the region is the physical and mental distress caused by lost jobs and social disruption. There is a high male death rate, explained in large part by alcohol abuse and related suicides. In Russia, approximately 7 million deaths per year are alcohol related. This is almost 100 times the number of similar deaths in the United States, and Russia's population is only half as large!

> Between 1990 and 2005, male life expectancy in Russia dropped from 63.9 to 58 years, the shortest in any industrialized country.

In addition, after 1991, many people in the region began to suffer nutritional deficiencies caused by sharply falling incomes and food scarcities, some of which were brought on by conflict in Russia's internal republics and in Caucasia.

Environmental pollution also plays a large, if poorly understood, role in untimely illness and death. A third or more of the population may be affected by noxious airborne

last fifteen years. To fully appreciate what these reforms mean, it is important to know something of the Soviet institutions that were previously in place.

The Former Command Economy

Although the long-term goal of the command economy was to achieve Communism, its shorter-term goal was to end the severe deprivation suffered under the czars. To some extent, this goal was met. The Soviet economy grew rapidly until the 1960s, and abject poverty was largely eradicated. Basic necessities such as housing, food, health care, education, and transportation were all provided for free or at very low cost—a remarkable accomplishment for any country.

Nonetheless, the Soviet command economy was less efficient than market economies in allocating resources. With bureaucrats setting production goals for the whole Soviet Union, even small miscalculations created massive problems. Scarcities of food and raw materials were common throughout the Soviet era.

A lack of competition in the civilian economy meant that producers had no incentive to use more efficient production methods. Each Soviet factory tended to have a monopoly on a particular product. Quality also suffered because hard work and innovation in the workplace was rarely rewarded. Promotions generally went to those with connections in the Communist Party. As a result, most consumer goods, such as cars or washing machines, were of poor quality and available only at high cost to a privileged few.

In science and military technology, however, the Soviet Union achieved feats that had the United States and its allies scrambling to keep up. Soviet scientists and engineers launched the world's first satellite in 1957 and the first manned spacecraft in 1961. The scientists and engineers of the region are still making significant breakthroughs in physics, biology, metallurgy, computer technology, and a wide variety of other fields. ▧

Soviet regional development schemes. A long-lasting legacy of Soviet central planning was the location of huge industrial projects in the farthest reaches of Soviet territory. This was done for a variety of reasons. Leaders wanted to buttress Russia's claims to distant territories by bringing the higher standards of living enjoyed in industrialized European Russia to all parts of the Soviet Union. Economic development projects also continued the pre-Communist Russian empire's policy of **Russification**—forcing non-Russian ethnic groups to conform to the state's goals by swamping them with large numbers of ethnic Russian migrants. Russians were given the best jobs and most powerful positions in the government. In this way, the possibility of revolt by remote ethnic minorities was minimized. Yet another concern was to disperse industrial centers throughout the country to make them safer from enemy attack.

These ambitious regional development schemes never really succeeded, in large part because of transport problems related to the region's huge size and challenging physical geography. While the region's rivers run mainly north-south, the primary transport needs created by the Soviet regional development schemes were east-west. This created a need for land transport systems such as railroads and highways.

The construction of land transport systems has long been held back by the region's climate. Long, harsh winters, during which it is difficult to build, give way to a period called the *rasputitsa* or "quagmire season," when melting permafrost turns many roads and construction sites into impassable mud pits. Huge distances and complex topography, especially in Siberia, make transportation even harder. As a result, few roads or railroads were ever built outside of European Russia.

Even today, this region more than two and a half times the size of the United States has less than one-sixth the number of hard-surface roads and virtually no multilane highways. No road and only one main rail line, the Trans-Siberian Railroad (Figure 5.17), runs the full east-west length, connecting Moscow with Vladivostok, the main port city on the Russian Pacific coast. By contrast, the road and rail network is fairly dense within European Russia, Belarus, and Ukraine.

Economic Reforms in the Post-Soviet Era

Russia's recent economic reforms have been ambitious, but haphazard. So far, the benefits have gone mainly to a small group of wealthy and politically connected individuals. The lives of the majority have become more difficult and unpredictable.

Privatization and lifting of price controls. The Soviet economy consisted almost entirely of industries owned and operated by the government. These have now been sold to private companies or individuals in a process called **privatization.** The hope is that they would be run more efficiently in a competitive free market setting. By 2000, approximately 70 percent of Russia's economy was in private hands, a significant change from the 100 percent state-owned economy of 1991.

Another major reform is that prices are no longer kept artificially low by the government to make goods affordable to all. Instead, they are determined by supply and demand. The lifting of price controls led to skyrocketing prices in the 1990s for the many goods that were in high demand but also in short supply. While a tiny few grew rich, many people had to use their savings to pay for basic necessities such as food. Eventually, as opportunities opened and competition developed, the supply of goods increased and prices fell. But in the interim, many people suffered.

Unemployment and loss of social services. Since being privatized, most formerly state-owned industries have cut many jobs in an attempt to compete against more efficient companies in the global free-market economy. Losing a job is especially devastating in a former Soviet country because one also loses the subsidized housing, health care, and other social

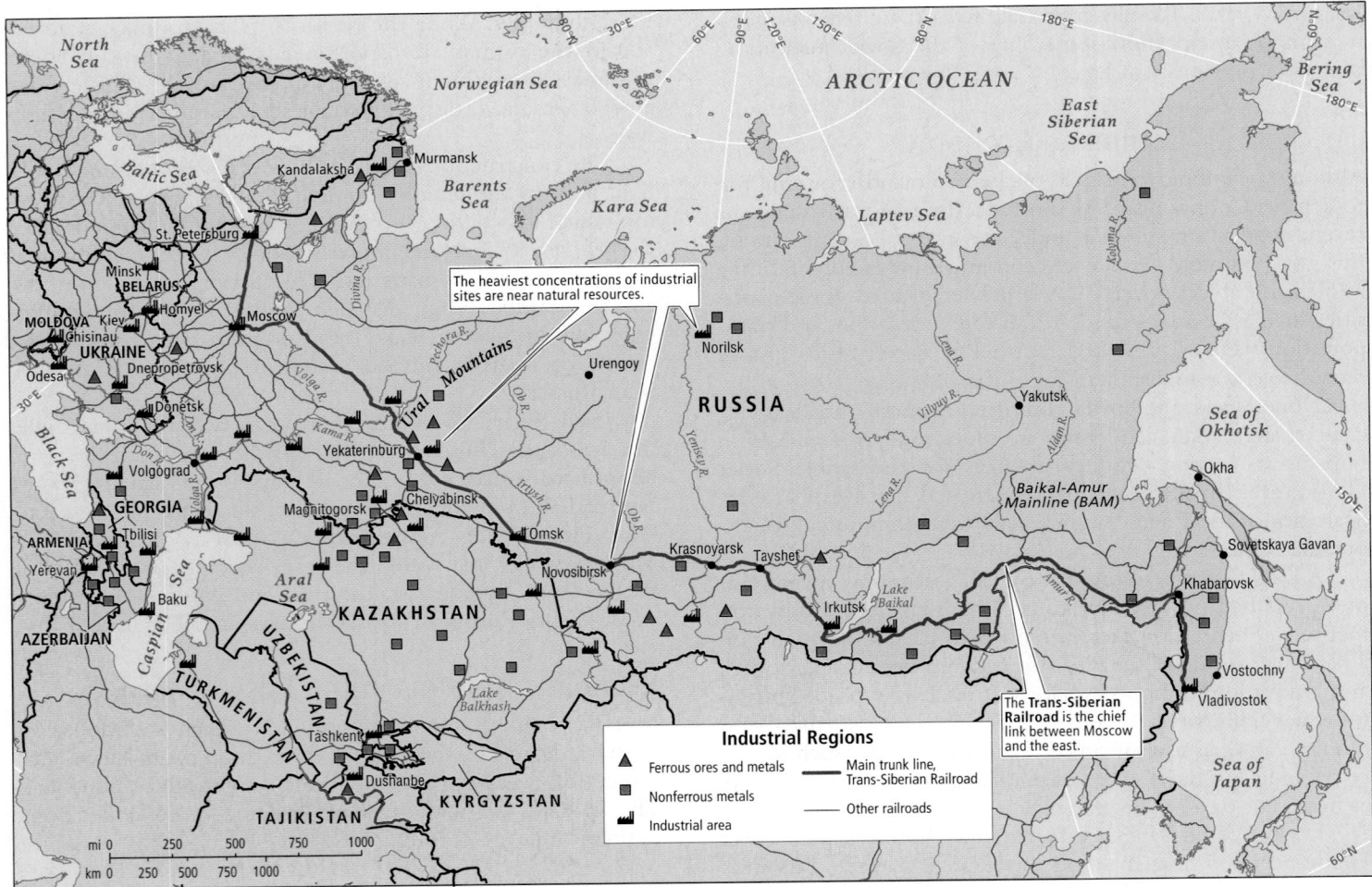

Figure 5.17 Principal industrial areas and land transport routes of Russia and the Newly Independent States. The industrial, mining, and transport infrastructure is concentrated in European Russia and adjacent areas. The main trunk of the Trans-Siberian Railroad and its spurs link industrial and mining centers all the way to the Pacific. Note that the frequency of industrial centers decreases with distance from the borders of European Russia. [Adapted from Robin Milner-Gulland with Nikolai Dejevsky, *Cultural Atlas of Russia and the Former Soviet Union*, rev. ed. (New York: Checkmark Books, 1998), pp. 186–187, 198–199, 204–205, 216–217]

services that were usually provided along with the job. Work was also the center of community life, so when a job ended, there was no social support group.

The new companies that have emerged rarely offer benefits to employees. There is little job security because most small private firms appear quickly and then fail. Discrimination is also a problem given the absence of equal opportunity laws. Job ads often contain wording such as "only attractive, skilled people under 30 need apply."

Widespread unemployment has been common since the end of the Soviet era. By 2006, official unemployment rates for the region ran from 1.9 percent in Belarus to 30 percent and higher in Caucasia and Central Asia. However, actual unemployment is probably much higher because many remaining state-owned firms cannot pay employees still listed as workers. The rate of **underemployment,** which measures people who are working too few hours to make a decent living or who are

highly trained yet working at menial jobs, may be even higher. It is estimated that in the mid-1990s, three-fifths of the Russian labor force was not being paid in full and on schedule.

The growing informal economy. To some extent, the new informal economy in the region is an extension of the old one that flourished under Communism. The black market of that time was based on currency exchange and the sale of hard-to-find luxuries. In the 1970s, for example, savvy Western tourists could enjoy a vacation on the Black Sea paid for by a pair or two of smuggled Levi's blue jeans and some Swiss chocolate bars. Today, many people who have lost stable jobs due to privatization now depend on the informal economy for their livelihood (Figure 5.18).

People often operate out of their homes, selling cooked foods, vodka made in their bathtubs, electronics they have assembled, or pirated computer software, to name just a few

Figure 5.18 The informal economy in Russia. Imported (probably smuggled) VCRs and TVs are sold in informal, non-taxpaying open markets like this one in Moscow. The markets are tended by men with connections to the Russian mafia. Police are paid to look the other way. [Gerd Ludwig/National Geographic Image Collection.]

products. So much of the economy is now in the informal sector that in many countries of the region, people may in fact be better off financially than official gross domestic product (GDP) per capita figures suggest.

Vignette Natasha is an engineer in Moscow. She has managed to keep her job and the benefits it carries, but in order to feed her family, she has to sell secondhand clothes in a street bazaar on the weekends. "Everyone is learning the ropes of this capitalism business," she laughs. "But it can get to be a heavy load. I've never worked so hard before!" Asked about her customers, Natasha says, "Many are former officials and high-level bureaucrats who just can't afford the basics for their families any longer." Some are older retired people whose pensions are so low that they resort to begging on Moscow's elegant shopping streets close to the parked Rolls-Royces of Moscow's rich. These often highly educated and only recently poor people buy used sweaters from Natasha and eat in nearby soup kitchens.

"But you know," Natasha continues, brightening and changing the subject, "the bales of used clothes I buy now come from the United States. A guy drives a carload from a container ship in Amsterdam harbor every two weeks. And there are lots of sturdy clothes, especially for children, some quite new with the price tags still on. But only occasionally is there a really fashionable item for a woman."

Source: This composite story is based on work by Alessandra Stanley, David Remnick, and David Lempert and Gregory Feifer. ■

Corruption and Organized Crime

The benefits of economic reforms in the post-Soviet era have been dampened by widespread corruption and the growth of organized crime. The process of privatization was rife with cor-

ruption. Officials, many of whom were no longer receiving a full salary, often allowed buildings, factories, and access to resources to be sold for a fraction of their real value in return for bribes. The buyers often got incredible deals and have profited immensely. Many are now known as *oligarchs* because of the political power their money brings.

Many oligarchs are closely connected to the so-called **Russian Mafia,** a highly organized criminal network dominated by former KGB (the soviet intelligence agency that was a counterpart to the CIA) and military personnel. This group has extended its influence into nearly every corner of the post-Soviet economy, but especially illegal activities and the arms trade. ◘ Since the fall of the Soviet Union, oligarchs and the Russian mafia have taken billions of dollars in profits to secret bank accounts outside the region.

The military and nuclear materials. In the post–9/11 world, many are concerned that corruption in the militaries of this region could put nuclear weapons from the former Soviet arsenal in the hands of terrorists. After huge funding cuts, weapons and even military rations were routinely sold on the black market. In 2007, Russian smugglers were caught trying to sell nuclear materials on the black market. However, recent years have witnessed some encouraging signs. Parts of the Russian military have received much more funding, and efforts to combat corruption are increasing. Moreover, all countries in the region are now cooperating with the International Atomic Energy Agency in controlling nuclear material. Many old nuclear weapons are being dismantled, and in 2006, the Central Asian republics signed a treaty creating a nuclear-free zone. ◘

Case study: Georgia's success in fighting corruption. The Caucasian republic of Georgia offers a hopeful contrast to the situation in the rest of this region. Since 2004, a new democratically-elected president has managed to dramatically reduce corruption that was once so widespread that it threatened to turn the country into a criminal enclave. For example, until 2003, Georgia's police were so poorly paid that they essentially had to collect their own salaries by shaking down the citizens. Civil servants had to take bribes just to feed their families. In 2004, however, Mikhail Saakashvili, a 35-year-old Georgian with a law degree from Columbia University in New York, was elected president. Once in office, Saakashvili fired many civil servants and police and replaced them with trained people whose pay was increased nearly twenty-fold to cover actual costs of living. Since then, calls to the police are up, not because of a crime wave, but because people can now report crimes and ask for assistance without fear of having to pay a bribe.

Oil and Gas Exports: Fueling Globalization

Crude oil and natural gas are the region's most lucrative exports and are dominating economies more than ever before. The struggle to control the region's oil and gas resources has become global and could be a source of conflict for years to come. However, the need for stable trading relationships in order to profit from the sale of oil and gas is also bringing greater stability.

Russia's energy-based economic recovery. Revenues from oil and gas have financed Russia's economic recovery, eliminating its once crippling foreign debt. As recently as 1998, Russia's debt was staggeringly high—90 percent of GDP—and a major impediment to the country's economic growth. Today, Russia has very low debt thanks to its status as the world's largest exporter of natural gas. The state-owned energy company **Gazprom** is the tenth-largest oil and gas entity in the world.

Gazprom and private oil companies account for more than half of Russia's federal tax receipts. Recently the central government has tightened control over the entire oil and gas sector by confiscating some facilities and strengthening rules to ensure that the industry pays all its taxes regularly. The increasing economic stability provided by Russia's energy sector has attracted foreign investment in all sectors of the economy. The influx of money is most visible in major cities like Moscow, where new skyscrapers are transforming the skyline (see Views of Russia and the Newly Independent States, page 176).

The struggle for Central Asia's oil. Many foreign investors are also interested in the energy resources of Central Asia. A tug-of-war has evolved between Russia and multinational energy companies over the rights to develop, transport, and sell these resources. Countries with designs on Central Asian oil include Russia, Turkey, Iraq, China, India, Pakistan, Iran, the United Kingdom, and the United States. The holders of these considerable oil reserves, the new Central Asian republics, are not powerful or experienced enough to stave off the covetous advances of world powers. Nor do they have the capital to develop the resources themselves. However, many in the Central Asian countries are reaping enormous financial rewards as deals are made.

A tug-of-war has evolved between Russia and multinational energy companies over the rights to develop, transport, and sell Central Asia's fossil fuel resources.

Contention over Central Asian oil and gas hinges primarily on pipeline routes. Russia, the United States, the European Union, Turkey, China, and India have all developed or proposed various routes (Figure 5.19). As of 2006, the United States had installed military bases to protect its interests in Georgia and Uzbekistan, both of which have pipelines. ▶️

Closer relations with the European Union and the world. Oil and gas exports have brought unprecedented levels of economic interdependence between the European Union and Russia. Gazprom, for example, earns 65 percent of its revenue from sales to the European Union, which buys 25 percent of its natural gas from Gazprom. Both the European Union and the United States want to ensure that Russia's new oil and gas wealth does not finance a return to the hostile relations of the cold war. Hence, Russia has been invited to join established global trading institutions, with entrance into the World Trade Organization scheduled for late 2008. Russia is also the newest member of the **Group of Eight (G8)**, an organization of the wealthiest and most highly industrialized nations.

These institutions may provide a forum for diffusing tensions that have developed in recent years as many of Russia's former allies have broken ties with it. Some are now members of the North Atlantic Treaty Organization (NATO), the old cold war Western military alliance against the Soviet Union. Further, many former Soviet allies have joined the European Union, and the newly independent states of Ukraine, Moldova, and Georgia are likely to join within the next 15 years. While actual membership in the European Union for Russia is unlikely any time soon because of its many environmental and political problems, closer relations are inevitable if Russia is to continue to provide so much of the European Union's fuel. With Russia's economy more strongly tied to the European Union and the global economy, greater cooperation will be possible on issues such as nuclear proliferation and global warming.

Food Production in the Post-Soviet Era

Across most of the region, agriculture is precarious at best, requiring expensive inputs of labor, water, and fertilizer. Only 10 percent of Russia's vast expanse is suitable for agriculture

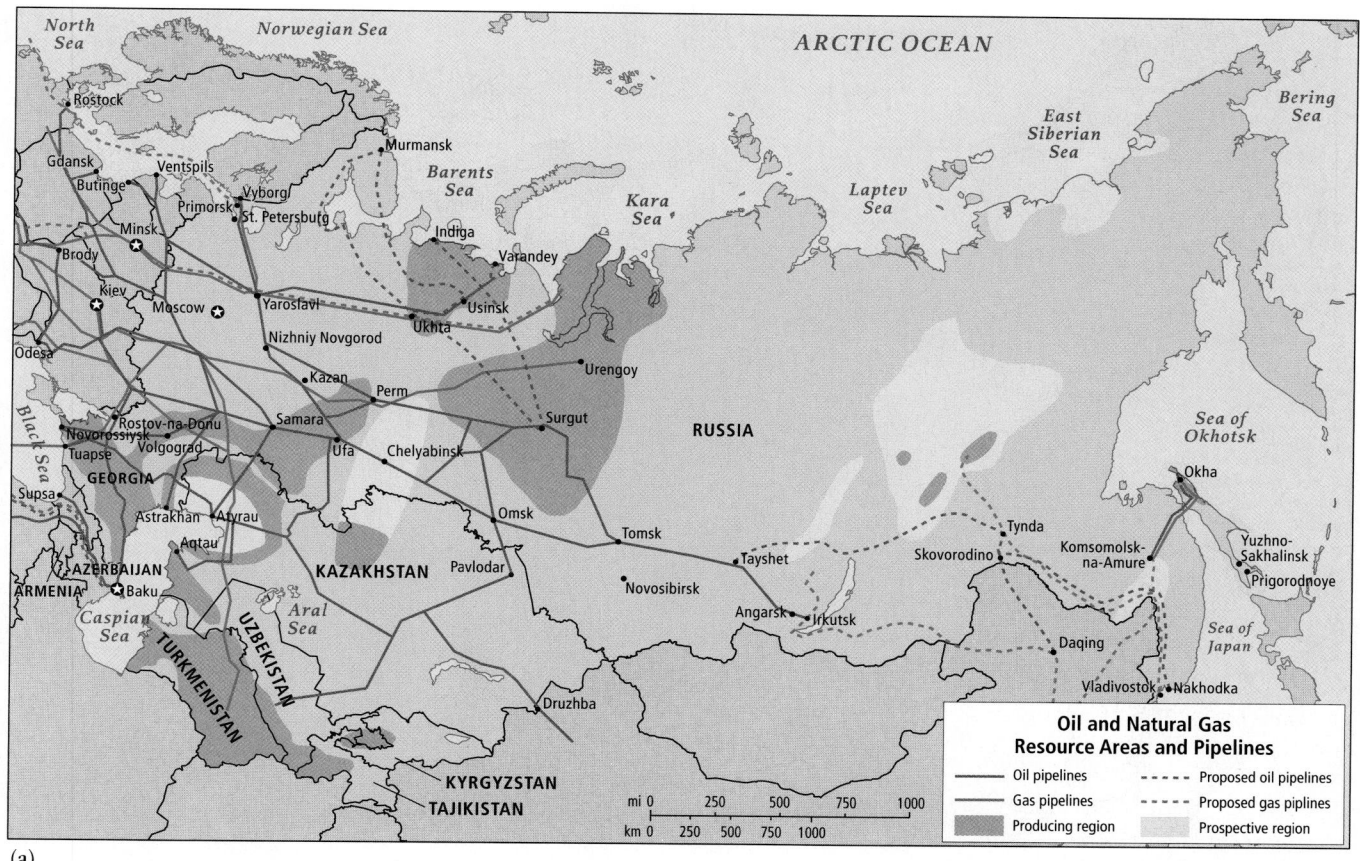

(a)

Oil Production, 2003	
Region or Basin	**1000 Barrels/day**
West Siberia	5,882
Volga-Urals	1,887
Precaspian	679
South Caspian	454
Timan-Pechora	373
Middle Caspian	261
South Turgay	209
Central Asia	161
North Caucasus	72
Far East	65
Azerbaijan Onshore	32
East Siberia	32
Baltic	—
Barents Sea	—
Total Region	**10,107**
Total World	**79,110**

(b)

Gas Production, 2003	
Region or Basin	**Billion Cubic Meters**
West Siberia	573.1
Central Asia	90.0
Precaspian	25.9
Volga-Urals	25.1
South Caspian	15.7
East Siberia	8.8
Timan-Pechora	3.6
Far East	1.9
Azerbaijan Onshore	0.4
Barents Sea	—
Total Region	**744.5**
Total World	**2,618.5**

(c)

Figure 5.19 Oil and natural gas: Resources and pipelines. [Adapted from "Russia Country Analysis Brief," May 2004, Energy Information Administration, U.S. Department of Energy, at http://www.eia.doe.gov/emeu/cabs/Russia/images/fsu_energymap.pdf.]

due to harsh climates and rugged landforms. The Caucasian mountain zones are some of the only areas in the region where rainfall adequate for agriculture coincides with a relatively warm climate and long growing seasons. Together with Ukraine and European Russia, this area is the agricultural backbone of the region (Figure 5.20). The best soils are in an area stretching from Moscow south toward the Black and Caspian seas, including much of Ukraine and Moldova. Irrigated agriculture is extensive in Central Asia, where long growing seasons support cotton, fruit and vegetables.

Agriculture went into a general decline in the 1990s, with yields dropping by 20 to 30 percent in most countries

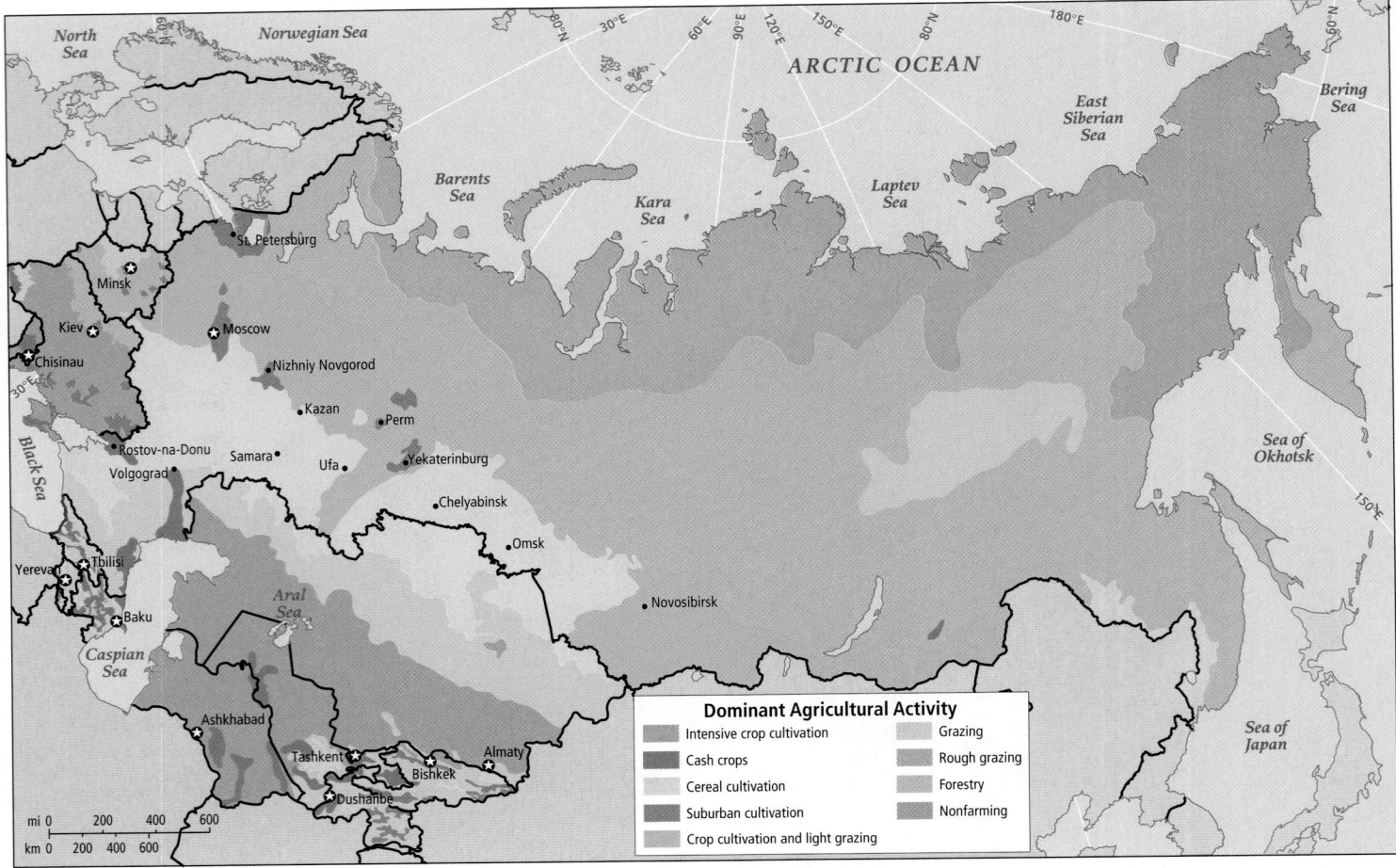

Figure 5.20 Agriculture in Russia and the Newly Independent States. Agriculture in this part of the world has always been a difficult proposition. The cold climate makes for short growing seasons, and soil fertility or lack of rainfall are problems in all but a few places (Ukraine, Moldova, and Caucasia). [Adapted from Robin Milner-Gulland with Nikolai Dejevsky, *Cultural Atlas of Russia and the Former Soviet Union*, rev. ed. (New York: Checkmark Books, 1998), pp. 186–187, 198–199, 204–205, 216–217.]

compared to Soviet production levels. This was due mostly to the collapse of the subsidies and trade arrangements of the Soviet era. Many large and highly mechanized farms suddenly found themselves without access to equipment, fuel, or fertilizers. Some went back to horse-drawn plows and animal fertilizers. Many of the newly independent states had to move away from the high degree of specialization that had been possible under the Soviet system of agricultural trade arrangements. For example, Central Asia could no longer be sure that Russia would buy its cotton exports and provide its grain imports. Therefore, Central Asia is now growing more of its own grain and exporting less cotton. All of these changes make it harder for farms to function, and agricultural productivity has decreased.

The massive collective farms of the Soviet era have only recently started to be privatized, but already many are seeing gains in efficiency from better management. Thousands of collective farms have been broken up into small tracts and sold to the people who worked on them. These independent farmers now have a much greater stake in their productivity and are making better decisions about what to grow, how to grow it, and where to sell it. At the same time, some are doing well enough to buy out other farmers, and large farms are slowly re-emerging throughout the region. They are expected to provide stiff competition to the remaining state-run collective farms.

> The massive collective farms of the Soviet era have only recently started to be privatized, but already many are seeing gains in efficiency from better management.

Ukraine and Georgia: feeding the European Union?

In the west of this region, Ukraine and Georgia are hoping to use their agricultural resources to gain closer association with the European Union. Ukraine is comparable to France in size, population, and agricultural output—agriculture accounts for one-fifth of the GDP and one-fourth of employment. Ukraine could give France some serious competition in the production of vegetables, fruits, and wines. To do so, however, Ukraine would have to make many adjustments to meet EU standards and continue to improve the efficiency of its farms, which are generally less efficient than farms in Europe.

While much of Russia and the Newly Independent States is arid or cold, Georgia is blessed with warm temperatures and abundant moisture from the Black and Caspian seas. In these favorable conditions, farmers can grow citrus fruits and even bananas. Before 1991, most of the Soviet Union's citrus and tea came from Georgia, as did much of its grapes and wine. In May 2006, Russia unexpectedly banned the importation of Georgian wine, claiming it was polluted with heavy metals. Trade experts suspect this was punishment for Georgia's resistance to Russian efforts to control the marketing of its wine and other products in the European Union.

Family gardens and urban space. Family gardens have long been an important source of nourishment in the region. In Soviet times, most cities were surrounded by many small parcels of government-held land on which urbanites maintained small garden plots. These provided them with a place to relax on weekends and to grow food that was cheaper and often less contaminated with pesticides than purchased food. These gardens are highly productive, accounting for 20 to 30 percent of agricultural produce in some countries on perhaps 1 percent of the agricultural land area.

Since the fall of the Soviet Union, family gardens have proven essential to the diets of millions across the region. During the economic crises of the 1990s, the cost of food suddenly took up as much as half of family budgets. Even those with access to gardens were eating fewer vegetables, less meat, eggs, and dairy products, and even less bread, long a staple of the region's diet. Today, entrepreneurial farmers (especially from Ukraine and Caucasia) supply attractive fresh food to the best city markets. However, even with the recent economic improvements, few ordinary people can afford these luxuries and many still grow much of their own produce.

As urban economies expand in this region, the growing market for real estate may displace family gardens. For example, since the transition to a market economy, middle-class and wealthy suburbs are beginning to ring the city of Moscow, surrounding the long-cultivated garden plots of inner-city residents (Figure 5.21). However, given the instability of jobs in Moscow and other large cities, garden plots are likely to remain important supplements to family nutrition that urban residents will fight hard to keep. Even so, there is a limit to what urban and suburban people can produce, given the short amount of time they have to cultivate and the small spaces available for gardens. As urban incomes rise, some families may give up gardening and instead buy more at the supermarkets.

Democratization

Democratization has not proceeded as fast as the introduction of a market economy. While several countries have held elections, many forms of authoritarian control remain. In several countries, elected representative assemblies often act as rubber stamps for very strong presidents and exercise only limited influence on policy. In Russia, Vladimir Putin exercised tight control when he was president of the country and is thought to have at least tacitly allowed the repression and possibly even execution of rivals, dissidents, and critics of his policies. In 2008, after his two terms as president ended, Putin tapped a former aid with no experience in elected office to succeed him as president. Putin then engineered his own election to the office of prime minister, which has no term limits. Many see these moves as an effort by Putin to hold on to power indefinitely. 🎥

Elsewhere in the region, progress toward true democracy is similarly limited. Throughout Central Asia, authoritarian leaders unaccustomed to criticism or sharing power are still the norm. In Belarus, elections were harshly criticized as unfair, with systematic intimidation of voters and arrests of the political opposition. However, in both Georgia and Ukraine similarly fraudulent elections resulted in sustained popular protest movements that forced authoritarian leaders to step down and let new leaders take office. 🎥

Figure 5.21 Gardening by city dwellers. Russians, like many Europeans, often journey on weekends to the urban fringe, where they cultivate small garden plots. Here, members of a Moscow cooperative gardening society are sowing potatoes. The small house in the background serves as storage for tools and as accommodation for gardeners who want to spend the night in order to extend their gardening hours. [TASS/Sovfoto.]

The media and political reform. In the Soviet era, all communications media were under government control. There was no free press, and public criticism of the government was punishable. Journalists, risking punishment by criticizing public officials and policies, were instrumental in bringing an end to the Soviet Union. Between 1991 and the early 2000s, the communications industry was a center of privatization, and several media tycoons emerged. Privately owned newspapers and television stations regularly criticized the policies of various leaders of Russia and the other republics. It appeared that a free press was developing. However, recent years have seen a return to state control of the media and limitations on free speech.

A turning point was Vladimir Putin's rise to power, after which the most critical newspapers and TV stations in Russia were shut down. Since then, critical analysis of the government has become rare throughout Russia. Journalists who have openly criticized Putin's policies have been treated to various forms of censorship and violence. Since 2000, more than a dozen have been killed. ◼◼

Closely related to the struggle to develop a free press is the general public availability of communications technology (Table 5.1 and Figure 5.22). Television sets are widely available, and many people receive European and other stations via satellite dishes. Access to telephone landlines is still limited, but mobile phones have increased dramatically, especially in Siberia and Central Asia, where they are essential to many new enterprises. Mobile phone service, however, remains expensive and unreliable, and it is not available to most people.

Mobile phone companies are hoping to help the Newly Independent States leapfrog into the global economy by facilitating Internet commerce. Throughout the region, however, personal computers are still rare—Russia has one-tenth the number of computers per capita that the United States has. Internet users are far fewer in Central Asia and Caucasia than elsewhere in the region. Nonetheless, information technology is increasingly available, and demand is growing.

Table 5.1 Increase in Internet use, 2000–2007

Country	Internet users as percentage of population, 2007	Percentage increase in Internet users since 2000
Armenia	5.1	400
Azerbaijan	8	5,556.7
Belarus	35	1,785.8
Georgia	4	778
Kazakhstan	2.7	471.4
Kyrgyzstan	5.2	442.6
Moldova	10.9	1,524
Russian Federation	16.5	664.5
Tajikistan	0.1	150
Turkmenistan	0.5	1,700
Ukraine	11.5	2,539.1
Uzbekistan	3.3	11,633.3
United States	69.4	114.7

Source: "Internet usage in Asia," at http://www.internetworldstats.com/stats3.htm; "Internet usage in Europe," at http://www.internetworldstats.com/stats4.htm; "Internet usage statistics for the Americas," at http://www.internetworldstats.com/stats2.htm#north.

Chechnya and Other Strains on the Russian Federation

Russia's long history of expansion into neighboring lands has left it with an exceptionally complex internal political geography. As the Russian czars and the Soviets pushed the borders of Russia eastward toward the Pacific Ocean over the past

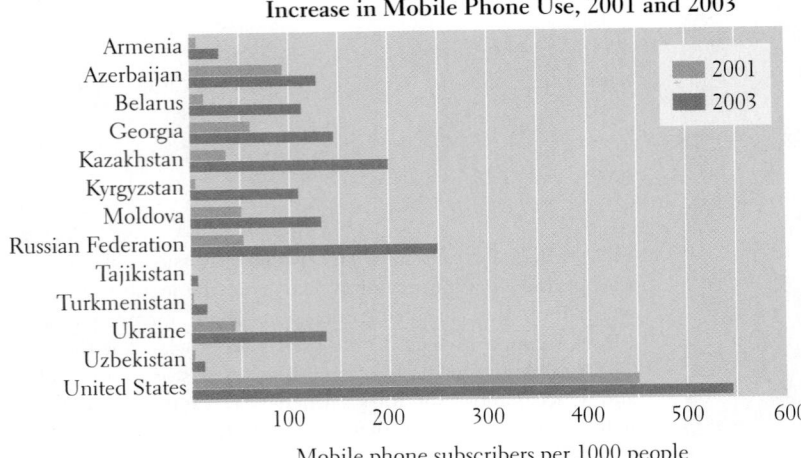

Increase in Mobile Phone Use, 2001 and 2003

Mobile phone subscribers per 1000 people

Figure 5.22 Access to mobile phones in Russia and the Newly Independent States. In general, access to information throughout the region is low compared with access in more developed countries. However, access to mobile phones has increased significantly since 2001. [*United Nations Human Development Report 2003* and *2005* (New York: United Nations Development Programme), Tables 11 and 13.]

Figure 5.23 Ethnic character of Russia and the Newly Independent States. The ethnic character of many former Soviet states and parts of Russia itself was changed by the policy of central planning as millions of Russians migrated to work on Soviet development schemes. Russians now form significant minority populations in many of the Newly Independent States and internal Russian republics where other ethnicities dominate. As of 2002, the ethnic makeup of Russia itself was 79.8 percent Russian, 3.8 percent Tatar, 2 percent Ukrainian, 1.2 percent Bashkir, 1.1 percent Chuvash, and 12.1 percent other. [Adapted from James H. Bater, *Russia and the Post-Soviet Scene* (London: Arnold, 1996), pp. 280–281; Graham Smith, *The Post Soviet States* (London: Arnold, 2000), p. 75; and *The World Factbook 2007*, at https://www.cia.gov/cia/publications/factbook/.]

500 years (see Figure 5.13 on page 186), they conquered a number of small non-Russian areas. These 30 republics and 10 autonomous regions now constitute 25 percent of Russia's land area (Figure 5.23). Many have significant ethnic minority populations that trace their origins to peoples who spoke such languages as German, Turkish, or Persian. In the more southerly republics, many are followers of Islam, the dominant religion in Central Asia. The peoples of most other republics are Christian, but some are also Buddhist, and many have animist beliefs that predate any of the organized religions.

Even during the Soviet period, but especially after its end, minorities organized to resist Russification and enhance local ethnic identities. Since the end of the Soviet Union, many ethnic Russians have returned to European Russia from the internal republics, Central Asia, and elsewhere.

Shortly after the breakup of the Soviet Union, several internal republics demanded greater autonomy, and two of them, Tatarstan and Chechnya, declared outright independence. Tatarstan has since been placated with greater economic and political autonomy. However, Chechnya's stronger resistance to Moscow's authority has led to the worst bloodshed of the post-Soviet era. For the time being, most of the other ethnic enclaves are content to stay closely associated with Russia. If they were to demand greater autonomy or independence, however, the territorial integrity of the vast Russian Federation would be threatened and its access to large, resource-rich areas diminished.

The conflict in Chechnya. Chechnya, a small Russian internal republic on the northern flanks of the Caucasus Mountains, has a fertile, hilly landscape that is home to 800,000 people. The Chechens converted to the Sunni branch of Islam in the 1700s, largely in response to Russian oppression. Since then, Islam has served as an important symbol of the Chechen identity. It was also an emblem of resistance against the Orthodox Christian Russians, who, following a long struggle, annexed Chechnya in the nineteenth century. After the Soviet Union was established, the Chechens and other groups in the Caucasus formed the Republic of Mountain Peoples, hoping to separate themselves from Russia, but the Soviets abolished the republic in 1924. In the 1930s, Chechens were forced onto collective farms, which supplied produce to distant Soviet cities.

In 1991, as the Soviet Union was dissolving, Chechnya declared itself an independent state. To the Russians, this declaration represented a dangerous precedent that could spark similar demands by other cultural enclaves throughout Russia. In addition, Russia wished to retain the region's agricultural and oil resources, and was planning to build pipelines across Chechen territory for transporting oil and gas to Europe from Central Asia. Since 1991, Chechen guerrillas seeking independence have repeatedly challenged Russian authority. The response has been bombing raids and other military operations that killed tens of thousands and created 250,000 refugees.

Chechens claim to be fighting simply for independence and the right to their own resources. Russia claims that Chechen factions have links to global terrorists and hence represent a threat to Russia's and the world's security. Russia's brutality throughout the conflict has raised doubts about its commitment to human rights and its overall ability to address internal political dissent. Fighting in Chechnya has died down in recent years, thanks in part to increased Russian investment in the republic's development. Still, sporadic clashes between Chechen militants and the Russian military took place in 2008.

> Fighting in Chechnya has died down in recent years, thanks in part to increased Russian investment in the republic's development.

Russia's involvement in Georgia's breakaway republics.

Just south of Chechnya in Georgia, conflicts between Russia and Georgia over the breakaway ethnic republics of South Ossetia and Abkhazia have worsened in the post Soviet era (Figure 5.24). Both republics are part of Georgia but have large ethnic Ossetian and Abkhazian populations that have long resisted Georgian control. Over the years, Russia has supported their agitation for secessions, even granting Russian citizenship to between 60 to 70 percent of these non-Georgian ethnic groups. This has been done in retaliation for Georgia's increasingly close relations with the United States and the EU, as evidenced by its candidacy for membership in NATO. A pipeline that links the oil fields of Azerbaijan with the Black Sea via Georgia has also been a source of contention. Russia would like to enhance its control over the oil resources of the Caspian Sea region with pipelines that go through Russian territory.

In August 2008, Georgia's military attempted to gain control over rebelling parts of South Ossetia, which had become major smuggling centers between Russia and Georgia, including the smuggling of nuclear materials discussed above. Russia responded with its much larger military, driving Georgia's forces out of South Ossetia. At the same time, fighters in Abkhazia drove Georgia's military out of that province, and Russian planes bombed a town near a Georgian pipeline. At present,

Figure 5.24 Areas of contention in Caucasia. [Adapted from European Center for Minority Issues, "Ethnopolitical map of the Caucasus," at http://info@ecmi.de/emap/m_caucasus.html.]

it remains uncertain whether Abkhazia and South Ossetia will break away and become independent countries, become provinces within the Russian Federation, or be satisfied by offers of greater autonomy within Georgia's federal structure. Regardless, Russia's relations with the EU and the United States have been severely strained.

SOCIOCULTURAL ISSUES

When the winds of change began to blow through the Soviet Union in the 1980s, few anticipated that the changes would happen so quickly and would so disrupt social stability. On one hand, new freedoms have encouraged self-expression and individual initiative, as well as a cultural and religious revival. On the other hand, women have faced new challenges in the post-Soviet era.

Religious Revival in the Post-Soviet Era

The official Soviet ideology was atheism, and religious practice and beliefs were seen as obstacles to revolutionary change. Now, religion is a major component of the general cultural revival occurring throughout the former Soviet Union. In European Russia (as well as in Georgia and Armenia), most people have some ancestral connection to Orthodox Christianity. Those with Jewish heritage form a sizable minority. Religious observance by both groups increased markedly in the 1990s, and many sanctuaries that had been destroyed or repurposed by the Soviets were rebuilt and restored.

A countertrend to the robust revival of Orthodox Christianity is the spread of evangelical Christian sects from the

United States (Southern Baptists, Adventists, and Pentecostals). The notion often promoted by this movement that with faith comes economic success may be particularly comforting both to those struggling with hardship and to those adjusting to new prosperity.

Vignette Valerii, age 35, once a government research scientist, now makes a comfortable living importing and exporting goods in the informal economy. Although he has to bribe officials and pay protection money to mobsters, his income has made his family much wealthier than their longtime friends. Vallerii's wife Nina is the only woman among them who does not work outside the home. In search of values that will guide them in these new circumstances, both have recently been baptized in an evangelical Christian sect. They say they chose this particular religious group because it promotes modesty, honesty, and commitment to hard work.

Source: Adapted from Timo Piirainen, Towards a New Social Order in Russia: Transforming Structures and Everyday Life (Aldershot, UK, and Brookfield, Vermont: Dartmouth Press, 1997), pp. 171–179. ■

Islam in Central Asia. Islam is now openly practiced and increasingly important politically in Central Asia, Azerbaijan, and some of Russia's internal ethnic republics. Especially in the Central Asian republics, however, the return to religious practices is often a subject of contention. Some local leaders still view traditional Muslim religious practices as obstacles to social and economic reform.

Many devout Muslims are wary of religious extremists from Iran and Saudi Arabia who may pose a security risk. Between 1992 and 1997, Tajikistan fought a civil war against Islamic insurgents with links to the Taliban in Afghanistan. In 2000, Uzbekistan and Kyrgyzstan joined forces to eliminate an extremist Islamic movement. But many religious leaders and human rights groups say that the fervor to eliminate radical insurgents has resulted in the persecution of ordinary devout Muslims, especially men. Human Rights Watch, an organization that monitors human rights abuses worldwide, reports that at least 4000 Muslim men have been arrested and detained in Uzbekistan alone. ▬◢

Gender: Challenges and Opportunities in the Post-Soviet Era

Soviet policy encouraged all women to work for wages outside the home. By the 1970s, 90 percent of able-bodied women in Russia were working full time, giving it the highest rate of female paid employment in the world. However, the traditional attitude that women are the keepers of the home persisted. The result was the double day for women. Unlike men, most women worked in a factory or

> By the 1970s, 90 percent of able-bodied women in Russia were working full time, giving that country the highest rate of female paid employment in the world.

office or on a farm for eight hours, and then returned home to cook, care for children, and do the housework (without the aid of household appliances). Because of shortages, they often had to stand in long lines to procure food and clothing for their families.

When market reforms reduced the number of jobs available to all citizens, President Gorbachev encouraged women to go home and leave the increasingly scarce jobs to men. Many women lost their jobs involuntarily, and by the late 1990s, 70 percent of the registered unemployed were women. But many, if not most, of these women were the sole support of their families due to illness, death, or divorce. Consequently, many have had to find new jobs. Often, the search is complicated by blatant gender bias in hiring and promotion practices. Job advertisements routinely specify gender, asking for an "attractive female receptionist" or a "male account executive."

The female labor force in Russia is now, on average, better educated than the male labor force. The same pattern is emerging in Belarus, Ukraine, Moldova, and parts of Muslim Central Asia. In Russia, the best-educated women commonly hold jobs as economists, accountants, scientists, and technicians, but they are unlikely to hold senior supervisory positions. As recently as 2005, the wages of women workers averaged 36 percent less than those of men.

The Trade in Women

Among the less savory entrepreneurial activities in the new market economies are those connected with the "marketing" of women. One part of this market is the Internet-based mail-order bride services aimed at men in western countries. Any viewer can find such services by going to Web sites related to Russia or other countries in the region. A woman in her late teens or early twenties, usually seeking to escape economic hardship, pays about $20 to be included in an agency's catalog of pictures and descriptions. (One Internet agency advertises 30,000 women listed.) She is then interviewed by the prospective groom, who travels—usually to Russia or Ukraine—to choose from the women he has selected from the catalog.

Kidnapping of women for sex work outside the region (Figure 5.25) has also increased. Precise numbers are hard to come by. However, in 2000, *The Economist* estimated that 300,000 such women were smuggled each year into the European Union, where the sex trade generates $9 billion annually. The business of supplying women is dominated by members of the Russian mafia, who take advantage of women desperately seeking jobs as domestic servants or waitresses in Europe. Upon arrival, they are forced to work as prostitutes or strippers. They may later be sold for a few thousand dollars to brothels in parts of the world where their European looks will be considered exotic. They are often unprotected from sexually transmitted diseases. Despite many national and global efforts to curtail the trade in women, help still reaches only a minority.

Figure 5.25 The trade in women. Seventeen-year-old Lyubov looks out a prison window as she awaits deportation from Israel as an illegal sex worker. Six months ago, Lyubov left her Russian coal mining city only to be unknowingly sold into prostitution for $9000. The importation of women is a big industry in Israel; women are recruited from the former Soviet Union and sent to brothels throughout the country. [Reuters.]

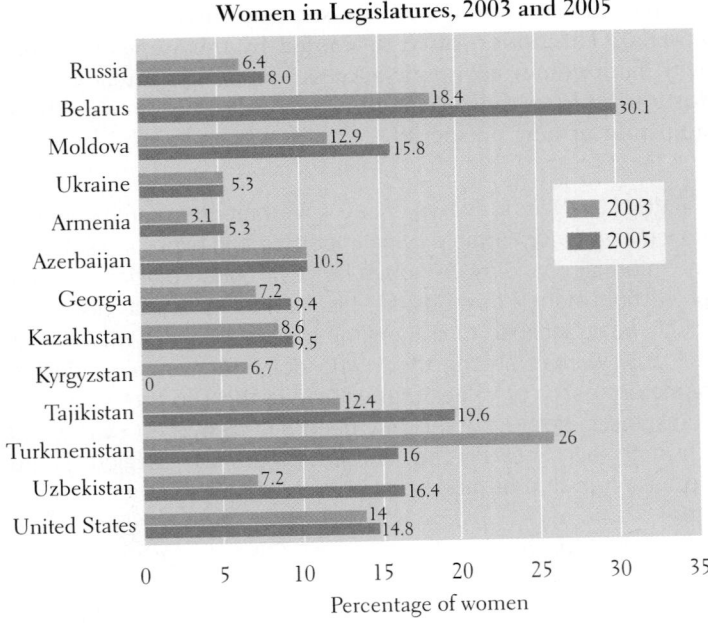

Women in Legislatures, 2003 and 2005

	2003	2005
Russia	6.4	8.0
Belarus	18.4	30.1
Moldova	12.9	15.8
Ukraine	5.3	
Armenia	3.1	5.3
Azerbaijan		10.5
Georgia	7.2	9.4
Kazakhstan	8.6	9.5
Kyrgyzstan	0	6.7
Tajikistan	12.4	19.6
Turkmenistan	26	16
Uzbekistan	7.2	16.4
United States	14	14.8

Percentage of women

Figure 5.26 Women legislators. This graph shows the percentage of legislators who are women in Russia and the Newly Independent States (with the United States statistics for comparison). Belarus and Turkmenistan stand out as having the most women lawmakers, but both are authoritarian societies in which true democratic participation is rare. [Adapted from *United Nations Human Development Report 2005* (New York: United Nations Development Programme), Gender Empowerment Measure, Table 26, and http://www.ipu.org/wmn-e/classif.htm.]

The Political Status of Women

Although they were granted equal rights in the Soviet constitution, women never held much power. In 1990, women accounted for 30 percent of Communist Party membership, but just 6 percent of the governing Central Committee. Since the fall of the Soviet Union, the political empowerment of women has advanced the least where democracy has developed the most, perhaps because of longstanding cultural bias against women in positions of power. Where governments are less democratic and more authoritarian, as in Belarus and Turkmenistan, women actually hold more legislative positions (Figure 5.26). However, they may have been promoted undemocratically by leaders who wanted to appear more progressive in the eyes of international donors.

Support among women for women's political movements is not widespread as many fear being seen as anti-male or against traditional feminine roles. Nonetheless, the influence of women promises to increase. Since the late 1990s, especially in Caucasia and the Central Asian republics, women have been working against gender discrimination through nongovernmental organizations and the United Nations Development Programme (UNDP). Hence, formal representation in parliament may no longer be the best measure of women's political activity and influence.

Reflections on Russia and the Newly Independent States

This is a region of dramatic change. As the Soviet Union, these countries amazed the world with their rapid industrialization and societal transformations. Since the collapse of the Soviet system, they have attempted an even more rapid transformation to a free market economy. Through it all, Russia has retained its leadership position, though its power and influence do not compare with that of the Soviet Union. Some former Soviet allies in eastern Europe are now closely connected with the increasingly powerful European Union, as are some of this region's most important agricultural producers (Ukraine and Georgia). The Central Asian republics are likewise drawn toward their neighbors in Southwest, South, and East Asia.

While oil wealth is transforming relationships with the rest of the world, the benefits within the region have been uneven. Many are still reeling from the loss of jobs and social services since the collapse of the Soviet system. Women in particular have suffered from high unemployment. Meanwhile, corruption and crime have exploded. Partially in response to all of these problems and the uncertain futures they create, birth rates are falling and populations shrinking in most parts of the region.

In the face of these challenges, it is not surprising that central governments have remained strong and democracy weak. While some countries are eager to learn from the world's wealthy democracies, there is also widespread resistance to rapid change and to excessive influence from abroad. Many see stronger, more authoritarian government control as necessary to maintain stability, even if it restricts freedom of speech and weakens democratic processes.

All parts of the region will be hampered for years to come by an aging and inefficient infrastructure and by severe environmental degradation. As modernization and privatization proceed, the rapid development of the region's rich resource base will probably lead to more pollution. Greenhouse gas emissions are likely to increase, as are water and air pollution. However, rising living standards may also create pressures for high environmental standards, especially in this region's many highly polluted but also increasingly wealthy urban areas. And if societies continue to open, there will be exhilarating opportunities for entrepreneurialism, self-expression, and political empowerment. Perhaps this region's experience with rapid transformation will serve it well in the years ahead.

Chapter Key Terms

Bolsheviks 186
capitalists 186
Caucasia 188
centrally planned or socialist economy 186
Cold War 187
communism 186
Communist Party 186
czar 186
Gazprom 194

glasnost 188
Group of Eight (G8) 194
Mongols 185
nomadic pastoralists 184
nonpoint sources of pollution 180
perestroika 188
permafrost 178
privatization 191
Russian Federation 178

Russian Mafia 193
Russification 191
Slavs 185
Soviet Union 174
steppes 178
taiga 178
tundra 178
underemployment 192

Critical Thinking Questions

1. What were the social circumstances that gave rise to the Communist Revolution? Explain why the revolutionaries adopted a form of communism grounded in rapid industrialization and central planning.

2. Discuss the tensions that led to the cold war from the perspective of a Soviet citizen who lived through World War II.

3. Describe the challenges posed to agriculture and transportation by the physical environments (landforms and climate) of the various parts of this region: the North European Plain, the Caucasus, the Central Asian steppes, Siberia, and the various mountain ranges (Urals, Altai, and Pamirs).

4. In general terms, discuss the ways in which the change from a centrally planned economy to a more market-based economy has affected career options and standards of living. Consider the elderly, young professionals, members of the military, women, and former government bureaucrats.

5. Discuss the reasons for Russia's diminishing sphere of influence after 1991. How has Russia's role in global politics changed recently?

6. Compare the ways in which the process of democratization has proceeded in the different countries of the region.

7. How has the development of the region's fossil fuels created both conflict and cooperation between Russia and the rest of the world?

8. Given the changing options for women in the region, do you anticipate that birth rates will remain low or increase? Explain your reasoning.

9. Why do so many people maintain urban family gardens in this region? What forces might lead to the decline of urban gardening?

10. How would different parts of this region be affected if a global effort to reduce greenhouse gas emissions resulted in dramatically less use of fossil fuels?

The map on the left shows the region of North Africa and Southwest Asia, including countries such as Russia, Kazakhstan, Uzbekistan, Turkmenistan, Kyrgyzstan, Tajikistan, Afghanistan, Iran, Iraq, Saudi Arabia, Yemen, Oman, United Arab Emirates, Qatar, Bahrain, Kuwait, Syria, Georgia, Armenia, Azerbaijan, Eritrea, Ethiopia, Somalia, Djibouti, Kenya, and surrounding bodies of water including the Caspian Sea, Persian Gulf, Gulf of Oman, Arabian Sea, Red Sea, Gulf of Aden, and Indian Ocean. Labeled features include Caucasus Mts., Mt. Ararat (elev. 16,946), Elburz Mts., Zagros Mts., Arabian Peninsula, Rub'al Khali, Ethiopian Highlands, Horn of Africa, and Somali Peninsula. Cities labeled include Volgograd, Tbilisi, Yerevan, Baku, Ashgabat, Tashkent, Dushanbe, Mashhad, Tehran, Tabas, Tabriz, Mosul, Kermanshah, Esfahan, Kerman, Baghdad, Abadan, Al Basrah, Shiraz, Bushehr, Bandar-e Abbas, Kuwait, Rafha, Ad Dammam, Bandar Beheshti, Al Manamah, Ad Dawhah, Abu Zaby, Muscat, Riyadh, Al Jawarah, Al Madinah (Medina), Jiddah, Makkah (Mecca), Port Sudan, Abha, San'a, Taizz, Aden, Asmera, Djibouti, Addis Ababa, Mogadishu, Nairobi, and Victoria (Seychelles).

CHAPTER 6

North Africa and Southwest Asia

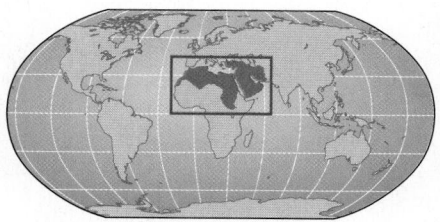

Global Patterns, Local Lives It is April 2003, during the early stages of the Iraq war, and a group of men in a coffee shop in Alexandria, Egypt, are visibly agitated as they watch a television broadcast from a neighborhood in Baghdad, Iraq. A grief-stricken man whose brother's house has just been destroyed by a U.S. bomb is screaming at the camera: "I don't want this freedom! I don't want this democracy! My brother and his children are dead!"

The channel they are watching is Al Jazeera, which with 50 million viewers is by far the most popular channel in Southwest Asia and North Africa (Figure 6.2). It is also the most controversial and criticized. Al Jazeera has built a reputation as a critic of leaders and governments throughout the region since its founding in 1996 by a group of journalists whose TV station was closed by the Saudi Arabian government. The network has been banned from time to time from operating inside Iran, Iraq, Jordan, and other countries due to its controversial reporting. It is now based in Qatar and is funded by that country's hereditary ruler.

Al Jazeera's broadcasts generally reflect popular sentiment among its viewers, such as the strong anti-U.S. sentiment that has followed the U.S. invasion of Iraq. But some of the station's key personnel have political opinions with which most U.S. citizens would agree. Senior Producer Samir Khader describes Al Jazeera's role as "to educate the Arab masses on something called democracy, respect of other opinions, free debate . . . and to try by using all these things to shake up these rigid societies, to awaken them."

Nevertheless, the U.S. government, which once praised Al Jazeera as a beacon of free speech for its coverage of corruption and human rights abuses, is now the station's most vocal critic. Former Secretary of Defense Donald Rumsfeld accused Al Jazeera of using actors to stage broadcasts like the one described above to deliberately enrage Iraqis against the U.S. presence in Iraq.

Figure 6.1 Regional map of North Africa and Southwest Asia.

Figure 6.2 Al Jazeera. News Anchor Muntaha al-Rumahi presents a panel discussion on Osama bin Laden from Kandahar, Afghanistan, in October 2001. [AFP/Getty Images.]

Al Jazeera is already hugely influential in the Arabic-speaking world for its willingness to criticize governments and expose corruption—a serious problem in the region. With the launching of Al Jazeera International, an English language news channel with broadcast centers throughout the world, the network is poised to become a global force. The station will cover issues from an international perspective. ▶

Adapted from "Control Room" an award-winning documentary about Al Jazeera; http://english.aljazeera.net/NR/exeres/55ABE840-AC30-41D2-BDC9-06BBE2A36665.htm;"Rumsfeld's Al Jazeera Outburst," The Sunday Times-Online, November 27, 2005: http://www.timesonline.co.uk/article/0, 2089-1892464,00.html; Lydia Pulsipher's conversations with Josh Rushing, University of Tennessee, April, 2006; http://www.alarabiya.net/english.htm#003. ■

Freedom of the press is just recently developing in North Africa and Southwest Asia (Figure 6.3). The 21 countries in the region (Figure 6.4) only achieved political independence in the twentieth century, and many of them not until after World War II. While a few countries were outright colonies of Europe, most were forced to submit to some type of control by European countries (and more recently by the United States). While Europe strongly influenced this region's economies and political systems, democratic institutions, such as a free press, were not encouraged. These institutions have been slow to develop in the independence era, a fact that makes the emergence of Al Jazeera all the more remarkable.

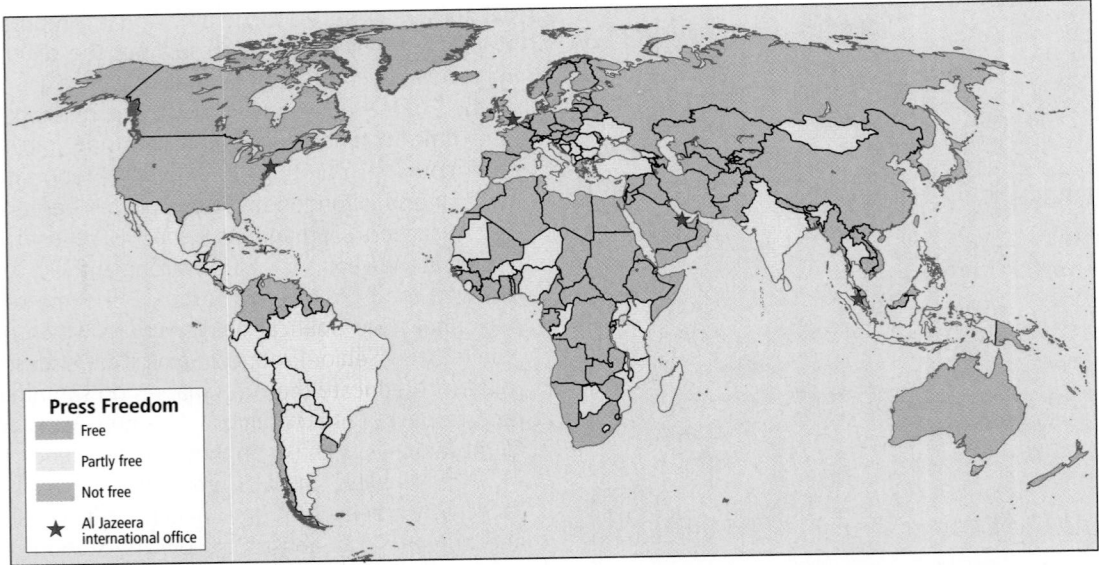

Figure 6.3 Freedom of the press worldwide, 2006. The three categories of "freedom of the press" shown on the map are based on several factors: the legal environment in which the media operate, the degree of independence from government ownership and influence, economic pressures on news content, and violations of press freedom ranging from murder of journalists to extralegal abuse and harassment. [Adapted from "Map of Press Freedom 2006," Freedom House, at http://www.freedomhouse.org/uploads/maps/fop_current.pdf.]

Press Freedom
- Free
- Partly free
- Not free
- ★ Al Jazeera international office

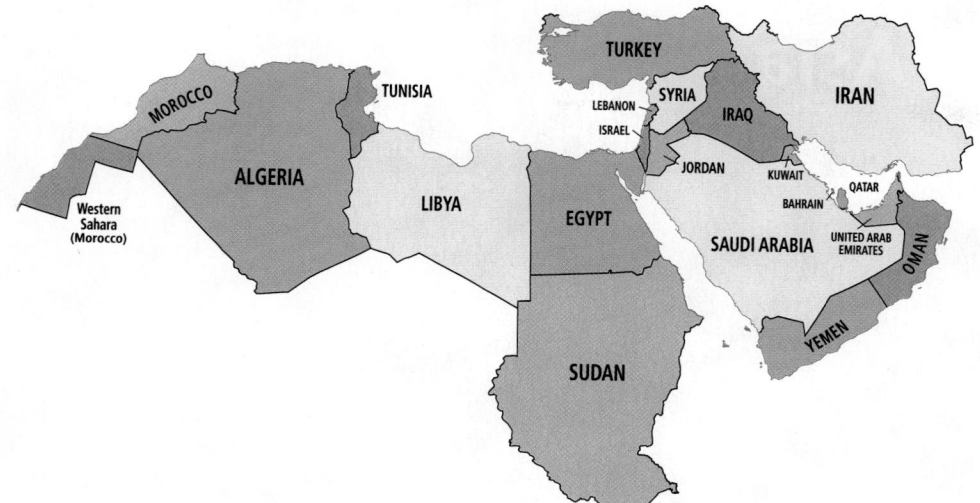

Figure 6.4 Political map of North Africa and Southwest Asia.

There is both wide variety and striking continuity in this region. Most countries share an arid climate and the vast majority of people practice **Islam,** a monotheistic religion that emerged in the seventh century C.E. when according to the Qur'an, the archangel Gabriel revealed the principles of the religion to the Prophet Muhammad. However, there is also great variety in Islam. Some practitioners are quite moderate in their thinking, while others are drawn to fundamentalist Islam, or **Islamism.** Islamist movements are grassroots religious revivals that seek political power to curb the non-Islamic influences that have become widespread throughout this region thanks to globalization. Islamist movements have been successful in some countries, but not in others.

Another source of variety is **fossil fuel** reserves. Formed from the fossilized remains of dead plants and animals, they are abundant in some countries but totally absent in others. In this region, they are found mainly around the Persian Gulf, where vast deposits of oil and natural gas exist. Fossil fuels are extracted from these reserves and exported throughout the world for tremendous profits.

There are many other sources of variety in this region. In some countries, nearly everyone lives in a city, while in others, life is more rural. Women may be educated and active in public life and government, or they may lead secluded domestic lives with few educational opportunities. Ethnicity also varies greatly within countries and across the region.

Questions to consider as you study North Africa and Southwest Asia. Also look at Views of North Africa and Southwest Asia, pages 208–209.

1. How will people get enough water to grow food? As the population of this dry region increases, obtaining enough water for agriculture will become harder. Will new technologies provide solutions, or will countries become even more dependent on imported food? How might global warming affect access to food?

2. How have the huge fossil fuel reserves of some countries in the region driven globalization? The vast energy resources of the Persian Gulf states have long attracted interest from European and North American companies eager to make profits, as well as from governments throughout the world in need of energy. How have the economies of the fossil fuel-rich countries in this region become linked to the global market for energy and to foreign sources of expertise?

3. How might democratization provide a path to peace between Islamists and the governments whose policies they often fervently oppose? Political movements based on Islamism have often been harshly repressed by undemocratic governments. Islamists have frequently responded violently with terrorism and armed resistance. Can democratization provide a path toward peace between opposing ideologies?

4. How does the low status of women contribute to this region's high population growth? This region has the second highest population growth rate after sub-Saharan Africa. Women are generally less educated than men and tend not to work outside the home. How do these factors contribute to high population growth?

5. How has globalization shaped different patterns of urbanization throughout the region? Some urban growth has been fueled by massive infusions of oil wealth. Elsewhere, economic reforms aimed at improving global competitiveness have resulted in massive migrations from rural areas. How have these contrasting forces produced dramatically different urban landscapes?

Views of North Africa and Southwest Asia

For a tour of these places, visit www.whfreeman.com/pulsipher
[Background image: Lori K. Perkins NASA/GSFC Code 610.3/587 Scientific Visualization Studio]

2. Population and gender. This region's high population growth rate will strain already scarce water and agricultural resources. Giving women more opportunities to study and work outside the home may reduce birth rates. This Palestinian woman and her three children live in the West Bank. See page 221–222 for more. Lat 31°46'32"N, Lon 35°15'16.43"E. [Armon]

1. Gender and Islam. Many conservative Muslims object to the liberalization of women's roles, preferring that they remain secluded within the home. Many women in this region are challenging such traditional views, as can be seen in the dress and somewhat defiant stance of this Egyptian woman as she speaks with Islamic students in Cairo. See page 225–227 for more. Lat 30°02'45"N, Lon 31°15'45"E. [DOD]

Cairo 1972

Cairo 2005

7. Globalization and urbanization: economic reform. Many cities have experienced an inflow of rural migrants since the implementation of economic reforms designed to reduce debt and increase global economic competitiveness. Cairo (shown right) is growing dramatically as rural migrants stream in to find work. The inset maps show Cairo in 1972, when its population was roughly six million, and 2005, when its population was about 16 million. See pages 222–223 for

3. Food and water. The greatest use of water in this dry region is for irrigated agriculture. However, traditional methods, such as trench irrigation (shown below in Iraq's Salah ad Din province) can result in the build up of salts in the soil which reduces fertility. Newer "drip irrigation" techniques (shown in the inset) can help solve this and other problems by drastically reducing the amount of water needed to grow food. See pages 211–213 for more. Lat 34°13'45"N, Lon 43°39'E. [Ben Barber/USAID; USDA]

4. Democratization in Iraq has created pressure for a rapid US Military withdrawal from the country. Iraqi parliamentarians, sensitive to Iraqi opinion polls showing strong opposition to any long-term US presence in Iraq, have demanded an end to the US occupation. Below is a demonstration against the US presence in Baghdad. See page 236 for more. Lat 33°18'49"N, Lon 44°23'26"E. [LCPL JENNIFER A. KRUSEN, USMC/DOD]

Salah ad Din

3

2

4

1&7 West Bank

Cairo

Baghdad

6

Saudi Arabia

5

Dubai

5. Globalization and urbanization: oil wealth. Over the past 30 years incredible amounts of money have flowed into the cities of the oil-rich Persian Gulf states. The city of Dubai (shown above) has used this wealth to create a modern diversified economy, and an urban landscape decorated with man-made islands (Palm Jumeirah is shown above) for the ultra-rich to live on, and dramatic new buildings, such as the Burj Al Arab Hotel (shown in the inset). See pages 222, 227–228, and 229 for more. Lat 25°07'N, Lon 55°08'E. [SvG; NASA]

6. Global warming and oil. A global shift away from fossil fuels would transform geopolitics in the region, as foreign powers like the United States and Europe would have less incentive to influence events here. Above, a US military plane flies over a natural gas well in Saudi Arabia. See page 216 for more. Lat 26°54'N, Lon 49°46'E. [Lt. Chuck Radosta, U.S. Navy/DOD]

The Geographic Setting

Terms to Be Aware Of

The common term *Middle East* is not used in this chapter because it describes the region from a European perspective. To someone in Japan, the region lies to the far west, and to a Russian, it lies to the south. The *Arab World* is also not used because not all people in the region are of the Arab ethnic group. In this book, the term used for this region is North Africa and Southwest Asia.

PHYSICAL PATTERNS

Climate

No other region in the world is as dry as North Africa and Southwest Asia (Figure 6.5). A belt of dry air that circles the planet between roughly 20° and 30° N creates desert climates in the Sahara of North Africa, the Arabian Peninsula, Iraq, and Iran.

The Sahara's size and southerly location make it a particularly hot desert region. In some places, temperatures can reach 130°F (54°C) in the shade at midday. With little water or vegetation to retain heat, nighttime temperatures can drop quickly to below freezing. Nevertheless, in even the driest zones,

humans survive at scattered oases, where they maintain groves of drought-resistant plants such as date palms. Desert inhabitants often wear light-colored, loose, flowing robes that reflect the sunlight during the day and provide warmth at night.

Landforms and Vegetation

The rolling landscapes of deserts and steppes cover most of North Africa and Saudi Arabia (see Figures 6.1 and 6.4). In a few places, mountains capture moisture, allowing plants, animals, and humans to flourish.

In northwestern Africa, the Atlas Mountains stretch from Morocco on the Atlantic coast to Tunisia on the Mediterranean coast. They block and lift damp winds from the Atlantic Ocean, creating rainfall of more than 50 inches (127 centimeters) per year in some places (see Figure 1.23 on page 27). In some Atlas Mountain locations, snowfall is sufficient to support a skiing industry.

Africa and Southwest Asia are separated by a rift formed between two tectonic plates—the African Plate and the Arabian Plate—that are moving away from each other (see Figure 1.21 on page 25). The rift, which began to form about 12 million years ago, is now filled by the Red Sea. The Ara-

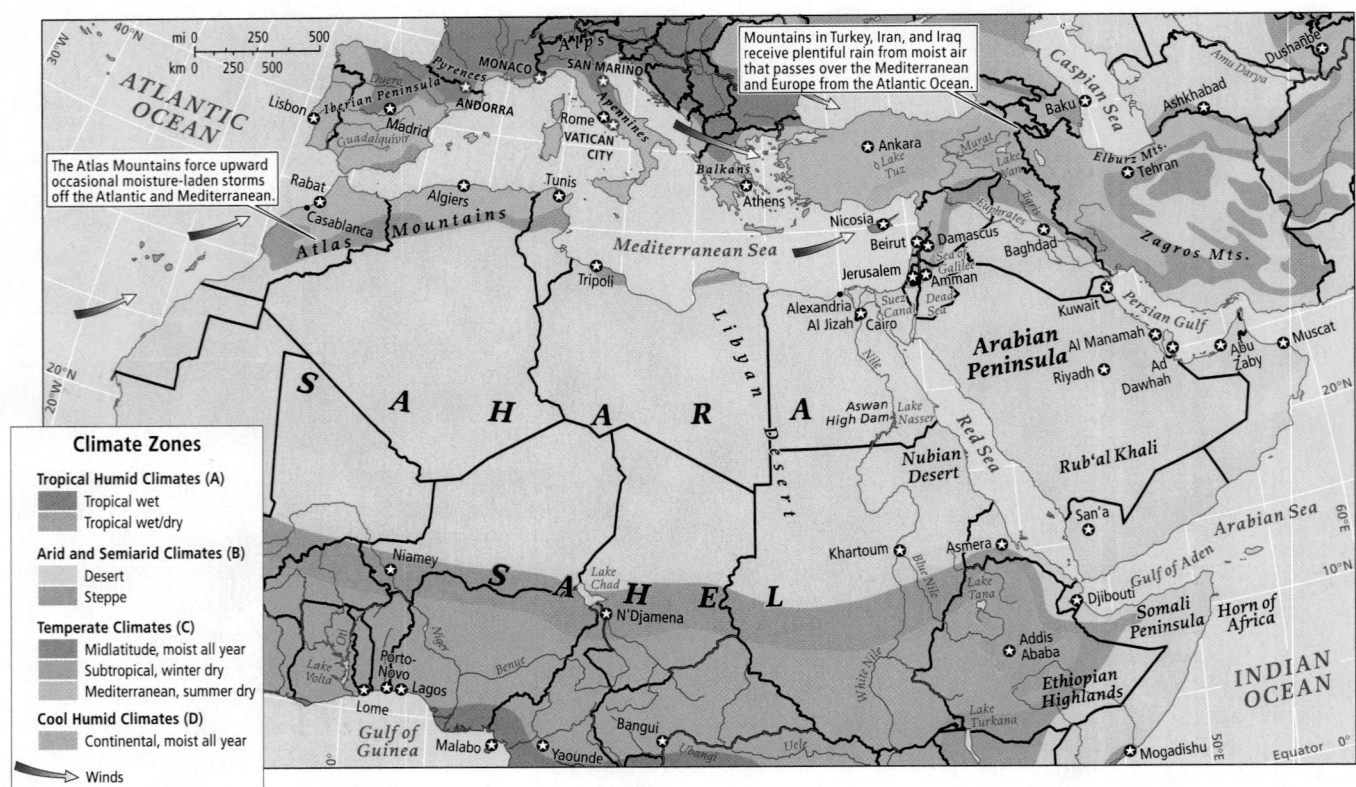

Figure 6.5 Climates of North Africa and Southwest Asia.

Figure 6.6 A wadi, or dry riverbed, in the Ahaggar Mountains of Algeria. [Victor Englebert/Photo Researchers.]

bian Peninsula lies to the east of this rift. There, mountains bordering the rift in the southwestern corner rise to 12,000 feet (3658 meters).

Behind these mountains to the east lies the great desert region of the Rub'al Khali. Like the Sahara, it has virtually no vegetation. The sand dunes of the Rub'al Khali, which are constantly moved by strong winds, are the world's largest, some reaching more than 2000 feet (610 meters).

The landforms of Southwest Asia are more complex. The Arabian Plate is colliding with the Eurasian Plate and pushing up the mountains and plateaus of Turkey and Iran (see Figure 1.21 on page 25). These same movements often result in earthquakes. Turkey's mountains lift damp air passing over Europe from the Atlantic, resulting in considerable rainfall. Only a little rain makes it over the mountains to the interior of Iran, which is very dry.

There are only three major rivers in the entire region, and all have attracted human settlement for thousands of years. The Nile flows north from the moist central East African highlands. It crosses arid Sudan and desert Egypt to a large delta on the Mediterranean. The Euphrates and Tigris rivers both begin in the mountains of Turkey, flowing southeast to the Persian Gulf.

A fourth and much smaller river, the Jordan, starts as snowmelt in the uplands of southern Lebanon and flows through the Sea of Galilee to the Dead Sea. Most other streams are dry riverbeds, or wadis, most of the year (Figure 6.6). They carry water only after the light rains that fall between November and April.

North Africa and Southwest Asia were home to some of the very earliest agricultural societies. However, agriculture is only possible in a few places. In spots along the Mediterranean coast, rain is sufficient to grow citrus fruits, grapes, olives, and many vegetables, though supplemental irrigation is often needed. In the valleys of the major rivers, seasonal flooding and irrigation support cotton, wheat, barley, and vegetables.

ENVIRONMENTAL ISSUES

Salam, the Arabic root of the word *Islam,* means peace and harmony. The **Qur'an** (or **Koran**), the holy book of Islam, requires Muslims to avoid spoiling or degrading human and natural environments and to share resources, especially water, with all forms of life. In practice, Muslims—the largest religious community in North Africa and Southwest Asia—and other residents of the region do conserve water better than most people in the world. Buildings are designed to conserve moisture by maximizing shade. Bathing often takes place in public baths, where water use is tightly controlled. Mountain snowmelt is captured and moved to dry fields and villages via underground water conduits called *qanats.*

However, this region's 450 million residents have such a limited water supply that these measures have not been enough to ensure an adequate supply. Growing populations virtually guarantee worsening water shortages in the future (Figure 6.7). By 2025, almost every country in the region will have less water than the UN considers necessary to support basic human development. Many approaches to water management have been attempted in this region, but all have run into obstacles that limit their effectiveness. Water shortage is one of the area's most significant environmental problems (Figure 6.8). ◣

Water and Food Production

The greatest use of water in North Africa and Southwest Asia is for irrigated agriculture, even though this activity does not contribute significantly to national economies. In Tunisia, for example, agriculture accounts for just 14 percent of GDP but 88 percent of all the water used. Only 9 percent of water is used in homes, and just 3 percent by industry. However, when irrigated agriculture is measured in terms of its value to diets, family budgets, and rural economies, it emerges as essential even in this water-stressed region. Even so, despite the fact

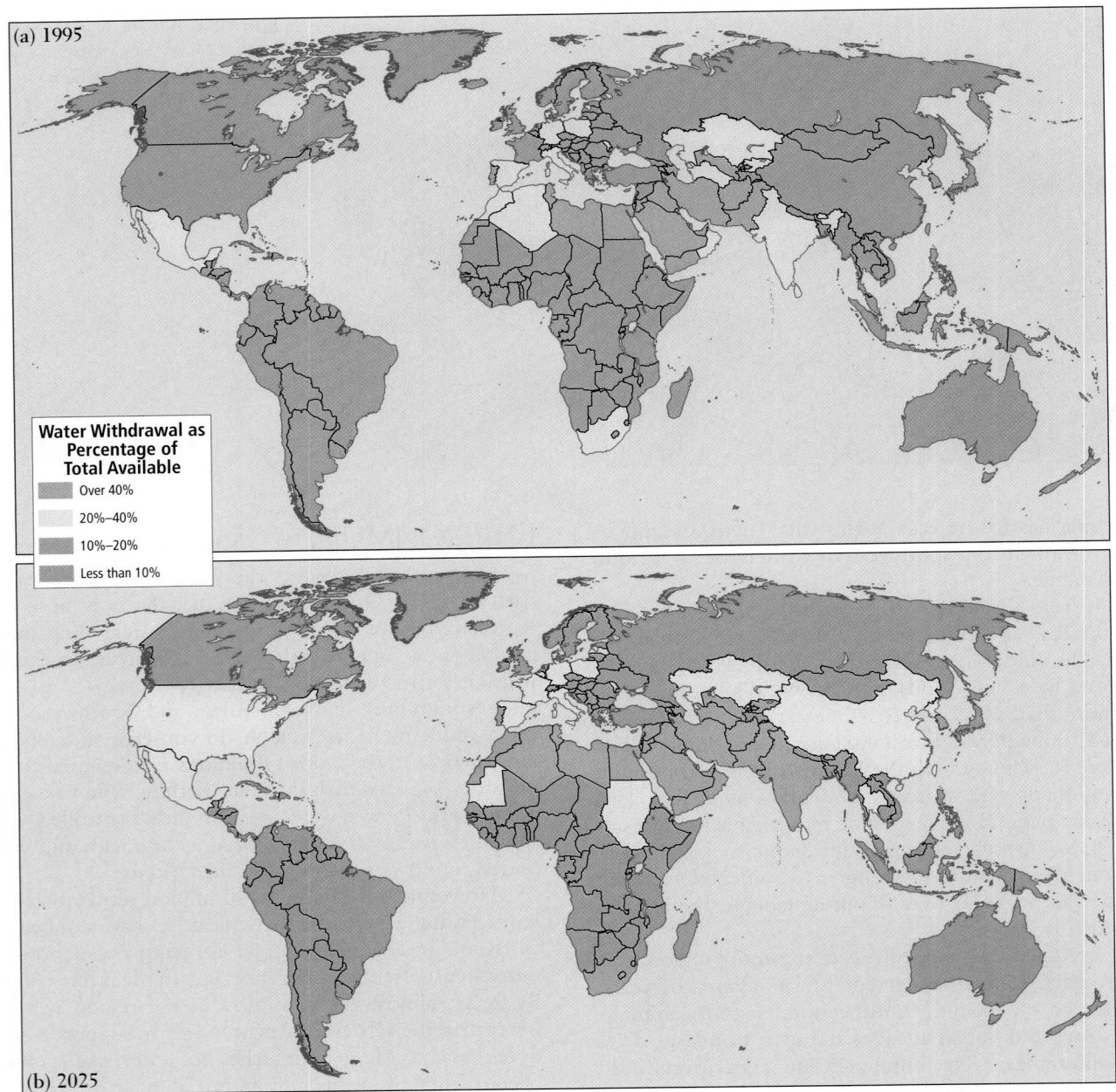

Figure 6.7 Global water availability, 1995 and 2025. Declining availability of potable water is a critical environmental issue for most countries in the region. An area experiences *water stress* when annual water supplies drop below 1700 cubic meters per person (in yellow on the map). *Water scarcity* occurs when supplies drop below 1000 cubic meters per person, the minimum needed for human health (orange on the map). By 2025, only Turkey will have sufficient water to maintain good health and meet development needs. [Adapted from "Freshwater Stress" map (New York: United Nations Environment Programme, 2002), at http://maps.grida.no/go/graphic/freshwater_stress.]

that most countries in the region subsidize agriculture in some way, almost all are highly dependent on imported food.

Until recently, agriculture has been confined to a few coastal and higher altitude zones where rain can support agriculture, and to river valleys (such as the Nile and Tigris and Euphrates valleys) where farms can be irrigated (see Figure 6.5). However, Libya, Egypt, Saudi Arabia, Turkey, Israel, and Iraq all have ambitious irrigation schemes that have expanded agriculture deep into the desert (Figures 6.9 and 6.10).

These and other irrigation projects have damaged soil fertility through **salinization.** This process occurs when large quantities of water are used to irrigate and a substantial amount of the water evaporates, leaving behind salts and other minerals. Most traditional irrigation methods—where trenches or furrows

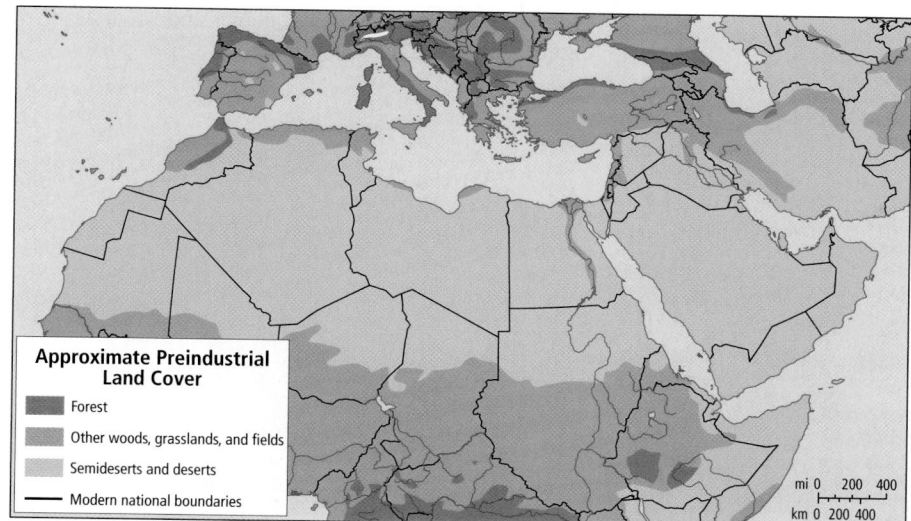

Figure 6.8 Land cover (preindustrial) and human impact (2002). In the preindustrial era **(a)**, human impact on the environments of this region was localized and light and not visible at this scale. By 2002 **(b)**, even the most remote desert locations bore some human impact. [Adapted from *United Nations Environment Programme, 2002, 2003, 2004, 2005, 2006* (New York: United Nations Development Programme), at http://maps.grida.no/go/graphic/human_impact_year_1700_approximately and http://maps.grida.no/go/graphic/human_impact_year_2002.]

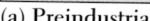

(a) Preindustrial

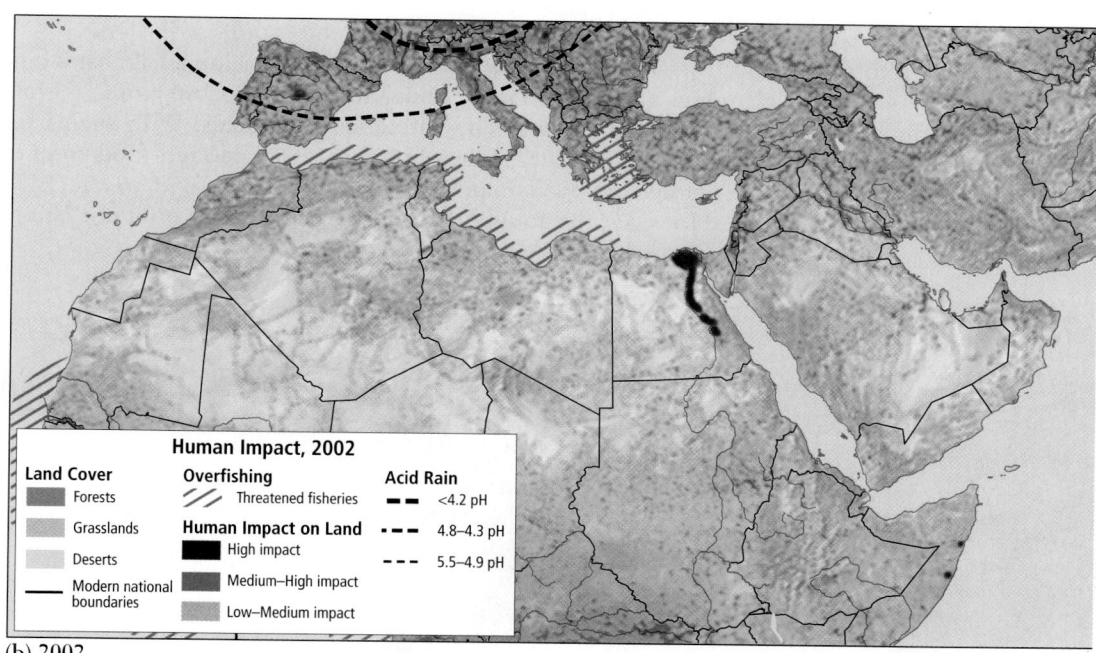

(b) 2002

> Most traditional irrigation methods, as well as modern irrigation systems that use large amounts of water, lead to soil salinization.

are dug and filled with water—and modern irrigation systems—which use large amounts of water—lead to salinization. The salts accumulate over time and inhibit plant growth. Salinization is particularly likely to occur in arid lands because there is little rain to wash away the salt.

Israel has developed more efficient techniques of *drip irrigation* that dramatically reduce the amount of water used, thereby limiting salinization and freeing up water for other uses. However, until very recently, poorer states have been unable to afford this somewhat complex technology, which

requires an extensive network of hoses and pipes to deliver water to each plant. Other countries are wary of depending on a technology developed by Israel, a country they deeply distrust.

Strategies for increasing water supply. Some strategies have been developed for increasing water supply, but they have their own problems. The fossil fuel–rich countries of the Persian Gulf have invested heavily in **seawater desalination** technologies that remove the salt from seawater, making it suitable for drinking or irrigating. Seventy percent of Saudi Arabia's drinking water is supplied by desalination plants, and some wheat fields there are irrigated this way. However, the process of

Figure 6.9 An irrigated wheat field in the Saudi Arabian desert. [Ray Ellis/Photo Researchers.]

desalination uses huge amounts of energy, making it too costly for most countries.

Many countries pump groundwater from underground aquifers to the surface for irrigation or drinking water. Libya has invested some of its earnings from fossil fuels in one of the world's largest groundwater pumping projects, known as the *Great Man-Made River*. This project supplies almost 2 billion gallons of water per day to Libya's coastal cities, as well as to 600 square miles of agricultural fields. With such irrigated agriculture, Libya hopes to grow enough food to end its current food imports, and even supply food to the EU. However, the aquifer that the project depends on is not being

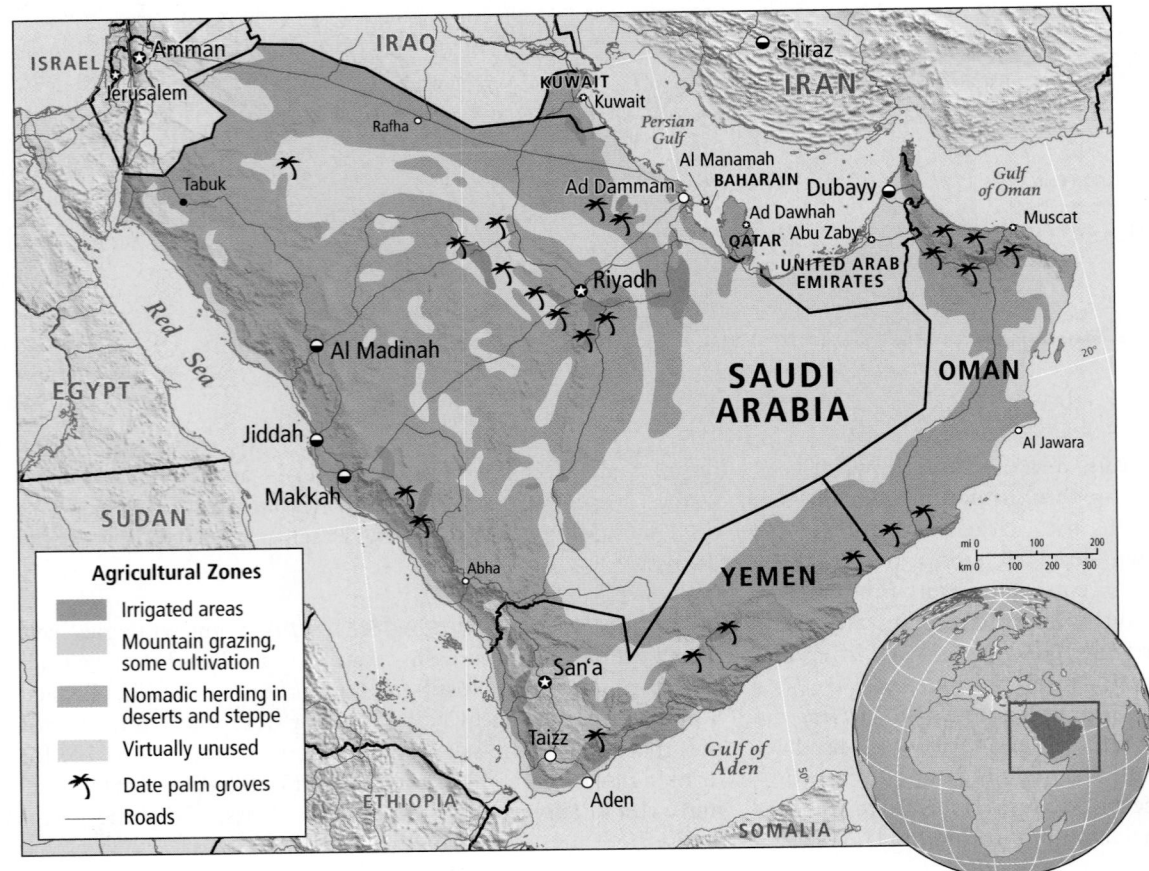

Agricultural Zones
- Irrigated areas
- Mountain grazing, some cultivation
- Nomadic herding in deserts and steppe
- Virtually unused
- Date palm groves
- Roads

Figure 6.10 Irrigated agriculture and other land uses on the Arabian Peninsula.

replenished and may run dry some day. Elsewhere in the region, groundwater pumping projects have caused land to sink. If too much water is taken from aquifers near the ocean, seawater can flow in, rendering them useless.

Dams and reservoirs have been built on the regions' major river systems to increase water supplies, but these projects have created new problems. In Egypt, for example, the natural cycles of the Nile River have been altered by the construction of the *Aswan Dams*. Downstream of these two dams, water flows have been reduced and floods no longer deposit fertility-enhancing silt on the land. As a result, expensive fertilizers are now necessary. Moreover, with less water and silt coming downstream, parts of the Nile delta are sinking into the sea. Upstream, the artificial reservoir created by the dams has flooded villages, fields, wildlife habitats, and historic sites.

Cross-border issues. Diplomatic relations within the region are strained by the need to share the water of rivers that flow across national boundaries. Turkey's *Southeastern Anatolia Project,* which involves the construction of several large dams on the Euphrates River, has reduced the flow of water to the downstream countries of Syria and Iraq. In negotiations over who should get Euphrates water, for example, Turkey argues that it should be allowed to keep more water behind its dams because the river starts in Turkey and most of its water originates there (Figure 6.11). Meanwhile, Iraq argues that Turkey has no right to diminish the water available downstream. Iraq points out that the Euphrates travels the longest distance in Iraq, and that the Sumerians (the ancient inhabitants of what is now Iraq) were the earliest major users of Euphrates water.

Desertification

A number of changes characterize the conversion of non-desert lands into deserts, or **desertification.** Soil moisture and vegetation are both reduced, and plants that are better adapted to dry conditions spread. Soil can become badly eroded by wind, and sand dunes can blow onto formerly vegetated land.

Desertification is occurring in many places. Globally, 23,000 square miles (60,000 square kilometers, or an area about the size of West Virginia) becomes new desert every year. It is not yet known just how much of this loss is due to natural climate variations and how much is due to human-induced climate change, though human activity clearly plays a major role.

A wide array of land use changes has contributed to the general drying of grasslands bordering the Sahara. As groundwater levels fall due to pumping for cities and irrigated agriculture, plant roots can no longer reach sources of moisture. International development agencies have encouraged settled cattle ranching of the type practiced in western North America. This has led to overgrazing of pastureland and excessive water use both by the cattle and by the people, many of whom have historically been nomadic herders accustomed to using very little water. When they are encouraged to settle and take up "modern" ways of life, their water use increases.

Figure 6.11 Dams on the Tigris and Euphrates. Turkey's construction of dams affects water availability throughout the Tigris and Euphrates river basins. The smaller map shows the full extent of Turkey's Southeastern Anatolia Project. [Adapted from "Turning the Tides" map, *Vital Water Graphics: Problems Related to Freshwater Resources,* United Nations Environment Programme, at http://maps.grida.no/go/graphic/regulation_of_the_tigris_and_euphrates_rivers]

Vulnerability to Global Warming: Water, Food, and Fossil Fuels

North Africa and Southwest Asia is especially vulnerable to both the effects of global warming and the world's attempts to reduce greenhouse gas emissions. Sea level rise of a few feet could severely impact the Nile delta, one of the poorest

Africa and Southwest Asia are especially vulnerable to both the effects of global warming and the world's attempts to reduce greenhouse gas emissions.

and most densely populated areas in the world. Elsewhere, shifting rainfall patterns resulting from changes in the regional climate could drastically reduce water availability or, conversely, increase damage from flooding.

Any of these changes would threaten this region's food security. Water shortages or flooding could reduce agricultural output, making countries even more dependent on food imports than they already are. This vulnerability has been made clear in recent years when food prices have skyrocketed on the global market, eroding the incomes of many poor people in this region.

Independent of any environmental changes, global efforts to reduce fossil fuel consumption could devastate economies and transform the region's geopolitics. Despite the enormous wealth that has flowed into the fossil fuel–rich countries over the past thirty years, few have undertaken significant economic diversification. Most remain heavily dependent on oil revenues and are worried about a future of reduced global fossil fuel consumption. Saudi Arabia, for example, argues that it must be compensated for all the oil that it will *not* be selling as the world moves to alternative energy sources. Relations with foreign powers such as the United States and Europe would also change, with the region's energy resources no longer providing an incentive for them to influence events here. However, the potential for global warming to create water and food shortages in the region could spark military conflicts between the region's countries, potentially drawing in the United States and Europe.

 *Ten key aspects of North Africa and Southwest Asia*

- By 2025, almost every country in the region will have less water than the UN considers necessary to support basic human development.
- Even though the vast majority of this region's water resources are devoted to irrigated agriculture, most countries are still highly dependent on imported food.
- While this region is nearly twice as large as the United States, most of its 450 million people are concentrated in the few areas that are useful for agriculture.
- More than 93 percent of the people in the region are followers of Islam, and political movements based on Islamism have transformed the region's politics in recent years.
- While many countries are still ruled by kings and ruling families, a region-wide shift toward democracy and greater political and personal freedoms is underway.
- Large reserves of fossil fuels have powerfully linked many urban and national economies to the global economy, bringing prosperity to a privileged elite and low-wage jobs to poor migrants from around the region and the world.

- Migration from rural to urban areas is increasing, with 58 percent of the population now living in urban areas.
- The Israeli–Palestinian conflict, which has spawned several major wars and innumerable skirmishes, is a persistent obstacle to political and economic cooperation in the region.
- While women are increasingly active as voters and politicians, they make up only 9 percent of national legislatures, the lowest of any world region and half the world average of 18 percent.
- Efforts to reduce global warming by shifting away from fossil fuels and toward alternative energy sources could devastate the wealthiest countries in this region whose economies are highly dependent on fossil fuel exports.

HUMAN PATTERNS OVER TIME

Important developments in agriculture, societal organization, and urbanization took place long ago in this part of the world. Three of the world's great religions were born here: Judaism, Christianity, and Islam. For many centuries, the region influenced Europe and the rest of the world through its advanced learning, trading strategies, and refined culture. More recently, especially since World War I, influences from Europe and the United States have transformed the region's politics and challenged traditional mores regarding religion, gender roles, and the role of the state in society.

Agriculture and the Development of Civilization

Between 10,000 and 8000 years ago, nomadic peoples founded some of the earliest known agricultural communities in the world. These communities were located in an arc formed by the uplands of the Tigris and Euphrates river systems (in modern Turkey and Iraq) and the Zagros Mountains of modern Iran. This zone is often called the **Fertile Crescent** (Figure 6.12) because of its plentiful fresh water, open forests and grasslands, abundant wild grains, and fish, goats, sheep, wild cattle, and other large animals.

The skills of these early people in domesticating plants and animals allowed them to build more elaborate settlements. The settlements eventually grew into societies based on widespread irrigated agriculture along the base of the mountains and along the two rivers. Nomadic herders living in adjacent grasslands traded animal products for the grain and manufactured goods produced in the settled areas.

Over the next several thousand years, agriculture spread to the Nile Valley, northwestern Africa, the eastern mountains of Persia (modern Iran), and ultimately worldwide. Eventually, the agricultural settlements took on urban qualities: dense settlement, specialized occupations, concentrations of wealth, and centralized government and bureaucracies. For example, the city of Sumer (in modern southern Iraq), which existed 5000 years ago, gradually extended its influence over the surrounding territory. The Sumerians developed wheeled vehicles, oar-driven ships, and irrigation technology.

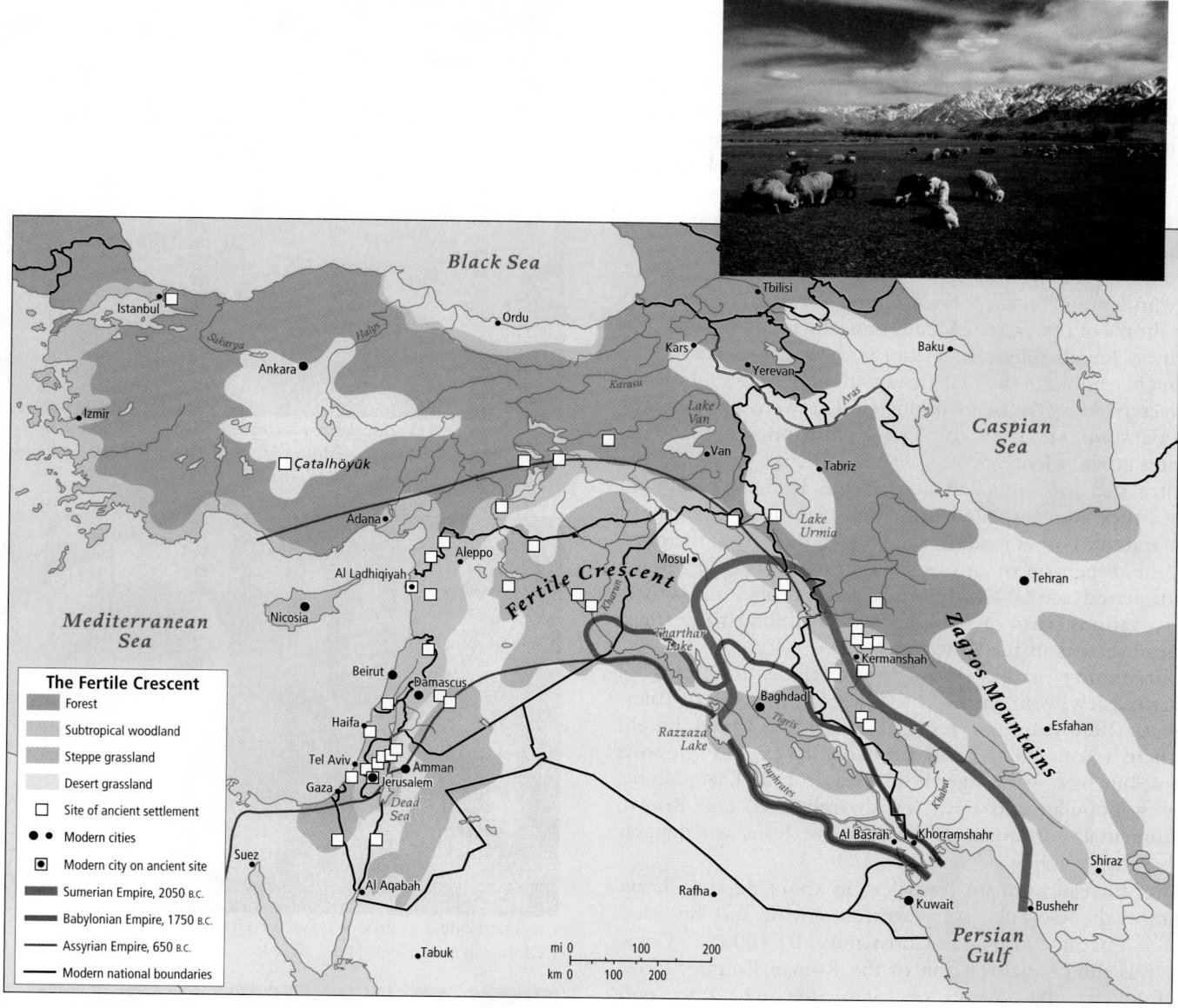

Figure 6.12 The Fertile Crescent, one of the earliest known agricultural sites. About 10,000 years ago, people in the Fertile Crescent began domesticating crops and animals. Three major empires developed successively in the eastern extent of the Fertile Crescent: the Sumerian, the Babylonian, and the Assyrian. [Map adapted from Bruce Smith, *The Emergence of Agriculture* (New York: Scientific American Library, 1995), p. 50; photo, Patrick Syder/Lonely Planet Images/Getty.]

At times, nomadic tribes would band together and sweep over settlements with devastating cavalry raids. They set themselves up as a ruling class, though they tended to adopt the cultures of the peoples they conquered.

Agriculture and Gender Roles

Some researchers think that the dawning of agriculture may mark the transition to markedly different roles for men and women. Archaeologist Ian Hodder reports that at the 9000-year-old site of Çatalhöyük, in south-central Turkey (see Figure 6.12), there is little evidence of gender differences. Families were small and men and women performed similar chores in daily life. Both had similar status and power, and both played key roles in social and religious life.

Scholars think that after the development of agriculture, as wealth and property became more important in human society, a concern with family lines of descent and inheritance emerged. This led in turn to the idea that women's bodies needed to be controlled so that a woman could not become pregnant by a man other than her husband and thus confuse lines of inheritance.

The Coming of Judaism, Christianity, and Islam

The very early religions of this region were based on a belief in many gods who controlled natural phenomena. Then, several thousand years ago, **monotheistic** belief systems—those based on one god—emerged. The three major monotheistic world religions—Judaism, Christianity, and Islam—all have their origins in the eastern Mediterranean (Figure 6.13).

Judaism was founded approximately 4000 years ago. According to tradition, it was begun by the patriarch Abraham, who led his followers from Mesopotamia (modern Iraq) to the shores of the eastern Mediterranean (modern Israel and Palestine). Jewish religious history is recorded in the Torah (the first five books of the Old Testament of the Bible). Judaism is characterized by the belief in one God, Yaweh, a strong ethical code summarized in the Ten Commandments, and an enduring ethnic identity.

After the Jews rebelled against the Roman Empire in 73 C.E., they were expelled from the eastern Mediterranean and migrated to other lands in a movement known as the **diaspora** (the dispersion of an originally localized people). Many Jews dispersed across North Africa and Europe, and others went to various parts of Asia. Jews were among the earliest European settlers in the Americas.

Christianity is based on the teachings of Jesus of Nazareth, a Jew, who gathered followers in the area of Palestine about 2000 years ago. Jesus, also known as Christ, taught that there is one God, who primarily loves and supports humans, but who will judge those who do evil. This philosophy grew popular, and both Jewish (religious) and Roman (governmental) authorities of the time saw Jesus as a dangerous challenge to their power.

After his execution in Jerusalem in about 32 C.E., Jesus' teachings (the Gospels) were written down, and his ideas spread and became known as Christianity. By 400 C.E., Christianity was the official religion of the Roman Empire. However, following the spread of Islam after 622 C.E., only remnants of Christianity remained in Southwest Asia and North Africa.

Islam is now the overwhelmingly dominant religion in the region. Islam emerged in the seventh century C.E. after the Prophet Muhammad transmitted the Qur'an to his followers through the spoken word. Born in about 570 C.E., Muhammad was a merchant and caravan manager in the small trading town of Makkah (Mecca) on the Arabian Peninsula near the Red Sea. Followers of Islam, called **Muslims,** believe that Muhammad was the final and most important in a long series of revered prophets, which included Abraham, Moses, and Jesus.

The religion of Islam has virtually no religious hierarchy or central administration. The world's one billion Muslims may communicate directly with God (Allah). A clerical intermediary is not necessary, though there are numerous mullahs (clerical leaders) who help their followers interpret the Qur'an. As a result, the interpretation of Islam varies widely within and among countries and from individual to individual.

(a) A bride and groom participate in a Jewish Israeli Mehndi (henna) ceremony prior to their wedding in 2005. [Ronen Zvulun/Reuters/Landov.]

(b) The bride and groom leave the church following an Arab Christian wedding in Jerusalem. [PonkaWonka.]

(c) A Muslim bride and groom are serenaded during wedding ceremonies in Tehran. [Abbas/Magnum Photos.]

Figure 6.13 Weddings in the three major religions of the region.

The Spread of Islam

The Bedouin—nomads of the Arabian Peninsula—were among the first converts to Islam. They were already spreading the faith and creating a vast Islamic sphere of influence by the time of Muhammad's death in 632 C.E.. Over the next century, Muslim armies built an Arab–Islamic empire over most of Southwest Asia, North Africa, and the Iberian Peninsula of Europe (Figure 6.14).

While Europe was stagnating during the medieval period (450–1500), the Arab–Islamic empire nurtured learning and economic development. Muslim scholars traveled throughout Asia and Africa, advancing the fields of history, mathematics, geography, and medicine. Centers of learning flourished from Baghdad (Iraq) to Toledo (Spain). The early Islamic development of banks, trusts, checks, and receipts fostered vibrant economies and wide-ranging trade.

By the end of the tenth century, the Arab–Islamic empire had begun to break apart. From the eleventh to the fifteenth centuries, Mongols from eastern Central Asia conquered parts of the Arab-controlled territory, forming the Mughal Empire. Meanwhile, beginning in the 1200s, nomadic herders in west-ern Anatolia (Turkey) had begun to forge the Ottoman Empire, the greatest Islamic empire the world has ever known.

By the fifteenth century, the Ottoman Muslims had defeated the Christian Byzantine Empire, which was the successor to the Roman Empire. They took over the Byzantine capital, Constantinople, and renamed it Istanbul. Very soon, the Ottomans controlled most of the eastern Mediterranean, Egypt, and Mesopotamia. By the late 1400s, they also controlled much of southeastern and central Europe. At about the same time, the Arab Muslims lost their control of the Iberian Peninsula to Christian kingdoms. Today, Islam still dominates in a huge area stretching from Morocco to western China and northern India, and from Malaysia to Indonesia.

Western Domination and State Formation

The Ottoman Empire ultimately withered in the face of a Europe made powerful by the Industrial Revolution. Throughout the nineteenth century, North Africa provided raw materials for Europe in a trading relationship dominated by European merchants. In 1830, France became the first European country to exercise direct control over a North African

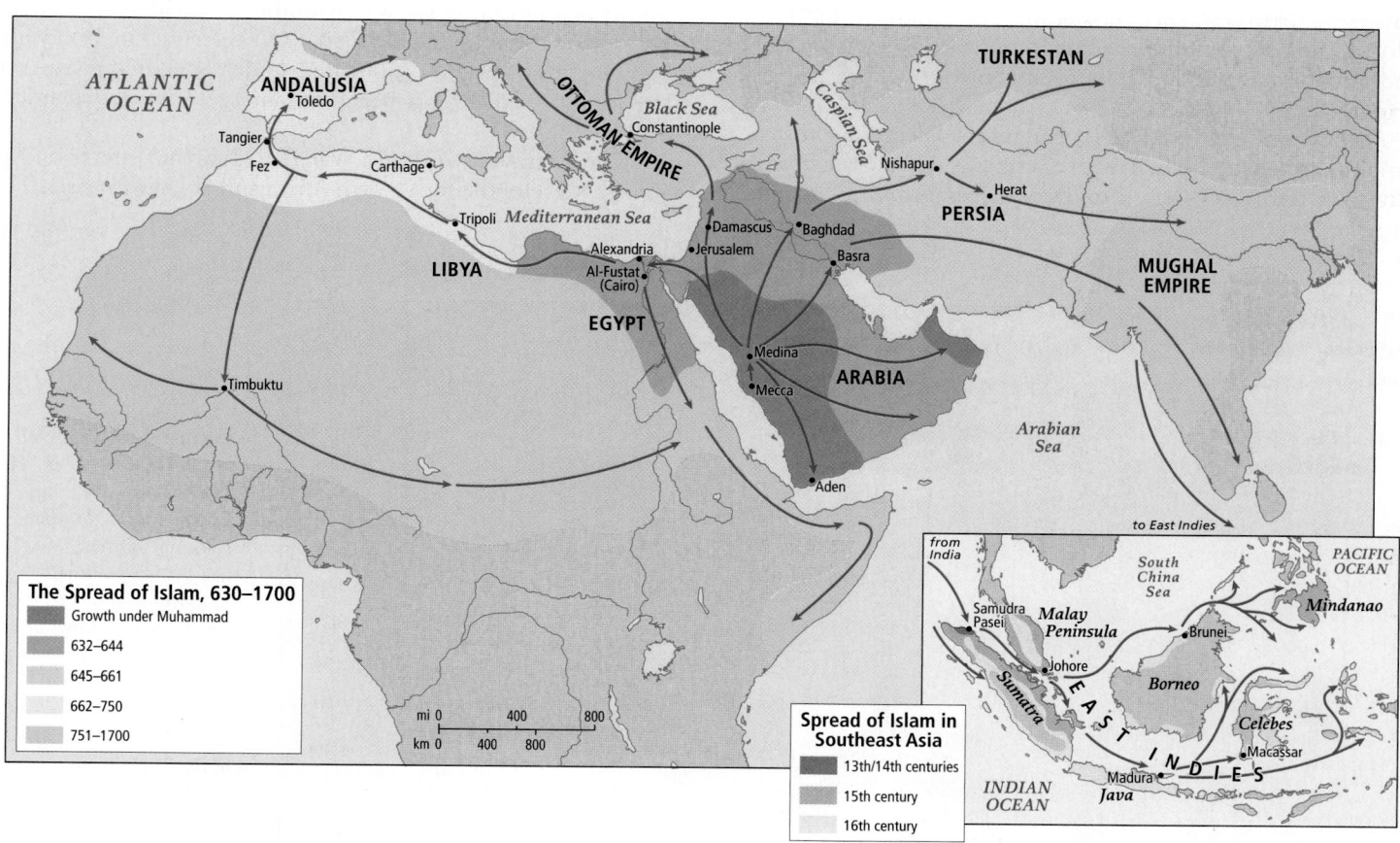

Figure 6.14 The spread of Islam, 630–1700. In the first 120 years following the death of the Prophet Muhammad in 632, Islam spread primarily by conquest. Over the next several centuries, Islam was carried to distant lands by both traders and armies. [Adapted from Richard Overy, ed., *The Times History of the World* (London: Times Books, 1999), pp. 98–99.]

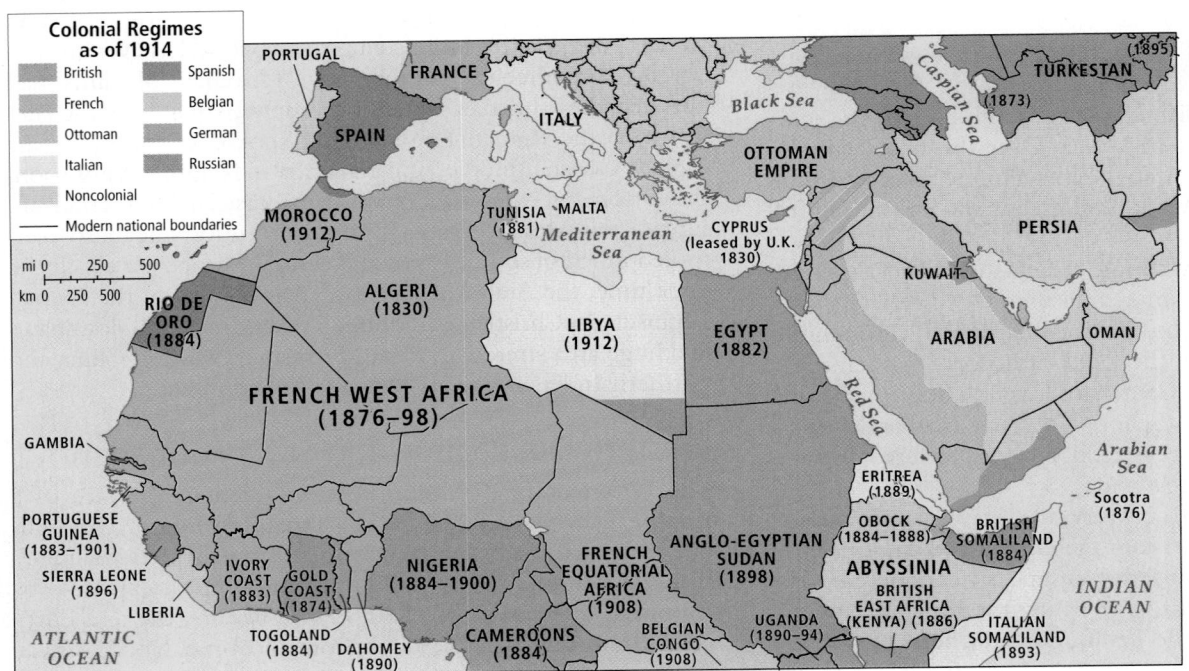

Figure 6.15 Colonial regimes in 1914. The dates on the map indicate when Europeans took control of each country. [Adapted from *Hammond Times Concise Atlas of World History* (Maplewood, N.J.: Hammond, 1994), pp. 100–101.]

territory (Algeria). France took control of Tunisia in 1881 and Morocco in 1912; Britain gained control of Egypt in 1882 and Sudan in 1898; and Italy took control of Libya in 1912 (Figure 6.15).

World War I (1914–1918) brought the fall of the Ottoman Empire, which had allied with Germany. Of all the former Ottoman territories, only Turkey was recognized as an inde-

pendent country after the war, with the rest divided up between France and Britain (Figure 6.16). On the Arabian Peninsula, Bedouin tribes were consolidated under Sheikh Ibn Saud in 1932, and Saudi Arabia began to emerge as an independent country.

The aftermath of World War II had further effects on the political development of North Africa and Southwest Asia. The

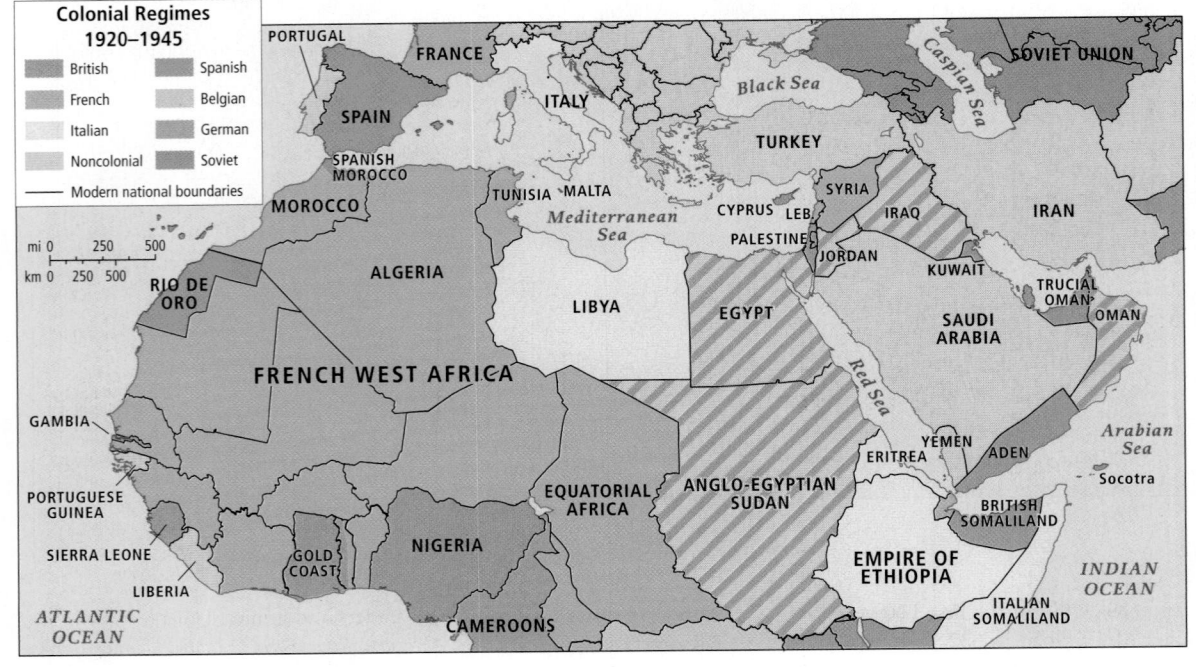

Figure 6.16 Colonial regimes, 1920–1945. The striped areas covering Egypt, Sudan, Jordan, Iraq, and Oman reflect the British colonial practice of allowing local rulers to govern while controlling many of their policies and actions. [Adapted from *Rand McNally Historical Atlas of the World* (Chicago: Rand McNally, 1965), pp. 36–37; and *Cultural Atlas of Africa* (New York: CheckMark Books, 1988), p. 59.]

United States supported those local leaders who were most sympathetic to U.S. cold war policies (as opposed to those of the Soviet Union) and most likely to maintain a friendly attitude toward U.S. business interests. In Iran and Saudi Arabia, where vast oil deposits became especially lucrative by mid-century, European and U.S. energy companies played a key role in deciding who ruled. The governments of these countries showed their loyalty to the foreign energy companies with low taxes on oil exports. Relatively few of these tax revenues were invested in creating opportunities for poor or middle class people, but a tiny ruling elite grew fabulously wealthy.

POPULATION PATTERNS

Although the region as a whole is nearly twice as large as the United States, most of the population is concentrated in the few areas that are useful for agriculture. Vast tracts of desert are virtually uninhabited, while 450 million people are packed into coastal zones, river valleys, and mountainous areas that capture orographic rainfall (Figure 6.17). Population densities in these areas can be quite high. For example, some of Egypt's urban neighborhoods have over 260,000 people per square mile (100,000 per square kilometer), a density ten times higher than that of New York City, the densest city in the United States.

While fertility rates have dropped significantly since the 1960s, at 3.5 children per woman, they are still higher than the world average of 2.7. Only sub-Saharan Africa, at 5.5 children per woman, is growing faster. At present growth rates, the population of the region will reach 540 million by 2025. Such population growth will severely strain supplies of fresh water and food.

> While fertility rates have dropped significantly since the 1960s, at 3.5 children per woman, they are still higher than the world average of 2.7.

Population Growth and Gender Roles

This region's high population growth rate can be explained in part by the relatively low status of women. As noted in many world regions, population growth rates are higher in societies where women are less educated and work less outside the home. In places where women have opportunities to work or study outside the home, they usually choose to have fewer children. Figure 6.18 shows that as of 2003, considerably less than 50 percent of women across the region (except in Israel and Turkey) worked outside the home at jobs other than farming. Moreover, only 64 percent of females can read, whereas 82 percent of males can.

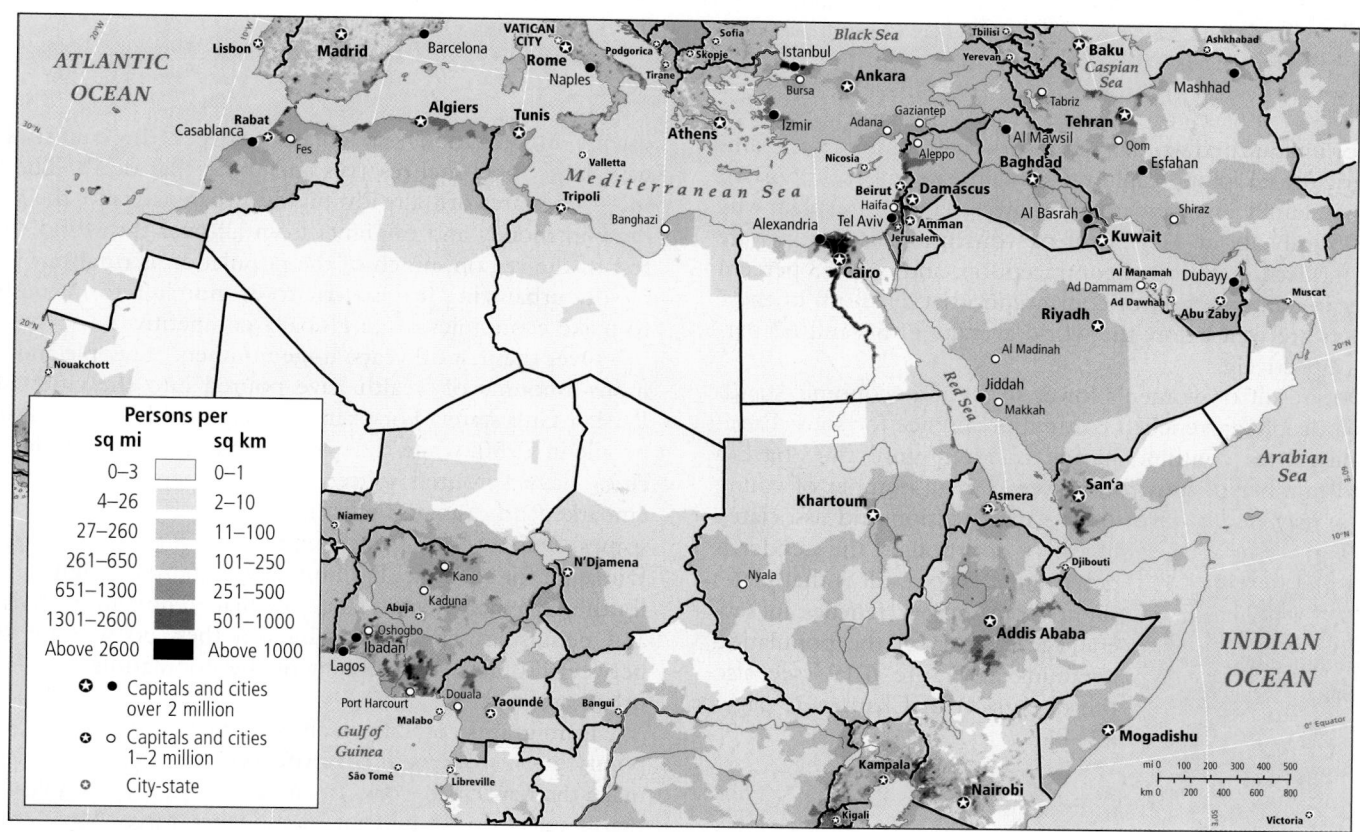

Figure 6.17 Population density in North Africa and Southwest Asia. [Data courtesy of Deborah Balk, Gregory Yetman, et al., Center for International Earth Science Information Network, Columbia University, at http://www.ciesin.columbia.edu.]

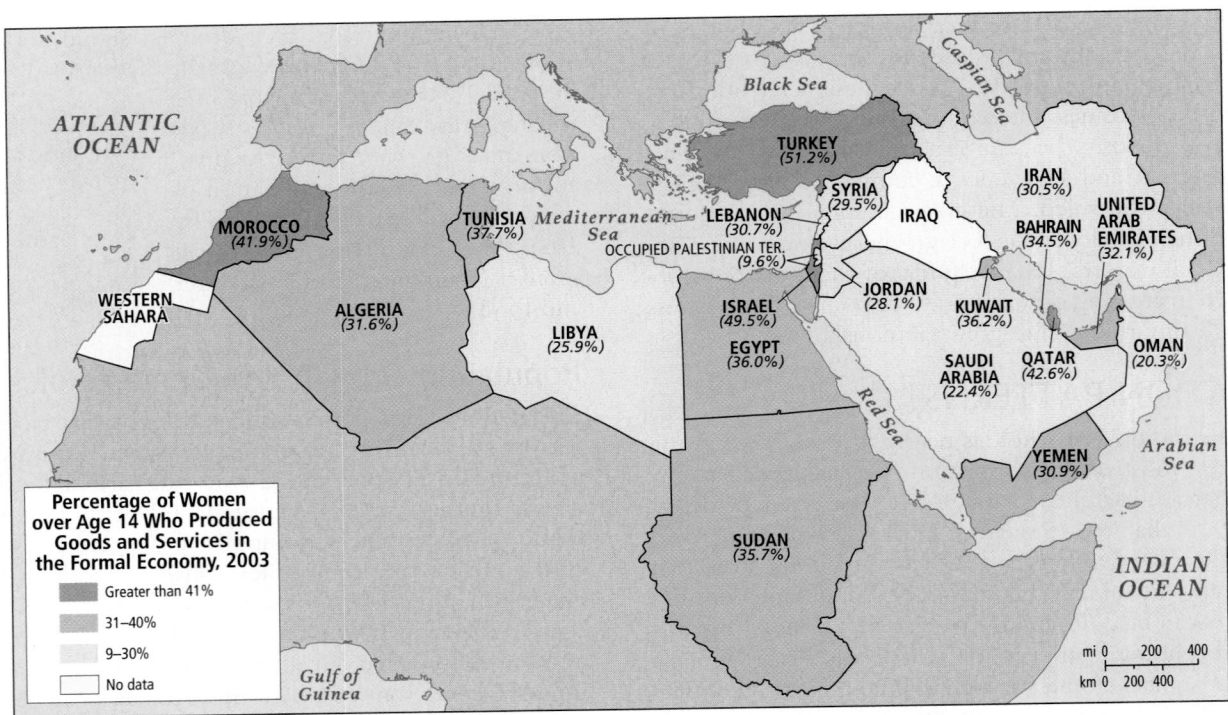

Figure 6.18 Percentage of women who are wage-earning workers in the region's countries. [Adapted from Joni Seager, *Women in the World: An International Atlas* (New York: Viking Penguin, 1997), pp. 66–67, and *United Nations Human Development Report 2005* (New York: United Nations Development Programme), Table 25.]

For uneducated women who work only at home or in family agricultural plots, children remain the most important source of family involvement and power. This may explain why in 2005 only about 41 percent of women in this region were using modern methods of contraception, and only 53 percent were using any method of contraception at all. Both of these numbers are well below the world average of 54 and 62 percent, respectively.

One result of women's lower social and economic standing is a deeply entrenched cultural preference for sons. Families sometimes continue having children until they have a desired number of sons. Moreover, a small number of young females may not survive due to malnutrition and associated illness or abortions if the gender is determined. The result is that males slightly outnumber females in several age cohorts of the population pyramids in Figure 6.19 (see also the discussion in Chapter 1, pages 42–44).

> One result of women's lower social and economic standing is a deeply entrenched cultural preference for sons.

Urbanization and Globalization

Urban populations have boomed in this region largely in response to global economic forces. However, the processes driving urbanization vary greatly between the countries that have large fossil fuel reserves and those that don't. The fossil fuel–rich states are already highly urbanized and their cities draw in money and migrants from all over the world. In the rest of the region, much of the population is rural but is now rapidly urbanizing in response to economic reforms designed to make economies more globally competitive.

Over the past 30 years, huge numbers of people and enormous amounts of wealth have poured into the cities of the Persian Gulf states. For example, Baghdad had only 580,000 people in 1950, while today it is home to seven million. Most cities have sprouted glittering skyscrapers, and some have embarked on dramatic schemes to beautify their urban landscapes. Dubai (one of the United Arab Emirates; also spelled Dubayy), for example, has built elaborate new neighborhoods for the extremely rich on fancifully shaped human-made islands and peninsulas off its coastline. In the Persian Gulf states, between 70 and 100 percent of the population now lives in urban areas.

Immigrants come from all over the world to work as temporary guest workers in construction and other industries that make the new cities work. In Saudi Arabia, for example, immigrants make up 88 percent of the labor force. Most employers prefer Muslim employees, and over the last two decades, several hundred thousand workers have arrived from Palestinian refugee camps in Lebanon and Syria. Other major labor

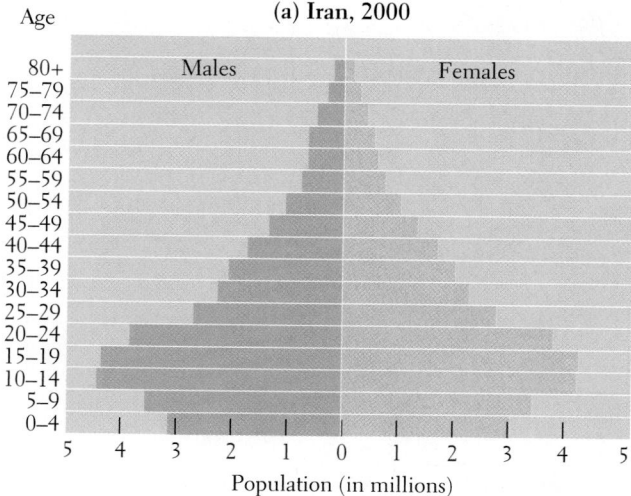

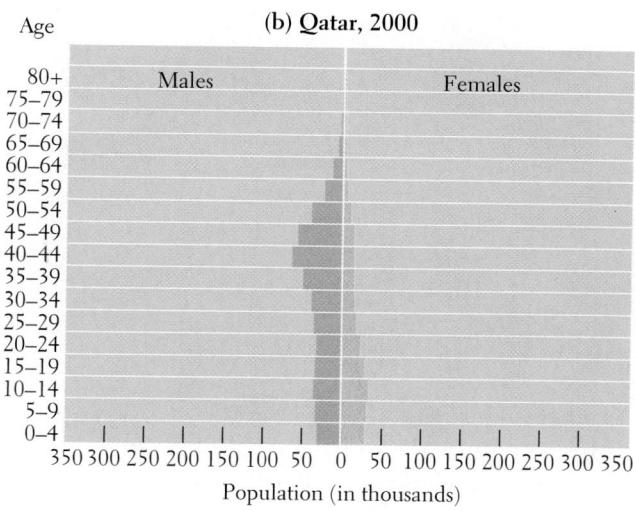

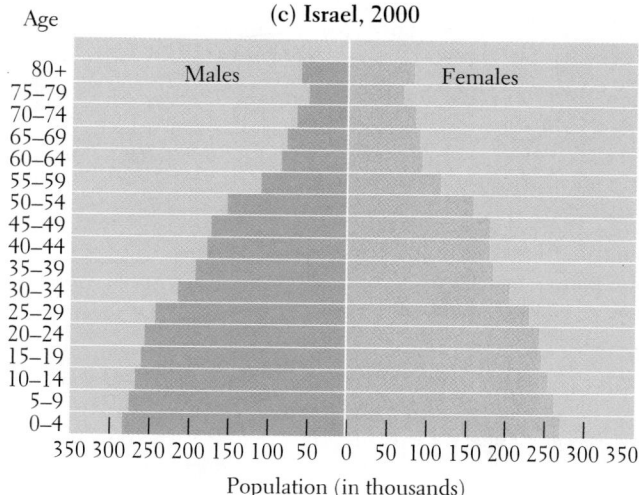

Figure 6.19 Population pyramids for Iran, Qatar, and Israel. The population pyramid for Iran is at a different scale (millions) from those for Israel and Qatar (thousands). The imbalance of Qatar's pyramid in the 25–54 age groups is caused by the presence of numerous male guest workers. Note too that all three pyramids show missing females in the younger age groups. (This is most easily observed by drawing lines from the ends of the male and female age bars to the scale at the bottom of the pyramid and comparing the numbers.) [Adapted from "Population Pyramids for Iran," "Population Pyramids for Qatar," and "Population Pyramids for Israel" (Washington, D.C.: U.S. Bureau of the Census, International Data Base, May 2000), at http://www.census.gov/ipc/www/idb/pyramids.html.]

source countries are Egypt, Pakistan, and India. Many female domestic workers come from Muslim countries in South and Southeast Asia. Immigrant workers on the Arabian Peninsula are only temporary residents with no job security, often living in squalid conditions alongside the opulent lifestyles of those enriched by oil and gas.

Significant migration from rural villages to urban areas has also occurred in response to economic reforms (discussed further on pages 229–230). Until recently, most people lived in small settlements. By 2007, however, 58 percent of the region's people lived in urban areas (Figure 6.20). There are now more than 120 cities with populations of at least 100,000, and there are 35 cities with more than one million people. The largest city in the region, and one of the largest in the world, with 18 million residents, is Cairo.

Poor urban migrants in this region often occupy the medieval interiors of old cities. Streets are narrow pedestrian pathways, and there is little plumbing, sewage disposal, or clean water. The ancient dwellings, dating back 500 years or more, are worthy of historic preservation, but the inhabitants are far too poor to provide even routine maintenance.

Emigration from the region. Many people come to the city only to find that jobs are almost impossible to come by, especially for workers with more than a basic education. Consequently, many millions emigrate to other parts of the world in search of work. Most of those leaving the region are young men because women typically do not travel widely, or on their own. Tens of millions of people from North Africa and Southwest Asia are guest workers in Europe, and the remittances they send home to their families significantly increase local standards of living.

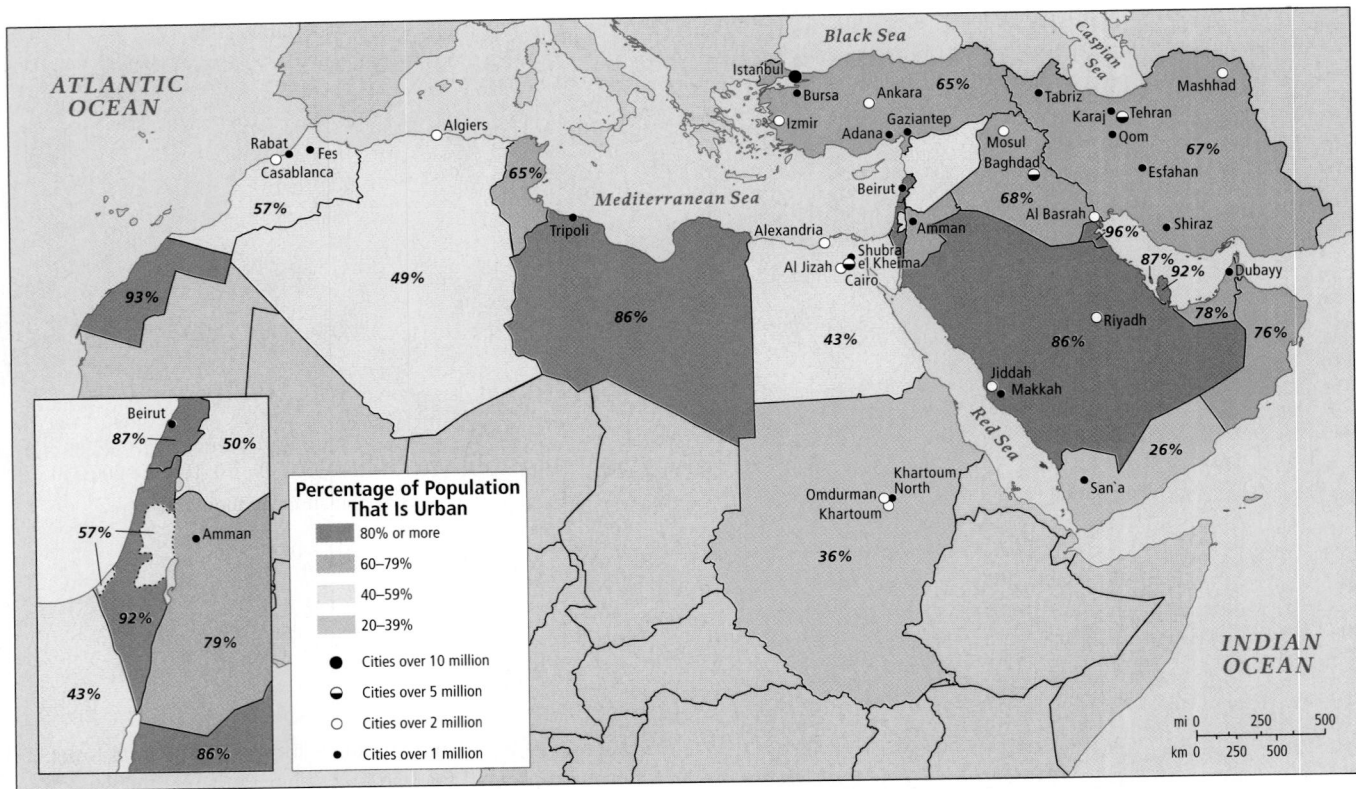

Figure 6.20 Percentage of the population living in cities in North Africa and Southwest Asia. [Data from *2005 World Population Data Sheet* [(Washington, D.C.: Population Reference Bureau) and *World Gazetteer,* at http://www.world-gazetteer.com/wg.php?x=&men=gcis&lng=en&dat=32&srt=npan&col=ahq&geo=-1.]

Current Geographic Issues

SOCIOCULTURAL ISSUES

In this section, we'll explore the basics of Islam and examine the broad social changes occurring in the region with regard to gender roles and restrictions on women.

Religion in Daily Life

More than 93 percent of the people in the region are followers of Islam. The Five Pillars of Islamic Practice embody the central teachings of Islam:

1. A testimony of belief in Allah as the only God and in Muhammad as his Messenger (Prophet).

2. Daily prayer at five designated times (daybreak, noon, midafternoon, sunset, and evening). Although prayer is an individual activity, Muslims are encouraged to pray in groups and in mosques.

3. Obligatory fasting (no food, drink, or smoking) during the daylight hours of the month of Ramadan, followed by a light celebratory family meal after sundown.

4. Obligatory almsgiving (*zakat*) in the form of a "tax" of at least 2.5 percent. The alms are given to Muslims in need. *Zakat* is based on the recognition of the injustice of economic inequity. Although it is usually an individual act, the practice of government-enforced *zakat* is returning in certain Islamic republics.

5. Pilgrimage **(hajj)** at least once in a lifetime to the Islamic holy places, especially Makkah (Mecca), during the twelfth month of the Islamic calendar.

Source: Carolyn Fluehr-Lobban, *Islamic Society in Practice* (Gainesville: University of Florida Press, 1994).

Saudi Arabia occupies a prestigious position in Islam, as it is the site of two of Islam's three holy shrines: Makkah, the birthplace of the Prophet Muhammad and of Islam, and Medina, the site of the prophet's mosque and his burial place. (The third holy shrine is in Jerusalem.) The fifth pillar of Islam has placed Makkah and Medina at the heart of Muslim religious geography. Each year, a large private sector service industry organizes and oversees the five- to seven-day hajj for more than 2.5 million foreign visitors (Figure 6.21).

Figure 6.21 The Grand Mosque in Makkah (Mecca), 2004. Devout Muslims pray as they circle the Kaaba (the black rectangular structure at the upper right) inside the Grand Mosque. [AP Photo.]

Beyond the Five Pillars, Islamic religious law, called **shari'a,** "the correct path," guides daily life according to the principles of the Qur'an. Some Muslims believe that no other legal code is necessary in an Islamic society as the shari'a provides guidance in all matters of life, including worship, finance, politics, marriage, diet, hygiene, war, and crime. Insofar as the interpretation of shari'a is concerned, the Islamic community is split into two major groups: **Sunni** Muslims, who today account for 85 percent of the world community of Islam, and **Shi'ite** (or **Shi'a**) Muslims, who live primarily in Iran but also in southern Iraq and southern Lebanon.

The Sunni–Shi'ite split dates from just after the death of Muhammad when divisions arose over who should succeed the Prophet and have the right to interpret the Qur'an for all Muslims. This division continues today. The original disagreements have been exacerbated by countless local disputes over land, resources, and philosophies. In Iraq, for example, conflict between Sunnis and Shi'ites has been intensified by the rivalry over political power and fossil fuel resources that followed the U.S. invasion.

> The Sunni Shi'ite split arose over who should succeed the Prophet and interpret the Qur'an.

Gender Roles and Gender Spaces

Carefully specified gender roles are common in many cultures, and there is often a spatial component to these roles. In both rural and urban settings, men and boys go forth into *public spaces*—the town square, shops, the market. Women have inhabited primarily *private spaces.*

Traditional family compounds included a courtyard that was usually a private, female space within the home; the only men who could enter were relatives. For the urban upper classes, female space was an upstairs set of rooms with latticework or shutters at the windows from which it was possible to look out at street life without being seen. Today, the majority of people in the region live in urban apartments, yet even here there is a demarcation of public and private space. One or two formal reception areas are reserved for nonfamily visitors and rooms deeper into the dwelling for family activities.

Now, women as well as men go out into public spaces, but how women enter these spaces remains an issue. Customs vary not only from country to country, but also from rural to urban settings and by social class.

In some parts of the region, particularly in Saudi Arabia, the requirement that women stay out of public view (also known as **female seclusion**), is strictly enforced. Women should not be in public spaces except when on important business and accompanied by a male relative. Elsewhere (in Egypt, for example), affluent urban women may observe seclusion more strictly than do rural women. Although rural women are traditional in their outlook, they have many tasks that they must perform outside the home: agricultural work, carrying water, gathering firewood, and marketing.

In the more secular Islamic countries—Morocco, Tunisia, Libya, Egypt, Turkey, Lebanon, and Iraq—women regularly engage in activities that place them in public spaces. Increasingly, female doctors, lawyers, teachers, and businesspeople are found in even the most conservative societies. Figure 6.22 compares the various levels of restrictions on women across the region.

Many women in this region use clothing as a form of private space. This is done with the many varieties of the **veil.** This may be a garment that totally covers the body and face, or just a scarf that covers the hair. The veil allows a devout Muslim woman to preserve a measure of seclusion when she enters a public space, thus increasing the space she may occupy with her honor preserved.

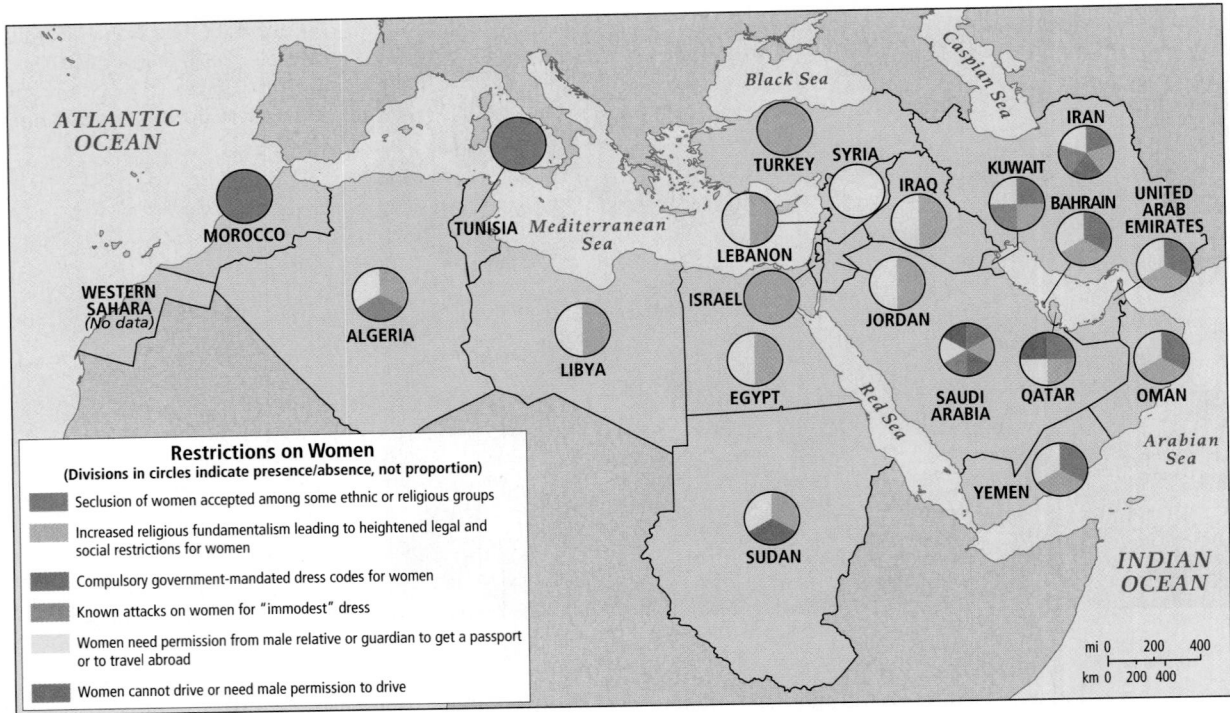

Figure 6.22 Variation in restrictions on women. [Adapted from Joni Seager, *The Penguin Atlas of Women in the World, Completely Revised and Updated* (New York: Viking Penguin, 2003), pp. 14–15.]

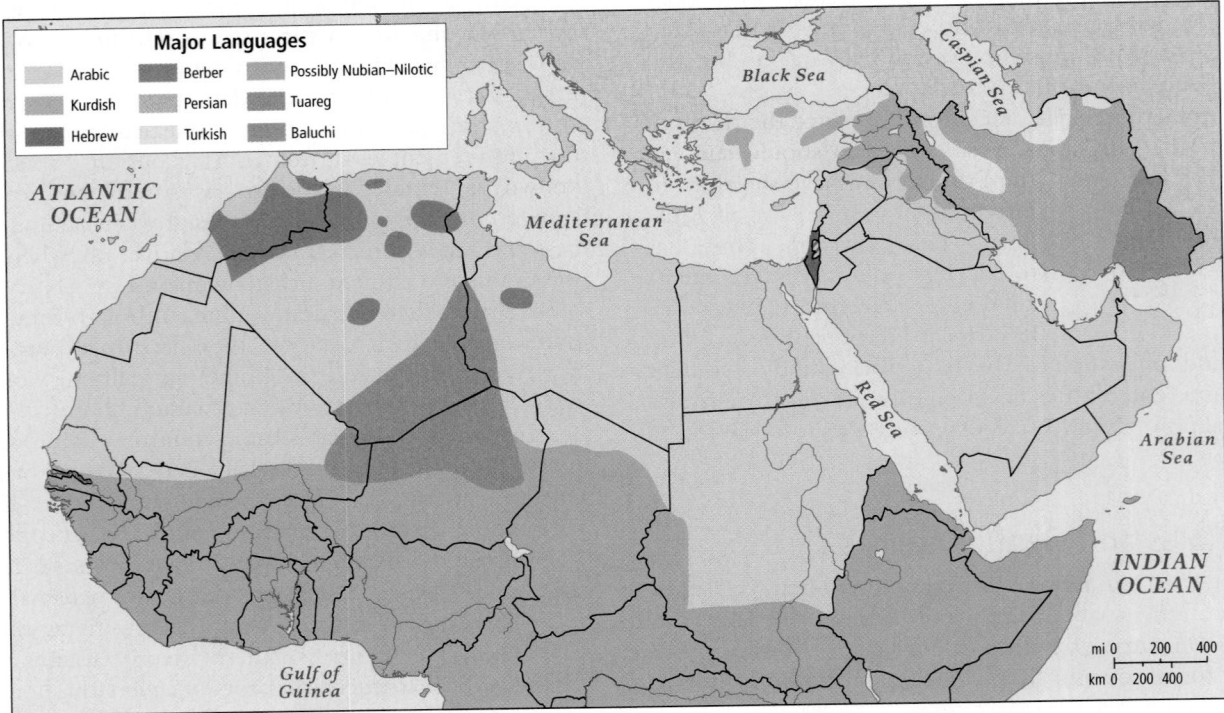

Figure 6.23 Major languages of North Africa and Southwest Asia. [Adapted from Charles Lindholm, *This Islamic Middle East: An Historical Anthropology* (Oxford: Blackwell, 1996), p. 9.]

There is considerable controversy over the origin and validity of female seclusion and veiling as Muslim customs. Scholars of Islam say that these ideas do not derive from the teachings of the Prophet Muhammad and predate Islam by thousands of years. In fact, Muhammad may have been reacting against such customs when he advocated equal treatment of males and females. Muhammad's first wife, Khadija, did not practice seclusion, and worked as an independent businesswoman whose counsel Muhammad often sought. 📹

The Qur'an allows a woman, married or not, to manage her own property and keep her wealth or a monthly salary for herself. By contrast, the custom of transferring the wealth a woman brings to a marriage to her husband's control was not legally ended in Britain until 1870 and in some parts of the United States until the 1960s.

One source of contention in this region is the custom of **polygyny,** the taking of more than one wife. In many countries in the region, polygyny is legally permitted for Muslim men. Although the Qur'an allows polygyny under certain conditions, it is not encouraged. Nor is it a common modern practice. It is estimated, for example, that less than 4 percent of males in North Africa have more than one wife. The Qur'an limits the number of wives to four and requires that all are treated as equals.

Urbanization and modernization are also important factors in polygynous practice. When agriculture was the main economic activity, multiple wives with several children each may have been more productive economically. Urban life, with its small living spaces and cash requirements, favors smaller families. Democratization has led to bans against polygyny in Tunisia, Lebanon, and Palestine.

Language and Diversity

The eastern Mediterranean is the ancient home of people who spoke Semitic languages, including the people of Assyria, Babylonia, and Phoenicia. Modern speakers of Semitic languages include the Arabs and Jews.

Arabic is now the official language in all countries of North Africa and Southwest Asia except Iran, Turkey, and Israel (where Farsi, Turkish, and Hebrew, respectively, are spoken). This uniformity of language masks considerable cultural diversity, however. Numerous minorities within the region, such as Berbers, Tuaregs, Nubians, Kurds, and Turkomans, have their own dialects and non-Arabic languages (Figure 6.23). French and English are also important second languages, especially in urban areas. The dominance of English on the Internet has contributed to its spread in recent years.

ECONOMIC AND POLITICAL ISSUES

There are major economic and political barriers to peace and prosperity within North Africa and Southwest Asia today. Wealth from fossil fuel exports remains in the hands of a few elites. Most people are low-wage urban workers or relatively poor farmers or herders. The economic base is unstable because the main resources, fossil fuels and agricultural commodities, are subject to wide price fluctuations on world markets. A more diverse range of industries is just beginning to emerge and is in need of investment. Meanwhile, in many poorer countries, large national debts are forcing governments to streamline production and cut jobs and social services.

> The economic base of the region is unstable because the main resources, oil and agricultural commodities, are subject to wide price fluctuations on world markets.

Political and economic cooperation in the region has been thwarted by a complex tangle of hostilities between neighboring countries. Many of these hostilities are the legacy of outside interference in regional politics. The Israeli–Palestinian conflict has profoundly affected politics throughout the region, especially since the massive migrations of Jews from Europe and the former Soviet Union that followed World War II. The Iran–Iraq war of 1980–1988 and the Gulf War of 1990–1991 were at least in part instigated by pressures from outside the region. The same is true of the Iraq war that began in the spring of 2003.

Globalization and Fossil Fuel Exports

Two-thirds of the world's known reserves of oil and natural gas are found in this region. Most reserves are located around the Persian Gulf (Figure 6.24) in the countries of Saudi Arabia, Kuwait, Iran, Iraq, Oman, Qatar, and the United Arab Emirates (UAE). Oil and gas are also found in the North African countries of Algeria, Tunisia, Libya, and Sudan.

Early in the twentieth century, European and North American companies were the first to exploit the region's fossil fuel reserves. These companies paid governments a small royalty for the right to extract oil (natural gas was not widely exploited in this region until the 1960s, though gas extraction has grown rapidly since then). Oil was processed at onsite refineries owned by the foreign companies and sold at very low prices, primarily to Europe and the United States, and eventually to other countries, such as Japan.

The governments of the region did not assume control of their oil and gas reserves until the 1970s, when they declared all fossil fuel resources and industries to be the property of the state. Even before this, however, in the 1960s, the oil-producing countries organized a **cartel,** a group of producers strong enough to control production and set prices for its products. The cartel is called **OPEC—the Organization of Petroleum Exporting Countries.** OPEC now includes all the states marked with a blue box on Figure 6.24, plus Venezuela, Indonesia, and Nigeria. OPEC members cooperate to periodically restrict or increase oil production, thereby significantly influencing the price of oil on world markets. A move to create an OPEC-like cartel for natural gas is currently underway.

Many factors outside of OPEC's control strongly influence world oil prices. Consumers affect prices by reducing or

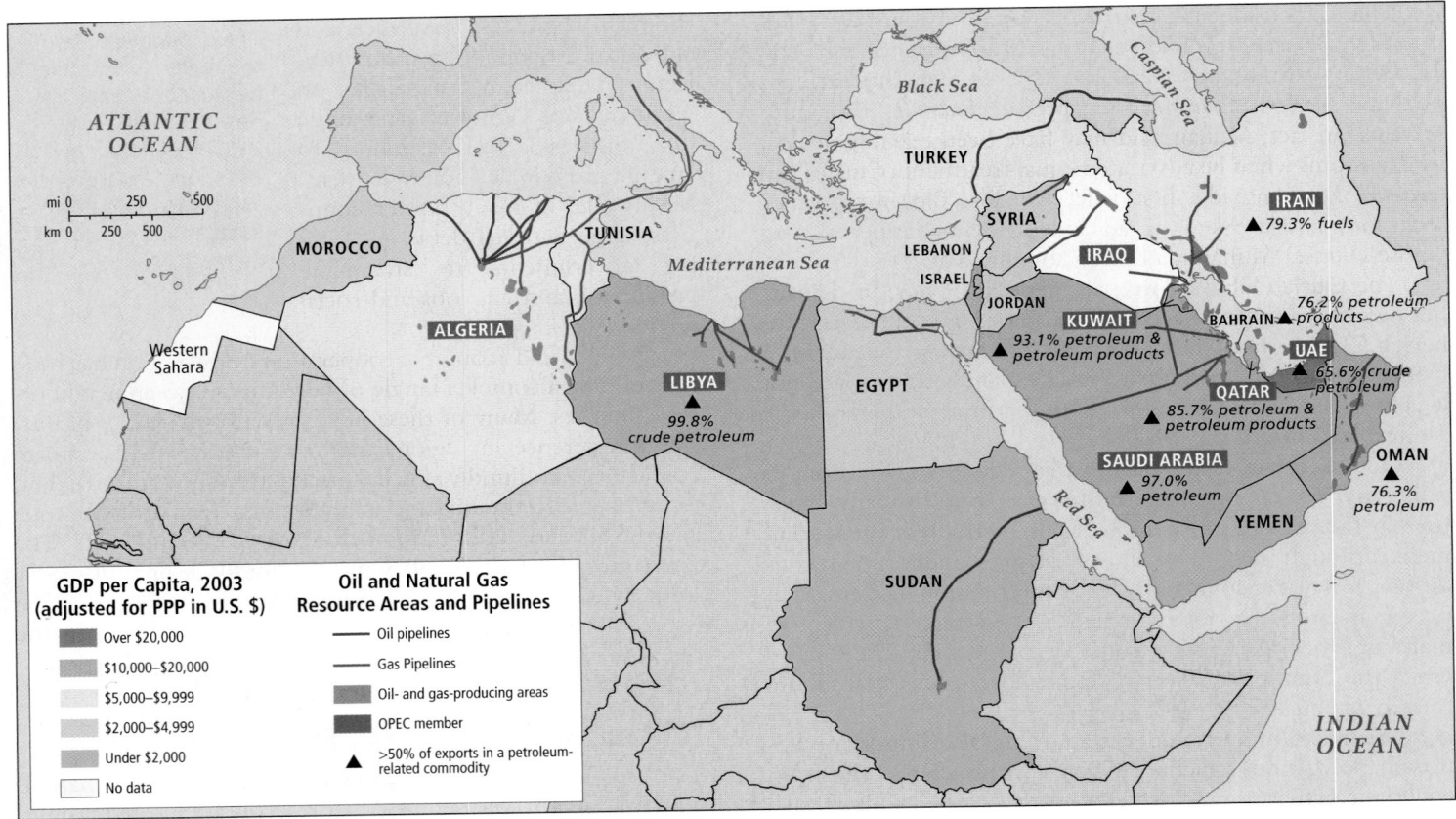

Figure 6.24 Oil and gas resources in North Africa and Southwest Asia. [Adapted from Rafic Boustani and Philippe Farques, *The Atlas of the Arab World—Geopolitics and Society* (New York: Facts on File, 1990), pp. 85, 88, 89; Richard Overy, ed., *The Times History of the World* (London: Times Books, 1999), p. 304; *Hammond Atlas of the Middle East*, revised (Union, N.J.: Hammond, 2001), pp. 8–9.

expanding demand. Geopolitical events, such as the September 11, 2001, attacks, also affect prices. In recent decades, a major factor in oil prices has been the increasing demand for oil in rapidly industrializing China and India. Figure 6.25 shows the flow of oil from the region.

The major oil-producing countries raised prices dramatically in 1973, and oil income in Saudi Arabia alone shot up from U.S.$2.7 billion in 1971 to U.S.$110 billion in 1981. Yet, because the OPEC countries did not invest much of their earnings at home in basic human resources, they have been slow to improve their economic bases. Like their poorer neighbors, they have remained highly dependent on the industrialized world for much of their technology, manufactured goods, skilled labor, and expertise.

Nevertheless, fossil fuel wealth has brought significant benefits to the region as a whole. Those countries that have the largest amounts (Saudi Arabia, the UAE, and Kuwait) now have good roads, airports, new cities, irrigated agriculture, and petrochemical industries. (Less attention has been given to education, social services, public housing, and health care.) The non–fossil fuel producing countries generally do not share directly in the wealth of the Persian Gulf states, though a few Islamic charities have benefited. They do, however, receive money sent home by their many citizens working in the Persian Gulf.

Over the next several decades, fossil fuel prices will probably rise as demand for energy rises. But eventually the depletion of fossil fuel reserves, rising costs, and the development of alternative energy sources will force many countries to find other ways of earning an income. Until then, this region will continue to supply much of the world with fossil fuels.

Economic Diversification and Growth

Greater **economic diversification**—expansion of an economy to include a wider array of activities—could have a significant impact on the region. It could bring economic growth and broader prosperity and limit the damage that a drop in the price of oil, gas, or other commodities on the world market would bring. For the most part, however, diversification has failed to occur for a number of reasons. Misguided economic development policies, a lack of private investment, political tensions, and war have all played a part.

Successes and failures in diversification. By far the most diverse economy is that of Israel, which has a large knowledge-based service economy and a particularly solid manufacturing

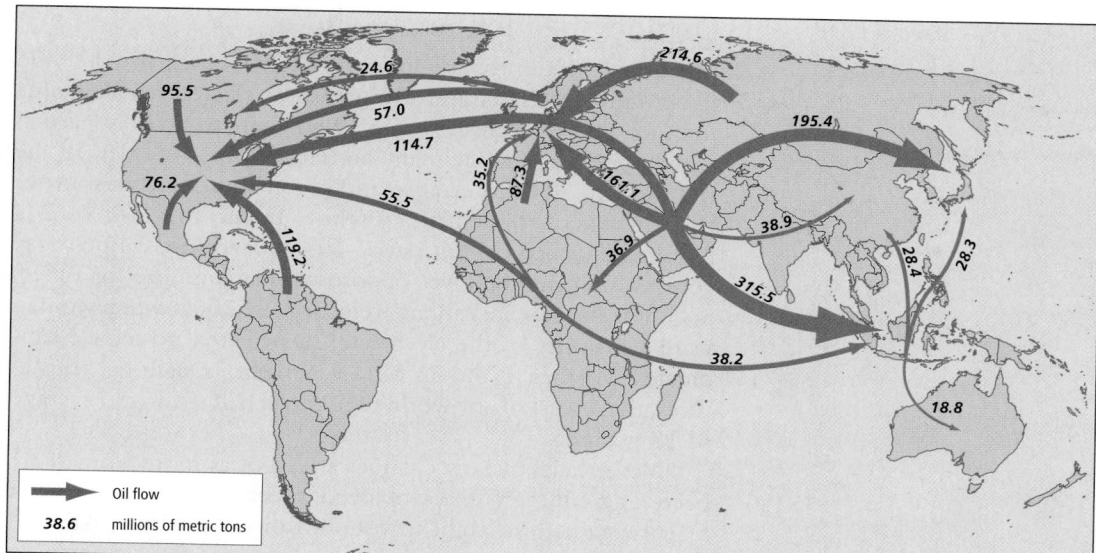

Figure 6.25 World oil flow in millions of metric tons, 2003. The world's largest flows of oil come from North Africa and Southwest Asia. Europe and the United States have many other sources, while Southeast Asia and Japan have few. [Adapted from "Flow of oil," *National Geographic Atlas of the World Eighth Edition* (Washington, D.C.: National Geographic Society, 2005), Plate 20.]

base. Israel's goods and services and the products of its modern agricultural sector are exported worldwide. Turkey is the next most diversified, with Egypt, Morocco, and Tunisia also starting to move into many new economic activities. Some fossil fuel–rich countries have tried to diversify into other industries. In Dubai, for example, only 6 percent of the economy is based on fossil fuels, with trade, manufacturing, and financial services now dominant. Even so, most Persian Gulf countries are still highly dependent on fossil fuels exports.

Aside from these few successes, diversification has been restricted by economic development policies based on *import substitution* (see Chapter 3, pages 118–119). Beginning in the 1950s, many governments, such as those of Turkey, Egypt, Iraq, Israel, Syria, Jordan, Tunisia, and Libya, established state-owned enterprises for the production of goods specifically for local consumption. Machinery and metal items, textiles, cement, processed food, paper, and printing were among the major products.

The countries then protected these enterprises from foreign competition with tariffs and other trade barriers. With only small local markets to cater to, profitability was low and the goods were relatively expensive. Without competitors, the products were also shoddy and unsuitable for sale in the global marketplace. Meanwhile, the extension of government control into so many parts of the economy nurtured corruption and bribery, which further hindered these projects.

Economic diversification and growth were also limited by a lack of financing from private investors within the region. Potential investors, such as members of the Saudi royal family, have generally preferred more profitable investments in Europe, North America, or Southeast Asia. Only recently have these investors started to keep more of their money in the region. Investment in tourism, particularly, has been growing rapidly in recent years.

Finally, both international and domestic military conflicts and the ensuing political tensions have resulted in some of the highest levels (proportionately) of military spending in the world. Four of the world's top five military spenders (as a percentage of GDP) are in this region. All the region's countries except Tunisia and Libya are above the global average of 2 percent (Figure 6.26). High levels of military spending divert funds from other types of development.

Globalization, Debt, and Structural Adjustment Programs

By the 1980s, the failures of import substitution had saddled many of the poorer governments of the region with crippling debt burdens. In response, the major international lending institutions, led by the International Monetary Fund, imposed structural adjustment policies (SAPs). Such programs shift governments away from state-led economic development and limit spending on social services (see Chapter 3, pages 119–122). Some countries have achieved the goals of SAPs, namely decreasing debt, increasing foreign investment, and attaining greater global economic competitiveness.

Overall, however, SAPs have not relieved debt burdens and have instead deepened poverty, eroded public health and education systems, and led to massive migration from rural areas to urban slums. In Egypt in the 1980s, for example, reductions in food and housing subsidies doubled poverty in rural areas and increased it by half in urban areas. Hundreds of thousands of men migrated to cities in the oil-rich Persian Gulf states to work under short-term construction labor contracts. Millions of other poor rural people were pushed into Egypt's already crowded cities. Because unemployment was already high, many sought work in the informal economy, selling food, clothing, and household items on the street. Wages

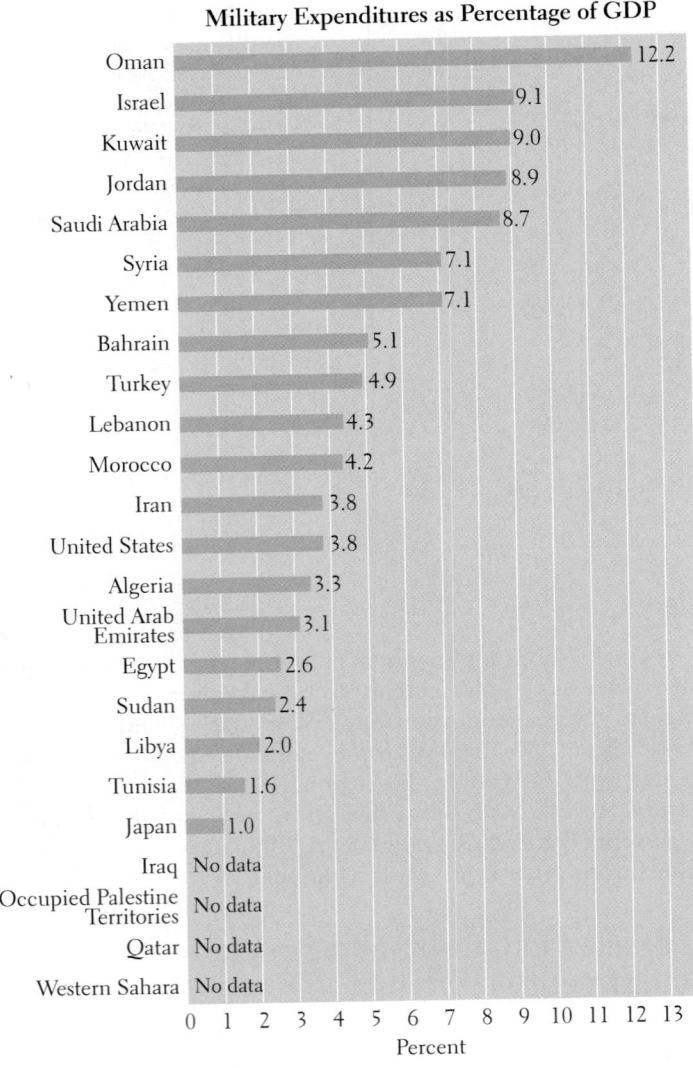

Military Expenditures as Percentage of GDP

Country	Percent
Oman	12.2
Israel	9.1
Kuwait	9.0
Jordan	8.9
Saudi Arabia	8.7
Syria	7.1
Yemen	7.1
Bahrain	5.1
Turkey	4.9
Lebanon	4.3
Morocco	4.2
Iran	3.8
United States	3.8
Algeria	3.3
United Arab Emirates	3.1
Egypt	2.6
Sudan	2.4
Libya	2.0
Tunisia	1.6
Japan	1.0
Iraq	No data
Occupied Palestine Territories	No data
Qatar	No data
Western Sahara	No data

Percent

Figure 6.26 Military expenditures as a percentage of GDP, 2003. North Africa and Southwest Asia lead the world in military spending as a percentage of GDP. The world average is 2 percent of GDP; Tunisia is the only country in the region below that figure. [Adapted from *United Nations Human Development Report 2005* (New York: United Nations Development Programme), Table 20.]

are low in the informal economy, and few of its enterprises pay any taxes, which are badly needed to support social services such as education and health care. Throughout the region, urban populations are depending more and more on nongovernmental organizations and Islamic charities to provide these services.

Such problems make it harder to attract private foreign investment that would contribute to economic diversification and growth. Nevertheless, beginning in the 1990s, Lebanon, Egypt, Turkey, Jordan, Saudi Arabia, Israel, and Kuwait all drew increasing levels of foreign investment.

Democratization and Gender

Most countries in this region now allow women to vote, and two countries (Israel and Turkey) have elected female prime ministers. Nevertheless, women who want to actively participate in politics still face many barriers. Women are denied the right to vote in Saudi Arabia, and in Lebanon they must prove that they went to elementary school before they can vote (a requirement men do not have). ◼ Sometimes, women may vote in high numbers yet oppose women running for office. Such was the case in Bahrain's elections of 2002, when women were a majority of the electorate, but elected no female candidates. Women make up only 9 percent of national legislatures, the lowest of any world region and half the world average of 18 percent.

Nevertheless, many changes are afoot as more women are becoming educated and employed outside the home. Women recently gained the right to vote in Kuwait and the UAE, and Saudi Arabia's government has pledged to allow women to vote in the next round of elections in 2009. However, change in this region's gender politics is often halting, as the following vignette illustrates.

Vignette Raufa Hassan al-Sharki is in her forties and holds a Ph.D. in Social Communications from the University of Paris. In 1996, she founded the Empirical Research and Women's Studies (ERWS) Center at San'a University in Yemen. As Yemen's most outspoken feminist, her overarching goal is to help women learn how to vote independently. Yemen's Islamist party, Islah, supports her efforts.

Despite having the right to vote, Yemeni women usually do not participate in the political process. Nearly 70 percent of the population lives in rural areas, where only 1 in 10 women can read. Typically, few girls go to school. Most work at home, herding cattle, grinding wheat, and carrying water. After marriage, the average woman bears six children. Dr. al-Sharki has found that husbands generally keep their wives' and daughters' voter registration certificates because both men and women believe the women would be likely to lose them. As a result, men often control whether and how a woman votes.

Reception to change is mixed, and Dr. al-Sharki always gets permission from the local **sheikh,** a patriarchal tribal leader, before she talks to the women in a village. But lately the sheikhs, too, seem to be changing their views. Some recognize that if they support women's right to vote and encourage them to do so, the sheikhs' own sons may profit in future elections when female voters are more numerous. Others may support women's education but not necessarily their political empowerment. Sheikh Ahmed Abdulrahman Jahaf, whose daughter is running for Parliament, believes that an educated woman is more attractive as a bride than as a public official.

For all her caution, Dr. al-Sharki eventually moved too fast for Yemen. In 1999, after she organized a successful international conference at ERWS, the government reorganized the center with an all-male staff and board of directors and removed al-Sharki as its director. She continues her work by explaining gender issues on the Arabian Peninsula to audiences in American universities.

Sources: Daniel Pearl, "Yemen steers a path toward democracy with some surprises," Wall Street Journal (March 28, 1997): A1, A11; Beloit College Web site: "Noted human rights activist named 2006 Weissberg Distinguished Professor in International Studies," http://www.beloit. edu/~pubaff/releases/05-06/0506weissberg_profile.htm ■

Islamism and Democracy in a Globalizing World

Especially since the attacks of September 11, 2001, Islamism (see page 207) has been seen as a great threat to the region's political stability and economic development. Islamist movements are grassroots religious revivals that seek political power to curb the non-Islamic influences that have become widespread throughout the Muslim world thanks to globalization. Many Islamists who seek to take control of secular governments have been severely repressed. In some instances, however, Islamists have been able to work productively within the democratic process.

The roots of Islamism. The roots of Islamism are complex. For many decades, and into the present in some cases, most governments in this region have been authoritarian and undemocratic. Free speech and the right to hold public meetings have been severely limited, as has freedom of the press to criticize governments or expose corruption. In many countries, the only public spaces in which people have been allowed to gather are mosques, and the only public discussions not subject to censorship by the government have been religious discourses. Hence, political discussions and the political movements that developed from them have been strongly shaped by the views of the region's Islamic leaders.

Islamism is also rooted in the mixture of governmental and religious authority that has characterized most of the region's empires and countries throughout their history. In several countries—Saudi Arabia, Yemen, the UAE, Oman, and Iran—the state enforces the religious principles of Islam. In these **theocratic states,** Islam is the official religion, political leaders must be Muslim, and they are considered divinely guided. The legal system is based on conservative interpretations of shari'a. ◻️

Other countries—Algeria, Egypt, Morocco, Iraq, Turkey, and Tunisia—have declared themselves to be **secular states.** In these countries, theoretically, there is no state religion and no direct influence of religion on affairs of state. In practice, however, religion plays a public role as most political leaders are at least nominally Muslim and Islamic ideas influence in government affairs.

Islamism, globalization, and culture. Many Islamists object to what they see as a modern global culture originating in Europe and North America that undermines Islamic values. Some object to the liberalization of women's roles, especially

Figure 6.27 An Iraqi woman listens to a question during the Facilities Protection Service (FPS) final exam. She will serve alongside male counterparts guarding buildings, dams, and other ministry facilities.

to their being educated, employed, and otherwise active outside the home (Figure 6.27). Many also lament what they see as the global spread of open sexuality, consumerism, and hedonism, transmitted in part by TV, movies, and popular music. They worry that such ways will lead to family instability, alcoholism, drug addiction, and juvenile street crime.

> Many Islamists object to what they see as a modern global culture, originating in Europe and North America, that undermines Islamic values.

These concerns are similar to those of other religious fundamentalists around the world, such as Hindu nationalists in India who worry about the erosion of traditional values and ethical systems. Most Islamists hark back to older interpretations of the Qur'an, favoring a simple, prayerful life focused on family and community, traditional gender roles, and respect for the elderly.

Differing Islamist perspectives. Islamist factions vary greatly in the perspectives and fervor they bring to their causes. While some groups, such as Al Qaeda, advocate violent resistance and drastic reforms, the majority seek simply to moderate non-Islamic influences. For example, most see the advantages of bringing people into the computer age, but they abhor Internet pornography. The proper role of the press is also hotly debated. Some Islamists believe that a free press is essential, while others think the press should not go beyond simple statements about events. Many Islamists also favor government policies based on the Qur'an and the shari'a.

Islamist movements have a strong popular base in the slums of the largest cities, where the economic situation of millions has been worsened by SAPs. Islamism also appeals to refugees from the region's military conflicts, whose suffering has led them to question the basic philosophy of the governments under which they live. 🎥 At the same time, many Islamist activists, especially the leaders, are recent university graduates who are frustrated by their inability to find employment or to participate in the political process. Some Islamist leaders are also respected religious men without much non-religious education who argue that all would be well if people rejected secularism and returned to fundamentalist versions of Islam.

The perils of repression. After decades of oppression at the hands of a secular, undemocratic, and highly corrupt government in Iran, an Islamist revolution led by Ayatollah Ruhollah Khomeini overthrew the government in 1979. Fearing a repeat of this scenario, secular governments in several countries—Egypt and Algeria, for example—attempted to forcibly repress Islamist movements. Both of the latter countries suffered from years of Islamist-sponsored terrorist campaigns as a result. In Algeria, violence has diminished only in the last decade following free and fair elections that incorporated Islamists into the government. In Egypt, where repression persists and elections are plagued by government-sponsored intimidation and fraud, Islamist terror and violence continues. 🎥

Even governments that have strong links with Islamists have difficulty controlling them without democratic reforms. For example, the Saud family in Saudi Arabia came to power and remains there by cooperating with the very conservative Wahhabi school of Islamist thinking. Today, however, Saudi Arabia faces a growing terrorist-based insurgency fomented by radical Islamist militants. Islamists have already forced the Saudi government to remove foreign soldiers from the country, mostly U.S. troops who had remained in the country since the first Gulf War. In a move to diffuse tensions, Saudi Arabia held its first democratic elections in 2005, though these were limited to local municipal elections. Similar elections have since been held in the United Arab Emirates. 🎥

Democratization: a path to peace with Islamists? In recent years, elections have brought Islamists to power in Iraq, Algeria, Jordan, and elsewhere. Many fear that their policies may worsen relations with other countries in the region as well as with the non-Muslim world. In Iran, for example, a government run by both appointed religious leaders and elected officials has sponsored Islamist terrorism in other countries, such

as Israel and Lebanon. However, other Islamist governments, such as Turkey's, do not sponsor terror. Moreover, the views of Islamists usually become more moderate once they achieve power and are faced with the complexities of governing. 🎥

Generally, governments that have allowed free elections in which Islamists can run for office have enjoyed greater stability. This may partially explain the region-wide shift toward democracy in recent years. While many countries are still ruled by kings and royal families, almost all countries also now have elected national legislatures and municipal councils. Some countries are also reducing censorship of the press (see the opening vignette on page 205) and expanding public forums in which political issues can be discussed.

> Generally, governments that have allowed free elections in which Islamists can run for office have seen greater stability.

Turkey has been the most successful at using the democratic process to constructively engage Islamist political parties. Turkey's government has been controlled by Islamists on and off for decades. Nevertheless, there have been no significant departures from the country's safeguards for freedom of religion or its steady improvement on women's rights, education, and freedom of the press. Nor has the country's long-term goal of becoming a modern, prosperous member of the European Union been compromised. 🎥

Demand for greater political freedoms is increasing throughout the region, and many public spaces outside of mosques are emerging. Extensive opinion polls of the citizens of Algeria, Lebanon, Jordan, and Palestine have revealed a broad consensus that governments need to protect freedom of speech, a free press, and the right to hold public meetings. If the growing demand for improved educational and employment opportunities is met, then some of the discontent from which Islamism feeds may also diminish. However, such changes will also require broad social reforms and support of the institutions that make up a *civil society* (see Chapter 1 page 47). 🎥

The Israeli–Palestinian Issue

The Israeli–Palestinian conflict has lasted longer than 50 years and spawned several major wars and innumerable skirmishes. The conflict is a persistent obstacle to political and economic cooperation in the region. Israel has by far the most modern and diversified economy. Since the 1950s, its development has been facilitated by the immigration of relatively well-educated middle class Jews from the United States, Europe, Russia, and South America. Israel's excellent technical and educational infrastructure, its diverse and prospering economy, and large aid contributions from the United States and private interests, have made the country one of the region's wealthiest and most militarily powerful countries. 🎥

The Palestinian people, by contrast, are severely impoverished and undereducated after years of repression and the loss of the bulk of their lands to Israel over the past 60 years (Table 6.1). Palestinians now live in two main areas—the West

Table 6.1 Circumstances and state of human well-being among Palestinians and Israelis, 2005

	Population (in millions)	Infant mortality (per 1000 live births)	Percentage of unemployed	Percentage of population in poverty, 2003
Palestinians	3.8	21	50	75
Israelis	7.1	5.1	8.9	18

Sources: Population Reference Bureau, *2005 World Population Data Sheet,* 2005; *United Nations Human Development Report 2005* (New York: United Nations Development Programme); *Arab Human Development Report* 2003.

Bank and the Gaza Strip, both of which are highly dependent on Israel's economy and subject to regular Israeli military actions in retaliation for Palestinian suicide bombings and rocket fire. 📹

Creation of the state of Israel. In the late nineteenth century, in response to centuries of persecution in Europe and Russia, a small group of Jews, known as **Zionists,** began to purchase land in what was then called Palestine. Most sellers were wealthy non-Palestinian Arabs and Ottoman Turks living outside of Palestine. These landowners had long leased their lands to Palestinian tenant farmers and herders or granted them the right to use the land. This right was negated by the

sales, and historians still debate whether or not the Palestinians were compensated adequately by either their former landlords or the Zionists.

On their new land, the Zionists established communal settlements called *kibbutzim,* which received a small flow of Jews from Europe and Russia. While Jewish and Palestinian populations intermingled in these early years, tensions rose as Zionist land purchases in the early twentieth century increasingly displaced Palestinians.

Efforts to secure a Jewish homeland continued and gained important political support from Britain, the ruling colonial power in Palestine (Figure 6.28). By 1946, following the death of 6 million Jews in Nazi death camps during World War II,

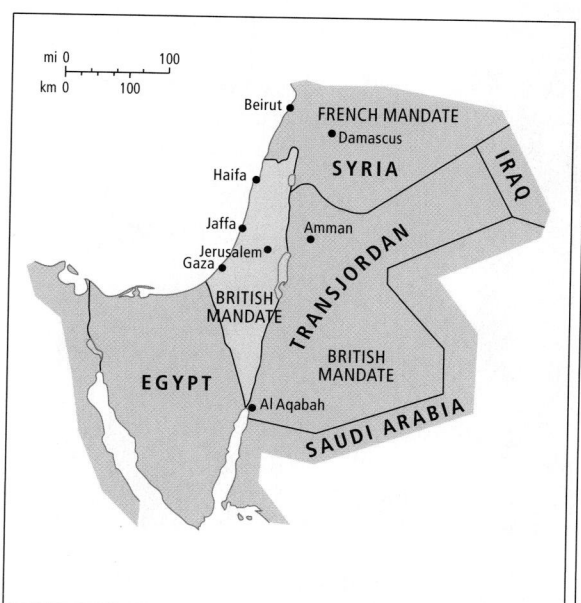

(a) Palestine, 1923
In the 1920s, Britain controlled what is now Israel and Jordan (Transjordan was the precursor to Jordan).

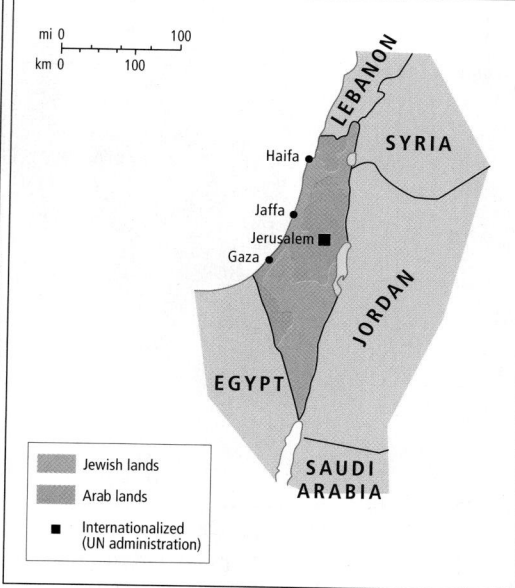

(b) UN Partition Plan, 1947
After World War II, the United Nations developed a plan for separate Jewish and Palestinian (Arab) states.

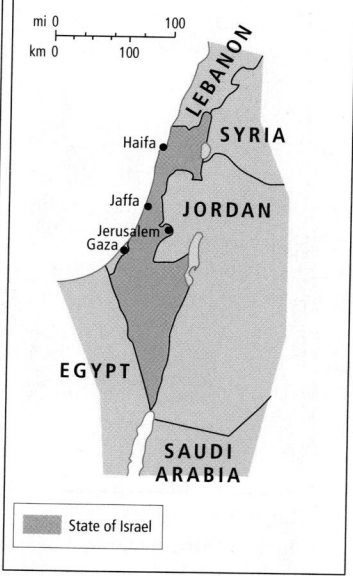

(c) Israel, 1949
The Jewish settlers did not agree to that plan; instead, they fought and won a war, creating the state of Israel. [Adapted from Colbert C. Held, *Middle East Patterns—Places, Peoples, and Politics* (Boulder, Colo.: Westview Press, 1994), p. 184.]

Figure 6.28 Israel and Palestine, 1923–1949.

strong sentiment had grown throughout the world in favor of a Jewish homeland. Hundreds of thousands of Jews migrated to Palestine, and many took up arms to convince Britain to support their goal of a Jewish state. Warfare between the Jews and Palestinians began in 1948 on the very day that the last British soldier left and the state of Israel was proclaimed.

Continuing conflict. The Palestinians and neighboring countries fiercely objected to the establishment of a Jewish state on formerly Palestinian land. They feared that all Palestinians would lose their remaining lands and become refugees. Then as now, the conflict between Zionists and Palestinian Arabs was less about religion than control of land and water.

In 1948, Israel was invaded by troops from Lebanon, Syria, Iraq, Egypt, and Jordan. Israel prevailed, and the Palestinians' land shrank yet further, with the remnants incorporated into Jordan and Egypt by 1949. In the repeated conflicts over the next decades—such as the Six Day War (1967) and the Yom Kippur War (1973)—Israel again defeated its larger Arab neighbors, expanding into territories formerly controlled by Egypt, Syria, and Jordan (Figure 6.29a)

The most recent of Israel's conflicts with its neighbors occurred in 2006. Hezbollah, an anti-Israeli militia based in southern Lebanon and financed largely by Iran, kidnapped two Israeli soldiers. Israel responded with a 34-day counter attack that destroyed most of Lebanon's infrastructure. However, it failed to defeat Hezbollah, which launched thousands of rockets into northern Israel. Hezbollah remains a powerful military and political force within southern Lebanon to this day.

Intifada. Since 1948, hundreds of thousands of Palestinians have fled the war zones, with many removed to refugee camps in nearby countries. Some Palestinians stayed inside Israel and became Israeli citizens, but they have not been treated as equal to Jewish Israelis by the state.

During this time Palestinians have mounted two prolonged uprisings, known as the first **intifada** (1987–1993) and the second intifada (2000–present). Both periods have been characterized by escalating violence. Palestinian suicide bombers have targeted Israeli civilians, killing hundreds, maiming thousands, and psychologically impacting all Israelis. In response, the Israeli military has used deadly force to quell demonstrations and to punish the families and communities of the suicide bombers.

Territorial disputes. When Israel occupied Palestinian lands in 1967, the United Nations Security Council passed a reso-

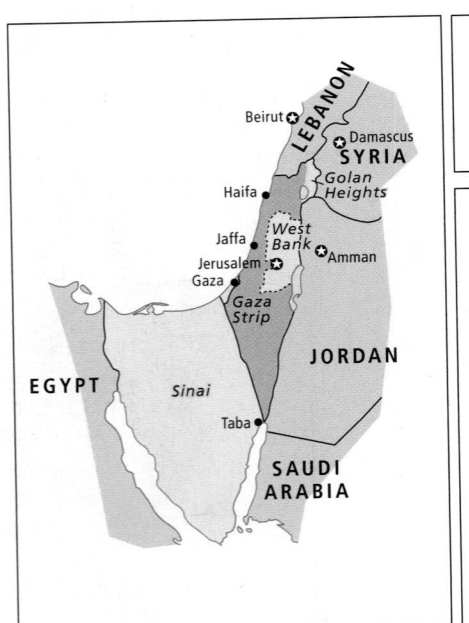

(a) 1967–1973
In 1967, Israel soundly defeated combined Arab forces and took control of the Sinai Peninsula, the Gaza Strip, the Golan Heights, and the West Bank.

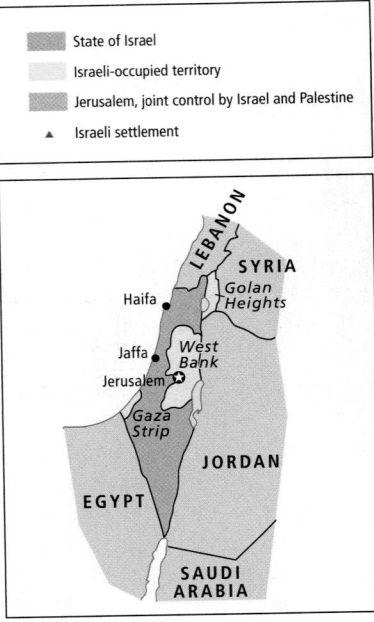

(b) 1993
In subsequent peace accords, the Sinai Peninsula was returned to Egypt, but Israel maintained control over the other territories it had occupied.

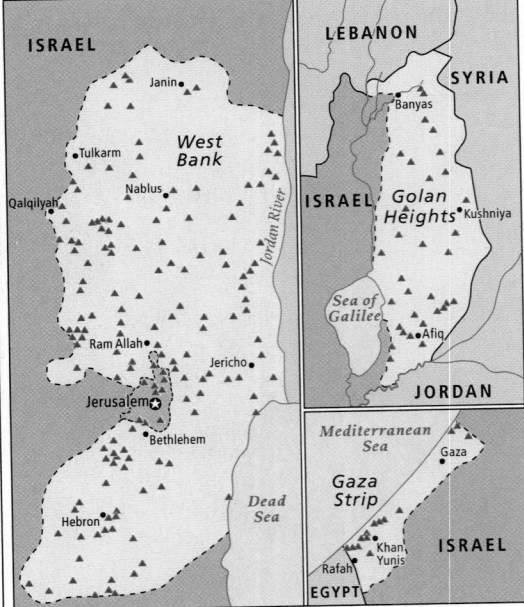

(c) 2002
Israel built Jewish settlements in the occupied territories, though those in Gaza were removed in 2005. In 2002, Israel began building a barrier (not shown on the map) around Palestinian territory on the West Bank. [Adapted from Colbert C. Held, *Middle East Patterns—Places, Peoples, and Politics* (Boulder, Colo.: Westview Press, 1994), p. 184.]

Figure 6.29 Israel after 1949. When the state of Israel was created in 1949, many of its Arab neighbors were opposed to a Jewish state.

Figure 6.30 Israeli soldiers entering Neve Dekalim settlement in the Gaza Strip, 2005. In 2005, Israel began pulling out of Jewish settlements in the Gaza Strip. Jewish settlers (seen here in the background) often attempted to block the pullout; in this case, they set tires on fire in protest. [AP Photo/Oded Balilty.]

lution requiring Israel to return those lands, known as the **occupied territories,** in exchange for peaceful relations between it and neighboring Arab states. This *land-for-peace formula,* which sets the stage for an independent Palestinian state, has been only partially fulfilled.

Since 1967, Israel has secured control over the land and water resources of the occupied territories with hundreds of Jewish settlements in the West Bank and the Gaza Strip (Figures 6.29b and 6.29c). Israel took a major step toward peace in 2005 when it removed all Jewish settlements from the Gaza Strip (Figure 6.30). However, in the eyes of Palestinians, this progress was negated by the construction of the **West Bank barrier,** which began in 2003.

The West Bank barrier, a 25-foot-high concrete wall in some places and a fence in others, now surrounds much of the West Bank and encompasses many of the remaining Jewish settlements there. The barrier separates around 30,000 Palestinian farmers from their fields and effectively annexes 6 to 8 percent of the West Bank to Israel. It also severely limits Palestinian access to much of the city of Jerusalem, most of which is now on the Israeli side of the barrier. The barrier has been declared illegal by the World Court in the Hague (Netherlands) and the UN, and it is opposed by the United States. However, it is very popular among Israelis because it has reduced the number of Palestinian suicide bombings. 🎥

Economic interdependence. Economic relations with Israel will continue to have a huge impact on Palestinian daily life with or without Palestinian independence. Israel is the largest trading partner of the West Bank and Gaza Strip and provides their currency (the Israeli shekel), electricity, and most imports. It employs tens of thousands of Palestinian workers in factories within Israel.

However, many Palestinians and some Israelis criticize this relationship as exploitive. Israel tightly controls Palestinian access to the world market and has historically discouraged industrial development in the occupied territories. Unequal economic relationships combined with the ongoing violence mean that most of the population of the occupied territories lives below the poverty line (70 percent in Gaza and 55 percent in the West Bank).

Economic relations improved somewhat after a peace agreement with Israel in 1993 (the Oslo Accords). This agreement resulted in the first substantive move toward an independent Palestinian state in the West Bank and Gaza Strip, the establishment of the Palestinian Authority. Though limited by Israeli control of its borders and regular Israeli military operations in its territory, the Palestinian Authority still functions as an independent governing body. One of its major achievements is increased economic cooperation with Israel. For example, the Palestinian Authority and the Israeli government have established several industrial parks in the West Bank and Gaza Strip in recent years. These developments have been dubbed "peace parks" because of their goal of overcoming conflict through economic development.

Israeli companies are by far the most enthusiastic investors in these ventures, more so even than Palestinians living abroad. However, the tensions of the second intifada have brought frequent border closures that constrain these and other economic development projects. The situation worsened in 2006 when Palestinian elections gave a parliamentary majority to Hamas, a political party and activist organization long associated with suicide bombings and other violence against Israel.

In response to the success of Hamas, the European Union, the United States, and Canada withdrew financial aid to Palestine. More importantly, Israel placed a blockade on trade that crippled the economies of the West Bank and the Gaza Strip. This financial pressure figured in the subsequent decision by Hamas to stop calling for the destruction of Israel—a major if somewhat ironic concession given Palestine's high level of economic dependence on Israel. The blockade was later eased in the West Bank but remains in effect in Gaza, where extremists regularly fire rockets into Israel. 📹

Local peace efforts. Little recognized in the international press or among world leaders is the fact that ordinary citizens, both Israeli and Palestinian, have separately and collaboratively been designing ways to end the conflict. These efforts acknowledge the national aspirations and the right to land of both parties.

Examples of these bottom-up peace initiatives include joint Israeli–Palestinian peace demonstrations and women's groups that have tried to end the Israeli occupation of the West Bank and Gaza Strip. Palestinian and Israeli Physicians for Human Rights have joined to address the medical problems of the overwhelmingly poor Palestinians. Groups from both sides hold youth camps so that Israeli and Palestinian children can break the cycle of hatred through personal friendship.

The Iraq War (2003–present) and Democratization

The U.S.-led invasion and occupation of Iraq has transformed the politics of this region. Anti-U.S. sentiment has exploded as has criticism of leaders and governments, such as Israel's, who are seen as supporting the goals of the United States. 📹 Meanwhile, the U.S. attempt to create a democratic Iraq has led to many unexpected consequences.

The origins of the Iraq war are complex. The United States has had a long-standing relationship with Iraq, driven in large part by that country's considerable oil reserves, the second largest in the world after Saudi Arabia's. The United States also supported Iraq in its war with Iran, whose revolutionary Islamist government has had a hostile relationship with the United States. U.S.–Iraq relations took a dramatic turn for the worse in 1990 when Saddam Hussein, who ruled Iraq as a dictator, invaded Kuwait. The United States forced Iraq's military out of Kuwait in the Gulf War of 1990–1991, and afterward placed the country under crippling economic sanctions.

In searching for an explanation for the terrorist attacks of September 11, 2001, the U.S. administration of George W. Bush focused on Iraq and its president Saddam Hussein. The Bush administration was convinced that Iraq had or was creating an arsenal of weapons of mass destruction. The United States declared war on March 20, 2003, with the goals of confiscating Iraq's weapons of mass destruction, removing Saddam from power, and turning Iraq into a democracy. Great Britain and a few other countries joined the United States as allies, but most of the world objected to the war.

On May 1, 2003, President Bush declared the war won. However, terrorist bombs and insurgent attacks by Iraqis continued to take the lives of U.S. and allied soldiers at the rate of more than two per day. By the end of 2008, over 4200 U.S. troops had been killed and 33,000 wounded. Furthermore, over 150,000 veterans of the conflict had been diagnosed with some form of serious mental disorder, including post-traumatic stress disorder (PTSD). The death toll for Iraqis, including civilians, has been much higher, 200,000 at minimum. Some statistics place the death toll as high as 650,000. Many Iraqi deaths were due to poor conditions created by the war rather than direct contact with foreign troops or insurgents. 📹

The war has had many unintended consequences. While Saddam has been removed from power and executed, weapons of mass destruction have not been found, nor have any links between the 9/11 attacks and Saddam's government been uncovered. Democratization is the main stated justification for continuing the war, but ongoing violence has inhibited the democratic process. Long-simmering tensions between Iraq's major religious and ethnic groups have exploded onto the political scene. Sunnis in the northwest dominated the country under Saddam although they constituted only 32 percent of the population. Shi'ites in the south, at 60 percent of the population, have dominated politics since the fall of Saddam. This has given Iran, whose population is mostly Shi'ite, greater influence in Iraq. Meanwhile, Kurds in the north, who were brutally suppressed under Saddam, have become increasingly defiant of the Iraqi national government in Baghdad. Saddam's treatment of the Kurds ultimately led to his conviction of mass murder and execution in 2006. 📹

Iraqi opinions. Recent studies of Iraqi public opinion indicate a number of important trends that will shape a more democratic Iraq. The United States has called for a weaker federal government in Iraq, with the Shi'ite south, Sunni northwest, and Kurdish north having greater independence (Figure 6.31). However, polls show that a majority of Iraqis want a strong central government that can protect them from violent insurgents and maintain control of the country's large fossil fuel reserves. Polls also show that the vast majority of Iraqis want locally based insurgents to stop fighting and all foreign insurgents to leave. Most Iraqis strongly oppose any long-term U.S. military presence in the country. This position has also been endorsed by the Iraqi parliament.

Refugees

There are millions of refugees in the region. Usually they are displaced by human conflict, but environmental disasters such as earthquakes or long-term drought also displace many people. When Israel was created, as many as 2 million Palestinians were displaced to refugee camps in Lebanon, Syria, Jordan, the West Bank, and the Gaza Strip. Palestinians still constitute

Figure 6.31 Iraq, Iran, and Turkey. The small triangle with Baghdad at one point shows the densest concentration of Sunni Muslims in Iraq, a center of intense resistance to the U.S. occupation. The area of Kurdish concentration overlaps three countries. Shi'ites are mostly southeast of Baghdad. [Adapted from Edgar O'Ballance, *The Kurdish Struggle 1920–1994.* (New York: St. Martin's Press, 1996), p. 235.]

the world's oldest and largest refugee population, numbering at least 5 million.

International organizations contribute some funds to support a basic level of services for refugees, but these displaced people are still a huge drain on the resources of their host countries. In Jordan, for example, refugees—including about 4 million Palestinians and over 1 million Iraqis—now account for more than half the total population of the country. This huge population influx has changed life for all Jordanians. Iran is sheltering more than 3 million Afghans and Iraqis. Across the region, even more people are refugees within their home countries. More than 2.4 million Iraqis are internal refugees, and in Sudan, between 5 and 6 million people, including those fleeing violence in Darfur, live in refugee camps. ▇▶

Throughout the region, refugee camps often become semi-permanent communities of stateless people in which whole generations are born, mature, and die. Although residents of these camps may show enormous ingenuity in creating a community and an informal economy under adverse conditions, the cost in social disorder is high. Tension and crowding create health problems. Children rarely receive enough schooling. Disillusionment is widespread. Years of hopelessness, extreme hardship, and lack of employment take their toll.

The Crisis in Darfur

The crisis in the Darfur region of Sudan (Figure 6.32) is a complex story of long-standing ethnic and tribal animosities, worsened by climate change and the recent discovery of oil in the area.

Officials from Sudan and neighboring Chad point to a 40 percent decline in rainfall over the past fifty years as one of the causes of the conflict. As nomadic Arab herders from the drier north stray into the wetter south in search of water for their animals, they come into increasing conflict with non-Arab sedentary farmers. The recent discovery of oil reserves has inflamed the situation further. The non-Arab farmers of Darfur have organized rebel militias who demand greater control over these oil reserves. Because Sudan's government is already fighting rebels in the southern part of the country over that area's oil reserves, its military could not afford to fight another war in Darfur. Therefore, the government turned to the Arab herders of northern Darfur, organizing them into militias known as *Janjaweed* and supporting them in attacks on the non-Arab farmers of southern Darfur.

The conflict in Darfur has been called the world's greatest current humanitarian disaster, with 450,000 dead and at least 2.5 million refugees. Even so, countries in need of oil,

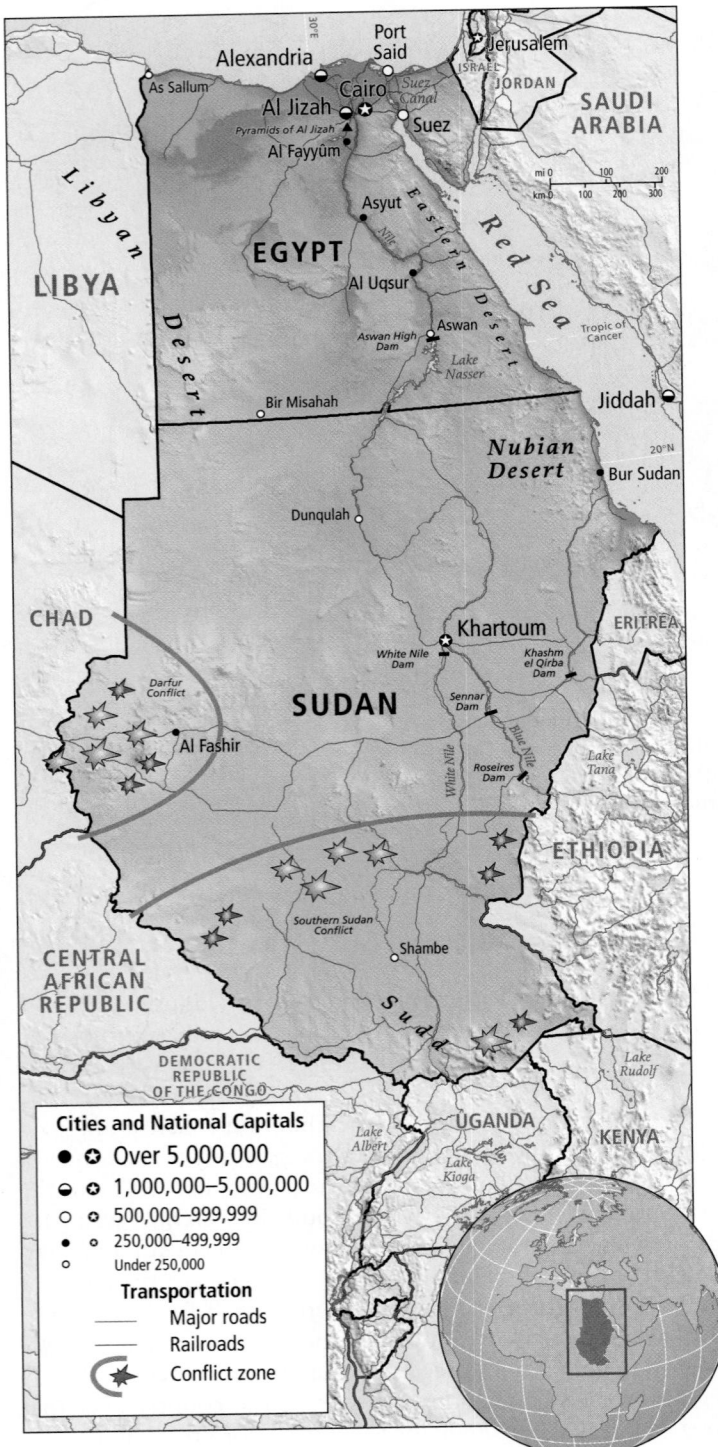

Figure 6.32 Conflict zones in Sudan. [Conflict zones adapted from USAID, http://www.USAID.gov/locations/sub-saharan_africa/sudan/sudan_bombjuly.pdf; http://www.lib.utexas.edu/maps/africa/darfur_villages_0802_2004.jpg]

such as China, are willing to overlook Sudan's humanitarian and environmental disasters and invest heavily in the country's energy industry.

Oil extraction in an intense conflict zone like Darfur often results in environmental disaster. Pipelines and derricks make easy targets for rebel insurgent groups, and in places like Iraq, they have been bombed repeatedly to deprive the government of oil revenues. This results in major inland oil spills and long-lasting fires that destroy habitat and threaten water and soil resources.

Reflections on North Africa and Southwest Asia

At first glance, North Africa and Southwest Asia may seem hopelessly mired in a morass of unsolvable problems. Violence, religious fundamentalism, economic inequities, growing urban slums, water shortages, extreme gender discrimination, expanding populations, and dependence on energy resources that the world may be turning away from—all suggest a gloomy future. Yet, many of the changes currently underway could provide a route through the daunting maze of issues this region faces.

The existing shift toward democracy and away from repressive authoritarian rule may diffuse the threat posed by fundamentalist Islam to the region's security and to relations with the rest of the world. Pressures for democracy are growing in the fossil fuel-rich countries, where an increasingly educated, informed, and globally aware citizenry is demanding greater freedom of speech, a larger voice in politics, and more equality for women. In the rest of the region, urbanization could also result in a more-educated citizenry, as it has in much of the rest of the world. But this will require economic growth, increased employment, and upgrades to educational systems that have suffered years of funding cuts.

Democracy and greater economic opportunities could also help diffuse some of the violent conflicts that currently block international cooperation. In Iraq, democracy may bring policies that result in greater cooperation and peaceful coexistence among sectarian, ethnic, and other factions currently in conflict. In Israel and Palestine, recognition of the economic interdependence of the two countries may also inspire better relations. An independent and democratic Palestinian state might prove crucial in relieving the hopelessness and deprivation that currently feed this conflict.

Improvements in education and economic growth could also aid the region's ability to provide its growing population with enough food and water. Water-saving agricultural technologies could stretch water supplies much further. But any significant shift in this direction would require farmers to become more educated and better able to invest in sophisticated and expensive technologies. If educational and employment opportunities were extended to women, then the region's high population growth rate could fall faster, relieving some of the pressure on food and water resources.

Global warming may provide additional motivation to make these changes. Water-saving technologies may become more necessary if rainfall patterns shift even slightly. Better-educated populations may find it easier to adapt to the complex economic disruptions created both by changing regional climates and a global shift away from fossil fuels. Similarly, citizens faced with the wide range of uncertainties that global warming may bring could demand a level of responsiveness and accountability that only democracies can deliver.

This may all be wishful thinking. But then again, it might not be.

Chapter Key Terms

cartel 227

Christianity 218

desertification 215

diaspora 218

economic diversification 228

female seclusion 225

Fertile Crescent 216

fossil fuel 207

hajj 224

intifada 234

Islam 207

Islamism 207

Judaism 218

monotheistic 218

Muslims 218

occupied territories 235

OPEC (Organization of Petroleum Exporting Countries) 227

polygyny 227

Qur'an (or Koran) 211

salinization 212

seawater desalination 213

secular states 231

shari'a 225

sheikh 230

Shi'ite (or Shi'a) 225

Sunni 225

theocratic states 231

veil 225

West Bank barrier 235

Zionists 233

Critical Thinking Questions

1. How have technological approaches to this region's water shortage created new problems? Can you devise a workable strategy for dealing with water shortages that does not depend on technological change?

2. Argue for or against government support of agriculture in this region. How do the potential impacts of global warming on the region's food production and on global food prices influence your position?

3. To what extent is the colonial history of North Africa and Southwest Asia visible in the present-day map of the region?

4. Consider the various factors that encourage relatively high human fertility in this region and design themes for a public education program that would effectively encourage lower birth rates. Which population groups would you target? How would you incorporate cultural sensitivity into your project?

5. Consider the new forces that are affecting urban landscapes: immigration and globalization. Compare how these forces differ in the cities of oil-rich states and those in the rest of the region.

6. Describe the role that oil has played in the region's economy. Who has benefited the most from oil development?

7. Compare and contrast the public debate over the proper role of religion in the politics of this region with similar debates in the United States.

8. How might democratization lead to liberalization of gender roles? What is the evidence that such changes might be slow in coming?

9. How are the forces that led to the formation of the state of Israel related to globalization?

10. How might the war in Iraq both aid and hinder the spread of democracy in this region?

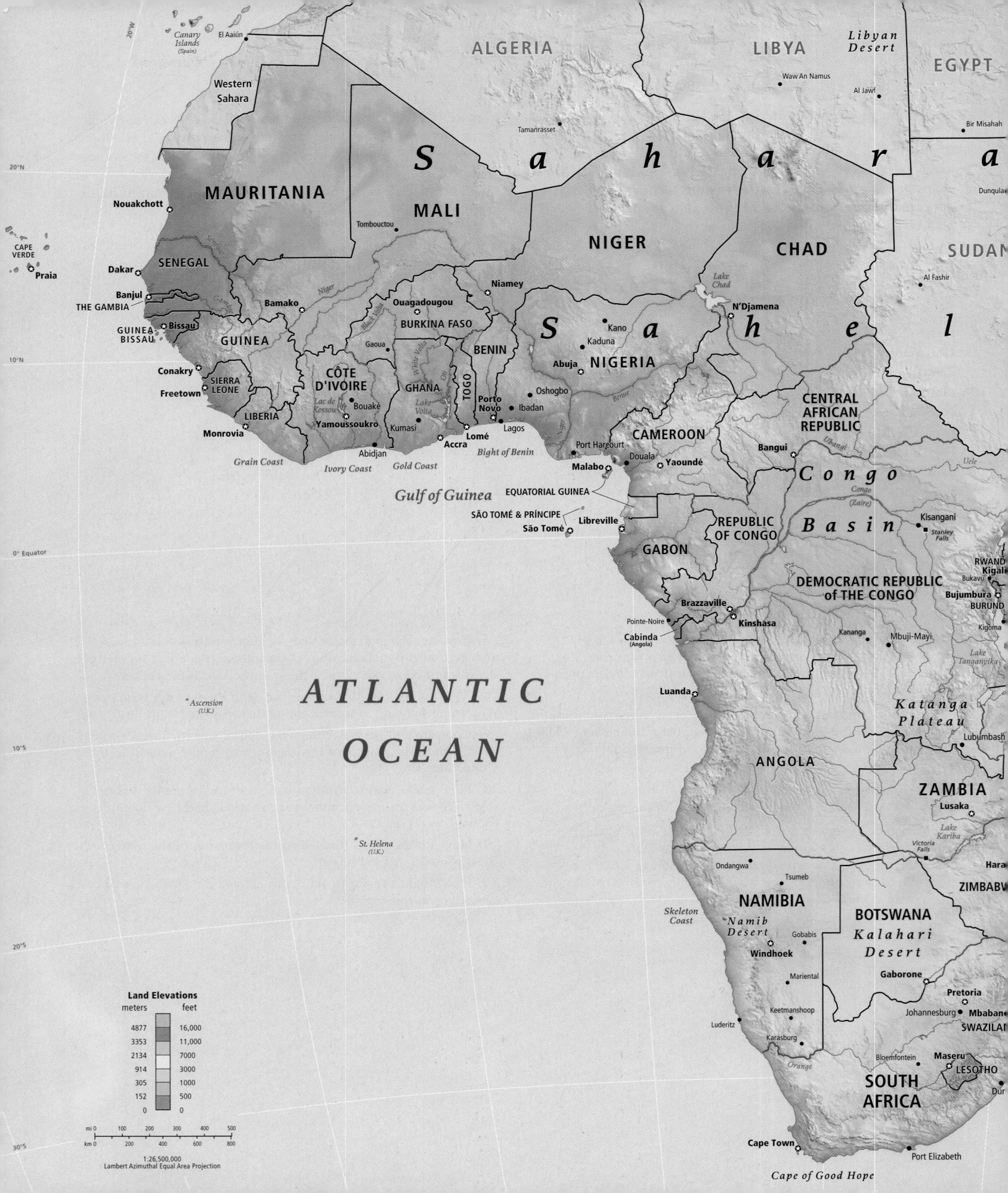

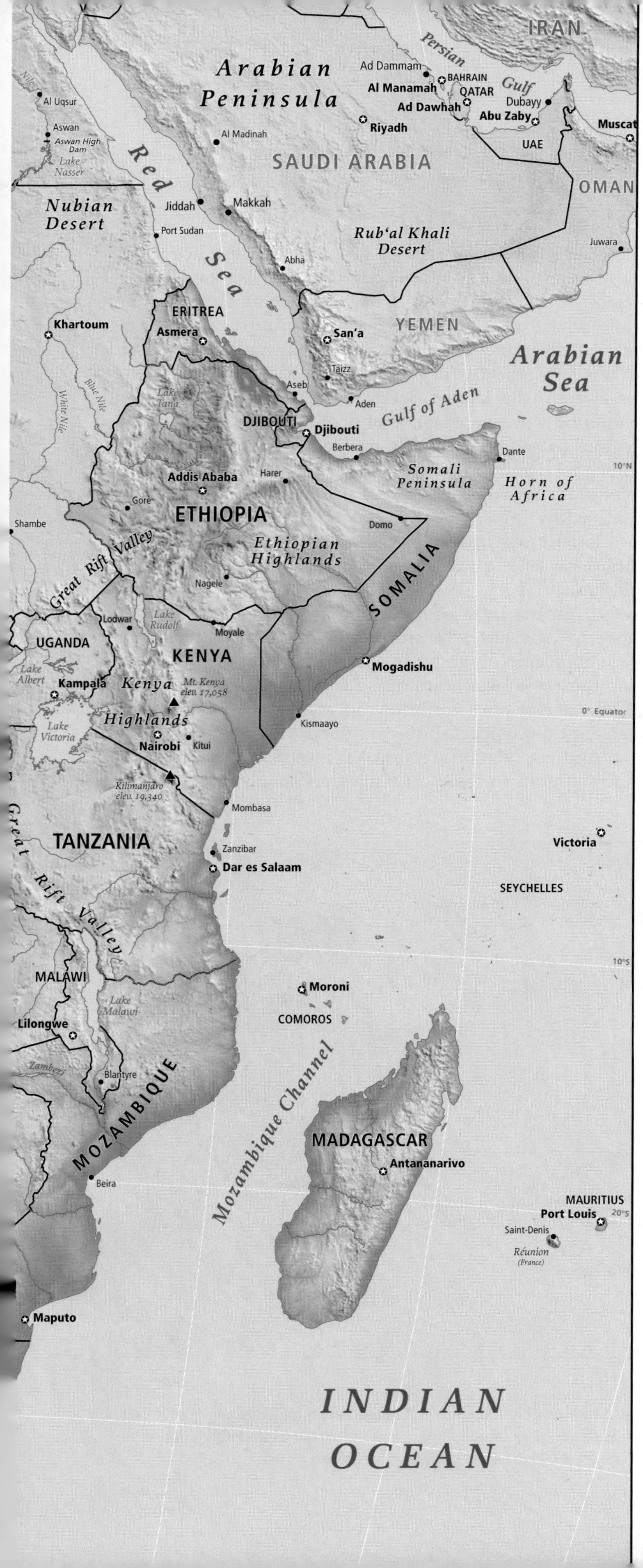

CHAPTER 7

Sub-Saharan Africa

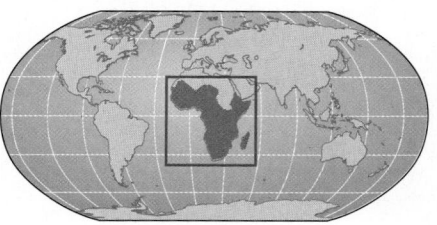

Global Patterns, Local Lives Liberian environmental activist Silas Siakor is an affable and unassuming fellow. But his casual style conceals a fierce dedication to his homeland and a remarkable ability at sleuthing.

At great personal risk, Siakor uncovered evidence that seventeen international logging companies were bribing Liberia's president, Charles Taylor, with cash and guns. In return, Taylor looked the other way while the companies illegally logged Liberia's forests, which are home to many endangered species such as forest elephants and chimpanzees. Taylor used the money and weapons he received to equip his own personal armies. Made up largely of kidnapped and tragically abused children, Taylor's armies fought those of other Liberian warlords in a 14-year civil war that took the lives of 150,000 civilians. The war also spilled over into neighboring Sierra Leone where another 75,000 died. ◼ Meanwhile, the logging companies—based in Europe, China, and the Middle East—reaped huge fortunes from the tropical forests.

Silas Siakor secretly prepared a clear, well-documented report substantiating the massive logging fraud (Figure 7.2). He showed the connections between the timber industry, illegal weapons trade, and the funds that gave Taylor the power to wage civil war. In response to Siakor's report, the United Nations Security Council voted to impose sanctions to stop the timber trade (Figure 7.3) and prosecute some of the people involved. Charles Taylor fled to Nigeria, but he was eventually turned over to face a war crimes tribunal in The Hague, Netherlands.

Democratic elections followed in Liberia, and Africa's first elected woman president, Ellen Johnson-Sirleaf, took office in January 2006. In a move that was bold given the poverty and political instability of her country, she cancelled all contracts with timber companies pending a revision of Liberian forestry law. Silas Siakor, who won the international Goldman Environmental Prize in April

FIGURE 7.1 Regional map of sub-Saharan Africa.

FIGURE 7.2 Silas Siakor at work. Siakor is documenting illegally taken logs that are being salvaged in Liberia's Rivercess County. [Goldman Environmental Foundation.]

2006, says the most rewarding outcome of his work is seeing that power for change can still lie with "the little fellow."

Adapted from "Silas Kapanan 'Ayoung Siakor: A voice for the forest and its people," Goldman Environmental Prize Web site, http://www. goldmanprize.org/node/442; Scott Simon, "Reflections of a Liberian

environmental activist," "Weekend Edition Saturday," National Public Radio, April 29, 2006, http://www.npr.org/templates/story/story.php? storyId=5370987.

The story of Silas Siakor and Liberia illustrates some of sub-Saharan Africa's challenges. Liberia, like much of this region, is still developing strong democratic governing institutions that can direct the country's resources to those who need them most. At the same time, Liberia is unique in that it was sub-Saharan Africa's first republic, founded by freed slaves from the United States in 1822. They hoped Liberia would be a shining example of democracy and prosperity in a region that was then mostly controlled by outsiders. Instead, rampant corruption severely limited Liberia's democracy and economic development, and large parts of the country were annexed by Britain and France.

Liberia also exemplifies the ways in which sub-Saharan Africa has failed to benefit substantially from economic globalization. Liberia has rich timber and mineral (diamond) resources, but most of the profits from these industries have gone to Liberians, who are already quite wealthy, and to foreign companies. Eighty percent of Liberians are unemployed, and those who work earn only a few hundred dollars per year. Adult literacy rates are below 56 percent. In 2007, the average life expectancy at birth was 45 years.

And yet, Silas Siakor's success illustrates how Africans can find solutions to these challenges, seemingly against all odds.

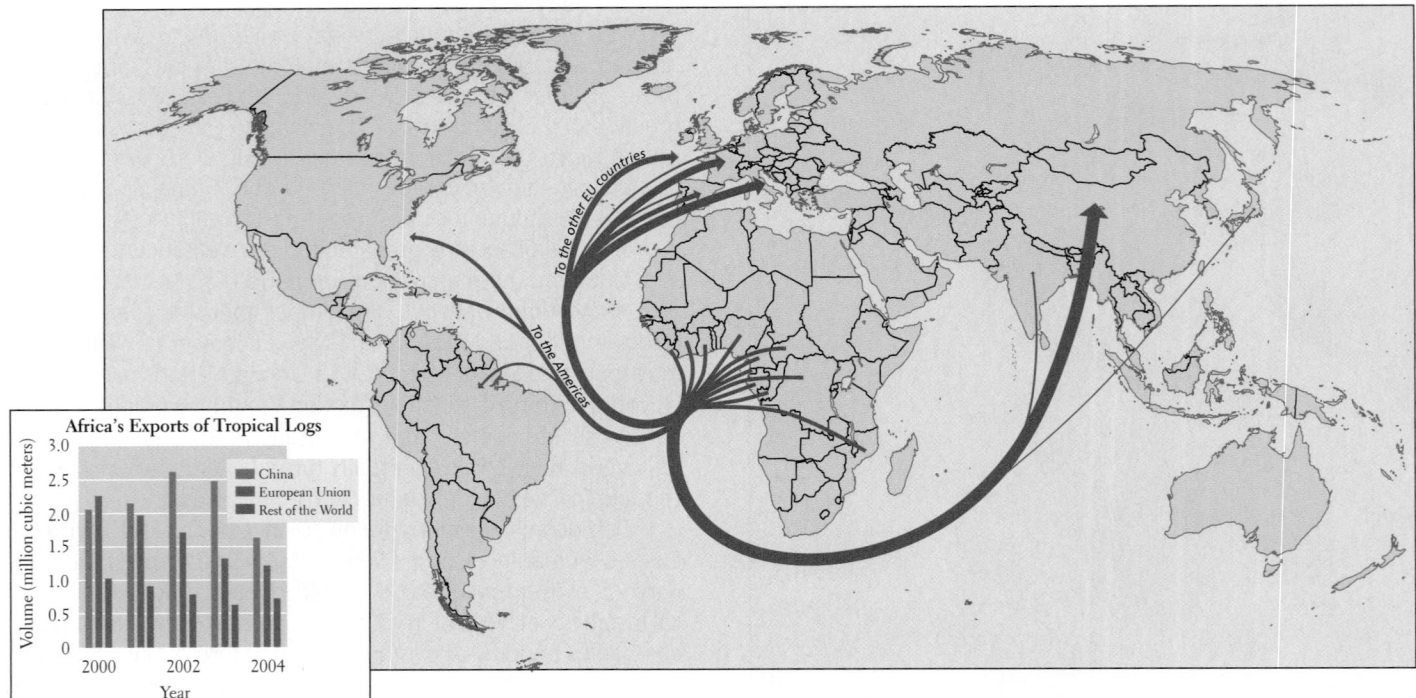

FIGURE 7.3 Destinations of Africa's exported tropical logs, 2000–2004. Most end-point consumers of Africa's tropical woods are in North America and Europe. Though China is the largest single recipient of exported logs, much of that wood ends up in furniture sold by firms like Ikea in the United States and Europe. [Data from "Exports by, and imports from, Africa" (Global Timber.Org.UK), at http://www.globaltimber.org.uk/africa.htm.]

It shows as well that African efforts toward achieving peace and prosperity can be both aided and thwarted by the rest of the world. Foreign logging companies may finance civil war in Liberia, but the international court in The Hague may help Liberians hold their leaders accountable for criminal acts.

Sub-Saharan Africa (see Figure 7.1) is home to about 750 million people. It contains several of the fastest-growing economies in the world, as well as some of the world's richest deposits of oil, gold, platinum, copper, and other strategic minerals. During the era of European colonialism (1850s–1950s), this massive wealth flowed out of Africa. Since the granting of political independence in the 1950s, 1960s, and 1970s, wealth is still flowing out of Africa. Globalization in an environment of rampant corruption and weak or non-existent democracy has enabled investors from the rich countries of the world to reap most of the wealth derived from Africa's mineral, forest, and agricultural resources. Charles Taylor's Liberia is just one example among many. Today, the average per capita income in sub-Saharan Africa is the lowest in the world. It is not surprising, then, that Africa is impoverished and often at war with itself.

But amid the turmoil, there are many hopeful signs that the conflicts can be resolved and the well-being of the majority enhanced. Democracy is expanding throughout the region, and despite ongoing violence and many unelected leaders, more Africans are able to participate in governing themselves. Since independence, African countries and societies have struggled to determine appropriate pathways to economic development. And although many African countries have not yet found an environmentally sustainable path to prosperity, a few have recently experienced significant economic growth.

Questions to consider as you study Sub-Saharan Africa (also see Views of Sub-Saharan Africa, pages 244–245)

1. Why is Africa particularly vulnerable to global warming? Many Africans depend on agriculture and herding for their livelihood. How are these occupations sensitive to changes in temperature and water availability that global warming is likely to bring? How do poverty and low access to cash influence the vulnerability of many Africans to global warming?

2. How has Africa been shaped by globalization? Europeans colonized Africa, setting up enduring structures that still leave many countries relying on exports of cheap raw materials to wealthier regions. How does this arrangement put Africa in a weak position in the global economy? How are Africa's many civil wars related to European colonialism and globalization?

3. What are the forces working for and against democracy in Africa? While a shift toward democracy is occurring throughout Africa, powerful forces still work against democracy in many countries. To what extent can the weakness of democracy in Africa be traced to institutions put in place during the colonial era? What role have elections played in both diffusing and inciting violence?

4. How is urbanization influencing population growth in Africa? Urbanization is slowing down population growth in Africa, but there are still large families even in the cities. What aspects of urban African life encourage large families? How has rapid urbanization contributed to the situation?

5. How are issues of gender influencing trends in both population growth and politics? Recent progress in political reform and the control of population growth has involved greater recognition of women's important roles in African societies. How does the education and empowerment of women contribute to lower population growth rates? How have national crises led to greater political empowerment of women?

The Geographic Setting

Terms to Be Aware Of

Although we name this region Sub-Saharan Africa (to recognize that North Africa is not included), we often refer to it simply as Africa for convenience. Sub-Saharan Africa is defined by the countries shown in Figure 7.4 (page 246). Notice that Sudan is not included. Sudan's location and strong Arab influence brings that country into closer association with North Africa and Southwest Asia, and it is discussed in Chapter 6.

The naming of African countries can often be confusing. There are two neighboring countries called Congo—the Democratic Republic of the Congo and the Republic of Congo. Because these designations are both lengthy and eas-

ily confused, we abbreviate them in this text. The Democratic Republic of the Congo (formerly Zaire) carries the name of its capital in parentheses: Congo (Kinshasa). The same is true for the Republic of Congo: Congo (Brazzaville). Check the regional map (see Figure 7.1) to note the locations of these countries and capitals.

PHYSICAL PATTERNS

The African continent is big—the second largest after Asia and three times the size of the United States. But Africa's great size is not matched by its surface complexity. More than one-fourth of the continent is covered by the Sahara, many

Views of Sub-Saharan Africa

For a tour of these places, visit www.whfreeman.com/pulsipher

[Background image: Trent Schindler NASA/GSFC Code 610.3/587 Scientific Visualization Studio]

1. Global warming, food and water. Because global warming is likely to bring changes in rainfall patterns, many are looking to irrigation to provide more stability to food production systems. This is especially important in dry areas like the Sahel, shown here. See pages 251–252 for more. Lat 10°51'6"N, Lon 0°52'10"W. [Euan Denholm/IRIN (UN Office for the Coordination of Humanitarian Affairs); Louis Stippel/USAID]

5. Democracy is expanding across Africa, but not without complications. More than 1 million refugees were created by violence following the disputed 2006 elections in Congo (Kinshasa), shown below. See pages 274–275 for more. Lat 4°18'25"S, Lon 15°18'31"E. [Tiggy Ridley/IRIN (UN Office for the Coordination of Humanitarian Affairs)]

(1) Sahel

(2) Lagos

Kinshasa

2. Urbanization is occurring faster in Africa than anywhere else in the world. Most migrants to Africa's cities end up living in the slums that house 72 percent of the urban population. Lagos, Nigeria is shown here. See pages 261–262 for more. Lat 6°29'47"N, Lon 3°23'36"E. [Dulue Mbachu/IRIN (UN Office for the Coordination of Humanitarian Affairs); Sarah Simpson/ IRIN (UN Office for the Coordination of Humanitarian Affairs)]

3. Population and gender. Gender issues, such as female literacy, are increasingly seen as central to efforts to reduce Africa's population growth rate, the highest in the world. These women are waiting outside a health clinic in Goma, Congo (Kinshasa). See pages 259–260 for more. Lat 1°36'28"S, Lon 29°08'24"E. [Sylvia Spring/IRIN (UN Office for the Coordination of Humanitarian Affairs)]

3
Goma

4
Limpopo

4. Globalization has increased Africa's dependence on the export of low value raw materials, such as the copper nugget shown in the center inset. Relatively expensive manufactured goods, such as the TV equipment in the South African mall (photo on the right), must be imported. Meanwhile copper miners in South Africa's Limpopo province (photo on the left) work long hours for little pay. See pages 265–271 for more. Lat 24°0'S, Lon 31°08'E.
[Kim Wylie/USAID; Gary Parent/http://commons.wikimedia.org/wiki/Image:Cuivre_natif1_(USA).jpg; Henry Trotter/http://commons.wikimedia.org/wiki/Image:Canal-Walk-Food-Court.jpg]

FIGURE 7.4 Political map of sub-Saharan Africa.

thousands of square miles of what, to the unpracticed eye, appears to be a homogeneous desert landscape (see Chapter 6). Africa has no major mountain ranges, but it does have several high peaks, including Mount Kilimanjaro (19,324 feet [5890 meters]) and Mount Kenya (17,057 feet [5199 meters]). At their peaks, both have permanent snow and ice, though these features have shrunk dramatically, possibly due to global warming and associated local environmental changes.

Landforms

The surface of the continent of Africa can be envisioned as a raised platform, or *plateau,* bordered by fairly narrow and uniform coastal lowlands. The platform slopes downward to the north and has an upland region with several high peaks in the southeast and lower areas in northwest. These landforms have obstructed transport and hindered connections to the outside world as routes from the interior plateau to the coast must negotiate steep *escarpments* (long cliffs) around the rim of the continent. Africa's long, uniform coastlines also give it few natural harbors.

Geologists usually place Africa at the center of the ancient supercontinent Pangaea (see Figure 1.21 on page 25). As several landmasses broke off from Africa and moved away—North America to the northwest, South America to the west, and India to the northeast—Africa readjusted its position only slightly. Hence, it did not pile up long linear mountain ranges as the other continents did (see pages 240–241).

Today, Africa is breaking up along its eastern flank. There, the Arabian Plate has split away and drifted to the northeast, leaving the Red Sea, which separates Africa and Asia. Another split, known as the Great Rift Valley, extends south

from the Red Sea more than 2000 miles (3200 kilometers) (see Figure 7.1). In the future, Africa is expected to break apart along these rifts.

Climate

Most of sub-Saharan Africa has a tropical climate (Figure 7.5). Average temperatures generally stay above 64°F (18°C) year-round everywhere except at the more temperate southern tip of the continent and in cooler upland zones (hills and plateaus). Seasonal climates in Africa differ more by the amount of rainfall than by temperature.

Most rainfall comes to Africa by way of the **intertropical convergence zone** (ITCZ), a band of atmospheric currents that circle the globe roughly around the equator (see inset, Figure 7.5). At the ITCZ, warm winds converge from both the north and south and push against each other. This causes the air to rise, cool, and release moisture in the form of rain. The rainfall produced by the ITCZ is most abundant in Africa near the equator. Here, dense tropical rainforests flourish in places such as the Congo Basin.

The ITCZ shifts north and south seasonally, generally following the area of Earth's surface that has the highest average temperature at any given time. Thus, during the height of summer in the Southern Hemisphere in January, the ITCZ might bring rain far enough south to water the dry grasslands, or steppes, of Botswana. During the height of summer in the Northern Hemisphere in August, the ITCZ brings rain as far north as the southern fringes of the Sahara—an area called the **Sahel,** where steppe and savanna grasses grow. Poleward of both of these extremes, a belt of descending dry air blocks the effects of the ITCZ. As a result, deserts are found in Africa

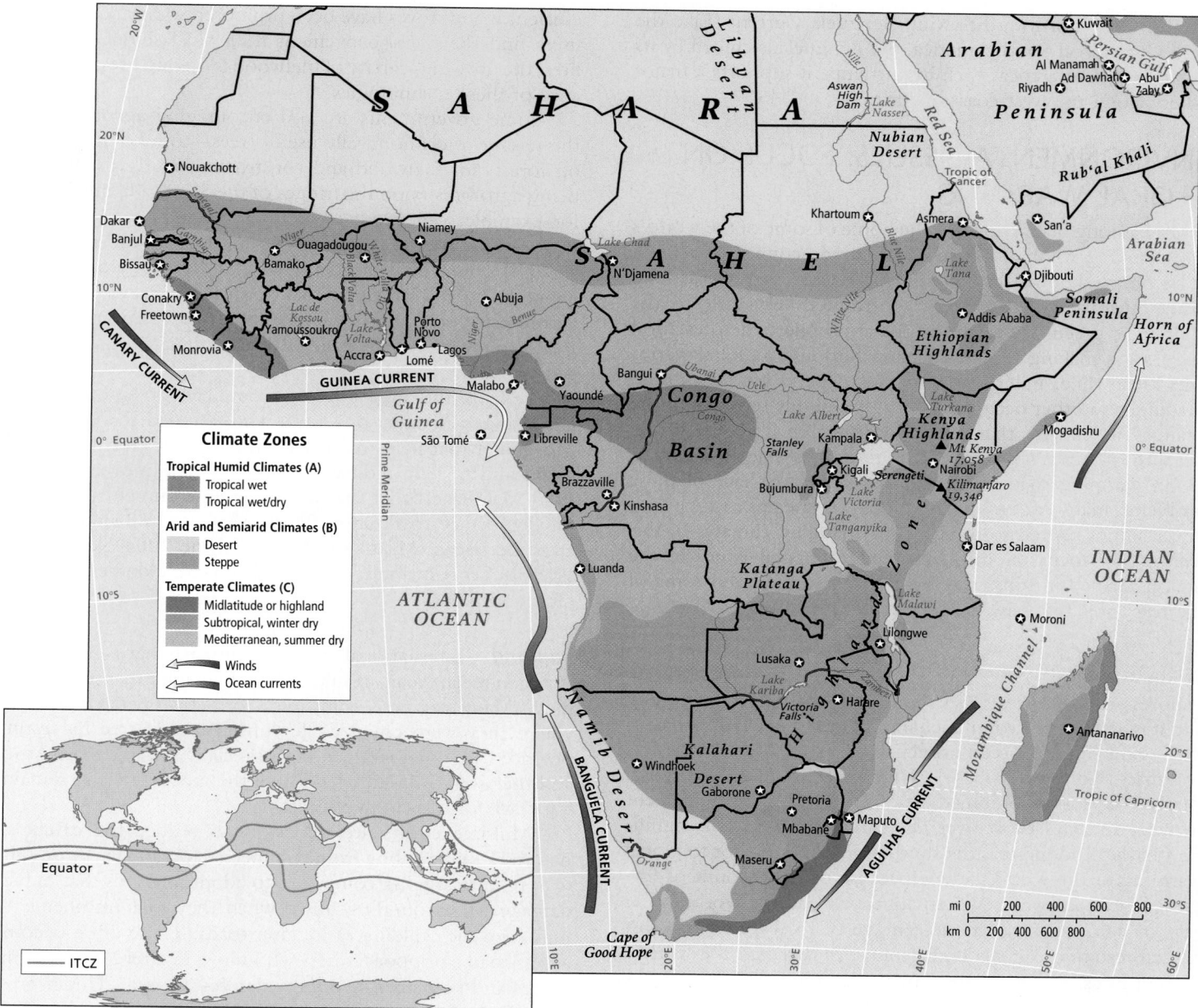

FIGURE 7.5 Climates of sub-Saharan Africa. The inset shows the position and range of the intertropical convergence zone (ITCZ). [Adapted from http://www.astrosurf.com/luxorion/meteo-tropicale.htm.]

(and on other continents) at roughly 30°N latitude (the Sahara) and 30°S latitude (the Namib and Kalahari deserts).

The tropical wet climates that support equatorial rain forests are bordered on the north, east, and south by seasonally wet/dry tropical woodlands. These give way to moist tropical savanna or steppe, where tall grasses and trees intermingle in a semiarid environment. These tropical wet and wet/dry climates have provided suitable land for agriculture for thousands of years. Farther to the north and south lie the true desert zones of the Sahara and the Namib and Kalahari. This

banded pattern of African ecosystems is modified in many areas by elevation and wind patterns.

Without mountain ranges to block them, wind patterns can have a strong effect on climate in Africa. Winds blowing north along the east coast keep ITCZ-related rainfall away from the **Horn of Africa,** the triangular peninsula that juts out from northeastern Africa below the Red Sea. As a consequence, the Horn of Africa is one of the driest parts of the continent. Along the coast of the Namib Desert, moist air from the Atlantic is blocked from moving over the desert by cold air above the northward-flowing Benguela Current. Like the Peru Current off South America, the Benguela is chilled by its passage past Antarctica. Rich in nutrients, it supports a major fishery along the west coast of Africa.

ENVIRONMENTAL ISSUES: FOCUS ON GLOBAL WARMING

Africa has long been known as the home of some of the world's most striking wildlife, but in the scientific community Africa is becoming known for its vulnerability to global warming. While Africans have generally contributed very little to the buildup of greenhouse gases in the atmosphere, deforestation in Africa is making significant contributions to global warming. Meanwhile, millions of Africans depend on agriculture, herding, and other occupations that are highly sensitive to the changes in climate that global warming is likely to bring. Because of Africa's poverty, it is much less able to adapt to global warming than other regions. Yet in the face of these problems, many Africans are developing strategies to cope with uncertainties related to the region's changing climate. In this section, we focus on the many environmental issues in the region, shaped in some way by global warming, and some of the responses that have been offered.

Deforestation and Global Warming

Deforestation is Africa's main contribution to global warming, and it is a major environmental impact of human activity across the region (Figure 7.6). Between 1990 and 2005, five African countries (Congo [Kinshasa]), Nigeria, Tanzania, Zambia, and Zimbabwe) lost a combined area of forest roughly equivalent to what Indonesia lost over the same period. This is significant in that Indonesia is the third-largest contributor to global warming in the world, largely because of the amount of forest it loses every year. African countries lead the world in the *rate* of deforestation, the percentage of total forest area lost to deforestation. Of the eight countries that had the world's highest rates of deforestation between in 1990 and 2005, six are African (Burundi, Togo, Nigeria, Benin, Uganda, and Ghana).

Most of Africa's deforestation is driven by the growing demand for farmland and fuelwood, though logging by international timber companies is also increasing. We look first at strategies for reducing fuelwood consumption and logging and then discuss agriculture.

Fuelwood. Africans use wood or charcoal to supply nearly all their domestic energy. Wood and charcoal remain the cheapest fuels available, in part because of African traditions that consider forests to be a free resource held in common. Even in Nigeria, which is a major oil producer, most people use fuelwood because they cannot afford petroleum products.

For a decade or more, many Africans have recognized the need to use fuelwood more sustainably. Solar ovens and fuel-efficient wood stoves have been promoted widely. While many users find them less convenient than old-fashioned cooking fires, the growing scarcity of fuelwood is driving greater acceptance of these technologies.

Some governments are also encouraging **agroforestry**—the raising of economically useful trees—to take the pressure off forests for fuelwood and construction materials. By practicing agroforestry on the fringes of the Sahel, a family in Mali, for example, can produce fuelwood, building, and fencing materials, medicinal products, and food all on the same piece of land. This would double what could be earned from single-crop agriculture or animal herding. However, critics of agroforestry note that sometimes the trees used are *invasive species* (see Chapter 11, page 413) that threaten many African ecosystems.

Perhaps the best known proponent of agroforestry is Dr. Wangari Maathai of Kenya, who founded the Green Belt Movement that helps rural women plant trees for use as fuelwood and to help reduce soil erosion. The movement grew from Maathai's belief that a healthy environment is essential for democracy to flourish. Thirty years and 30 million trees after she began, Maathai was awarded the 2004 Nobel Peace Prize for her contributions to sustainable development, democracy, and peace.

Logging. International timber companies have moved into Africa in recent years, resulting in a dramatic expansion of logging. Many of these companies formerly operated in Asia, where they were criticized for unsustainable tree harvesting practices. Logging is expanding especially fast in central Africa and the Congo Basin, which have the most extensive remaining tracts of African rain forest.

Multiple efforts are underway to reduce the extent of deforestation resulting from commercial logging in Africa. One is to require logging companies to adopt methods that reduce damage to surrounding forest when they log. Another is to minimize the building of logging roads. These often become access roads for poor farmers who move in after logging companies and permanently occupy deforested areas. Under pressure from consumers in the EU, many European logging companies are using these and other more sustainable practices. However, the Asian logging companies, which are expanding rapidly in Africa, are generally not adopting such practices.

Food Production and Water Resources

Africa's food production systems have both advantages and disadvantages in helping the region adapt to global warming. Scientists predict that global warming will cause growing seasons to become shorter in Africa because temperatures will be too high for optimal plant growth. Water will also become more scarce as higher temperatures increase evaporation rates and shift prevailing winds, leading to changes in rainfall patterns. Some widely used African agricultural techniques may

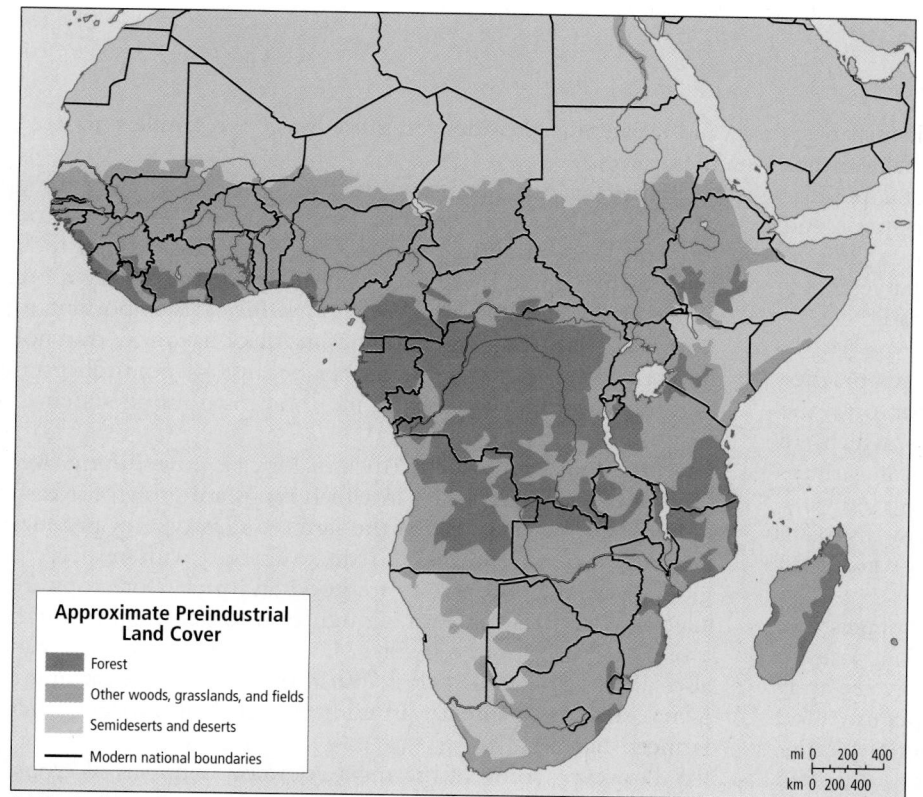

(a) Preindustrial

Approximate Preindustrial Land Cover

- Forest
- Other woods, grasslands, and fields
- Semideserts and deserts
- Modern national boundaries

mi 0 200 400
km 0 200 400

FIGURE 7.6 Land cover (preindustrial) and human impact (2002). Although human impact was significant in the preindustrial era (a), its effects are not visible at this scale. By 2002 (b), however, human impact was pervasive. Notice especially the impact on the forests of Congo (Kinshasa) and Cameroon. [Adapted from *United Nations Environment Programme,* 2002, 2003, 2004, 2005, 2006 (New York: United Nations Development Programme), at http://maps.grida.no/go/graphic/human_impact_year_1700_approximately and http://maps.grida.no/go/graphic/human_impact_year_2002.]

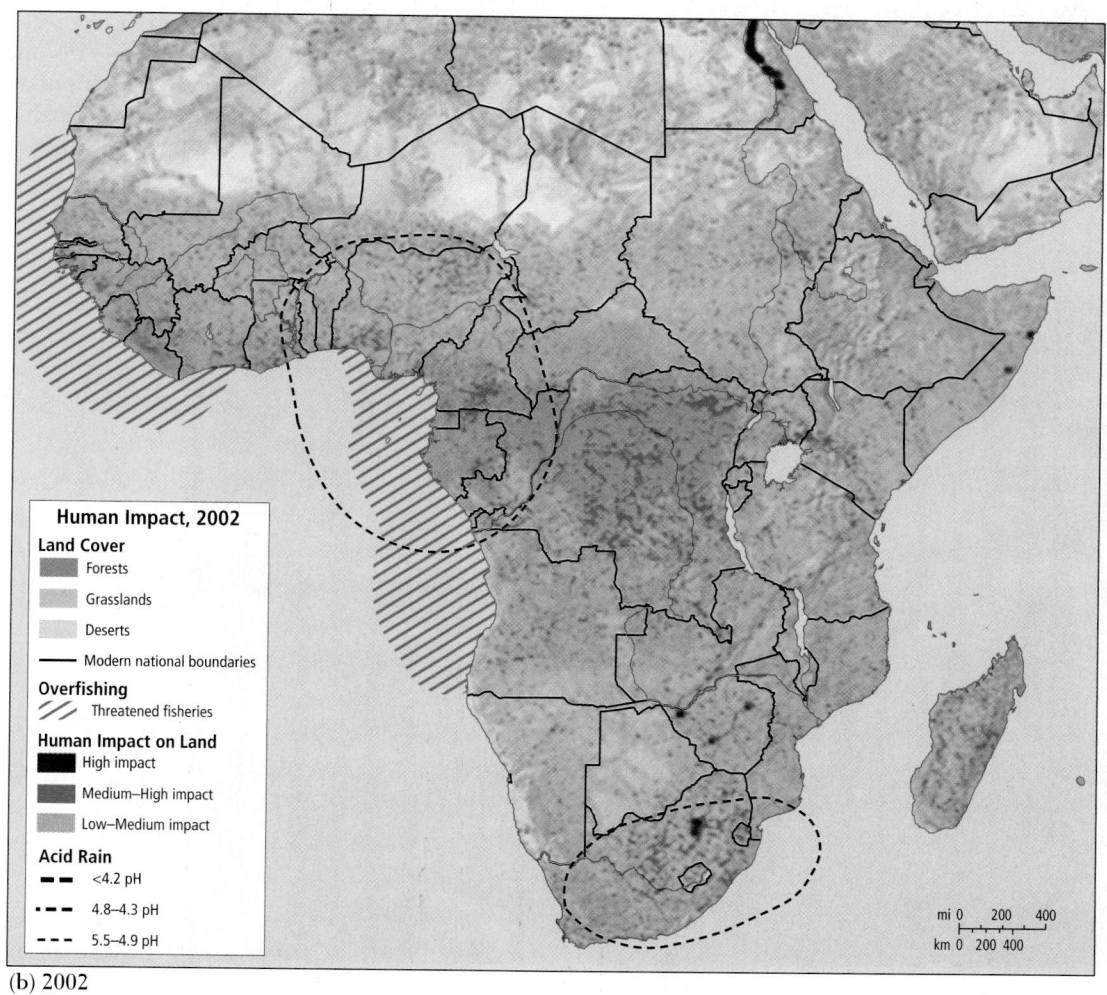

(b) 2002

Human Impact, 2002

Land Cover
- Forests
- Grasslands
- Deserts
- Modern national boundaries

Overfishing
- Threatened fisheries

Human Impact on Land
- High impact
- Medium–High impact
- Low–Medium impact

Acid Rain
- <4.2 pH
- 4.8–4.3 pH
- 5.5–4.9 pH

mi 0 200 400
km 0 200 400

Scientists predict that global warming will cause growing seasons to become shorter in Africa because temperatures will be too high for optimal plant growth.

be adaptable to these disturbances. Nevertheless, many rural Africans (66 percent of the total population) remain extremely poor and hence limited in their ability to cope with the reduced harvests that global warming may bring.

Subsistence and mixed agriculture. Most Africans practice **subsistence agriculture,** which provides food for only the farmer's family and is usually done on small farms, which are about 2 to 10 acres (1 to 4 hectares). Most African farmers also practice **mixed agriculture** (Figure 7.7), raising a diverse array of crops and a few animals as livestock. Many also fish, hunt, and gather some of their food from forest or grassland areas.

These widely practiced techniques have advantages and disadvantages in a world of global warming. Mixed agriculture combined with hunting and gathering provide a diverse array of strategies for coping with the changes in temperature and rainfall that global warming may bring. For example, traditional Nigerian farmers grow complex tropical gardens, often with 50 or more species of plants at one time. Some of the plants can handle drought, while others can withstand intense rain or heat. Hence the diversity of the gardens allows sustainable production of more than enough food for their families under a variety of climatic conditions. Hunting, fishing,

and gathering provide yet more ways for families to feed themselves.

However, the subsistence nature of most African farming can also leave families without much cash. If harvests are too low and hunting and gathering fails to provide enough food to feed the family, there won't be much money to buy food. While such situations can lead to *famine,* it is important to note that the most serious famines in Africa have occurred not because of low harvests but rather because of political instability that disrupts economies and food distribution systems.

Commercial agriculture. Much of Africa is now shifting over to commercial agriculture, in which crops are grown for cash rather than just as food for the farmer. This type of production also has advantages and disadvantages with respect to global warming. If harvests are good and prices for crops are adequate, farmers can earn enough cash to get them through a year or two of poor harvests. Having some cash income can also allow poor families to invest in other means of earning a living, such as opening a grinding mill to help process other farmers' harvests.

However, some of the most common commercial crops, such as peanuts, cacao beans (used for making chocolate), rice, or coffee, are often less well adapted to environments outside their native range. This makes them more likely to suffer reduced yields if temperatures increase or water becomes scarce. Moreover, to maximize profits, these crops are often

FIGURE 7.7 Small farm plots surrounding a Kikuyu village near Mount Kenya. The foothills around Mt. Kenya north of Nairobi, settled by the Kikuyu about 400 years ago, are today one of the most intensively farmed areas in the country. Most of the plots are small, but the farmers have moved beyond bare subsistence and are producing salable surpluses. [Russell Middleton/University of Wisconsin–Madison, African Studies Program.]

grown in large fields of only a single plant species. This can leave farmers with few options if the harvest fails.

The potential for commercial agriculture to help farmers adapt to the uncertainties of global warming is also limited by the instability of prices for commercial crops. This is because commercial agriculture is often part of a globalized production system, where prices for agricultural products are set by the global market. Prices can rise or fall dramatically from year to year because of overproduction or crop failures both in Africa and abroad. Hence, even in a year of ideal climatic conditions and large harvests, farmers might not earn significant income if global prices for their crops are too low.

Another downside of commercial production is that it is usually based on the permanent conversion of forests to fields. This can put a greater strain on soil and water resources, leaving agricultural systems more vulnerable to changes in the regional climate. Many African soils rapidly lose their fertility when cultivated. This is especially true in the warmer, wetter areas, where the organic matter in the soil decays rapidly. To maintain soil quality in such a climate, African subsistence farmers have long used *shifting cultivation,* in which patches of forest are cleared for a short time and then left to regrow (see Chapter 3, page 109). After a few decades, the soil is ready to cultivate again.

However, if fields are cleared permanently, as many commercial agriculture systems demand, soil fertility declines and crops become more likely to fail even under ideal climatic conditions. Chemical fertilizers can compensate for this loss, but these can be unaffordable for many farmers. Moreover, rains almost always wash much of the fertilizer into nearby waterways, thus polluting them. This may ultimately hurt farmers by reducing the quantity of available fish, which many rural people depend on as a source of protein.

These aspects of commercial agricultural systems make them poorly adapted even to ideal climatic conditions in Africa, let alone the hotter and more drought-prone climates that global warming may bring. Part of the problem lies in ignorance of local conditions and cultivation techniques on the part of foreign agricultural "experts" who first introduced many commercial systems. For example, in Nigeria, it is women who grow most of the food for family consumption and who are guardians of the knowledge that makes Nigeria's complex farming systems possible. However, women were rarely included or even consulted by the British experts who promoted commercial agricultural development projects. At best, women were employed as field laborers. In consequence, diverse subsistence agricultural systems based on numerous plant species were replaced by less stable commercial systems based on a single plant species.

Many agricultural scientists now recognize this loss and are trying to incorporate the knowledge of African farmers, male and female, into more diverse commercial agriculture systems. For example, scientists at Nigeria's International Institute for Tropical Agriculture are developing a variety of cultivation systems, using many different plants that help each other cope with varying climatic conditions. Most of these systems are designed for both subsistence and commercial agriculture, and hence can give families both a stable food supply and cash to help them ride out crop failures. The end result is reduced vulnerability to global warming for poor rural farmers.

Irrigation and water management: alternatives in the Niger River basin. Because global warming is likely to result in changing rainfall patterns, many are looking to irrigation to provide more stability to agricultural systems. However, the massive ecological changes created by large-scale irrigation projects have made some Africans more appreciative of irrigation on a smaller scale.

The Niger is one of Africa's most important rivers. It carries summer floodwaters northeast into the normally arid lowlands of Mali and Niger (Figure 7.8). There the waters spread out into lakes and streams that nourish wetlands. For a few months of the year (June through September), the wetlands ensure the livelihoods of millions of fishers, farmers, and pastoralists. These people share the territory in carefully synchronized patterns of land use that have survived for millennia. Wetlands along the Niger produce eight times more plant matter per acre than the average wheat field, provide seasonal pasture for millions of domesticated animals, and are an important habitat for wildlife.

The governments of Mali and Niger now want to dam the Niger River and channel its water into irrigated agriculture

FIGURE 7.8 Threatened seasonal wetlands. Every rainy season, the Bani River, a major tributary of the Niger, floods and turns the city of Djénné, Mali, into a series of islands. The annual flood revitalizes the surrounding soil and the river's marine life and permits the population to feed itself. A proposal to build a dam upstream threatens the city's self-sufficiency. [Sarah Leen/Matrix.]

FIGURE 7.9 Market gardens along the Sénégal River in Mali. The river is an important source of moisture in this Sahel region, providing water for nearly 10 million people. Women transport water in containers up the steps from the river to their gardens, where they water plants individually so as to conserve moisture. This high-altitude photo shows the tiny walled garden plots. [Yann Bertrand-Artaud/ALTITUDE.]

projects that will help to feed the more than 26 million people in the two countries. Rising food prices in global markets, mounting population pressure, and the threat of changing rainfall patterns due to global warming are driving the two governments to undertake the massive project. However, the dams will forever change the seasonal rise and fall of the river. The irrigation systems may also pose a threat to human health. Most such systems lose a great deal of water through leakage, and the standing pools of water they create often breed mosquitoes that spread tropical diseases such as malaria and schistosomiasis.

Many smaller-scale alternatives are available. In some parts of Mali, for example, farmers are using hand- or foot-powered pumps to bring water from rivers or ponds directly to the individual plants that need them. This is in some ways a more modern version of traditional African irrigation practices whereby water is delivered directly to the roots of their plants by human water brigades (Figure 7.9). Smaller-scale projects provide the same protection against drought that the larger systems offer, but are much cheaper and simpler to operate.

They also avoid the large-scale ecological disruption of larger projects, which will help reduce the overall ecological damage that may come with global warming.

Herding and desertification. Herding, or **pastoralism,** is practiced by millions of Africans, primarily in savannas, on desert margins, or in the mixture of grass and shrubs called open bush. Herders use the milk, meat, and hides of their animals, and they typically trade with settled farmers for grain, vegetables, and other necessities.

Many traditional herding areas in Africa are now experiencing desertification, the process by which arid conditions spread to areas that were previously moist. While herding may be partially to blame, some scientists think desertification may be caused in part from human-induced global warming (see Chapter 6, page 215).

The effects of desertification are most dramatic in the region called the Sahel (Arabic for "shore" of the desert). This band of arid grassland, 200 to 400 miles (320 to 640 kilometers) wide, runs east-west along the southern edge of the Sahara (see Figure 7.5 on page 247). Over the last century, desertification has shifted the Sahel to the south. For example, the *World Geographic Atlas* in 1953 showed Lake Chad situated in a forest well south of the southern edge of the Sahel. By 1998, the Sahara itself was encroaching on Lake Chad.

The Sahel and other dry ecosystems in Africa are fragile environments where water and soil resources are barely sufficient for the needs of native grasses. Rainfall is already low in the Sahel, at 10 to 20 inches (25 to 50 centimeters) per year, and there are only low levels of organic matter in the soil to provide nutrients. Consequently, any further stress, such as fire, plowing, or intensive grazing, may cause grasses to die off. Rain evaporates more quickly from the barren land, and soon the dry soil is blown away by the wind. The remaining sand piles up into dunes as the grassland becomes more like a desert. In the Ethiopian highlands, an area that is home to 85 percent of Ethiopia's population and 75 percent of its livestock, desertification has damaged 80 percent of the land. Some areas are so severely damaged that food can no longer be produced there.

Indigenous animal herders are often blamed for desertification, and their actions may speed up natural cycles or increase the area affected. Economic development efforts that fail to carefully consider environmental impacts are also to blame. For example, the World Bank once promoted cattle raising in the Sahel as a more profitable alternative to the traditional animals, such as camels and goats. But cattle worsened desertification by putting more stress on native grasslands than traditional herding animals do. Other projects aimed at expanding irrigated agriculture into dry grasslands have also worsened desertification. ▶

Scientists are now exploring the possibility that reducing the intensity of herding and agriculture in the Sahel may not only reduce desertification, but also help combat global warming. This would happen as a result of **carbon sequestration,** the removal and storage of carbon taken from the atmosphere. When left undisturbed, many native grasses and other plants absorb atmospheric carbon and store it in extensive root systems that reach deep underground. The vast extent of the Sahel and other dry grassland zones raises the possibility that these areas, if properly managed, could remove enough carbon from the atmosphere to at least partially counteract global warming.

Wildlife and Global Warming

Africa's world renowned wildlife faces multiple threats, from both humans and natural forces, all of which could become more severe due to global warming. Managers of African national parks have had some success in reducing these threats, but many challenges remain.

Wildlife management. Wildlife is especially vulnerable to global warming. Increases in average temperature or rainfall patterns can dramatically affect food availability, potentially requiring species to migrate to new areas to survive. Even if animals don't have to move to entirely new territory, changing environmental conditions can create new threats to their survival. The Intergovernmental Panel on Climate Change estimates that 25 to 40 percent of the species in Africa's national parks may become endangered as a result of global warming.

Wildlife managers will have to develop new management techniques to help animals survive. For example, the annual 1800-mile-long migration of more than a million wildebeest, zebras, and gazelles in Kenya's Masai Mara game reserve, one of the greatest wildlife spectacles on the planet, took a tragic turn in 2007. The migration requires animals to traverse the Mara River (Figure 7.10), a difficult feat that usually results in the drowning of a thousand or so animals. In the past, the park's managers have taken a hands-off approach to the migration, considering the losses natural. However, extremely heavy rains in the area in 2007, which some scientists think are related to global warming, swelled the Mara River to record levels. When the animals tried to cross, 15,000 drowned. Park managers are now considering taking a more active role in helping the migrating animals cope with unusual climatic conditions that may worsen with global warming. This may involve stopping animals from attempting a river crossing where many have already drowned and directing them to a safer crossing.

Bush meat. The dependence of farmers on hunting wild game (often called *bush meat*) for part of their food and income needs is already a major threat to wildlife in much of Africa. For example, endangered species such as gorillas and chimpanzees are being killed in record numbers in the Congo basin, often by farmers who need food or extra income. If crop harvests are diminished by global warming, many farmers will become even more dependent on bush meat. The threat to many wild populations has led to calls to expand protected areas for wildlife and establish new ones. However, Africa's existing parks are already struggling to deal with poaching (illegal hunting) within the parks by members of surrounding communities. Some park managers have reduced poaching by promoting alternative livelihoods in the communities where poachers come from.

Community-based ecotourism. With one-third of the world's preserved national park land, Africa has tremendous potential for ecotourism (see Chapter 3, pages 107–108). Some parks are now using money earned from ecotourism to sponsor economic development in nearby communities that once lived in part off poaching. For example, in 1985 Kasanka National Park in Zambia was plagued by poaching that threatened its wildlife populations. Park managers decided to redirect some of the funds generated by tourist visits toward economic development and education in surrounding communities. Locals were encouraged to sell products to tourists and were employed to build tourist lodges and wildlife-viewing infrastructure in the park. Funding was also provided to local schools and clinics,

FIGURE 7.10 Wildebeest crossing the Mara River. More than a million wildebeest, zebras, and gazelles migrate 1800 miles each year, primarily within the Masai Mara game reserve in Kenya and the Serengeti National Park in Tanzania. [Danita Delimont/Alamy]

and students were included in research projects within the park. Assistance was given to improve local farming techniques and to encourage alternative livelihoods, such as bee-keeping and agroforestry. Today, poaching in Kasanka is very low, its wildlife populations are booming, tourism is growing, and local communities have an ongoing stake in the park's success.

Ten Key Aspects of Sub-Saharan Africa

- Scientists predict that global warming will cause growing seasons to become shorter and rainfall less predictable throughout Africa.
- The main objectives of European colonial administrations in Africa were to extract as many raw materials as possible and to create markets in Africa for European manufactured goods.
- In fewer than 50 years, Africa's population has more than tripled, growing from around 200 million in 1960 to 750 million in 2007.
- The average annual growth rate of African cities is the highest in the world at more than 5 percent.
- Africa suffers more than any other world region from infectious diseases, with the world's worst epidemics of malaria, schistosomiasis, and HIV-AIDS.

- Africa's status as the poorest region of the world is related in part to its weak position in the global economy as an exporter of raw materials to wealthier regions.
- In recent years, Africa has become a new frontier for Asia's large and growing economies, especially those of China and India.
- Democratic elections have provided part of the solution to ending Africa's civil wars by offering former combatants the possibility of becoming non-violent elected leaders.
- Women across Africa are assuming positions of power in the wake of national crises that have undermined faith in traditional male leadership and highlighted the contributions of women.
- More than a thousand languages, falling into more than a hundred language groups, are spoken in Africa.

HUMAN PATTERNS OVER TIME

Africa's rich past has often been misunderstood and dismissed by people from outside the region. European slave traders and colonizers called Africa the "Dark Continent" and assumed it was a place where little of significance in human history had occurred. The substantial and elegantly planned cities of Benin in western Africa and Loango in the Congo Basin, which European explorers encountered in the 1500s, never became part of Europe's image of Africa. Even today, most people outside the continent are unaware of Africa's internal history or its contributions to world civilization.

The Peopling of Africa and Beyond

Africa is the original home of humans. It was in eastern Africa that the first human species evolved more than 2 million years ago, although they differed anatomically from humans today. These early, tool-making humans ventured as far as the Caspian Sea as early as 1.8 million years ago. Anatomically, modern humans (*Homo sapiens*) evolved separately in eastern Africa about 200,000 years ago, and by about 90,000 years ago, they had reached the eastern Mediterranean. Like earlier humans, modern humans radiated out of Africa, spreading across mainland and island Asia and eventually into Europe.

Early Agriculture, Industry, and Trade in Africa

In Africa, people began to cultivate plants as early as 7000 years ago in the Sahel and the highlands of present-day Sudan and Ethiopia. Agriculture was brought to equatorial Africa 2500 years ago and to southern Africa about 1500 years ago. Trade routes spanned the African continent, extending north to Egypt and Rome, and east to India and China. Gold, elephant tusks, and timber from tropical Africa were exchanged for a wide variety of goods.

Some of the first people in the world to learn how to smelt iron and make steel did so in northeastern Africa about 3400 years ago. By 700 C.E., when Europe was still recovering from the collapse of the Roman Empire, a remarkable civilization with advanced agriculture, iron production, and gold-mining technology had developed in the highlands of southeastern Africa in what is now Zimbabwe. Known now as Great Zimbabwe, this empire traded with merchants from Arabia, India, Southeast Asia, and China, exchanging the products of its mines and foundries for silk, fine porcelain, and exotic jewelry. The Great Zimbabwe empire collapsed around 1500 for reasons as yet little understood.

> The Great Zimbabwe empire traded with merchants from Arabia, India, Southeast Asia, and China, exchanging the products of its mines and foundries for silk, fine porcelain, and exotic jewelry.

Complex and varied social and economic systems existed in many parts of Africa well before the modern era. Several influential centers made up of dozens of linked communities developed in the forest and the savanna of the western Sahel. There powerful kingdoms and empires, such as that of Ghana (700–1000 C.E.), rose and fell. Some rulers converted to Islam, periodically sending large entourages on pilgrimages to Makkah (Mecca), where their opulence was a source of wonder.

Africans also traded slaves. Long-standing customs of enslaving people captured during war fueled this trade. The treatment of slaves within Africa was sometimes brutal and sometimes reasonably humane. Long before the beginning of Islam, a slave trade with Arab and Asian lands to the east developed (Figure 7.11). After the spread of Islam began around 700, close to 9 million African slaves were exported to parts of Southwest, South and, Southeast Asia. Protections afforded slaves in Africa were lacking when slaves were traded to non-Africans. For example, male slaves were often castrated before they were offered for sale to Muslim traders.

Europeans and the Slave Trade

The course of African history shifted dramatically in the mid-1400s, when Portuguese sailing ships began to appear off Africa's west coast. The names given to stretches of this coast by the Portuguese and other early European maritime powers reflected their interest in Africa's resources: the Gold Coast, the Ivory Coast, the Pepper Coast, and the Slave Coast.

By the 1530s, the Portuguese had organized a slave trade with the Americas. The trading of slaves by the Portuguese, and then by the British, Dutch, and French, was more widespread and brutal than any trade of African slaves that preceded it. African slaves became part of the elaborate production systems supplying the raw materials and money that fueled Europe's Industrial Revolution.

To acquire slaves, the Europeans established forts on Africa's west coast and paid nearby African kingdoms with weapons, trade goods, and money to make slave raids into the interior. Some slaves were taken from enemy kingdoms in battle. Many more were kidnapped from their homes and villages in the forests and savannas. Most slaves traded to Europeans were male because the raiding kingdoms preferred to keep captured women for their reproductive capacities. Between 1600 and 1865, about 12 million captives were packed aboard cramped and filthy ships and sent to the Americas. Up to one-quarter of them died at sea. Those who arrived in the Americas went primarily to plantations in South America and the Caribbean. About one-fourth were sent to the southeastern United States (see Figure 7.11).

The European slave trade severely drained the African interior of human resources and set in motion a host of damaging social responses within Africa that are not well understood even today. The trade enriched those African kingdoms that could successfully conquer their neighbors and force some if not all of them into slavery. It also made the slave-trading kingdoms dependent on European trade goods and technologies, especially guns and other European weaponry. The slave trade impoverished the many smaller, less powerful kingdoms and other communities that were repeatedly terrorized and robbed of men, women, and children.

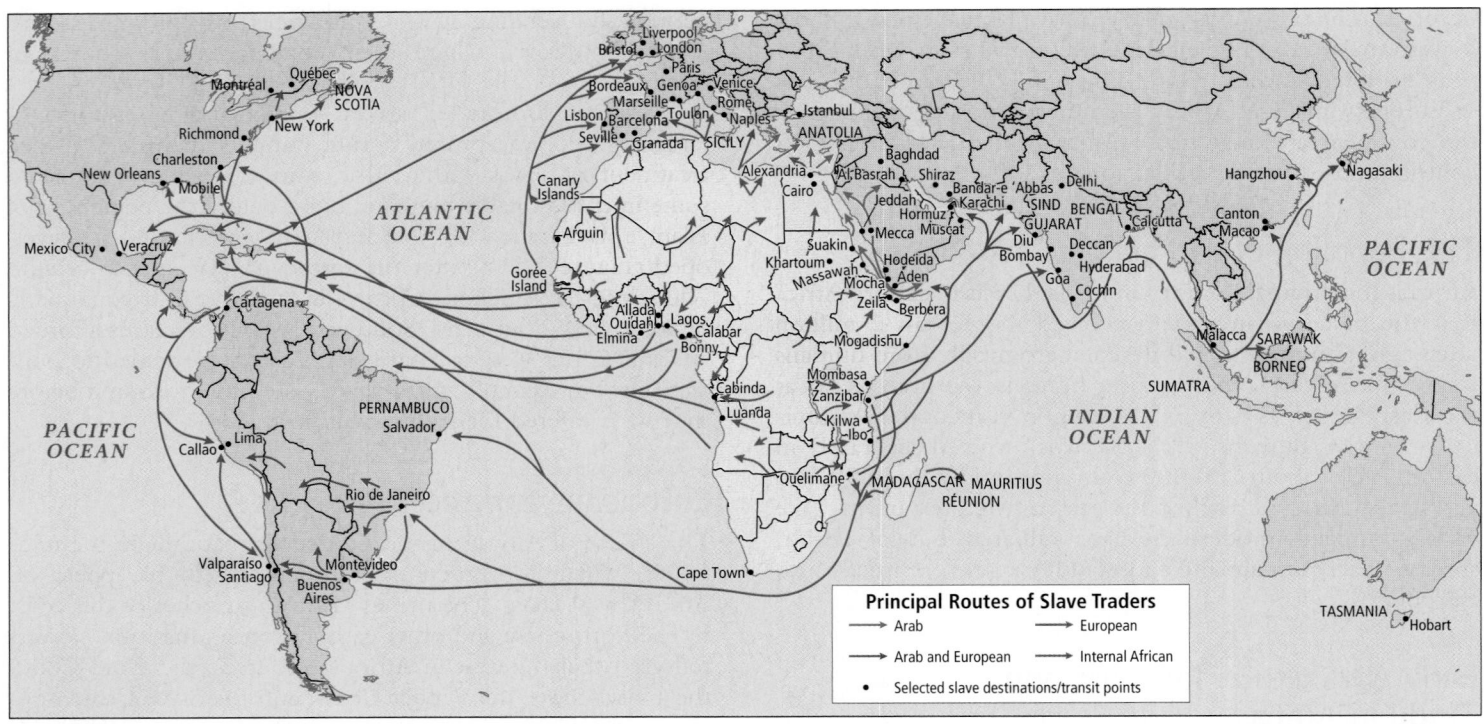

FIGURE 7.11 African slave trade. [Adapted from the work of Joseph E. Harris, in Monica Blackmun Visona et al., eds., *A History of Art in Africa* (New York: Harry N. Abrams, 2001), pp. 502–503.]

Slavery today. Slavery persists in modern Africa and is a growing problem that some argue dwarfs the trans-Atlantic slave trade of the past. Today, slavery is most common in the Sahel region, where several countries have made the practice officially illegal only in the past few years. People may become enslaved during war, are sold by their parents or relatives to pay off debts, or are forced into slavery when trying to migrate or find a job in a city. Slaves often work as domestic servants or as prostitutes, and they are increasingly used in commercial agriculture, mining, and war. It is hard to know exactly how many Africans are currently enslaved, but estimates range from several million to tens of millions.

The Scramble to Colonize Africa

The European slave trade wound down by about the mid-nineteenth century, as Europeans found it more profitable to use African labor within Africa to extract raw materials for Europe's growing industries. European interests extended from the coasts inland toward fertile agricultural zones, areas of mineral wealth, and places with large populations that could serve as sources of labor.

European colonial powers competed avidly for territory and resources, and by World War I, only two areas in Africa were still independent (Figure 7.12). Liberia, on the west coast, was populated by former slaves from the United States. Ethiopia (then called Abyssinia), in East Africa, managed to defeat early Italian attempts to colonize it. Otto von Bismarck, the German chancellor who convened the 1884 Berlin Con-ference at which the competing powers first partitioned Africa, declared: "My map of Africa lies in Europe." With some notable exceptions, the boundaries of most African countries today derive from the colonial boundaries set up between 1884 and 1916 by European treaties. These territorial divisions lie at the root of many of Africa's current problems.

The main objectives of European colonial administrations in Africa were to extract as many raw materials as possible and to create markets in Africa for European manufactured goods. The case of South Africa provides insights into European expropriation of African lands and the subjugation of African peoples. In this case, these aims led to the infamous system of racial segregation known as **apartheid.**

Case Study: The Colonization of South Africa

In the 1650s, the Dutch took possession of the Cape of Good Hope, which is located at the southern tip of Africa, from the Portuguese. Dutch farmers, called Boers, expanded into the interior, bringing with them herding and farming techniques that used large tracts of land and depended on the labor of enslaved Africans. The British were also interested in the wealth of South Africa, and in 1795 they gained control of areas around the Cape of Good Hope. Slavery was outlawed throughout the British Empire in 1834, prompting large numbers of slave-owning Boers to migrate to the northeast. There, in what became known as the Orange Free State and the Transvaal, the Boers often came into intense and violent conflict with African populations.

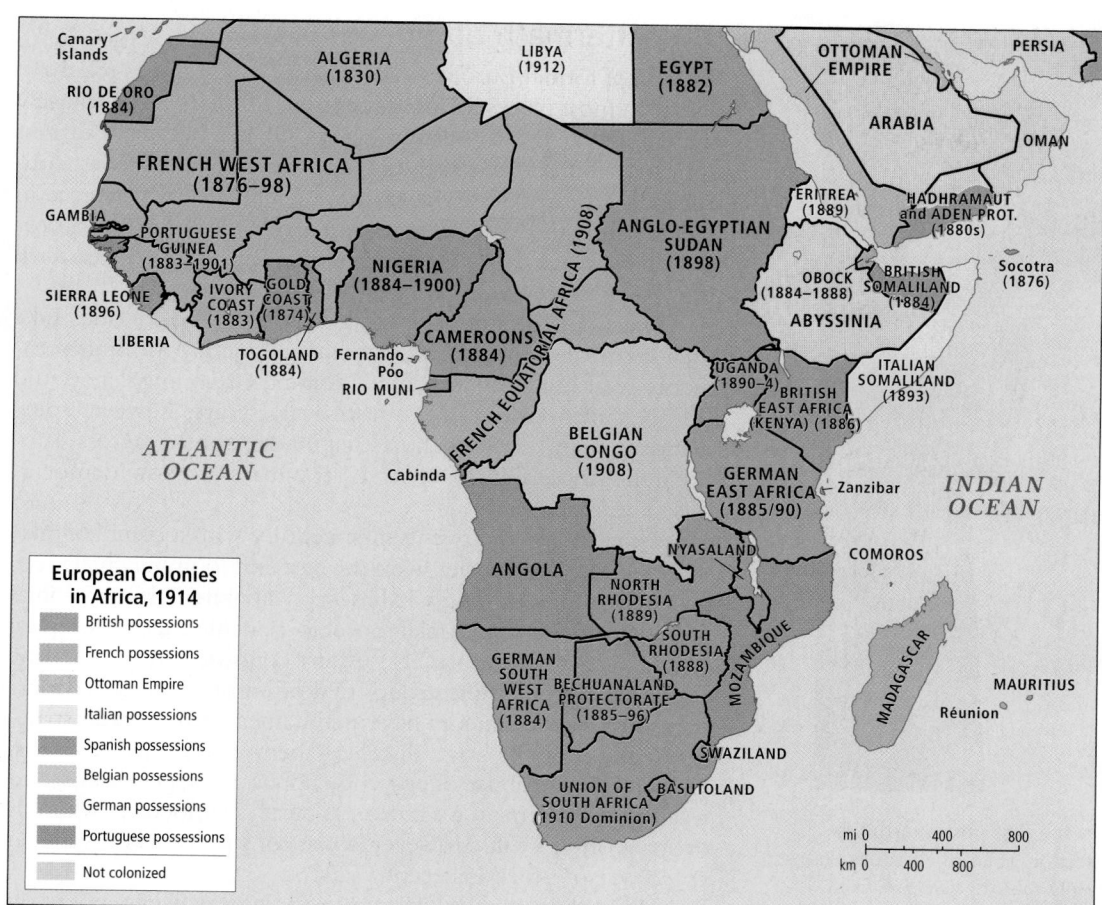

FIGURE 7.12 The European colonies in Africa in 1914. The dates on the map indicate the beginning of officially recognized control by the European colonizing powers. Countries without dates were informally occupied by colonial powers for a few centuries. [Adapted from Alan Thomas, *Third World Atlas* (Washington, D.C.: Taylor & Francis, 1994), p. 43.]

In the 1860s, extremely rich deposits of diamonds and gold were unearthed in these areas, securing the Boers' economic future. Africans were forced to work in the diamond and gold mines under extremely unsafe conditions and for minimal wages. They lived in unsanitary compounds that travelers of the time compared to large cages.

Britain, eager to claim the wealth of the mines, invaded the Orange Free State and the Transvaal in 1899, waging the bloody Boer War. The war gave them control of the mines briefly, until resistance by Boer nationalists forced the British to grant independence to South Africa in 1910. This independence, however, applied to only a small minority of whites: the Boers (who have since been known as Afrikaners) and some British who chose to remain. Black South Africans, more than 80 percent of the population, lacked legal political rights until 1994.

In 1948, the long-standing segregation of South African society was reinforced by apartheid, a system of laws that required everyone except whites to carry identification papers at all times and to live in racially segregated areas. Eighty percent of the land was reserved for the use of white South Africans, who at that time made up just 15 percent of the population. Blacks were assigned to ethnicity-based "homelands." The homelands were considered independent enclaves within the borders of, but not legally part of, South Africa. Never-

theless, the South African government exerted strong influence in them. Democracy theoretically existed throughout South Africa, but blacks were only allowed to vote in the homelands.

The African National Congress (ANC) was the first and most important organization fighting to end racial discrimination in South Africa. Formed in 1912 to work nonviolently for civil rights for all South Africans, the ANC grew into a movement with millions of supporters. Its members endured decades of brutal repression by the tiny minority of white South Africans.

Violence increased throughout South Africa until the late 1980s, when it threatened to engulf the country in civil war. The difficulties of maintaining order, combined with international pressure, forced the white-dominated South African government to initiate reforms that would end apartheid. A key reform was the dismantling of the homelands and a range of laws that dictated where people could live and vote. Finally, in 1994, the first national elections in which Black South Africans could participate took place. Nelson Mandela, an ANC leader who had been jailed by the South African government for 27 years, was elected the country's president (Figure 7.13). Today in South Africa, the long process of dismantling systems of racial discrimination still continues.

FIGURE 7.13 Nelson Mandela votes in the first post-apartheid election. On April 27, 1994, the first election open to all adults was held in South Africa. Nelson Mandela, only recently released from prison, was elected president. It was also the first time he was able to vote. [David Turnley/CORBIS.]

The Aftermath of Independence

The era of formal European colonialism in Africa was relatively short. In most places, it lasted for about 80 years, from roughly the 1880s to the 1960s. In 1957, Ghana became the first African colonial state to achieve its independence. The last sub-Saharan African country to gain independence was Eritrea in 1993, although Eritrea won its independence not from a European power but from its neighbor Ethiopia after a 3-year civil war.

The road to nation building has been a rocky one. Like their colonial predecessors, most independent African governments were undemocratic and dominated by privileged and Europeanized elites. In the last several years, however, pro-democracy movements have begun to spread across sub-Saharan Africa, and 23 of its 47 countries are now democracies (Figure 7.14).

Africa enters the twenty-first century with a complex mixture of enduring legacies from the past and looming challenges for the future. Although it has been liberated from colonial domination, most colonial borders remain intact (compare Figure 7.12 with Figure 7.4). Many countries are plagued by conflict between ethnic groups that often turns violent. Moreover, most governments have maintained bureaucratic structures and policies that distance them from their citizens. Corruption and abuse of power by bureaucrats, politicians, and wealthy elites remain a serious problem. Democracy, where it exists, is often weak. However, some corrupt leaders have been voted out of office in recent years.

Meanwhile, many economies remain heavily dependent on their agricultural and mineral exports, the prices of which are

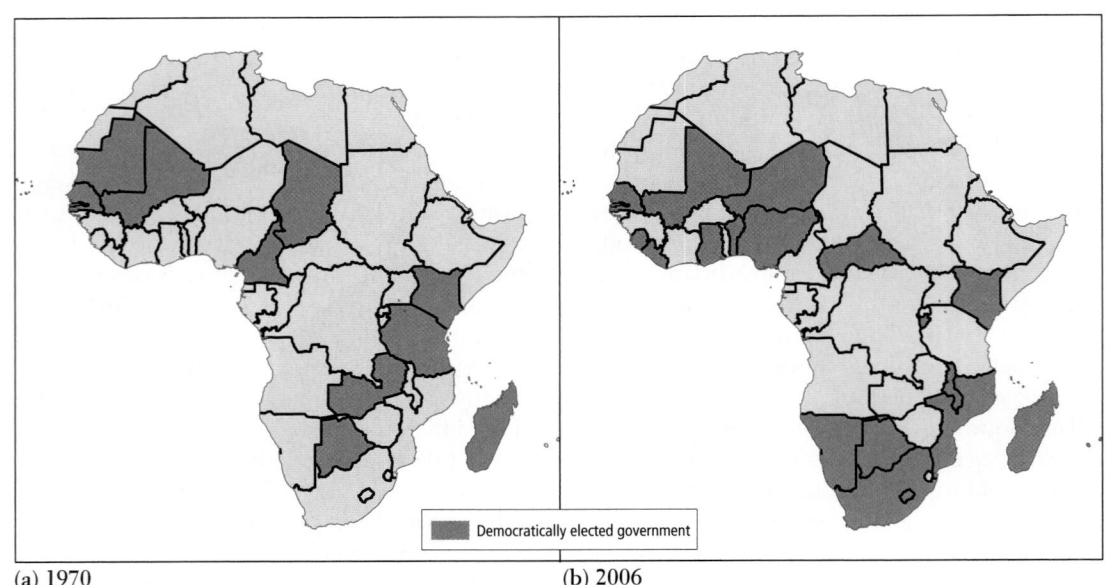

(a) 1970   (b) 2006

Democratically elected government

FIGURE 7.14 Democratically elected governments in sub-Saharan Africa. Between 1970 and 2006, the number of democratically elected governments in sub-Saharan Africa rose and fell, but by 2006, there were 23. It is important to note, however, that the criteria for democratic elections have been strengthened significantly since 1970. [Adapted from Barbara McDade, "Freedom in Africa today," in *Freedom in the World 2006*, at http://www.freedomhouse.org/uploads/special_report/36.pdf; and from http://en.wikipedia.org/wiki/Image: Freedom_House_electoral_democracies_2006.png.]

highly unstable on global markets. Most economies are also dependent on imports of relatively expensive manufactured goods. Hence, Africa finds itself at a disadvantage in a world of increasing globalization. Many countries remain economically dependent on their former European colonizers.

Solutions to Africa's economic problems have been slow in coming, and poverty has increased rapidly. In the recent past, Africa has faced declining economic productivity and rising debt. The region also faces the challenge of having a very young and fast-growing population. Perhaps the brightest spot in Africa's future is the willingness to consider innovative alternatives to conventional solutions.

POPULATION PATTERNS

While overall population densities in Sub-Saharan Africa are relatively low (32 people per square kilometer as opposed to the global average of 49 people per square kilometer), this can be misleading. First, populations are distributed very unevenly across the region (Figure 7.15). While some of the drier rural

areas are almost devoid of humans, other areas, especially Africa's growing cities, are some of the most crowded places on earth. Second, African populations are growing faster than any others on earth. In fewer than 50 years, Africa's population has more than tripled, growing from around 200 million in 1960 to 750 million in 2007. By 2050, Africa's population will be 1.7 billion. ■► So many places that are relatively uncrowded now may change dramatically over the next few decades. Nevertheless, population growth is slowing throughout Africa due to a number of factors.

Population Growth and the Demographic Transition

The geographer Ezekiel Kalipeni has found that many Africans are not yet choosing to have smaller families because they view children as both an economic advantage and a spiritual link between the past and the future. Childlessness is considered a tragedy. Not only do children ensure a family's genetic and spiritual survival, they still do much of the work on family

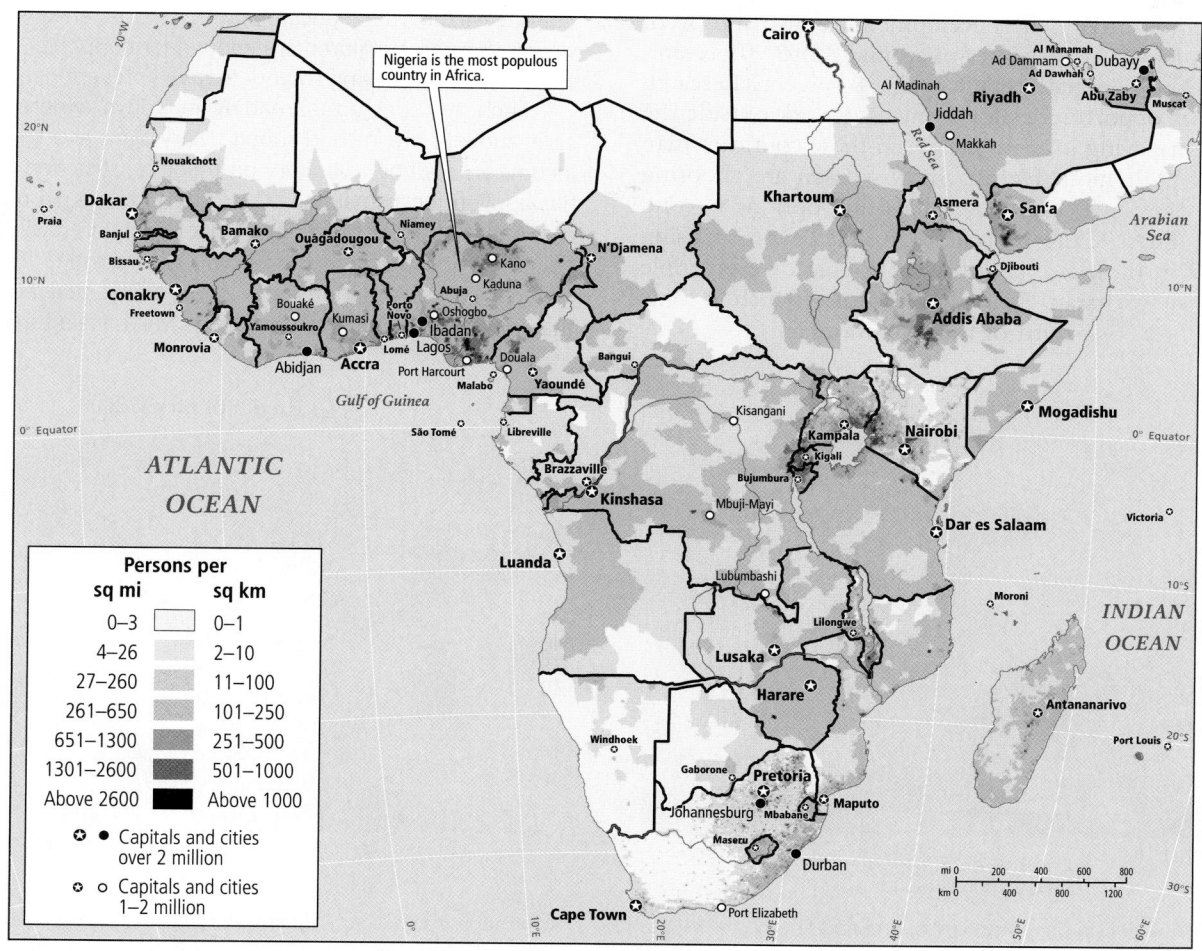

FIGURE 7.15 Population density in sub-Saharan Africa. [Data courtesy of Deborah Balk, Gregory Yetman, et al., Center for International Earth Science Information Network, Columbia University, at http://www.ciesin.columbia.edu.]

farms. Moreover, in this region of high infant mortality, parents have many children in the hope of raising a few to maturity. In all but a few countries, the *demographic transition*—the sharp decline in births and deaths that usually accompanies economic development (see Figure 1.37 on page 45)—has barely begun.

Five countries further along in the demographic transition.

Despite the overall high growth rate for Africa, fertility rates have declined significantly in a handful of countries that are further along in the demographic transition. All have seen higher levels of economic development, better health care systems, and more educational and work opportunities, especially for women. In South Africa, Botswana, Seychelles, Réunion, and Mauritius (the last three being tiny island countries off Africa's east coast), circumstances have changed sufficiently to make smaller families desirable.

In all five countries, per capita incomes are five to ten times the sub-Saharan African average of U.S.$2000. Advances in health care have cut the infant mortality rate to about half the regional average of 92 infant deaths per 1000 live births. Hence, parents can have only a few children and expect most to live to adulthood. Women's development has also improved, as reflected in female literacy rates of around 80 to 90 percent, compared to the regional average of 54.4 percent. Research also shows that opportunities for women to work outside the home at decent-paying jobs are greater in these countries than in the rest of the region. Hence, many women are choosing to use contraception because they have life options beyond motherhood. Indeed, the percent of married women using contraception in these five countries is double or even triple the rate for sub-Saharan Africa as a whole, which is only 22 percent (about one-third of the world average).

Contrasts in population growth.

The population pyramids in Figure 7.16 demonstrate the contrast between countries experiencing rapid growth (such as Nigeria, Africa's most populous country) and countries just entering the demographic transition, such as South Africa. South Africa's pyramid has contracted at the bottom because its birth rate has dropped from 35 per 1000 to 23 per 1000 since 1990. This decrease may be the consequence of economic and educational improvements and social changes that have come about since the end of apartheid in the early 1990s. But it may also be a consequence of the spread of HIV among young adults and the babies they bear. HIV-AIDS is now a major cause of the slowing of population growth rates in Africa. In any case, contraception is increasingly being used throughout Africa as more opportunities open for women.

Vignette Mary, a Kenyan farmer, has just had her third son. The father of Mary's children works in a distant city and visits the family only several weeks per year. He supports himself with his earnings and buys occasional nonessentials for the family. Mary owns only one cow and a small piece of land that can't be further divided, so all she can provide for her children is an education. Mary says that she can afford to educate only three children. She tells an interviewer that three children are enough for happiness, and so today, at age 29, she is having surgery that will prevent conception.

Such attitudes are spreading in Kenya, where food, health care, and jobs are in short supply. Mary plans to augment her farm income by starting a sanitary pit toilet construction business. She has applied for a small loan (U.S.$150) for this purpose. The success of her business could mean that her children will become well educated and that she herself will gain prestige. Studies of the effects of these small business loans, or *microcredit*, have shown that women who receive them tend to increase their use of birth

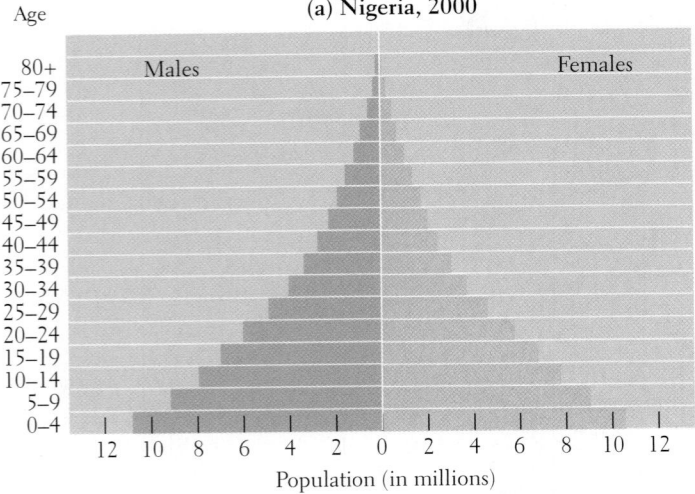

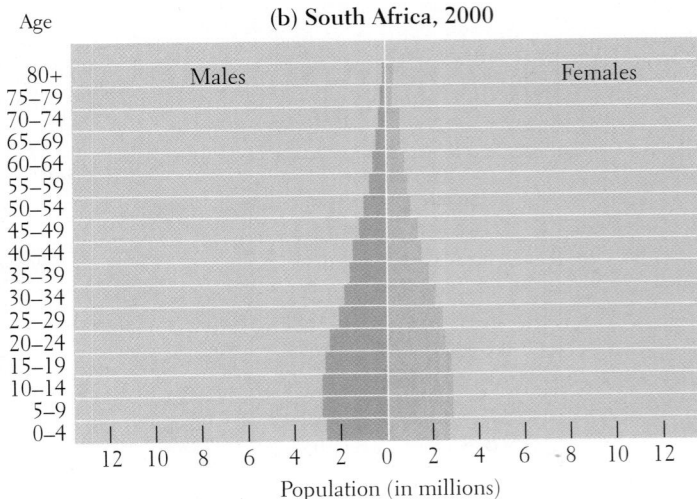

FIGURE 7.16 Population pyramids for Nigeria and South Africa. Nigeria, with a total population of 144 million, has a population growth rate of 2.8 percent. South Africa, with a population of 48 million, has a growth rate of 0.9 percent. [Adapted from

"Population pyramids for Nigeria" and "Population pyramids for South Africa" (Washington, D.C.: U.S. Census Bureau, International Data Base), at http://www.census.gov/ipc/www/idb/pyramids.html.]

control markedly. In addition, when women like Mary accomplish their goals, they become role models for other women, who then limit their families so that they can also become self-sufficient owners of small businesses.

Sources: Jeffrey Goldberg, New York Times Magazine *(March 2, 1997): 39; World Resources, 1996–1997 (New York: Oxford University Press, 1996), p. 5. Information on the role of microcredit in fertility patterns is from Fiona Steele, Sajeda Amin, and Ruchira T. Naved, "The impact of an integrated micro-credit program on women's empowerment and fertility behavior in rural Bangladesh," Policy Research Division Working Paper no. 115 (New York: Population Council), 1998.* ■

Urbanization and Population Growth

Africa is experiencing a massive shift in population from rural to urban areas, with major implications for population growth. On average, rural African women give birth to about 6.6 children, while urban African women give birth to 4.7. The decline in fertility rates has much to do with the demographic transition. All the factors that influence the demographic transition (increased economic development, better health care, and more educational opportunities) are strengthened by urbanization. However, even in urban Africa, the demographic transition has only just begun, as Africa's urban fertility rate of 4.7 children per woman is still almost double the world average (for rural and urban areas) of 2.7.

Part of the reason that urban fertility rates remain high is that Africa's cities have been unable to deliver as much improvement in living standards, especially access to clean water and sanitation, as cities in other regions. With poverty and disease still widespread in Africa's cities, urban families, much like rural families, have more children to ensure that some will survive into adulthood.

Exploding urban growth. The average annual growth rate of African cities is the highest in the world at more than 5 percent. In the 1960s, only 15 percent of Africans lived in cities; now about 34 percent do (Figure 7.17). By 2030, Africa's urban population will have doubled to about 530 million

> The average annual growth rate of African cities is the highest in the world at more than 5 percent.

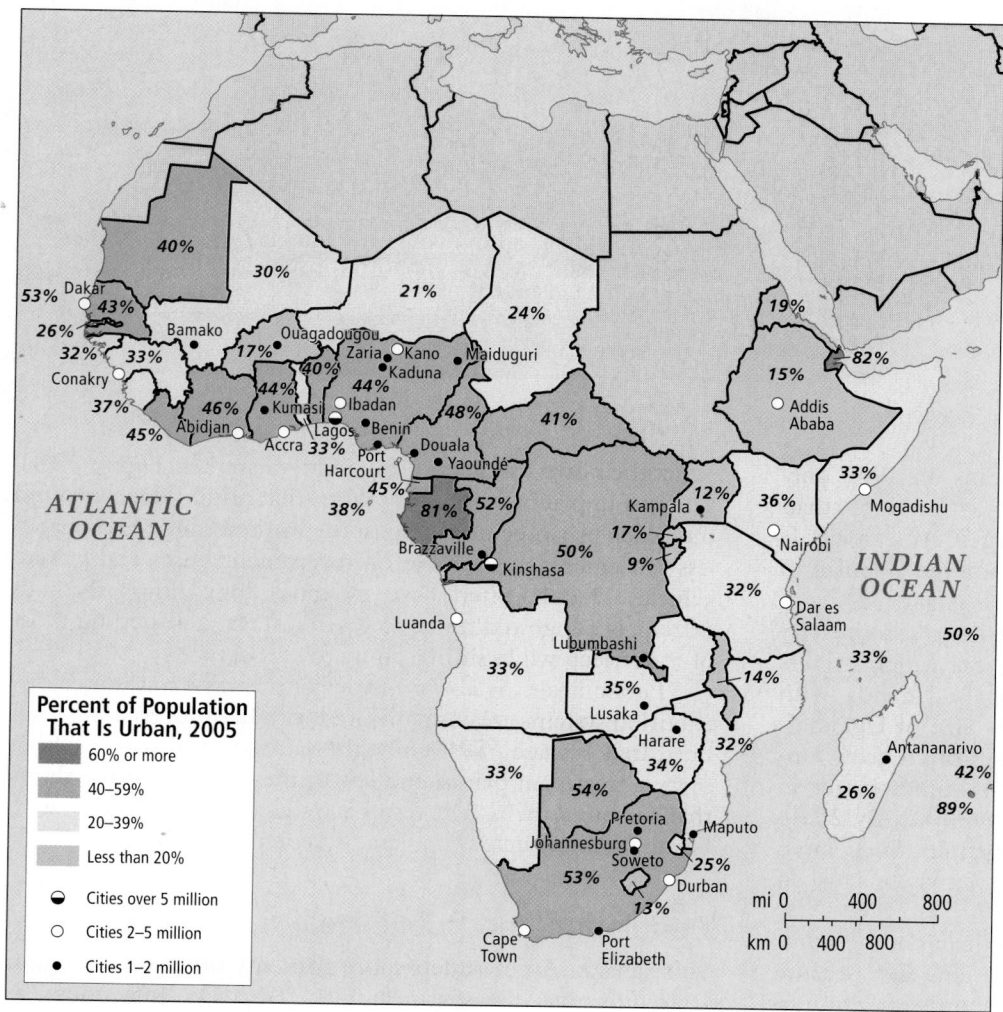

FIGURE 7.17 Percentage of the population living in cities in sub-Saharan Africa, 2005. Slightly more than 36 percent of the region's population is urban, but there is great variation from place to place. Thirty-six sub-Saharan African cities now have populations greater than 1 million, and seven of them have more than 3 million people. [Data from *2007 World Population Data Sheet* (Washington, D.C.: Population Reference Bureau); and World Gazetteer, at http://www.worldgazetteer.com/wg.php?x= &men=gcis&lng=en&dat=32&srt=npan &col=ahq&geo=-1.]

Percent of Population That Is Urban, 2005
- 60% or more
- 40–59%
- 20–39%
- Less than 20%

◐ Cities over 5 million
○ Cities 2–5 million
● Cities 1–2 million

FIGURE 7.18 Urban Africa. African urban landscapes vary radically, ranging from oppressive shantytowns to modern mid-rise office buildings to single-family homes and apartment towers to elegant villas set in landscaped grounds. Shown here is Lagos, Nigeria. [L. Gilbert/ CORBIS SYGMA.]

and will account for about half of all Africans. In 1960 only one African city—Johannesburg, South Africa—had more than one million people; today, 40 do. The largest African city is Lagos, Nigeria, where various estimates put the population between 8 and 11 million.

Most African countries have one very large *primate city* (see Chapter 3, page 113), usually the capital, which attracts virtually all migration. For example, Kampala, Uganda, with 1.4 million people, is almost ten times the size of Uganda's next largest city, Gulu. These primate cities often grow rapidly. For instance, Lagos, Nigeria, was 48 times larger in 2000 (11 million) than it was in 1950 (230,000). By 2020, Lagos will have more than 20 million people. Such rapid and concentrated urbanization usually overwhelms urban infrastructure.

Most migrants to Africa's cities end up living in the slums that house 72 percent of the urban population. ◻◼ Because governments and private investors have paid little attention to the need for affordable housing, most migrants have to con-

struct their own dwellings using found materials (Figure 7.18). The vast unplanned one-story slums that result often surround the older urban centers. Transport in these huge and shapeless settlements is a jumble of government buses and private vehicles. People often have to travel long hours through extremely congested traffic to reach distant jobs, getting most of their sleep while sitting on a crowded bus.

Public health is also a major concern as many water distribution systems are contaminated with harmful bacteria from untreated sewage. ◻◼ Only the largest African cities have sewage treatment plants, and few of these extend to the slums that surround them. The result is frequent outbreaks of waterborne diseases such as cholera, dysentery, and typhoid.

Population and Public Health

Sub-Saharan Africa suffers more than any other world region from infectious diseases, including HIV-AIDS. Infectious diseases are by far the largest killers in Africa, responsible for

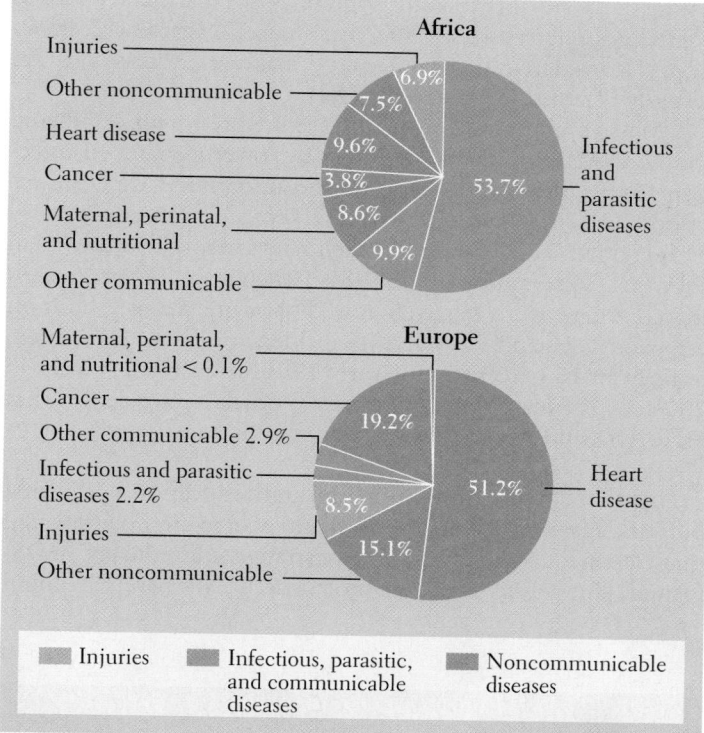

Leading Causes of Death, 2002

Africa

Injuries — 6.9%
Other noncommunicable — 7.5%
Heart disease — 9.6%
Cancer — 3.8%
Maternal, perinatal, and nutritional — 8.6%
Other communicable — 9.9%
Infectious and parasitic diseases — 53.7%

Europe

Maternal, perinatal, and nutritional < 0.1%
Cancer — 19.2%
Other communicable 2.9%
Infectious and parasitic diseases 2.2%
Injuries — 8.5%
Other noncommunicable — 15.1%
Heart disease — 51.2%

Injuries | Infectious, parasitic, and communicable diseases | Noncommunicable diseases

FIGURE 7.19 Leading causes of death in Africa and Europe, 2002. [Adapted from *Global Burden of Disease Estimates, 2002* (Geneva: World Health Organization).]

about 50 percent of all deaths (Figure 7.19). Some are linked to particular ecological zones. For example, people living between the 15th parallels north and south of the equator are most likely to be exposed to sleeping sickness (trypanosomiasis), which is spread among people and cattle by the bites of tsetse flies. The disease attacks the central nervous system and, if untreated, results in death. Several hundred thousand Africans suffer from sleeping sickness, and most of them are not treated because they cannot afford the expensive drug therapy.

Africa's most common chronic tropical diseases, *schistosomiasis* and *malaria*, are linked to standing fresh water. Thus, their incidence has increased with the construction of dams and irrigation schemes. Schistosomiasis is a debilitating, though rarely fatal, disease affecting about 170 million sub-Saharan Africans. It develops when a parasite carried by a particular freshwater snail enters the skin of a person standing in water. Malaria, which is spread by the anopheles mosquito (which lays its eggs in standing water) is even more costly. The disease kills at least one million sub-Saharan Africans annually, most of them children under age five. Malaria also infects millions of adult Africans who are left feverish, lethargic, and unable to work efficiently because of the disease.

Until recently, relatively little funding was devoted to controlling the most common chronic tropical diseases. That is now changing, with major international donors funding research in Africa and elsewhere. More than 60 research groups in Africa are working on a vaccine that will prevent malaria in most people.

HIV-AIDS in Africa

The leading cause of death in Africa is acquired immunodeficiency syndrome (AIDS), caused by the human immunodeficiency virus (HIV; Figure 7.20), As of 2008, more than 15 million sub-Saharan Africans had already died of the disease, and as many as 80 million more AIDS-related deaths are expected by 2025. In 2008, sub-Saharan Africa had 66 percent of the estimated worldwide total of 33 million deaths. While the epidemic appears to have stabilized, in that the number of new infections each year is generally not increasing, levels of infection are still very high. Indeed, infection rates for adults in Africa are the world's highest, at 6.1 percent. Rates are highest in southern Africa, which accounted for 38 percent of global HIV-AIDS deaths in 2007. The epidemic has

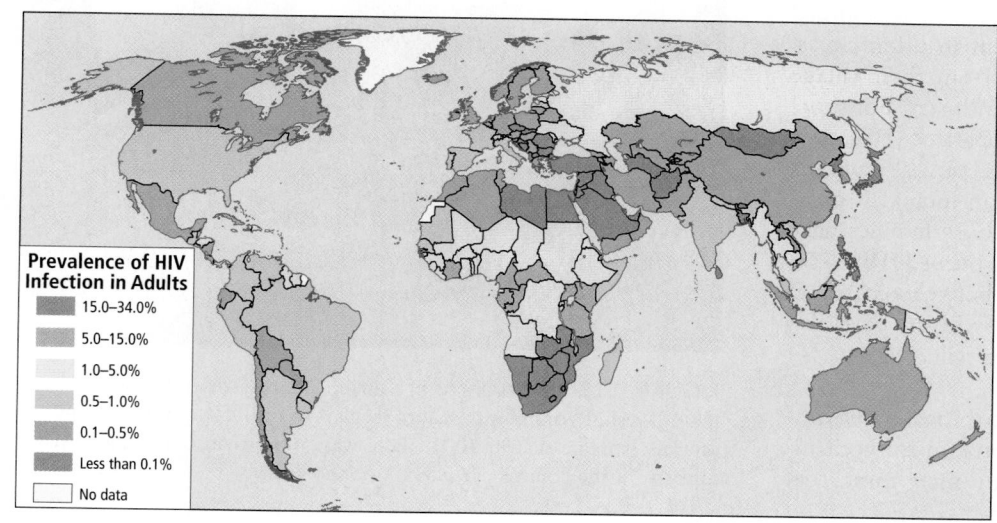

Prevalence of HIV Infection in Adults
15.0–34.0%
5.0–15.0%
1.0–5.0%
0.5–1.0%
0.1–0.5%
Less than 0.1%
No data

FIGURE 7.20 Global prevalence of HIV-AIDS. [Adapted from *UNAIDS 2006 Report,* Chapter 2, at http://www.unaids.org/en/KnowledgeCentre/HIVData/GlobalReport/2008/]

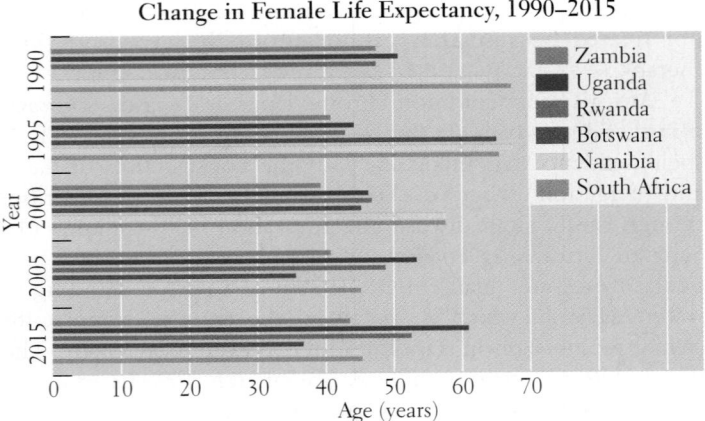

Change in Female Life Expectancy, 1990–2015

FIGURE 7.21 Current and projected effects of HIV-AIDS on female life expectancy in selected countries. As the effects of HIV-AIDS began to be felt in sub-Saharan Africa, life expectancy for women dropped dramatically in several countries (after 1990 in Zambia, Uganda, and Rwanda; and by 2000 in Botswana, Namibia, and South Africa; there are no 1990 data for Botswana). Expanded sex education programs, medical research, and more affordable drugs may increase female life expectancy again. Recovery is likely to be slow, however, with Uganda and Rwanda recovering the fastest (notice the lengths of these bars in 2005 and 2015). [Adapted from U.S. Census Bureau, International Programs Center, unpublished tables, Slide 23, at http://www.census.gov/ipc/www/slideshows/hiv-aids/TextOnly/Slide23.html.]

significantly constrained economic development in southern Africa.

In Africa, HIV-AIDS affects women more than men: women make up 59 percent of the region's HIV-infected adults. The reasons for this pattern are related to the social status of women. Women are often infected by their husbands, who may visit sex workers when they travel for work or business. Wives have little power to insist that their husbands use condoms. Figure 7.21 shows the dramatic effect of HIV-AIDS on female life expectancy, especially in Botswana and South Africa.

The rapid urbanization of Africa has speeded the spread of HIV-AIDS, which is much more prevalent in urban areas. Many poor new urban migrants, removed from their village support systems, become involved in the sex industry. For some poor urban women, occasional sex work is part of what they do to survive economically. In some cities, virtually all sex workers are infected. Meanwhile, many urban men visit prostitutes on a regular basis, especially those whose families have remained in rural villages. These men often bring HIV-AIDS back to their rural homes. As transportation between cities and the countryside has improved, bus and truck drivers have also become major carriers of the disease to rural villages.

Myths, taboos, and education. Education is the key to combating the myths and social taboos that have made controlling HIV-AIDS so difficult in Africa. Many men think that

only sex with a mature woman can cause the disease, so very young girls are increasingly sought as sex partners. (This is sometimes referred to as "the virgin cure.") Elsewhere, infection is considered such a disgrace that even those who are severely ill refuse to get tested. ◾

Massive education programs are credited with stabilizing the HIV-AIDS epidemic in Africa by lowering rates of infection among those who can read and understand explanations of how HIV is spread. For example, levels of infection have so far remained low in Senegal, which made major investments in HIV-AIDS prevention in the 1980s, when infection rates across the continent were relatively low. Following major education campaigns, Uganda lowered its incidence of new HIV infections from 15 percent in the early 1990s to just 4.1 percent in 2004. By contrast, in areas where top officials have denied that HIV-AIDS was a problem, infection rates have soared.

Treatment. While prevention through education is the first bulwark against HIV-AIDS, so many people are now infected that treatment is the most effective strategy for reducing AIDS-related illness and death. Unfortunately, the overwhelming

FIGURE 7.22 AIDS orphans in Kenya. The majority of these children in a makeshift community school in Nairobi are HIV-positive. Kenyan officials estimate that in 2005, there were more than one million AIDS orphans in the country. [Reuters/Corinne Dufka.]

majority of Africans cannot afford the costly combination of drugs that keeps victims alive in such places as North America, Europe, and Japan. The antiretroviral "drug cocktail" produced by American and European drug companies costs about U.S.$10,000 per year per patient. In 2000, in an attempt to lower this cost, a global movement began to challenge the patents of the drug companies. By the end of that year, drug companies in Cuba and India began to offer a regimen that cost just $1.00 a day—but even that price is too high for many Africans. As of 2008, barely 10 percent of African AIDS patients had access to antiretroviral drugs, though some governments have plans to subsidize treatment. ▱

Consequences. Across the continent, the consequences of the HIV-AIDS epidemic are enormous. The disease has severely strained the health care systems of most countries.

Demand for treatment is exploding, while many health care workers themselves are infected. Because so many young people are dying of AIDS, decades of progress in improving the life expectancy of Africans have been erased. Without HIV-AIDS, life expectancy in sub-Saharan Africa would now be about 62 years. Instead, it was 47 years in 2008.

Nelson Mandela has said that he now views HIV-AIDS as a challenge for South Africa that is greater than apartheid ever was. Millions of parents, teachers, skilled farmers, craftspeople, and trained professionals have been lost. More than 12 million children have been orphaned, many without any family left to care for them or to pass on vital knowledge and skills (Figure 7.22). ▱

> Nelson Mandela now views HIV-AIDS as a challenge for South Africa that is greater than apartheid ever was.

Current Geographic Issues

Most countries of sub-Saharan Africa (see Figure 7.4 on page 246) have been independent of colonial rule for about 50 years. While many countries are still struggling with poverty and other effects of the colonial era, others are achieving higher standards of living for their citizens.

ECONOMIC AND POLITICAL ISSUES

Today, Africa is the poorest region of the world and exhibits widespread political turbulence. Much of Africa's situation relates to patterns of globalization that have increased its dependence on the export of raw materials to generate economic growth.

In this section, we examine the economic and political impacts of three waves of globalization in Africa. We also look at the current pressures moving some African countries away from raw materials exports and toward new types of economic development. We conclude by examining the forces pushing African countries toward more democratic political systems.

Commodities and Globalization in Africa

For centuries, Africa's primary role in the global economy has been as a producer of raw materials, usually produced through agriculture or mining. The raw materials are exported to wealthier countries that process them into more valuable manufactured goods. When traded, raw materials are called **commodities.** Two qualities of commodities limit their potential for lifting poor countries out of poverty: their prices are unstable, and they are of relatively low value.

Because they are considered to be of generally uniform quality, regardless of where they are produced, commodities can be traded on a global basis. For example, on the global

major commodities exchanges, a pound of copper may sell for U.S.$3 to 5, regardless of whether it is mined in South Africa, Chile, or Papua New Guinea. The global scale of this trade makes the prices of commodities highly responsive to changes anywhere on the globe that might influence the supply or demand.

As discussed earlier in the chapter, changes in the prices of globally traded agricultural commodities can wreak havoc with farmers' incomes. The many African governments that depend heavily on taxes collected on commodity exports can suffer similar disruption. The result can be nation-wide economic chaos that limits the ability of governments to manage their economic growth over the long term. In times when prices are high, earnings are often spent unwisely. When prices are low, cuts have to be made to essential government-provided services, such as electricity, road maintenance, or health care.

Another problem with commodities is that they are much less valuable than the more sophisticated manufactured goods into which they are processed. For example, copper might be worth U.S.$4 per pound on a global commodity exchange. But a pound of fine copper wire, used in stereo systems, might sell for U.S.$30. Hence, exports of commodities provide much lower profits than exports of finished manufactured goods. Very few countries have achieved broad prosperity by exporting commodities.

Three Waves of Globalization

Three waves of globalization have transformed Africa over the past several centuries. Two have increased Africa's dependence on exports of commodities, while the current wave could result in the development of some African manufacturing industries.

1. The colonial and early independence era. The modern system of African commodity production, and the political arrangements that made it possible, date largely from the era when European powers colonized almost all of Africa.

2. The structural adjustment era. Economic crises swept through most of Africa in the 1980s, partially because the export of commodities did not bring about broad prosperity. At the same time, massive debts that had accumulated over previous decades forced countries into even greater dependence on commodities exports.

3. The era of diverse globalization. The current wave of globalization is experiencing new sources of investment coming into Africa from across the globe. There is new potential for funding the manufacturing industries that could lift many African countries out of poverty and dependence on commodities exports, but many barriers remain.

Colonial and Early Independence Era of Globalization: Commodity Dependence

The European colonial powers that ruled Africa during the nineteenth and early twentieth centuries were interested mainly in extracting mineral and agricultural resources. As in British India, many early colonial governments in Africa evolved directly out of private corporations whose primary goal was to extract as many tons of copper, cotton, coffee, or other commodities as possible.

The welfare of Africans and their future development was a relatively low priority for most colonial administrators. Education and health care for Africans were generally neglected. Instead, profits from commodity exports were directed toward the colonial governments themselves—the "mother country" (Britain, France, Germany, Belgium)—and the European corporations that owned most commodity-producing ventures in Africa. In this sense, these were *mercantilist* governments (see Chapter 3, page 117). The Africans who worked on European-owned mines or plantations were very poorly paid, as were those who grew crops on their own lands. Moreover, the governments of colonial Africa discouraged any other economic activities. Hence, it was impossible for African economies to make a transition to the more profitable manufacturing-based industries that were transforming Europe and North America.

Corruption and inefficiency in the independence era.

When African nations started gaining independence in the 1950s and 1960s, the government institutions developed by the Europeans generally stayed in place. Therefore, economies remained focused on the production of commodities for export (Figure 7.23). While slightly more of the profits stayed within African countries, they were generally not wisely invested. Many leaders and bureaucrats considered it their right, much as their colonial predecessors did, to enrich themselves and their associates with government tax revenues derived from

commodity exports. Many governments also grew larger and more inefficient, while contributing minimally to economic development. Enormous sums also went to fund government-run economic development projects that were plagued by corruption and incompetence. Most farmers and miners remained poorly paid. As a result, even in Africa's largest, most prosperous economies, there is still widespread poverty today.

South Africa: an example and a warning.

Only one African country, South Africa, has a strong base in manufacturing. Early on, profits from its commodities exports (mainly minerals) were reinvested into related manufacturing industries serving the mining sector. Today, South Africa is a world leader in, among other things, the manufacture of mining and railway equipment. The country also has a well-developed service sector, with particular strengths in finance and communications that developed in part to support the mining and manufacturing industries. With only 6 percent of Africa's population, South Africa today produces 30 percent of the region's economic output.

However, South Africa has a dual economy. For centuries, it has had a well-off minority European population, whose skills and external connections fostered economic development within South Africa. After independence in 1910, the Dutch (Afrikaners) and British who remained in the country continued to dominate the economy. Even though the labor of black South Africans was essential to the country's prosperity, most remained poor. Under the apartheid system, 84 percent of black South Africans lived at a bare subsistence level. This pattern began to change at the end of apartheid, as black South Africans benefited from government-backed loans and other programs designed to encourage entrepreneurship. In 2007, however, 50 percent of black South Africans still lived below the poverty line (compared with 7 percent of white South Africans), and only 22 percent of black South Africans had finished high school (compared with 70 percent of white South Africans). The average income for a white South African is still more than 5 times that of a black South African. These inequalities contribute to high rates of crime and violence that have encouraged many skilled South Africans of all races to emigrate to Europe or North America.

Hence, South Africa provides both an example and a warning to the rest of Africa. Its experience shows what is possible when profits from raw materials exports are kept within Africa and are wisely directed toward industrialization. However, it also reveals the consequences when economic gains are not evenly spread throughout the population.

The Structural Adjustment Era of Globalization

By the 1980s, economic and political problems (we discuss the latter on pages 271–274) had left most African countries poor and dependent on their volatile and relatively low-value

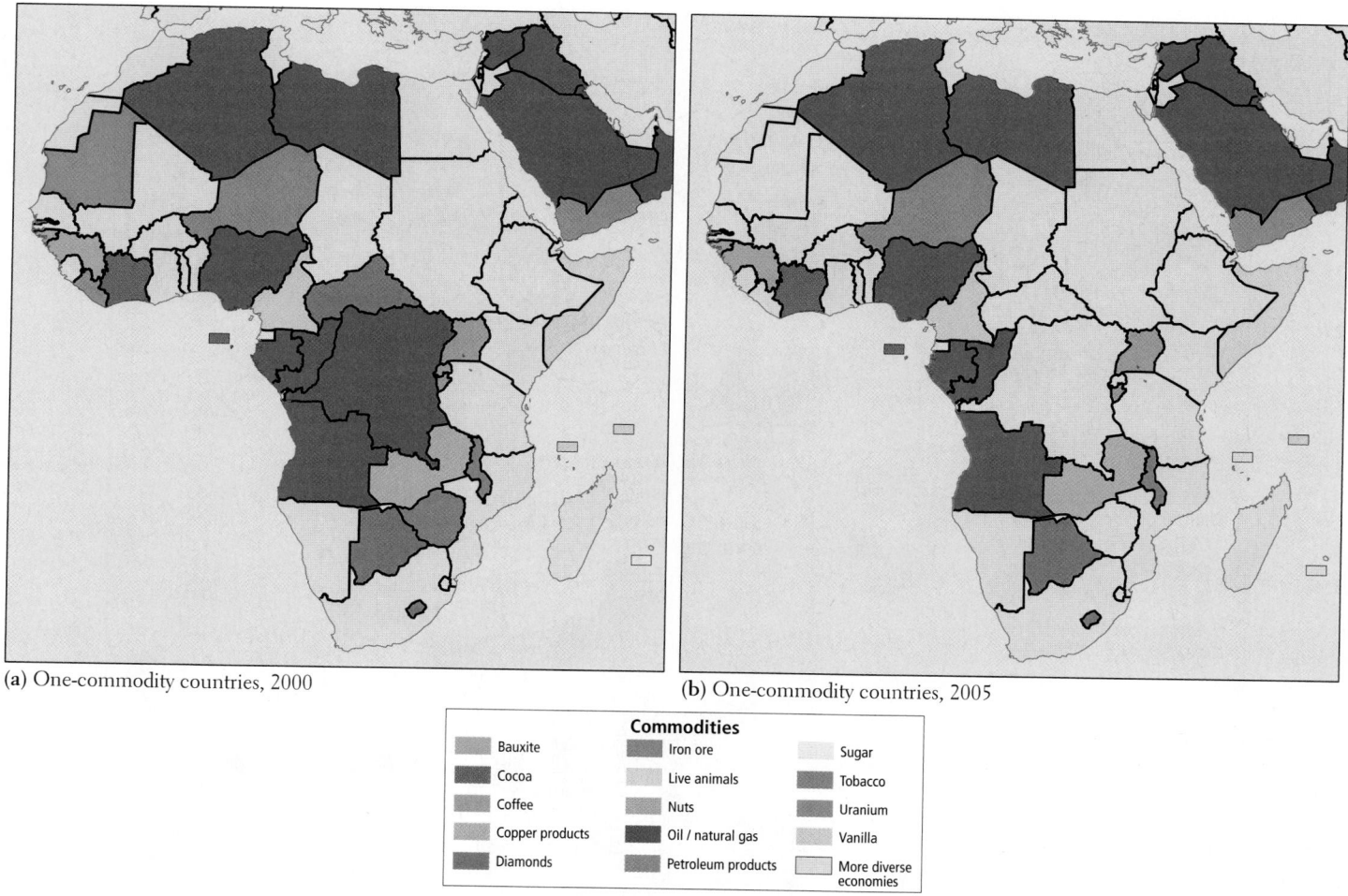

(a) One-commodity countries, 2000

(b) One-commodity countries, 2005

Commodities

Bauxite	Iron ore	Sugar
Cocoa	Live animals	Tobacco
Coffee	Nuts	Uranium
Copper products	Oil / natural gas	Vanilla
Diamonds	Petroleum products	More diverse economies

FIGURE 7.23 One-commodity countries. (a) African countries that depended on only one commodity for more than 50 percent of their export earnings in 2000. (b) Five years later, ten more African countries had diversified their economies sufficiently to no longer depend on only one main commodity. [Adapted from George Kurian, ed., *Atlas of the Third World* (New York: Facts on File, 1992), p. 76; data from *United Nations Human Development Indicators 2000* (New York: United Nations Development Programme); 2005 data from the *CIA World Factbook, 2006,* at https://www.cia.gov/cia/publications/factbook/index.html; *People Daily Online*, at http://english.people.com.cn/200509/20/eng20050920_209525.html.]

commodities exports. In the 1960s and 1970s, governments had tried to invest in manufacturing industries, but these projects generally failed due to a number of factors, discussed below. Worse yet, most African countries had taken out massive loans for these projects and were struggling to repay them. A breaking point came in the early 1980s when an economic crisis swept through Africa and much of the developing world, leaving most countries unable to repay their debts. In response the IMF and the World Bank designed *structural adjustment programs* or SAPs (see Chapter 3, pages 119–122) to help countries repay their loans. These programs and their successors had many unintended consequences, and ironically, they failed at their primary objective—reducing debt (Figure 7.24). In fact, they actually increased most countries' dependence on commodities exports, making it harder for them to move into more profitable manufacturing industries.

Why did investments in manufacturing fail? While corruption and civil war have ruined many investments in African manufacturing, tariffs in developed countries have also played a crucial role. The story of a shoe factory in Tanzania provides a case in point.

In the 1980s, the World Bank gave a loan to the government of Tanzania to develop an export-oriented shoe factory. Tanzania's large supply of animal hides was to be used to manufacture high-fashion shoes for the European market, using expensive imported machinery. However, due to EU tariffs that protected shoemakers in Europe, the Tanzanian factory never managed to export shoes to Europe. Ideally, the factory would have stayed in business by producing shoes for the African market. However, its machinery could produce only fancy dress shoes, not the practical working shoes most Africans need. Years went by and, as the unused factory deteriorated,

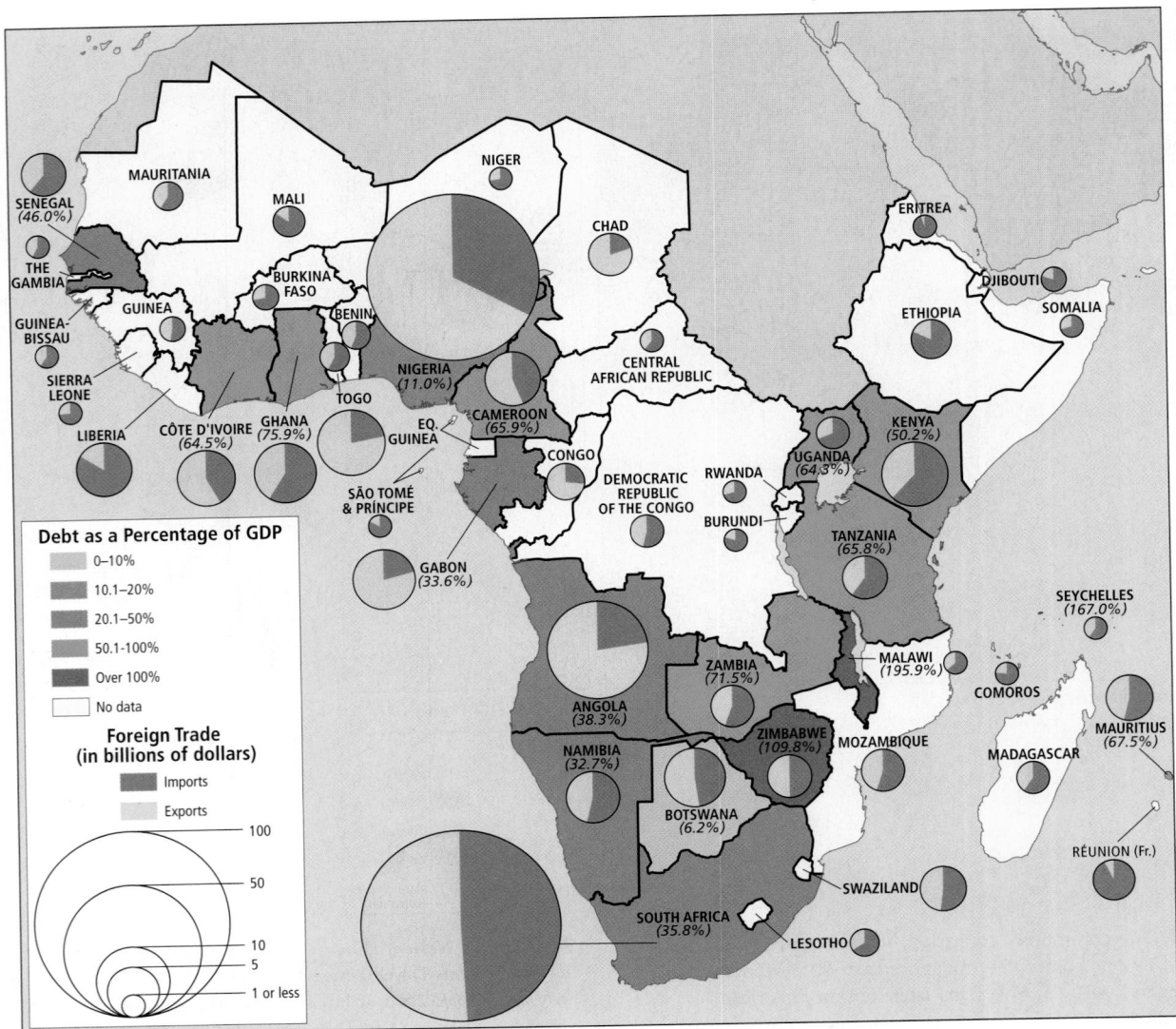

FIGURE 7.24 Economic issues: Public debt, imports, and exports. In most sub-Saharan African countries, imports exceed exports—even in South Africa, which has a large, diverse economy. As these countries borrow money for development, their public debt increases. Even in countries where exports are greater than imports (except for Nigeria), public debt is at least one-third of the GDP. [Debt and trade data from the *CIA World Factbook, 2006*, at https://www.cia.gov/cia/publications/factbook/index.html.]

Tanzania struggled to repay its loan to the World Bank. Such experiences, repeated hundreds of times over, left African countries impoverished, highly indebted, and unable to repay their loans by the 1980s.

What did structural adjustment programs achieve in Africa? In order to force countries to repay their loans, SAPs required governments to reduce their involvement in the economy by selling off government-owned enterprises. Government payrolls, along with many social and agricultural programs, were slashed so that tax revenues could be devoted to loan repayment. If countries refused to implement SAPs, the international banks cut off any future lending for economic development.

In Africa, the SAPs did accomplish some good. They tightened bookkeeping procedures and thereby curtailed corruption and waste in bureaucracies. They closed some corrupt state-owned monopolies in industries and services, and opened some sectors of the economy to medium- and small-scale business entrepreneurs. They also made tax collection more efficient.

At the same time, SAPs created some major problems. They failed to reduce the debt burden. On the contrary, debt continued to grow, despite the fact that SAPs forced Africa as a whole to spend more on debt payments than on health care and education combined. One reason debt persists is that SAPs have generally eliminated more jobs than they have created. Large state bureaucracies have been significantly reduced, and

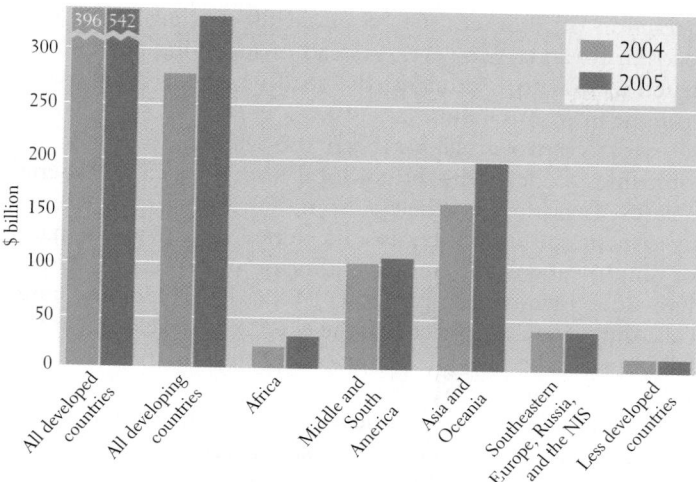

FIGURE 7.25 Foreign direct investment (FDI) around the world. Both developed and developing countries around the world had increases in FDI in 2005. FDI inflows to the 50 least developed countries rose to U.S.$10 billion, but were still marginal compared with the growth of FDI in the developing countries of Asia, Oceania, Latin America, and the Caribbean. [Adapted from *World Investment Report 2006* (New York: UN Conference on Trade and Development), pp. 39–40, at http://www.unctad.org/en/docs/wir2006ch2_en.pdf.]

state-owned businesses have been sold to the private sector. This has resulted in the firing of many employees in order to cut costs. Meanwhile, few new jobs have been created.

SAPs were supposed to create jobs by attracting a greater share of the world's foreign direct investment (FDI) to Africa because of all the reductions in corruption and mismanagement. This has not happened, however. As of 2005, foreign direct investment in Africa was less than 1.5 percent of the world total (Figure 7.25). Prospective investors in sub-Saharan Africa have been discouraged by problems that SAPs either ignored or made worse. Loss of public funds for schools perpetuated an under-skilled workforce. As unemployment rose, so did political instability. Deteriorating infrastructure reduced the quality of health care, transport, and financial services, all of which scared away investors. Thus, SAPs have also made it harder for the poor majority to make a decent living and stay healthy.

> Prospective investors in sub-Saharan Africa have been discouraged by problems that structural adjustment programs either ignored or worsened.

Food production and SAPs.
Structural adjustment programs reduced the availability of food for Africa's own consumption by shifting resources toward the production of cash crops for export. Between 1961 and 2005, per capita food production in Africa actually decreased by 14 percent, making it the only region on earth where people are eating less well now than in the past (see Figure 1.19 on page 23).

SAPs tipped the scales in favor of export-oriented commercial agriculture by encouraging African countries to lower the value of their currencies relative to the currencies issued by other countries. Such **currency devaluation** promotes the sale of African export crops by making them cheaper on the world market. But at the same time, it makes all imports more expensive. Therefore, farmers who grow food for the African market must spend much more on seeds, fertilizers, pesticides, and farm equipment, few of which are produced within the region. The shift to export crops also left farmers with less time and space to grow food for local markets. Increasingly, Africans must pay for expensive imported food.

Even export crops failed to earn as much as expected. SAPs were implemented in so many countries throughout the world that they produced an oversupply of export crops. Hence, prices for these crops have fallen globally.

Industry and SAPs.
Several aspects of SAPs have worked against export-oriented manufacturing industries. SAPs forced countries to earn as much money from exports as quickly as possible in order to pay off debts. This meant encouraging most countries to remain focused on commodity exports. Any further investments in manufacturing industries were postponed indefinitely. SAPs also forced African countries to remove their own tariffs on textiles from other countries, resulting in a flood of imported cloth from China. Meanwhile, currency devaluations made imported equipment and spare parts for local industries prohibitively expensive. The overall results were factory closings and job losses in African textile industries.

The informal economy and SAPs.
For many Africans, the informal economy has provided a partial escape from the hardships brought on by SAPs. Most people who work in the informal economy perform useful and productive tasks, such as selling garden produce, prepared food, or craft items. Others make a living by smuggling scarce or illegal items, such as drugs, weapons, endangered animals, or ivory.

In most African cities, informal trade once supplied perhaps one-third to one-half of all employment; now it often provides more than two-thirds. This creates problems for governments because the informal economy is very difficult to tax, so less money is available to pay for government services or to repay debts. Moreover, the profits of many informal businesses have declined over time as more people compete to sell goods and services to people with less disposable income. And when large numbers of men lost their jobs in factories or the civil service, they crowded into the streets and bazaars as vendors, displacing the women and young people who formerly dominated there.

Often, women are left with much worse jobs, working long hours for little pay and using hazardous chemicals or techniques. Some women have turned to sex work—a growing sector in the informal economy—putting themselves at high risk of contracting HIV-AIDS. In addition, more children from disintegrating families must fend for themselves on the streets. Cities such as Nairobi, Kenya, which had very few street children before SAPs, now have thousands.

SAP successors and debt relief. In response to the now widely recognized failures of SAPs, the IMF and the World Bank replaced them with "Poverty Reduction Strategy Papers," or PRSPs, in 2000. While these policies are essentially the same as SAPs, they differ by including the possibility that a country may have all or most of its debt "forgiven" (paid off by the IMF and the World Bank) if they follow the rules of PRSPs. A few countries in Africa have benefited from debt reduction and economic improvements as a result of following the PRSPs, but most remain highly indebted.

The Era of Diverse Globalization

The current wave of globalization is resulting in new sources of investment as well as new pressures on the prices of Africa's export commodities. While Europe and the United States are still the largest source of investment in Africa, Asia's influence on African economies is increasing dramatically. At the same time, Africans working and living abroad, primarily in Europe and North America, are sending home more money (*remittances*) than Africa is receiving from all other sources of foreign investment.

Asia's growing influences: China and India rushing in. In recent years, Africa has become a new frontier for Asia's large and growing economies, especially those of China and India. Through their demand for Africa's export commodities, through direct investment in Africa, and through the sale of their manufactured goods in Africa, China and India now exert a more powerful influence on Africa than ever before. Of the two, China's influence is by far the greater.

Both China and India are consuming ever larger amounts of Africa's commodity exports. Though these two countries consume only about 15 percent of Africa's exports, their share is growing twice as fast as any of Africa's other trading partners. In the coming decades, China is likely to become Africa's largest trading partner, eclipsing both the EU and the United States.

However, increasing Chinese and Indian demand for resources is a double-edged sword. It has brought higher and more stable prices for Africa's export commodities in recent years. However, increasing demand for food in China, India, and other parts of Asia has also helped drive up global food prices. This has been particularly painful for the many poor African countries that are highly dependent on food imports.

China's huge investments in Africa have become somewhat controversial. Much like Europe during the colonial era, China has invested mainly in commodity exports, especially minerals, timber, and oil. These investments have brought dramatic improvements in infrastructure but also fewer jobs for Africans than were hoped for. For example, to facilitate the extraction of resources, China paved 80 percent of the main roads in Rwanda. However, instead of supplying needed construction jobs to Rwandans, the road was built with laborers brought in from China.

Most controversial has been the willingness of Chinese companies to deal with brutal local leaders, such as Liberia's Charles Taylor. Nevertheless, some Africans praise Chinese investment specifically because it comes with fewer demands for improvement on human-rights or environmental issues than does investment from the EU and the United States. This, they argue, helps achieve the economic development that will improve the lives of Africans more quickly than any labor or environmental standards.

Chinese and Indian influences on African manufacturing cut both ways. Some companies, like the TATA Group, a large industrial conglomerate in India, are helping to boost this crucial sector with investments in automobile production and other types of manufacturing. However, this is occurring mainly in South Africa's already well-developed manufacturing sector. Meanwhile, Chinese- and Indian-manufactured goods have flooded Africa. Some Africans praise these goods because they are much cheaper than what was previously available from Japan, Europe, or the United States. However, others complain of a de-industrialization that is occurring in their countries as a result of Chinese and Indian investment. For example, a textile mill in Zambia closed after it was bought by a Chinese company. Now, the Zambian-grown raw cotton that was once processed by the mill is shipped to China where it is made into clothing. Nevertheless, both China and India could have a much more positive influence over the long term because both allow African-manufactured goods much greater access to their markets than do the EU or the United States.

Investment from Africans living abroad and returning home. Perhaps the most hopeful sign for the African economy is that Africans living outside of Africa are becoming a major source of investment in Africa. Since 2000, money sent home by Africans working in Europe and elsewhere has amounted to more than the total foreign direct investment in the region. This money is used to build houses for their families, to start small businesses, to fund education for children, and to help the needy in their communities. Remittances are a more stable source of investment than foreign direct investment and are much more likely to reach poorer communities. One disadvantage is that they tend to come in amounts too small to start anything but small businesses.

More substantial enterprises are developing out of the efforts of Africans who have received education and training abroad and brought their skills home. This is especially true in the information technology sector. In 1994, for example,

> Since 2000, money sent home by Africans working in Europe and elsewhere has amounted to more than total foreign direct investment in the region.

Africa Online, now the continent's largest Internet service provider (ISP), was founded by three young Kenyans who returned from the United States after graduating from college. Although Internet service is still largely confined to the main cities, all 54 countries on the African continent have had local full-service dial-up ISPs since 2000.

Regional and Local Economic Development

Seeking alternatives to past development strategies, many African governments are focusing on regional economic integration along the lines of the European Union. Local-scale and locally designed grassroots development is also being pursued.

Regional economic integration. For the many African countries, such as Togo or the Gambia, that are simply too small to function efficiently in the world economy, regional integration is crucial. The national markets of such countries have not nurtured a significant industrial base, and many remain heavily dependent on trade with former colonial powers in Europe. In 2002, only 11 percent of the total trade of sub-Saharan Africa was conducted between African countries.

Establishing regional trade links is hard because the necessary transport and communication networks are not in place. Air travel, and even long-distance telephone calls, from one African country to another must often be routed through Europe. Roads are often poor, even between major cities. Moreover, each of the 47 countries of sub-Saharan Africa has its own tangle of bureaucratic regulations for business, trade, and work permits. All of these impediments make doing business in Africa about 50 percent more expensive than doing business in Asia, so even many Africans choose to go elsewhere with their investment money.

A number of organizations are working toward economic integration (Figure 7.26). These organizations share several goals: reducing tariffs between members, forming common currencies, reestablishing peace in war-torn areas, upgrading transportation and communication infrastructure, and building regional industrial capacity. Full-scale economic union along the lines of the European Union is a long-term goal.

Grassroots economic development. This development strategy aims at providing sustainable livelihoods in rural and urban areas. One approach is **self-reliant development,** which consists of small-scale self-help projects, primarily in rural areas. These projects use local skills to create products or services for local consumption. Crucially, local control is maintained so that participants retain a sense of ownership. One district in Kenya has more than 500 such self-help groups. Most members are women who terrace land, build water tanks, and plant trees. They also build schools and form credit societies.

The issue of rural transport illustrates how a focus on local African needs can generate unique solutions. When non-Africans learn that transport facilities in Africa are in need of

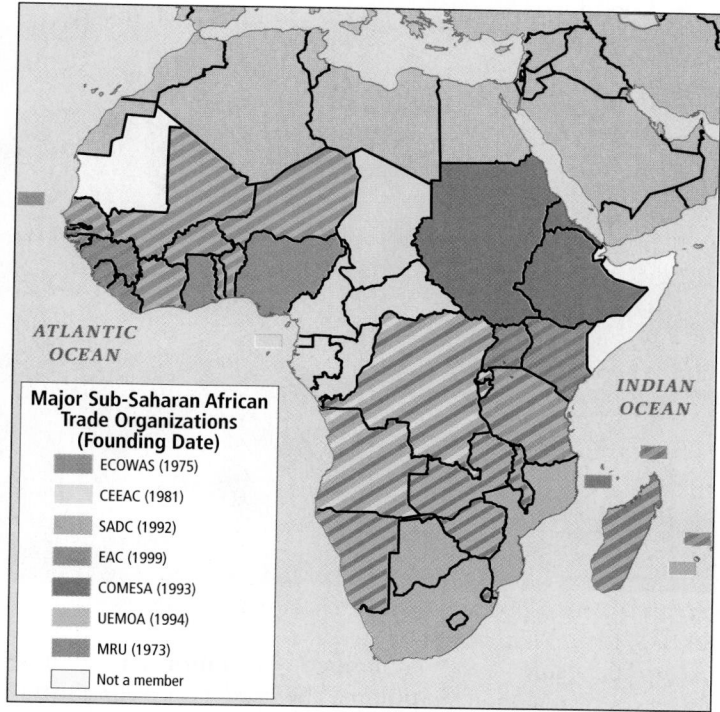

Major Sub-Saharan African Trade Organizations (Founding Date)
- ECOWAS (1975)
- CEEAC (1981)
- SADC (1992)
- EAC (1999)
- COMESA (1993)
- UEMOA (1994)
- MRU (1973)
- Not a member

FIGURE 7.26 Principal trade organizations in sub-Saharan Africa. Observe that some countries belong to more than one organization.

development, they usually imagine building and repairing roads for cars and trucks. But a recent study that analyzed village transport on a local level discovered that 87 percent of the goods moved are carried via narrow footpaths *on the heads of women!* Women "head up" firewood from the forests, crops from the fields, and water from wells (Figure 7.27). An average adult woman spends about 1.5 hours each day moving the equivalent of 44 pounds (20 kilograms) more than 1.25 miles (2 kilometers).

Yet the often dilapidated footpaths trod by Africa's load-bearing women have been virtually ignored by African governments and international development agencies, which have instead spent hundreds of millions on roads for motorized vehicles. A few grassroots-oriented non-governmental organizations are starting to make much less expensive but much more necessary improvements to Africa's footpaths. Meanwhile, women have been provided with bicycles, donkeys, and even motorcycles that can travel on the footpaths. This gives them more time for education and income generation.

Conflict and Colonial Legacies: Divide and Rule

Africa's economic development has been held back by frequent civil wars that are in many ways the legacy of colonial era policies of **divide and rule.** Divisions and conflicts between communities (often ethnic or religious groups) were deliberately

FIGURE 7.27 A common method of transport in Africa. These women are returning from the market at Kisangani, Congo (Kinshasa). Women throughout Africa are responsible for the transport of a large percentage of trade goods. [Tom Friedmann/Photo Researchers.]

> The borders and administrative units of the African colonies were designed so that different and sometimes hostile groups would be under the same jurisdiction.

intensified by European colonial powers. The borders and administrative units of the African colonies were designed so that different and sometimes hostile groups would be put together under the same jurisdiction (Figure 7.28). This made it harder for African groups to unite to overthrow their foreign rulers, and easier for Europeans to maintain their control over the flow of commodities out of Africa.

The resulting conflicts facilitated rule by foreigners who could appear as impartial governors with no ethnic loyalties, capable of benevolent intervention to settle disputes. After independence, however, rule by Africans was more difficult. Because Africans inevitably belonged to one local ethnic group or another, they could not be seen as impartial. The result has been years of carnage as ethnic and other hostilities developed into civil wars.

Case study: conflict in Nigeria. Nigeria was and remains a creation of British divide-and-rule imperialism. Many disparate groups—speaking 395 indigenous languages—have been joined into one unusually diverse country.

In present-day Nigeria, there are four main ethnic groups. The Hausa (21 percent of the population) and Fulani (9 percent) are both predominantly Muslim. Until recently, they lived mostly as herders in the northern grasslands and semi-desert. Hausa and Fulani elites collaborated extensively with the British during the colonial period, and since independence in 1960, they have dominated the high levels of Nigeria's government and military. The Yoruba (20 percent), who practice animism and Christianity, live in southwestern Nigeria. Once nearly all Yoruba were settled farmers, but many now live in urban areas as laborers, tradespeople, and professionals. The Igbo (17 percent) are primarily farmers, centered in the southeast. Originally animists (see definition below), many Igbo are now also Christian. Hundreds of other ethnic groups live among these larger groups.

The colonial administration reinforced a north-south dichotomy. Among the Hausa and Fulani, the British ruled via local Muslim leaders who did not encourage public education. In the south, among the Yoruba and Igbo, Christian missionary schools open to the public were common. At independence, the south had more than ten times as many primary and secondary school students as the north. The south was more prosperous, and southerners also held most government civil service positions. Yet, the northern Hausa dominated the top political posts. Over the years, bitter and often violent disputes have erupted between the southern Yoruba-Igbo and northern Hausa-Fulani regarding the distribution of economic development funds, jobs, and oil revenues, as well as over environmental damage from oil extraction. ◼

In 1966, a 3-year civil war erupted after an Igbo-led military coup. A Hausa-led countercoup resulted in the slaughter of 30,000 Igbos. In response, the Igbo portion of southeastern Nigeria tried unsuccessfully to secede. Violence and ensuing food shortages resulted in more than 200,000 deaths.

Geographic strategies have often been used to reduce tensions in Nigeria. One approach has been to create more political states (Nigeria now has 30) and thereby reallocate power to smaller local units with fewer ethnic and religious divisions. Recently, large, wealthy states have been subdivided to spread oil profits more evenly. Although dividing the country into more states has increased administrative costs, it seems to have eased ethnic and religious hostilities like those described in the following vignette.

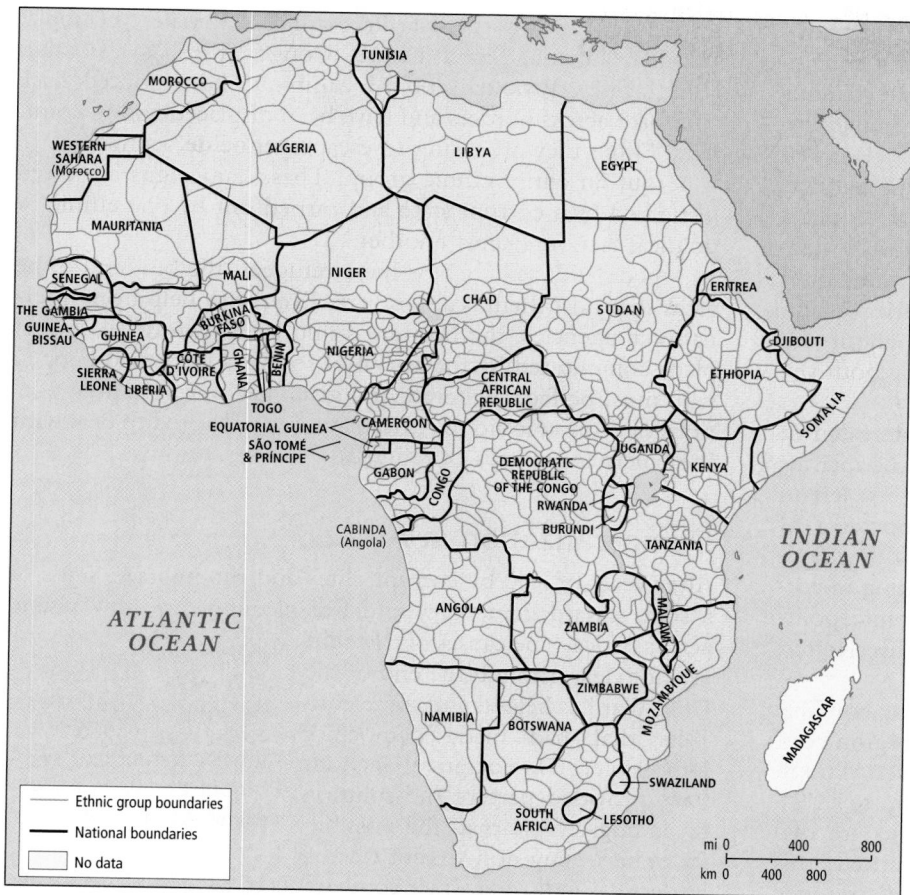

FIGURE 7.28 Ethnic groups in sub-Saharan Africa. This map indicates the large number of ethnic groups spread across the continent of Africa. Superimposed on this pattern are the present national boundaries, which were, for the most part, imposed by European colonizers. Very rarely do ethnic group and national boundaries match. [Adapted from James M. Rubenstein, *An Introduction to Human Geography* (Upper Saddle River, N.J.: Prentice Hall, 1999), p. 246.]

Vignette *It is a vicious circle—you cannot get investment while there is violence and killing and disturbance and then if you don't get investment which [would create] employment, then the killing, the violence will go on. So it is a very, very bad vicious circle.*

—President Olusegun Obasanjo of Nigeria,
interview with the BBC, 2002

In 2002, Amina Lawal, then pregnant, was sentenced to death by stoning for adultery. She lived in a Muslim village in northern Nigeria, where shari'a had replaced the corrupt secular legal system 3 years before, when Nigeria's military dictatorship ended. The democratically elected government in Abuja, in the Christian south, declared she would never be executed. Her case exacerbated tensions between the different ethnic groups in the north and south.

This simmering conflict erupted during preparations for the Miss World pageant, which was to be held in Abuja that year. When a Christian fashion writer suggested that the Prophet Muhammad would surely have chosen one of the beauty contestants for his wife, the northern Muslims were offended. Riots between Muslims and Christians broke out in the northern city of Kaduna, and the pageant, which Christian Nigerian officials had

hoped would help to improve Nigeria's global image, packed up and left.

Amina Ladan-Baki, a banker and world rights activist, believes that the discord between ethnic and religious groups in Nigeria is manipulated to suit the agendas of politicians. "The politicians mislead people. They use religion. They use diverse cultures to unite or disunite. 'Vote for me because I am of your stock.'"

In February 2004, a shari'a court of appeals overturned Amina Lawal's conviction. However, other women have been charged with adultery in the 12 Nigerian states that follow shari'a, and they face the possible sentence of death by stoning. Meanwhile, continuing political and economic tensions between the north and south make it likely that, as a matter of pride and identity, the north will retain its conservative approach to Islam.

To learn more about Amina Lawal and the Miss World pageant, watch the FRONTLINE/World video "The Road North: What the Miss World Riots Reveal About a Divided Country." ■

Conflict and the Cold War in Africa. The conflicts that grew out of divide-and-rule policies were deepened and prolonged by the Cold War between the United States and the former Soviet Union (see Chapter 1, pages 46–47 and Chapter 5, page 187). After independence, some African governments turned to socialist models of economic development, often receiving economic and military aid from the Soviet Union. Other governments became allies of the United States, receiving equally generous aid (see Figure 5.14, page 187). Both the United States and the USSR tried to undermine each other's allies by arming and financing rebel groups.

In the 1970s and 1980s, southern Africa became a major area of tension. The United States aided South Africa's apartheid government in military interventions against Soviet-allied governments in Namibia, Angola, and Mozambique. Another area of cold war tension was the Horn of Africa, where Ethiopia and Somalia fought intermittently throughout the 1960s, 1970s, and 1980s. Each side was funded by both the Soviets and Americans at different times.

It is important to note that many Africans interested in solving their countries' problems are moving beyond criticisms of European colonial powers and the structures they left in place. Increasingly, they fault the greed and incompetence of their own leaders and government bureaucrats, and they fault their own people for tolerating and often participating in corruption. Many are focusing on new modes of economic, political, and social development discussed elsewhere in this chapter.

Conflict and refugees. Sub-Saharan Africa's many conflicts have resulted in huge numbers of refugees. This region contains about 19 percent of the world's refugees even though it has only 11 percent of the world's population (Figure 7.29). The United Nations estimates that if people who are displaced within their home countries are also counted, this region would have about 28 percent of the world's refugee population. Three-fourths of Africa's refugees are women and children.

Throughout the last decade of the twentieth century, for example, people in Chad, Sudan (see Chapter 6, pages 237–238,

for an account of the Darfur conflict), Somalia, Ethiopia, Uganda, Liberia, Sierra Leone, Congo (Kinshasa), Congo (Brazzaville), Rwanda, and Mozambique have poured back and forth across borders and have been displaced within countries. Often they are trying to escape **genocide,** campaigns to wipe out an entire ethnic group. These campaigns are often instigated by a corrupt state and carried out by one ethnic or political faction against another.

As difficult as life is for these refugees, the burden on the areas that host them is also severe. Even with help from international agencies, the host areas find their own development plans complicated by the arrival of so many distressed people, who must be fed, sheltered, and given health care. Large portions of economic aid to Africa have been diverted to deal with the emergency needs of refugees.

Democratization in Africa

After years of rule by corrupt elites and the military, signs of a shift toward democracy and free elections are now visible across Africa. The causes of this shift are not clear, though the end of the Cold War has helped end some conflicts that had made democracy impossible. Democratic elections have provided part of the solution to ending civil wars as the possibility of becoming non-violent elected leaders has induced former combatants to lay down their arms. In other cases, public outrage against corrupt ruling elites has resulted in elections that brought a change of leadership.

> After years of rule by corrupt elites and the military, signs of a shift toward democracy and free elections are now visible across Africa.

The holding of elections has increased dramatically. In 1970, only 11 states had held elections since independence. By 2008, 23 out of 44 sub-Saharan African states had held open, multiparty, secret-ballot elections with universal suffrage. The trend toward democracy could lead to governments that are more responsive to the needs of their people.

Nevertheless, there are still many states that have never held truly free elections. Indeed, for most of the independence era, the institutions that make up most African governments have been **authoritarian** in that they emphasize the authority of the government over the rights of the people to elect leaders or freely express their opinions in public or through the media. Intimidation and abuse by national militaries, which have often assumed control of governments, has also been widespread. Often, these tendencies are a legacy of the divide-and-rule tactics discussed earlier.

In many cases, elections have been so flawed by corruption that they have sparked massive violence. Such was the case in Kenya in 2008, where more than 1000 people died and 600,000 were dislocated by mobs of voters enraged by rigged elections. ◪ A similar sequence of events occurred in 2006 in Congo (Kinshasa) when the first elections held in 46 years resulted in violence that left more than 1 million Congolese as refugees within their own country.

FIGURE 7.29 Refugees in Africa. These Bantu people from Somalia are waiting in Kenya to participate in a resettlement program that will bring them to the United States. [AP/Wide World Photos.]

In other countries, elected leaders have become as corrupt and authoritarian as the leaders they replaced. In the 1960s and 1970s, Robert Mugabe of Zimbabwe became a hero to many people for his successful guerilla campaign against what was then called Rhodesia. This highly discriminatory white minority government was allied with apartheid South Africa but not formally recognized by any other country. Mugabe was elected president in 1980, following free, multiparty elections. Over the years, however, his authoritarian policies alienated more and more Zimbabweans. In the 1990s, he implemented a highly controversial land-redistribution program that resulted in his supporters gaining control of the country's best farmland, much of which had been in the hands of white Zimbabweans. This contributed to a chronic food shortage and a massive economic crisis that left 80 percent of Zimbabweans unemployed. Resulting political violence has created 3 to 4 million refugees, most of whom have fled to neighboring South Africa. Nevertheless, Mugabe held onto power through brutal repression of the political opposition and rigged elections in 2002 and 2008. 📹

African women in politics and civil society. Women across Africa are assuming positions of power in the wake of national crises that have undermined faith in traditional male leadership and highlighted the contributions of women. The opening of this chapter describes the changes that President Ellen Johnson-Sirleaf of Liberia has been making. In Rwanda, which was devastated by genocide and mass rape in the 1990s, women now make up 40 percent of the mayors and hold 56 percent of the parliamentary seats. 📹 Rwanda is the first country in the world where women make up a majority of the national legislature. In Mozambique, 34.8 percent of the parliament is female, and in South Africa, 32.8 percent. All told there are 11 African countries where the percentage of women in national legislatures is above the world average of 18.2 percent.

In many cases, these figures reflect government policies as much as the willingness of Africans to be led by women. All of the countries cited above have quotas that guarantee a certain percentage of legislative seats for women, and fewer women are elected in the countries without quotas. The quotas are often a response to national crises, especially civil wars, in which women fought alongside men in battle or suffered disproportionately from the chaos and destruction of war. 📹 Quotas are usually a reflection of a larger commitment toward the empowerment of women, sometimes written into a country's constitution, that encourages female participation in politics and civil society. It is important to note, however, that many female leaders in Africa, such as Ellen Johnson-Sirleaf and at least half of Rwanda's female parliamentarians, were elected without quotas.

Beyond the arena of electoral politics, businesswomen are joining with women who run education and health NGOs to challenge all types of gender discrimination throughout the region.

Vignette Charity Kaluki Ngilu will never forget the day she became a professional politician. She was washing dishes in her kitchen in Kitui, east of Nairobi, Kenya, when she saw a group of women who had worked with her on community health projects approaching her back door. Mrs. Ngilu answered the knock on the door, drying her hands on an apron. The women said they wanted her to run for parliament in Kenya's first multiparty elections. She assumed they were joking.

That was in 1992. Ngilu beat the governing party's incumbent, then became a major advocate for women's issues. In 1997, she was the first woman to run for president in Kenya. She didn't win, but men were among her strongest supporters because they believed she was capable of making bigger changes than a man could. By 2004, she was minister of health, in charge of Kenya's greatest challenge: responding to the HIV-AIDS epidemic.

Sources: James C. McKinley, Jr., "A woman to run in Kenya? One says, 'Why not?'" New York Times (August 3, 1997): 3; Kennedy Graham, ed., The Planetary Interest: A New Concept for the Global Age (London: Rutgers University Press, 1999); "Women taking control of power in Africa," "Talk of the Nation," National Public Radio, April 20, 2006. ■

SOCIOCULTURAL ISSUES

To the casual observer, it may appear that a majority of sub-Saharan Africans live traditional lives in rural villages. A closer look, however, reveals the imprint of other cultures, particularly European, on such aspects of traditional culture as religion, gender roles, health, and community life.

Religion

Africa's rich and complex religious traditions derive from three main sources: indigenous African belief systems, Islam, and Christianity.

Indigenous belief systems. Traditional African religions are among the most ancient on earth and are found in every part of the continent. Figure 7.30a highlights the countries in which traditional beliefs remain particularly strong. Traditional African beliefs and rituals often seek to bring departed ancestors into contact with people now alive—people who in turn are the connecting links in a timeless spiritual community that stretches into the future. The future is reached only if present family members procreate and perpetuate the family heritage through storytelling, coming-of-age rituals, and offerings to the ancestral spirits. Religious leaders are often powerful men (and occasionally women) who act as both politicians and spiritual leaders.

Most traditional African beliefs can be considered **animist** in that spirits, including those of the deceased, are considered to exist everywhere—in trees, streams, hills, and art objects, for example (Figure 7.31). In return for respect (expressed through ritual), these spirits offer protection from sickness, accidents, and the ill will of others. Many African religions are

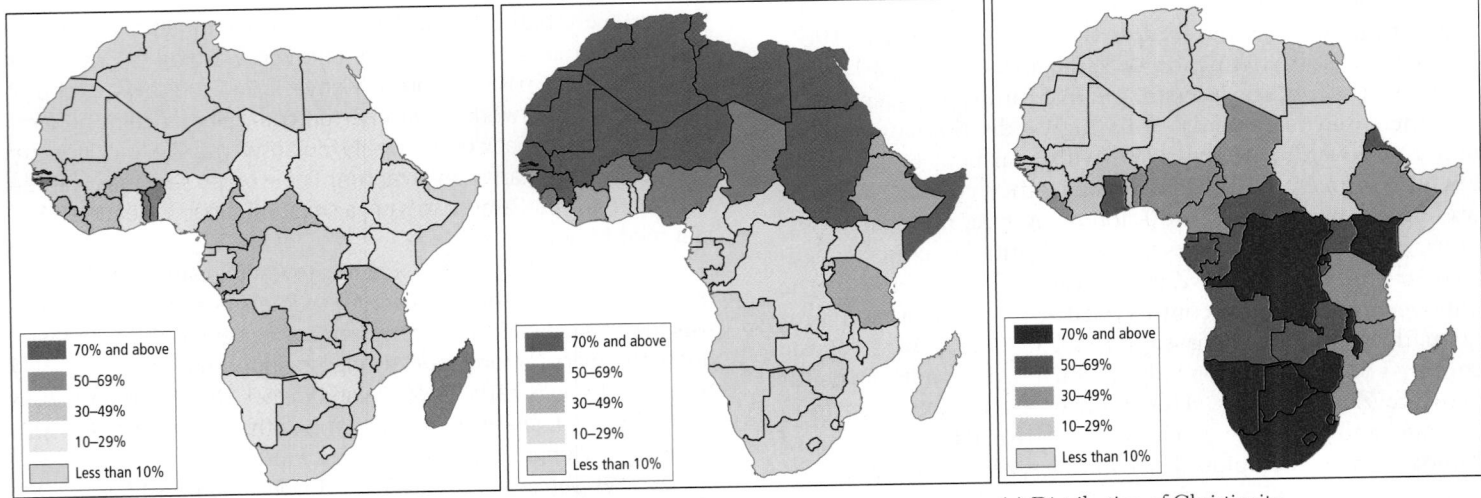

(a) Distribution of traditional religions   (b) Distribution of Islam   (c) Distribution of Christianity

FIGURE 7.30 Religions in Africa. Notice that the various religions in Africa overlap in distribution; in many countries, one is dominant but others are present. [Adapted from Matthew White, "Religions in Africa," in *Historical Atlas of the Twentieth Century* (October 1998), at http://users.erols.com/mwhite28/afrorelg.htm; revised with new data from the *CIA World Factbook, 2006*, at https://www.cia.gov/cia/publications/factbook/index.html.]

fluid and adaptable to changing circumstances. For example, Osun, the god of water, traditionally credited with healing powers, is now also invoked for those suffering economic woes.

Religious beliefs in Africa, as elsewhere, are not static but evolve continually as new influences are encountered. If Africans convert to Islam or Christianity, they commonly retain aspects of their indigenous religious heritage. The three maps in Figure 7.30 show a spatial overlap of belief systems, but they do not convey the philosophical blending of two or more faiths, which is widespread. In the Americas, the African diaspora has influenced the creation of new belief systems developed from the fusion of Roman Catholicism and African beliefs. Voodoo in Haiti or Santeria in Cuba are examples of this fusion. In the 1990s, African versions of evangelical Christianity began to influence U.S. evangelicals when African missionaries were sent to places such as Atlanta and Birmingham.

Islam and Christianity. Islam began to extend south of the Sahara soon after Muhammad's death in 632, and today, about one-third of sub-Saharan Africans are Muslim. As Figure 7.30b shows, Islam is now the predominant religion throughout the Sahel, where Muslim traders from Southwest Asia and North Africa brought the religion. Powerful Islamic empires have arisen here since the ninth century. The latest of them challenged French domination of the region in the late nineteenth century. Muslim traders also spread Islam along Africa's eastern coast.

Today about half of Africans are Christian. Christianity first came to the region via Ethiopia in the fourth century, well before it spread throughout Europe. However, Christianity didn't come to the rest of Africa until nineteenth-century missionaries from Europe and North America became active along the west coast (Figure 7.30c). Many Christian missionaries provided the education and health services that colonial administrators had neglected. In the 1980s, old-line established churches began to gain adherents in Africa. The Anglican Church (Church of England), for example, grew so rapidly in Kenya, Uganda, and Nigeria that by 2000, there were more Anglicans in these countries than in the United Kingdom. The

FIGURE 7.31 Coffin art in Ghana. Ghana's coffin carvers accomplish an apparently seamless interweaving of religion, art, and modern life. They are experts at interpreting the life of the deceased artistically: a bus for a bus driver, a Mercedes-Benz for a chauffeur, a plane for a pilot. Both rich and poor prize the coffin carver's art. People may commission a coffin long before death comes, often spending their life savings to buy a niche in the cultural memory of the community. [Carol Beckwith and Angela Fisher/Robert Estall Photo Library, UK.]

FIGURE 7.32 Worshipers at the Miracle Center in Kinshasa, Democratic Republic of the Congo. [Robert Grossman/CORBIS SYGMA.]

Anglican Church in Africa attracts the educated urban middle class, whereas modern evangelical versions of Christianity appeal to the less educated, more recent urban migrants—the most rapidly growing African populations. Further, the independent, evangelical sects made more room for folk traditions that were often forbidden by the European missionaries.

Evangelical Christianity and the gospel of success. In urban Africa, certain interpretations of Christianity are being combined with African beliefs in the importance of sacrificial gifts to ancestors and the power of miracles. The result is one of the world's fastest growing Christian movements (Figure 7.32). Preachers at the Miracle Center in Kinshasa describe the new gospel of success forcefully: "The Bible says that God will materially aid those who give to Him . . . We are not only a church, we are an enterprise. In our traditional culture you have to make a sacrifice to powerful forces if you want to get results. It is the same here." Generous gifts to churches are promoted as a way to bring divine intervention to alleviate miseries, whether physical or spiritual. Practitioners donate food, television sets, clothing, and money; one woman gave 3 months' salary in the hope that God would find her a new husband.

   Like all religious belief systems, the gospel of success is best understood within its cultural context. Many of the believers, new to the city, feel isolated and are seeking a supportive community to replace the one they left behind. People view their material contributions to the church as similar to the labor and goods they previously donated to maintain their standing in their home village. In return for these dues and volunteer services, members receive social acceptance and community assistance in times of need.

Gender Relationships

Long-standing African traditions dictate a fairly strict division of labor and responsibilities between men and women. In general, women are responsible for domestic activities, including rearing the children, tending the sick and elderly, and maintaining the house. Women collect water and firewood and prepare nearly all the food. Men are usually responsible for preparing land for cultivation. In the fields intended to produce food for family use, women sow, weed, and tend the crops as well as process them. In the fields where cash crops are grown, men perform most of the work.

   When husbands in search of cash income migrate to work in the mines or in urban jobs, women take over nearly all agricultural work. They tend to work with simple hand tools in the fields and in the home, and during their reproductive years, they often do field labor with a child strapped on their backs. When there are small agricultural surpluses or handcrafted items to trade, it is women who transport and sell them in the market. Throughout Africa, married couples tend to keep separate accounts and manage their earnings as individuals. When a wife sells her husband's produce at the market, she usually gives the proceeds to him.

   In rural areas, African men on average do not have as many responsibilities as women, nor do they work as hard or for as many hours. The retreat of men from virtually all tasks directly related to supporting domestic life seems to have started with European colonialism. The prevailing European attitude at the time was that women were lesser creatures who should remain in the home and let their husbands deal with the outside world. This attitude was reflected in the colonial policy of recruiting men for cultivating cash crops. At first, men took on such activities to meet the colonial administration's requirement that taxes be paid in cash. Now men work for cash to pay for children's school fees, basic electricity, and certain consumer goods. And many men migrate to the cities or the mines to work, at least seasonally. Women are left to shoulder all the domestic work as well as what was formerly the shared work of subsistence agriculture.

   In the precolonial past, there were social controls that tempered gender relationships. Most marriages were social alliances between families; therefore, husbands and wives spent most of their time doing their tasks with family members of their own sex rather than with each other. For most of the day, women were influenced primarily by other women.

   Traditional gender relationships were modified by Islam and Christianity. In most cases, men gained power and women lost freedoms. Muslim women were restricted to domestic spaces, no longer able to move about at will, trade in the markets, or engage in public activities. While Christian women operate in the public sphere more than Muslim women do, they too are socialized to restrict their activities to the home. It is important to note that having multiple wives—the practice of **polygyny** (Figure 7.33)—is more common in sub-Saharan Africa (where it has ancient pre-Muslim roots) than

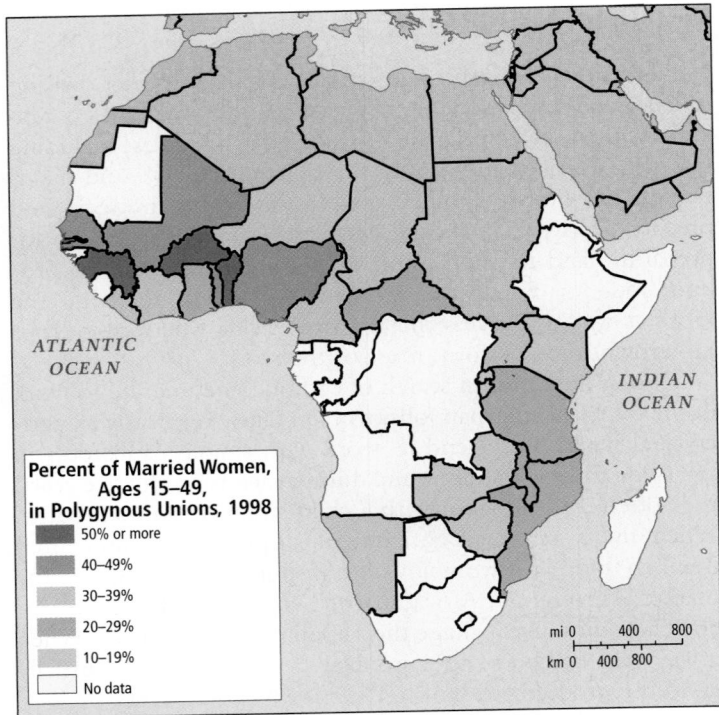

FIGURE 7.33 Polygyny in sub-Saharan Africa. The practice of polygyny (men having multiple wives) is more common in sub-Saharan Africa than anywhere else in the world. Nevertheless, only a small minority of Africans are in polygynous marriages today. [Adapted from *The World's Women, 2000: Trends and Statistics* (New York: United Nations, Department of Economic and Social Affairs, 2000), pp. 27–28.]

in Muslim North Africa. Nevertheless, only a small minority of Africans are in polygynous marriages today.

Female Circumcision

A practice known as **female circumcision,** popularly called *cutting,* is widespread in at least 27 countries throughout the central portion of the African continent (Figure 7.34). In the procedure, which is usually performed without anesthesia, parts of the labia and the entire clitoris are removed. In the most extreme cases (called *infibulation*), the vulva is stitched nearly shut. Female circumcision eliminates any possibility of sexual stimulation for the woman and makes urination and menstruation difficult. Intercourse is painful and childbirth is particularly devastating because the scarred flesh is inelastic. A 2006 medical study conducted with the help of 28,000 women in six African countries showed that women who had undergone circumcision were 50 percent more likely to die during childbirth, and their babies were at similarly high risk. The practice also leaves women exceptionally susceptible to HIV infection.

Often mistakenly thought to be a Muslim practice, this custom is actually older than Islam. It is now practiced by Muslims, Christians, and followers of ancient traditional African

religions. The practice is probably intended to ensure that a female is a virgin at marriage and that she thereafter has a low interest in intercourse other than for procreation. While the practice is in decline today, it is still widespread among some groups. Among the Kikuyu of Kenya, nearly all females would have been circumcised 40 years ago, but today only about 40 percent are. About 130 million girls and women in Africa have been circumcised, with about 2 million more circumcised each year.

Because female circumcision results in the removal of healthy parts of the body, many people, including physicians, refer to it as *female genital mutilation.* Many African and world leaders have concluded that the practice is an extreme human rights abuse, and it

> Because female circumcision results in the removal of healthy parts of the body, many people refer to it as female genital mutilation.

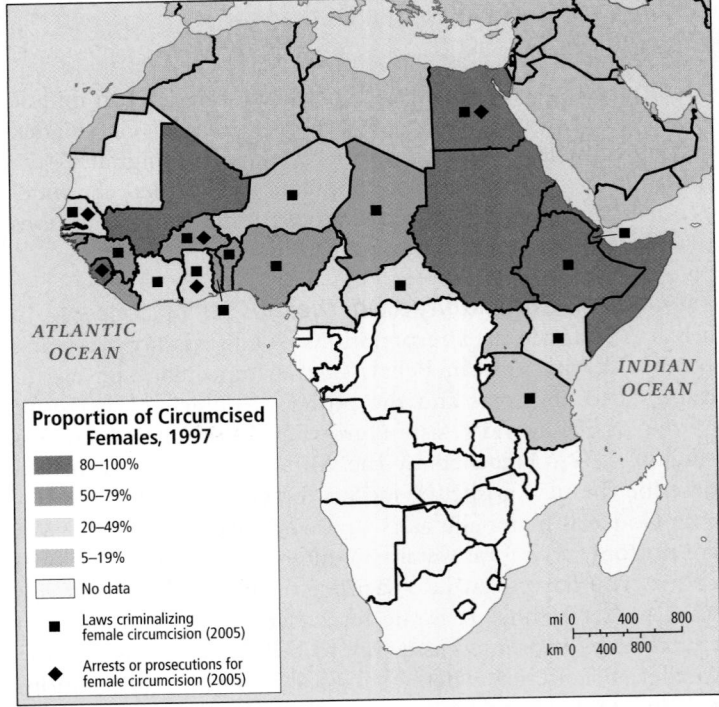

FIGURE 7.34 Female circumcision in Africa, 1997–2005: Prevalence and legality. This practice, mapped here using 1997 data, occurs all across the center of the African continent in spite of government policies against it. Female circumcision has been declared illegal in the countries indicated, but enforcement is lax in many. [Adapted from Joni Seager, *The State of Women in the World: An International Atlas* (London: Penguin, 1997), p. 53; additional data from Amnesty International, at http://www.amnesty.org/ailib/intcam/femgen/fgm9.htm; http://www.crlp.org/pub_fac_fgmicpd.html; *The World's Women, 2000: Trends and Statistics* (New York: United Nations, Department of Economic and Social Affairs, 2000), pp. 159–161.]

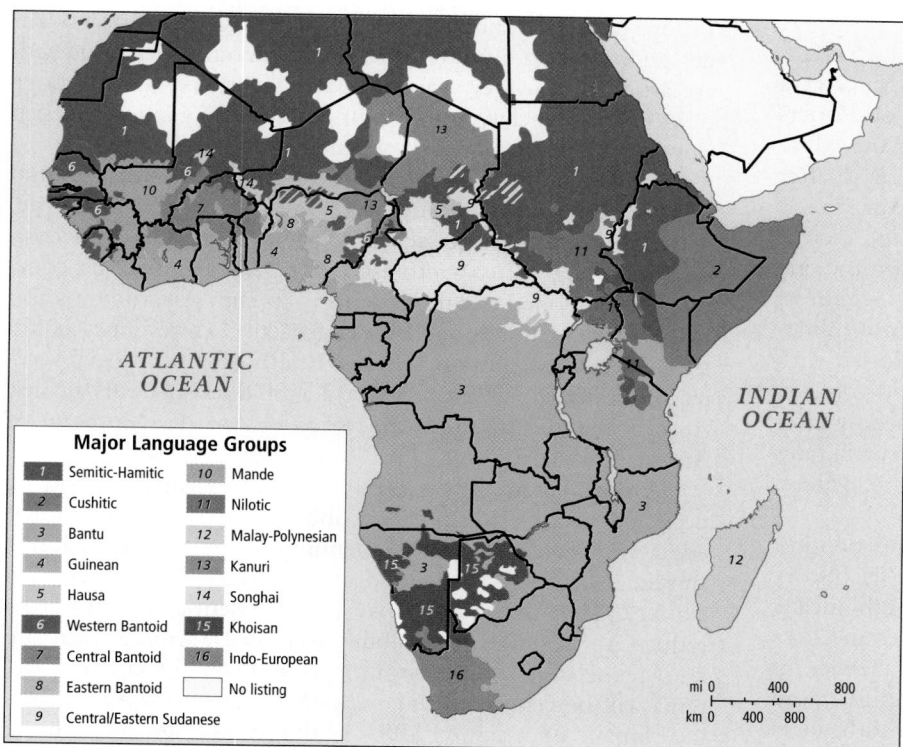

FIGURE 7.35 Major language groups of Africa. [Adapted from Edward F. Bergman and William H. Renwick, *Introduction to Geography—People, Places, and Environment* (Englewood Cliffs, N.J.: Prentice Hall, 1999), p. 256; and Titus Didactica, Frankfurt, Germany, http://titus.uni-frankfurt.de/didact/karten/afr/afrikam.htm.]

is now against the law in 16 countries. However, because it is so deeply ingrained in some value systems, the most successful eradication campaigns are those that emphasize the threat circumcision poses to a woman's health.

Ethnicity and Language

Ethnicity, as we have seen throughout this book, refers to the shared language, cultural traditions, and political and economic institutions of a group. The map of Africa presented in Figure 7.35 shows a rich and complex mosaic of languages. Yet despite its complexity, this map does not adequately depict Africa's cultural diversity (compare it with Figure 7.28 on page 273).

Most ethnic groups have a core territory in which they have traditionally lived, but very rarely do groups occupy discrete and exclusive spaces. Often several groups share a space, practicing different but complementary ways of life and using different resources. For example, one ethnic group might be subsistence cultivators, another might herd animals on adjacent grasslands, and a third might be craft specialists working as weavers or metalsmiths.

People may also be very similar culturally and occupy overlapping spaces but identify themselves as being from different ethnic groups. Hutu and Tutsi cattle farmers in Rwanda share an occupation, similar languages, and similar ways of life. However, European colonial policies exaggerated ethnic differences,

and Hutu and Tutsi now think of themselves as having very different ethnicities. In the 1990s and again in 2004, the Hutu and Tutsi people engaged in several devastating episodes of mutual genocide.

While some African countries have only a few ethnic groups, others have hundreds. Cameroon, sometimes referred to as a microcosm of Africa because of its ethnic complexity, has 250 different ethnic groups. Different groups often have extremely different values and practices, making the development of cohesive national policies difficult. Nonetheless, the vast majority of African ethnic groups have peaceful and supportive relationships with one another.

To a large extent, language correlates with ethnicity. More than one thousand languages, falling into more than a hundred language groups, are spoken in Africa. Most Africans speak their native tongue and a **lingua franca** (language of trade). Some languages are spoken by only a few dozen people, while others, such as Hausa, are spoken by millions of people from Côte d'Ivoire to Cameroon. Some African languages are now dying out and being replaced by languages that better suit people's needs or that have become politically dominant. Increasingly, a few lingua francas, such as Hausa, Arabic, and Swahili, are taking over in different areas. Former colonial languages such as English, French, and Portuguese (all classed as Indo-European languages in Figure 7.35) are also widely used in commerce, politics, education, and on the Internet.

Reflections on Sub-Saharan Africa

Late one evening in a restaurant in Central Europe, after a lengthy conversation that touched on some of the world's perplexing problems, a friend leaned across the elegant white tablecloth and asked, "But don't you think that Africa is, after all, better off for having been colonized by Europeans?"

How does one reply to such a question? Africa, the ancient home of the human species, is, of all regions on earth, the one in greatest need of attention. Africa is the poorest region of the world, yet the reasons for Africa's poverty are not immediately apparent to the casual observer. Africa is blessed with many kinds of resources—agricultural, mineral, and forest—but the greatest value is added to these resources only after they leave Africa. Africa is not densely occupied, and birth rates are dropping, but population growth, especially in urban areas, is hindering efforts to improve standards of living.

The reasons for Africa's poverty and social and political instability emerge only through an exploration of its history over the last several centuries. Colonialism methodically removed Africans from control of their own societies and lives and forced Africa's people and resources into the service of Europe. Even today, with colonialism officially dead for more than three decades, many Africans are struggling with borders developed by Europeans and economies still focused on cheap and unstable raw materials exports to Europe and elsewhere. In view of all this, it is hard to believe that anyone would think Africa is better off for having been colonized.

Africa is changing, however. Africans are beginning to devise new economic development strategies and political institutions to replace the ones imposed by outsiders. African leaders, an increasing number of whom are democratically elected, are articulating wider visions. In particular, the contributions of women to society, and the importance of gender issues in slowing population growth are increasingly recognized.

It is tempting to suggest that the rest of the world should just "leave Africa to the Africans," but that view is unrealistic. Three more likely strategies are currently emerging as a consensus among Africa specialists. First, SAPs should be completely abandoned and more debts to foreign governments and international lending agencies cancelled. Tax revenues could then once again support schools, health care, and social services. Second, the developed world should lower tariffs against African manufactured products to foster development of African industries.

A third suggestion is that future aid to Africa should be designed to take advantage of indigenous skills and knowledge, and that development planning and aid money address should local needs as defined and managed by local African experts. This perspective may prove essential to efforts to reduce Africa's high vulnerability to global warming. Intimate knowledge of African environments and food production systems will be central to this region's effort to adapt to growing climatic uncertainty. Much of this knowledge lies with the poor rural Africans who have arguably gained the least from Africa's "development" so far. Perhaps efforts to adapt to global warming will bring greater benefits to this group.

Chapter Key Terms

agroforestry 248

animist 275

apartheid 256

authoritarian 274

carbon sequestration 253

commodities 265

currency devaluation 269

divide and rule 271

female circumcision 278

genocide 274

Horn of Africa 247

intertropical convergence zone (ITCZ) 246

lingua franca 279

mixed agriculture 250

pastoralism 252

polygyny 277

Sahel 246

self-reliant development 271

subsistence agriculture 250

Critical Thinking Questions

1. How do humans affect the vulnerability of African wildlife populations to global warming?

2. In what ways might traditional African subsistence food cultivation techniques be less vulnerable to global warming than commercial agricultural systems? In what ways might they be more vulnerable?

3. How are education and economic opportunity for women linked to Africa's rapid population growth?

4. How does urbanization contribute to the spread of HIV-AIDS in Africa?

5. Why does being an exporter of commodities place a country at a disadvantage in the global economy? What would be some ways to amend the situation?

6. How might some of Africa's post-independence civil wars be related to the colonial past?

7. How are Africa's trading relationships with China and India different from Africa's trading relationships with Europe and the United States?

8. How is the recent increase in female elected leaders in Africa in part a response to crisis?

9. Give an example of how traditional African religions have influenced Christianity in this region.

10. If you were designated to make a speech to the Rotary Club in your town on hopeful signs in Africa, what would you include in your talk? What pictures would you show?

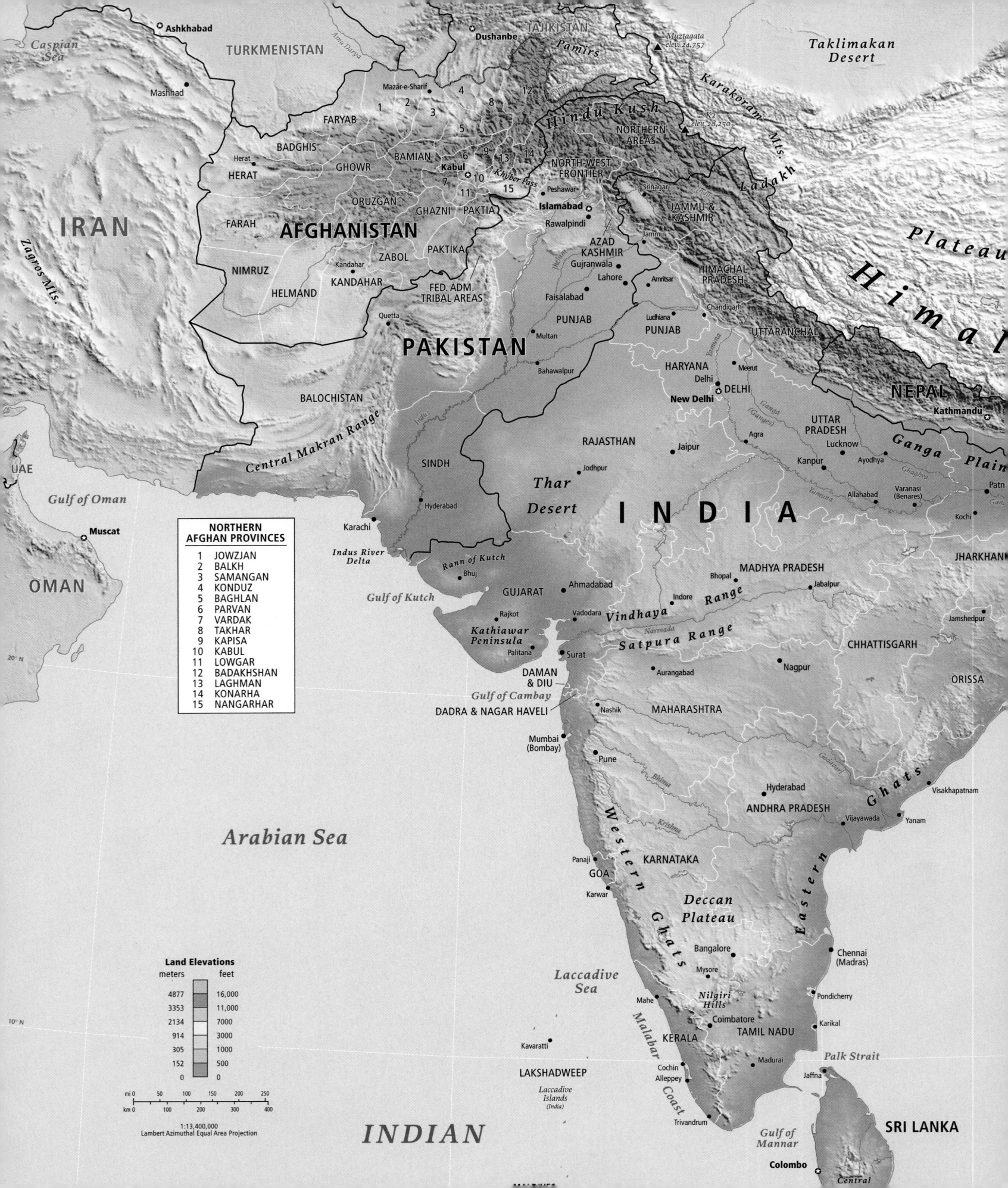

CHAPTER 8

South Asia

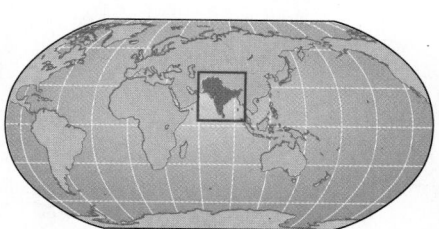

Global Patterns, Local Lives It is April 16, 2006. Narendra Modi, the governor of the state of Gujarat in India, draped in garlands by well-wishers, is embarking on a hunger strike. He is protesting a decision by India's national government in New Delhi to limit the height of the Sardar Sarovar Dam on the Narmada River in the neighboring state of Madhya Pradesh. Damming the river has become the centerpiece of Gujarat's efforts to deal with the periodic droughts that affect as many as 50 million of its citizens. Hence, Modi sees limiting the height of the dam as threatening his state's many irrigation programs.

Far away in New Delhi, Medha Patkar, the leader of the "Save the Narmada" movement, is in day 18 of her hunger strike in protest against the same dam. As water rises in the dam's reservoir, 320,000 farmers and fishers are being forced to relocate. While Indian law requires that these people be given land or cash to compensate them for what they have lost, so far only a fraction have received any compensation. Much of the land and cash has actually been pocketed by corrupt officials. In response, some of the farmers forfeited their right to compensation and refused to move, even as the rising waters of the reservoir consumed their homes. Forcibly removed by Indian police, many have since relocated to crowded urban slums.

Environmental problems have been at the heart of the Sardar Sarovar controversy since the project began in 1961 (Figure 8.2). The Narmada River was once a placid, slow-moving river—and one of India's most sacred. Now its natural cycles have been disturbed, causing massive die-offs of aquatic life and high unemployment among fishers.

The benefits of the dam have also been called into question. A major justification for the dam was that it would provide irrigation waters to drought-prone areas in Gujarat. However, 80 percent of the areas in Gujarat most vulnerable to drought will get no water from the project. Concerned that the economic benefits would be small and easily negated by the environmental costs, the World Bank withdrew its funding of the dam in the mid-1990s. Ecologists say that far less costly water-management strategies,

Figure 8.1 Regional map of South Asia.

283

Figure 8.2 The Sardar Sarovar Dam on the Narmada River, south of the Indian city of Ahmadabad. This dam is held up as a model for the development of modern India. But it is also reviled as a symbol of how the rights of the poor are trampled as they are forced to relocate with little or no compensation. [Dave Amit/Landov/Reuters.]

such as rainwater harvesting, groundwater recharge, and watershed management, would be better options for the drought-stricken farmers of Gujarat.

Less than a day after Gujarat's governor began his hunger strike, the Indian Supreme Court ruled that the Sardar Sarovar dam could be raised higher, so Modi ended the hunger strike. The following day, Medha Patkar ended her fast as well, because in the same decision the Supreme Court ruled that all people displaced by the dam must be adequately relocated. Furthermore, the decision confirmed that human impact studies are required for dam projects. No such study was provided for the Sardar Sarovar Dam.

Adapted from "Modi goes on fast over Narmada dam," India eNews, April 16, 2006, http://www.indiaenews.com/politics/20060416/4531.htm; "Dam protester's health gets worse," BBC News Online, April 4, 2006, http://news.bbc.co.uk/2/hi/south_asia/4876110.stm; Rahul Kumar, "Medha Patkar ends fast after court order on rehabilitation," One World South Asia, http://southasia.oneworld.net/article/view/131077/1/2220; "Narmada's revenge," Frontline, India's National Magazine 22(9) (2005), http://www.hinduonnet.com/fline/fl2209/stories/20050506002913300.htm; "Water harvesting, addressing the problem of drinking water," http://www.narmada.org/ALTERNATIVES/water.harvesting.html. ∎

The recent history of water management in the Narmada River valley highlights some key issues now facing South Asia and other regions with developing economies. Across the world, large and poor populations are depending on increasingly overtaxed environments. Improving their standard of living nearly always requires more water and energy. Efforts to meet these urgent needs often make neither economic nor environmental sense, but are driven to completion by political and social pressures. In the case of the Sardar Sarovar dam,

the wealthier, more numerous, and more politically influential farmers of Gujarat have tipped the scales in favor of a project that is creating more problems than it is solving. But the same can be said for dam projects discussed elsewhere in this book (Figure 8.3). Rarely do the backers of these and other economic development projects ask the fundamental question, "Development for whom?"

The countries that make up the South Asia region are Afghanistan and Pakistan in the northwest; the Himalayan states of Nepal and Bhutan; Bangladesh in the northeast; India; the island country of Sri Lanka; and several groups of islands in the Arabian Sea and the Bay of Bengal (Figure 8.4; also see Figure 8.1).

Questions to consider as you study South Asia (also see Views of South Asia, pages 286–287)

1. In what ways have recent changes in South Asian food production systems contributed to urbanization? Farming has been made both more productive and more expensive with the introduction of new seeds, fertilizers, pesticides, and equipment. While some farmers have become wealthier, many have found themselves unable to compete and have moved to cities. What kind of jobs and housing do these people usually find in urban areas?

2. How have issues related to gender transformed South Asian populations in recent decades? A strong preference for sons has produced a gender imbalance throughout most of South Asia. Why are sons preferred over daughters? Where in South Asia is this preference the weakest?

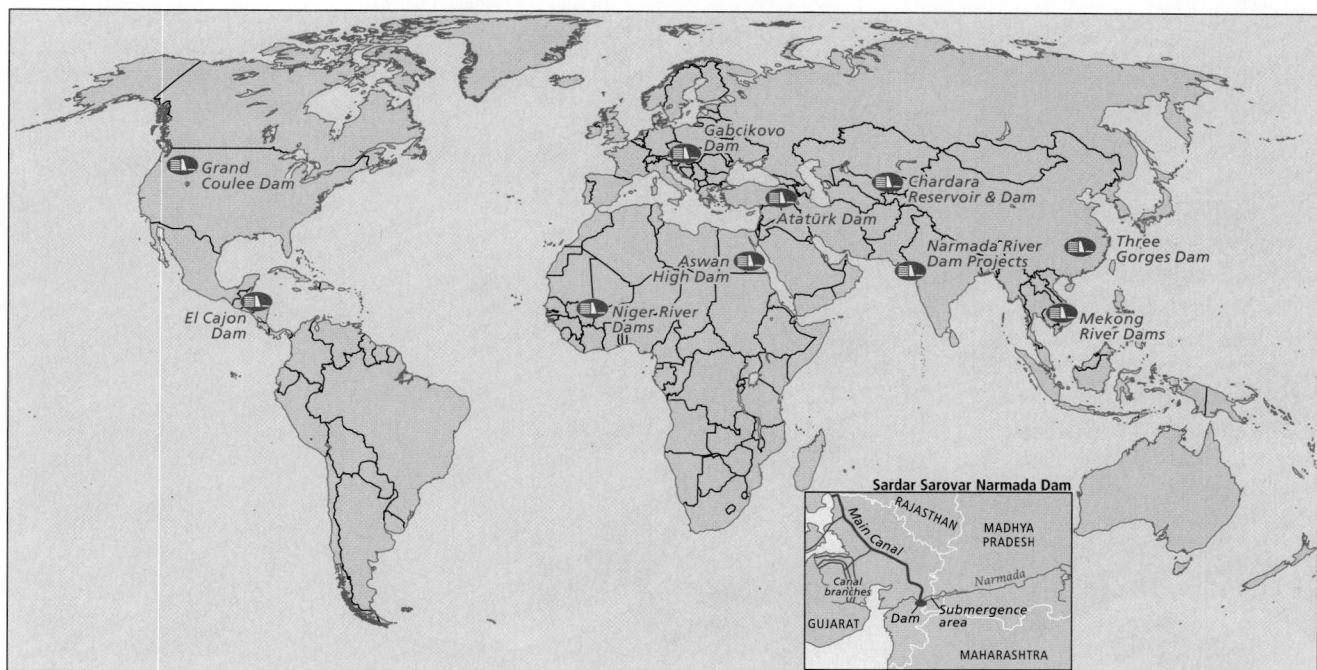

Figure 8.3 Major dams around the world. This map shows the location of dams featured or mentioned in this book. All were built with the intention of improving human circumstances. However, all have also resulted in considerable damage to the environment, and dislocation and hardship for people.

3. **How has globalization benefited some Indian workers in recent years?** Well-educated and technically-skilled workers have gained from a recent boom in foreign investment. Many international companies are taking advantage of India's large, college-educated workforce that is low cost relative to Europe or the United States. How are these jobs linked to the global economy?

4. **What role has the democratic process played in South Asia's many violent conflicts?** Many conflicts have been made worse by an unwillingness on the part of governments and warring parties to recognize the results of elections, or even to allow people to vote. Some conflicts have been diffused, at least for the short term, by incorporating former combatants into the democratic process. How has democracy, or a lack thereof, contributed to violent conflict in South Asia?

5. **In what ways is South Asia vulnerable to water problems associated with global warming?** Low-lying, densely populated areas such as coastal Bangladesh are particularly vulnerable to sea level rise. However, many more people will be impacted by the melting of glaciers in the Himalayas that feed the region's major rivers during the dry season. What other vulnerabilities to global warming does this region have? What solutions to global warming are being explored in South Asia?

Figure 8.4 Political map of South Asia.

Views of South Asia

For a tour of these places, visit www.whfreeman.com/pulsipher

[Background image: Trent Schindler NASA/GSFC Code 610.3/587 Scientific Visualization Studio]

2. Globalization. Information technology and television have brought global influences to even the most remote parts of South Asia. These rural Afghan's are learning how to use computers in Kabul. See page 315–317 for more. Lat 34°31'03"N, Lon 69°10'46"E. [USAID]

1. Gender and democracy. Women have been elected to lead South Asia's four largest countries. However, the status of women lags behind that of men throughout South Asia. The late Benazir Bhutto of Pakistan is shown here in Karachi. See page 312 for more. Lat 24°51'N, Lon 67°E.
[SRA Gerald B. Johnson/DOD]

5. Urbanization and food production. Changes in agriculture have forced many poor farmers into the cities. Often they find only low paying work, such as pulling a rickshaw. Many live in slums, like Mumbai's Dharavi (shown on the lower right). See page 302–304 for more. Lat 19°02'44"N, Lon 72°51'29"E.
[Kaukab Jhumra Smith/USAID; http://commons.wikimedia.org/wiki/Image:Calcutta_rickshaw.jpg; http://commons.wikimedia.org/wiki/Image:Straatbewoners_Bombay.jpg]

2

Kabul

1 Karachi

5

Mumbai

1780 **1849**
1900
1935
1956
1964
1971

2001

Gangotri Glacier

Himalaya Mountains

3. Global warming and water. South Asia's three largest rivers, including the Brahmaputra (shown here flooding during the summer monsoon), are fed by glaciers that will be gone in 50 years. See page 291 for more. Lat 29°N, Lon 94°55'E. [NASA; Staff Sgt. Val Gempis/DOD]

3

Brahmaputra

4

Bangladesh

4. Population. While Population growth is slowing, population density is already extremely high in most of South Asia. These rural Bangladeshi's were flooded out of their homes by a hurricane (cyclone) in 2007. See page 300–302 for more. Lat 23°11'N, Lon 90°40'E. [Mohammad Shahidul Islam/USAID; Sue McIntyre/USAID]

The Geographic Setting

Terms to Be Aware Of

Because its clear physical boundaries set it apart from the rest of the Asian continent, the term **subcontinent** is often used to refer to the entire Indian peninsula, including Nepal, Bhutan, India, Pakistan, and Bangladesh.

Many cities in South Asia have recently adopted new names to replace the names given them during British colonial rule. The city of Bombay, for example, is now officially *Mumbai*, Madras is *Chennai*, Calcutta is *Kolkata*, Benares is *Varanasi*, and the Ganges River is the *Ganga River*.

PHYSICAL PATTERNS

Many of the landforms, and even the climates, of South Asia are the result of huge tectonic forces. These forces have positioned the Indian subcontinent along the southern edge of the Eurasian continent, where the warm Indian Ocean surrounds it and the massive mountains of the Himalayas shield it from cold air flows from the north.

Landforms

The Indian subcontinent and the territory surrounding it dramatically illustrate what can happen when two tectonic plates collide. Millions of years ago, the Indian-Australian Plate, which carries India, broke free from the eastern edge of the African continent and drifted to the northeast (see Figure 1.21, page 25). As it began to collide with the Eurasian Plate about 60 million years ago, India became a giant peninsula jutting into the Indian Ocean. As the relentless pushing from the south continued, both the leading (northern) edge of India and the southern edge of Eurasia crumpled and buckled. The result is the world's highest mountains—the Himalayas, which rise more than 29,000 feet (8800 meters)—as well as other very high mountain ranges to the east and west. The continuous compression also lifted up the Plateau of Tibet, which rose up behind the Himalayas to an elevation of more than 15,000 feet (4500 meters) in some places.

South and southwest of the Himalayas are the Indus and Ganga river basins, also called the Indo-Gangetic Plain. Still farther south is the Deccan Plateau, an area of modest uplands 1000–2000 feet (300–600 meters) in elevation, interspersed with river valleys. This upland region is bounded on the east and west by two moderately high mountain ranges, the Eastern and Western Ghats. These mountains descend to long but narrow coastlines interrupted by extensive river deltas and floodplains. The river valleys and coastal zones are densely occupied; the uplands only slightly less so.

Because of its high degree of tectonic activity and deep crustal fractures, South Asia is prone to devastating earth-quakes, such as the magnitude 7.7 quake that shook the state of Gujarat in western India in 2001 and the 7.6 quake that hit the India–Pakistan border region in 2005. Coastal areas are also vulnerable to tidal waves or *tsunamis* that are cause by undersea earthquakes. A massive tsunami wrecked much of coastal Sri Lanka and southern India in 2004, killing tens of thousands of people there. ▶

Climate

The end of the dry [winter] season [April and May] is cruel in South Asia. It marks the beginning of a brief lull that is soon overtaken by the annual monsoon rains. In the lowlands of eastern India and Bangladesh, temperatures in the shade are routinely above a hundred degrees; the heat causes dirt roads to become so parched that they are soon covered in several inches of loose dirt and sand. Tornadoes wreak havoc, killing hundreds and flattening entire villages. . . .

This is also a time of hunger, as with each passing day thousands of rural families consume the last of their household stock of grain from the previous harvest and join the millions of others who must buy their food. Each new entrant into the market nudges the price of grain up a little more, pushing millions from two meals a day to one.

ALEX COUNTS, *GIVE US CREDIT* (NEW YORK: TIMES BOOKS, 1996), P. 69

From mid-June to the end of October [summer] is the time of the river. Not only are the rivers full to bursting, but the rains pour down so relentlessly and the clouds are so close to village roofs that all the earth smells damp and mildewed, and green and yellow moss creeps up every wall and tree . . . Cattle and goats become aquatic, chickens are placed in baskets on roofs, and boats are loaded with valuables and tied to houses . . . As the floods rise villages become tiny islands, . . . self-sustaining outpost[s] cut off from civilization . . . for most of three months of the year.

JAMES NOVAK, *BANGLADESH: REFLECTIONS ON THE WATER* (BLOOMINGTON: INDIANA UNIVERSITY PRESS, 1993), PP. 24–25

These two passages highlight the contrasts between South Asia's dominant winter and summer wind patterns, known as **monsoons** (Figure 8.5). In winter, cool, dry air flows from the Eurasian continent to the ocean. In summer, warm, moisture-laden air flows from the Indian Ocean over the Indian subcontinent, bringing with it heavy rains. The abundance of this rainfall is amplified by the intertropical convergence zone (ITCZ). Air masses moving south from the Northern Hemisphere converge near the equator with those moving north from the Southern Hemisphere. As the air rises and cools,

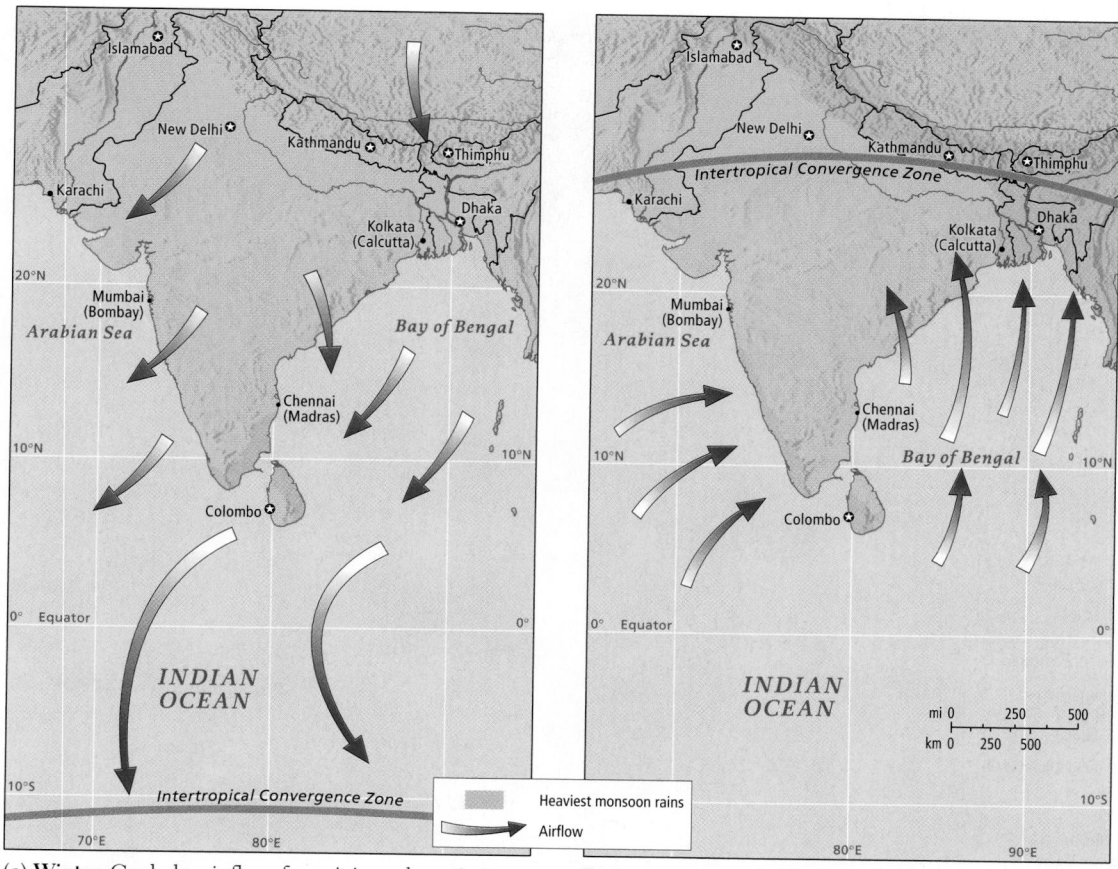

(a) Winter. Cool, dry air flows from Asian subcontinent.

(b) Summer. Warm, moist air flows to Asian subcontinent.

Figure 8.5 Winter and summer monsoons in South Asia. **(a)** In the winter, cool, dry air blows from the Eurasian continent south across India toward the ITCZ, which in winter lies far to the south. **(b)** In the summer, the ITCZ moves north across India, picking up huge amounts of moisture from the ocean, which are then deposited over India.

copious precipitation is produced. As described in Chapter 7 (see pages 246–247), the ITCZ shifts north and south seasonally. The intense rains of South Asia's summer monsoon are likely caused by the ITCZ being sucked onto the land by a vacuum created when huge volumes of air over the Eurasian landmass heat up and rise into the upper atmosphere.

The monsoons are major influences on South Asia's climate. In early June, the warm, moist ITCZ air of the summer monsoon first reaches the mountainous Western Ghats. The rising air mass cools as it moves over the mountains, releasing rain that nurtures dense tropical rain forests and tropical crops in the central uplands. Once on the other side of India, the monsoon gathers additional moisture and power in its northward sweep up the Bay of Bengal, sometimes turning into tropical cyclones.

As the monsoon system reaches the hot plains of West Bengal and Bangladesh in late June, columns of warm rising air create massive, thunderous cumulonimbus clouds that drench the parched countryside. Precipitation is especially intense in the foothills of the Himalayas, which hold the world record for annual rainfall—about 35 feet, even though there is no rain at all for half the year. Monsoon rains run in a band parallel to the Himalayas that reaches across northern India all the way to northern Pakistan by July. These patterns of rainfall are reflected in the varying climate zones (Figure 8.6) and agricultural zones (see Figure 8.29 on page 313) of South Asia.

By November, the cooling Eurasian landmass sends cooler, drier air over South Asia. This heavier air from the north pushes the warm, wet air back south to the Indian Ocean. Very little rain falls in most of the region during this winter monsoon. However, parts of southeastern India and Sri Lanka receive winter rains as the ITCZ drops moisture picked up on its now southward pass over the Bay of Bengal.

The monsoon rains deposit large amounts of moisture over the Himalayas, much of it in the form of snow and ice that creates huge glaciers. Meltwater from these glaciers feeds the headwaters of the three river systems that figure prominently in the region: the Indus, the Ganga, and the Brahmaputra. All

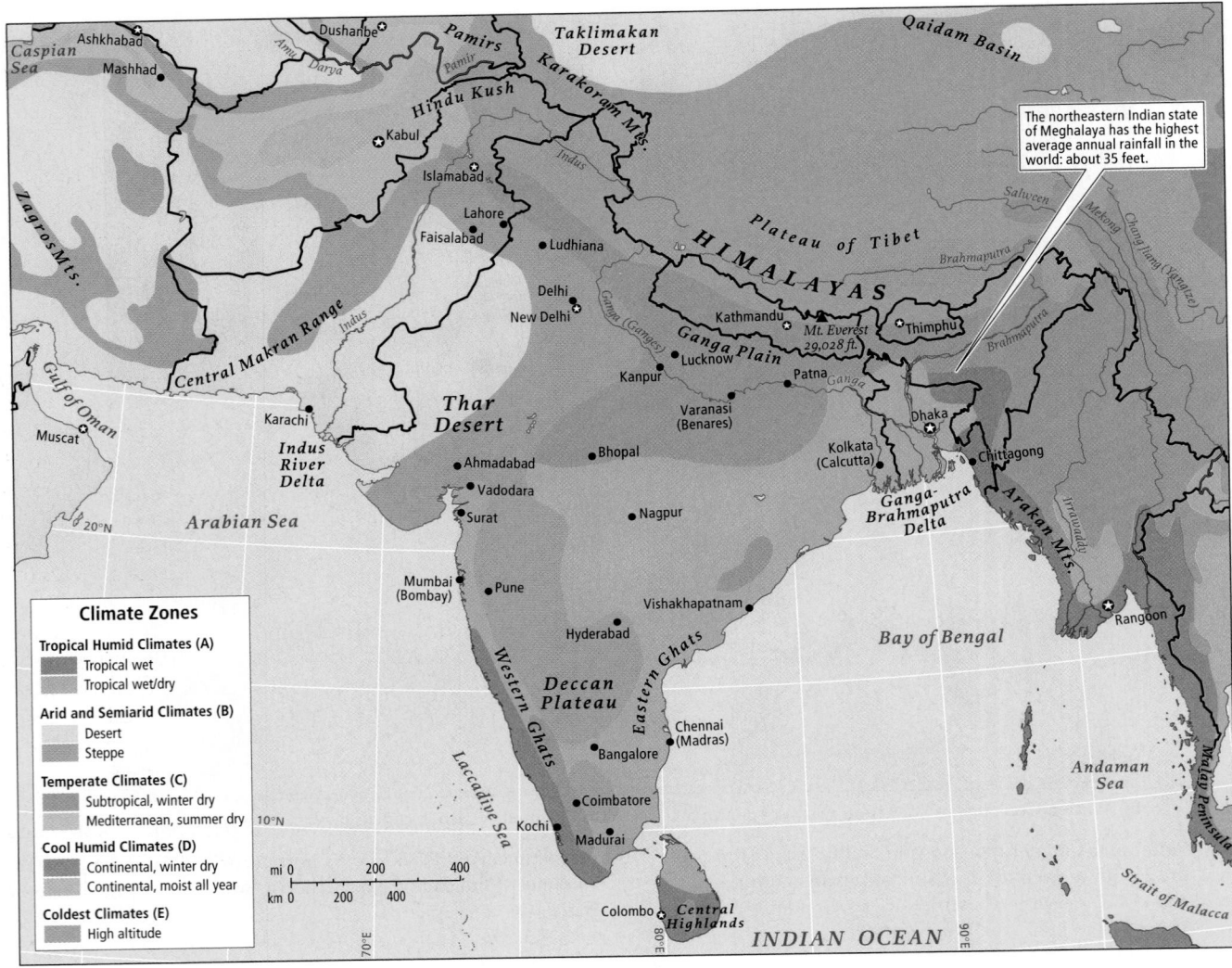

Figure 8.6 Climates of South Asia.

three rivers begin within 100 miles (160 kilometers) of one another in the Himalayan highlands near the Tibet–Nepal–India borders (see Figure 8.1).

These rivers, and many of the tributaries that feed them, are actively wearing down the surface of the Himalayas. They carry enormous loads of sediment, especially during the rainy season. Their velocity slows when they reach the lowlands, and much of the sediment settles out as silt. It is then repeatedly picked up and deposited by successive floods. As illustrated in the diagram of the Brahmaputra River in Figure 8.7, the seasonally replenished silt nourishes much of the agricultural production in the densely occupied plains of Bangladesh. The same is true on the Ganga and Indus plains.

ENVIRONMENTAL ISSUES

South Asia has been occupied by people for millennia, but as recently as 1700 (just prior to British colonization), population density and human environmental impacts were much lower than today (Figure 8.8a). By the beginning of the twenty-first century, population density and human impacts had vastly grown (Figure 8.8b). Today, South Asia has a range of serious environmental problems, including potentially devastating vulnerability to global warming, deforestation, water shortages, water pollution, and industrial pollution. However, the region is implementing a broad array of solutions, some of which mix modern environmental research and technologies with South Asia's deep cultural roots.

South Asia's Vulnerability to Global Warming

In South Asia, global warming puts more lives at risk due to sea level rise, water shortage, and crop failure than in any other region in the world.

Sea level rise. Bangladesh, where tens of millions of poor farmers and fishers live near sea level, is more vulnerable to *sea level rise* (see pages 17–20 in Chapter 1) than any other

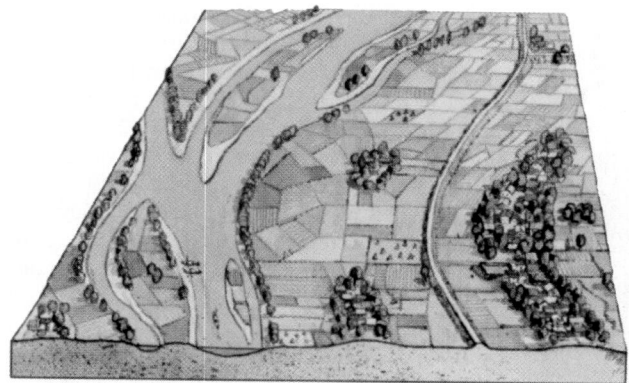

(a) **Premonsoon Stage:** The river flows in multiple channels across the flat plain.

(b) **Peak Flood Stage:** During peak flood stage, the great volume of water overflows the banks and spreads across fields, towns, and roads. It carves new channels, leaving some places cut off from the mainland.

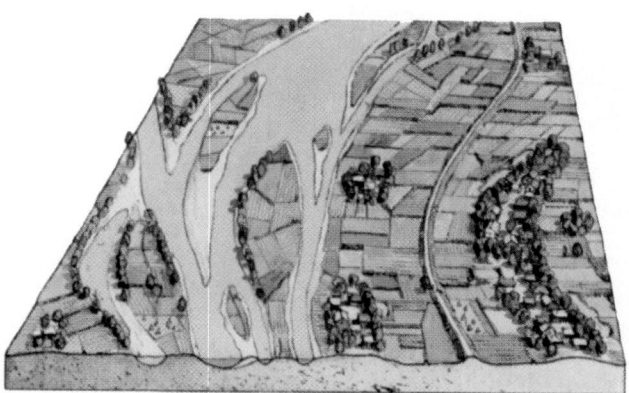

(c) **Postmonsoon Stage:** The river returns to its banks, but some of the new channels persist, changing the lay of the land. As the river recedes, it leaves behind silt and algae that nourish the soil. New ponds and lakes form and fill with fish.

Figure 8.7 The Brahmaputra River in Bangladesh at various seasonal stages. People who live along the river have learned to adapt their farms to a changing landscape. Along much of the river, farmers are able to produce rice and vegetables nearly year-round. [Adapted from *National Geographic* (June 1993): 125.]

country on earth. With 149 million people already squeezed into a country the size of Iowa, as many as 17 million people might have to find new homes if sea levels rise by five feet. The greatest economic impacts of sea level rise, however, could come from the submergence of parts of South Asia's wealthiest city, Mumbai, and the many other large, coastal urban centers.

> Bangladesh is more vulnerable to sea level rise than any other country on earth.

Water shortage. South Asia's three largest rivers are fed by glaciers high in the Himalayas that are now melting fast due to global warming. Seven hundred three million people, almost half of the region's population, depend on these rivers for irrigation and drinking water. The immediate effect of glacial melting may be flooding. However, as glaciers shrink and provide less and less water to rivers each year during the dry season, the long-term effect will be water shortage. Unless dramatic action against global warming is successful, most Himalayan glaciers will disappear within 50 years. As a result, South Asia's largest rivers may run dry during the winter monsoon.

Crop disruption. Shifting rainfall patterns and rising temperatures could also result in lower yields for some of South Asia's most important food crops such as rice, maize, and millet. Avoiding these effects may require the development of new crop varieties and better irrigation techniques.

Responses to global warming. South Asia is pioneering some innovative responses to global warming. India is home to the world's largest producer of plug-in electric cars, the Reva company, which sells its cars in India and the UK for about U.S.$8000. Even factoring in emissions from the plants that generate the electricity used to charge the cars, electric cars contribute substantially less to global warming than do gasoline- or diesel-powered cars. Many South Asian countries are also investing in solar energy, which could work better there than in many other regions because of the many cloudless days and proximity to the equator. Wind power is another area of potential growth. India is already the world's fourth largest user of wind power, though wind supplies only around 3 percent of the country's electricity.

With regard to water, the melting of Himalayan glaciers may require both water conservation and increased water storage to last through the dry winter monsoon. Fortunately, South Asia has much experience with such technologies. India and Pakistan have both pioneered methods for increasing the rate at which water deposited during the summer monsoon percolates through the soil and into underground aquifers. More water is thus available for irrigation during the dry season. Nevertheless, drip irrigation technology (see page 211 in Chapter 6) would aid water conservation in agriculture, which uses the most water of any human activity. All South Asian countries have experience with drip irrigation, though its relatively high cost has held back widespread implementation.

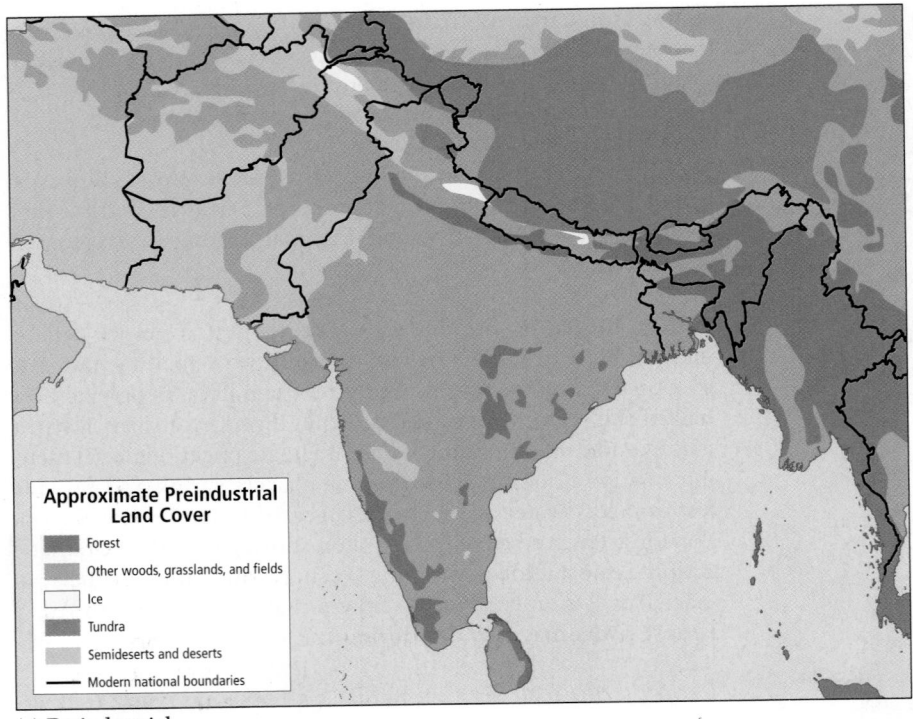

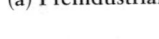

(a) Preindustrial

Figure 8.8 Land cover (preindustrial) and human impact (2002). South Asia has been occupied by humans for thousands of years. Part **(a)** shows land cover in the preindustrial era, when much of the forest had already been removed. By 2002 **(b)**, the human impact on land, water, and air was extensive throughout the region, but variable in intensity. Remote and sparsely occupied parts of Afghanistan and Pakistan were impacted by war. [Adapted from *United Nations Environment Programme, 2002, 2003, 2004, 2005, 2006* (New York: United Nations Development Programme), at http://maps.grida.no/go/graphic/human_impact_year_1700_approximately and http://maps.grida.no/go/graphic/human_impact_year_2002.]

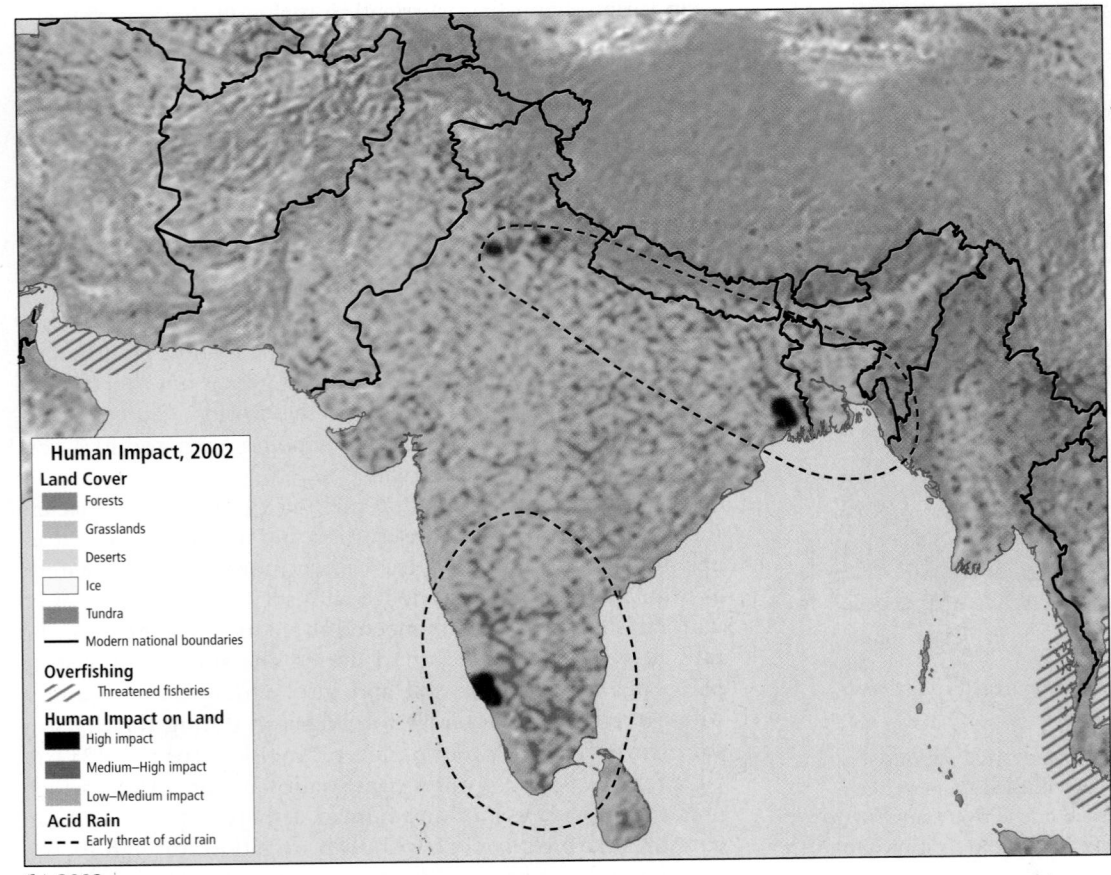

(b) 2002

Deforestation

Deforestation has been going on in South Asia since the first agricultural civilizations developed more than five thousand years ago. Ecological historians have shown that, as the forests vanished, the western regions of the subcontinent (from India to Afghanistan) became drier and drier. The pace of deforestation increased dramatically over the past two hundred years. By the mid-nineteenth century, perhaps a million trees a year were felled for use in building railroads alone.

In the twenty-first century, South Asia's forests are still shrinking due to commercial logging and expanding village populations. Many of South Asia's remaining forests are in mountainous or hilly areas, where forest clearing dramatically increases erosion during the rainy season. One result is massive landslides that can destroy villages and close roads. With fewer trees and less soil to hold onto the water, rivers and streams become clogged with mud and debris. The effects can reach far downstream so that increased flooding in the plains of Bangladesh is linked to deforestation in the Himalayas.

Unlike China and many other nations facing similar problems, the countries of South Asia have a healthy and vibrant culture of environmental activism that has brought the consequences of deforestation to the attention of the public. In 1973, for example, in the Himalayan district of Uttar Pradesh, India, a sporting-goods manufacturer planned to cut down a grove of ash trees so that his factory, in the distant city of Allahabad, could use the wood to make tennis racquets. The trees were sacred to nearby villagers, however, and when their protests were ignored, a group of local women took dramatic action. When the loggers came, they found the women hugging the trees and refusing to let go until the manufacturer decided to locate another grove.

The women's action grew into the Chipko (literally, "hugging") or Social Forestry Movement. The movement has spread to other forest areas, slowing deforestation and increasing ecological awareness. Proponents of the movement argue that management of forest resources should be turned over to local communities. They say that people living at the edges of forests possess complex local knowledge of those ecosystems gained over generations—knowledge about which plants are useful for building materials, food, medicinal uses, and fuel. These people have the incentive to manage forests carefully because they want their descendants to benefit from forests for generations to come. These and other activist movements focused on saving forests are a reaction to a pattern found throughout South Asia, in which the resources of rural areas are channeled to urban industries without consideration of the needs of local rural people. ▶️

Despite the success of the Chipko movement, it is often difficult to convince impoverished people to conserve forest resources. Growing local populations contribute to deforestation as people gather firewood for cooking and harvest tree foliage as fodder for their animals. Often their need for income leads them to collaborate with poachers of rare forest plants and animals. Moreover, the powerful industrial and government interests that currently control many forest reserves are not likely to yield control to local people easily. Despite these problems, local residents are gaining increasing control of some forest reserves in India, Nepal, Sri Lanka, and Pakistan.

Case Study: Nature Preserves in the Nilgiri Hills

The Mudumalai Wildlife Sanctuary and neighboring national parks in the Nilgiri Hills (part of the Western Ghats) harbor some of the last scraps of forest in southern India. Here, in an area of about 600 square miles, live a few of India's last wild tigers (Figure 8.9) and a dozen or more other rare species, such as sloth bears and barking deer. ▶️

Figure 8.9 Endangered wildlife. This photo was taken during an unexpected encounter with a Bengal tiger in the Mudumalai Wildlife Sanctuary in the Nilgiri Hills in southwestern India. The Bengal tiger is a rare and endangered species. [Alex Pulsipher.]

Even much smaller forest reserves play important roles in conservation. At 287 acres, Longwood Shola is a tiny remnant of the ancient tropical forests that once covered the Nilgiris. This patch of forest protects three streams that supply fresh water to 16 downstream villages, and it harbors 13 mammal species, 52 bird species (including an ancestor of the domestic chicken), and 118 plant species, many of them found only in the Nilgiris. Longwood is managed by two local naturalists, members of ethnic minorities native to the Nilgiri Hills. Among the projects the naturalists have initiated is a reforestation effort for which they themselves are growing the seedlings from local native species.

Phillip Mulley, a naturalist, Christian minister, and leader of the Badaga ethnic group, points out that the indigenous peoples of the Nilgiris must now compete for space with a growing tourist industry (1.7 million visitors in 2005). In addition, huge tea plantations were recently cut out of forestlands by the state government to provide employment for Tamil refugees from the ongoing conflict in Sri Lanka. So while the forestry department and citizen naturalists are trying to preserve forestlands, the social welfare department, faced with a huge refugee population, is cutting them down.

Sources: Lydia and Alex Pulsipher's field notes, Nilgiri Hills, June 2000; *Tamil Nadu Human Development Report*, 2003, at http://data.undp.org.in/shdr/tn/TN%20HDR%20final.pdf.

Water

One of the most controversial environmental issues in South Asia today is the use of water. South Asia has more than 20 percent of the world's population, but only 4 percent of its fresh water. It is not surprising, then, that India and Bangladesh have disputes over access to the waters of the Ganga River.

Conflicts over Ganga River water.

In recent years, during the dry season, India has diverted 60 percent of the Ganga's flow to Kolkata to flush out channels where silt is accumulating and hampering river traffic. However, the diversions deprive Bangladesh of normal freshwater flow. Less water in the Ganga delta allows salt water from the Bay of Bengal to penetrate inland, ruining agricultural fields. The diversion has also caused major alterations in Bangladesh's coastline, damaging its small-scale fishing industry. Thus, to serve the needs of Kolkata's 15 million people, the livelihoods of 40 million rural Bangladeshis have been put at risk, triggering protests in Bangladesh.

In the late 1990s, India signed a treaty promising a fairer distribution of water, but as of 2008, Bangladesh was still receiving a considerably reduced flow. As with other major environmental problems, a solution has been hard to achieve

> South Asia has more than 20 percent of the world's population, but only 4 percent of its fresh water.

through established democratic processes. Not only is the adversely affected population poor and rural, but it is located in a different region—in this case, in a different country—from the politicians and bureaucrats who are making decisions.

Similar water use conflicts occur between states within India and between the wealthier and poorer sectors of the population. ▶ For example, the state of Haryana's diversion of water from the Ganga deprives farmers downstream in Uttar Pradesh of the means to irrigate their crops. And in Delhi, just 17 five-star hotels use about 210,000 gallons (800,000 liters) of water daily, which would be enough to serve the needs of 1.3 million slum dwellers.

Water purity of the Ganga River.

Water purity has become an issue in historic religious pilgrimage towns such as Varanasi, where each year millions of Hindus come to die, be cremated, and have their ashes scattered over the Ganga River. As the number of such final pilgrimages has increased, wood for cremation fires has become scarce. As a result, incompletely cremated bodies are being dumped into the river, where they pollute water used for drinking, cooking, and bathing. In an attempt to deal with this problem, the government recently installed an electric crematorium on the riverbank. It is attracting quite a bit of business, as a cremation here costs 30 times less than a traditional funeral pyre.

Of greater concern now is the large amount of industrial waste and sewage dumped into the river (Figure 8.10). Most sewage enters the river in raw form because Varanasi's sewage system (built by the British early in the twentieth century) long ago exceeded its capacity. Pumps have been installed to move the sewage up to a new and expensive sewage processing plant, but the plant is so overwhelmed by the volume of water during the rainy season that it can process only a small fraction of the city's sewage.

Veer Bhadra Mishra, a Brahmin priest and professor of hydraulic engineering at Banaras Hindu University in Varanasi, is on a mission to clean up the Ganga using unconventional methods. He is working with engineers from the United States to build a series of processing ponds that will use India's heat and monsoon rains to clean the river at half the cost of other methods. In addition, he preaches a contemporary religious message to the thousands who visit his temple on a bank of the sacred Ganga. The belief that the Ganga is a goddess who purifies all she touches leads many Hindus to think that it is impossible to damage this magnificent river. Mishra reminds them that because the Ganga is their symbolic mother, it would be a travesty to smear her with sewage and industrial waste.

Industrial Pollution

In many parts of South Asia, the air as well as the water is endangered by industrial activity. Emissions from vehicles and coal-burning power plants are so bad that breathing Delhi's air is equivalent to smoking 20 cigarettes a day. The acid rain caused by industries up and down the Yamuna and Ganga

Figure 8.10 Pollution of the Ganga. The city of Varanasi dumps waste into the Ganga River through many sewage pipes like this one. The Ganga is India's most sacred yet most polluted river. [AP Photo/John McConnico.]

rivers is destroying good farmland and such great monuments as the Taj Mahal.

M. C. Mehta, a Delhi-based lawyer, became an environmental activist partly in response to the condition of the Taj Mahal. For more than 20 years, he has successfully promoted environmental legislation that has removed hundreds of the most polluting factories from India's river valleys. His efforts are also a response to a horrible event that took place in central India in 1984. At that time, an explosion at a pesticide plant in Bhopal produced a gas cloud that killed at least 15,000 people and severely damaged the lungs of 50,000 more. The explosion was largely the result of negligence on the part of the U.S.-based Union Carbide Corporation, which owned the plant, and the local Indian employees who ran it. In response to the tragedy, the Indian government launched an ambitious campaign to clean up poorly regulated factories.

 Ten Key Aspects of South Asia

- Precipitation is especially intense in the foothills of the Himalayas, which hold the world record for annual rainfall—about 35 feet, even though there is no rain at all for half the year.
- South Asia's three largest rivers are fed by glaciers high in the Himalayas that are now melting fast due to global warming.
- The British controlled most of South Asia from the 1830s through 1947, profoundly influencing the region politically, socially, and economically.
- Many rural people are pushed into urban migration by agricultural modernization that can make farming unaffordable for the poor.

- A significant gender imbalance is occurring in South Asian populations due to cultural customs that make sons more likely to contribute to a family's wealth than daughters.
- Less than 5 percent of registered marriages among Indian Hindus cross caste lines.
- Mumbai's annual per capita GDP is three times that of the next wealthiest city, India's capital, Delhi.
- Seventy percent of South Asians live in rural villages.
- To take advantage of India's large, college-educated, low-cost workforces, companies in North America and Europe are "outsourcing" an increasing number of jobs to Indian cities like Bangalore, Mumbai, and Ahmadabad.
- Many violent conflicts in South Asia center around the right of people to run their own affairs through local democratic processes.

HUMAN PATTERNS OVER TIME

A variety of groups have migrated into South Asia over the millennia, many of them as invaders who conquered peoples already there. Despite much blending over time, the continued co-existence and interaction of many of these groups make South Asia both a richly diverse and extremely contentious place.

The Indus Valley Civilization

There are indications of early humans in South Asia as long as 200,000 years ago, but the first evidence of modern humans in the region is about 38,000 years old. The first large agricultural communities, known as the **Indus Valley civilization** (or **Harappa** culture), appeared about 4500 years ago along the Indus River in modern-day Pakistan and northwest India. Their architecture and urban design were quite advanced for the time. Homes featured piped water and sewage disposal. Towns were well planned, with wide, tree-lined boulevards laid out in a grid. There is also evidence of a trade network that extended to Mesopotamia and eastern Africa.

Much of the Indus Valley civilization's agricultural system survives to this day, including techniques for storing monsoon rainfall to be used for irrigation in dry times; methods for cultivating wheat, barley, and other crops adapted to arid conditions; and the use of wooden plows drawn by oxen. Vestiges of its language, and possibly biological traits as well, survive today among the Dravidian peoples of southern India, who originally migrated from the Indus region.

The reasons for the decline of the Indus Valley civilization after about 800 years (3700 years ago) are debated. Some scholars believe that complex geologic and ecological changes brought about a gradual demise. Others argue that foreign invaders brought a swift collapse. In any case, most agree that aspects of Harappa culture blended with subsequent foreign influences to form the foundation of many modern South Asian cultural traditions.

A Series of Invasions

The first recorded invaders of South Asia, the Aryans, moved into the rich Indus Valley and Punjab from Central and Southwest Asia 3500 years ago. Many scholars believe that the Aryans, in conjunction with the Harappa and other indigenous cultures, instituted some of the early elements of classi-cal Hinduism, the major religion of India. One of those elements was the still influential caste system (discussed on pages 306–308), which divides society into hereditary hierarchical categories.

In addition to the Aryans, invaders from greater Central Asia included the Persians, the armies of the Greek general Alexander the Great, and numerous Turkic and Mongolian peoples. Starting about 1000 years ago, Arab traders and religious mystics came by sea to the coasts of southwestern India and Sri Lanka, introducing Islam into the region.

The invasion in 1526 by the **Mughals**, a group of Turkic Persian people from Central Asia, intensified the spread of Islam. The Mughals reached the height of their power and influence in the seventeenth century, controlling the north-central plains of South Asia. The last great Mughal ruler (Aurangzeb) died in 1707, but the legacy of the Mughals has remained. One aspect of that legacy is the more than 420 million Muslims now living in South Asia. The Mughals also left a unique heritage of architecture, art, and literature that includes the Taj Mahal, the fortress at Agra (Figure 8.11), miniature painting, and the tradition of lyric poetry. The Mughals helped to produce the Hindi language, which became the language of trade of the northern subcontinent and is still used by more than 400 million people.

After the end of Mughal rule, a number of regional states and kingdoms rose and competed with one another (Figure 8.12). The absence of one strong power created an opening for yet another invasion. Several European trading companies competed to gain a foothold in the region. Of these, Britain's was the most successful. By 1857, the British East India Company, acting as an extension of the British government, had become the dominant power in the region.

Figure 8.11 The great fortress at Agra. Built by the Mughals in 1565–1571, the fortress has walls 72 feet (22 meters) high enclosing an area about 1.5 miles (2.4 kilometers) in circumference. [J. H. C. Wilson/Robert Harding Picture Library.]

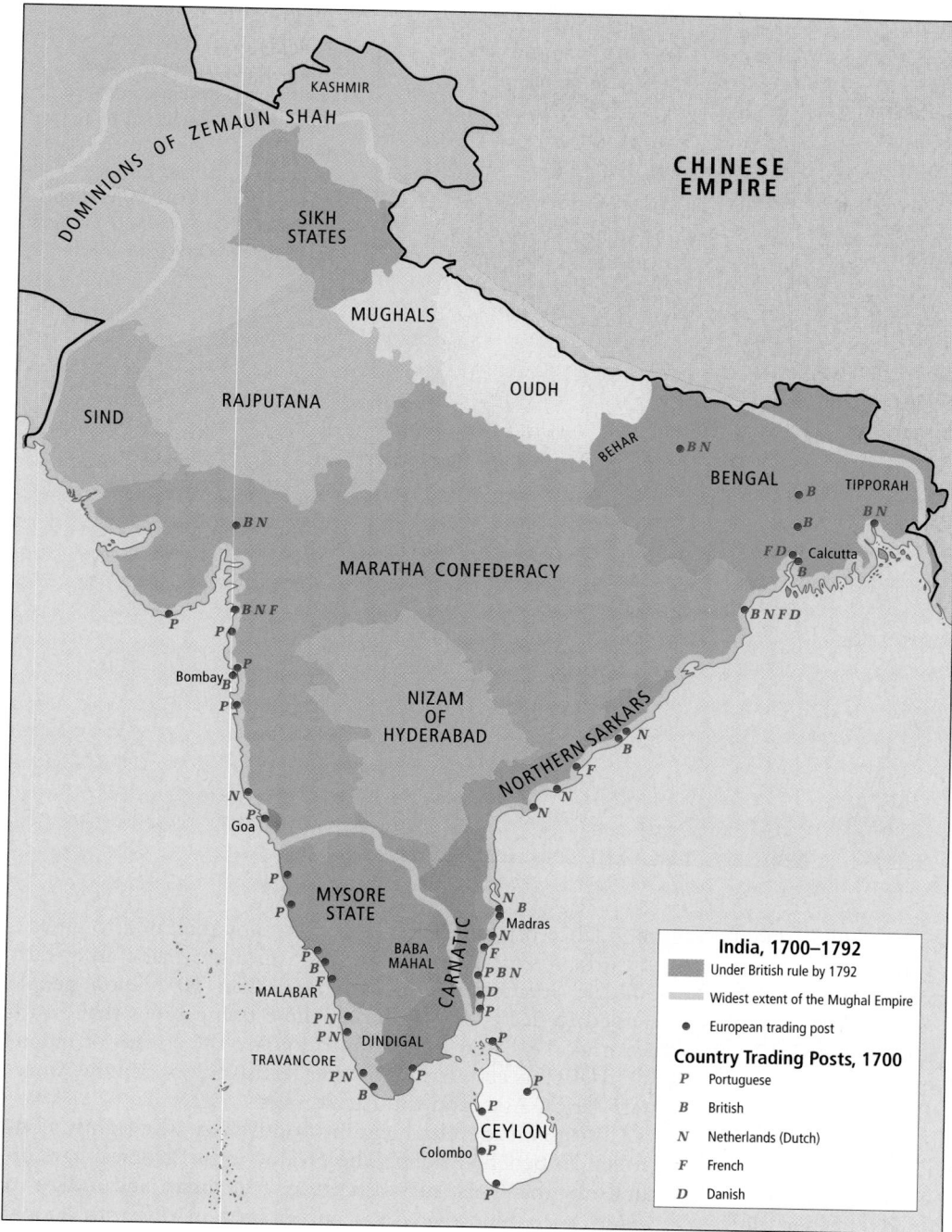

Figure 8.12 Precolonial South Asia. By 1700, several European nations had established trading posts along the coast of India and Ceylon. With the Mughal empire in decline at that time, a number of regional states emerged. Rivalry between these states paved the way for British conquest by the end of the eighteenth century. [Adapted from William R. Shepherd, *The Historical Atlas* (New York: Henry Holt, 1923–1926), p. 137; and Gordon Johnson, *Cultural Atlas of India* (New York: Facts on File, 1996), p. 111.]

The Legacies of Colonial Rule

The British controlled most of South Asia from the 1830s through 1947 (Figure 8.13), profoundly influencing the region politically, socially, and economically. Even areas not directly ruled by the British felt the influence of colonialism. Afghanistan repelled British attempts at military conquest, but the British continued to meddle there, attempting to make Afghanistan a "buffer state" between British India and Russia's expanding empire. Nepal remained only nominally independent during the colonial period, and Bhutan became a protectorate of the British Indian government.

The deindustrialization of South Asia. As in their other colonies, the British used South Asia's resources primarily for their own benefit, often with disastrous results for South Asians. One example was the fate of the textile industry in Bengal (modern-day Bangladesh and the Indian state of West Bengal).

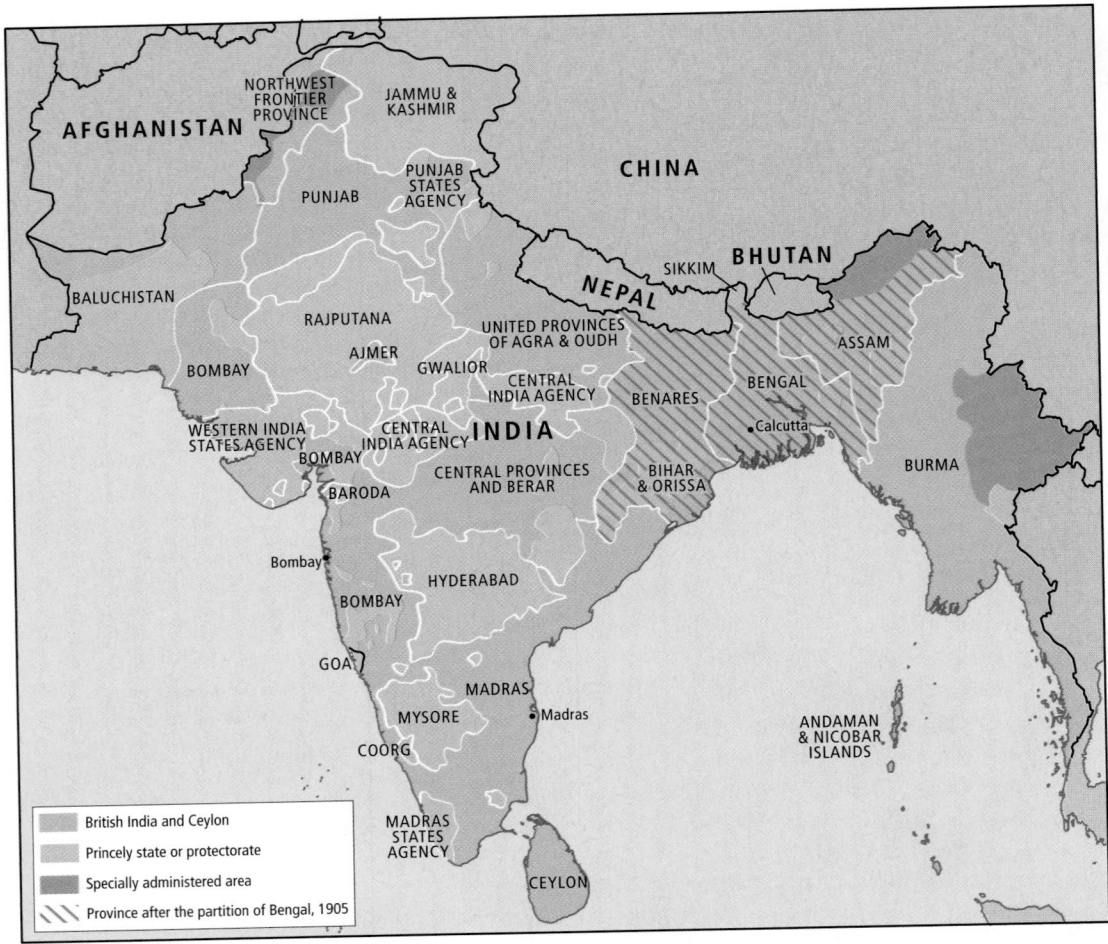

Figure 8.13 British Indian Empire, 1860–1920. After winning control of much of South Asia, Britain controlled lands from Baluchistan to Burma, including Ceylon and the islands between India and Burma. [Adapted from Gordon Johnson, *Cultural Atlas of India* (New York: Facts on File, 1996), p. 158.]

Bengali weavers, long known for their high-quality muslin cotton cloth, initially benefited from the greater access British traders gave them to overseas markets in Asia, the Americas, and Europe. By 1750, South Asia had an advanced manufacturing economy, which produced 12 to 14 times more cotton cloth than Britain alone and more than all of Europe combined. However, during the second half of the eighteenth century, Britain's own highly mechanized textile industry developed cheaper cloth that replaced Bengali muslin, first in world markets and eventually throughout South Asia. The British textile industry was further aided by the British East India Company, which began severely punishing Bengalis who continued to run their own looms. Thus, as one British colonial official put it, while the mills of Yorkshire prospered, "the bones of Bengali weavers bleached the plains of India."

> As one British colonial official put it, while the mills of Yorkshire prospered, "the bones of Bengali weavers bleached the plains of India."

Many people who were pushed out of their traditional livelihood in textile manufacturing were compelled to find work as landless laborers. But rural South Asia already had an abundance of agricultural labor, so many migrated to emerging urban centers. In the 1830s, a drought made an already difficult situation worse, and more than 10 million people starved to death. It was during these trying times that South Asian workers were pressed into joining the stream of indentured laborers emigrating to other British colonies in the Americas, Africa, Asia, and the Pacific. 📹

Economic development in South Asia was tightly controlled to benefit Britain. The production of tropical agricultural raw materials, such as cotton, tea, sugar, and indigo (a widely used blue dye), was encouraged in order to supply Britain and its other colonies. Industrial development, which might have competed with Britain's own industries, was discouraged.

Nonetheless, British rule did bring some benefits. Trade with the rest of Britain's empire brought prosperity to a few areas, especially the large British-built cities on the coast, such as Bombay (now Mumbai), Calcutta (Kolkata), and Madras (Chennai). The British built a railroad system that boosted trade within South Asia and greatly eased the burden of personal transport. In addition, English became a common language for South Asians of widely differing backgrounds, assisting both trade and cross-cultural understanding.

Democratic and administrative legacies. Contemporary South Asian governments retain institutions put in place by the British to administer their vast empire. These governments inherited many of the shortcomings of their colonial forebears, such as highly bureaucratic procedures, resistance to change, and a tendency to remain aloof from the people they govern. Nonetheless, the governments have proved functional in most cases. In particular, though democratic government was not instituted on a large scale until the final days of the empire, it has given people an outlet for voicing their concerns and has enabled many peaceful transitions of elected governments since 1947.

Independence and partition. The tremendous changes brought by the British inspired many resistance movements among South Asians. Some of these were militant movements intent on pushing the British out by force, while others focused on political agitation for greater democracy as a route to South Asian political independence. Although both types of action were brutally repressed, the democracy movements achieved world-wide attention and, after decades of struggle, success.

In the early twentieth century Mohandas Gandhi, a young lawyer from Gujarat, emerged as a central political leader of South Asia's independence movement. His tactics of **civil disobedience** focused on the non-violent violation of laws imposed by the British that discriminated against South Asians. Gathering a large group of peaceful protesters, he would notify the government that the group was about to break a discriminatory law—for example, the law that made it illegal for South Asians to produce salt, thus requiring its importation from Britain. If the authorities ignored the act, the demonstrators would have made their point and the law would be rendered moot. On the other hand, if the government used force against the peaceful demonstrators, it would lose the respect of the masses. Throughout the 1930s and 1940s, this technique was used to slowly but surely undermine British authority across South Asia. The hunger strikes described in the chapter-opening vignette reflect Gandhi's legacy of non-violent political protest.

In 1947, independence was granted to British India, which was divided into two independent countries: predominantly Hindu India and Muslim Pakistan (Figure 8.14) (Afghanistan, Bhutan, and Nepal were never officially British colonies; Ceylon [now Sri Lanka] became independent in 1948). The **Partition**, which Gandhi greatly lamented, was perhaps the most enduring and damaging outcome of colonial rule.

The idea of two nations was first suggested by Muslim political leaders concerned about the fate of a minority Muslim population in a united India with a Hindu majority. Though partition was highly controversial, it became part of the independence agreement between the British and the Indian National Congress (India's principal nationalist party). Northwestern and northeastern India, where the population was predominantly Muslim, became a single country consisting of two parts, known as West and East Pakistan, separated

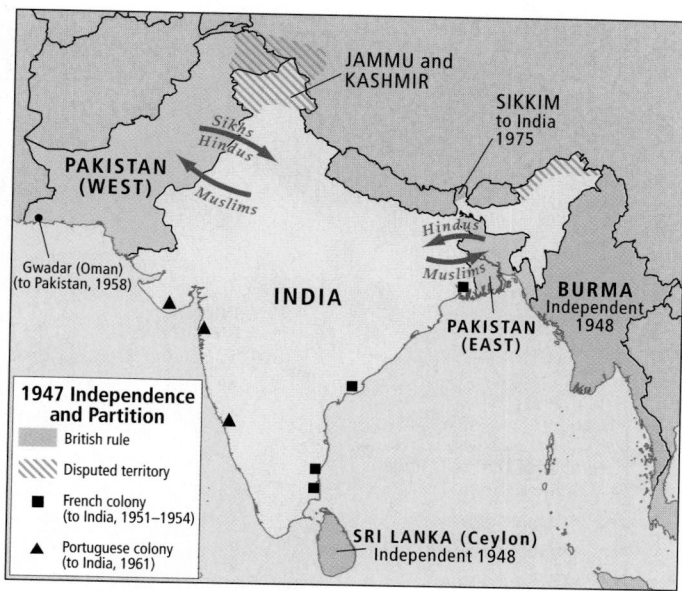

Figure 8.14 Independence and Partition. India became independent of Britain in 1947. By 1948, the old territory of British India was partitioned into the independent states of India and Pakistan (East and West). The Jammu and Kashmir region was contested space, and remains so today. Sikkim went to India, and both Burma and Sri Lanka became independent. Following additional civil strife, East Pakistan became the independent country of Bangladesh in 1971. [Adapted from *National Geographic* (May 1997): 18.]

by India. Although both India and Pakistan maintained secular constitutions, with no official religious affiliation, the general understanding was that Pakistan would have a Muslim majority and India a Hindu majority. Fearing that they would be persecuted if they did not move, more than 7 million Hindus and Sikhs migrated to India from their ancestral homes in what had become Pakistan. A similar number of Muslims left their homes in India for Pakistan. In the process, families and communities were divided, looting and rape were widespread, and between 1 and 3.4 million people were killed.

Partition was the tragic culmination of "divide-and-rule" tactics the British used throughout the colonial era (see Chapter 7, pages 271–274). This approach heightened tensions between South Asian Muslims and Hindus, thus creating a role for the British as indispensable and benevolent mediators. The legacy of these tactics includes not only the Partition, but also the repeated wars and skirmishes, strained relations, and ongoing arms race between India and Pakistan.

After independence. In the more than 60 years since the departure of the British, South Asians have experienced both progress and setbacks. Democracy has expanded steadily, albeit somewhat slowly. India is now the world's most populous democracy and is gradually dismantling age-old traditions that hold back poor, low-caste Hindus, women, and other disadvantaged groups.

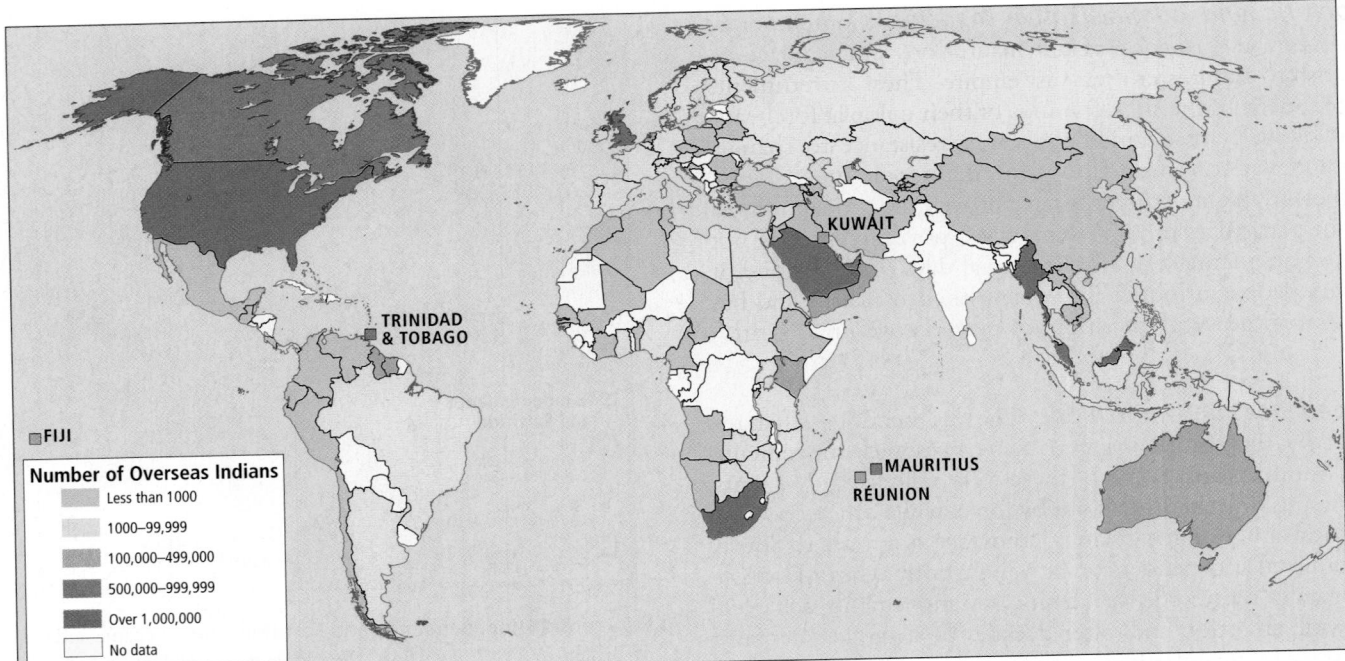

Figure 8.15 Indian diaspora communities around the world. This map makes clear that people of South Asian origin now live virtually everywhere on earth. They engage in many occupations, but are found particularly in the professions of technology development, commerce, and agriculture. [Data from *Report of the High Level Committee on The Indian Diaspora*, at http://indiandiaspora.nic.in/contents.htm, specifically "Estimated size of overseas community: Countrywise," at http://indiandiaspora.nic.in/diasporapdf/part1-est.pdf.]

Changes in agriculture have brought prosperity to some South Asians, and forced others to migrate to the cities. In most countries, urban-based industrial and service economies now constitute a far larger share of GDP than agriculture, with the information technology (IT) sector growing especially fast in India. Nonetheless, as of 2007, the South Asian countries had a collective annual GDP per capita of only about U.S.$3500, adjusted for PPP (only sub-Saharan Africa's was lower). Poverty persists for the majority in both rural and urban areas throughout the region.

Although recent years have experienced rapid economic growth throughout South Asia, many highly skilled people have left the region to find better jobs elsewhere (Figure 8.15), contributing to an ongoing *brain drain* (See Chapter 3, page 115). Some who leave are fleeing South Asia's many violent conflicts, one of which has resulted in the emergence of a new nation: in 1971, a bloody civil war led to the division of Pakistan into Bangladesh (formerly East Pakistan) and Pakistan (formerly West Pakistan).

POPULATION PATTERNS

South Asia is one of the most densely populated regions in the world (Figure 8.16). The region already has more people (1.54 billion) than China (1.33 billion), which has almost twice the land area of South Asia. By 2050, India alone is expected to overtake China.

Population Growth Factors in South Asia

Many factors have encouraged South Asia's high population growth over the past century. In rural areas, there is a strong incentive for large families because children contribute their labor to farming, and hence contribute to family income, from an early age. ◼▸ Large families are also a response to poor access to health care, which means that many babies die in infancy. As a result, couples often choose to have many children to ensure that at least some will survive to adulthood. A further incentive for large families is the fact that children are the only retirement plan that most South Asians will ever have. Hence, it is imperative that at least one child, and preferably more, reach maturity and be able to care for their elderly parents. Over the short term, the youthfulness of South Asia's population will ensure growth even if radical population reduction efforts are undertaken.

The long-term picture shows population growth slowing. While South Asian nations have been trying to reduce population growth since 1952, this goal was not achieved until the 1970s. Today, population growth rates are still declining in most countries. India spends more than one billion dollars a year on population control programs. Fertility rates (the average number of children per woman) have indeed declined significantly everywhere in South Asia except for Afghanistan. (Figure 8.17). By some estimates, the region's population could level off at around 2.4 billion in 2050.

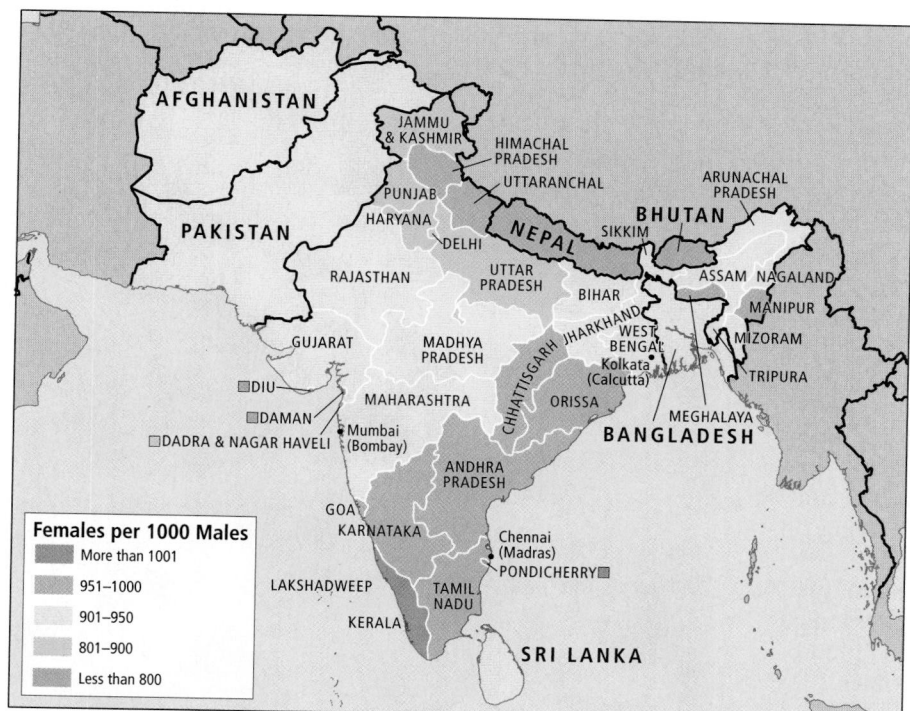

Figure 8.19 Ratio of females to males in the South Asia region, 2000–2001. The green, khaki, yellow, and light orange colors show where females are unusually underrepresented in the population. Only Kerala has a somewhat larger number of females than males. The text offers some explanations for these patterns. [Adapted from "Sex ratio map of India—2001," at http://www.mapsofindia.com/census2001/sexratio/sexratio-india.htm.]

Females per 1000 Males
- More than 1001
- 951–1000
- 901–950
- 801–900
- Less than 800

South Asia's current urban population of around 460 million could expand to as much as 712 million by 2025.

Mumbai has 19 million; Kolkata, 15 million; Delhi, 16 million; and Dhaka, 13 million. All of these cities are growing fast, and South Asia's current urban population of around 460 million could expand to as much as 712 million by 2025.

Many middle class South Asians move to cities for education, training, or business opportunities. Meanwhile, many poorer rural people are pushed into urban migration by agricultural modernization that can make farming unaffordable. Other urban migrants are refugees who have left drought-stricken or flooded landscapes. As much as one-quarter of South Asia's urban population is extremely poor and forced to live in vast slums where living standards may be only slightly better than in the most impoverished rural areas.

Vignette One consequence of South Asia's rapid urban growth is a lack of affordable housing. Many people simply live on the streets. In a National Public Radio interview several years ago, an Indian journalist asked a bicycle rickshaw driver (Figure 8.20) about himself as he peddled her through Delhi. He replied that his belongings—a second set of clothes, a bowl, and a sleeping mat—were under the seat where she was sitting. He had come to Delhi from the countryside 14 years before and had never found a home. He knew virtually no one, had few friends and no family, and no one had ever inquired about him before. He worked virtually around the clock and slept here and there for 2 hours at a time.

Source: Gagan Gill, "Weekend Edition," National Public Radio (August 16, 1997).

Case study: wealth and poverty in Mumbai. Bombay is the name by which most Westerners know South Asia's wealthiest city, but since 1995 its official name has been Mumbai, after the Hindu goddess Mumbadevi. Mumbai is the largest deepwater harbor on India's west coast. Its metropolitan area, with close to 20 million people, hosts India's largest stock exchange and the nation's central bank. It pays about a third of the taxes collected in the entire country and brings in nearly

Figure 8.20 A bicycle rickshaw and driver in Delhi. [Lindsay Hebberd/Woodfin Camp & Associates.]

Figure 8.21 Mumbai's Dharavi section, Asia's largest slum. Women who must wash clothes in the watercourse flowing through Dharavi have a difficult time because garbage and raw sewage pollute the water daily. [Steve McCurry/National Geographic Image Collection.]

40 percent of India's trade revenue. Its annual per capita GDP is three times that of India's next wealthiest city, the capital Delhi. Mumbai's wealth also extends into the realm of culture through the city's flourishing creative arts industries, including the internationally known Bollywood film industry. 🎥

Mumbai's wealth is most evident when one looks up at the elegant high-rise condominiums built for the city's rapidly growing middle class. But at street level, the urban landscape is dominated by the large numbers of people living on the sidewalks, in narrow spaces between buildings, or in large, rambling shantytowns. The largest of these communities, Dharavi, houses more than 600,000 people on less than 1 square mile. Because most of those people have no plumbing, Dharavi and slums like it pose major health problems for their residents and those downstream (Figure 8.21). 🎥

And yet, Dharavi is known for its inventive entrepreneurs. One young man, for example, collects and sells aluminum cans that once held ghee, a popular cooking oil in India. He says he makes about 15,000 rupees a month (U.S.$480), nearly twice the salary of the average college professor in India and much more than he made as a truck driver. Hence, despite widespread poverty, Mumbai has more than a few success stories. 🎥

Current Geographic Issues

This section will explore daily life in a village. The cultural characteristics that touch all lives yet vary greatly in practice, such as ethnicity, religious beliefs, social stratification, and gender roles, will also be discussed. Finally, we will take a look at the region's complex and interrelated economic and political issues.

SOCIOCULTURAL ISSUES

Many general aspects of life in South Asia have been described, but it is in more intimate settings that relationships among individuals, and within and between groups, are most easily discerned (Figure 8.22). We begin by visiting a village to absorb something of the rhythm of life.

Village Life

The writer Richard Critchfield, who has studied village life in more than a dozen countries, writes that the village of Joypur (Bangladesh) in the Ganga-Brahmaputra delta is set in "an unexpectedly beautiful land, with a soft languor and gentle rhythm of its own." In the heat of the day, the village is sleepy: naked children play in the dust, women meet to talk softly in the seclusion of courtyards, and chickens peck for seeds. Here and there, under the trees, people ply their various trades: a tinker mends pots; the village tailor sews school uniforms on his hand-cranked sewing machine.

In the early evening, mist rises above the rice paddies and hangs there "like steam over a vat." It is then that the village comes to life, at least for the men. The men and boys return

Figure 8.22 Several generations of a northern Indian family from the village of Dharamkot in Himachal Pradesh province. [David Morgan.]

from the fields, and after a meal in their home courtyards, the men come "to settle in groups before one of the open pavilions in the village center and talk—rich, warm Bengali talk, argumentative and humorous, fervent and excited in gossip, protest and indignation" as they discuss their crops, an upcoming marriage, or national politics.

The vast majority—about 70 percent—of South Asians live in hundreds of thousands of villages like Joypur. Even many of those now living in South Asia's giant cities were born in a village or visit an ancestral rural community on occasion.

Language and Ethnicity

There are many distinct ethnic groups in South Asia, each with its own language or dialect. In India alone, 18 languages are officially recognized, but hundreds of distinct languages actually exist. This complexity results partly from the region's history of multiple invasions from outside. However, long periods of isolation experienced by particular groups were also a factor. In Figure 8.23, number 21 indicates some of the most ancient culture groups in the region. Remnants of Austro-Asiatic languages in these groups that were once more widely distributed were left as isolated pockets when sweeping cultural changes were brought by invaders. These languages are distantly akin to others found farther east in Southeast Asia. The languages represented by numbers 1–12 are linked to various groups of Aryan people who entered South Asia from Central Asia during prehistory.

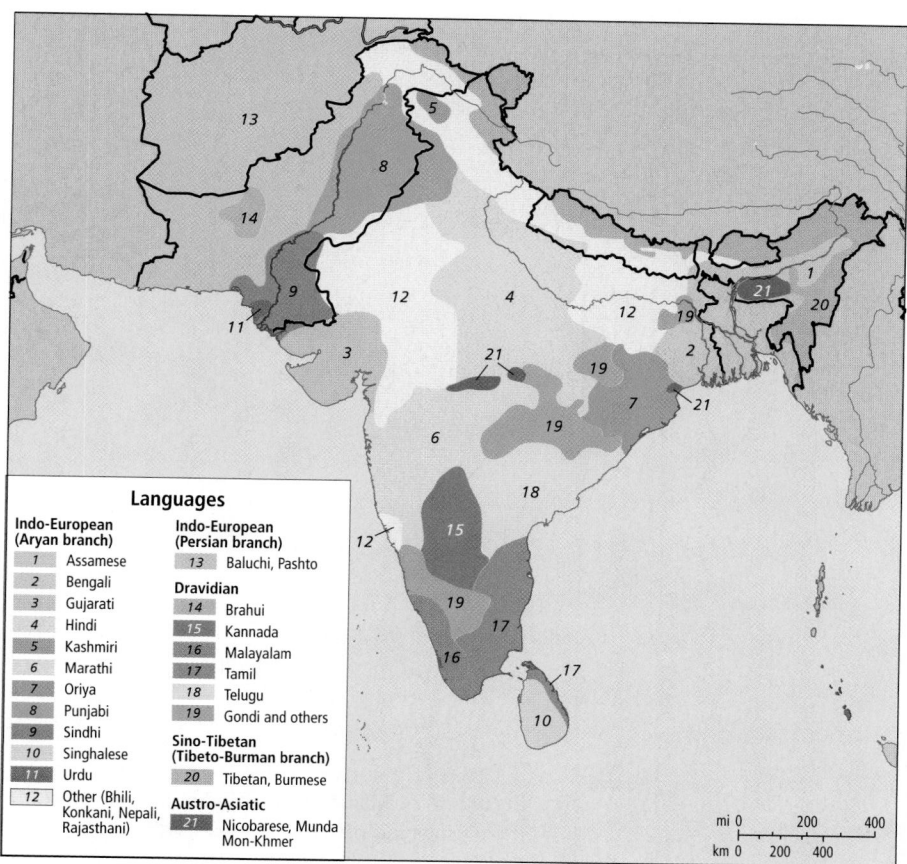

Languages

Indo-European (Aryan branch)
1	Assamese
2	Bengali
3	Gujarati
4	Hindi
5	Kashmiri
6	Marathi
7	Oriya
8	Punjabi
9	Sindhi
10	Singhalese
11	Urdu
12	Other (Bhili, Konkani, Nepali, Rajasthani)

Indo-European (Persian branch)
13	Baluchi, Pashto

Dravidian
14	Brahui
15	Kannada
16	Malayalam
17	Tamil
18	Telugu
19	Gondi and others

Sino-Tibetan (Tibeto-Burman branch)
20	Tibetan, Burmese

Austro-Asiatic
21	Nicobarese, Munda Mon-Khmer

mi 0   200   400
km 0   200   400

Figure 8.23 Major language groups of South Asia. The modern pattern of language distribution in South Asia is a testimony to the fact that this region has long been a cultural crossroads. [Adapted from Alisdair Rogers, ed., *Peoples and Cultures* (New York: Oxford University Press, 1992), p. 204.]

The Dravidian language-culture group, represented by numbers 14–19, is another ancient group that predates the Aryan invasions by a thousand years or more. Today, Dravidian languages are found mostly in southern India, but a small remnant of the extensive Dravidian past can still be found in the Indus Valley in south-central Pakistan.

By the time of British colonization, Hindi—an amalgam of Persian and Sanskrit-based northern Indian languages—was the *lingua franca* of northern India and what is today Pakistan. Today, variants of Hindi serve as national languages for both India and Pakistan, though it is the first, or native, language of only a minority. Only Chinese and English are spoken by more people worldwide. English is a common second language throughout the region. For years, it was the language of the colonial bureaucracy, and it remains a language used at work by professionals. Between 10 and 15 percent of South Asians speak, read, and write English.

Religion: Hinduism and Caste

The main religious traditions of South Asia are Hinduism, Buddhism, Sikhism, Jainism, Islam, and Christianity (Figure 8.24). (For a discussion of Islam and Christianity, refer to Chapter 6, pages 216–217.)

Hinduism. **Hinduism** is a major world religion practiced by approximately 900 million people, 800 million of whom live in India. It is a complex belief system, with roots both in highly localized folk traditions (known as the Little Tradition) as well as in a broader system based on literary texts (known as the Great Tradition).

A major tenet of classical Hindu philosophy, as described in the 4000-year-old scriptures called the Vedas, is that all gods are merely illusory manifestations of the ultimate divinity, which is formless and infinite. Some devout Hindus worship no gods at all, but may engage in meditation, yoga, and other spiritual practices. Such practices are designed to liberate them from illusions and bring them closer to the ultimate reality, described as infinite consciousness. The average person, however, is thought to need the help of personified divinities in the form of gods and goddesses. While some deities are recognized by all Hindus, many are found only in one region, one village, or even one family.

There are some things that almost all Hindus have in common. One is the belief in reincarnation, the idea that any living thing that desires the illusory pleasures (and pains) of life will be reborn after it dies. A reverence for cows, which are seen as only slightly less spiritually advanced than humans, also binds all Hindus together. This attitude, along with the Hindu

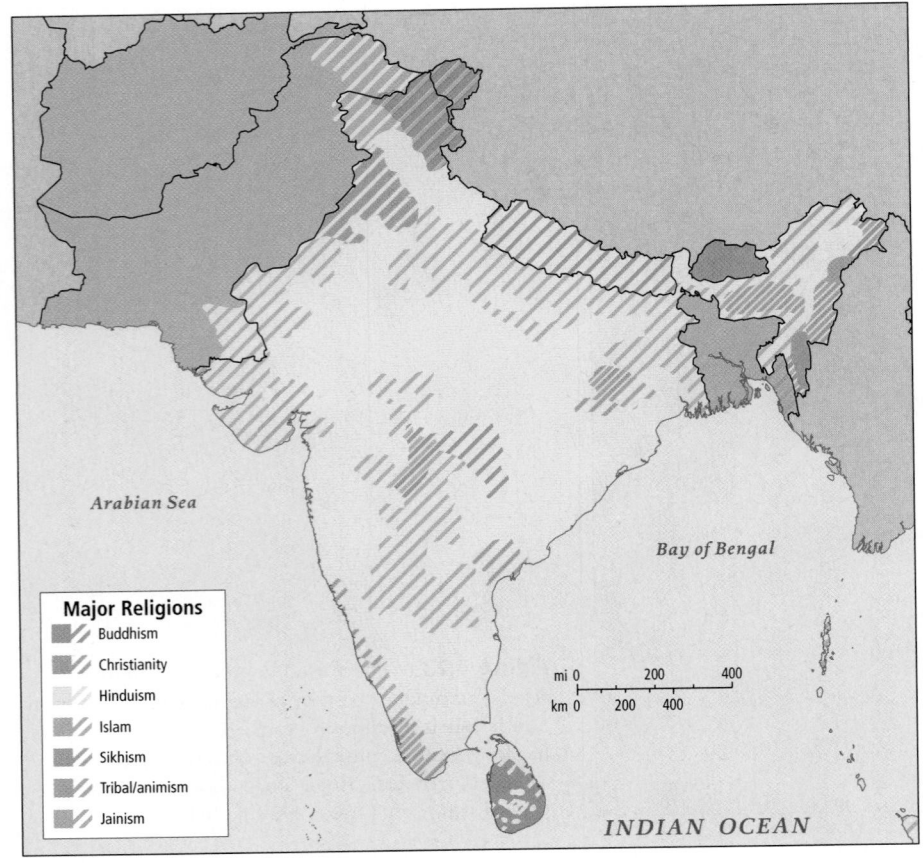

Figure 8.24 Major religions in South Asia. Notice the overlapping patterns in many parts of the region. [Adapted from Gordon Johnson, *Cultural Atlas of India* (New York: Facts on File, 1996), p. 56.]

Major Religions
- Buddhism
- Christianity
- Hinduism
- Islam
- Sikhism
- Tribal/animism
- Jainism

prohibition on eating beef, may stem from the fact that cattle have been tremendously valuable in rural economies as the primary source of transport, field labor, dairy products, fertilizer, and fuel (animal dung is often burned).

Caste. Hinduism includes the **caste** system, a complex and ancient system of dividing society into hereditary hierarchical categories. One is born into a given subcaste, or community (called a *jati*). Traditionally, that happenstance largely defined one's experience for a lifetime—where one would live, where and what one could eat and drink, with whom one would associate, one's marriage partner, and often one's livelihood. The classical caste system has four main divisions or tiers, called *varna*, within which are many hundreds of *jati*s and sub-*jati*s, which vary from place to place.

Brahmins, members of the priestly caste, are the most privileged in ritual status. Thus, they must conform to those behaviors that are considered most ritually pure (for example, strict vegetarianism and abstention from alcohol). Then, in descending rank, are Kshatriyas, who are warriors and rulers; Vaishyas, who are landowning farmers and merchants; and Sudras, who are low-status laborers and artisans. A fifth group, the Dalits—"the oppressed" or untouchables—is actually considered to be so lowly as to have no caste. Dalits perform those tasks that caste Hindus consider the most despicable and ritually polluting: killing animals, tanning hides, cleaning, and disposing of refuse. A sixth group, also outside the caste system, is the Adivasis, who are thought to be descendants of the region's ancient original inhabitants.

Although *jati*s are associated with specific subcategories of occupations, in modern economies this aspect of caste is more symbolic than real. Members of a particular *jati* do, however, follow the same social and cultural customs, dress in a similar manner, speak the same dialect, and tend to live in particular neighborhoods or villages. This spatial separation arises from the higher-caste communities' fears of ritual pollution through physical contact or sharing of water or food with lower castes. When one stays in the familiar space of one's own *jati*, one is enclosed in a comfortable circle of families and friends that becomes a mutual aid society in times of trouble. The social and spatial cohesion of *jati*s helps to explain the persistence of a system that seems to put a burden of shame and poverty on the lower ranks.

It is important to note that caste and class are not the same thing. Class refers to economic status, and there are class differences within caste groups because of differences in wealth. Historically, upper-caste groups (Brahmins and Kshatriyas) owned or controlled most of the land, and lower-caste groups (Sudras) were the laborers, so caste and class tended to coincide. There were many exceptions, however. Today, as a result of expanding educational and economic opportunities, caste and class status are less connected. Some Vaishyas and Sudras have become large landowners and extraordinarily wealthy businesspeople, while some Brahmin families struggle to achieve a middle-class standard of living. By and large, however, Dalits remain very poor.

Caste, politics, and culture. In the twentieth century, Mohandas Gandhi began an organized effort to eliminate discrimination against "untouchables." As a result, India's constitution bans caste discrimination. However, in recognition that caste is still hugely influential in society, upon independence from Britain, India began an affirmative action program. The program reserves a portion of government jobs, places in higher education, and parliamentary seats for Dalits and Adivasis. Together, the two groups now constitute approximately 23 percent of the Indian population and are guaranteed 22.5 percent of government jobs. In 1990, this program was extended to include other socially and educationally "Backward Castes" (such as very disadvantaged *jati*s of the Sudras caste), reserving an additional 27 percent of government jobs for these groups. Reserving half of government jobs in this way has resulted in considerable controversy, however (Figure 8.25).

Figure 8.25 Caste still matters. In May 2006, medical students and many younger doctors across India protested government plans to increase enrollment quotas in elite institutions for lower-status castes and tribal groups. Students argued that admission should be based only on merit. In August 2006, the India Supreme Court ruled against the government (in favor of the students' position). [AP Photo/Bikas Das.]

At the local level, most political parties design their vote-getting strategies to appeal to subcaste loyalties. They often secure the votes of entire *jati* communities with such political favors as new roads, schools, or development projects. These arrangements fly in the face of the official ideologies of the major political parties, which deny any caste loyalties, and of Indian government policies, which actively work to undermine discrimination on the basis of caste. Currently, the role of caste in politics seems to be increasing, as several new political parties that explicitly support the interests of low castes have emerged. This assertion of political rights by low castes has met with a backlash from upper-caste groups, resulting in a number of violent clashes in recent years. Effectively, caste has been woven into the political system in ways that create and maintain tension and conflict.

Nevertheless, among educated people in urban areas, the campaign to eradicate discrimination on the basis of caste has been remarkably successful. Some Dalits throughout the country are now powerful officials. Members of high and low castes now ride city buses side by side, eat together in restaurants, and attend the same schools and universities. For some urban Indians—especially educated professionals who meet in the workplace—caste is disappearing as the crucial factor in finding a marriage partner. Nonetheless, it would be incorrect to conclude that caste is now irrelevant in India. Less than 5 percent of registered marriages cross even *jati* lines. Nearly everyone notices social clues that reveal an individual's caste, and in rural areas, where the majority of Indians still reside, the divisions of caste remain prevalent.

> For some urban Indians—especially educated professionals who meet in the workplace—caste is disappearing as the crucial factor in finding a marriage partner.

Geographic Pattern in Religious Beliefs

There is an uneven and overlapping distribution pattern of religions in South Asia, as Figure 8.24 shows. Hindus, as we have observed, are the most numerous, and they are found mostly in India. Other religions are important in various parts of the region.

The 420 million Muslims in the region form the majority in Afghanistan, Pakistan, Bangladesh, and the Maldives. Muslims are a large and important minority in India, where at 120 million, they form about 12 percent of the population. They live mostly in the northwestern and central Ganga Plain.

Sikhism was founded in the fifteenth century as a challenge to both Hindu and Islamic systems. Sikhs espouse belief in one God, high ethical standards, and meditation. Philosophically, Sikhism accepts the Hindu idea of reincarnation but rejects the idea of caste. (In everyday life, however, caste plays a role in Sikh identity.) The 18 million Sikhs in the region live mainly in Punjab, in northwestern India. Their influence in India is greater than their numbers because many Sikhs hold positions throughout India in the government, military, and police.

Buddhism began about 2600 years ago as a reform and reinterpretation of Hinduism. Its origins are in northern India, where it flourished early in its history before spreading eastward to East and Southeast Asia. Only 1 percent of South Asia's population—about 10 million people—are Buddhists. They are a majority in Bhutan and Sri Lanka.

Jainism, like Buddhism, originated as a reformist movement within Hinduism more than 2000 years ago. Jains (about 6 million people, or 0.6 percent of the region's population) are found mainly in cities and in western India. They are known for their educational achievements, nonviolence, and strict vegetarianism.

The first Christians in the region are thought to have arrived in the far southern Indian state of Kerala with St. Thomas, the Apostle of Christ, in the first century C.E. Today, Christians are an important minority along the west coast of India, on the Deccan Plateau, and in northeastern India.

Animism is practiced throughout South Asia, especially in central and northeastern India, where there are indigenous people whose occupation of the area is so ancient that they are considered aboriginal inhabitants. (For a discussion of animism see Chapter 7, pages 275–276.)

The Hindu-Muslim Relationship

The great Indian independence leaders Mohandas Gandhi and Jawaharlal Nehru both emphasized the common cause that once united Muslim and Hindu Indians: throwing off British colonial rule. Since independence, members of the Muslim upper class have been prominent in Indian national government and the military. Muslim generals have served India willingly, even in its wars with Pakistan after the Partition. Hindus and Muslims often interact amicably, and they occasionally marry each other.

But there is a darker side to the Hindu-Muslim relationship. Throughout South Asia, relations between the region's two largest religious groups are often quite tense. Especially in Indian villages, some Hindus regard Muslims as members of low castes. Religious rules about food are often the source of discord because dietary habits are a primary means of distinguishing caste. Because Hindus forbid killing cows, the consumption of beef and the processing of cow hides into leather is permitted for only the lowest castes. Muslims, on the other hand, run slaughterhouses and tanneries (though discreetly), eat beef, and use cowhide to make shoes and other items. To some Hindus, therefore, some Muslims appear to have offensive customs. Also fueling this perception is the occasional conversion to Islam of entire low-caste or tribal Hindu villages seeking to escape the hardships of being members of a disadvantaged social category.

The Hindu-Muslim relationship is no less complex in Bangladesh. After partition in 1947, some Hindu landowners remained. In some Bangladeshi villages today, while Muslims may be a majority, Hindus are often somewhat wealthier. Although the two groups may coexist amicably for many years, they view themselves differently, and conflict resulting from religious differences—euphemistically called **communal conflict**—can erupt over seemingly trivial events.

Vignette The sociologist Beth Roy, who studies communal conflict in South Asia, recounts an incident in the village of Panipur (a pseudonym), Bangladesh (Figure 8.26). The incident started when a Muslim farmer either carelessly or provocatively allowed one of his cows to graze in the lentil field of a Hindu. The Hindu complained, and when the Muslim reacted complacently, the Hindu seized the offending cow. By nightfall, Hindus had allied themselves with the lentil farmer and Muslims with the owner of

the cow. More Muslims and Hindus converged from the surrounding area, and soon there were thousands of potential combatants lined up facing each other. Fights broke out. The police were called. In the end, a few people died when the police fired into the crowd of rioters. Relationships in the village were deeply affected by the incident. In the words of Roy, the dispute "delineat[ed] distinctions of caste, class, and [religious] culture so complex they intertwinë like columbines climbing on an ancient wall."

Source: Beth Roy, Some Trouble with Cows—Making Sense of Social Conflict *(Berkeley: University of California Press, 1994), pp. 18–19.*

Geographic and Social Patterns in the Status of Women

South Asia has a number of women in very high positions of power, and young women today have greater educational and employment opportunities than those of a generation ago. However, the overall status of women in the region is notably lower than the status of men. Women's status and welfare are lowest in the belt that stretches from the northwest in Afghanistan across Pakistan, western India, and the Ganga Plain into Bangladesh. Women fare better in eastern, central, and southern India and in Sri Lanka, where different marriage and inheritance practices give women greater access to education and resources (Figure 8.27).

Urban women generally enjoy greater individual freedom than rural women, with many now pursuing professional careers and some becoming involved in politics. In rural areas, middle- and upper-caste women are more restricted in their movements than are lower-caste women. Lower-caste women, however, must contend with sexual harassment and exploitation from upper-caste men.

The socioeconomic status of Muslim women is notably lower than that of their Hindu and Christian counterparts. A recent national survey in India reported that Muslims on the whole have an average standard of living well below that of most Hindus. This disparity translates into educational levels for Muslim women that are significantly below the national average. In India, Muslim women's workforce participation rates also tend to be the lowest. Low rates of education and workforce participation for women also prevail in Muslim-dominated countries such as Pakistan and Afghanistan.

Women and Afghanistan's Taliban. Women in Afghanistan have suffered sometimes brutal repression since a conservative Islamist movement, the **Taliban**, gained control of the government there in the mid-1990s. The Taliban supported strict and distorted interpretations of Islamic law. Females, even urban professional women, had to live in seclusion. Girls and women were not allowed to work outside the home or attend school, and they had to wear a heavy, completely concealing garment, called a *burqa*, whenever they were out of the house. (Men also had to follow a dress code, though a less restrictive one.) The Taliban even decreed that women must whisper and not make noise as they walked, because the sound of their

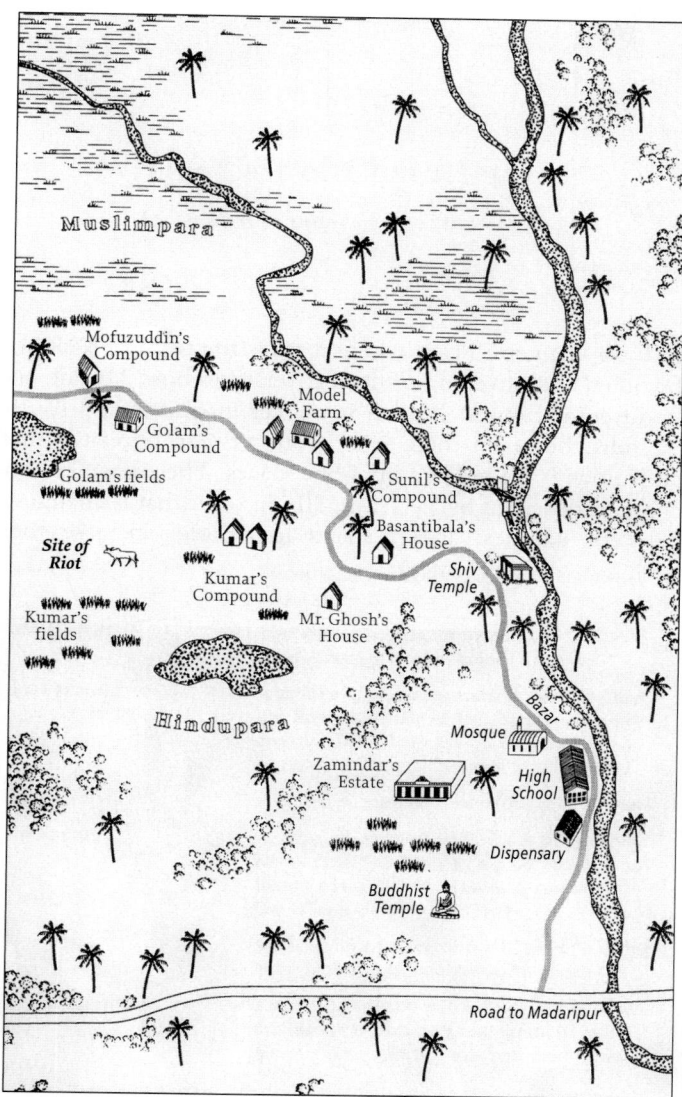

Figure 8.26 Some trouble with cows. This map of the village of Panipur (a pseudonym) illustrates how intimately the separate Muslim and Hindu communities were connected. The map shows Muslim and Hindu areas, the area where the riot took place, and numerous other features of village life. [Courtesy of the University of California Press.]

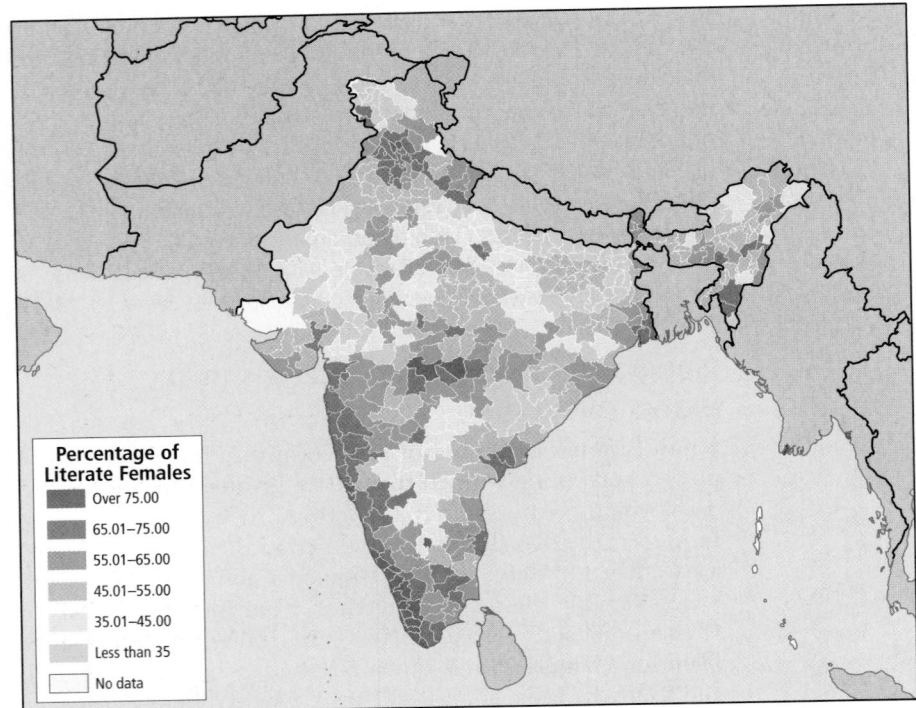

Figure 8.27 Female literacy in India, 2005. Female literacy lags behind male literacy in nearly all parts of the country. Female literacy is crucial to improving the lives of children: women who can read often seize opportunities to earn some income and nearly always use this income to help their children. The white lines indicate district boundaries. [Adapted from "India female literacy districtwise," *Maps of India*, at http://www.mapsofindia.com/census2001/female literacydistrictwise.htm.]

Percentage of Literate Females
- Over 75.00
- 65.01–75.00
- 55.01–65.00
- 45.01–55.00
- 35.01–45.00
- Less than 35
- No data

footsteps was distracting and potentially erotic to men. Although the Taliban were driven from power in November of 2001, cultural and religious conservatism continues to adversely affect Afghan women. Concerted efforts to change their lives for the better are under way. ▣

Vignette From behind her microphone at Radio Sahar (Dawn), Nurbegum Sa'idi speaks to a female audience on a wide range of topics. Located in the city of Herat, Radio Sahar is the latest in a network of independent women's community radio stations which have sprung up in Afghanistan since early 2003. Radio Sahar provides 2 hours of daily programming consisting of educational items that address cultural, social, and humanitarian matters as well as music and entertainment.

Given the high percentage of illiterate women with little or no access to education, stations like Radio Sahar and others (Figure 8.28) provide one of the most powerful ways to reach and educate women. The broadcasts allow them to connect with one another in this conservative, male-dominated society. As Sa'idi attests, "It's great when you feel you can bring about change. The feedback we have been getting from listeners tells us that Sahar is providing new hope for the women in Herat."

Source: United Nations Office for the Coordination of Humanitarian Affairs Integrated Regional Information Network, "Afghanistan: New radio station to tackle women's problems," November 20, 2003.

Purdah. The practice of concealing women from the eyes of non-family men, especially during their reproductive years, is known as **purdah.** It is observed in various ways across the region. The practice is strongest in the northwest and across the Indo-Gangetic Plain, where it takes the form of seclusion of women and of veiling or head covering in both Muslim and Hindu communities. Purdah is weaker in central and southern India, but even there separation between unrelated men and women is maintained in public spaces. The custom is generally not observed by low-caste Hindus, but that is changing. In recent decades, as low-status households increase their

Figure 8.28 Women's community radio in Afghanistan. Radio journalist Parwin of Radio Quyaash, a sister station of Radio Sahar, interviews the Afghan head of Women's Affairs on Children's Day. [© Leslie Knott.]

economic standing, their ability to seclude women signals surplus wealth and increased ritual purity. Unfortunately for women, the effect of this trend is to limit their economic independence and autonomy in the long run.

Vignette Night is falling in Ahraura, a village in the state of Uttar Pradesh in north-central India. In the enclosed women's quarters of a walled compound, Mishri is finishing her day by the dying cooking fire as her one year-old son tunnels his way into her sari to nurse himself to sleep. Mishri, who is 27, lives in a tiny world bounded by the walls of the courtyard she shares with her husband, five children, and several of her husband's kin. Like many villages in northern India, her village observes the practice of purdah, in which women keep themselves apart from men.

That Mishri can observe purdah is a mark of status because it shows she need not help her husband in the fields. Within the compound, she works from sunup to sundown, only chatting for moments with two women who cover their faces and scurry from their own courtyards to hers for the short visit. Mishri is devoted to her husband, who was chosen for her by her family when she was 10; out of respect, she never says his name aloud.

Source: Adapted from Alex Pulsipher's field notes, 2000.

Dowry, marriage, and widowhood.
Throughout South Asia, most women are partners in marriages arranged for them. Especially in wealthier, better educated families, the wishes of the bride and groom are considered, but in some cases they are not. This is especially true of child marriage, when a young girl (often as young as 12) is married off to a much older man. In most cases, the bride's family pays the groom's family a sum of money called a **dowry** at the time of the marriage. Dowry originated as an exchange of wealth between land-owning, high-caste families. With her ability to work reduced by purdah, an upper-caste female was considered a liability for the family that took her in.

Usually, a bride goes to live in her husband's family compound, where she becomes a source of domestic labor for her mother-in-law. Most brides work at domestic tasks for many years until they have produced enough children to have their own crew of small helpers, at which point they gain some prestige and a measure of autonomy.

Motherhood in South Asia determines much about a woman's life. A woman's power and mobility increase when she has grown children and becomes a mother-in-law herself. On the other hand, in some communities, the death of a husband, no matter what the cause, is a disgrace to a woman and can completely deprive her of all support and even of her home, children, and reputation. Widows may be ritually scorned and blamed for their husbands' deaths. Widows of higher caste rarely remarry, and in some areas, they become bound to their in-laws as household labor or may be asked to leave the family home.

Female infanticide and bride burning.
Changing dowry customs appear to be a cause of the growing incidence in India of both female infanticide and bride burning. Until the last several decades, it was the custom among the lower castes to pay a *bride price*, not a dowry: the groom paid the family of the bride a relatively small sum that symbolized the loss of their daughter's work to her family's economy and the gain of her labor by his.

Increasing education for males and increasing family affluence have reinforced the custom of dowry. Young men came to feel that their diplomas increased their worth as husbands and gave them the power to demand larger and larger dowries. Soon, the practice spread through lower-caste families wanting to upgrade their status. Now poor families are crippled by the dowries they must pay to get their daughters married. A village proverb captures this dilemma: "When you raise a daughter, you are watering another man's plant." Some poor families view the birth of a daughter as such a calamity that they are led to the desperate act of female infanticide—killing second and third daughters soon after birth.

The reinstitution of dowry also appears to be a cause of the practice known as bride burning, or dowry killing. In this phenomenon, a husband and his relatives stage an "accident," such as a kitchen fire, that kills his wife. The wife's death enables the widower to marry again and collect a new dowry. Bride burning has been reported for many years, especially in India; the National Crime Records Bureau of India reported that 7026 such deaths occurred in 2005.

> The reinstitution of dowry also appears to be a cause of the practice known as bride burning, or dowry killing.

Earning power, population, and the status of women.
Efforts to increase the status of women are gaining momentum in South Asia because it is believed this will slow population growth and increase economic well-being. If women channel less of their energy into reproduction and more into economic activities, their families will benefit from better health and increased educational opportunities.

In Mazār-e-Sharif, Afghanistan, the state governor has supported the recent establishment of a women's bazaar, where all the shops are owned and operated by women, but the customers are of both sexes. So far, only four women have been brave enough to withstand the harassment of men, who resent the bazaar. Fariba Majid, the Women's Ministry representative who founded the bazaar, hopes for at least 20 shopkeepers in time. She remarks that, for the bazaar to survive, the women must quickly turn a profit; otherwise, their husbands will put a stop to it. But for now, one such entrepreneur, Ms. Barmaki, who runs her handicraft shop out of a cargo container, has won the consent of her husband to employ their 12-year-old daughter.

A minority of upper- and middle-class urban professional women in India, Pakistan, Sri Lanka, and Bangladesh may have more in common with their counterparts in Europe and America than they do with village women in their own countries. In the region's major cities, there are growing numbers of highly successful businesswomen, female directors of companies, highly qualified female technicians, high-ranking female academics, and women who serve prominently in government. The daughters of these women are free to pursue careers, and in some cases may choose their own spouse.

Gender and democratization. The status of women has been advanced throughout South Asia by moves toward greater democracy. With the exception of the Maldives, Nepal, Afghanistan, and Bhutan, all South Asian countries have had female heads of state. However, it is important to note that all of them were either wives or daughters of previous heads of state. ◼

At lower levels of government, progress has been mixed. In the 1980s, Indian Prime Minister Rajiv Gandhi introduced *panchayati raj* (village governing councils) to encourage gender equality in village life. Thirty percent of the seats on these local councils must be reserved for women during a given election cycle. At the parliamentary level, however, Indian women remain very poorly represented. In 2006, only 9 percent of the Indian Parliament was female. At that time, a confederation of Muslim and Hindu women's groups asked the prime minister for legislation that would reserve one-third of the seats in the lower house of Parliament and in state assemblies for women for a 15-year trial period. Such quotas are already in place in Pakistan (22 percent), Nepal (33 percent), and Bangladesh (14 percent). It is hoped that these temporary quotas will afford women sufficient political experience to win elections without the aid of quotas.

ECONOMIC ISSUES

South Asia is a region of startling economic contrasts. India, for example, is home to hundreds of millions of desperately poor people, but it is also a global leader in the computer software industry and has a space program. The British colonial system deepened the extent of South Asia's poverty and widened the gap between rich and poor. However, the current wealth disparities in the region have resulted mostly from economic policies favored by post-independence leaders. Despite India's celebrated democratic traditions, its poor have often been left out of the political process and bypassed or hurt by economic reforms.

Agriculture remains the basis of South Asian economies, but rapid industrialization and self-sufficiency have been the dream since independence. Recent measures to encourage economic development include an emphasis on information technology in India and innovative financing strategies pioneered in Bangladesh. Such approaches reflect the strength of the service sector, which has expanded more rapidly than either agriculture or industry in all South Asian countries over the past decade.

Food Production and the Green Revolution

Although 60 percent of the region's population is engaged in agricultural labor, the contribution of agriculture to most national economies hovers below 25 percent of the GDP. Production per unit of land has increased dramatically over the past 50 years, agriculture is the least efficient economic sector, meaning that it gets the lowest return on investments of land, labor, and cash. Figure 8.29 shows the distribution of agricultural zones in South Asia.

Until the 1960s, agriculture across South Asia was based largely on traditional small-scale systems that managed to feed families in good years, but often left them hungry in years of drought or flooding. Moreover, these systems did not produce sufficient surpluses for the region's growing cities. Even now cities must import food from outside the region. Much of South Asia's agricultural land is still cultivated by hand, and for decades, agricultural development was neglected in favor of industrial development, especially in India. Nonetheless, by the 1970s, important gains in agricultural production had begun.

Beginning in the late 1960s, a so-called **green revolution** boosted grain harvests dramatically through the use of new agricultural tools and techniques. Such innovations included seeds selected for high yield and for resistance to disease and wind damage, fertilizers, mechanized equipment, irrigation, pesticides, herbicides, and double-cropping. To a lesser extent, an increase in the amount of land under cultivation also contributed to a greater yield. Where the new techniques were used, yield per unit of farmland improved by more than 30 percent between 1947 and 1979, and both India and Pakistan became food exporters. ◼

The benefits of the green revolution have been very uneven. Some Indian states, such as Punjab and Haryana, which have extensive irrigation networks, have gained tremendously. Others have lagged behind. Many poor farmers who were unable to afford the special seeds, fertilizers, pesticides, and new equipment could no longer compete and had to give up farming. Most migrated to the cities, where their skills matched only the lowest paying jobs.

> Many poor farmers who were unable to afford special seeds, fertilizers, pesticides, and new equipment could no longer compete and had to give up farming.

Green revolution technologies have also inadvertently reduced the utility of many crops for the rural poor, especially women. The new varieties of rice and wheat, for example, yield more grain. However, they produce fewer of the other plant components previously used by women, such as the wheat straw used to thatch roofs, make brooms and mats, and feed livestock. Moreover, women's already low status in agricultural communities often erodes further as their labor is replaced by new technologies. Equipment such as small tractors and mechanized grain threshers are usually controlled by male members of the family.

The need for green revolution alternatives. Many observers are concerned about South Asia's long-term ability to maintain current levels of food production. The green revolution's chemical fertilizers, pesticides, and high levels of irrigation all

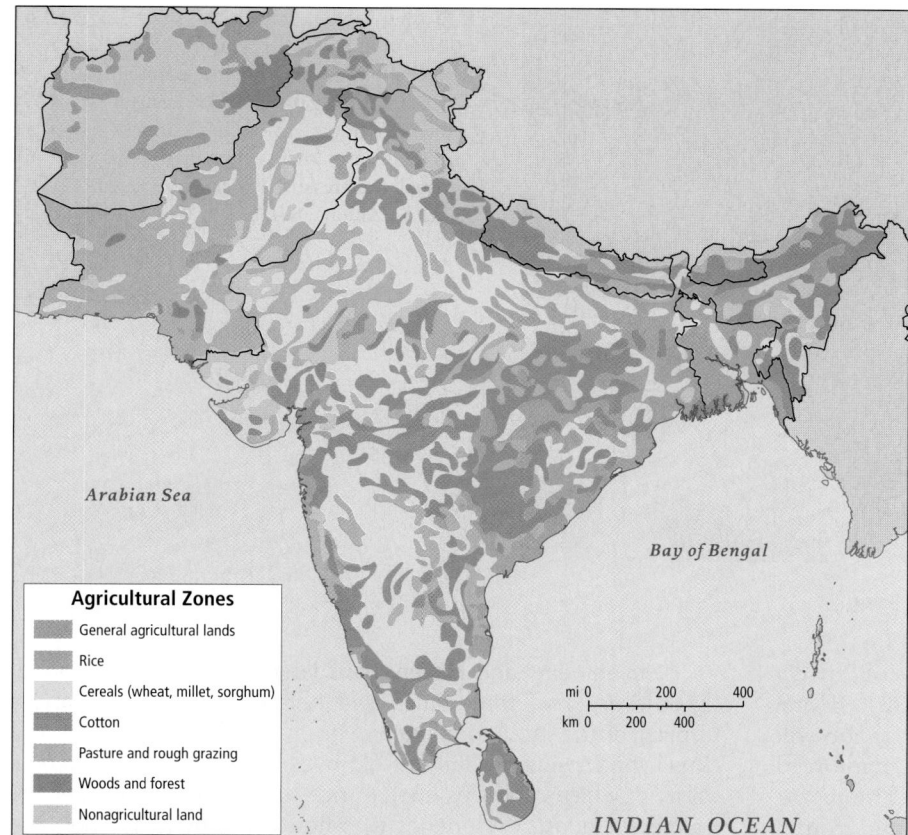

Figure 8.29 Agricultural zones in South Asia. Agricultural zones in this region form a complex pattern influenced by landforms, climate, cultural customs, and recent development theories. [Adapted from Gordon Johnson, *Cultural Atlas of India* (New York: Facts on File, 1996), p. 34; "Afghan Economy (map)," SESRTCIC (Statistical, Economic and Social Research and Training Centre for Islamic Countries) InfoBase, at http://www.lib.utexas.edu/maps/middle_east_and_asia/afghanistan_econ_1982.jpg.]

contribute to the loss of soil fertility over time through the buildup of salt in soils. This *salinization* (see page 210 in Chapter 6) is already reducing yields in many areas, such as the Pakistani Punjab, that country's most productive, but highly irrigated, agricultural zone.

A potential remedy for some of the failings of the green revolution style of agriculture is **agroecology.** Traditional methods are used to fertilize crops, such as animal manure and inter-cropping of legumes, and natural predators are used to control pests. Unlike green revolution techniques, the methods of agroecology are not disadvantageous to poor farmers because the necessary resources are readily available in most rural areas. To participate, farmers do not need access to cash but to knowledge, which can be taught orally to small groups and over the radio. Studies in southern India that compared agroecology techniques with those of the green revolution found that their productivity and profitability were equal. However, agroecology techniques reduced soil erosion and loss of soil fertility.

Malnutrition. Increased food supplies have reduced, but not eliminated, hunger and malnutrition in the region. Between 1970 and 2001, the amount of food produced per capita in South Asia increased 18 percent, and the proportion of undernourished people dropped from 33 percent of the population to 22 percent. Still, this represents roughly half of the world's undernourished population. Because of corruption and social discrimination, government programs to provide food to the poor have generally failed to reach those most in need. For example, in India, despite massive resources devoted to child malnutrition, about half of children show signs of malnutrition.

Microcredit: A South Asian Innovation for the Poor

Over the past four decades, a highly effective strategy for helping lift people out of extreme poverty has been pioneered in South Asia. **Microcredit** makes very small loans (generally under U.S.$100) available to poor would-be business owners. Throughout South Asia, as in most of the world, banks are generally not interested in administering the small loans that poor people need. Instead, they must rely on small-scale moneylenders, who often charge interest rates as high as 30 percent or more *per month*. In the late 1970s, Muhammad Yunus, an economics professor in Bangladesh, started the Grameen Bank, or "People's Bank," which makes small loans, mostly to people in rural villages who wish to start businesses.

The loans often pay for the start-up costs of small enterprises such as cell phone–based services, chicken raising, small-scale egg production, or construction of pit toilets. Potential

Figure 8.30 Microcredit for poor entrepreneurs. In this village in Bangladesh, women borrowers and participants in the local Grameen Bank gather weekly, usually at someone's house, to discuss business and pay their loan installments. For many, this is an unusual and treasured social outing. [IFAD Photo by Anwar Hossain.]

borrowers are organized into small groups that are collectively responsible for paying back the loans (Figure 8.30). If one member fails to repay a loan, then everyone in the group will be denied loans until the loan is repaid. This system, reinforced with weekly meetings, creates incentives to repay. The repayment rate on the loans is extremely high, averaging around 98 percent, much higher than most banks achieve.

So far, the Grameen Bank has been an enormous success in Bangladesh, where it has loaned over U.S.$5.6 billion to more than 6 million borrowers. Similar microcredit projects have been established in India and Pakistan and throughout Africa, Middle and South America, North America, and Europe. In 2006, Dr. Yunus was awarded the Nobel Peace Prize for his work in microcredit. ▱

Vignette In a small hamlet in Bangladesh, not too far from the Indian border, is the house of Mosamad Shonabhan, a 32-year-old married woman whose life has been changed by her 11-year participation in the Grameen Bank. Everyone agreed that she had been the smartest of her family's children. However, because her father earned only 50 cents a day as a farm laborer, she could not go to school, and was instead married at the age of 14 to a young barber. For a year, she lived in her father-in-law's house, but then financial problems forced her to move back into her father's house. There she faced increasingly dire circumstances as her father's health deteriorated.

After a few years, a local political leader suggested that Mosamad join the Grameen Bank's lending program. She was afraid to go, as she had never handled money and had heard a local rumor that the bank's real purpose was to convert people to Christianity. Nevertheless, she went. Eventually, she took out a loan for $40.00 that would allow her to set up a small rice-husking operation in her father's backyard.

Eleven years and eleven loans later, Mosamad earns about $1.50 every day—three times what her father had made—and is a pillar of the local community. Her main source of income is a small shop inside her father's old house. She also leases an acre of land, which produces enough rice to feed her family and the numerous guests and friends who now come by to see her. She plans to open another shop that her husband will run. She is being taught to read by her 15-year-old daughter, whom she plans to send to university.

Source: Adapted from Alex Pulsipher's field notes, 2000.

Industry over Agriculture: A Vision of Self-Sufficiency

After independence from Britain in 1947, South Asia's new leaders favored industrial development over agriculture. Influenced by socialist ideas, they believed that government involvement in industrialization was necessary to ensure the levels of job creation that would cure poverty. Another of their goals was to reduce the need to import manufactured goods from the industrialized world (see Chapter 3, pages 118–119 for a discussion of *import substitution industrialization*). To accomplish this, governments took over the industries they believed to be the linchpins of a strong economy: steel, coal, transport, communications, and a wide range of manufacturing and processing industries.

South Asian industrial policies in the decades after independence generally failed to meet their goals. The emphasis on industrial self-sufficiency was ill suited to countries that had such large agricultural populations. In India, for example, governments invested huge amounts of money in a relatively small industrial sector—even today industry employs only 12 percent of the population compared with the 60 percent employed in agriculture. Since such a small portion of the population directly benefited from this investment, industrialization failed to significantly increase South Asia's overall prosperity.

Another problem was that the measures intended to boost employment often contributed to inefficiency. One policy encouraged industries to employ as many people as possible, even if they were not needed. So, for example, until recently, it took more than 30 Indian workers to produce the same amount of steel as one Japanese worker. Consequently, for years, Indian steel was not competitive in the world market. In addition, as in the former Soviet Union, decisions about which products should be produced were made by ill-informed government bureaucrats rather than being driven by consumer demand. Until the 1980s, items that would improve daily life for the poor majority, such as cheap cooking pots or simple tools, were produced in only small quantities. At the same time, there was a relative abundance of kitchen appliances and cars that only a tiny few could afford.

Economic Reform: Globalization and Competitiveness

During the 1990s, much of South Asia began to undergo a wave of economic reforms. In other regions structural adjustment programs (SAPs; see Chapter 3, pages 119–122) were mandated by the International Monetary Fund and World Bank. By contrast, India's economic reforms were initiated by the Indian government itself in response to a financial crisis that emerged in the 1980s. Although privatization of India's public sector industries and banks has proceeded slowly, marketization of the economy has been significant. International investment has poured into the country in recent years, and a wide range of foreign consumer goods are now available. The result has been rapid economic growth over the past decade, with recent years seeing India register high rates of yearly economic growth (9.6 percent in 2007), comparable to China's (11.4 percent in 2007).

India's manufacturing boom. Freed from a maze of regulations, both foreign and Indian companies are investing heavily in manufacturing and other industries (Figure 8.31). Drawn by India's large and cheap workforce, its large domestic market for manufacturers of all sorts, as well as its excellent educational infrastructure, many companies are setting up global manufacturing headquarters in major Indian cities. For example, nearly every major global automobile company is currently establishing significant manufacturing facilities somewhere in India. So many global manufacturers have flocked to India in recent years that the country may soon challenge China as an exporter of manufactured goods. Whether or not that happens, many

> India's current manufacturing boom is benefiting from a previous boom in offshore outsourcing that started in the 1990s.

of India's urban poor should see rising incomes in the coming years as manufacturing jobs grow.

India's offshore outsourcing revolution. India's current manufacturing boom is benefiting from a previous boom in offshore outsourcing that started in the 1990s. **Offshore outsourcing** occurs when a company contracts to have some of its business functions performed in a country other than the one where its products or services are actually developed, manufactured, and sold. To take advantage of India's large, college-educated, and relatively low-cost workforce, companies in North America and Europe are outsourcing an increasing number of jobs to cities such as Bangalore, Mumbai, and Ahmadabad. A wide array of jobs are being outsourced, including information technology (IT), engineering, telephone support, pharmaceutical research, and "back office" work.

Major global finance firms are increasingly seeking out the highly skilled workers of Mumbai's "Wall Street"—Dalal Street—to provide finance and accounting services. Stiff global competition makes cost cutting imperative for finance firms, and India's relatively low salaries provide a solution. While a junior analyst from an Ivy League school costs $150,000 a year in the United States, a graduate of a top Indian business school costs only $35,000 a year in India. Yet, for that Indian employee, this salary buys a much higher standard of living than the U.S. employee would enjoy. ◄▪

Service sector growth. India's outsourcing growth reflects a broad pattern of steady expansion in the service sector throughout South Asia. Between 20 and 40 percent of South Asian workers now work in services, which contribute well over 50 percent of GDP in India, Pakistan, Sri Lanka, and Bangladesh. Within this sector, facilities that engage in trade, transport, storage, and communication (including information technology) show the most growth. Finance, insurance, real estate, business services, and tourism have also grown quickly. All these activities are connected in some way with international commerce and benefit from India's success at developing information technology (Figure 8.32).

Impact of India's reforms on wealth distribution. The impact of the economic reforms on India's wealth distribution is the subject of much debate. The highly skilled and educated urban middle and upper classes have gained the most by far. India's middle class now stands at 50 million and is likely to grow dramatically in the near future. Like SAPs in Middle and South America and Africa, the new economic policies may ultimately produce wider disparities in income. Though many of India's more than 300 million poor urban people are gaining from recent growth in manufacturing and other industries, such growth will only indirectly benefit the 72 percent of India's population still living in rural areas, many of whom remain very poor.

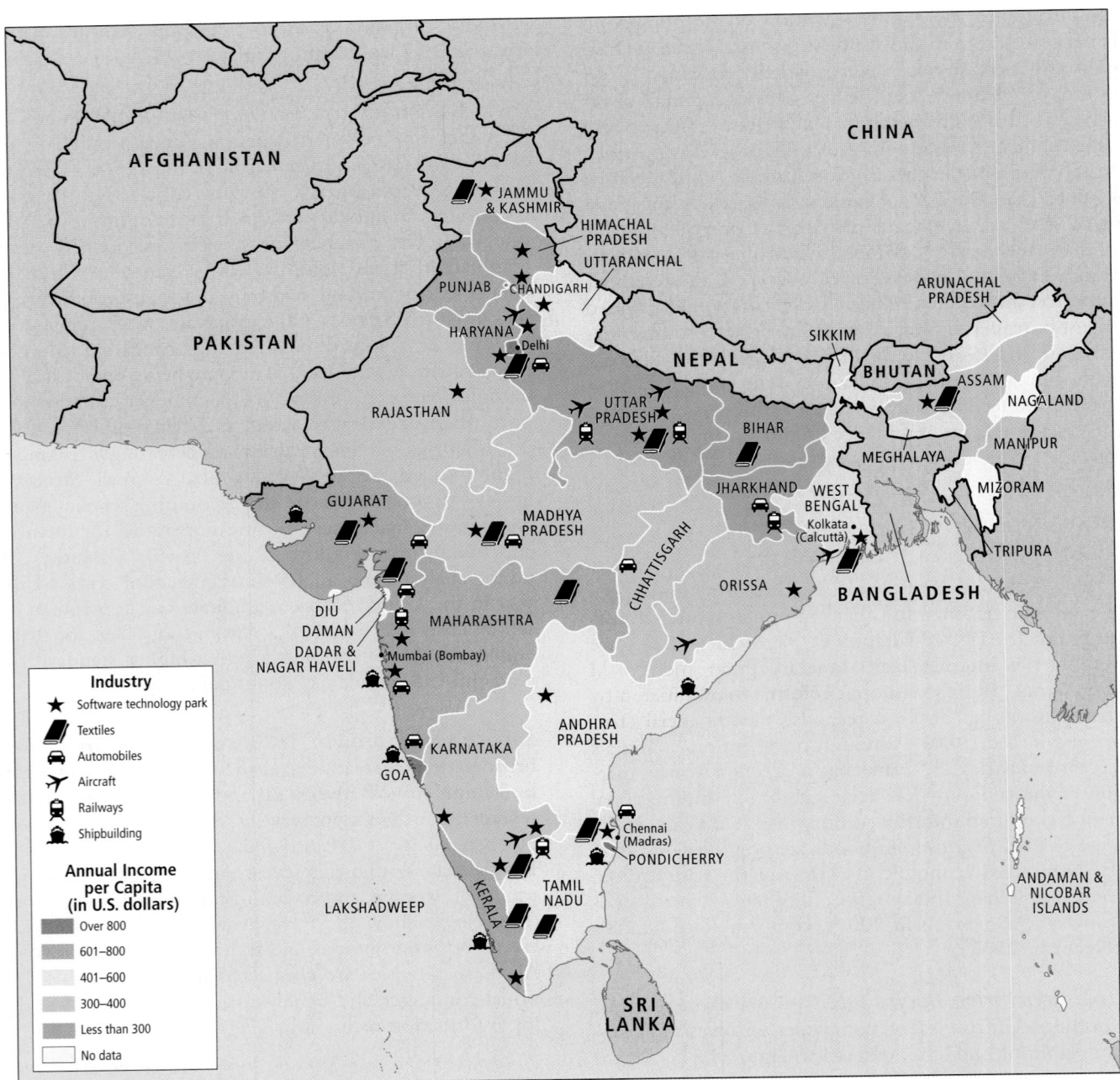

Figure 8.31 Per capita annual income and industrial and information technology centers in India. The presence of industry is often associated with higher incomes, so planners may seek to bring industry to low-income places. In some areas, poverty may be so great that even fairly intensive industry is able to raise average incomes only slowly. [Data from Directorate of Economics & Statistics of the respective state governments, 2006.]

Concerns about globalization. Though globalization has brought a number of improvements to South Asia, conservative leaders fear that access to global flows of information and products will cause a breakdown of traditional social and economic relationships, resulting in a flood of Westernization. In addition to the Internet, television is a major agent of globalization that is having a dramatic cultural impact, as the following vignette illustrates. ◪

Vignette The tiny kingdom of Bhutan, couched in the Himalayas between China and India, had for centuries been more or less secluded from the rest of the world. That all changed in June 1999, when a royal decree legalized television, making Bhutan the last country in the world to "plug in." Within a short time, several entrepreneurs, such as Rinzy Dorji, whom the Bhutanese call "The Cable Guy," were in business. For U.S.$5 a month, the price of a bag of chilis, Rinzy provides Bhutanese

Figure 8.32 Training employees for India's service economy. An instructor in a Cochin outsourcing firm teaches other employees the skills necessary to provide technical support to U.S. clients. [© Stephen Voss.]

households with 45 cable TV channels—everything from the BBC to *Baywatch*.

Although Rinzy's business is booming, not everyone welcomes the new technology. As Kinley Dorji, editor of Bhutan's only newspaper, describes, "Soon after television started, we started getting letters . . . from children, children who seemed very [anxious]. The letters . . . specifically asked about this World Wrestling Federation program, 'Why are these big men standing there and hitting each other? What is the purpose of it?' They didn't understand. . . . Now, a few months later my son jumps on me one morning and says, 'I am Triple H, and you can be Rock.' And, suddenly we are fighting. Suddenly these [TV wrestlers] are new heroes for our children."

Foreign Minister Lyonpo Jigma Thinley thoughtfully observes, "People have suddenly realized that there are so many things they desire, which they were not even aware of before. The truth is that most of these television channels are commercially driven. And some of the Bhutanese people are driven toward consumerism. And that is inevitable. It's unfortunate, but inevitable." On the other hand, he has also heard people saying, "My God! We didn't know that we are living in a peaceful country. There seems to be violence and crime everywhere in the world." He concludes, "So, in a way, the positive thing is that people realize how good a life they are living in this country."

To learn more about the impact of satellite and cable TV in Bhutan, see the FRONTLINE/World Video "Bhutan: The Last Place."

POLITICAL ISSUES

Since independence in 1947, South Asian countries have peacefully resolved many conflicts, smoothed numerous potentially bloody transfers of power, and nurtured vibrant public debate over the issues of the day. India is often cited as a bastion of democracy that serves as an example of political enlightenment for the rest of the developing world. However, South Asians have also missed many opportunities to resolve conflicts through democratic means, favoring the use of force instead. Indeed, there are signs that corruption, religious nationalism, and violent conflict are an increasing threat to democracy throughout South Asia. However, there are also some signs that democracy is strengthening.

Religious Nationalism

Increasingly, people frustrated by government inefficiency, corruption, caste politics, and the failure of governments to deliver on their promises of broad-based prosperity are joining religious nationalist movements (Figure 8.33). **Religious nationalism** is the association of a particular religion with a political unit—be it a neighborhood, a city, or an entire country. Political control over a given territory is often the ultimate goal of such movements.

Although both India and Pakistan were formally created as secular states, religious nationalism has long been a reality, shaping relations between people and their governments in those countries. India is increasingly thought of as a Hindu state, and Pakistan and Bangladesh as Muslim states. Many people in the dominant religious group strongly associate their religion with their national identity.

Hindu nationalism in India is supported predominantly by urban men from middle- and upper-caste groups. Its proponents fear the erosion of their castes' political influence and particularly resent the extension of the quota system for government jobs and seats in universities to lower-caste groups (see pages 307–308). This sentiment has been fueled by

Figure 8.33 Religious nationalism. Indian holy men participate in the eleventh religious parliament in the Hindu temple town of Vadtal, near Ahmadabad, in 2006. The parliament was organized by the Hindu nationalist group VHP, which takes extreme positions against Muslims. The painting behind the group shows the Hindu goddess Durga killing Pakistani president Pervez Musharraf (on the left), and the Hindu god Lord Shiva dancing on Osama bin Laden (on the right). [AP Photo/Ajit Solanki.]

alliances among politically mobilized lower castes who are no longer willing to follow the dictates of the dominant castes.

Political parties based on religious nationalism have gained popularity throughout South Asia, often fueling conflicts between religious majorities and minorities. Although their members think of these parties as forces that will purge their country of corruption and violence, they are usually no less corrupt or violent than other parties.

Babar's Mosque or Ram's Temple? The geography of religious nationalism. Proponents of religious nationalism often try to gain mass support through political campaigns and acts of terrorism that interweave with South Asian history, mythology, and landscapes. In late 1992 and early 1993, a series of deadly riots occurred throughout South Asia. The events were the culmination of a long campaign supported by India's leading Hindu nationalist political party, the Bharatiya Janata Party (BJP) and its allies. The riots were triggered by the destruction of a Muslim mosque by some 300,000 Hindus brought to the town of Ayodhya on the Ganga Plain by BJP allies. The Mughal emperor Babar had built the mosque in 1528–1529, supposedly on the ruins of a Hindu temple marking the birthplace of the Hindu god Ram.

After the mosque's demolition, mobs commanded by urban Hindu nationalist organizations (many of them funded by Hindus living abroad) burned and looted selected Muslim businesses and homes throughout India. The police often acted in complicity with the rioters. Nearly 5000 people died.

Despite the violence, the BJP turned the incident into a symbol of Hindu power, going on to take control of the national government of India from 1998–2004.

Violence erupted again in 2002, when a train full of Hindus returning to Gujarat from the disputed site at Ayodhya was attacked and burned by a Muslim mob. Ensuing riots killed between 2000 and 2500 people, most of whom were Muslims. Then in 2005, five Muslim terrorists from Pakistan-administered Kashmir attacked the site of the mosque, now occupied by a makeshift Hindu temple. All of the terrorists were killed in the attack, and a few people died in the ensuing riots.

The selection of the Ayodhya mosque-temple site as a locus of protest was calculated to elicit violent reactions from both Muslims and Hindus. The locations and tactics chosen reflected the urban base of Hindu nationalist parties and the violent gangster-style tactics still common in local South Asian politics.

Hindu nationalism, corruption, and multi-party democracy. Despite the violence and hatred promoted by its political campaigns, the emergence of the BJP was an important step in the strengthening of India's democracy. The BJP's popularity was fueled in part by widespread opposition to corruption which had become rampant over the nearly fifty years during which the Indian National Congress (INC) had controlled the central government. Unchecked by any major political opposition, INC leaders became notorious for abusing their power for their own financial gain. The BJP ultimately proved similarly corrupt and was voted out of office after only six years in power. Since then, however, it has remained a strong, well-organized opposition party, ready to expose the corruption of the Indian National Congress or any other party that may come to power.

Regional Political Conflicts and Democracy

The most intense armed conflicts in South Asia today are **regional conflicts,** in which nations dispute territorial boundaries or a minority actively resists the authority of a national or state government (Figure 8.34). Most represent failed opportunities to resolve problems through the democratic processes.

Conflict in Punjab. Punjab is the ancestral home of the Sikh community. When India and Pakistan were partitioned in 1947, the state of Punjab was divided between the two countries. Large numbers of Sikhs chose to live in the Indian part of Punjab, where they thought their unique identity would be better protected. Since that time, however, Sikhs have felt alienated from the rest of India, despite their wealth and influence in many cities outside the Punjab. In the 1970s, moderate Sikh elected officials issued demands for more water rights and greater respect for local democratic processes. The demands were generally ignored by India's national government.

In the early 1980s, militant Sikh extremists barricaded themselves in the holiest Sikh shrine, the Golden Temple in Amritsar. They then used the shrine as a base of operations to agitate for an independent Sikh nation. In 1984, government forces attacked the shrine, damaging the temple and killing numerous innocent pilgrims as well as the militants. Shortly thereafter, Indian Prime Minister Indira Gandhi, who had called for the attack, was assassinated by two of her Sikh bodyguards. Riots and organized mob violence spread throughout India over the next few days, resulting in the deaths of more than 2700 Sikhs.

Since then, agreements acceding to many Sikh demands for water rights and control of religious sites were signed but not implemented. This alienated Sikhs yet further and led to more political violence, resulting in the deaths of 25,000 people. Further violence has been avoided for the time being, but the situation in Punjab remains sensitive.

Conflict in Kashmir. Since 1947, between 60,000 and 100,000 people have been killed in violence in Kashmir that could likely have been resolved democratically.

Kashmir has long been a Muslim-dominated area, and in 1947, some Kashmiris and all of Pakistan's leaders believed that it should be turned over to Pakistan. However, the maharaja (king) of Kashmir at the time, a Hindu, wanted Kashmir to remain independent. The most popular Kashmiri political leader and significant portions of the populace, on the other hand, favored joining India. They preferred India's stated ideals of secular, non-religiously based government to Pakistan's less robust safeguards for secularism. When Pakistan-sponsored raiders invaded western Kashmir in 1947, the maharaja quickly agreed to join India. A brief war between Pakistan and India resulted in a cease-fire line that became a tenuous boundary.

A popular vote for or against joining India was never held, due to India's resistance. Pakistan attempted another invasion of Kashmir in 1965, but was defeated. The two countries are technically still waiting for a UN decision on where the final border will be. In the meantime, Pakistan effectively controls the thinly populated mountain areas north and west of the densely populated valley known as the Vale of Kashmir. India holds nearly all the rest of the territory, where it maintains more than 500,000 troops. The Ladakh region of Kashmir (see Figure 8.34) is the object of a more limited border dispute between India and China.

> After years of military occupation, many Kashmiris support independence from both India and Pakistan, but neither country is willing to hold a vote on the matter.

After years of military occupation, many Kashmiris support independence from both India and Pakistan. However, neither country is willing to hold a vote on the matter. As in Punjab, much of the conflict in Kashmir has centered around the right of people to run their own affairs through local democratic processes. India's national leaders have often appointed their own favorites in an attempt to maintain strong central control. Anti-Indian Kashmiri guerrilla groups, equipped with weapons and training from Pakistan, have carried out many bombings and assassinations of these appointees. Blunt counterattacks launched by the Indian government have killed large numbers of civilians and alienated most Kashmiris. Meanwhile, sporadic fighting between India and Pakistan continues along the boundary line, in the world's highest battle zone, at an altitude of 20,000 feet (6000 meters).

Another complication in the Kashmir dispute is the fact that both India and Pakistan—which came close to war in 1999 and again in 2002—have nuclear weapons. Many see the conflict in Kashmir as more likely to result in the use of nuclear weapons than any other conflict in the world because of the nationalistic fervor of the protagonists.

War and Reconstruction in Afghanistan

In the 1970s, political debate in Afghanistan became polarized. On one side were urban elites, who favored industrialization and democratic reforms. Opposing them were rural conservative religious leaders, whose positions as landholders and ethnic leaders were threatened by the proposed reforms. Some urban elites allied themselves with the Soviet Union, which supported some of their goals.

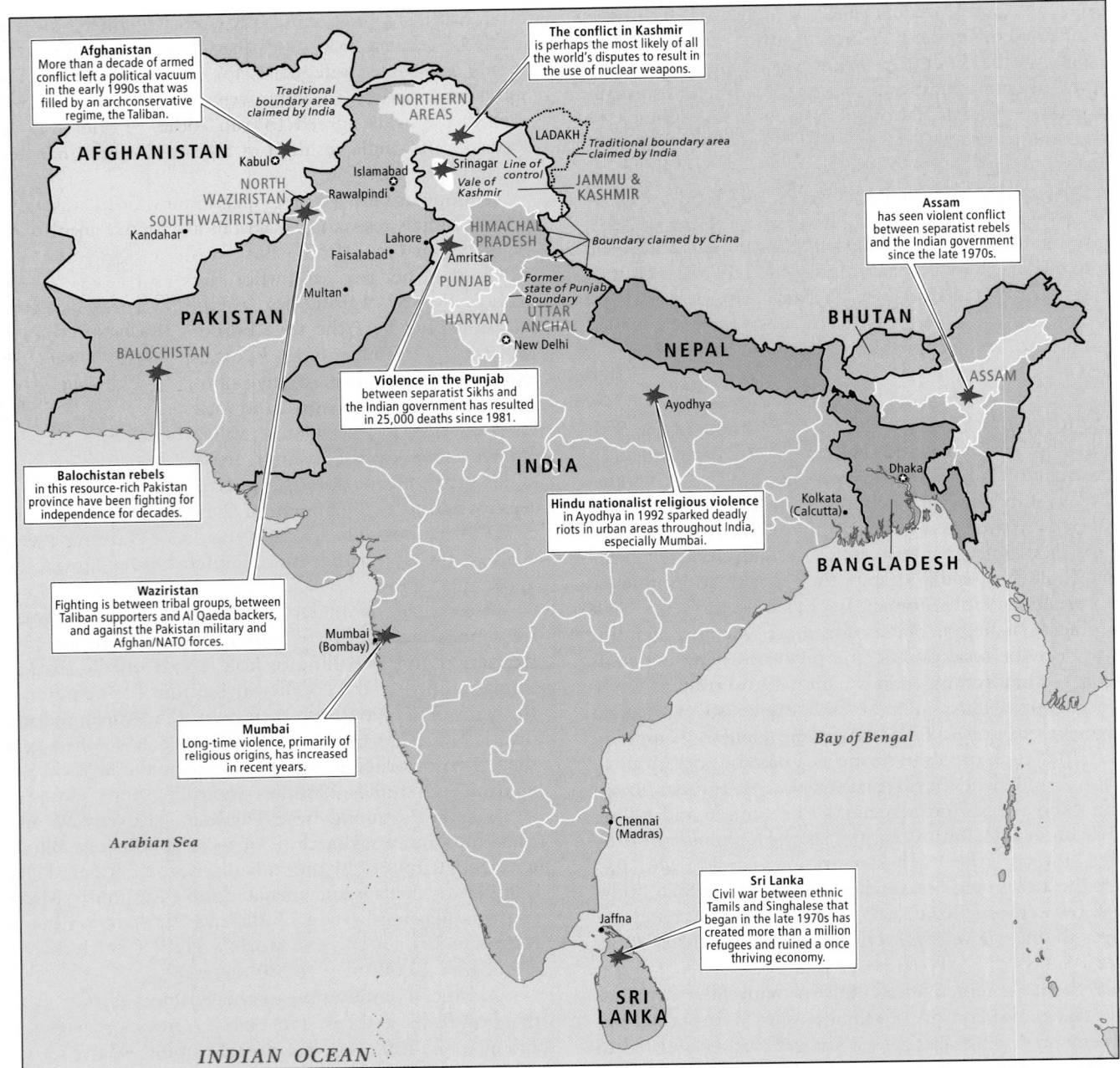

Figure 8.34 Conflict in South Asia. Some of the best known violent conflicts in South Asia, such as those over Kashmir and Punjab, are rooted in disputes that have remained unsettled since the Partition of India and Pakistan. Some conflicts, such as those in Afghanistan and Pakistan, are linked to international disputes and some to religious differences; nearly all are in some way related to disputes over land and resources.

Fearing that a civil war in Afghanistan would destabilize neighboring Soviet republics in Central Asia, the Soviet Union invaded Afghanistan in 1979. An anti-Soviet resistance group, the mujahedeen, was formed by the rural conservative leaders (often erroneously labeled "warlords") and their followers. As resistance to Soviet domination increased, the mujahedeen became ever more strongly influenced by militant Islamist thought. The United States and its regional ally Pakistan overlooked this development because it was still the cold war era, and they gave considerable support to any anti-Soviet movement. The mujahedeen proved to be tenacious fighters, and in 1989, after heavy losses, the Soviets gave up and left the country. Anarchy prevailed for a time as the Afghan factions fought one another, but the rural conservatives eventually defeated the reformist urban elites.

The Rise of the Taliban. In the early 1990s, the radical religious-political-military movement called the Taliban emerged from among the mujahedeen. For the most part, the Taliban are illiterate young men from remote villages, led by students from the *talib*s (Islamist schools of philosophy and law). The Taliban wanted to control corruption and crime and minimize Western ways introduced or reinforced by the Russian occupation and seen as licentious by rural Afghans. They also wanted to strictly enforce shari'a, the Islamic social and penal code (see Chapter 6, page 223). Efforts by the Taliban to purge Afghan society of non-Muslim influences included greatly restricting women (see pages 309–310), promoting only fundamentalist Islamic education, and banning the production of opium, to which many Afghan men had become addicted. By 2001, the Taliban controlled 95 percent of the country, including the capital, Kabul.

The events of September 11, 2001, focused the United States and its allies on removing the Taliban, who were giving shelter to Osama bin Laden and his Al Qaeda network. By late 2001, the Taliban were overpowered by an alliance of Afghans, supported heavily by the United States and the United Kingdom. The United Nations stepped in to help establish an interim coalition government. A national assembly was convened to designate a new national government and appoint a head of state in 2002, and to ratify a new constitution in 2004.

Seven years after the Taliban were ousted, they were back again, effectively thwarting the ability of Afghanistan's new government to ensure security and to meet the needs of people outside Kabul. 📹 Based in rural areas in both Pakistan and Afghanistan, the Taliban are aided by widespread distrust of both the government in Kabul, which is seen as corrupt, and the international military forces and private security personnel stationed in Afghanistan since 2001. Meanwhile, the vast majority of the country remains desperately poor despite the increased attention from international donors since 2001. Most people favor democratic government based on Islam, but feel that their concerns have not been adequately addressed.

Sri Lanka's Civil War

Violence between Sri Lanka's majority Singhalese and minority Tamil communities has left 68,000 people dead, produced more than one million refugees, and severely impeded Sri Lanka's economic development. Though both groups coexisted peacefully for centuries, beginning in the 1970s, antidemocratic government policies that favored Singhalese over Tamils resulted in violence.

The Singhalese have dominated Sri Lanka since their migration from Northern India several thousand years ago. Today, they make up about 74 percent of Sri Lanka's population of 20 million. Most Singhalese are Buddhist. Tamils, an ethnic group from South India, make up about 18 percent of the total population of Sri Lanka. About half of these Tamils have been in Sri Lanka since the thirteenth century, when a Tamil Hindu kingdom was established in the northern par the island (see Figure 8.23). The other half was brought over by the British in the nineteenth century to work on tea, coffee, and rubber plantations. Some Tamils have done well, especially in urban areas, where they dominate the commercial sectors of the economy. However, many others have remained poor laborers isolated on rural plantations.

Upon its independence in 1948, Sri Lanka had a thriving economy, led by a vibrant agricultural sector and a government that made significant investments in health care and education. It was poised to become one of Asia's most developed economies. But Singhalese nationalism was already alienating many Tamils. Singhalese was made the only official language, and Tamil plantation workers were denied the right to vote. Efforts were also made to deport hundreds of thousands of Tamils to India. Protests against these moves were brutally repressed by the government. Conditions in rural areas took a drastic turn for the worse in the 1960s when global prices for the country's chief agricultural exports declined. In response, the government shifted investment away from agricultural development and toward urban manufacturing and textile industries, which were dominated by Singhalese.

After years of skirmishes, full scale civil war broke out in 1983 between the Sri Lankan government and a variety of Tamil groups led by a guerilla army known as the Tamil Tigers. India intervened in 1987 at the request of the Sri Lankan government, but this failed to end the violence. Violence between the Tamil Tigers and the government has flared intermittently, often with great intensity, ever since. While the main battleground is in the north, the entire island has been subjected to terrorist bombings and kidnappings. 📹

In spite of the violence, economic growth is surprisingly robust in Sri Lanka. Driven by strong growth in food processing, textiles, and garment making, Sri Lanka is today the wealthiest nation in South Asia on a per capita basis.

Nepal's Rebels Become Elected Leaders

In Nepal, a dramatic expansion of democracy has paved the way toward a tentative peaceful reconciliation between warring parties. Until 2008, Nepal was governed by a royal family that paid only superficial respect to the country's elected legislature and multiparty democracy, which were introduced in 1990. In 1996, revolutionaries inspired by the ideals of the late Chinese leader Mao Zedong (but with no support from China) waged a "people's war" against the Nepalese monarchy. After a decade of civil war, during which 13,000 Nepalese died, the Maoists had both military control of much of the countryside and strong political support from most Nepalese, who objected to the dictatorial rule of the latest monarch, King Gyanendra. Massive protests forced Gyanendra to step down in 2006, and soon thereafter the Maoists declared a ceasefire with the government. In 2008, the Maoists expanded their power with sweeping electoral victories that gave them a majority in parliament and made their former rebel leader prime

...if the Maoists will be able to maintain ...ow they have been peacefully integrated ...nainstream.

Democracy in South Asia

With so man..., ...tical hot spots in South Asia, conditions vary considerably throughout the region, but in general democracy seems to be expanding. In India a more competitive multi-party system has emerged alongside efforts to reduce corruption. Pakistan has recently returned to democratic government after almost a decade of military dictatorship, during which the human rights of its citizens were repeatedly violated. Similarly in Bangladesh, after years of military dictatorship, democratic elections have occurred with some regularity, though government corruption is a recurring cause for public protest. Following years of civil war and authoritarian rule, both Afghanistan and Nepal are now democracies, though the political stability of both countries remains in question. Even the tiny Himalayan kingdom of Bhutan granted its people the right to elect local representatives in 2002 and national leaders in 2008. Sri Lanka, however, remains an open wound, with no signs yet on the horizon of reconciliation through the democratic process or any other political means.

Reflections on South Asia

Despite the number, scale, and complexity of problems facing South Asia, there are many reasons for optimism when looking toward the future. This region's enormous population is straining resources, but population growth is also slowing as women gain education and economic empowerment. Changes in agriculture are creating environmental problems and forcing more rural people into the cities, but urbanization may increase access to education, jobs, and health care, and may help slow population growth in the region as a whole.

Poverty and malnutrition are widespread but decreasing. South Asian innovations such as micro-credit have brought millions out of extreme poverty, and a recent boom in manufacturing promises to transform the incomes of India's vast population of urban poor. Globalization has already boosted the ranks of India's middle class with sustained growth in off-shore outsourcing.

South Asia's long history of democratic politics is threatened by rampant corruption, religious nationalism, and violent conflict. And yet, democracy has expanded significantly in recent decades. Several countries have turned away from various forms of dictatorship, and India's multi-party democracy has become more vital.

Of any world region, South Asia has the largest populations vulnerable to the effects of global warming. Yet its prospects are not entirely dim. South Asia is pursuing many important solutions to the climate crisis in the form of electric vehicles, alternative energy, and water conservation.

In this time of tumultuous change and expanding global connections, South Asia's problems are indeed daunting. But they are no more daunting than those faced by Mohandas Gandhi when he said "We must be the change we wish to see." The solutions emerging from South Asia today suggest that many have taken these words to heart.

Chapter Key Terms

agroecology 313
Buddhism 308
caste 307
civil disobedience 299
communal conflict 309
dowry 311
green revolution 312
Harappa culture 296

Hinduism 306
Indus Valley civilization 296
Jainism 308
jati 307
microcredit 313
monsoon 288
Mughals 296
offshore outsourcing 315

Partition 299
purdah 310
regional conflict 319
religious nationalism 317
Sikhism 308
subcontinent 288
Taliban 309
varna 307

Critical Thinking Questions

1. Dams and the reservoirs they create are sources of irrigation water and generators of electricity. What are the factors that have rendered so many places in South Asia, as well as elsewhere around the world, in need of markedly more water and electricity?

2. Explain why some would say that India has been part of globalization for thousands of years. Tie this history to what is happening in the present.

3. What are some of the lingering features of the British colonial era in South Asia? To what extent is globalization reinforcing or erasing these features?

4. How are changes in agriculture resulting in urbanization?

5. Describe microcredit and its impact on extreme poverty in South Asia.

6. Describe how South Asia is vulnerable to global warming. What are some responses to the climate crisis that are emerging from this region?

7. What are the main challenges to democracy in South Asia? In what ways is South Asian democracy strengthening?

8. What factors have led to a recent boom in manufacturing industries in South Asia?

9. What factors have encouraged high population growth in South Asia in the past? What factors may encourage slower population growth in the future?

10. What are the main factors creating South Asia's gender imbalance?

RUSSIA

KAZAKHSTAN

Astana

Lake Balkhash

Almaty

Bishkek

KYRGYZSTAN

UZB.

Naryn

Victory Peak
elev. 24,700 ▲

Tian Shan

Kashi

TAJ.

Muztagata
elev. 24,757 ▲

PAKISTAN

K2
(Godwin Austen)
elev. 28,250 ▲

Yining

Junggar Basin

Urumqi

Kuqa

Bosten Hu

Turfan Depression

Tarim

Taklimakan Desert

Tarim Basin

XINJIANG UYGUR

Yutian

Kunlun

Lop Nor

Yumen

W. Sayan Mts.

Yenisey

Tannu Ola

Uvs Nuur

Hovd

Ulungur

Dzavhan

Hangayn Mts.

Han Us Nuur

Hovsgol Nuur

Selenge

Orhon

Ulan-Ude

Lake Baikal

Shilka

Ulan Bator ✪

MONGOLIA

Dzamin Uud

Choybalsan

Kerulan

Hentiyn Nuruu

Argun

Hulun Nur

Buir Nur

Yablonovyy Range

Oron

Greater Khingan Range

**NEI MONGOL
(INNER MONGOLIA)**

Gobi   Desert

Liao

Old Beds

Huang He (Yellow)

Baotou

Hohhot

Zhangjiakou

Datong

Beijing

Tangshan

Yongding

Baoding

TIANJIN

Tianjin

HEBEI

Bo He

Qaidam Basin

Golmud

QINGHAI

Qinghai Hu

Xining

Huanghe

**NINGXIA
HUIZU**

Yinchuan

Lanzhou

GANSU

Ordos Desert

Loess Plateau

SHANXI

Luliang Shan

Taiyuan

Taihang Shan

Shijiazhuang

Handan

Jinan

Zibo

SHANDONG

**TIBET
(XIZANG)**

Plateau   of   Tibet

Xigaze

Lhasa

Gyangze

Tsangpo

Nu (Salween)

Mekong

CHINA

Xianyang

Xian

SHAANXI
Qin Ling

Luoyang

Zhengzhou

HENAN

Xuzhou

JIANGS

Honze Hu

Bengbu

Xinyang

Hiainan

Zhenjiang

Hefei

ANHUI

Nan

Delhi

New Delhi

Himalayas

Kanpur

Lucknow

NEPAL

Kathmandu

Mt. Everest
elev. 29,028 ▲

Ganga Plain

Patna

Ganga (Ganges)

BHUTAN

Thimphu

Brahmaputra

INDIA

Kolkata
(Calcutta)

BANGLADESH

Dhaka

Mouths of the Ganga

Chittagong

Eastern Ghats

Gongga Shan
elev. 24,790 ▲

Chengdu

SICHUAN

Sichuan Basin

Chongqing

Qiaotou

Three Gorges Dam

Yichang

Wuhan

HUBEI

Daba Shan

Dabie Shan

Changjiang

Han

Dongting Hu

HUNAN

Changsha

Pingxiang

Nanchang

JIANGXI

Poyang Hu

Gan

Wuyi Shan

FUJIAN

Dalou Shan

Wuliang Shan

Yunnan-Guizhou Plateau

Kunming

Dian Chi

YUNNAN

Guiyang

GUIZHOU

Xiang

Wu

Nanning

**GUANGXI
ZHUANGZU**

Nan Ling

GUANGDONG

Guangzhou (Canton)

Zhu Jiang (Pearl)

Rongcheng

Shantou

Hong Kong
(China S.A.R.)

Macao
(China S.A.R.)

Zhanjiang

Haikou

HAINAN

BURMA

Arakan Yoma

Salween

Rangoon ✪

Gulf of Martaban

LAOS

Vientiane ✪

Hanoi ✪

Gulf of Tonkin

South China Sea

THAILAND

Bangkok ✪

CAMBODIA

VIETNAM

Bay of Bengal

Preparis North Channel

Andaman Sea

10°N

CHAPTER 9

East Asia

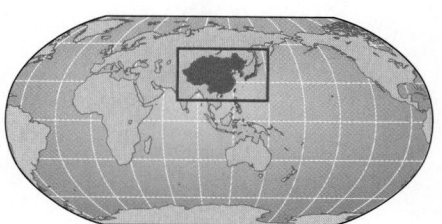

Global Patterns, Local Lives In 2000, at age 18, Li Xia (Li is her family name) left her farming village in China's Sichuan Province for the city of Dongguan in the southern province of Guangdong. She was accompanied by two friends. A few months earlier, the government had taken their families' farmland for an urban real estate project, paying compensation of just U.S.$2000 per family. The three young women accepted an offer to work in a Dongguan toy factory so they could send money back to their families, who now had to buy food and housing. Like so many migrants, they did not have the necessary official permission to leave their home territory. Despite this, they were excited. On the 4-day bus trip, they entertained dreams of one day shopping in the Wal-Mart supercenter located in Dongguan.

When the young women arrived in the city, they joined 5 million other recent internal migrants (Figure 9.2). Sixty percent of these were illegal migrants without residency rights, and hence dependent on their employers for housing. Like many others, Li Xia and her friends soon found that the labor recruiters had lied about their wages. They would be paid U.S.$30, not U.S.$45, per month. In addition, they would work 12-hour shifts in 100°F heat, receive no overtime pay, and have just one day off a month. But there was no point in protesting. Their families in Sichuan needed the money they would send home, and the recruiters purposely brought in thousands of extra workers to replace any complainers.

Xia felt better when she saw that the toy factory was a clean, modern building known locally as "Palace of Girls" (nearly all 3500 employees were female). Within a day, she had completed her training, signed a 3-year contract, and mastered her task of putting eyes on stuffed animals destined for toddlers in the United States and Europe. Her enthusiasm faded, however, when she learned that she and her friends would be spending much of their money on the expensive but low-quality food provided by the company, and would be sharing one small room and a tiny bath with eight other women and a rat or two. Her hopes of one day becoming a factory supervisor were dashed when she discovered that to obtain such a position she might have to grant sexual favors to the plant manager.

FIGURE 9.1 Regional map of East Asia.

FIGURE 9.2 Workers in the new Chinese economy. Chinese women produce Santa Claus dolls at the Fly Ocean Toy factory on the outskirts of Guangzhou (Canton) in Guangdong Province. This factory is similar to the factory in Dongguan where Lee Xia worked. [Guang Niu/Getty Images.]

In less than 6 months, Xia broke her contract and returned home to her village in Sichuan. There, using ideas and assertiveness she had gained from her time in Dongguan, she opened a snack stand. In only one month, she made ten times her investment of U.S.$12. But Xia yearned to return to the southern coast

to try again. A few months later, she returned to Dongguan with her sister, leaving the stand to her sister's husband (who also cared for the couple's baby).

Since her first arrival, the city had grown by 20 percent and now had 1400 foreign companies trying to hire thousands of workers. Through connections, Xia and her sister easily found jobs in a Taiwanese-owned factory at twice the wages Xia had originally earned at the other factory. Only one year after her first trip to Dongguan, Xia was making about U.S.$100 a month, enough to live relatively comfortably with only three roommates. She was sending money home and once again saving to open a bar in her home village.

Adapted from Kathy Chen, "Boom-town bound," Wall Street Journal (October 29, 1996): 1, A6; "Life lessons," Wall Street Journal (July 9, 1997); Peter S. Goodman, "A turn from factories, China's labor pool is shifting," Washington Post National Weekly Edition (October 11–17, 2004): 19; Louisa Lim, "The end of agriculture in China," Reporter's Notebook, National Public Radio, May 19, 2006. ■

Li Xia's experiences illustrate how urbanization, globalization, and changes in gender roles are transforming East Asia. The needs of rural areas are being subverted to the needs of China's burgeoning cities. Rural land-grabs by developers have increased fifteen-fold since the mid-1990s, and farmers rarely get a fair deal. Meanwhile, millions of rural young adults have flocked to work in factories in East Asia's coastal cities, producing goods for sale on global markets (Figure 9.3).

In China, most rural to urban migration is illegal because the *hukou* (household registration) **system,** begun in the Maoist era and only now under revision, effectively tied rural

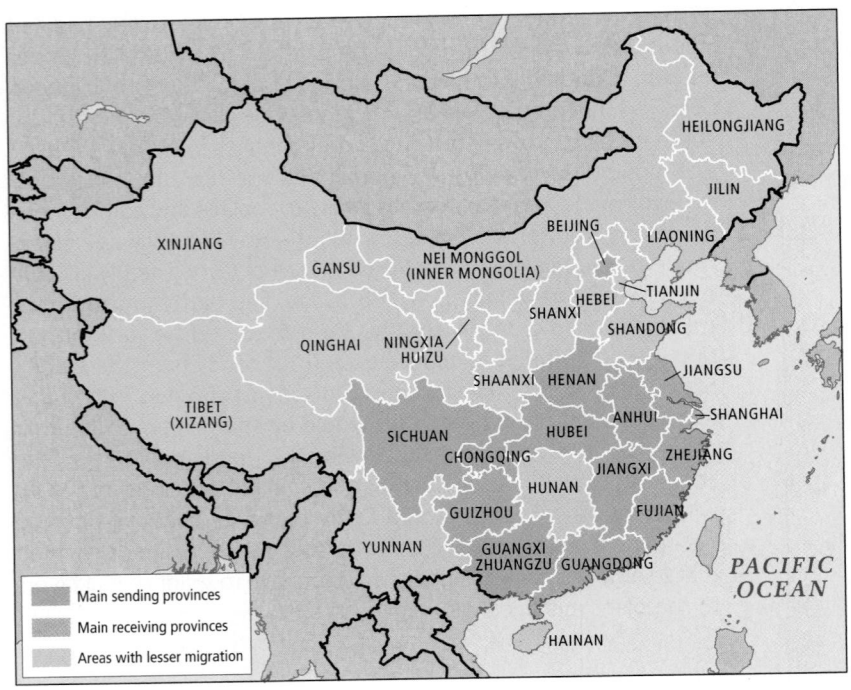

FIGURE 9.3 Rural-to-urban migration in China. By 2003, some 114 million migrant workers had moved from rural areas to urban jobs in China. About one-third of these migrants were women. The map identifies the interior provinces from which most of the rural workers came, and the primarily coastal areas where they found work. [Adapted from Zhan Shaohua, "Rural labour migration in China: Changes for policies," *Policy Papers/10* (United Nations Economic, Social and Cultural Organization, 2005), p. 14, at http://unesdoc.unesco.org/images/0014/001402/140242e.pdf.]

people to the place of their birth. In a desperate search for work, over 100 million people like Li Xia are ignoring the *hukou* system, but by doing so, they became part of what is called the **floating population,** which has no rights to housing, schools, or health care. These urban migrants generally work in menial, low-wage jobs and make agonizing sacrifices to send money home to children and spouses living in rural areas. While the remittances help, migration deprives poor rural locations of the energy, initiative, and leadership of young adults. This contributes to the growing disparity in wealth between East Asia's poor rural locations and booming, over-crowded urban centers.

Li Xia's story also hints at a surprising change in the labor market that is underway in China: the labor surplus is disappearing! Turnover of low-paid, ill-treated workers is very high, and firms are beginning to offer better wages and other amenities to retain workers. The next time we encounter Li Xia, in a personal vignette in the Economic and Political Issues section, the continuation of her story will reveal how improving circumstances for workers are changing China and will probably bring higher prices to consumers worldwide.

China is part of the region of East Asia, home to nearly one-fourth of humanity. This vast territory stretches from the Taklimakan Desert in far western China to Japan's rainy Pacific coast line, and from the frigid mountains of Mongolia in the north to the subtropical forests of China's southeastern coastal provinces (see Figure 9.1). East Asia (Figure 9.4) includes the countries of China, Mongolia, North Korea, South Korea, Japan, and Taiwan (the last has been independent since World War II but has never been recognized by China). These countries are grouped together because of their cultural and historical links, many of which stem from China. Because of China's great size, enormous population, and huge economy, we give it particular emphasis in this chapter. Japan, whose large and prosperous economy make it a major player on the world stage, is also emphasized.

Questions to consider as you study East Asia (also see Views of East Asia, pages 328–329)

1. How is China's spectacular urban growth, and the massive increase in pollution that has come with it, linked to processes of globalization set in motion three decades ago? China's cities have grown by 450 million since 1980, when China's economy was opened to the global economy. Rapid urban economic growth has made China's cities some of the most polluted on the planet. How did globalization and China's response to it lead the country's cities to where they are today?

2. How might global warming result in changes in water availability? East Asia has long suffered from enormously destructive droughts and devastating floods. Now China's highest glaciers are melting. How might this effect the country's rivers? In what ways is China more vulnerable to global warming than Japan, the Koreas, or Taiwan?

3. In what ways has China's embrace of globalization created pressures that might eventually lead to democratization? The growing gap in wealth between China's globalizing cities and its much poorer rural areas has created a reservoir of discontent, yet the most recent shifts toward democracy are happening in urban areas. What changes in urban areas have led to these new developments?

4. How has East Asia's ability to feed itself been transformed by globalization? Why might the region's food systems be criticized for a lack of sustainability? East Asian countries have become more able to buy food on the global market and less able to produce the food they need at home. What kinds of problems might this create both for East Asia and for the rest of the world?

5. Why are East Asia's populations aging so rapidly and why does this matter? All countries in East Asia either have already gone through the *demographic transition* or are in the midst of it now. As birth rates slow and life expectancies increase, more retirees must be supported by fewer younger workers. Why is this significant? How is Japan responding to this challenge?

6. How has the interaction between the government's population policy and Chinese culture resulted in a shortage of women in China? China's one-child policy has left most couples with only one chance to give their families an heir. Why do most families want a male child? What problems have been created by the resulting shortage of women?

FIGURE 9.4 Political map of East Asia.

Views of East Asia

For a tour of these places, visit www.whfreeman.com/pulsipher

[Background image: [Trent Schindler NASA/GSFC Code 610.3/587 Scientific Visualization Studio]

1. Urbanization. Shanghai (shown below) and its surrounding suburbs and nearby towns now account for over 25% of China's GDP. See pages 350–353 for more. Lat 31°14'12"N, Lon 121°30'E. [Brian Kell]

2. Population and gender. The poster below (seen on a building in Beijing) promotes the "one child policy", which has helped control China's population growth, but also contributed to a shortage of women. See pages 344–346 for more. Lat 39°54'13"N, Lon 116°23'15"E. [Zhou Yuwei]

1987   2006

. Water. China's Three Gorges Dam as transformed the Chang Jiang river, egrading it as a habitat for endangered sh and forcing the relocation of illions of people. The pictures above ompare the un-dammed river in 1987, o the dammed river in 2006. See page 35–336 for more. Lat 30°49'N, on 111°E. [NASA/Goddard Space Flight enter Scientific Visualization Studio/USGS]

6. Globalization, Guangdong. China is now the worlds largest exporter of manufactured goods, These Chinese-made cargo cranes are en route from Guangdong to Germany. See pages 349–354 for more. Lat 22°39'N, Lon 113°40'50"E. [http://commons.wikimedia.org/wiki/Image:Containerb%C3%BCcken_f%C3%BCr_Eurogate_009.jpg]

3. Global warming and air pollution. China now rivals the US as the world's largest producer of greenhouse gasses. Especially around Beijing, industries and vehicles create a deadly haze of air pollution (shown below). China is responding with stricter emissions standards than the US has. See pages 332 and 336–337 for more. Lat 39°53'N, Lon 116°32'E. [Image courtesy the SeaWiFS Project, NASA/Goddard Space Flight Center, and ORBIMAGE ID: GL-2002-001716 NASA Goddard Space Flight Center; http://commons.wikimedia.org/wiki/Image:Beijing_traffic_jam.JPG]

4. Food. As incomes have risen so has demand for meat and fish. Japan is a major consumer of fish and has been criticized by environmentalists throughout the world for overfishing tuna (shown here snaged in a net). Tokyo, shown on the map, has the largest fish market in the world. See pages 334–335 for more. Lat 35°39'40"N, Lon 139°46'11"E. [Courtesy of United Nations Food and Agriculture Organization]

2&3
Beijing

4
Tokyo

1
Shanghai

7
Three Gorges Dam

5
Taipei

6
Guangdong

5. Democracy, Taipei. The right to protest is well established in Taiwan, South Korea, and Japan, and is emerging in Mongolia, but is not recognized yet in China and North Korea. These women are marching in Taipei, Taiwan for the rights of lesbians to adopt. See pages 354–356 for more. Lat 25°02'05"N, Lon 121°31'18"E. [http://commons.wikimedia.org/wiki/Image:Taiwan_Pride_2006_COSWAS.JPG]

The Geographic Setting

Terms to Be Aware Of

East Asian place-names can be very confusing to English-speaking readers. We give place-names in English transliterations of the appropriate Asian language, taking care to avoid redundancies. For example, *he* and *jiang* are both Chinese words for *river*. Hence the Yellow River is the Huang He, and the Long River is the Chang Jiang (also called the Yangtze); it is redundant to add the term *river* to either name. The word *shan* appears in many place-names and usually means "mountain."

Pinyin (a spelling system based on Chinese sounds) versions of Chinese place-names are now commonplace. For example, the city once called Peking in English is now Beijing, and Canton is Guangzhou.

The region popularly known as Manchuria is here referred to as China's Far Northeast to emphasize its geographical location. Although China refers to Tibet as Xizang, people around the world who support the idea of Tibetan self-government avoid using that name. This text uses Tibet for the region (with Xizang in parentheses), and Tibetans for the people who live there.

PHYSICAL PATTERNS

A quick look at the regional map of East Asia (see Figure 9.1) reveals that its topography is perhaps the most rugged in the world. East Asia's varied climates result from a dynamic interaction between huge warm and cool air masses and the land and oceans. Its rapidly expanding human populations have affected the variety of ecosystems that have evolved there over the millennia and that still contain many important and unique habitats.

Landforms

The complex topography of East Asia is partially the result of the slow-motion collision of the Indian subcontinent with the southern edge of Eurasia over the past 60 million years. This tremendous force created the Himalayas and lifted up the Plateau of Tibet (depicted in gray and gold in Figure 9.1), which can be considered the highest of four descending steps that define the landforms of mainland East Asia moving roughly west to east.

The second step down is a broad arc of basins, plateaus, and low mountain ranges (depicted in yellowish tan in Figure 9.1). These landforms include the broad, rolling highland grasslands and deep, dry basins and deserts of Western China and Mongolia. This step also includes the rugged Yunnan–Guizhou Plateau, which is dominated by a system of deeply folded mountains and valleys that bend south through the Southeast Asian peninsula.

The third step, directly east of this upland zone, consists mainly of broad coastal plains and the deltas of China's great rivers (shown in shades of green in Figure 9.1). Starting from the south, this step is defined by three large lowland river basins: the Zhu Jiang (Pearl River) basin, the massive Chang Jiang basin, and the Huang He lowland basin on the North China Plain. Low mountains and hills (shown in light brown) separate these river basins. China's Far Northeast and the Korean Peninsula are also part of this third step.

The fourth step consists of the continental shelf, covered by the waters of the Yellow Sea, the East China Sea, and the South China Sea. Numerous islands—including Hong Kong, Hainan, and Taiwan—are anchored on this continental shelf; all are part of the Asian landmass.

The islands of Japan have a different geological origin: they are volcanic rather than being part of the continental shelf. They rise out of the waters of the northwestern Pacific in the highly unstable zone where the Pacific, Philippine, and Eurasian plates grind against one another. Lying along a portion of the Pacific Ring of Fire (see Figure 1.22 on page 26), the entire Japanese island chain is particularly vulnerable to disastrous eruptions, earthquakes, and **tsunamis** (seismic sea waves). The volcanic Mount Fuji, perhaps Japan's most recognizable symbol (Figure 9.5), last erupted in 1707. In 2001, however, there were deep internal rumblings, suggesting that the mountain's period of dormancy may be ending.

The East Asian landmass has few flat portions, and many of them are very dry or very cold. Consequently, the large numbers of people who occupy the region have had to be particularly inventive in creating spaces for agriculture. They have

FIGURE 9.5 Mount Fuji. Japan's highest peak, at 12,388 feet (3776 meters), provides a stately distant backdrop to a woman working in tea fields. At one time considered a sacred mountain, Mount Fuji today attracts some 200,000 climbers annually, 30 percent of them foreigners. [Chad Ehlers/Alamy.]

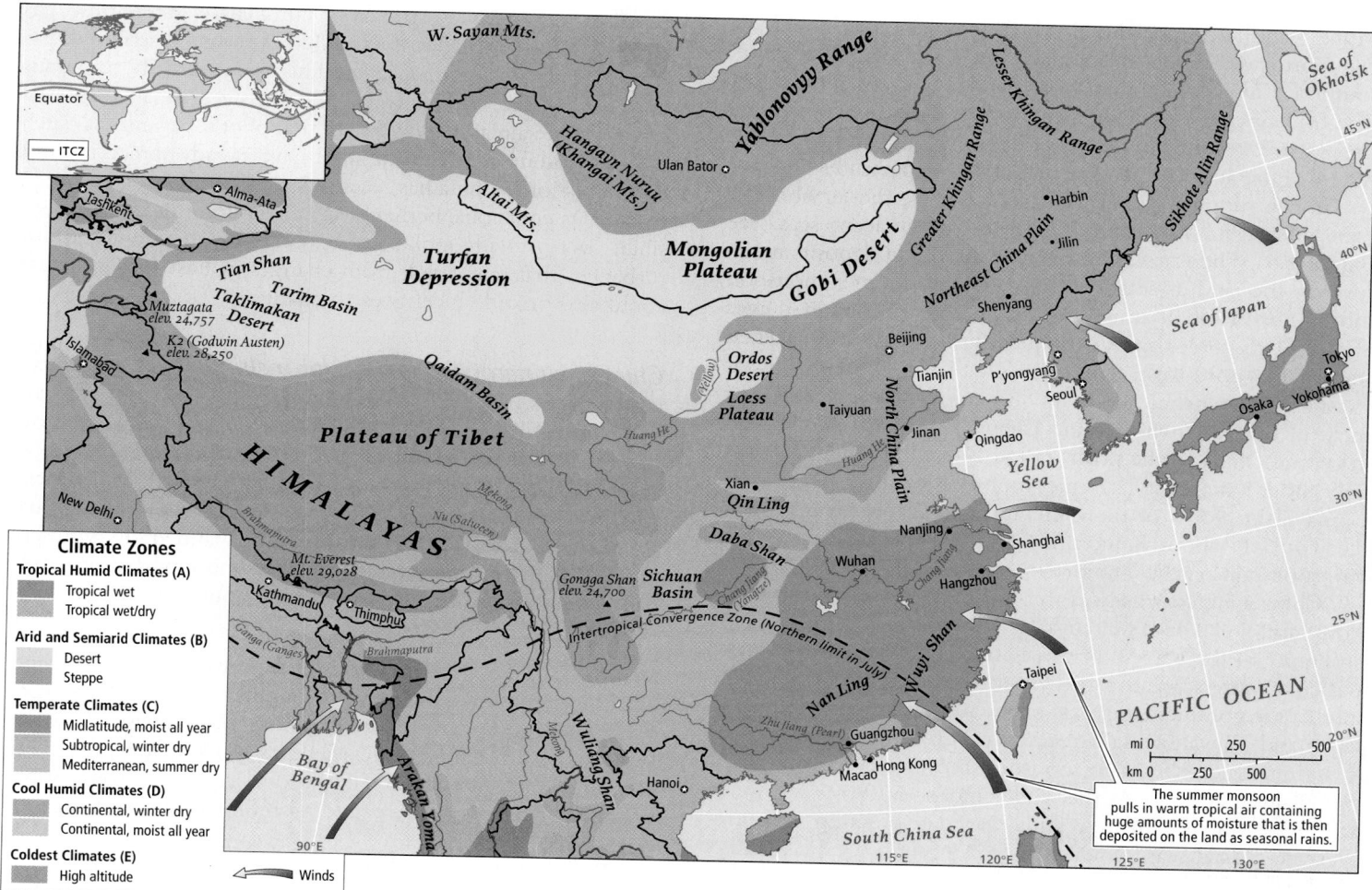

FIGURE 9.6 Climates of East Asia.

cleared and terraced entire mountain ranges using only simple hand tools. They have irrigated drylands with water from melted snow, drained wetlands using elaborate levees and dams, and applied their complex knowledge of horticulture and animal husbandry to help plants and animals flourish in difficult conditions.

Climate

East Asia has two contrasting climate zones (Figure 9.6): the dry interior west and the wet (monsoon) east. Recall from Chapter 8 that the term *monsoon* refers to the seasonal reversal of surface winds that flow from the Eurasian continent to the surrounding oceans during winter and from the oceans inland during summer.

The dry interior. Because land heats up and cools off more rapidly than water, the interiors of large landmasses in the midlatitudes tend to experience intense cold in winter and powerful heat in summer. Western East Asia, roughly corre-

sponding to the first two topographic steps described above, is an extreme example of such a midlatitude continental climate because it is very dry. With little vegetation or cloud cover to retain the warmth of the sun after nightfall, summer daytime and nighttime temperatures may vary by as much as 100°F (55°C).

Grasslands and deserts of several types cover most of the land in this dry region. Only scattered forests grow on the few relatively well-watered mountain slopes and in protected valleys supplied with water by snowmelt. In all of East Asia, humans and their effects are least conspicuous in the large, uninhabited portions of the deserts of Tibet (Xizang), the Tarim Basin in Xinjiang, and the Mongolian Plateau.

The monsoon east. The monsoon climates of the east are influenced by the extremely cold conditions of the huge Eurasian landmass in the winter and the warm temperatures of the surrounding seas and oceans in the summer. During the dry winter monsoon, descending frigid air sweeps south and east through East Asia, producing long, bitter winters on the

Mongolian Plateau, on the North China Plain, and in China's Far Northeast. While occasional freezes may reach as far as southern China, winters are shorter and less severe in that area. The cold air of the dry winter monsoon is partially blocked by the east-west mountain ranges of the Qin Ling, and the warm waters of the South China Sea moderate temperatures on land.

In the summer, as the continent warms, the air above it rises, pulling in wet tropical air from the adjacent seas (see Figure 9.6). The warm, wet air from the ocean deposits moisture on the land as seasonal rains. As the summer monsoon moves northwest, it must cross numerous mountain ranges and displace cooler air. Consequently, its effect is weakened toward the northwest. Thus the Zhu Jiang basin in the far southeast is drenched with rain and enjoys warm weather for most of the year, whereas the Chang Jiang basin, which lies in central China to the north of the Nan Ling range, receives only about 5 months of summer monsoon weather. The North China Plain, north of the Qin Ling and Dabie Shan ranges, receives only about 3 months of monsoon rain. Very little monsoon rain reaches the dry interior.

China's Far Northeast is wet in summer, and neighboring Korea and Japan have wet climates year-round because of their proximity to the sea. All of these areas still have hot summers and cold winters because of their northerly location and exposure to the continental effects of the huge Eurasian landmass. Japan and Taiwan actually receive monsoon rains twice: once in spring, when the main monsoon moves toward the land, and again in autumn, as the winter monsoon forces warm air off the continent. This retreating warm air picks up moisture over the coastal seas, which is then deposited on the islands. Much of Japan's autumn precipitation falls as snow.

Natural hazards. The entire coastal zone of East Asia is intermittently subjected to **typhoons** (tropical cyclones). Japan's location along the northwestern edge of the Pacific Ring of Fire means that it has many volcanoes, and it also experiences earthquakes and tsunamis. These natural hazards are a constant threat in Japan; the heavily populated zone from Tokyo southwest through the Inland Sea is particularly endangered. Earthquakes are also a serious natural hazard in Taiwan and in China's mountainous interior. 🎥

ENVIRONMENTAL ISSUES

Today, East Asia's worst environmental problems result from high population density combined with rapid urbanization and environmentally insensitive economic development (Figure 9.7). Here we focus on global warming, water issues, food production, and air pollution.

Global Warming: Emissions and Vulnerability in East Asia

Global warming is a growing concern in East Asia. China now rivals the United States as the world's largest producer of greenhouse gases (see Chapter 1, page 19), and Japan is also a major emitter. New studies of cumulative greenhouse gas emissions since 1850 show that China's recent greenhouse gas emissions are so large that its total historical contribution trails only that of the United States and Russia. China's historical emissions are already greater than that of other countries (such as Japan and most of Europe) that have already agreed to limit their emissions. China has recently given attention to the need to limit its emissions, perhaps out of recognition of its responsibility as a growing global leader. China's vulnerability to disturbances that may result from changes in East Asia's climate could also provide incentives to change.

China's vulnerability to glacial melting. Two of China's largest rivers, the Huang He and the Chang Jiang, are partially fed by glaciers in the Himalayas. These glaciers are now melting so fast that scientists predict they will be gone in fifty years. One result could be significantly lower flows in these two great rivers during the dry winter monsoon. Both have already begun to run low during winter, and trade is suffering as many river boats and barges become stranded on sandbars. Irrigation for dry season farming could also suffer.

Droughts. Triggered by periods of abnormally low rainfall and abnormally high temperatures, droughts occur in China every year. They often cause more suffering and damage than any other natural hazard. Many scientists think that higher temperatures and shifting rainfall patterns resulting from global warming may have contributed to recent droughts throughout China.

Droughts have also been worsened by human activity on a local or regional scale. When people begin to live or farm in dry environments, as many millions have done in China and Mongolia during the twentieth century, *desertification* (see Chapter 6, page 215) can result. Natural vegetation is cleared to grow crops, which are irrigated with water pumped to the surface from underground aquifers. However, many dry areas in China are subject to strong winds that can blow away topsoil. Further, irrigated crops are often less able to hold the soil than the natural vegetation. As a result, huge dust storms have plagued China in recent years. High dunes of dirt and sand have appeared almost overnight in some areas bordering the desert, threatening crops and homes. 🎥

Water shortages have been particularly intense in the North China Plain, which produces half of China's wheat and a third of its corn. Here the water table is falling more than 10 feet a year due to increased use of groundwater for irrigation and urban needs. Meanwhile, withdrawals of water from the Huang He often make its lower sections run completely dry during the winter and spring. As a result, the rapidly growing urban areas of the North China Plain, such as China's capital, Beijing, are facing severe water shortages that constrain industrial expansion and economic growth. In the coming years, more water will have to be diverted from agricultural purposes to support Beijing, whose demand for water is projected to double in the next decade.

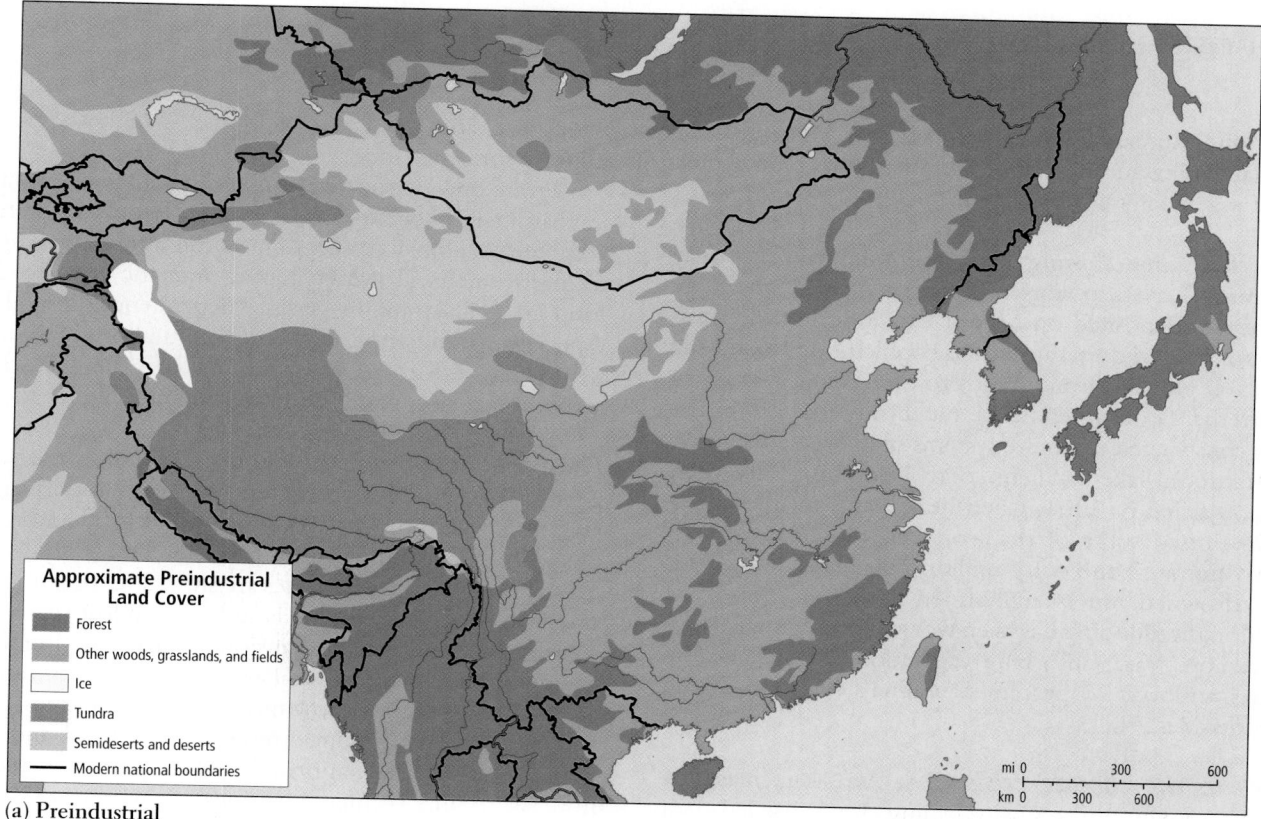

(a) Preindustrial

Approximate Preindustrial Land Cover

- Forest
- Other woods, grasslands, and fields
- Ice
- Tundra
- Semideserts and deserts
- Modern national boundaries

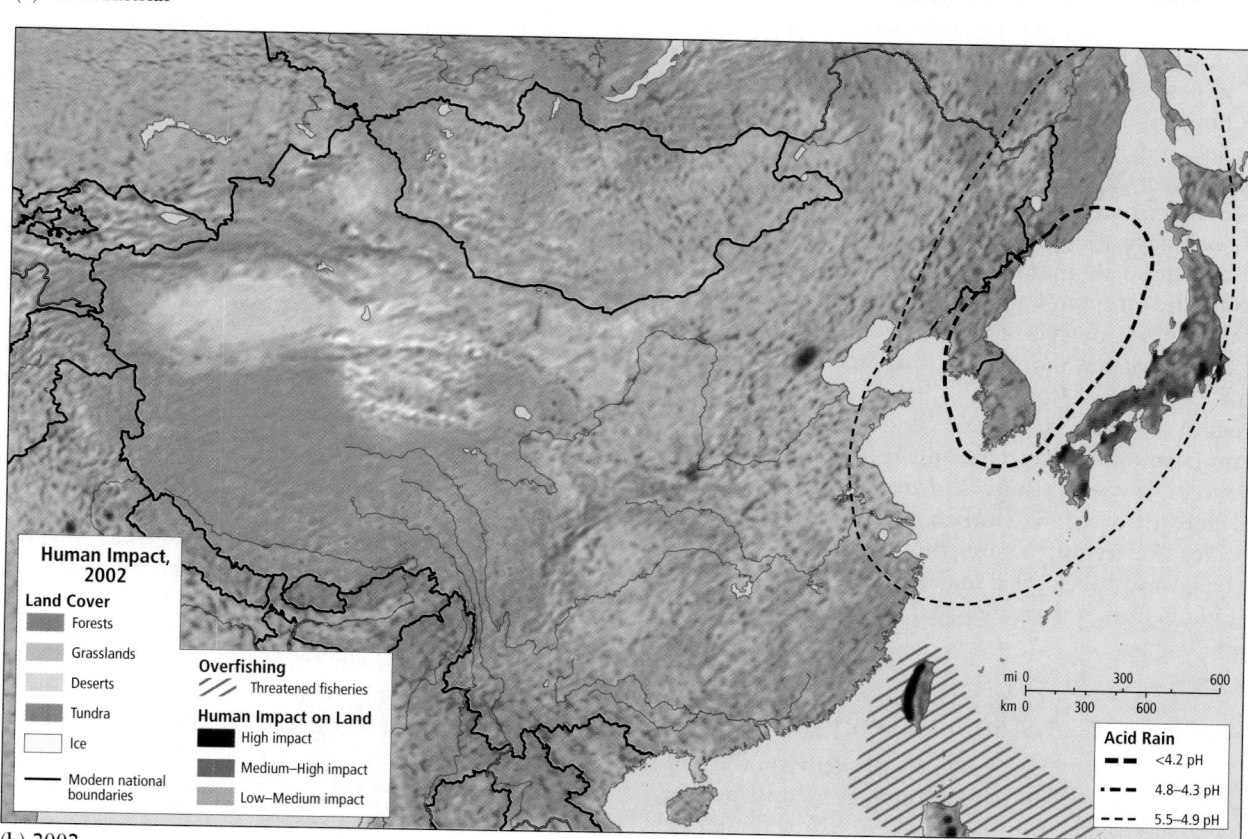

(b) 2002

Human Impact, 2002

Land Cover
- Forests
- Grasslands
- Deserts
- Tundra
- Ice
- Modern national boundaries

Overfishing
- Threatened fisheries

Human Impact on Land
- High impact
- Medium–High impact
- Low–Medium impact

Acid Rain
- <4.2 pH
- 4.8–4.3 pH
- 5.5–4.9 pH

FIGURE 9.7 Land cover (preindustrial) and human impact (2002). East Asians had already changed land cover extensively by 1000 years ago (a), and by 2002 (b), human impact had significantly intensified. Nearly every location was affected in one way or another, and in some places the impact was extreme. [Adapted from *United Nations Environment Programme, 2002, 2003, 2004, 2005, 2006* (New York: United Nations Development Programme), at http://maps.grida.no/go/graphic/human_impact_year_1700_approximately and http://maps.grida.no/go/graphic/human_impact_year_2002.]

Japan, the Koreas, and Taiwan are less vulnerable to drought than China due to their proximity to the ocean, which gives them a generally wetter climate.

Flooding. The same shifting patterns of rainfall that may worsen droughts can also worsen flooding. Already, the huge amounts of rain deposited on eastern China during the summer monsoon periodically cause catastrophic floods along the major rivers. If global warming leads to even slight changes in rainfall patterns, flooding could be much more severe. Engineers have constructed elaborate systems of dikes, dams, reservoirs, and artificial lakes to help control flooding. However, these systems failed in 2004, when heavy rains in the Chang Jiang basin caused some of the worst flooding in two centuries. Two hundred and forty million people were affected, with 3656 drowned and 14 million left homeless.

Like drought, flooding has been worsened by human activities that remove water-absorbing vegetation and soil. Deforestation has removed 25 percent of China's forests over the past three decades.

Responses to the climate crisis. East Asia is increasingly focused on responding to global warming. Japan has led such efforts for decades. In 1997 the Japanese city of Kyoto hosted the meeting in which countries first committed to reduce their greenhouse gas emissions. Japanese auto-makers such as Toyota were among the first to develop and sell hybrid gas-electric vehicles, and Japan is now second only to Germany in installed solar power generating capacity. However, Japan has seen little actual reduction of its greenhouse gas emissions. China may soon take the lead in efforts to reduce emissions because of its potentially massive markets and access to huge flows of foreign investment. China's government has set a target of generating 10 percent of its energy from renewable sources by 2010, a much more ambitious target than Japan's 1.35 percent.

With regard to water shortages, China has more limited options than Japan, which receives plentiful rain. With few untapped sources of water, China is focusing on conservation. Already, 30 percent of China's urban water is now recycled, and many cities are trying to raise this number. Supporting these conservationist policies is a major new national effort to remove pollutants from wastewater discharged by industry and farming. ◼▸

Food Security and Sustainability in East Asia

> About three-quarters of the food consumed in Japan, South Korea, and Taiwan is imported.

East Asia's **food security,** the ability of a country to feed its people, is increasingly linked to the global economy. The wealthiest countries in the region are already heavily dependent on food imports from around the globe. About three-quarters of the food consumed in Japan, South Korea, and Taiwan is imported. China is currently self-sufficient for basic necessities such as rice, but it is highly dependent on imports for other foods.

From one perspective, high food imports are a sign of increasing food security in East Asia because countries with competitive globalized economies can afford to bring in food from elsewhere. However, recent dramatic increases in global food prices illustrate the perils of dependence on food imports, especially for the poor. ◼▸ If agricultural production declines worldwide due to global warming or other causes, people across the region could be hit with much higher food prices.

Food production and sustainability. Only a relatively small portion of East Asia's vast territory can support agriculture (Figure 9.8). In much of this area, food production has been pushed well beyond what can be sustained over the long term; hence, East Asia's fertile zones are shrinking. In China, roughly one-fifth of the agricultural land has been lost since 1949 due to urban and industrial expansion and to agricultural mismanagement that has created soil erosion and desertification. Also, as urban populations grow more affluent, their taste for meat and other animal products requires more land and resources than the plant-based national diet of the past. As a result, China now has to import huge quantities of soybeans (41 percent of the world's total in 2005) for use as animal feed.

Rice cultivation. The region's most important grain is rice, and its cultivation has dramatically transformed landscapes throughout southern China, Japan, Korea, and Taiwan. In these areas, rainfall is sufficient to sustain **wet rice cultivation,** which is capable of very high productivity. Elaborate systems of water management are required, as the roots of the plants must be submerged in water early in the growing season. Centuries of painstaking human effort have channeled rivers into intricate irrigation systems, and whole mountainsides have been transformed into cascading terraces that evenly distribute the water. Geographer Chiao-Min Hsieh describes how "[e]verywhere one can hear water gurgling like music as it brings life and growth to the farms." However, these same cultivation techniques, combined with extensive forestry and mining, have led to the loss of most natural habitat in all but the most mountainous, dry, or remote areas. It is unlikely that the area under wet rice cultivation can be expanded.

Japan and the sea. Many East Asians depend heavily on fishing for food. The Japanese in particular have had a huge impact on the seas not only of this region but of the entire world. With only 2 percent of the world's population, Japan consumes 15 percent of the global fish catch. There are some 4000 coastal fishing villages in Japan, sending out tens of thousands of small crafts to work nearby waters each day. Japan also sends out large fishing ships, complete with onboard canneries and freezers, to harvest oceans around the world. Japan has long been criticized by environmentalists for overfishing tuna around the globe and for its consumption of endangered marine species, especially whales. With more than 75 percent

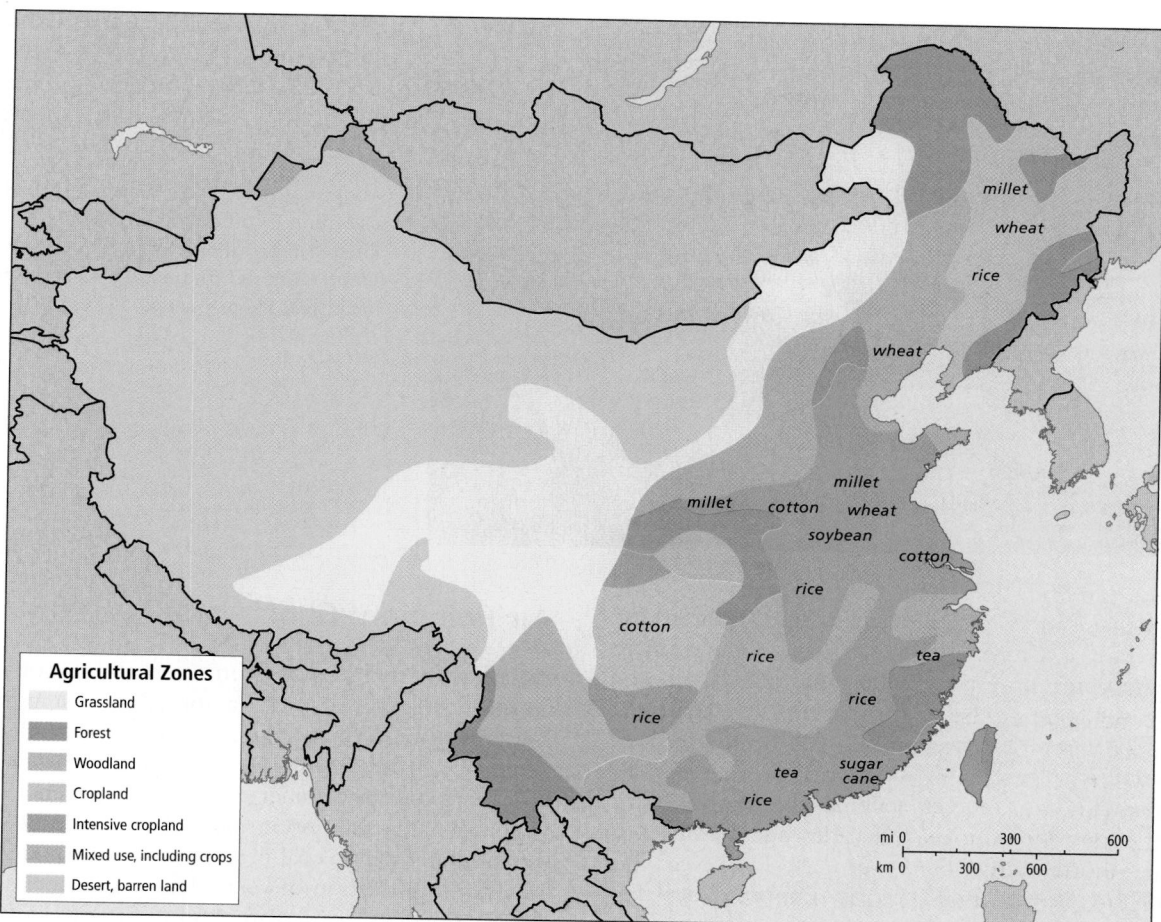

FIGURE 9.8 China's agricultural zones. As part of the economic reforms instituted in recent years in China, greater regional specialization in agricultural products is taking place. [Adapted from "World agriculture," *National Geographic Atlas of the World, Eighth Edition* (Washington, D.C.: National Geographic Society, 2005), p. 19; and "China—Economic, minerals," *Goode's World Atlas 21st Edition* (New York: Rand McNally, 2005) p. 207.]

of the world's fisheries either fully exploited or in decline, there is little room for Japan to expand its fish intake.

Three Gorges Dam: The Power of Water

The Three Gorges Dam (Figure 9.9) is the largest engineering project in history and the largest dam in the world at 600 feet (183 meters) high and 1.4 miles (2.3 kilometers) wide. It is designed to improve navigation on the Chang Jiang (Yangtze), control flooding, and generate electricity equivalent to about 3 percent of the national total. The scheduled completion date is 2009.

Many qualified experts involved with the project see serious flaws. As many as 80 cracks have appeared in the dam, which suggests shoddy construction. Similar defects led to the failure of China's much smaller Banqiao Dam in 1975, which was responsible for 171,000 deaths due to flooding and ensuing famine. Another worry is that earthquakes could seriously weaken the dam, which sits above a seismic fault. The enormous weight of water in the 370-mile (600-kilometer) long reservoir behind the dam could lubricate the fault and trigger an earthquake. Even if the dam holds, its power generation potential will probably be reduced by the buildup of silt behind the dam due to soil erosion upstream.

Any failure of the dam would be a financial as well as human disaster. Construction costs have already grown to $25 billion, but the real costs may be three times this figure, due in part to massive theft from the project by corrupt officials.

Then there are the incalculable costs associated with relocating the 1.9 million people who once lived where the dam now forms a reservoir. Thirteen major cities, 140 large towns, hundreds of small villages, 1600 factories, and 62,000 acres (25,000 hectares) of farmland have been submerged. The reservoir has destroyed important archaeological sites, as well as some of China's most spectacular natural scenery. There are significant environmental costs as well. The giant sturgeon, for

FIGURE 9.9 The Three Gorges Dam. This 2006 photo shows a distant view of the world's largest hydropower dam project under construction, which is located on the Chang Jiang (Yangtze River). At the time it was taken, less than 3000 cubic meters of concrete remained to be placed before the dam would be completed—9 months ahead of schedule according to official Chinese sources. The dam is situated near Xiling Gorge, the easternmost of the Three Gorges in the middle reaches of the Chang Jiang. [AP Photo/Xinhua, Du Huaju.]

example, a fish that can weigh as much as three-quarters of a ton and is as rare as China's giant panda, may go extinct. Sturgeon used to swim more than 1000 miles (1600 kilometers) up the Chang Jiang past the location of the dam to spawn. Now the sturgeon's reproductive process has been irretrievably interrupted.

International funding sources such as the World Bank withdrew their support of the dam decades ago because of these and other shortcomings of the project. However, the construction of *Da Ba* (the Big Dam) is supported by Chinese industrialists who need the energy, construction companies that have prospered from building the dam, and government officials eager to impress the world and leave their mark on China with this mega-project. 📹

Air Pollution: Choking on Success

Although air pollution is often severe throughout East Asia, the air quality in China's cities is among the worst in the world. Coal burning is the primary cause. China is the world's largest consumer and producer of coal, accounting for 40 percent of all the coal burned on the planet each year. Between 1975 and 2005, China's coal consumption more than quadrupled. By 2006, it was bringing a new coal-fired power plant online every week.

The combustion of coal releases high levels of two pollutants, suspended particulates and sulfur dioxide, both of which can cause respiratory ailments. In Chinese cities, these emissions can be ten times higher than World Health Organization guidelines (Figure 9.10). China's government estimates

Ambient Concentrations of Air Pollutants

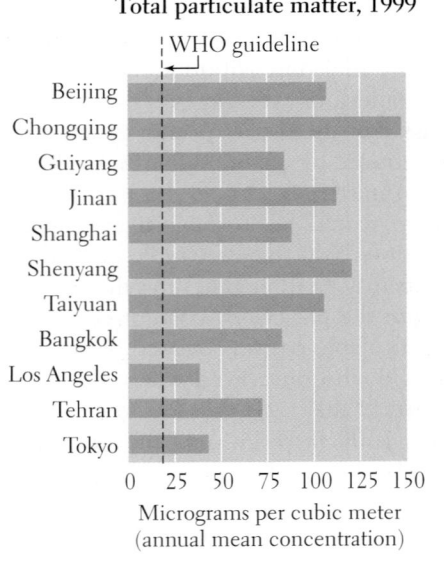

Total particulate matter, 1999

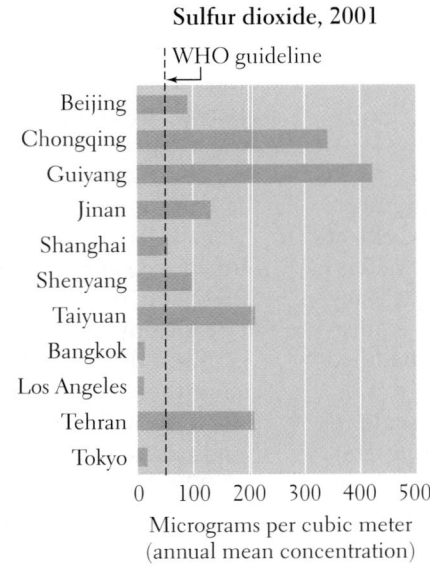

Sulfur dioxide, 2001

FIGURE 9.10 Air pollutant concentrations in selected Chinese and other large cities, 1999 and 2001. The vertical dashed lines indicate World Health Organization guidelines for safe concentration levels. [Adapted from http://devdata.worldbank.org/wdi2005/Table3_13.htm#about.]

that emissions from coal burning cause roughly 656,000 premature deaths each year and cost the country about 10 percent of its GDP due to health problems and lost productivity.

Air pollution occurs primarily in urban areas, where industries are concentrated and where there are large numbers of homes to be heated. The worst pollution is in northeastern China, where homes and industries that both depend heavily on coal for fuel are in close proximity. Sulfur dioxide from coal burning also contributes to acid rain, which is displaced to the east by prevailing winds, reaching as far as Korea, Japan, Taiwan, and beyond. Particulates from China's coal burning are transported globally by high-altitude west-to-east flowing jet streams, thus affecting air quality in North America and Europe.

Air pollution from vehicles is also severe even though the use of cars for personal transport in China has hardly begun. This is because China's vehicles have traditionally had very high rates of lead and carbon dioxide emissions. For example, Beijing has much worse air quality than Los Angeles even though it has only one-tenth the vehicles. Fortunately, the government is taking action. As of 2008, new Chinese cars have to conform to higher mileage and stricter emissions standards than U.S. cars. Even so, to provide decent air quality during the 2008 summer Olympics, China had to remove half of Beijing's cars from the road.

FIGURE 9.11 Pollution in Taiwan. A family masked against air pollution goes for a drive in a park in Kaohsiung, southwestern Taiwan. [Jodi Cobb/National Geographic Image Collection.]

Air pollution outside China. Public health risks related to air pollution are serious in the largest cities of Japan (Tokyo and Osaka), Taiwan (Taipei), and South Korea (Seoul), and in adjacent industrial zones. Antipollution legislation is being passed, and enforcement is increasing, but high population densities and rising expectations for better standards of living make it difficult to improve environmental quality. Taiwan is a case in point.

Taiwan has some of the dirtiest air on earth (Figure 9.11). Primary causes are the island's extreme population density of 1600 people per square mile (615 per square kilometer) and its high rate of industrialization. In the far north, around the metropolitan area of Taipei, seven cities house a total of 3.26 million people. Densities can rise to 5000 people or more per square mile (1930 per square kilometer). As standards of living have increased, so has per capita energy use. There are now four motor vehicles (cars or motorcycles) for every five residents, which adds up to more than 16.5 million exhaust-producing vehicles on this small island. In addition, there are nearly eight registered factories for every square mile (three per square kilometer), all emitting waste gases. As a result, Taiwan's air is acknowledged by the government to be six times dirtier than that of the United States or Europe.

Ten Key Aspects of East Asia

- From 1100 to 1600, China was the world's most developed region: it had the wealthiest economy, the largest cities, the highest living standards, and the most advanced technologies.
- After three decades of rapid economic growth since opening up to the global economy, China has the third largest economy in the world, following the European Union and the United States.
- China's urban population has grown by 450 million since 1980, due mainly to massive migration from poor rural areas to booming and globalizing coastal cities.
- In response to growing public protests in urban areas, China's government is increasingly allowing elections to be held at the local level.
- Overall population growth has slowed as government policies, urbanization, and changing gender roles have given East Asia the lowest rate of natural increase of all world regions except Europe.
- A cultural preference for sons over daughters has produced gender imbalance in all East Asian populations, with males outnumbering females substantially in many age groups.
- China now rivals the United States as the world's largest producer of greenhouse gases.
- One-third of China's population does not have safe drinking water, and only 24 percent has access to adequate sanitation.
- One-fifth of China's agricultural land has been lost since 1949 due to soil erosion, desertification, and urban, and industrial expansion.

- Female workers earn about two-thirds of what male workers earn in China, and about half of what men earn in South Korea and Japan.

HUMAN PATTERNS OVER TIME

East Asia is home to some of the most ancient civilizations on earth. Settled agricultural societies have flourished in China for more than 7000 years. A little more than 2000 years ago, the basic institutions of government that still exist across the region were established in eastern China. Until the twentieth century, China was the main source of wealth, technology, and culture across East Asia.

Chinese civilization evolved from several hearths, including the North China Plain, the Sichuan Basin, and the lands of interior Asia that were inhabited by Mongolian nomadic pastoralists. On East Asia's eastern fringe, the Korean Peninsula and the islands of Japan and Taiwan were profoundly influenced by the culture of China, but they were isolated enough that each developed a distinctive culture and maintained political independence most of the time. In the early twentieth century, Japan industrialized rapidly by integrating European influences that China disdained.

Bureaucracy and Imperial China

Although humans and their ancestors have lived in East Asia for hundreds of thousands of years, the region's earliest complex civilizations appeared in various parts of China about 4000 years ago. Written records exist only from the civilization that was located in north-central China. There, a small,

militarized, feudal aristocracy controlled vast estates on which the majority of the population lived and worked as semi-enslaved farmers and laborers. The landowners usually owed allegiance to one of the petty kingdoms that dotted northern China. These kingdoms were relatively self-sufficient and well defended with private armies.

Between 400 and 221 B.C.E., a long period of war ensued among these petty kingdoms, after which a single dominant kingdom emerged. What became known as the Qin empire prevailed mainly because it used a trained and salaried bureaucracy in combination with a strong military to extend the monarch's authority into the countryside (Figure 9.12).

The Qin system proved more efficient than the old feudal allegiance system it replaced. The estates of the aristocracy were divided into small units and sold to the previously semi-enslaved farmers. The empire's agricultural output increased because the people worked harder to farm land they now owned. In addition, the salaried bureaucrats were more responsible than the aristocrats they replaced, especially about building and maintaining levees, reservoirs, and other tax-supported public works that reduced the threat of flood, drought, and other natural disasters. Although the Qin empire was short-lived, subsequent empires maintained Qin bureaucratic ruling methods, which have proved essential in governing a united China.

Confucianism, Order, and Continuity

Closely related to China's bureaucratic ruling tradition is the philosophy of **Confucianism.** Confucius, who lived about 2500 years ago, was an idealist interested in reforming government and eliminating violence from society. He thought

FIGURE 9.12 An emperor's army. After unifying China in 221 B.C.E., Qin Shi Huang became its emperor. One of his many public works was the creation of a life-sized terra-cotta army of thousands of lifelike soldiers, each with a distinctive face and body, in a vast underground chamber that became his mausoleum. Since 1974, archaeologists have been excavating, restoring, and preserving this site, which is located near modern-day Xian. It attracts more than 2 million visitors per year. [Courtesy of Mark Samols.]

human relationships should involve a set of defined roles and mutual obligations. If each person understood his or her proper role and acted accordingly, a stable, uniform, and enduring society would result. Confucian values include courtesy, knowledge, integrity, and respect for and loyalty to parents and government officials. These values are still widely shared throughout East Asia today.

The model for Confucian philosophy was the patriarchal extended family. The oldest male held the seat of authority and was responsible for the well-being of everyone in the family. All other family members were aligned under the patriarch according to age and gender. Beyond the family, Confucianism held that commoners must obey imperial bureaucrats, and that everyone must obey the emperor, who in turn was obliged to ensure the welfare of society. As the supreme human being, the emperor was seen as the source of all order and civilization—the grand patriarch of all China.

Over the centuries, Confucian philosophy penetrated all aspects of Chinese society. Concerning the ideal woman, for example, a student of Confucius wrote: "A woman's duties are to cook the five grains, heat the wine, look after her parents-in-law, make clothes, and that is all! When she is young, she must submit to her parents. After her marriage, she must submit to her husband. When she is widowed, she must submit to her son."

Bias against merchants. For thousands of years, Confucian ideals were used to maintain the power and position of emperors and their bureaucratic administrators at the expense of merchants. In parable and folklore, merchants were characterized as a necessary evil, greedy and disruptive to the social order. At times, high taxes left merchants unable to invest in new industries or trade networks. At other times, however, the anti-merchant aspect of Confucianism was less influential, and trade and entrepreneurship flourished. Confucianism was never as strong in southern coastal China, which had a vibrant maritime society in which trade was important. Some scholars of China argue that this is reflected today in southern China's dynamic economy.

Cycles of expansion, decline, and recovery. Although the Confucian bureaucracy allowed Chinese empires to expand (Figure 9.13), its resistance to change often led to periods of decline. The heavy tax burden on farmers led to periodic revolts that weakened imperial control. Invasions by nomadic peoples from what is today Mongolia and western China dealt the final blow to several such weakened Chinese empires despite massive defenses, such as the Great Wall built along China's northern border. Nevertheless, the Confucian bureaucracy always recovered, and China's cultural and eco-

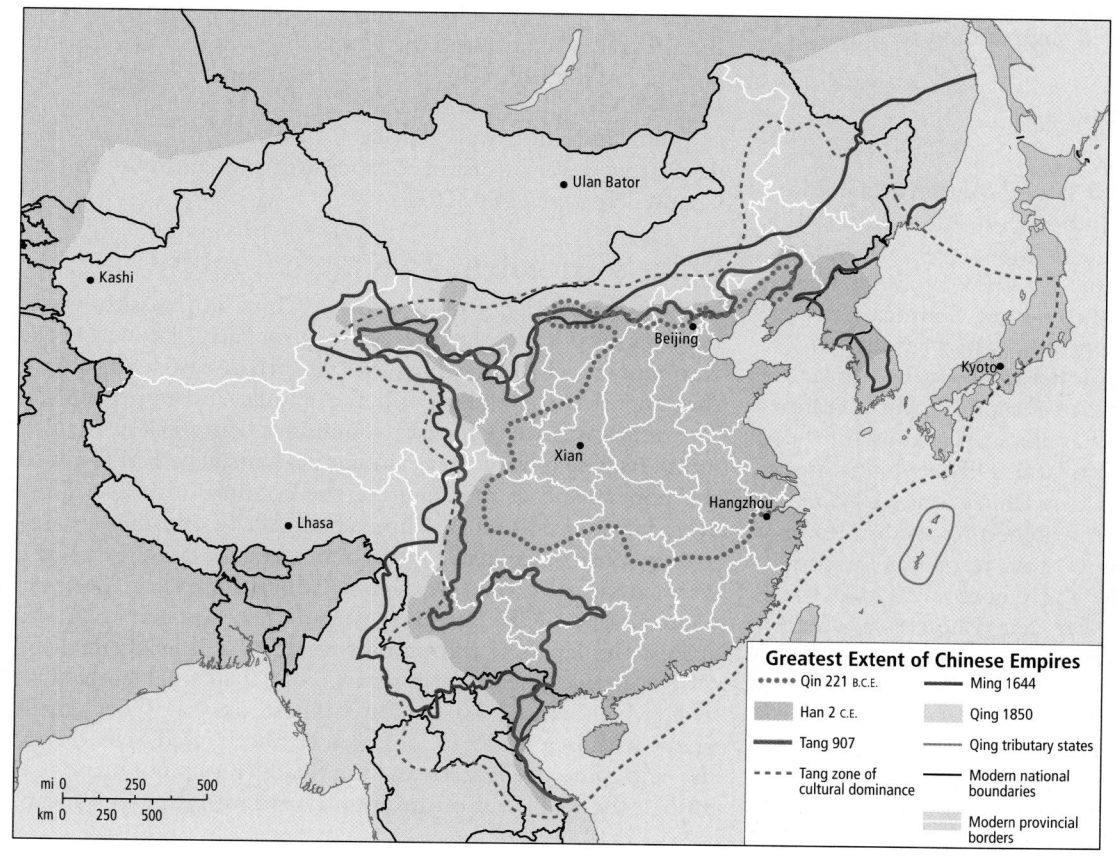

Greatest Extent of Chinese Empires

- • • • • Qin 221 B.C.E.
- Han 2 C.E.
- Tang 907
- - - - Tang zone of cultural dominance
- Ming 1644
- Qing 1850
- Qing tributary states
- Modern national boundaries
- Modern provincial borders

FIGURE 9.13 The extent of Chinese empires, 221 B.C.E.–1850 C.E. The Chinese state has expanded and contracted throughout its history. [Adapted from *Hammond Times Concise Atlas of World History*, 1994.]

nomic sophistication usually overwhelmed the invaders. After a few generations, the nomads were indistinguishable from the Chinese.

One nomadic invasion did result in important contacts between China and the rest of the world. In the 1200s, the Mongolian military leader Genghis Khan and his descendants were able to conquer all of China. They then pushed west across Asia as far as Hungary and Poland (see Chapter 5, page 185). It was during this empire (the Yuan, or Mongol) that traders such as the Venetian Marco Polo made the first direct contacts between China and Europe. These contacts proved much more significant for Europe, which was dazzled by China's wealth and technologies, than for China, which saw Europe as backward and barbaric.

Indeed, from 1100 to 1600, China remained the world's most developed region, despite enduring several cycles of imperial expansion, decline, and recovery. It had the largest economy, the highest living standards, the most advanced technologies, and the most magnificent cities. Improved strains of rice allowed dense farming populations to expand throughout southern China and supported large urban industrial populations. As early as 1078, people in northern China were producing twice as much iron as people in England did 700 years later. Nor was innovation lacking: Chinese inventions included papermaking, printing, paper currency, gunpowder, and improved shipbuilding techniques.

> As early as 1078, northern China was producing twice as much iron as England did 700 years later.

Why Didn't China Colonize an Overseas Empire?

From 1405 to 1433, Zheng He, a Chinese Muslim admiral in the emperor's navy, ventured far into the outside world, sailing in ships that were bigger and more technologically advanced than those of Columbus. In fleets of more than 250 ships, Zheng He ventured throughout Southeast Asia, across the Indian Ocean, and along the coasts of India, continuing into the Persian Gulf and down the coast of Africa.

These lavish voyages were funded for almost thirty years, but they never resulted in an overseas empire like those of the European countries a century or two later. Other regions simply didn't have many trade goods that China wanted. Moreover, the empire was continually threatened by the armies of nomads from Mongolia, so there were no resources for conquering new territory. Eventually the emperor decided the explorations weren't worth the effort, and China as a whole turned inward. In the years following Zheng He's voyages, contacts with the rest of the world were minimized as the emperor focused on upgrading the Great Wall to defend against nomads from Mongolia. As a result, the pace of technological change slowed, leaving China ill prepared to respond to growing challenges from Europe after 1600.

European and Japanese Imperialism

By the mid-1500s, Spanish and Portuguese traders interested in acquiring China's silks, spices, and ceramics had found their way to East Asian ports. In exchange, they brought a number of new food crops from the Americas such as corn, peppers, peanuts, and potatoes. These new sources of nourishment initiated a spurt of economic expansion and population growth and by the mid-1800s, China's population was more than 400 million.

By the nineteenth century, European influence had increased markedly as European merchants gained access to Chinese markets. In exchange for Chinese silks and ceramics, British merchants supplied opium from India, which was one of the few things that Chinese merchants would trade for. The emperor attempted to crack down on this trade because of its debilitating effects on Chinese society. The result was the Opium Wars (1839–1860), in which Britain badly defeated China. Hong Kong became a British possession, and British trade expanded throughout China.

The final blow to China's early preeminence in East Asia came in 1895, when a rapidly modernizing Japan won a spectacular naval victory over China in the Sino-Japanese War. After its defeat by the Japanese, the Qing empire made only halfhearted attempts at modernization, and in 1912, it collapsed after a coup d'état. During the decline of the Qing empire until China's Communist Party took control in 1949, much of the country was governed by provincial rulers in rural areas and by a mixture of Chinese, Japanese, and European administrative agencies in the major cities (Figure 9.14). During this era, radical ideologies gained popularity in new Western-style universities as intellectuals searched for a new basis of political authority to replace Confucianism. Of particular interest were various forms of socialism and communism (see Chapter 5, pages 186–187).

China's Turbulent Twentieth Century

Two rival reformist groups arose in China in the early twentieth century. One was the Nationalist Party, known as the Kuomintang (KMT), which was an urban-based movement that appealed to workers and the middle and upper classes. The other was the Chinese Communist Party (CCP), which found its base in rural areas among peasants. The KMT gained the upper hand at first, uniting the country in 1924. However, Japan's invasion of China changed the dynamic.

By 1937, Japan had control of most major cities. The KMT were confined to the few deep interior cities not in Japanese control. The CCP, however, waged a constant guerilla war against the Japanese throughout rural China. This gained the CCP a heroic status as a resistance to Japan's brutal occupation, which caused 10 million Chinese deaths. When Japan finally withdrew in 1945, defeated by the United States, Russia, and other Allied forces, the vastly more popular CCP pushed the KMT out of the country and into exile in Taiwan.

FIGURE 9.14 "Twelfth-Night Cake for Kings and Emperors." This French cartoon lithograph, made in 1898, shows Britain, Germany, Russia, and Japan imperialistically carving up China as France watches. A stereotypical Chinese official throws up his hands to show he is powerless to stop them. [The Granger Collection, New York.]

In 1949, the CCP, led by Mao Zedong, proclaimed the People's Republic of China, with Mao as president.

Mao's Communist revolution. Mao Zedong's revolutionary government became the most powerful China had ever had. It dominated all the outlying areas of China's Far Northeast, Inner Mongolia, and western China (Xinjiang Uygur), and launched a brutal occupation of Tibet (Xizang). The People's Republic of China was in many ways similar to a traditional Chinese empire. The Confucian bureaucracy was replaced by the Chinese Communist Party, and Mao Zedong became a sort of emperor with unquestioned authority.

The chief early beneficiaries of the revolution were the masses of Chinese farmers and landless laborers. On the eve of the revolution, huge numbers lived in abject poverty. Famines were frequent, infant mortality was high, and life expectancy was low. The vast majority of women and girls held low social status and spent their lives in unrelenting servitude.

The revolution drastically changed this picture. All aspects of economic and social life became subject to central planning by the Communist Party. Land and wealth were reallocated, often resulting in an improved standard of living for those who needed it most. Heroic efforts were made to improve agricultural production and to reduce the severity of floods and droughts. The masses, regardless of age, class, or gender, were mobilized to construct huge public works projects (roads,

dams, canals, mountains terraced into fields) almost entirely by hand. "Barefoot doctors" with rudimentary medical training dispensed basic medical care, midwife services, and nutritional advice to the remotest locations. Schools were built in the smallest of villages. Opportunities for women became available, and some of the worst abuses against them—such as the crippling binding of women's feet to make them small and childlike—stopped. Most Chinese who are old enough to have witnessed these changes say that the revolution did indeed improve overall living standards for the majority.

Nonetheless, this progress came at enormous human and environmental costs. During the **Great Leap Forward** (a government-sponsored program of massive economic reform initiated in the 1950s), 30 million people died from famine brought on by poorly planned development objectives. Meanwhile, deforestation, soil degradation, and agricultural mismanagement became widespread. In the aftermath, some Communist Party leaders tried to correct the inefficiencies of the centrally planned economy only to be demoted or jailed as Mao Zedong attempted to stay in power.

In 1966, the **Cultural Revolution,** a series of highly politicized and destructive mass campaigns, enforced support for Mao and punished dissenters. Everyone was required to study the "Little Red Book" of Mao's sayings (Figure 9.15). Educated people and intellectuals were a main target of the

FIGURE 9.15 China's Cultural Revolution. Taken in the late 1960s, this photo shows a group of Chinese children in front of a picture of Chairman Mao Zedong. The children each hold a copy of Mao's "Little Red Book." [Hulton Archive/Getty Images.]

Cultural Revolution because they were thought to instigate dangerously critical evaluations of Mao and Communist Party central planning. Tens of millions of Chinese scientists, scholars, and students were sent out of the cities, to labor in mines and industries, or to jail, where as many as one million died. Children were encouraged to turn in their parents. Petty traders were punished for being capitalists, as were those who adhered to any type of organized religion. The Cultural Revolution so disrupted Chinese society that by Mao's death in 1976, the communists had been thoroughly discredited.

Changes after Mao.

Two years after Mao's death, a new leadership formed around Deng Xiaoping. Limited market reforms were instituted, but the Communist Party retained tight political control. Now, after thirty years of reform and remarkable levels of economic growth, China's economy is the third largest in the world. However, human rights are still often abused and political activity remains tightly controlled.

Japan Becomes a World Leader

Although China is once again the behemoth of East Asia, for much of the twentieth century, Japan, with only one-tenth the population and 5 percent of the land area of China, dominated East Asia economically and politically. Japan's rise as a modern global power results largely from its response to challenges from Europe and North America.

Modern Japanese populations are descended from migrants from the Asian mainland, the Korean Peninsula, and the Pacific islands. Ideas and material culture imported from these places include Buddhism, Confucian bureaucratic organization, architecture, the Chinese system of writing, the arts, and agricultural technology. Beginning in the mid-sixteenth century, active trade with Portugal brought new ideas and technology that strengthened Japan's wealthier feudal lords (*shoguns*), allowing them to unify the country under a military bureaucracy. However, the shoguns monopolized contact with the outside world, allowing no Japanese to leave the islands on penalty of death.

A second period of radical change developed in response to the 1853 arrival in Tokyo Bay of a small fleet of U.S. naval vessels. The foreigners, carrying military technology far in advance of Japan's, forced its government to open the economy to international trade and political relations. In response, a group of reformers (the Meiji) seized control of the Japanese government, setting the country on a crash course of modernization and industrial development. They sent Japanese students abroad and recruited experts from around the world, especially from Western nations, to teach everything from foreign languages to modern military technology.

Between 1895 and 1945, Japan fueled its economy with resources from a vast colonial empire. Equipped with imported European and North American military technology, its armies occupied first Korea, then Taiwan, then eastern and coastal China, and eventually much of Southeast Asia as well

(Figure 9.16). Many of these areas still harbor resentment for the brutality they suffered at Japanese hands. Japan's imperial ambitions ended with its defeat in World War II and subsequent occupation by U.S. military forces until 1952. ▄▀

The U.S. government imposed many social and economic reforms. Japan was required to create a democratic constitution and reduce the emperor to symbolic status. Its military was reduced dramatically, forcing it to rely on U.S. forces to protect it from attack. With U.S. support, Japan rebuilt rapidly after World War II, and it eventually became a giant in industry and global business, exporting automobiles, electronic goods, and many other products. Japan's economy is still among the world's largest and most technologically advanced, though China has replaced it as the next largest economy after the EU and the United States.

Chinese and Japanese Influences on Korea, Taiwan, and Mongolia

The history of the entire region is largely grounded in what transpired in China and Japan. All East Asian countries, for example, have been affected by ancient Confucian philosophy, the relics of which can be found in present-day political and social policy. North Korea, and for a while Mongolia, joined China in the communist experiment. Taiwan and South Korea have chosen to model themselves after Japan's state-aided market economy (discussed on pages 347–348).

The Korean War and its aftermath.

Korea was a unified country until 1945. At the end of World War II, the Soviet Union declared war against Japan and invaded Manchuria and the northern part of Korea. To prevent all of Korea from coming under Soviet control, the United States proposed dividing the country at the 38th parallel. The United States took control of the southern half of the peninsula, where it instituted reforms similar to those in Japan. The Soviet Union took the northern half, where it implemented communist development strategies. After the United States withdrew its troops in the late 1940s, North Korea attacked South Korea. The United States returned to defend the south, leading a 3-year war against North Korea and its allies, the Soviet Union and China.

After great loss of life on both sides and devastation of the peninsula's infrastructure, the Korean War ended in a truce and the establishment of a demilitarized zone (DMZ) at the 38th parallel in 1953. North Korea closed itself off from the rest of the world, and to this day it remains impoverished and defensive, occasionally gaining international attention by rattling its nuclear arsenal. Meanwhile, South Korea has developed into one of the world's most prosperous and technologically sophisticated market economies. ▄▀

Taiwan.

Taiwan was for a long time a poor agricultural island on the periphery of China, and it was part of Japan's empire between 1895 and 1945. In 1949, the Chinese nationalists

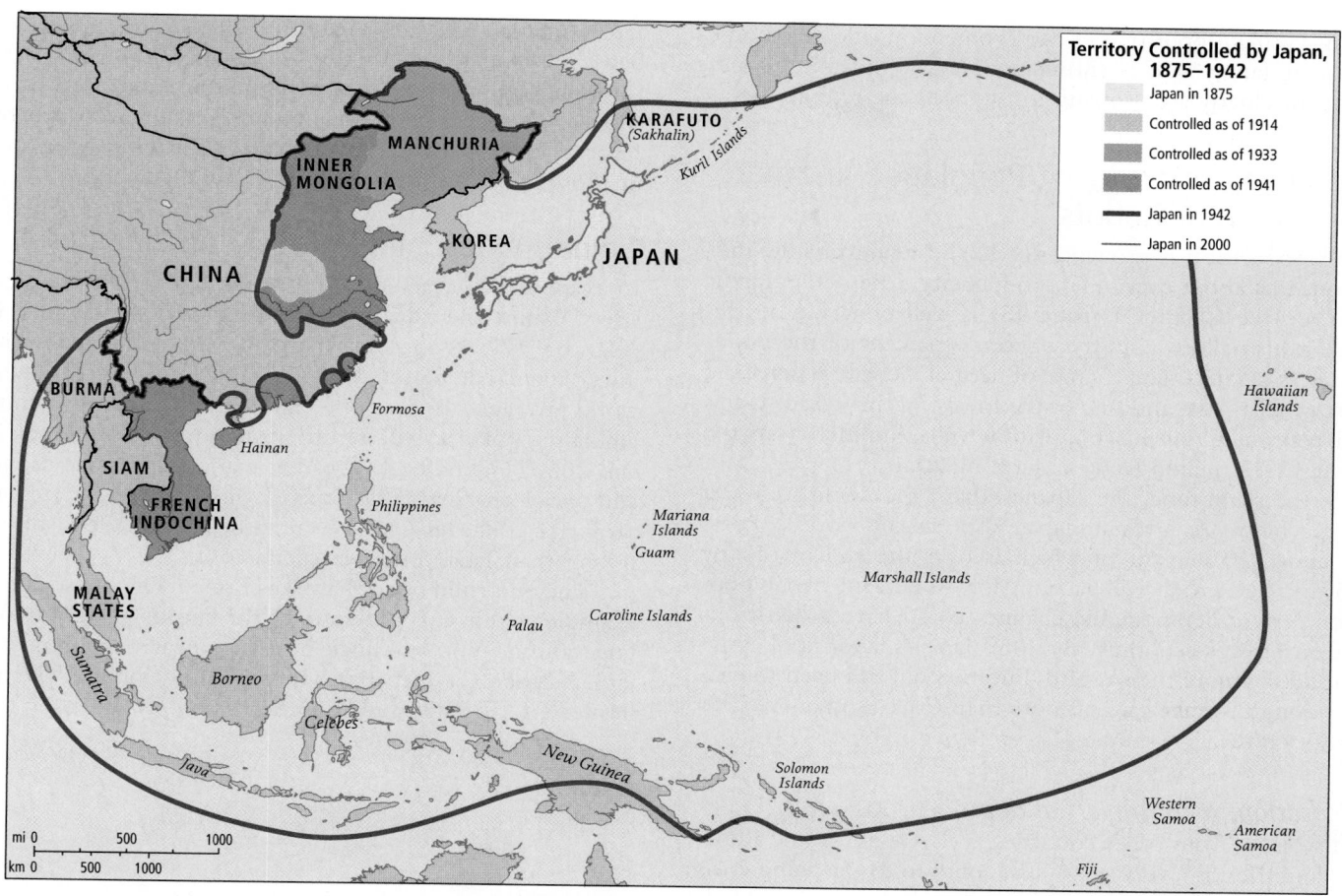

FIGURE 9.16 Japan's expansions, 1875–1945. Japan colonized Korea, Taiwan (Formosa), Manchuria, China, parts of Southeast Asia, and several Pacific islands to further its program of economic modernization and to fend off European imperialism in the early twentieth century. [Adapted from *Hammond Times Concise Atlas of World History,* 1994.]

(the KMT), pushed out of mainland China, set up an anti-communist government in Taiwan, naming it the Republic of China (ROC). For the next 50 years, with U.S. aid and encouragement, it became a modern industrialized economy whose value dwarfed that of China until the 1990s. Today, Taiwan is still an economic powerhouse and has invested heavily in the rapid transformation of East Asian and Southeast Asian economies. Throughout this time, China has never relinquished claim to Taiwan and considers it a rebellious province deserving occasional threats of invasion and military occupation.

Mongolia. In the distant past, Mongolia's nomadic horsemen periodically posed a threat to China, and the Mongolian ruler Genghis Khan (ca. 1162–1227) and his successors eventually conquered most of China. As a consequence, China became obsessed with controlling its northern neighbor. China did control Mongolia from 1691 until the 1920s. A communist revolution began soon thereafter, and Mongolia continued as

an independent communist country under Soviet guidance until the breakup of the Soviet Union in 1989. Since then, it has been on a difficult road to a free market economy.

POPULATION PATTERNS

While East Asia is the most populous world region, families are having many fewer children than in the past. Of all world regions, only Europe has a lower rate of natural increase (– 0.1 percent increase per year compared to East Asia's 0.5 percent). In China this is due to government policy that harshly penalizes families for having more than one child. Elsewhere, urbanization and changing gender roles are encouraging smaller families. Only in Mongolia (with 2.6 million people) and North Korea (with 22.9 million) are women still averaging two or more children each. But even in these two countries, urbanization

> Of all world regions, only Europe has a lower rate of natural increase than East Asia.

and increasing opportunities for women outside the home mean that family size is shrinking. Hence, the numbers of dependent elderly are growing throughout the region. 🎥

Responding to an Aging Population in Japan: Immigration or Robots?

Japan's population is growing slowly and aging rapidly, raising concerns about economic productivity. The demographic transition (see Chapter 1, page 45) is well underway in this highly industrialized country where 79 percent of the population lives in cities. Japan's rate of natural increase is zero, the lowest in East Asia, and one of the lowest in the world. If this trend continues, Japan's population will plummet from the current 127.7 million to 95 million by 2050.

At the same time, the Japanese have the world's longest life expectancy. As a result, Japan also has the world's oldest population, 20 percent of which is over the age of 65. By 2055, this age group will account for 40 percent of the population. These demographic changes could have a disastrous effect on Japan's economy. By 2050 Japan's labor pool could be shrunk by more than a third, but it would still need to produce enough to take care of more than twice as many retirees as it does now.

Immigration: a controversial approach. Recruiting immigrant workers from other countries, as is done in some European countries, is a very unpopular solution to the aging crisis in Japan. Many Japanese consider foreigners a source of "pollution," and the few small minority populations with cultural connections to China or Korea have long faced discrimination. The children of foreigners born in Japan are not granted citizenship, and some communities that have been in the country for generations are still thought of as foreign. Today, immigrants are fingerprinted and photographed upon entering Japan and must carry an "alien registration card" at all times.

Nonetheless, foreign workers are dribbling into Japan in a multitude of legal and illegal guises. Many are "guest workers" from South Asia brought in to fill the most dangerous and low-paying jobs with the understanding that they will eventually leave. Others are the descendants of Japanese who once migrated to South America. Regardless, Japan's foreign population remains tiny, making up only 1.2 percent of the total population (approximately 1.5 million people). A recent United Nations report estimates that Japan would have to import over 640,000 immigrants per year just to maintain its present workforce and avoid a 6.7 percent annual drop in its GDP.

Robots. In a novel approach to Japan's demographic changes, its government has invested enormous sums of money in robotics over the past decade. Robots are already widespread in Japanese industries such as auto manufacturing. Now they are being developed to care for the elderly, to guide patients through hospitals, to look after children, and even to make sushi. The government plans to replace up to 15 percent of Japan's workforce with robots by 2025, though they are still unaffordable for most businesses and individuals.

China's One-Child Policy

In response to fears of overpopulation and environmental stress, China has had a one-child-per-family policy since 1979 (Figure 9.17). As a result, its rate of natural increase is slightly lower than that of the United States, and roughly half the world average. If the one-child family pattern continues, China's population will start to shrink sometime between 2025 and 2050 (Figure 9.18), creating many of the same economic and social problems that Japan is now facing. In fact, these problems could hit China even harder because it is still much poorer than Japan on a per capita basis.

The one-child policy has transformed Chinese society. For example, within two generations, the kinship categories of sibling, cousin, aunt, and uncle have disappeared from most families. Further, the sometimes brutal enforcement of the policy has led to shocking human rights violations. In many places,

FIGURE 9.17 China's one-child-per-family policy. A man walks past a sculpture in Beijing which promotes China's one-child-per-family policy. Enforcement of the policy, introduced in 1979 to control China's population growth, has relaxed a bit in recent years. [AP Photo/Greg Baker.]

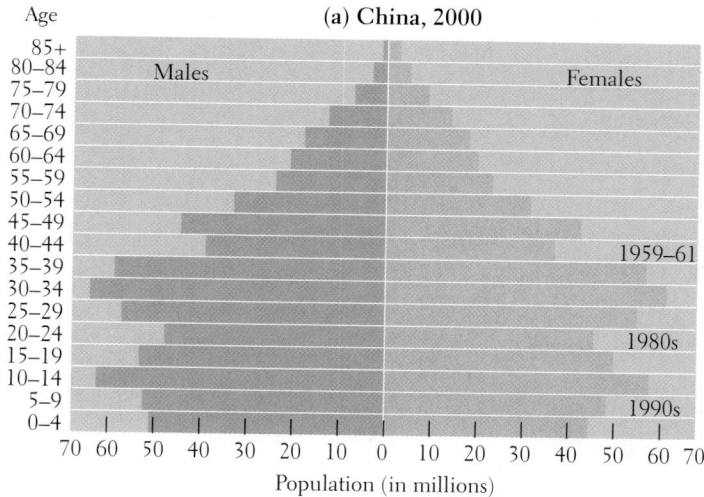

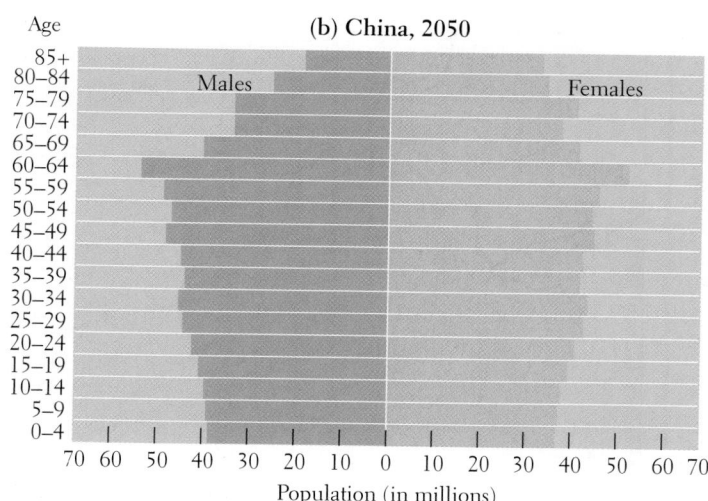

FIGURE 9.18 Population pyramids for China, 2000 and 2050 (projected). [Adapted from U.S. Census Bureau, International Data Base (Washington, D.C.), at http://www.census.gov/cgi-bin/ipc/idbpyrs.pl?cty=CH&out=s&ymax=250&submit=Submit.]

the government has waged a campaign of forced sterilizations and even forced abortions for mothers who already have one child.

Cultural preference for sons. The prospect of a couple's only child being a daughter, without the possibility of having a son in the future, is the aspect of the one-child policy that has caused families the most despair. For years, the makers of Chinese social policy have sought to eliminate the preference for males by empowering women economically and socially. They believe that female children will be just as desirable as males when it is clear that well-trained, powerful daughters can bring honor to the family name and earn sufficient income for their families. Nevertheless, the preference for sons remains strong.

East Asia's Missing Females

The pyramid in Figure 9.18a illustrates how the preference for sons has created a gender imbalance in China's population. Up to age 30 and older, the number of males exceeds the number of females by small but significant amounts. In 2000, there were roughly 44 million girls but over 50 million boys aged 0–4, indicating a deficit of about 6 million girls. Viewed another way, this is a ratio of 114 boys to every 100 girls. (The normal sex ratio at birth is 105 boys to 100 girls.) What happened to the missing girls?

There are several possible answers. Given the preference for male children, the births of these girls may simply have gone unreported as families hoped to conceal their daughter and try again for a son. There are many anecdotes of girls being raised secretly or even disguised as boys. Adoption records indicate that girls are given up for adoption much more

often than boys. Or, the girls may have died in early infancy, either through neglect or infanticide. Finally, some parents have access to medical tests that can identify the sex of a fetus. There is evidence that in China, as elsewhere around the world, some of these parents choose to abort a female fetus.

The cultural preference for sons persists elsewhere in East Asia as well. A deficit of girls appears on the 2000 population pyramids for Japan, the Koreas, Mongolia, and Taiwan. Nonetheless, evidence shows that attitudes may be changing. In Japan, South Korea, and Mongolia, the percentage of women receiving secondary education equals or exceeds that of men. Comparable data are not available for North Korea, China, and Taiwan. In the latter two countries, however, educated young women are increasingly recruited for middle management positions, suggesting that soon they could have incomes equal to those of young men.

A shortage of brides. A side effect of the preference for sons is that there is now a growing shortage of women of marriageable age throughout East Asia. China alone had an estimated deficit of 10 million women aged 20–35 in 2000. Females are also effectively "missing" from the marriage rolls because many educated young women are too busy with career success to meet eligible young men. Some are even too busy to attend the singles parties arranged to bring them together with males of compatible education and income, so they send their parents! The mother of one such daughter away on a business trip recently attended a singles gathering in Shanghai where she patiently recited to interested males

> A side effect of the cultural preference for sons is a growing shortage of women of marriageable age throughout East Asia.

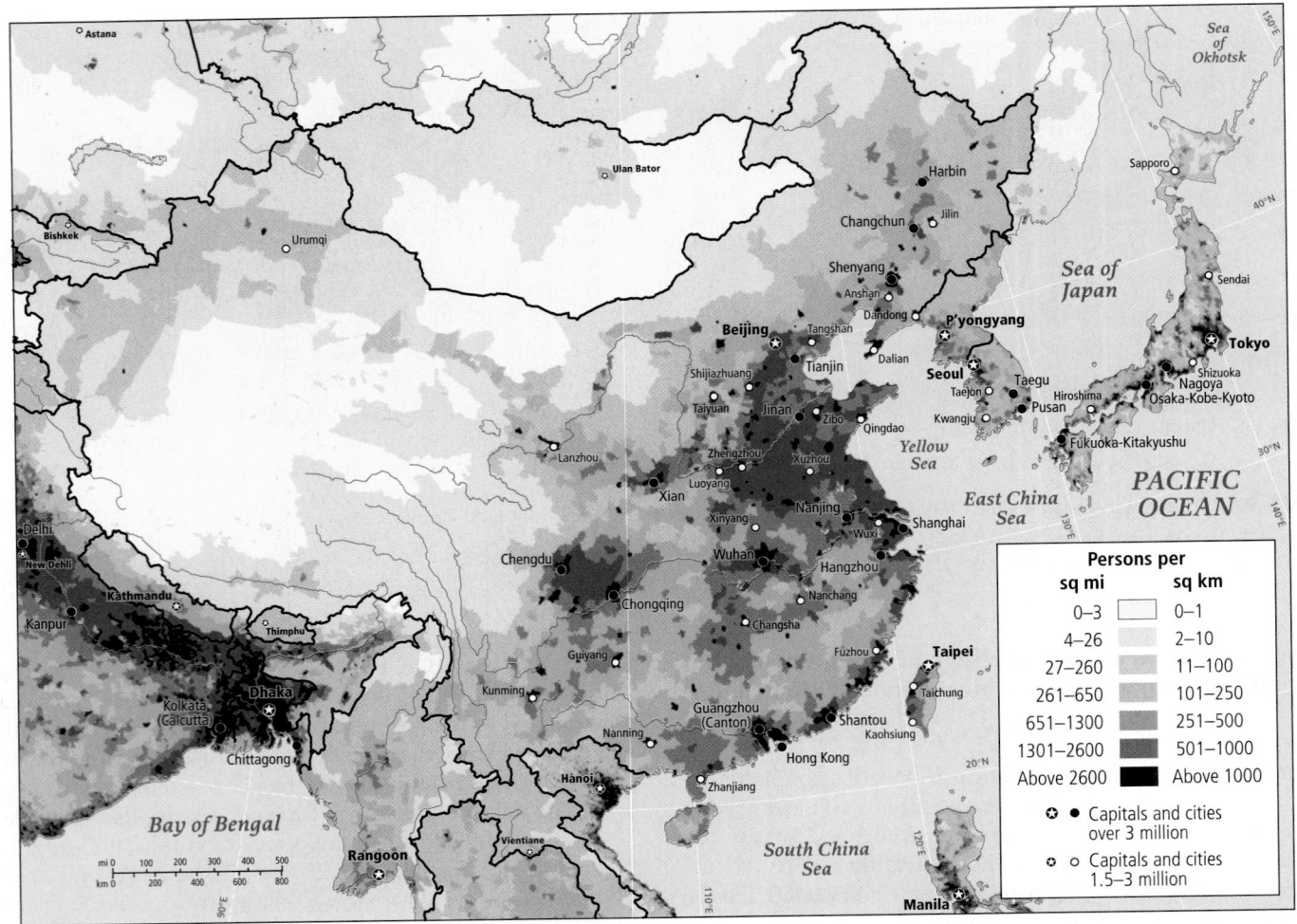

FIGURE 9.19 Population density in East Asia. [Data courtesy of Deborah Balk, Gregory Yetman et al., Center for International Earth Science Information Network, Columbia University, at http://www.ciesin.columbia.edu.]

the vital statistics of her daughter, including her age, height, weight, hobbies, salary, and career ambitions. No mention was made of future children.

A darker side of the shortage of women in China is increasing kidnapping and forced prostitution of young girls and women. ■▶ Further, China's growing millions of single young men are emerging as a major threat to civic order. Research has shown that societies with large numbers of unattached males often experience a surge in drug abuse, violent crime, and even terrorism.

Population Distribution

As you can see in Figure 9.19, people are not evenly distributed on the land. China, with 1.3 billion people, has more than one-fifth of the world's population. However, 90 percent

of these people are clustered on the roughly one-sixth of the total land area that is suitable for agriculture. People are concentrated especially densely in the eastern third of China: in the North China Plain, the Sichuan Basin, the middle and lower Chang Jiang (Yangtze) basin, and the delta of the Zhu Jiang (Pearl River) in the southeast.

The west and south of the Korean Peninsula are also densely settled, as are northern and western Taiwan. In Japan, settlement is concentrated in a band stretching from the cities of Tokyo and Yokohama on Honshu island south through the coastal zones of the Inland Sea to the islands of Shikoku and Kyushu. This urbanized region is one of the most extensive and heavily populated metropolitan zones in the world, accommodating well over half of Japan's total population. The rest of Japan is mountainous and more lightly settled.

Current Geographic Issues

Although the countries of East Asia adopted new economic systems only after World War II, most of them are making progress in creating a better life for their citizens. In fact, Japan, Taiwan, South Korea, and China's Hong Kong Special Administrative Unit (discussed on pages 351–352) have among the highest standards of living in the world. The great challenge in this part of the world will be to achieve and maintain improved living standards for large numbers of people in a manner that does not place too much stress on the environment.

ECONOMIC AND POLITICAL ISSUES

After World War II, the countries of East Asia established two basic types of economic systems. The communist regimes of China, Mongolia, and North Korea relied on central planning by the government to set production goals and to distribute goods among their citizens. In contrast, Japan, Taiwan, and South Korea established **state-aided market economies** with the assistance and support of the United States and Europe. In this type of economic system, market forces, such as supply and demand and competition for customers, determine many economic decisions. However, the government intervenes strategically, especially in the financial sector, to make sure that certain economic sectors develop. Investment in the country by foreigners is also limited so that the government can retain greater control over the direction of the economy. In the case of Japan, South Korea, and Taiwan, government intervention was designed to enable **export-led growth.** This economic development strategy relies heavily on the production of manufactured goods destined for sale abroad, primarily to the large economies of North America and Europe.

More recently, the differences among East Asian countries have diminished as China and Mongolia have set aside strict central planning and adopted reforms that rely more on market forces. China now also relies heavily on exports of its manufactured goods to North America and Europe. Politically, however, contrasts remain stark. While Japan, South Korea, Taiwan, and Mongolia have become democracies, unelected governments in China and North Korea maintain a tight grip on politics and the media, though some experimentation with local level democracy is occurring in China.

The Japanese Miracle

Throughout the nineteenth century, the economies of Japan, Korea, and Taiwan were minuscule compared with China's. Then, during the twentieth century, all three grew tremendously. Credit belongs in large part to the Japanese model of a state-aided market economy and export-driven economic growth.

Japan rises from the ashes. Japan's recovery after its crippling defeat at the end of World War II is one of the most remarkable tales in modern history. Except for Kyoto, which was spared because of its historical and architectural significance, all of Japan's major cities were destroyed by the United States—most notably Hiroshima and Nagasaki—with nuclear weapons and Tokyo with incendiary bombs. ▰◀

Key to Japan's rapid recovery was government guidance of private investors in creating new export-based industries. The government provided financing and protected new industries from foreign competition. Trade deals with the United States and Europe were negotiated by the Japanese government, giving Japan large and wealthy foreign markets in which to sell its products. Japan's government also negotiated deals with labor unions that guaranteed lifetime employment by a single company for most workers in return for relatively modest pay.

> Key to Japan's rapid economic recovery after World War II was government guidance of private investors in creating new export-based industries.

All of this produced explosive economic growth of about 10 percent per year between 1950 and the 1970s. The leading sectors were export-oriented automobile and electronics manufacturing. Their products could sell at much higher volumes in the large prosperous economies of North America and Europe than in the relatively small Japanese and nearby Asian economies. Japanese brand names such as Sony, Panasonic, Nikon, and Toyota became household words around the world. Although growth has slowed considerably, Japan's "economic miracle" continues to have an immense worldwide effect. Resources from all parts of the world are shipped to Japan, and Japanese purchases support local economies that employ millions around the globe.

Since 1992, employment has shifted significantly from manufacturing to services, such that the service sector now employs 68 percent of the workforce and constitutes 72 percent of Japan's GDP. Banking and financial services are especially important industries that have contributed to Tokyo's status as a world city (Figure 9.20).

Productivity-boosting innovations. Over the years, Japan has made two key innovations in manufacturing that have boosted its productivity. The **kanban,** or "just in time," **system** clusters companies that are part of the same production process close together so that they can deliver parts to each other when they are needed. For example, factories that make automobile parts are clustered around the final assembly plant, delivering parts literally minutes before they will be used. This saves money by speeding production and reducing the need for warehouses.

FIGURE 9.20 Modern Tokyo at twilight. The Shinjuku business and shopping district in western Tokyo is known for its over-the-top sight and sound experiences. Every conceivable product or service is available. [Peter Adams/CORBIS.]

A related innovation is the **kaizen,** or "continuous improvement system." This system ensures that fewer defective parts are produced because production lines are constantly surveyed for errors. Production lines are also constantly adjusted and improved so as to save time and energy. Both *kanban* and *kaizen* systems have been imitated by companies around the world and have been taken overseas by Japanese companies investing abroad. For example, Toyota uses *kaizen* and *kanban* in all its U.S. plants (see Chapter 2, page 81).

Is the party over? Since the 1980s, Japan's economy has slowed considerably, as have East Asia's other state-aided market economies. Some analysts have faulted the long-standing close relationship between government and industry for nurturing favoritism, corruption, and inefficiency. For example, government protection from global competition for outdated sectors of the economy, such as farming, has burdened consumers with the highest food prices on earth. A single cantaloupe can cost as much as U.S.$10, a head of lettuce U.S.$5, and a pound of fine beef more than U.S.$50. Another problem was a poorly timed expansion of industrial productive capacity. Between 1988 and 1992, Japan increased its productive capacity by an amount equal to the entire economy of France. Unfortunately, this happened at a time when the demand for its products was beginning to diminish at home and abroad.

Some argue that Japan could give its economy a boost by paying its workers more, enabling them to buy more of the high-quality items that Japan exports to North America and Europe. However, others point out that slow growth rates are common in similarly rich countries.

Mainland Economies: Communists in Command

Economic development proceeded on a dramatically different course on East Asia's mainland after World War II. Here, communist economic systems transformed China, Mongolia, and North Korea. Private property was abolished, and the state took full control of the economy, loosely following the example of the Soviet command economy (see Chapter 5, page 191). These sweeping changes ultimately proved less successful than was hoped.

By design, most people in the communist economies were not allowed to consume more than the bare necessities. On the other hand, the "iron rice bowl" policy guaranteed nearly everyone a job for life, enough food, and housing that was better than what they had before. Nevertheless, overall productivity remained low.

The Commune system When the Communist Party first came to power in China in 1949, its top priority was to make great improvements in both agricultural and industrial production. Similar early goals were held by the communist regimes in North Korea and Mongolia, though they had much smaller populations and resource bases to work with. In China, the first strategy was land reform. The initial effort was to take large, unproductive tracts out of the hands of landlords and put them into the hands of the millions of landless farmers. By the early 1950s, much of China's agricultural land was divided into tiny plots.

It soon became clear that these small plots were not going to produce enough to feed the people who were leaving agriculture to work in the industrial cities. In response, communist leaders joined small landholders together into cooperatives so that they could pool their labor and resources to increase agricultural production. In time, the cooperatives became full-scale communes, with an average of 1600 households each. The communes, at least in theory, took care of all aspects of life. They provided health care and education, and built rural industries to supply themselves with such items as fertilizers and small machinery. The rural communes also had to fulfill the ambitious expectations of the leaders in Beijing for better flood control and expanded irrigation systems.

The Chinese commune system had several difficulties. Farmers had too little time to farm because they were required to spend so much time building roads, levees, and drainage ditches or working in the new rural industries. Local Communist Party administrators often compounded the problem by overstating harvests to impress their superiors in Beijing. The leaders in Beijing responded by requiring larger food shipments to the cities, which created food shortages in the countryside.

Focus on heavy industry. The communist leadership in China, North Korea, and Mongolia believed that investment in heavy industry would dramatically raise living standards. Massive investments were made in the mining of coal and other minerals and in the production of iron and steel. Heavy machinery was produced to build roads, railways, dams, and other infrastructural improvements that would increase overall economic productivity. Funds for heavy industry were to come from the already strained agricultural sector. Farmers were required to sell their products to the state at artificially low prices, and the state directed the profits gained in reselling those products toward industrialization. Other funds for industry came from profits in mining and forestry.

However, the focus on heavy industry didn't have the desired effect. Much as in India (see Chapter 8, pages 314–315), the vast majority of the population remained poor rural laborers, receiving little benefit from industries that created jobs mainly in urban areas. Little attention was paid to producing consumer goods (such as cheap pots, pans, and other household items) that would have driven economic growth and improved living standards for the rural poor. Even in the urban areas growth remained sluggish because, much as in the Soviet command economy, small miscalculations by bureaucrats resulted in massive shortages and production bottlenecks that constrained economic growth.

Communist Struggles with Regional Disparity

For centuries, China's interior west has been poorer than its coastal east. The interior west has been locked into herding and agricultural economies, while even in the most restricted times, the economies of the east have benefited from trade and industry. This spatially uneven development has continued to plague the Chinese in modern times. Right after the revolu-

tion, economic policy focused on **regional self-sufficiency,** with each region encouraged to develop independently. Agricultural and industrial sectors were both developed by the state in the hope of creating jobs and evening out the national distribution of income. Communist leaders also believed that dispersed industrial development and regional self-sufficiency would foil an invading enemy's effort to destroy China's productive capacity.

Government funds were used to set up industries in nearly every province, regardless of practicality. For example, a steel industry was established in Inner Mongolia at great expense, although other wealthier provinces already had steel industries that could have been improved with a far greater payoff. Similarly, huge, mechanized grain farms were developed on cleared forestlands in China's Far Northeast. The same effort elsewhere would have yielded better results and allowed the forests to be saved for other purposes.

Globalization and Market Reforms in China

By the late 1970s, it was clear that China's command economy was not going to make dramatic improvements in living standards. In the 1980s, China's leaders enacted market reforms that changed the country's economy in four ways. First, economic decision making was decentralized. Second, farmers and small businesses were permitted to sell their produce and goods in competitive markets. Third, regional specialization, rather than regional self-sufficiency, was encouraged (Figure 9.21). Finally, the government allowed foreign investment in Chinese export-oriented enterprises and the sale of foreign products in China. This four-part shift to a market-based economy dramatically improved the efficiency with which food and goods were produced and distributed.

China's market reforms have transformed the region and indeed the whole world. Today, China is the world's largest producer of manufactured goods. Equally important, it now represents a market of more than 1.3 billion potential consumers. Nearly every major company in the world is eager to sell its goods to customers in China. Meanwhile, Mongolia has participated in this revolution only modestly and North Korea, not at all.

> China is the world's largest producer of manufactured goods and has a market of more than 1.3 billion potential consumers.

Regional specialization. Decentralizing decision making encouraged regional specialization. Managers of many state-owned enterprises now set production levels and prices for goods and services according to the demands of consumers in the open market.

In response, managers and entrepreneurs (sometimes with assistance from the state) are taking advantage of the different resources and opportunities offered by different areas of the country. The old colonial city of Shanghai, for example, has once again become a center of trade and finance. And the

FIGURE 9.21 Regional specialization in agriculture. Nearly half of China's crop production takes place along the fertile banks of the Chang Jiang (Yangtze River). Among the crops grown are rice, wheat, barley, corn (maize), beans, cotton, and hemp. Note the terrace-style fields throughout the landscape. [Michael S. Yamashita/ CORBIS.]

Zhu Jiang (Pearl River) delta has evolved into a massive industrial center for the production of export goods. Meanwhile, without support from the government, many old state-run enterprises, especially in China's Far Northeast and in remote interior areas, have collapsed. While their inefficiency is not missed, the employment they once offered to millions of skilled workers is.

Market reforms in rural China. One of the most remarkable developments resulting from China's decentralization of economic control is the rapid growth of new rural enterprises. These now constitute one-quarter of the Chinese economy, produce 40 percent of its exports, and employ more than 128 million people, more than the Chinese government itself. Rural enterprises provide a wide variety of goods and services—from operating mines to making cooking pots to assembling electronic equipment. Such enterprises may still be village collectives or communes, but increasingly they are privately owned. 📹 All of them leave major decisions to managers and price their products according to market demand.

Despite the importance of rural enterprises in the economy, their growth has been accompanied by some problems. They are significant contributors to environmental stress. In the mountains west of Beijing, for example, industrial pollution exceeds that of the city. Paper mills pollute waterways, and trucks linking rural industries with urban and overseas markets contaminate the air. Corruption is also a problem. Some managers steal funds from their own enterprises, evade taxes, and pay officials to look the other way when they violate environmental regulations.

Urbanization and Globalization in China

While reform in rural areas has been significant, urban areas have witnessed even more dramatic transformation. China's reforms have lead to the most massive urbanization in the world's history. Since 1980, when reforms were initiated, the urban population has more than quadrupled, growing by 450 million. This explosive growth is fueled largely by the highly globalized economies of the big coastal cities, whose urban factories supply consumers throughout the world. While the 600 million urban Chinese still represent only 44 percent of the country's total population, they reflect a massive shift of resources away from rural areas. Despite the success of some enterprises, only a few rural provinces have raised their per capita GDP significantly (Figure 9.22).

China's cities are among the fastest growing on earth, in terms of both population and economic value. Shanghai and its surrounding suburbs and cities now account for more than 25 percent of China's GDP. As a result of this expansion, urban China has a large and growing middle class, some of whom can afford lavish homes and vacation travel through the country. Their globalized incomes have transformed urban landscapes, and many old neighborhoods are being bulldozed to make way for row after row of high-rise apartment complexes. 📹

International trade and special economic zones. In the early 1980s, China's government was wary of the disruption that could result from a rapid opening of the economy to international trade. Consequently, it selected five coastal cities to function as free trade zones, or **special economic zones (SEZs)** (Figure 9.23, on page 352). Industries in these cities were allowed to recruit foreign investors and use capitalist management methods that had not yet been permitted in the rest of the country. In the late 1990s, the program was expanded to 32 other cities. These new locations were designated *economic and technology development zones (ETDZs;* Figure 9.23). Like SEZs, the ETDZs provide footholds for international investors and multinational companies eager to establish operations in the country.

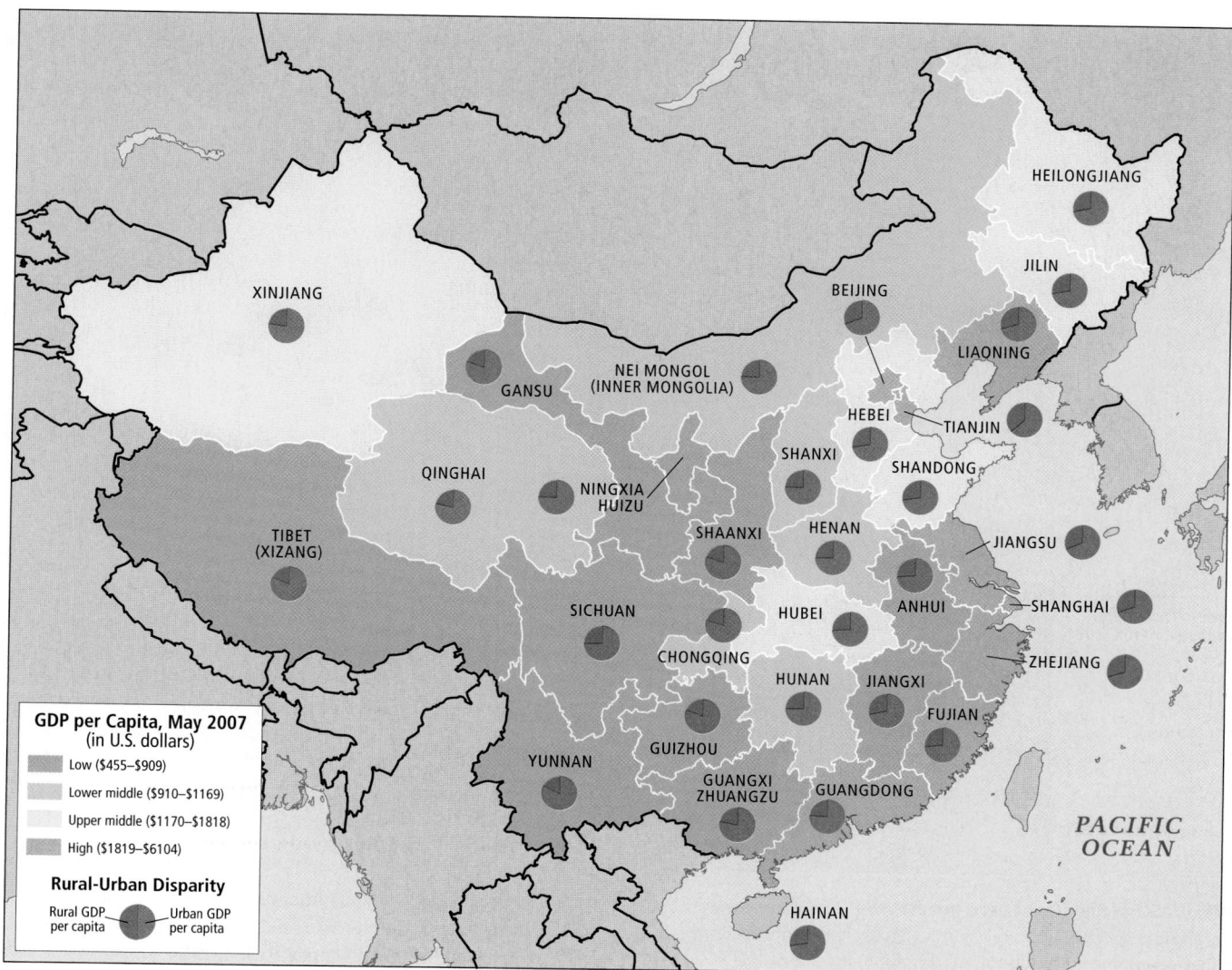

FIGURE 9.22 China's regional and rural-urban GDP disparities, 2004. The colors show that the average range of GDP per capita at the provincial level generally decreases from high (orange), along the coast, to low (green), in the interior. Within the provinces there is also a disparity between urban (higher) and rural (lower) GDP per capita, as shown by the pie diagrams. Notice that on the coast there are fewer rural-urban disparities than in the interior, where the rural (dark green) wedges are a smaller portion of the pie. [Adapted from "Exhibit 4. Classification of provinces by per-capita GDP," *China Human Development Report 2005* (Beijing: United Nations Development Programme China, 2006), p. 148, at http://www.undp.org.cn/modules.php?op=modload& name=News&file=article&catid=18&topic=40&sid=242&mode= thread&order=0&thold=0; province data from http://www.china.org.cn/ english/features/ProvinceView/148894.htm.]

Today the SEZs and ETDZs are China's greatest **growth poles,** meaning that their success is drawing yet more investment and migration (Figure 9.24 on page 352). In only 25 years, many coastal cities, including Dongguan (the city Li Xia migrated to in the chapter opening vignette) have grown from medium-sized towns or even villages into some of the largest urban areas in the world.

Hong Kong's special role. Hong Kong is one of the most densely populated cities on earth, and its residents have China's highest per capita income. In 2007, its annual GDP per capita (adjusted for PPP) was more than U.S.$42,000. Hong Kong was a British crown colony until July 1997, when Britain's 99-year lease ran out and Hong Kong became a special administrative region (SAR) of China. Many wealthy citizens fled, worried that China would absorb Hong Kong and no longer allow it economic and political freedom.

While Hong Kong's democracy has been curtailed, its role as China's unofficial link to the global economy has continued.

> Hong Kong is one of the most densely populated cities on earth, and its residents have China's highest per capita income.

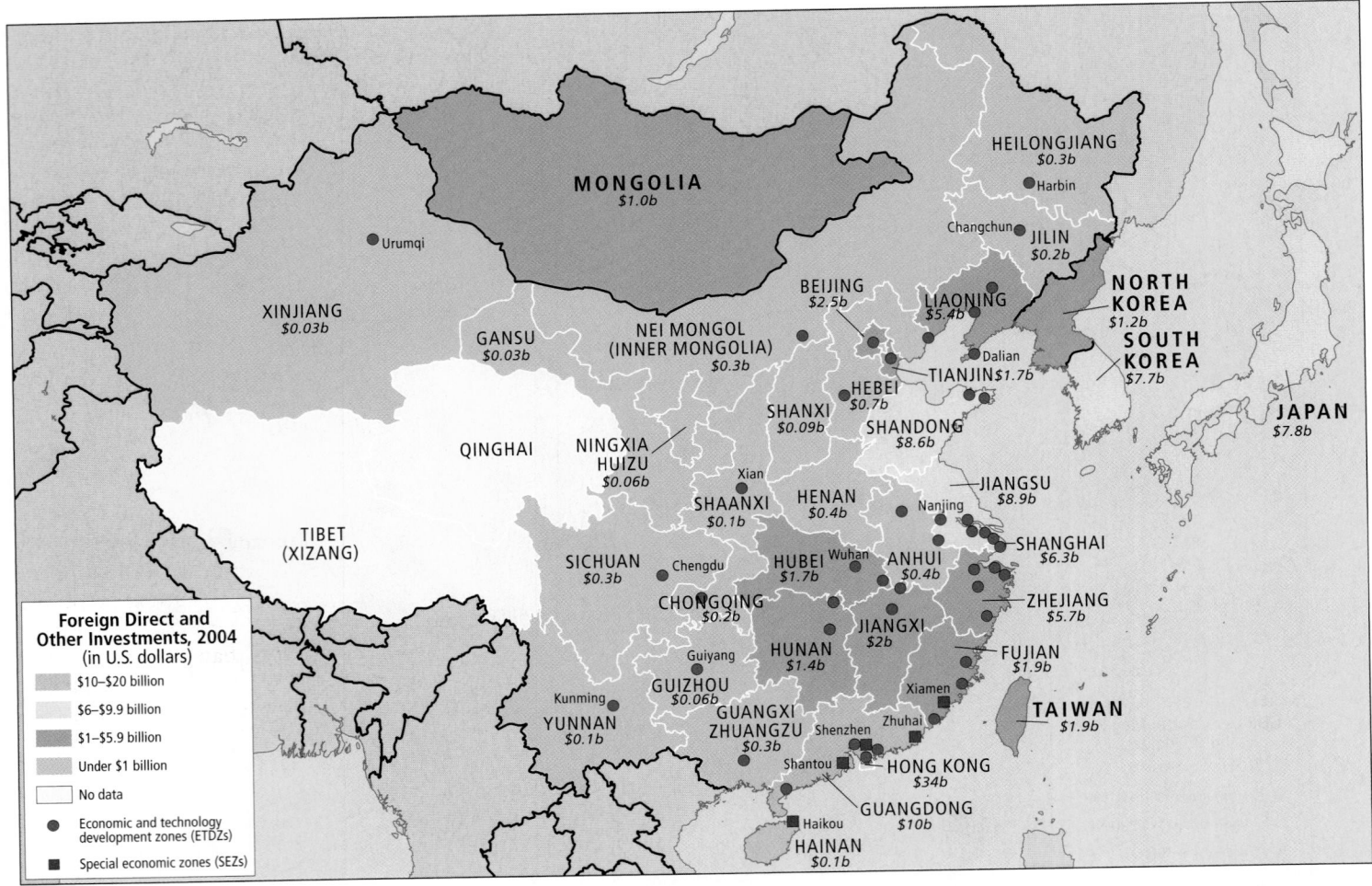

FIGURE 9.23 Foreign investment in East Asia. The map shows China's original special economic zones (SEZs) and more recently designated economic and technology development zones (ETDZs). The colors on the map reflect levels of foreign investment (direct and otherwise) in each of the countries of the region and in each of China's provinces. [FDI data from FDI Invest in China, at http://www.fdi.gov.cn/common/info.jsp?id=ABC00000000000022787. Specific Web site no longer available without registering at http://www.fdi.gov.cn/pub/FDI_EN/Statistics/default.htm.]

Before 1997, some 60 percent of foreign investment in China (Figure 9.25) was funneled through Hong Kong, and since then it has remained the financial hub for China's booming southeastern coast. Because so much of this investment comes from Japan, Korea, and Taiwan, Hong Kong is an important regional financial hub as well. ▰

Resistance to increasing rural-urban disparity. Many poor and rural inhabitants in China resent the growing income gap between the globalizing coastal cities and the rest of the country. The government has responded to protests and complaints by lowering taxes on farms and farm products and increasing farm subsidies. In 2006, rural incomes overall grew 7 percent faster than urban incomes, but because rural incomes started so much lower, overall regional disparities remained. Still, the possibility of making a living on the land has drawn some urban workers back to the farm.

Case study: Shanghai's inequitable transformation. Shanghai has a long history as a trendsetter. Its opening to Western trade in the early nineteenth century spawned a period of phenomenal economic growth and cultural development that led it to be called "the Paris of the East." As a result of China's recent reentry into the global economy, Shanghai is undergoing another boom, which has enriched some people and dislocated others.

In less than a decade, the city's urban landscape has been remade by the construction of more than one thousand business and residential skyscrapers, subway lines and stations, highway overpasses, bridges, and tunnels. The changes, however, have occurred too fast for planners to keep up with them. For hundreds of miles into the countryside, suburban development linked to Shanghai's economic boom is gobbling up farmland, and displaced farmers have rioted. The excessive opulence of Shanghai is causing discontent even among those who

FIGURE 9.24 DaimlerChrysler in Beijing. DaimlerChrysler AG formally opened its first factory in China in September 2006. The factory, in suburban Beijing, makes Mercedes-Benz and Chrysler sedans for the upscale market. With its U.S.$1.9 billion investment, DaimlerChrysler is joining a rush of foreign automakers for a share of the booming Chinese car market. [AP Photo/Elizabeth Dalziel.]

live there. Mortgage payments on new apartments leave many middle-class workers with little money for anything else.

Pudong, the new financial center for the city, sits across the Huangpu River from Shanghai's famous Bund—an elegant row of big brownstone buildings that served as the financial capital of China until half a century ago. Previously a maze of dirt paths and sprawling neighborhoods of simple tile-roofed houses, Pudong's former residents were pushed out to make way for soaring high-rises (Figure 9.26). The grandest is Jin Mao Tower, the fifth-tallest building in the world. Atop the tower, Hyatt Corporation has opened its first grand hotel in China, catering to wealthy investors from abroad.

Labor shortages in China? China's spectacular urban growth was based on hundreds of millions of new urban migrants who were willing to put up with almost any abuse to earn a little cash. However, so many factories, businesses, and shopping malls have been built or are under construction that experienced and skilled workers of all types are increasingly in short supply. Those with management skills are especially scarce.

To attract employees, some factory owners are offering higher pay, better working conditions, shorter workdays, and increasing time off. The extra costs these changes impose mean that China is no longer the cheapest place to manufacture products. Some factories are already moving to Vietnam, the Philippines, and countries in Africa with yet cheaper labor.

Origins of Investment in China, 2005

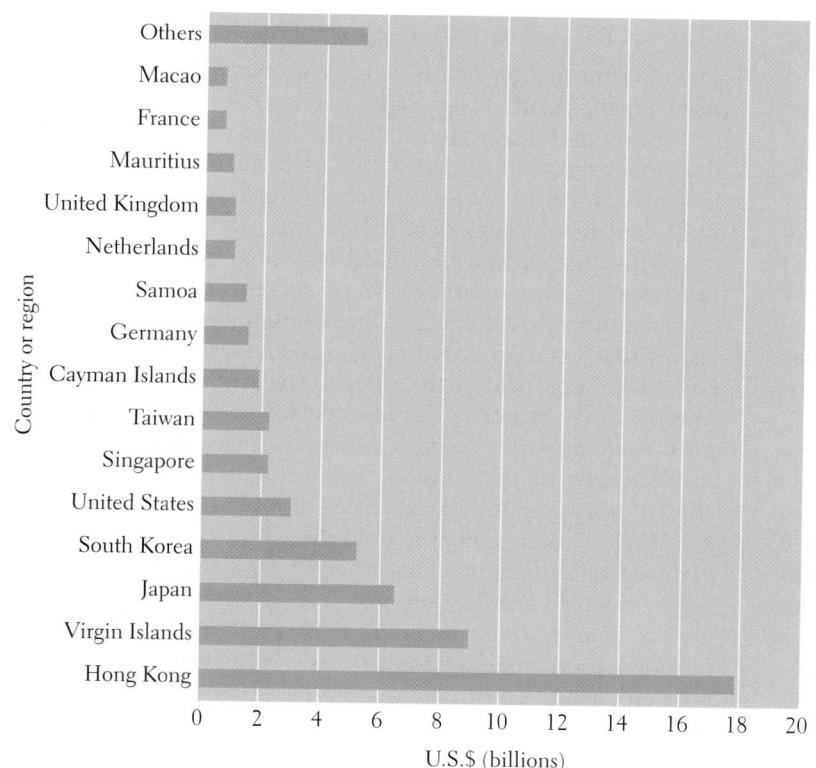

Figure 9.25 Top origins of investment in China in 2005. Total foreign direct investment in China in 2005 was U.S.$60.3 billion. The graph depicts the origin of the investors by country or locale and the amounts invested. Although investment is definitely high, the figures may be inflated by "round-tripping," which refers to capital that originates in China, is sent to a foreign tax haven or to Macao or Hong Kong, and then returns as "foreign" investment. [FDI data from FDI Invest in China, at http://www.fdi.gov.cn/common/info.jsp?id=ABC0000000000003075. Specific Web site no longer available without registering at http://www.fdi.gov.cn/pub/FDI_EN/Statistics/default.htm.]

(a)

(b)

FIGURE 9.26 Shanghai, many cities in one. Shanghai is China's largest and most modern urban area, yet it still retains many characteristics of the old city. (**a**) In one of the oldest parts of Shanghai, hundreds of traditional shops and stalls exhibit modern symbols. (**b**) Nearby stands the Jin Mao Tower, currently China's tallest building, ranking fifth highest in the world. [(**a**) © M. Lines; (**b**) © Khai Y. Chan.]

Others are producing more technologically sophisticated items, such as cell phones and computers, that sell for more money and hence allow for higher wages for workers. China's labor shortages have added a new twist to the story of Li Xia, whom we met in the chapter opening vignette.

> **Vignette** Li Xia and her sister returned home to rural Sichuan a second time. With the money she had saved from her second job in Dongguan, Xia tried to open a bar in the front room of her parents' house. She hoped to introduce the popular custom of karaoke singing she had enjoyed in the city, but people in her village couldn't stand the noise, and family tensions rose. Xia's sister, delighted to be home with her husband and baby, quickly found a job in a small new fruit processing factory. Her husband began farming again to fill the new demand for organic vegetables among the middle class in the Sichuan city of Chongqing. The government offered him a subsidy to learn new cultivation techniques if he would guarantee regular delivery of salad greens to a central purchasing depot.
>
> Amid all this success and her bar's failure, news that training was now available in Dongguan for skilled electronics assemblers convinced Xia to try again. Word was that the plant would be air-conditioned and that this time she could eventually earn U.S.$400 a month—enough to afford some serious shopping and perhaps one of the stylish new apartments built by the electronics firm for skilled workers.
>
> *Sources: Paul Wiseman, "Chinese factories struggle to hire," USA Today, April 11, 2005, http://www.usatoday.com/money/world/2005-04-11-china-labor_x.htm; Louisa Lim, "The end of agriculture," "Reporter's Notebook," National Public Radio, May 19, 2006, http://www.npr.org/templates/story/story.php?storyId=5411325; Mei* *Fong, "A Chinese puzzle: Surprising shortage of workers," Wall Street Journal (August 16, 2004): B1; David Barboza, "Labor shortage in China may lead to a trade shift," New York Times (April 3, 2006), business section, 1.* ■

Democratization in East Asia

Pressures for greater democracy are growing throughout East Asia. Japan, South Korea, and Taiwan have well-established democratic political systems, and Mongolia has experienced a dramatic expansion of democracy since abandoning socialism in 1992. North Korea and China, however, remain under the tight control of undemocratic regimes.

With China now a globalized economy, many wonder how much longer the Communist Party can remain in control without instituting democracy. The Communist Party officially claims that China is a democracy, and indeed some elections have long been held at the village level and within the Communist Party. However, representatives to the *National People's Congress*, the country's highest legislative body, are appointed by the Communist Party elite, who maintain tight control throughout all levels of government. 📹

While radical change is unlikely in the near future, most China experts agree that a steady shift toward greater democracy is underway and possibly inevitable. Demands for political change built to a crescendo less than a decade after market reforms began, culminating in a series of pro-democracy protests that drew

> While radical change is unlikely in the near future, most China experts agree that a steady shift toward greater democracy is underway and possibly inevitable.

hundreds of thousands to Beijing's *Tiananmen Square* in 1989. These protests were brutally repressed, with thousands (the precise number is uncertain) of students and labor leaders massacred by the military. Since then, pressure for change has continued to mount both within China and internationally.

Urbanization, protest, and local democracy.
The shift of hundreds of millions of people to cities has created new pressures for democracy in China. Crime has increased in urban areas, as have protests by workers for better pay and living conditions. Farmers and urban dwellers displaced by new real estate development have rioted, as have unemployed workers in older cities where state run industries have closed in recent years. In 2003, China's government reported that there were 58,000 such public protests in the country. This number rose to 74,000 in 2004, and 87,000 in 2005. Since then, the government has stopped issuing complete statistics on protests.

In an effort to maintain control over China's increasingly rebellious urban populations, the government has allowed elections to be held for "urban residents committees." The idea is to provide a peaceful outlet for voicing frustrations and achieving limited change at the local level. Many of these elections are hardly democratic, with candidates selected by the Communist Party. However, in cities where unemployment is high and where protests have been particularly intense, elections tend to be more free, open, and truly democratic.

International pressures.
China's admission to the World Trade Organization in 2001 brought with it pressures for political change. Much of China's growth has been built on environmentally destructive activities and abuses of workers, both of which effectively lower production costs so that goods can sell at a lower price. Informed consumers and environmentalists in developed countries have long criticized China for its "no holds barred" pursuit of economic growth. Now many developing countries are also becoming critical. Increasingly, they see Chinese goods as unfairly competing with their industries and are calling for the WTO to require China to improve its record on human rights and the environment.

The 2008 Olympics in Beijing highlighted a number of issues that have also created pressures for greater democracy. Before the games, the international media shined its spotlight on human rights abuses of workers, protesters, prisoners, ethnic minorities, and religious groups such as Falun Gong. Particular attention was given to protests in Tibet and elsewhere around the world against China's ongoing repression of Tibetan Buddhist monasteries and the mass importation of non-Tibetans into the province. ▣

The 2008 Beijing games provide an interesting contrast with the 1988 Olympic games in Seoul, South Korea. In 1987, pressures for democracy were growing in South Korea after decades of undemocratic military rule. Fearful that massive protests would break out before the Olympics and that the games might be moved somewhere else as a result, South Korea's military leader stepped down and democratic elections were held.

China chose a different approach for the 2008 Olympic games, cracking down on demonstrators and limiting their access to the international media. The attention of the world was then focused on the lavish sports facilities and spectacular opening and closing ceremonies of the games (Figure 9.27). China spent $U.S.44 billion for the games as a whole, more than had been spent on the previous ten summer Olympic games *combined*. This extravagance became a subject of criticism by the global media. Many journalists pointed out that no democracy would ever be able to devote so many tax dollars to such an event, especially not in a country with as many pressing human and environmental problems as China has.

FIGURE 9.27 The 2008 Olympic Games in Beijing. China spent more than $U.S.44 billion on the 2008 Olympic Games, mostly on infrastructure (including sports facilities such as the "bird's nest" stadium (a), energy, transportation, and water supply projects in the city of Beijing. The opening ceremonies alone (b) cost $U.S.300 million. [(a) Alex Needham (b) U.S. Army photo by Tim/Hipps]

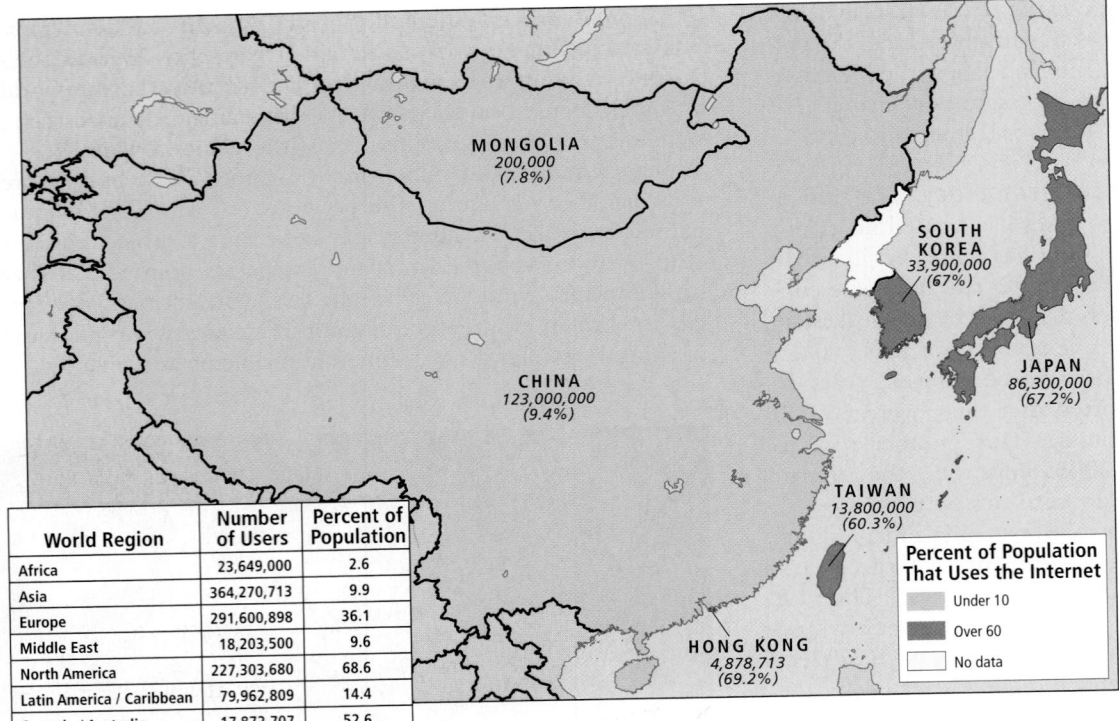

FIGURE 9.28 Internet use in East Asia, 2006. Some parts of East Asia have very high Internet use (rust color), while China and Mongolia have relatively low rates. Nonetheless, in these two countries, diffusion of information coming from the Internet has significant influence on public awareness. The government tries to control it, not always successfully. [Data from http://www.internetworld stats.com/stats3.htm.]

World Region	Number of Users	Percent of Population
Africa	23,649,000	2.6
Asia	364,270,713	9.9
Europe	291,600,898	36.1
Middle East	18,203,500	9.6
North America	227,303,680	68.6
Latin America / Caribbean	79,962,809	14.4
Oceania / Australia	17,872,707	52.6

Attention to China's human rights abuses adds to long-standing arguments by the United States and other countries that China should reduce the use of repressive political control and shift toward greater democracy. China has responded with an elaborate public relations and investment campaign emphasizing its **soft power.** Emissaries are sent to assure governments that China will be a trading partner and investor that will not meddle in their internal affairs or make demands that impinge on their sovereignty. Human rights and democratization are represented as purely internal affairs. However, China's willingness to invest in countries with human rights abuses even worse than its own, such as Sudan and Liberia, has earned it yet more criticism from the international community. Nor has "soft power" diplomacy distracted the world from China's military buildup in recent years, which has been considerable. ▄►

Information technology and democracy. The spread of information via the Internet has increased pressures for democratization from within China. Since 1949, China's central government has controlled the news media in the country. By the late 1990s, however, the expanding use of electronic communication devices was loosening central control over information.

By 2001, 23 million people were connected to the Internet in China, and by 2006, 123 million were connected (Figure 9.28). Just a few years ago, telephones were very rare, but now anyone can buy temporary cell phone access without showing identification. As a result, millions of Chinese have anonymous access to an international network of information. It is now much more difficult for the government to give inaccurate explanations for problems caused by inefficiency and corruption. Reporters can easily check the accuracy of government explanations by calling witnesses or the principal actors directly on cell phones and then posting the explanations on the Internet. Analysts both inside and outside China see the availability of the Internet to ordinary Chinese citizens as a watershed event that supports democracy.

Nevertheless, the Internet in China is not the open forum that it is in most Western countries. For example, if people writing blogs in China use the words "democracy," "freedom," or "human rights," they may get the following reminder: "The title must not contain prohibited language, such as profanity. Please type a different title." Such censorship has been aided by U.S. technology firms such as Google, Yahoo, and Microsoft allowing the Chinese government to use software that blocks access to certain Web sites for users in China. Nevertheless, the massive amount of information available online and the difficulties of controlling access to it still make information technology a powerful force for change within China. ▄►

SOCIOCULTURAL ISSUES

East Asia's economic progress has led to social and cultural change throughout the region. Modernized economies are changing work patterns and family structures. Meanwhile, cultural homogenization, discrimination, and political persecution are threatening East Asia's rich cultural diversity.

Gender, Family, and Work in Urban East Asia

Life in a small city apartment is very different from life among a host of relatives in a farming village. Nonetheless, throughout East Asia, urban wives still perform most domestic duties, looking after husbands and children and often elderly parents or parents-in-law as well. Although there is a small but growing group of trained young women who seek careers and are as ambitious as men, even many university-educated married women say they wish to earn only supplementary income for the household.

The changing work ethic. One reason that many women have not been able to take on outside work as well as family duties is that jobs in East Asian industrial economies are particularly demanding. Workdays are long, and commuting often adds more than 3 hours to the time away from home. Even more important is the "culture of work," which in Japan, Taiwan, and South Korea (less so in China) is based on male camaraderie and demands an especially high level of loyalty to the firm, often at the expense of family. In Japan, for example, after-work leisure time often must be spent with business colleagues if one is pursuing a promotion. It is considered disloyal to refuse overtime, and employees are discouraged from taking time off to spend with a spouse or to help care for children or elderly relatives. When hired, women usually support men as secretaries and assistants rather than participating as full members of corporate teams.

> Japanese women, when hired, are usually there to support men as secretaries and assistants, not to participate as full members of corporate teams.

By 2000, the East Asian urban work ethic and family structure were being publicly challenged, not least by the male workers themselves. In Tokyo, for example, a group calling itself Men Concerned About Child Care meets regularly to discuss ways to participate more fully in home and community life. Another group has formed to lobby for 4-hour workdays for both men and women, so that both can spend time with the family. This group has filed lawsuits against employers who routinely require employees to accept job transfers to distant locations where they cannot take their families. It is particularly significant that urban men are the ones challenging the extremes of the East Asian work ethic. They are the group that has been most regimented and most deprived of personal time and family life, but they are also the group that has the most power to change the system.

Inequality by the numbers. East Asia's highly gendered patterns of work and home life are reflected in statistics on education, employment, and pay. In China in 2005, 14 percent of women obtained college or technical training, compared with 16 percent of men. Female Chinese workers earn about two-thirds of what male workers earn. In Mongolia, almost twice as many women as men seek education beyond high school, and yet they earn only two-thirds of what men earn.

In the industrialized societies of Japan and South Korea, nearly all men attend high school, as do all Japanese women. Only 88 percent of South Korean women attend high school, but more women (61 percent) go to college or technical school than in Japan (47 percent). Japanese and South Korean women earn only about half of what men earn.

Indigenous Minorities

Cultural diversity exists throughout East Asia even though most countries have one dominant ethnic group. In China, for example, 93 percent of Chinese citizens call themselves "people of the Han." The name harks back about 2000 years to the Han empire, but it gained currency only in the early twentieth century, when nationalist leaders were trying to create a mass Chinese identity. The term *Han* does not denote an actual ethnic group but rather connotes people who share a general way of life, pride in Chinese culture, and a sense of superiority to ethnic minorities and outsiders. The main language spoken by the Han is Mandarin, although it is just one of many Chinese dialects. (All Chinese dialects use the same writing system.)

China's non-Han minorities number about 117 million people in more than 55 different ethnic or culture groups scattered across the country. Most live outside the Han heartland of eastern China (Figure 9.29). Some of these areas have been

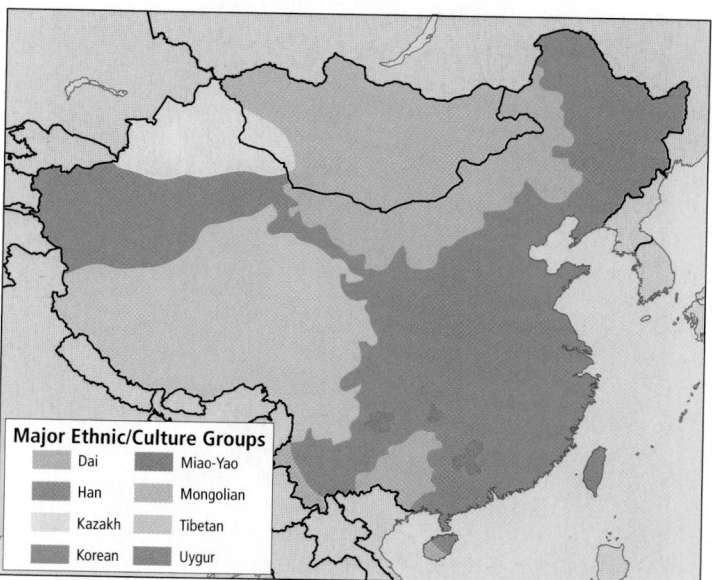

FIGURE 9.29 Major ethnic groups of China. This map shows the areas traditionally occupied by the Han and the ethnic minorities. It does not show the recent resettling of Han in Xinjiang and Tibet, nor does it show the Hui, people from many ethnic groups whose ancestors converted to Islam and who are found in disparate locations along the old Silk Road and in coastal southeastern China. [Adapted from Chiao-min Hsieh and Jean Kan Hsieh, *China: A Provincial Atlas* (New York: Macmillan, 1995), p. 12. The Web site http://www.index-china.com/minority/minority-english.htm supplies a comprehensive survey of minorities in China.]

designated autonomous regions, where minorities theoretically manage their own affairs. In practice, however, the Han-dominated Communist Party in Beijing controls the fate of the minorities, especially those considered to be security risks or who have resources of economic value. We profile just a few of China's ethnic groups here.

Western China's Muslims. One autonomous region is Xinjiang Uygur in the far northwest (see Figure 9.1), where Turkic-speaking Muslims, such as the Uygurs and Kazakhs, live. Historically, these peoples were nomadic herders, and a very few still are. Trade with similar peoples in Central Asia has revived since China's market reforms began (Figure 9.30). Because the Beijing government wishes to claim this area's oil and other mineral resources for national development, it has sent troops and hundreds of thousands of Han settlers to Xinjiang. The Han settlers fill most managerial jobs in mineral extraction, the military, and power generation. An important secondary role of the Han is to dilute the power of Uygurs and Kazakhs within their own lands.

Uygurs in particular are concerned about Han prejudices against Uygur ethnicity and religion. For example, one young

FIGURE 9.30 A Silk Road market. At the market in Kashi, you can still buy a camel, a fine Oriental rug, the original bagel, or the skin of an endangered animal. You can also get a haircut or get your teeth fixed, without a painkiller. [Reza/National Geographic Image Collection.]

Uygur man in Urumqi, Xinjiang's capital, writes of his discontent: "I am a strong man, and well-educated. But [Han] Chinese firms won't give me a job. Yet go down to the railroad station and you can see all the [Han] Chinese who've just arrived. They'll get jobs. It's a policy to swamp us."

Many people of Xinjiang now express their resistance to Han dominance through Islamic culture. Islamic prayers are increasingly heard publicly, more Muslim women are wearing Islamic dress, and Islamic architectural traditions are being revived. The Beijing government has responded by harshly punishing those suspected of Islamic fundamentalism.

Prosperous Muslim traders: the Hui. The original Hui people were descended from ancient Turkic Muslim traders who traveled the Silk Road across Central Asia from Europe to Kashgar to Xian (see Figure 5.11 on page 185). Though most Hui today are descended from non-Turkic peoples who converted to Islam, they are still culturally distinct. Groups of Hui, some numbering one million or more, live throughout northern, western, and southwestern China. The Hui have a long tradition of commercial activity, and they have been particularly successful in China's new free market economy. They are using their money not only for consumption of luxury goods, but also to revive religious instruction and to fund their mosques, which are now more obvious in the landscape.

The Tibetans. In contrast to the prosperous Hui are the Tibetans, an impoverished ethnic minority of nearly 5 million individuals scattered thinly over a huge, high mountainous region in western China. Their homeland is widely known as Tibet, but since Chinese troops invaded in 1950, the Chinese government has referred to it as the Xizang Autonomous Region. The Chinese government suppressed the Tibetan Buddhist religion, long the mainstay of most Tibetans' daily lives, by destroying thousands of temples and monasteries and massacring many thousands of monks and nuns. The spiritual and political leader of Tibet, the Dalai Lama, was forced into exile in India along with thousands of his followers. ▶

To dismantle traditional Tibetan society further, hundreds of thousands of Han Chinese were resettled in Tibet. They control the economy and the major cities, exploit mineral and forest resources, and force native Tibetans to adopt Chinese ways. A new railway link connecting Tibet more conveniently to the rest of China was completed in 2006. The Han in Tibet see the railway as a public service that will promote Tibetan development, but Tibetan activists see it as a conveyor belt for more Han dominance.

Women in Tibet have always had a relatively high position in society. Among the nomadic herders, they were free to have more than one husband, just as men were free to have more than one wife. In addition, the custom of the husband joining the wife's family allowed Tibetan women to attain a higher status than Han women. Although Buddhism introduced patriarchal attitudes from outside Tibet, it encouraged female independence. At any given time, up to one-third of the male population was living a short-term monastic life, so Tibetan Buddhist women have become particularly self-sufficient. Chinese culture has typically regarded the women of the western minorities as barbarian precisely because their roles were not circumscribed: they were not secluded, they rode horses, they worked alongside the men in herding and agriculture, and they were assertive.

Indigenous diversity in southern China. In Yunnan Province in southern China, more than 20 groups of ancient native peoples live in remote areas of the deeply folded mountains that stretch into Southeast Asia. These groups speak many different languages, and many have cultural and language connections to the indigenous people of Tibet, Burma, Thailand, or Cambodia. Women and men are treated more equally in this area than among the Han. A crucial difference may be that among several groups, most notably the Dai, the husband moves in with the wife's family at marriage and provides her family with labor or income. A husband inherits from his wife's family rather than from his birth family, and female children are valued just as highly as males.

Taiwan's many minorities. In Taiwan and the adjacent islands, the Han account for 95 percent of the population, but Taiwan is also home to 60 indigenous minorities. Some have cultural characteristics—languages, crafts, and agricultural and hunting customs—that indicate a strong connection to ancient cultures in far Southeast Asia and the Pacific. The mountain dwellers among these groups have resisted assimilation more than the plains peoples. Both groups may live on reservations set aside for indigenous minorities if they choose, but most are now being absorbed into mainstream urbanized Taiwanese life.

Japan's Ainu. There are several indigenous minorities in Japan and most have suffered considerable discrimination. A small and distinctive minority group is the **Ainu.** Now numbering only about 16,000, the Ainu are a racially and culturally distinct group who are thought to have migrated many thousands of years ago from the northern Asian steppes. They once occupied Hokkaido and northern Honshu, living by hunting, fishing, and some cultivation, but they are now being displaced by forestry and other development activities. Few full-blooded Ainu remain because, despite prejudice, they have been steadily assimilated into the mainstream Japanese population. Some now make a living in Ainu living history villages by demonstrating to tourists how to make traditional crafts. (Figure 9.31). Many are attempting to revive the "Ainu spirit" by teaching their children the Ainu language and traditional ways.

East Asia's Largest Cultural Export: The Overseas Chinese

China has had an impact on the rest of the world not only through its global trade, but also through the migration of its people to nearly all corners of the world. The first recorded emigration by the Chinese took place over 2200 years ago. Since then, China's contacts spread eastward to Korea and Japan, westward into Central and Southwest Asia via the Silk Road, and by the fifteenth century, to Southeast Asia, coastal India, Arabia, and even Africa.

FIGURE 9.31 Ethnic minorities in Japan. This Ainu chief lives and works in a traditional Ainu living-history village in the Akan National Park in eastern Hokkaido, Japan. He is working with wood that he carves to sell to visitors. [© F. Staud/www.photo travels.net.]

Trade was probably the first impetus for Chinese emigration. The early merchants, artisans, sailors, and laborers came mainly from China's southeastern coastal provinces. Taking their families with them, some settled permanently on the peninsulas and islands of what are now Indonesia, Thailand, Malaysia, and the Philippines. Today they form a prosperous urban commercial class known as the "Overseas Chinese."

In the nineteenth century, economic hardship in China and a growing international demand for labor spawned the migration of as many as 10 million Chinese to countries all over the world. By the mid-twentieth century, they were joined by many others fleeing the repression of China's communist revolution. As a result, "Chinatowns" are present in places as widely scattered as Singapore, London, São Paulo, and Toronto.

Reflections on East Asia

Just as the astounding economic rise of Japan, South Korea, and Taiwan changed the world over the past fifty years, the opening of China to the global economy will continue to transform East Asia and the world for the next fifty. The changes underway in China today are immense. Incomes have risen and cities have boomed, but at the cost of some of the worst air and water pollution in the world. Greenhouse gas emissions have increased, but so has the ability to use new technologies to create cleaner energy economies. China's ability to feed its people with its own agricultural resources is declining, but its ability to pay for food imported from abroad is increasing.

Wealth disparities in China have increased dramatically, with millions of incomes hurt by the collapse of the communist command economy. However, many have been helped by new economic opportunities created by the forces of free market globalization that have swept through the country. The shortages of skilled labor are bringing better wages and

working conditions, and may also hasten a shift toward the production of more technologically sophisticated and higher-quality goods.

A major question remains regarding China's growing and future prosperity. Where will China find the natural resources to supply increasing living standards for so many people? China's massive importation of food, minerals and oil is already affecting many regions of the world (see Chapter 3, page 106, Chapter 7, pages 241–242 and 270, and Chapter 10, page 373). These areas may not be willing or able to sustain such massive resource exports for long.

Throughout East Asia, as birth rates decline and populations age, some economies may suffer if they don't either import labor or increase mechanization. This shift is already occurring in Japan, South Korea, and Taiwan, whose existing broad prosperity has given them more breathing room. China's economy may take a bigger hit because the one-child-per-family policy has reduced population growth drastically *before* broader prosperity has been achieved. This, combined with the country's gender imbalance, may produce considerable social instability. And as more women throughout East Asia pursue careers instead of family life, birth rates will continue to fall, bringing further social and economic changes.

The combined impact of all of these transformations may eventually force China to adopt the democratic systems that already prevail among East Asia's wealthier countries. Such a shift could improve China's awful human rights record, and would no doubt be celebrated by the millions who have suffered from the tragic failures of judgment of the Communist Party political elite. Still, many of China's problems will remain even when strong democratic institutions are in place. Democracy has not solved the problems of an aging population in Japan, Taiwan, and South Korea, or of severe air pollution in Taiwan. In this ever more globalized world, the ways in which China and East Asia as a whole solve these problems will influence all of our lives.

Chapter Key Terms

Ainu 359
Confucianism 338
Cultural Revolution 341
export-led growth 347
floating population 327
food security 334

Great Leap Forward 341
growth poles 351
hukou system 326
kaizen system 348
kanban system 347
regional self-sufficiency 349

soft power 356
special economic zones (SEZs) 350
state-aided market economy 347
tsunami 330
typhoon 332
wet rice cultivation 334

Critical Thinking Questions

1. Why is the interior west of continental East Asia (western China and Mongolia) so dry and subject to extremes in temperature?

2. In what ways is China more vulnerable to global warming than South Korea, Japan, and Taiwan?

3. China used to be afflicted with recurring famines. What are the present concerns about its food security?

4. How might a more democratic political system have resulted in changes in the overall conception and design of the Three Gorges Dam?

5. Why might a Confucian praise or criticize China's current political system and its approach to economic and political change?

6. How might globalization have taken a different course if the voyages of Admiral Zheng He had inspired China to conquer a vast colonial empire like that of Britain or the Netherlands?

7. Contrast Japan's pre–World War II policies in East Asia with its present role in the region. What are the principal similarities or differences?

8. In what ways has the one-child-per-family policy helped maintain stability in China? How might it destabilize China in the future?

9. What problems related to urbanization and globalization might lead to greater political changes than have occurred already?

10. What forces are likely to change gender roles in East Asian countries in the future?

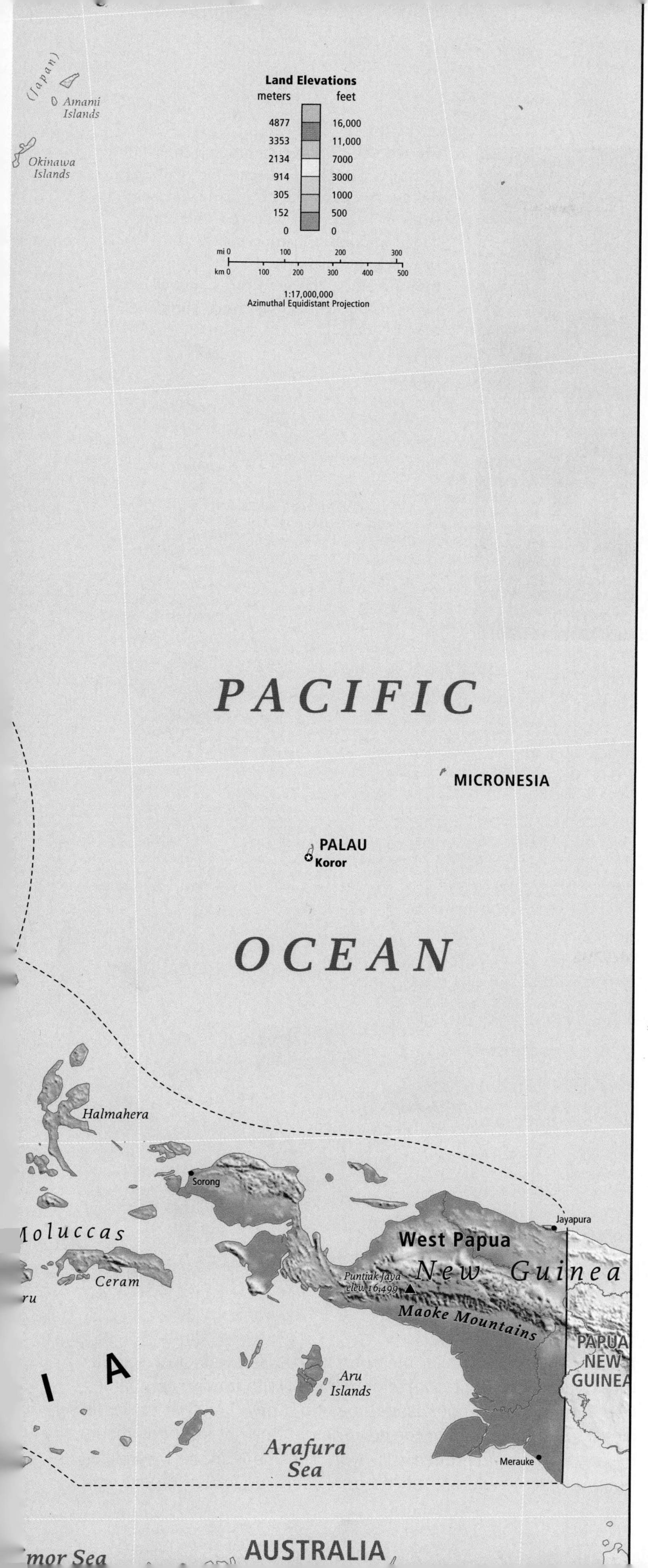

Land Elevations

meters	feet
4877	16,000
3353	11,000
2134	7000
914	3000
305	1000
152	500
0	0

mi 0   100   200   300

km 0   100   200   300   400   500

1:17,000,000
Azimuthal Equidistant Projection

Amami Islands

Okinawa Islands

(Japan)

PACIFIC

OCEAN

MICRONESIA

PALAU
Koror

Halmahera

Moluccas

Ceram

Sorong

Jayapura

West Papua
New Guinea

*Puntjak Jaya
elev. 16,499*

Maoke Mountains

PAPUA NEW GUINEA

Aru Islands

Arafura Sea

Merauke

AUSTRALIA

mor Sea

CHAPTER 10

Southeast Asia

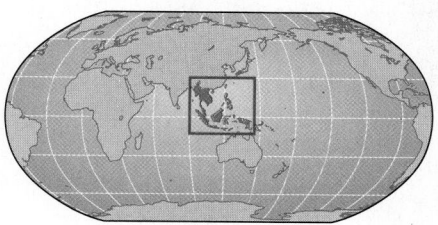

Global Patterns, Local Lives In December 2005, a group of indigenous people in the Maylasian state of Sarawak, on the island of Borneo, attended a public meeting wearing orangutan masks. They carried signs informing onlookers that, although the government protects orangutans, it ignores the basic right of indigenous people to live on their own ancestral lands.

Over the years, Sarawak forest dwellers have tried many tactics to save their lands from deforestation by logging companies and expanding oil palm plantations. The state government began giving logging companies licenses to cut down forests occupied by indigenous peoples in the mid-twentieth century. By the 1980s, 90 percent of Sarawak's lowland forests had been degraded and 30 percent had been clear-cut, with much of the cleared area replaced by oil palm plantations. As a consequence, indigenous people found that their hunts declined. Meanwhile, streams and rivers became polluted by fertilizers and the pesticides applied to the palm trees.

An early tactic indigenous communities used to stop the deforestation was to block logging roads with felled trees and their own bodies. But with government support, the logging companies removed the blockades and had the protesters arrested. Although news stories were widely circulated, the outside world paid little attention. In the 1990s, however, a group of citizens in Berkeley, California, who were concerned about news reports of the deforestation in Sarawak organized the Borneo Project. They offered to become a "sister city" to one indigenous community in Sarawak called Uma Bawang, giving help wherever it was needed. It soon became evident that an inventory of how the Uma Bawang used the forest could be crucial in validating their claim to their land and perhaps could keep timber extractors out.

With the help of the Borneo Project, Uma Bawang began a community-based mapping project in 1995. Using rudimentary compass and tape techniques, they began mapping both the extent and content of their forest home (Figure 10.2). Since then, indigenous people from across Sarawak have learned how to use global positioning systems (GPS), geographic information systems (GIS),

Figure 10.1 Regional map of Southeast Asia.

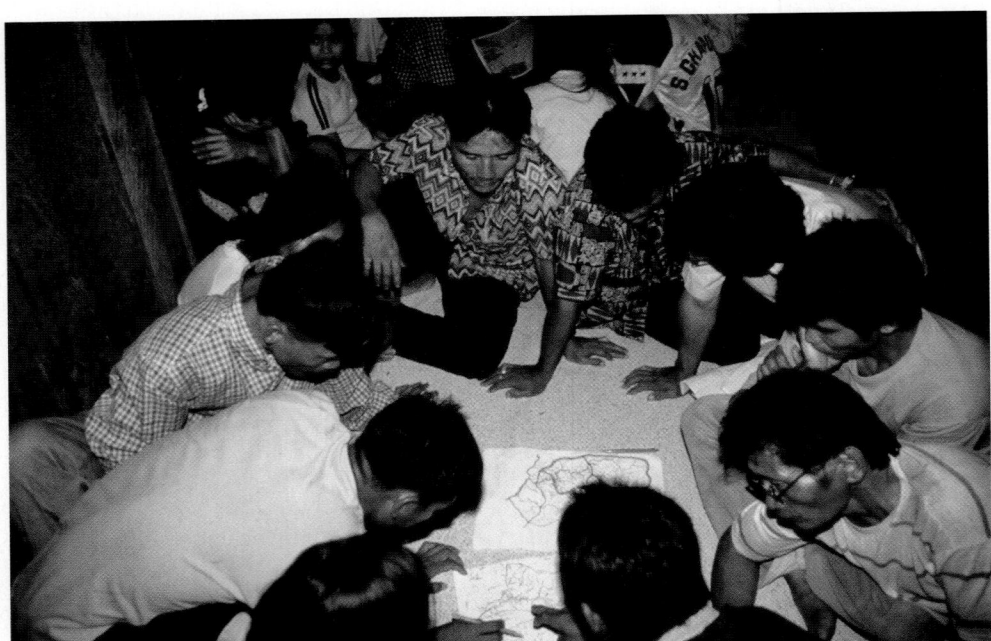

(a)

Figure 10.2 Villagers mapping their home territory. **(a)** These villagers first gathered to discuss and draw the boundaries of their lands in Penan, Sarawak. **(b)** Then they added important features such as huts (some of which are abandoned), sago palm trees, hunting areas, and graveyards, as well as physical features such as rivers. [Bruno Manser Fonds.]

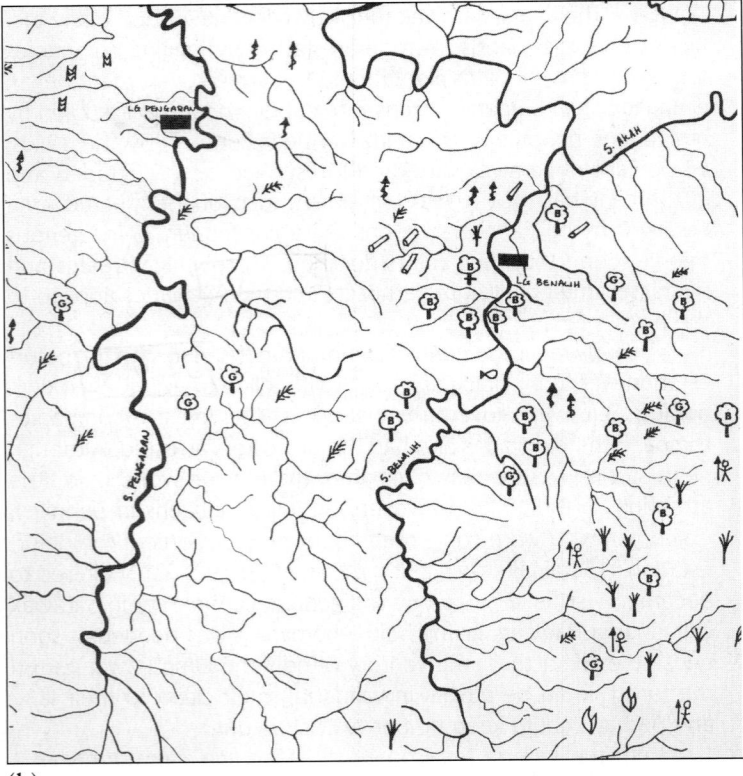

(b)

and satellite imagery to make more sophisticated maps. The power of these maps was demonstrated in 2001, when they helped the Iban people of central Sarawak win a precedent-setting court case that protected their lands from an encroaching oil palm plantation.

Despite their efforts, the ruling in favor of the Iban was over-turned in 2005; a final decision is still pending at the Malaysian federal level. Meanwhile, members of the Sarawak state legislature (who often profit from joint ventures with foreign companies operating on indigenous lands), passed a new law that makes it illegal to submit a map to the government that has not been made by a government-certified surveyor. The maps made by indigenous peoples could still be used in court, but their impact was much reduced. Meanwhile, no government-certified surveyors would agree to map indigenous land claims.

Though the legality of indigenous mapping may still be in question in Malaysia, the relevance of indigenous land claims has taken on global significance. Efforts to combat global warming are increasingly focused on combating deforestation (see Chapter 1, page 18). The UN and other organizations are starting programs that pay landowners to keep their forests intact. In response, indigenous groups all over the world have coordinated in a campaign to have their land claims recognized at the global level by the UN, thus making them eligible for payments. The Borneo Project, and thousands of other allies of indigenous groups, have supported this lobbying effort with online campaigns and carefully crafted statements and position papers.

By 2008, these efforts paid off. For the first time, representatives from indigenous forest communities around the world found themselves at the negotiating table in Washington, D.C., with timber and palm oil executives, national environment ministers, and other global warming stakeholders as they tried to figure out how to help stabilize the world's climate by keeping remaining forests intact.

Adapted from The Borneo Wire, Spring 2006, Jessica Lawrence, editor, Newsletter of The Borneo Project, 1771 Alcatraz Avenue, Berkeley, CA 94703, http://Borneoproject.org; "Rainforest dwellers successfully maintain logging road blockade in one of Malaysia's last virgin jungle areas," Bruno Manser Fonds, Society for the Peoples of the Rainforest. August 15, 2006, http://www.bmf.ch/en/en_index.html; Mark Bujang, "A community initiative: Mapping Dayak's customary lands in Sarawak," presented at the Regional Community Mapping Network Workshop, Nov. 8–10, 2004, Diliman, Quezon City, Philippines.

This account of the shifting tactics indigenous groups are using to secure their rights to ancestral lands highlights the extent to which local or national issues are becoming global issues. As problems like global warming become better understood, more issues are being addressed at the global level. This presents both challenges and opportunities for the world's indigenous people (Figure 10.3). If, like Uma Bawang, they

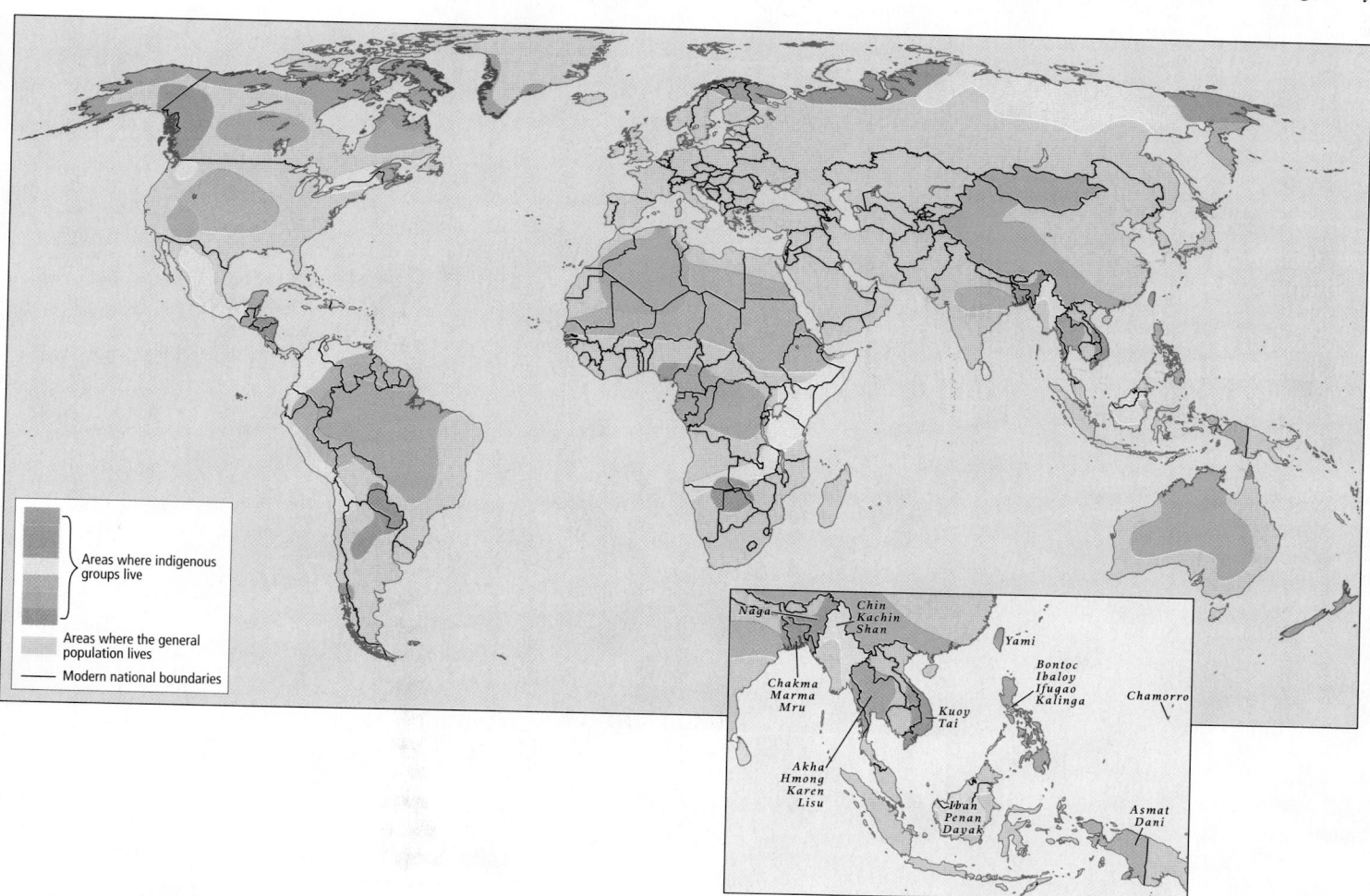

Figure 10.3 Indigenous groups worldwide. There are about 5000 distinct indigenous groups in the world. On this map, each color represents one or more of these groups that are related by language, culture, or an affinity to a geographic location. Many of these peoples have participated in community mapping projects, similar to those of the Sarawak forest dwellers, in order to identify and protect their rights to their traditional lands. [Adapted from "Struggling cultures," *National Geographic Atlas of the World, Eighth Edition* (Washington, D.C.: National Geographic Society, 2005), p. 15.]

Views of Southeast Asia

For a tour of these places, visit www.whfreeman.com/pulsipher
[Background image: Trent Schindler NASA/GSFC Code 610.3/587 Scientific Visualization Studio]

2. Urbanization. Agricultural jobs have declined in recent decades, pushing millions of rural people into cities such as Bangkok. Here many reside in slums such as these along the Chao Praya river. See pages 383–384 for more. Lat 13°46'45"N, Lon 100°30'08"E. [Pdtnc/Dreamstime.com]

1. Globalization is a driving force in the region's economies, especially in Singapore, shown here. The island city-state's global shipping facilities, export oriented manufacturing industries, and international financial sector have made it the region's wealthiest country. See pages 385–387 for more. Lat 1°16'N, Lon 103°50'E. [Rifleman 82/ http://commons.wikimedia.org/ wiki/File:Singapore_River_top_view.jpg]

8. Gender and sex tourism. Southeast Asia has become a global center for sex tourism. It is estimated that in 2008 70 percent of Thailand's 14 million visitors were looking for sex. This young woman is a paid dancer at a bar in Phuket, Thailand. See page 395–396 for more. Lat 7°53'47"N, Lon 98°17'43"E. [Image by ChrisO/http://commons.wikimedia. org/wiki/File:Gogo_dancers.jpg]

7. Population. While fertility rates have been declining, Southeast Asia's young population will still grow for several decades. This young family lives in Sumatra. See page 380–382 for more. Lat 3°35'N, Lon 98°40'E. [U.S. Navy photo by Photographer's Mate 3rd Class Jacob J. Kirk]

3 — Lao Cai

2 — Bangkok

Phuket 8

7

Sumatra

1 — Singapore

6 — Jakarta

3. Food production. Wet rice cultivation has transformed Southeast Asian landscapes. Even steep hillsides, such as these in Lao Cai, Vietnam have been terraced into fields. See page 374 for more. Lat 22°31'N, Lon 103°54'15"E. [Tttrung]

4. Water and global warming. Sea temperatures are rising due to global warming, threatening the coral reefs that sustain much of the region's fishing and tourism. Shown here are two reefs in Tubbataha National Marine Park, Philippines. The coral on the right is experiencing "bleaching." See pages 374–376 for more. Lat 8°54'N, Lon 119°56'E. [© 2005 Richard Ling <richard@research.canon.com.au>http://commons.wikimedia.org/wiki/File:Bicolor; Pacific Islands Fisheries Science Center/NOAA]

Tubbataha Reef

Sarawak

5. Global warming and deforestation. The expansion of oil palm plantations in Indonesia and Malaysia (Sarawak is shown here) has contributed significantly to both global warming and biodiversity loss. Deforestation is also driving orangutans (shown in the inset) toward extinction. See pages 373–374 for more. Lat 2°51'N, Lon 112°13'E. [Craig Antweiler; Mariana Ruiz Villarreal]

6. Democratization. Protests in Jakarta during the economic crisis of the late 1990s hastened the resignation of Suharto who had ruled in a semi-dictatorial style for 31 years. A new era of expanded democracy has followed. See page 390–392 for more. Lat 6°10'32"S, Lon 106°49'37"E. [Jonathan McIntosh]

develop partnerships that help them participate in global-scale lobbying efforts, they could see their land claims validated. Otherwise, their claims may go unrecognized. Either outcome will influence their ability to negotiate at the local and national scale.

In Southeast Asia, current trends toward the globalization of local and national issues can be better understood in the context of previous waves of globalization that have swept over the region. Conquered by European powers more than five hundred years ago, Southeast Asian nations gained independence after World War II and became key links in the global economy. Pressures on the region's resources intensified greatly when Southeast Asia entered an era of accelerated globalization in the 1970s. Governments and local investors began to aggressively market the minerals, forests, and cheap manufactured goods of the region on a global scale. By the 1990s, the forests of Thailand and the Philippines were largely depleted, and logging was in full swing in Malaysia and Indonesia. Timber flowed out of the region's forests toward the port of Singapore and on to building sites throughout the world.

Many rural farmers and forest inhabitants were forced or enticed into towns and cities, where they found work in booming manufacturing industries. Although their wages were low, they were enough to raise standards of living and provide the next generation with new educational opportunities. By the early 1990s, Southeast Asia was regarded as a model of rapid development that other regions might emulate. As we will see, that model has since become somewhat tarnished, both by economic crisis and serious environmental problems. Nevertheless, today Southeast Asia is again moving forward, thanks in large part to expanding manufacturing industries, which are using ever more sophisticated technologies.

Questions to consider as you study Southeast Asia (also see views of Southeast Asia, pages 366–367)

1. How is global warming linked to deforestation, food production, and water resources in Southeast Asia? Tropical rain forests are being cleared at an alarming rate in this region. To what extent is agriculture, both for export and local consumption, driving this process? How do these activities contribute to global warming? How is global warming likely to impact this region's water resources?

2. How are changes in agriculture and industry driving urbanization? Small family farms are being replaced by large mechanized corporate farms oriented toward commercial production. How has this change combined with the expansion of manufacturing and other industries to produce rapid urbanization? How are these changes related to globalization?

3. How has globalization produced both spectacular successes and tragic failures in the economies of this region? This region experienced rapid economic growth from the mid-1980s to the late 1990s. Beginning in 1997, however, a disastrous financial crisis shook the region for several years. How were both of these events related to globalization?

4. What factors are resulting in slower rates of population growth? In some countries, government policies have had a strong effect on the number of children that families are having. How have other factors, such as urbanization and economic growth, influenced population growth?

5. Can democratization reduce violence and corruption? While significant barriers to democracy remain, some countries have experienced dramatic expansions of democracy in recent years. How has the expansion of democracy affected corruption in this region? What is the evidence that democracy can reduce political violence associated with separatist movements?

6. How is urbanization transforming gender roles in Southeast Asian families? Although traditional family structures are patriarchal, they have some unique features that give women more power and freedom than they have in many other world regions. How are gender roles changing as families move to the city?

The Geographic Setting

Terms to Be Aware Of

Many governments in Southeast Asia (Figure 10.4) choose to dispense with place-names originating in their colonial past. However, when the governments that make these changes earn broad disrespect in the international community, usually by violating the human rights of their citizens, their chosen name is sometimes not acknowledged. Instead, the old name is used as a political statement to show disapproval for the government's actions. Such is the case with Burma, where a military government seized control in a coup d'état in 1990, changing the country's name to Myanmar. In this text, we use Burma instead of Myanmar to acknowledge the repression of the people of that country by a government they had no role in choosing.

Another potential point of confusion is Borneo, a large island that is shared by three countries. The part of the island known as Kalimantan is part of Indonesia; Sarawak and Sabah

Figure 10.4 Political map of Southeast Asia.

are part of Malaysia; and Brunei is a very small, independent, oil-rich country.

PHYSICAL PATTERNS

The physical patterns of Southeast Asia have a continuity that is not immediately obvious on a map of the region. On a map, one sees a unified mainland region that is part of the Eurasian continent and a vast and complex series of islands arranged in chains and groups. These landforms are actually related in origin. Climate is another source of continuity, with most of the region tropical or subtropical.

Landforms

Southeast Asia is a region of peninsulas and islands (see Figure 10.1). Although the region stretches over an area larger than the continental United States, most of that space is ocean; the area of all the region's land amounts to less than half that of the contiguous United States. The large mainland peninsula, sometimes called Indochina, that extends to the south of China is occupied by Burma, Thailand, Laos, Cambodia, and Vietnam. This peninsula itself sprouts a long, thin peninsular appendage that is shared by outlying parts of Burma and Thailand, a part of Malaysia, and the city-state of Singapore, which is built on a series of islands at the southern tip. The **archipelago** (a series of large and small islands) that fans out to the south and east of the mainland is grouped into the countries of Malaysia, Indonesia, the Philippines, Brunei, and Timor-Leste (East Timor). Indonesia alone has some 17,000 islands, and the Philippines has 7000.

The irregular shapes and landforms of the Southeast Asian mainland and archipelago are the result of the same tectonic

forces that were unleashed when India split off from the African Plate and crashed into Eurasia (see page 288). As a result of this collision, which is still under way, the mountainous folds of the Plateau of Tibet, which reach heights of almost 20,000 feet (6100 meters), bend out of the high plateau and turn south into Southeast Asia. There, they descend rapidly and then fan out to become the Indochina peninsula. The gorges widen into valleys that stretch toward the sea, each containing hills of 2000 to 3000 feet (600 to 900 meters) and a river or two flowing from the mountains of China to the north. The major rivers of the peninsula are the Irrawaddy and the Salween in Burma; the Chao Phraya in Thailand; the Mekong, which flows through Laos, Cambodia, and Vietnam; and the Black and Red rivers of northern Vietnam.

The curve formed by Sumatra, Java, the Lesser Sunda Islands (from Bali to Timor), and New Guinea conforms approximately to the shape of the Eurasian Plate's leading edge (see Figure 1.21 on page 25). As the Indian-Australian Plate plunges beneath the Eurasian Plate along this curve, hundreds of earthquakes and volcanoes occur, especially on the islands of Sumatra and Java. Volcanoes and earthquakes also occur in the Philippines, where the Philippine Plate is pushing against the eastern edge of the Eurasian Plate. The volcanoes of the Philippines are part of the Pacific Ring of Fire (see Figure 1.22 on page 26).

Volcanic eruptions, and the mudflows and landslides that occur in their aftermath, endanger and complicate the lives of many Southeast Asians. Over the long run, though, the volcanic material creates new land and provides minerals that enrich the soil for farmers. Earthquakes are especially problematic because of the tsunamis they can set off. The tsunami of December 2004, triggered by a giant earthquake just north of Sumatra, swept east and west across the Indian Ocean, taking the lives of 230,000 people and injuring as many more. It is thought to be one of the deadliest natural disasters in recorded history (Figure 10.5).

The now-submerged shelf of the Eurasian continent that extends under the Southeast Asian peninsulas and islands was above sea level during the recurring ice ages of the Pleistocene epoch, during which much of the world's water was frozen in glaciers. The exposed shelf, known as Sundaland (Figure 10.6), allowed ancient people and Asian land animals (such as elephants, tigers, rhinoceroses, and orangutans) to travel south to what became the islands of Southeast Asia.

Climate

The tropical climate of Southeast Asia is distinguished by continuous warm temperatures in the lowlands—consistently above 65°F (18°C)—and heavy rain (Figure 10.7). The rainfall is the result of two major processes: the monsoons (seasonally shifting winds) and the inter-tropical convergence zone (ITCZ), the band of rising warm air that circles Earth roughly around the equator (see pages 246–247). The wet summer season extends from May to October, when the warming of the

Figure 10.5 Banda Aceh, Indonesia, before and after the December 2004 tsunami. When the tsunami hit, this community at the western end of Sumatra was totally destroyed. [Digital Globe.]

Eurasian landmass sucks in moist air from the surrounding seas. Between November and April, there is a long dry season on the mainland, when the seasonal cooling of Eurasia causes dry air from the interior continent to flow out toward the sea. On the many islands, however, the winter can also be wet because the air that flows from the continent picks up moisture as it passes south and east over the seas. The air releases its moisture as rain after ascending high enough to cool. With rains coming from both the monsoon and the ITCZ, the island part of Southeast Asia is one of the wettest areas of the world.

Irregularly every 2 to 7 years, the normal patterns of rainfall are interrupted, especially in the islands, by the El Niño

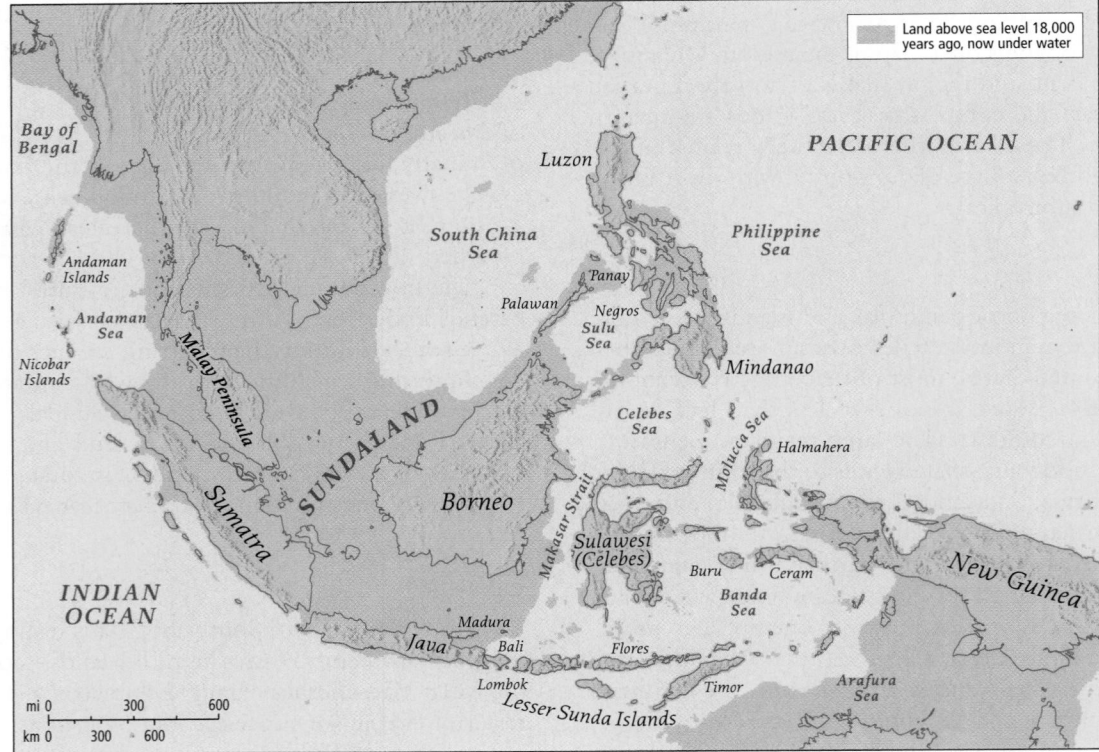

Figure 10.6 Sundaland 18,000 years ago, at the height of the last ice age. The now-submerged shelf of the Eurasian continent that extends under Southeast Asia's peninsulas and islands was exposed during the last ice age and remained above sea level until about 16,000 years ago, when that ice age was ending. [Adapted from Victor T. King, *The Peoples of Borneo* (Oxford: Blackwell, 1993), p. 63.]

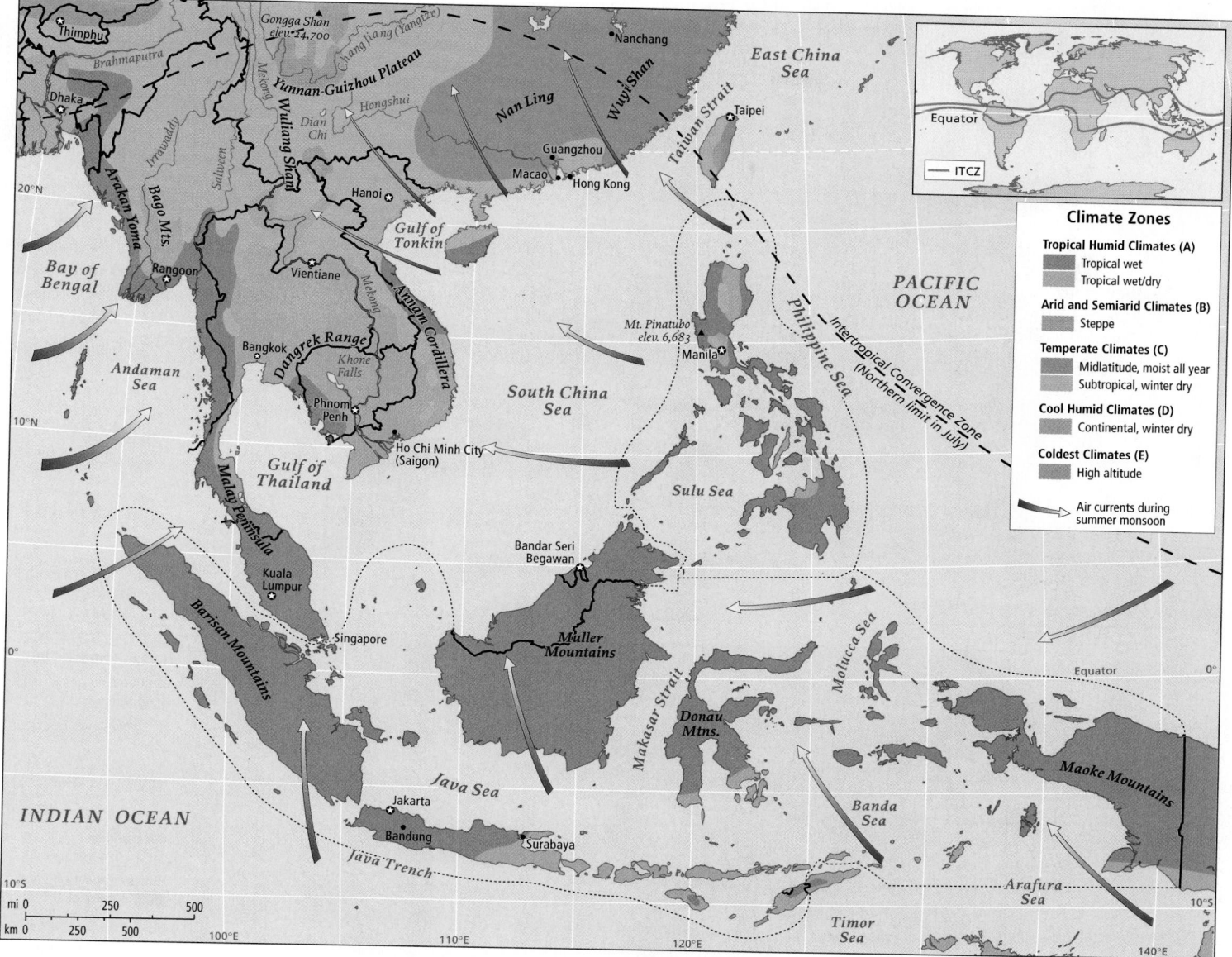

Figure 10.7 Climates of Southeast Asia. The inset shows the intertropical convergence zone (ITCZ), which is partly responsible for making Southeast Asia one of the wettest regions of the world.

phenomenon (see Figure 11.8 on page 409). In an El Niño event, the usual patterns of air and water circulation in the Pacific are reversed. Ocean temperatures are cooler than usual in the western Pacific near Southeast Asia. Instead of warm, wet air rising and condensing as rainfall, cool, dry air sits on the ocean surface. The result is severe drought, with often catastrophic results for farmers.

The soils in Southeast Asia are typical of the tropics. Although not particularly fertile, they will support dense and prolific vegetation when left undisturbed for long periods. The warm temperatures and damp conditions promote the rapid decay of **detritus** (dead organic material) and the quick release of useful minerals. These minerals are taken up directly by the

living forest rather than enriching the soil. Some of the world's most impressive rain forests still thrive in this region.

ENVIRONMENTAL ISSUES: FOCUS ON GLOBAL WARMING

This section focuses on the many environmental issues in Southeast Asia that are in some way related to global warming. As we saw in the chapter introduction, major environmental issues like deforestation have many impacts in this region beyond global warming (Figure 10.8). One is loss of living space for indigenous forest inhabitants, but orangutans, the Sumatran tiger, and the Sumatran rhinoceros are losing

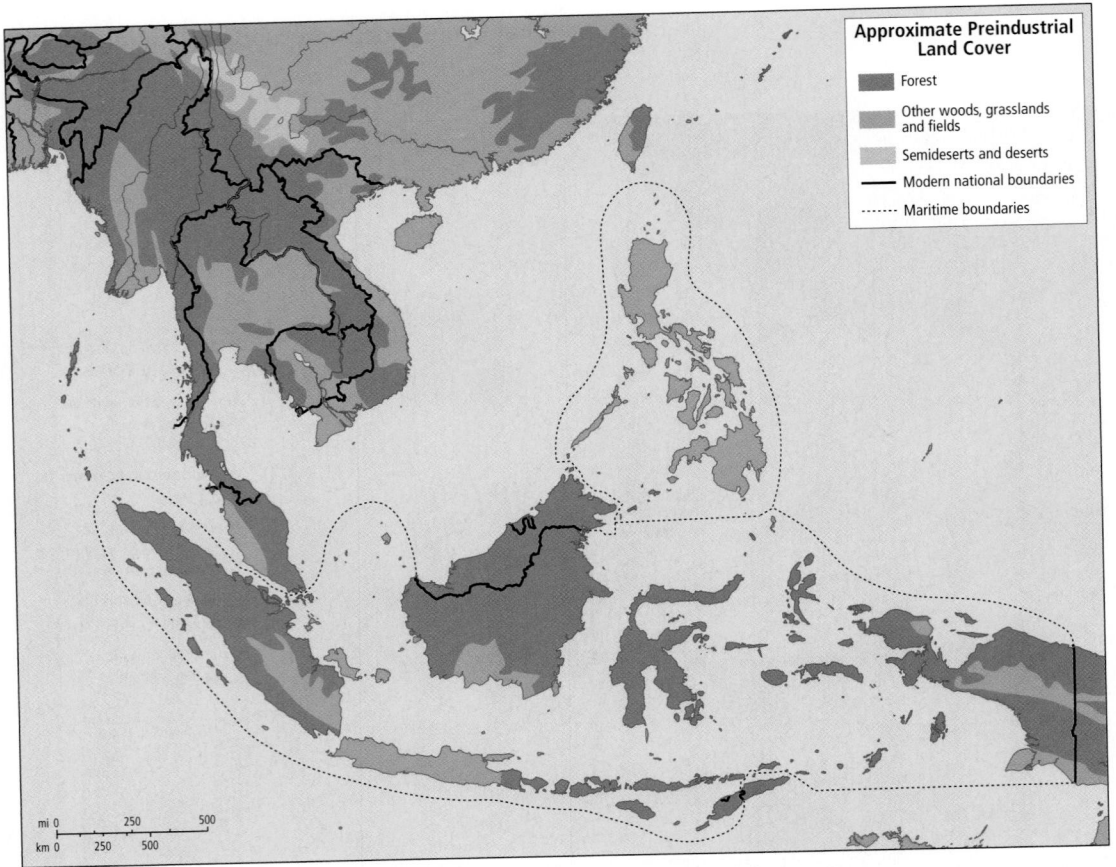

Approximate Preindustrial Land Cover
Forest
Other woods, grasslands and fields
Semideserts and deserts
Modern national boundaries
Maritime boundaries
mi 0 250 500
km 0 250 500

(a) Preindustrial

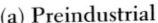

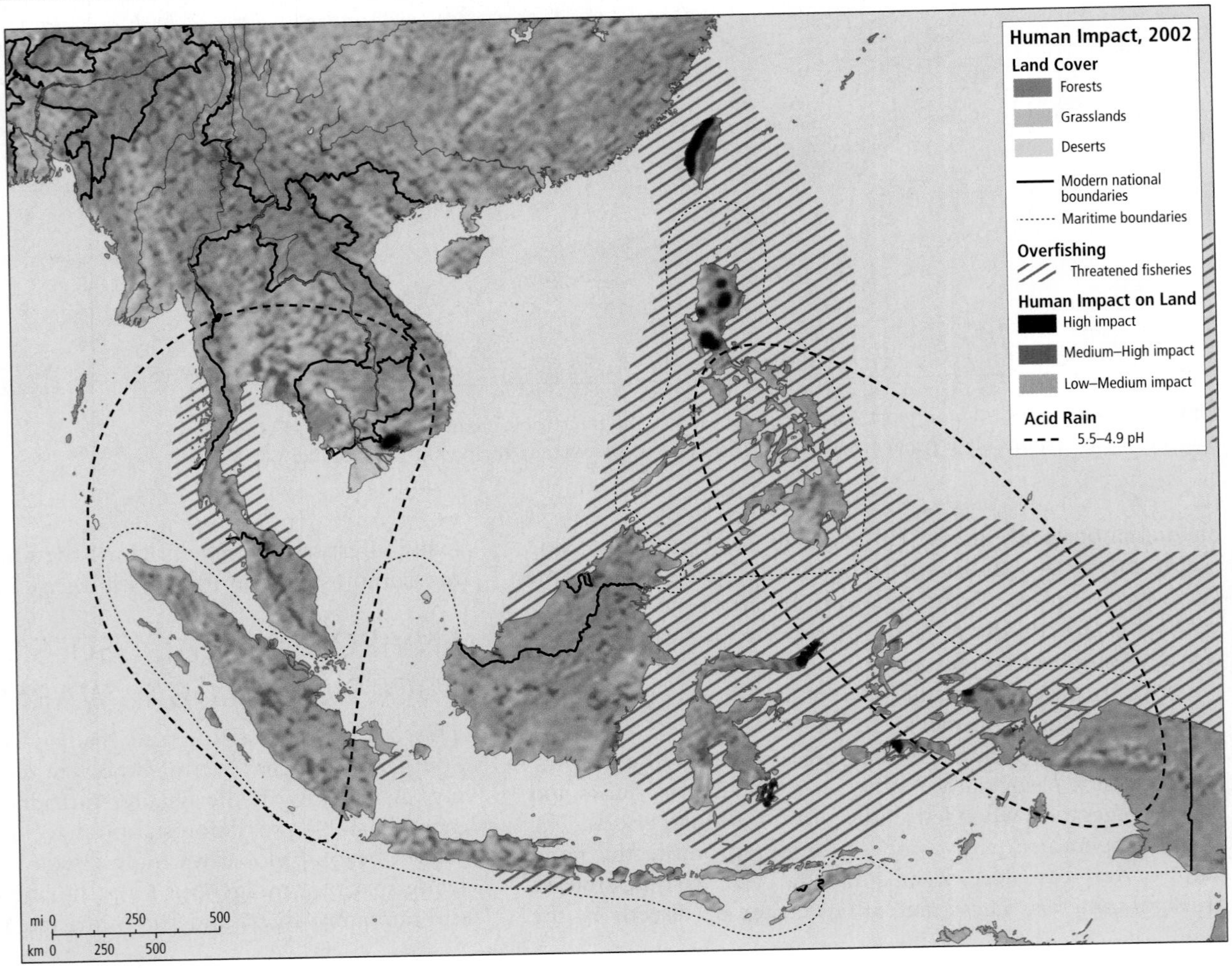

Human Impact, 2002
Land Cover
Forests
Grasslands
Deserts
Modern national boundaries
Maritime boundaries
Overfishing
Threatened fisheries
Human Impact on Land
High impact
Medium–High impact
Low–Medium impact
Acid Rain
5.5–4.9 pH
mi 0 250 500
km 0 250 500

(b) 2002

Figure 10.8 Land cover (preindustrial) and human impact (2002). Southeast Asia has been occupied by humans for hundreds of thousands of years. Though the land cover was modified by forest clearing and agricultural terracing in the preindustrial era **(a),** much of the human impact is not apparent at this scale. By 2002 **(b),** human impact was intensive and extensive, affecting land, water, and air. [Adapted from *United Nations Environment Programme, 2002, 2003, 2004, 2005, 2006* (New York: United Nations Development Programme), at http://maps.grida.no/go/collection/globio-geo-3.]

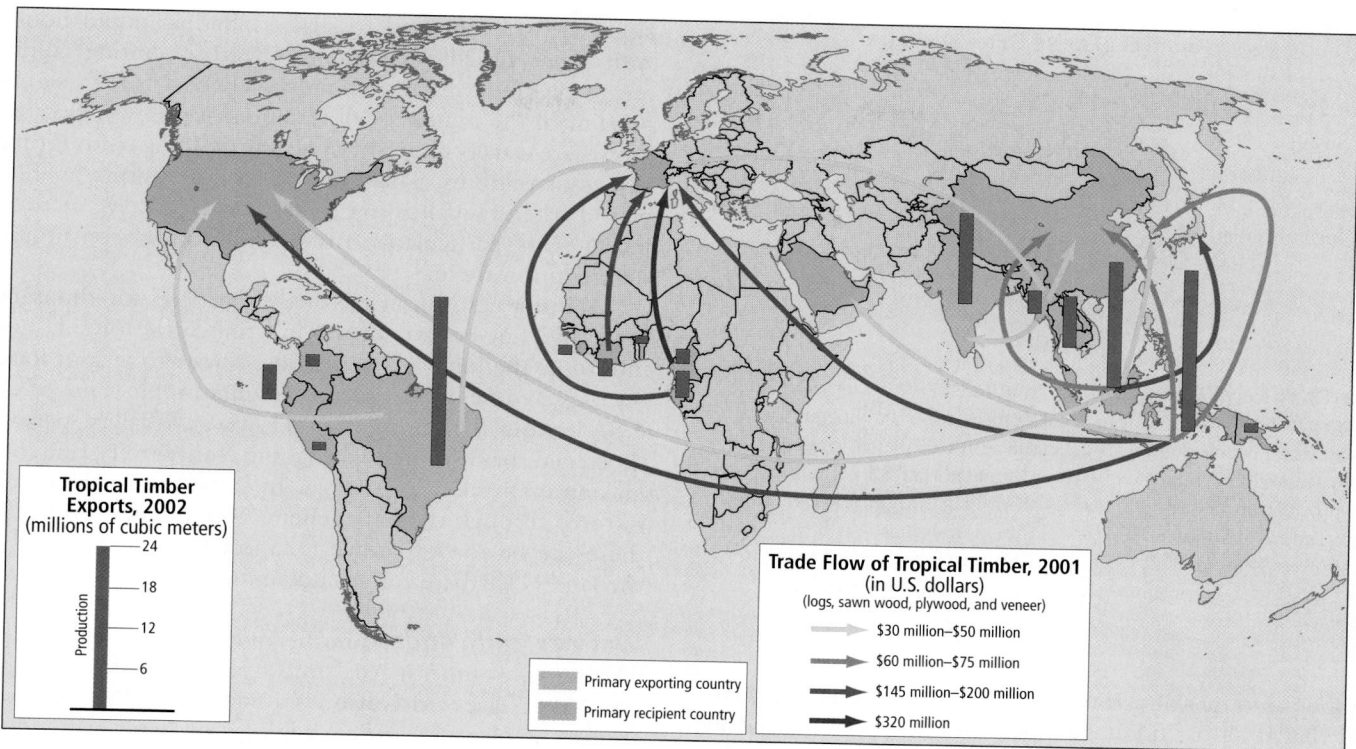

Figure 10.9 Trade in tropical timber. Most of the consumer demand for tropical timber is in North America, Europe, and Japan. Large quantities of tropical timber sold to China are made into furniture, plywood, and flooring, which are then sold to consumers in the developed world. The trade shown here is mostly legal, but much of the wood fueling China's wood-processing industries is harvested illegally in Southeast Asia, North Korea, Russia, and Africa. [Adapted from United Nations Conference on Trade and Development (New York: United Nations, 2004), pp. 44–45, at http://www.unctad.org/en/docs/ditccom20041ch22_en.pdf; and Peter S. Goodman and Peter Finn, "A corrupt timber trade," *Washington Post National Weekly Edition*, April 9–15, 2007, pp. 6–9.]

habitat as well. Tens of thousands of other less spectacular species may also be lost if rain forests are converted to agricultural uses. 🎥

However, efforts to reduce global warming and greenhouse gas emissions, are at the heart of many important environmental issues in Southeast Asia. Long driven by booming global demand for wood and agricultural products (Figure 10.9), deforestation in Southeast Asia results in significant emissions of greenhouse gases. Today, this system of land use is being modified by global efforts to reduce greenhouse gas emissions. Global warming is also likely to affect food production and access to water in much of this region. Meanwhile, marine ecosystems, which supply much of the fish eaten in Southeast Asia, may already be showing signs of degradation due to global warming.

Global Warming and Deforestation

Deforestation is a major contributor to global warming. Enormous amounts of CO_2 are released when forests are burned, and fewer trees mean that less carbon dioxide is absorbed from the atmosphere. Southeast Asia has the world's second highest rate of deforestation after sub-Saharan Africa. Every day, 13 to 19 square miles (34 to 50 square kilometers) of Southeast Asia's rain forests are destroyed. Much of this takes place in Indonesia where deforestation and burning associated with the conversion of forests into oil palm plantations occurs on a massive scale. These activities make Indonesia the world's third-largest contributor to global warming after China and the United States. Figure 10.10 shows the activities that result in deforestation across the region.

> Southeast Asia has the world's second highest rate of deforestation after sub-Saharan Africa.

Oil palm plantations. Palm oil, which is used for cooking and in food products throughout the world, is at the center of debates over global warming in Southeast Asia. Indonesia and Malaysia are the world's largest producers, and the expansion of their oil palm plantations has resulted in extensive deforestation. Much of this has occurred in lowland peat swamps, which store huge amounts of carbon in their soils. The swamps are drained and most of their vegetation burned to create plantations. Often, the fires spread underground to the peat beneath the forests, smoldering for years after the surface fires have been extinguished. These sub-surface fires release

Sources of Legal Deforestation

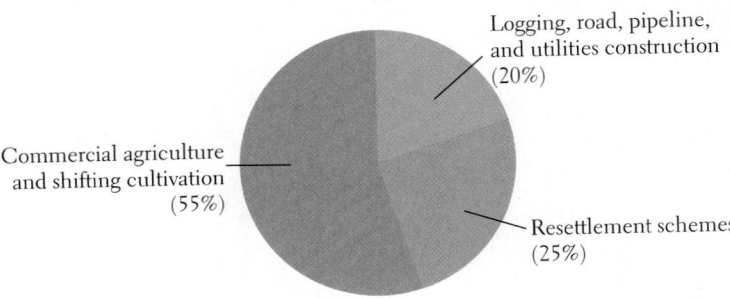

Logging, road, pipeline, and utilities construction (20%)

Commercial agriculture and shifting cultivation (55%)

Resettlement schemes (25%)

Figure 10.10 Legal deforestation in Southeast Asia, 2004. This pie diagram is an estimate of various legal deforestation activities in Southeast Asia. Much deforestation is the result of illegal logging. The 2004 World Wildlife Fund for Nature estimates that 83 percent of timber production in Indonesia stems from illegal logging. [Data from "Forestry issues—deforestation: Tropical forests in decline," at http://www.fao.org/forestry/site/33592/en; http://www.panda.org/about_wwf/what_we_do/forests/problems/forest_illegal_logging/in dex.cfm.]

enormous amounts of carbon into the atmosphere. It is estimated that if all of Indonesia's peat swamps were burned in this way, the equivalent of 12 years of global carbon emissions would be released into the atmosphere. Smoke from burning forests has covered much of the region in recent years, dramatically reducing air quality. At times the smoke is so bad that airplanes have difficulty landing in the cities, and even in rural areas, people are urged to wear masks.

Now, oil palms are being promoted as one potential solution to global warming because their oil can be converted into fuel for automobiles. Because the trees absorb carbon dioxide from the atmosphere when they are growing, palm oil can be considered a "carbon neutral" fuel. However, critics point out that if forests are burned to clear land for oil palm plantations, it can take decades to counteract the initial carbon emissions of deforestation. If a peat swamp is cleared for an oil palm plantation, more carbon is released through the burning of sub-surface peat than can ever be counteracted by carbon dioxide absorbed by the growing oil palms.

Global Warming and Food Production

Food production contributes to global warming in two ways. First, it is a major contributor to deforestation, and second, some types of cultivation actually produce significant greenhouse gases.

Shifting cultivation. Where population densities are relatively low, poor subsistence farmers practice the agricultural technique called shifting cultivation (described in Chapter 3, page 109). In the hills and uplands of mainland Southeast Asia and in many parts of the islands, farmers move their fields every 3 years or so. This is done to maintain soil fertility, which can require decades to recover after a few years of farming. Hence,

larger areas are needed to support human populations than with other agricultural systems. Because shifting cultivation means clearing land at each move, it accounts for a significant portion of the region's deforestation. Moreover, because the forests are usually cleared by burning, shifting cultivation often results in wildfires. This is especially true during an El Niño period when rainfall is low. These wildfires have increased in recent years, particularly in Indonesia, further contributing to deforestation there.

Shifting cultivation has been practiced for thousands of years and can be maintained indefinitely as long as nothing interrupts the long fallow periods. The regrowth of forest on once-cleared fields also absorbs significant amounts of carbon dioxide from the atmosphere. However, if fallow periods are shortened, or disrupted by logging, soil fertility can collapse, making cultivation impossible. In some cases, fertility can be restored through the use of chemical or organic fertilizers, but these can be too expensive for most farmers. In most cases, the land is left bare for an indefinite period.

Wet rice cultivation. Another major contributor to global warming is Southeast Asia's most productive form of agriculture. **Wet rice cultivation** (sometimes called paddy rice) entails the planting of rice seedlings by hand in flooded terraced fields that are first cultivated with hand-guided plows pulled by water buffalo. Wet rice cultivation has transformed landscapes throughout Southeast Asia. It is practiced throughout this generally well-watered region, but especially on rich volcanic soils and in places where rivers and streams bring a yearly supply of silt.

The flooding of rice fields also results in the production of methane, a powerful greenhouse gas responsible for about 20 percent of global warming. It is estimated that up to one-third of the world's methane is released from flooded rice fields where organic matter in soil undergoes fermentation as oxygen supplies are cut off. While wet rice has been cultivated for thousands of years, growing human populations have driven a 17 percent expansion of the area devoted to this crop in the last 25 years. Figure 10.11 shows patterns of agriculture throughout the region, including rice and other crops.

Global Warming and Water

Many of Southeast Asia's potential vulnerabilities to global warming are related to water resources. Here, we touch on two areas of vulnerability that may affect the region's economy and food supply: glacial melting and coral reef bleaching.

Melting glaciers. Like much of Asia, mainland Southeast Asia's largest rivers (the Irrawaddy, Salween, Mekong, and Red rivers) are fed during the dry season (November–March) by glaciers high in the Himalayas. These glaciers are now melting so rapidly that they may be gone in fifty years, introducing innumerable complex changes that scientists are only beginning to understand. However, some impact is a certainty,

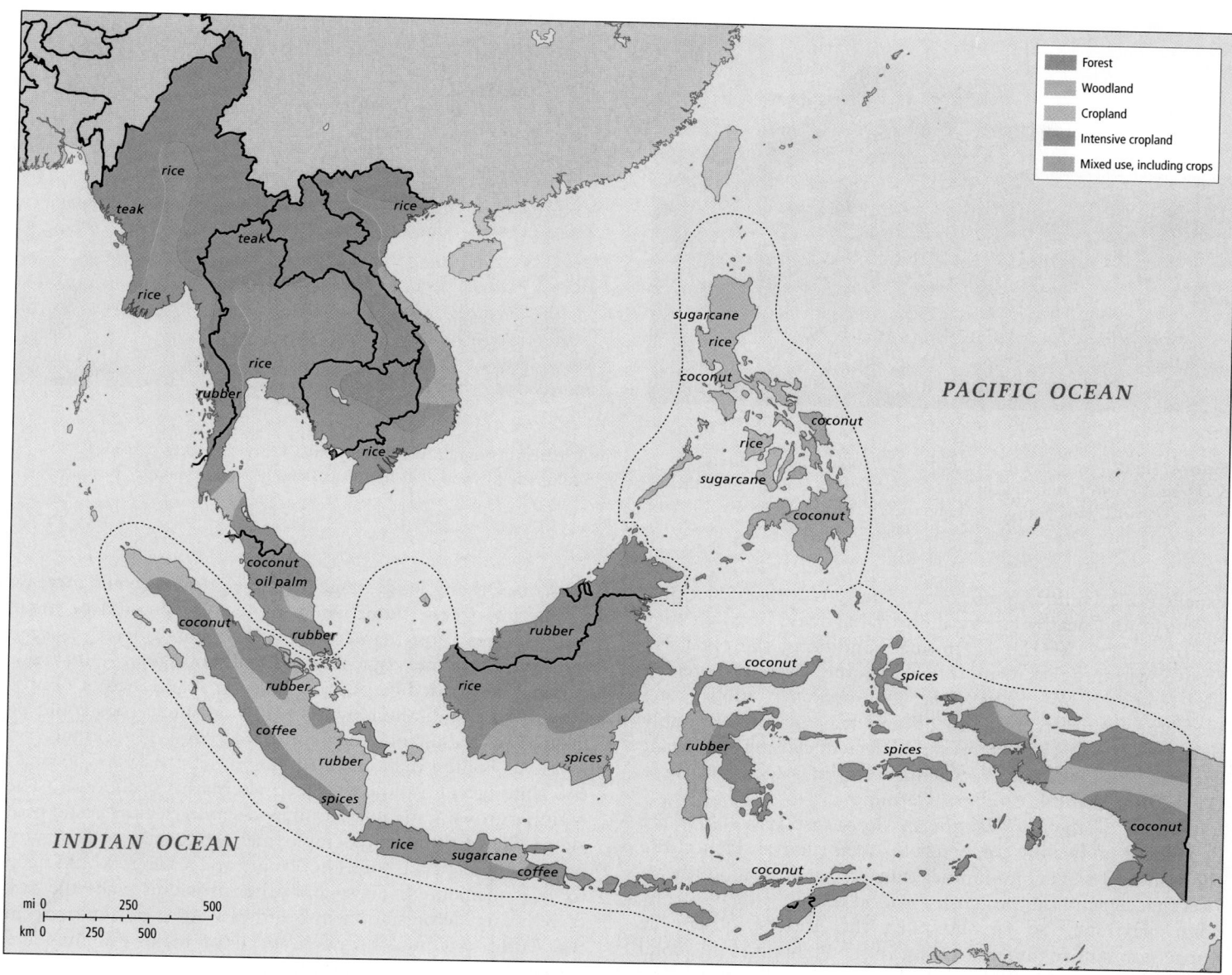

Figure 10.11 Agricultural patterns in Southeast Asia. Tropical forests and crops, rice production, and shifting cultivation dominate the agricultural patterns of Southeast Asia. [Adapted from *Hammond Citation World Atlas* (Maplewood, N.J.: Hammond, 1996,) pp. 74, 83, 84.]

especially for the 150 million people who live in the basins of these rivers.

As glacial melting accelerates, the immediate risk is flooding, which is already dramatic in some areas (Figure 10.12). The longer term concern is reduced dry season flows in the rivers as glaciers disappear. As much as 15 percent of this region's rice harvest depends on the dry season flows of the major rivers. The loss of these harvests would strain many farmers' incomes.

> Much of this region's rice harvest depends on the dry season flows of the major rivers of mainland Southeast Asia, which are threatened by glacial melting.

In addition, it could possibly lead to food shortages in cities. Coming on top of a global rise in food prices in recent years, this would place further strain on the incomes of poor people throughout the region.

Coral reef bleaching. Global warming is expected to increase sea temperatures in Southeast Asia, threatening the coral reefs that sustain much of the region's fishing and tourism. A coral reef is an intricate structure composed of the calcium-rich skeletons of millions of tiny living creatures called coral polyps. The polyps are subject to **coral bleaching,** or color loss, which results when photosynthetic algae that live in the corals are

(a)

(b)

Figure 10.12 Living in the wetlands. **(a)** People who live along the Mekong River in Cambodia and southern Vietnam usually build their houses on stilts. **(b)** During the rainy season, the river can rise 10 feet (3 meters) or more, broadening into a lazy flow 2 to 3 miles (3 to 5 kilometers) wide. [Michael Yamashita/Woodfin Camp & Associates.]

expelled. Bleaching may also occur when corals experience pollution from a nearby city or industrial activity, or when they are overfished. ◢▮ Under normal conditions, the coral will recover within weeks or months. However, severe or repeated bleaching can cause corals to die. Unprecedented global coral bleaching events occurred in 1998 and again in 2002. Roughly half of the world's coral reefs were affected, causing significant coral die-offs in some areas. Scientists generally agree that these events are the result of global warming.

Many of the fish caught in Southeast Asia's seas are dependent on healthy coral reefs for their survival. Hence, the thousands of rural communities throughout coastal Southeast Asia that depend on these fish for food are also threatened by coral bleaching. So far, however, the greatest observable impacts on humans are in tourism. In the Philippines, the coral bleaching event of 1998 brought a dramatic decline in tourists who come to dive the country's usually spectacular reefs, resulting in a loss of about 30 million dollars to the economy.

Responses to Global Warming

Southeast Asia is beginning to try to reduce its greenhouse gas emissions and to limit its vulnerability to the changes in regional climate that global warming could bring. As we saw in the opening vignette, the UN and other international organizations are favoring incentives to reduce deforestation on a global scale. According to their calculations, other global scale changes that could yield similarly large reductions in greenhouse gas emissions, such as restricting emissions from industry or transport, would cost more and require complex technological changes.

Critics point out that efforts to reduce deforestation must also attempt to reduce the demand for wood products and palm oil. Otherwise, shortages of these products could increase their prices to the point where more money could be made from deforestation than from the payments offered to keep forests intact. Some argue that a global shortage of wood products can be avoided by planting trees on abandoned or unproductive farmland. Similarly, oil palms or other crops could be planted on these lands to prevent a cooking oil shortage.

Even so, the millions of rural Southeast Asians who practice shifting cultivation will need alternative livelihoods. The region's growing manufacturing and service sectors could provide some of the needed employment, but massive programs promoting alternative livelihoods in rural settings would need to be implemented. These and other problems make the task of reducing deforestation in Southeast Asia, or anywhere in the world, a complex one in need of major funding and extremely competent administration.

All the region's governments have plans to reduce fossil fuel consumption, and some are delivering on those goals. Both the Philippines and Indonesia have significant potential for generating electricity from *geothermal energy* (heat stored in the earth's crust). This energy is particularly accessible near active volcanoes, which both countries have in abundance. Already, the Philippines generates 27 percent of its electricity from geothermal energy, and is second only to the United States in the amount of geothermal power it generates. Indonesia, too, is expanding its geothermal capacity, which by some estimates could provide all of its energy needs. Solar energy is another attractive option given that the entire region lies near the equator, the part of the earth that receives the most solar energy. For most countries, however, wind is the most cost-effective option, especially in Laos and Vietnam, where many population centers are in high wind areas.

Ten Key Aspects of Southeast Asia

- High rates of deforestation and associated burning make Indonesia the world's third largest contributor to global warming after China and the United States.
- Over the last five centuries, several European countries, and later the United States and Japan, established colonies or quasi-colonies that covered almost all of Southeast Asia.
- From the 1960s to the 1990s, Southeast Asian countries pursued export-led growth strategies, investing heavily in industries that manufactured products for export, primarily to developed countries.
- About 39 percent of the region's population is urban, and this proportion is increasing, thanks to changes in agriculture and industry.
- By 2050, 766 million people will be packed into Southeast Asia's land area, which is roughly half the size of the United States.
- The economic crisis of the late 1990s exposed corruption in Thailand and Indonesia, leading to widespread protest movements that expanded democracy in both countries.
- Women's wages average only about one-half to two-thirds those of men in Southeast Asia.
- Glaciers that feed mainland Southeast Asia's four largest rivers are melting so rapidly that they may be gone in 50 years,
- With the exception of the animist belief systems of the indigenous peoples, the major religious traditions of Southeast Asia all originated outside the region.

- Repression by Burma's military government has resulted in at least 680,000 refugees, with 500,000 displaced within Burma and 180,000 having fled to Thailand.

HUMAN PATTERNS OVER TIME

First settled in prehistory by migrants from the Eurasian continent, Southeast Asia was later influenced by Chinese, Indian, and Arab traders. Later still, it was colonized by Europe (1500s to early 1900s), the United States (1898 to 1946 in the Philippines), and Japan (during World War II). By the late twentieth century, colonial domination had ended and the region was profiting from the sale of manufactured goods to its former colonizers.

The Peopling of Southeast Asia

The modern indigenous populations of Southeast Asia arose from two migrations widely separated in time. In the first migration, **Australo-Melanesians,** a group of hunters and gatherers from the present northern Indian and Burman parts of southern Eurasia, moved into the exposed landmass of Sundaland about 40,000 to 60,000 years ago. Their descendants still live in Indonesia's easternmost islands and in small pockets on other islands and the Malay Peninsula.

In the second migration (about 10,000 years ago, at the end of the last ice age), people from southern China began moving into Southeast Asia (Figure 10.13). Their migration gained momentum about 5000 years ago, when a culture of skilled farmers and seafarers from southern China, the

Figure 10.13 Hmong women returning to their village. One of the last groups to move from China to Southeast Asia was the Hmong, who probably left China during the nineteenth century. Although many Hmong were displaced by the Vietnam War, some still live in Sapa, in northwestern Vietnam, among other Southeast Asian locales. [QT Luong/terragalleria.com.]

Austronesians, migrated first to Taiwan, then to the Philippines, and then into island Southeast Asia and the Malay Peninsula. Some of these sea travelers eventually moved westward to southern India and to Madagascar (off the east coast of Africa), and eastward to the far reaches of the Pacific islands (see Chapter 11).

Diverse Cultural Influences

Southeast Asia has been and continues to be shaped by a steady stream of cultural influences, both internal and external. Overland trade routes and the surrounding seas brought traders, religious teachers, and sometimes even invading armies from China, India, and Southwest Asia. These newcomers brought religions, trade goods (such as cotton textiles) and food plants (such as mangoes and tamarinds) deep into the Indonesian and Philippine archipelagos and throughout the mainland. Access by merchant ships from South Asia and the Persian Gulf was facilitated by alternating monsoon winds, which blew from the west in the spring and summer, and from the east in the autumn and winter. These winds carried people, spices, bananas, sugarcane, silks, and other Southeast Asian items to the wider world.

Religious legacies. Spatial patterns of religion in Southeast Asia reveal an island-mainland division that reflects the history of influences from India, China, Southwest Asia, and Europe. Both Hinduism and Buddhism arrived thousands of years ago via Indian monks and traders traveling by sea and along overland trade routes that connected India and China through Burma. Many early Southeast Asian kingdoms and empires switched back and forth between Hinduism and Buddhism as their principal religion. Spectacular ruins of these Hindu-Buddhist empires are scattered across the region, the most famous being the city of Angkor in present-day Cambodia. At its zenith in the 1100s, Angkor was among the largest cities in the world, and its ruins are now a World Heritage Site (see Figure 10.23 on page 390). Today, Buddhism dominates mainland Southeast Asia, while Hinduism is dominant only on the Indonesian islands of Bali and Lombok.

In Vietnam, people practice a mix of Buddhist, Confucian, and Taoist beliefs that reflect the one thousand years (ending in 938 CE) when it was part of various Chinese empires. China's traders and laborers also brought cultural influences to scattered coastal zones throughout Southeast Asia.

Islam and Christianity dominate in Island Southeast Asia. Islam came mainly through South Asia after India fell to Muslim (Mughal) conquerors in the fifteenth century. Muslim mystics and traders converted many formerly Hindu-Buddhist kingdoms in Indonesia, Malaysia, and parts of the southern Philippines, where Islam is still dominant. Roman Catholicism is the predominant religion in Timor-Leste, which was colonized by Portugal, and in most of the Philippines, which was colonized by Spain.

Colonization

Over the last five centuries, several European countries established colonies or quasi-colonies in Southeast Asia (Figure 10.14). Drawn by the region's fabled spice trade, the Portuguese established the first permanent European settlement in Southeast Asia at the port of Malacca, Malaysia, in 1511. Although better ships and weapons gave the Portuguese an advantage, their anti-Islamic and pro-Catholic policies provoked strong resistance in Southeast Asia. Only in Timor-Leste (East Timor) did the Portuguese establish Catholicism as the dominant religion.

By 1540, the Spanish had established trade links across the Pacific between the Philippines and their colonies in the Americas. Like the Portuguese, they practiced a style of colonial domination grounded in Catholicism, but they met less resistance because of their greater tolerance of non-Christians. The Spanish ruled the Philippines for more than 350 years, and as a result, the Philippines is the most deeply Westernized and certainly the most Catholic part of Southeast Asia.

The Dutch were the most economically successful of the European colonial powers in Southeast Asia. From the sixteenth to the nineteenth centuries, they extended their control of trade over most of what is today called Indonesia. The Dutch became interested in growing cash crops for export. Between 1830 and 1870, they forced indigenous farmers to leave their own fields and work part time without pay in Dutch coffee, sugar, and indigo plantations. The resulting disruption of local food production systems caused severe famines and provoked resistance that often took the form of Islamic religious movements. Such movements hastened the spread of Islam throughout Indonesia, where the Dutch had made little effort to spread Christianity.

Beginning in the late eighteenth century, the British established colonies at key ports on the Malay Peninsula. They held these ports both for their trade value and to protect the Strait of Malacca, the passage for sea trade between China and Britain's empire in India. In the nineteenth century, Britain extended its rule over the rest of modern Malaysia to benefit from its tin mines and plantations. Britain also added Burma to its empire, which provided access to forest resources and overland trade routes to China.

The French first entered Southeast Asia as Catholic missionaries in the early seventeenth century. They worked mostly in the eastern mainland area in the modern states of Vietnam, Cambodia, and Laos. In the late nineteenth century, spurred by rivalry with Britain and other European powers for greater access to the markets of nearby China, the French colonized the area.

In all of Southeast Asia, the only country not to be colonized was Thailand (then known as Siam). Like Japan, it protected its sovereignty through both diplomacy and a vigorous drive toward European-style modernization.

Struggles for Independence

Agitation against colonial rule began in the late nineteenth century when Filipinos fought first against Spain. They then fought against the United States, which had taken control of

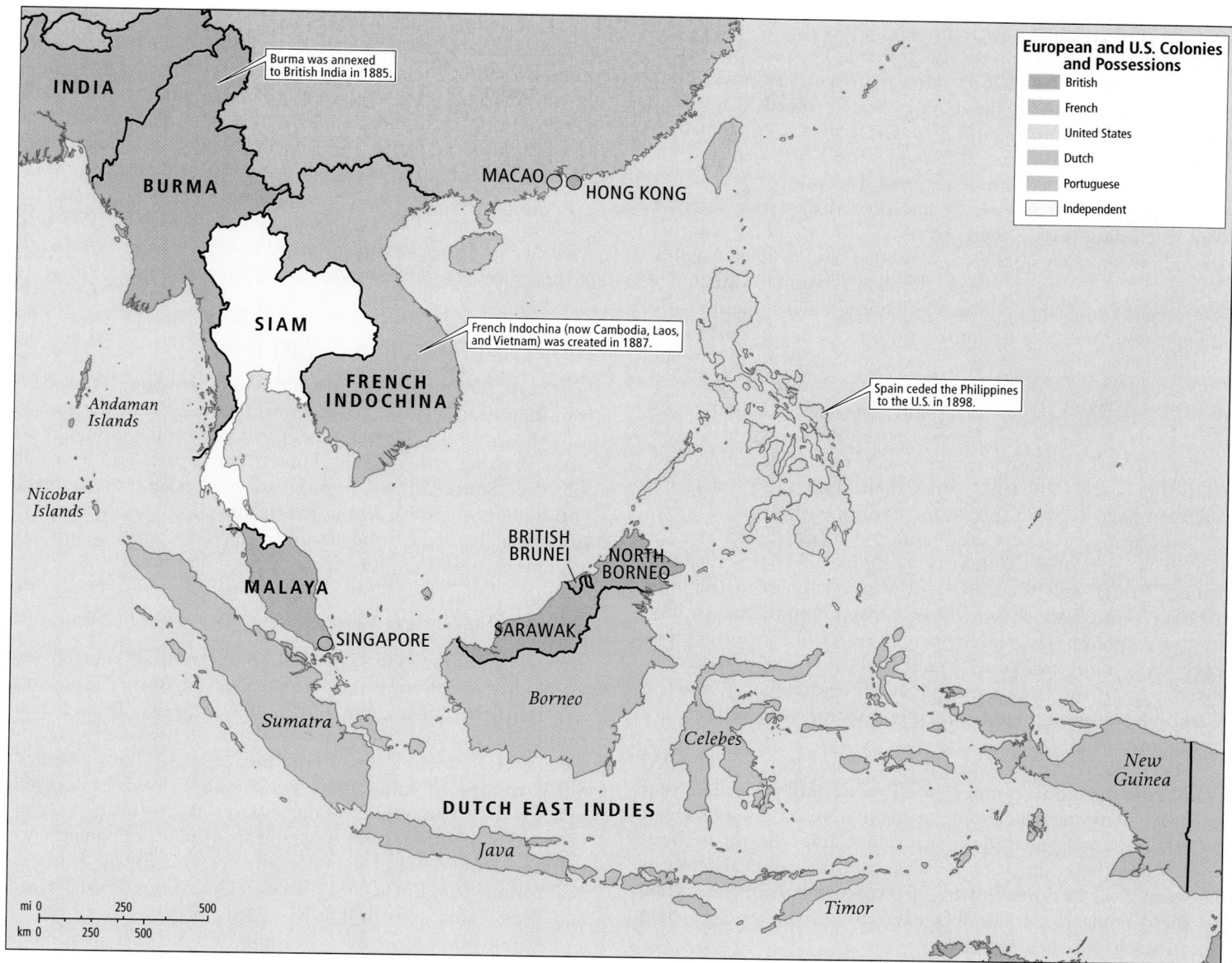

Figure 10.14 European and U.S. colonies in Southeast Asia, 1914. Of the present-day countries in Southeast Asia, only Thailand (formerly called Siam) was never colonized. [Adapted from *Hammond Times Concise Atlas of World History* (Maplewood, N.J.: Hammond, 1994), p. 101.]

the Philippines in 1898 after the Spanish-American War. However, independence for the Philippines and the rest of Southeast Asia was not won until the end of World War II. By then, Europe's ability to administer its colonies had been weakened, partly because its attention was diverted by the devastation of the war. Further, Japan had conquered most of European-held Southeast Asia and held it until its defeat by the United States several years later (see Figure 9.16 on page 343). By the mid-1950s, the colonial powers had granted self-government to most of the region.

The Vietnam War. The most bitter battle for independence took place in the French-controlled territories of Vietnam, Laos, and Cambodia (collectively called Indochina). Although

all three became nominally independent in 1949, France retained political and economic power. Various nationalist leaders, most notably Vietnam's Ho Chi Minh, headed resistance movements against continued French domination. The resistance leaders accepted military assistance from Communist China and the Soviet Union, even though they did not begin as communists and despite ancient antipathies toward China for its previous millennia of domination. In this way, the cold war was brought to mainland Southeast Asia.

In 1954, the French were defeated by Ho Chi Minh at Dien Bien Phu, in northern Vietnam. The United States, increasingly worried about the spread of international communism, stepped in. The Vietnamese resistance, which controlled the northern half of the country, attempted to wrest control of the southern half from the United States and a U.S.-supported South Vietnamese government. The pace of the war accelerated in the mid-1960s. After many years of brutal conflict, public opinion in the United States forced U.S. withdrawal from the conflict in 1973. The civil war continued in Vietnam, finally ending in 1975, when the North defeated the South and established a new national government.

More than 4.5 million people died during the Vietnam War, including more than 58,000 U.S. soldiers. Another 4.5 million on both sides were wounded, and bombs, napalm,

> More than 4.5 million people died during the Vietnam War, including more than 58,000 U.S. soldiers.

and defoliants ruined much of the Vietnamese environment. Land mines continue to be a hazard to this day. The withdrawal from Vietnam in 1973 ranks as one of the most profound defeats in United States history. After the war, the United States crippled Vietnam's recovery with severe economic sanctions that lasted until 1993. Since then, the United States and Vietnam have become significant trading partners.

In Cambodia, where the Vietnam War had spilled over the border, a particularly violent revolutionary faction called the Khmer Rouge seized control of the government in the mid-1970s. Inspired by the vision of a rural communist society, they attempted to destroy virtually all traces of European influence. They targeted Western-educated urbanites in particular, forcing them into labor camps, where more than 2 million Cambodians—one-quarter of the population—starved or were executed. ◼

In 1978, Vietnam deposed the Khmer Rouge and ruled Cambodia through a puppet government until 1989. A 2-year civil war then ensued. Despite a major United Nations effort to establish multiparty democracy in Cambodia throughout the 1990s, the country remains plagued by political tensions between rival factions and by government corruption. None of the Khmer Rouge leaders have ever been tried or held accountable for their actions.

Vignette Former Khmer Rouge village chief Choch states flatly: "It's not true. If I had done those things, how could I live here now?"

In Cambodia, victims of the Khmer Rouge reign of terror now often live as close neighbors to those who tortured and killed their loved ones. The perpetrators have never had to stand trial or face punishment because little forensic work has been done on the murders and massacres. Those responsible can easily deny having been involved.

One night in 1977, soldiers came to the home of Samrith Phum and took her husband away. She thought he was just going to a meeting, but he never came home. Samrith was then only 20 years old and had three young children, one a newborn. With her infant in her arms, she went to talk to Choch, the Khmer Rouge village chief, and asked him, "Brother, do you know where my husband is?" The village chief told her not to worry about other people's business. Says Samrith, "I didn't ask him any more after that. I was

hopeless. I knew my husband was dead." A short while later, Choch appeared at Samrith's door and said he would take her to see her husband. Instead, he drove to the nearby prison and locked her up with her baby. She was released a year later, after the Vietnamese drove the Khmer Rouge from power. Today, Samrith still lives just down the street from Choch. For his part, Choch denies any involvement in the killings or even ever being at the prison.

To learn more about the aftermath of the Khmer Rouge, watch the FRONTLINE/World ◼ *"Cambodia: Pol Pot's shadow" and read Amanda Pike's "Reporter's diary: In search of justice," at http://www.pbs.org/frontlineworld/stories/cambodia/diary03.html.* ◼

POPULATION PATTERNS

Southeast Asia's population is large and growing, but urbanization and population control efforts are slowing the rate of growth. Around 574 million people occupy the region, almost twice as many as live in the United States in a land area half the size. Southeast Asia's population is projected to reach 766 million by 2050, and much of this population will live in cities. At the same time, many other Southeast Asians are migrating to find employment outside the region.

Population Distribution

The population map in Figure 10.15 reveals that relatively few people live in the rugged upland reaches of Burma, Thailand, and northern Laos, or in much of Cambodia. In Malaysia, Indonesia, and the Philippines, wetlands, dense forests, mountains, and geographically remote areas are also lightly settled. Small groups of indigenous people have lived in forested uplands for thousands of years, supported by shifting cultivation and by hunting, gathering, and small-plot permanent agriculture. Despite resettlement schemes, much of Sumatra, Kalimantan (on Borneo), Sulawesi (Celebes), the Moluccas, and West Papua remain lightly settled.

About 60 percent of the people of Southeast Asia live in patches of particularly dense rural settlement along coastlines, on the floodplains of major rivers, and in the river deltas of the mainland. On the islands, settlement is most concentrated on Luzon (in the northern Philippines) and on Java. These places are attractive because the rich and well-watered volcanic soils allow intensive agriculture.

Population Dynamics

While Southeast Asia's young population will ensure steady growth for several decades, government policies have had a strong effect on population growth in several countries. This produces a highly variable geography of population growth that doesn't follow the usual rules of the *demographic transition* (see Chapter 1, page 45). Usually, rich countries grow more slowly than poor ones. However, in the absence of effective efforts at population control, Malaysia, one of Southeast Asia's wealthiest countries, has a population pyramid similar

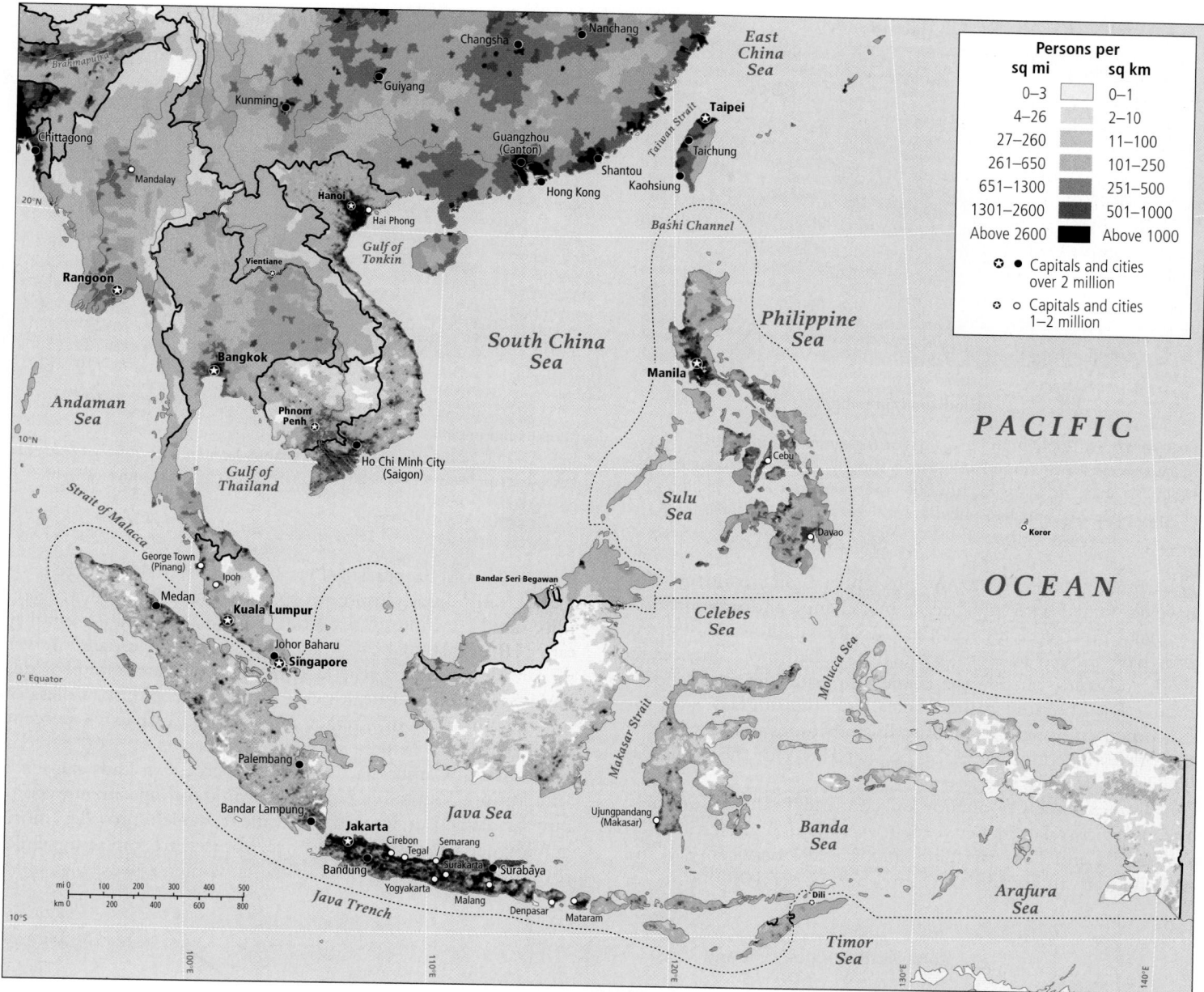

Figure 10.15 Population density in Southeast Asia. [Data courtesy of Deborah Balk, Gregory Yetman et al., Center for International Earth Science Information Network, Columbia University, at http://www.ciesin. columbia.edu.]

to much poorer Indonesia (Figure 10.16), and a slightly higher fertility rate. The average adult woman in Malaysia bears 2.9 children, as opposed to 2.4 children in Indonesia.

Despite this variability, fertility rates have declined sharply in all countries since the 1960s (Figure 10.17) due largely to economic growth and urbanization. For example, in 1960, the average adult woman in Singapore bore 4.9 children. By 2007, after decades of economic growth and rising living standards, Singapore's highly urbanized population was well into the *demographic transition*. The average woman bore only 1.3 children, which is below the replacement level (2.1 children per

adult woman). The Singapore government is now so concerned about the low fertility rate that it offers young couples various incentives for marrying and procreating. A greater source of population growth for Singapore is the steady stream of highly skilled immigrants that its vibrant economy attracts.

Thailand has the next lowest fertility rate in the region: 1.7 children per adult woman. Highly effective

Government-sponsored campaigns encouraging contraception have combined with rapid economic change and urbanization to make most Thai couples feel that smaller families are best.

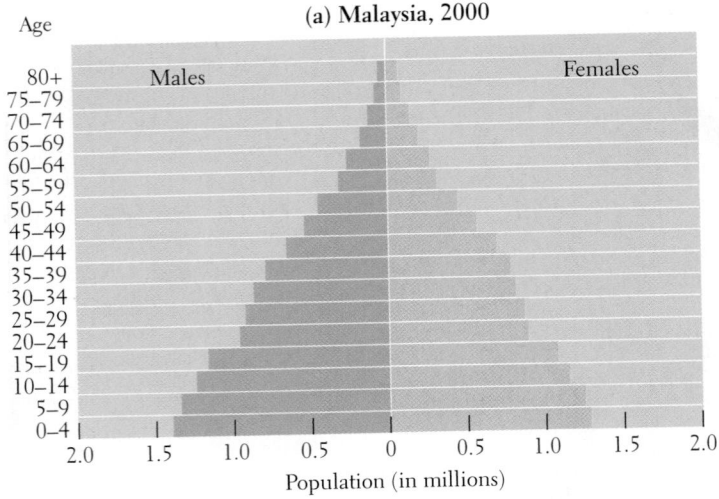

(a) Malaysia, 2000

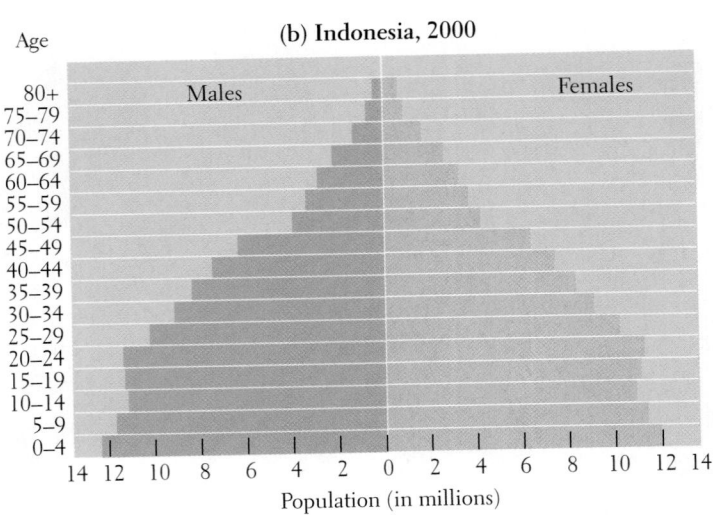

(b) Indonesia, 2000

Figure 10.16 Population pyramids for Malaysia and Indonesia. Note that the scale for the population axis is roughly eight times greater for Indonesia than for Malaysia. In 2000, Indonesia had 206.1 million people, while Malaysia had only 22.7 million people. [Adapted from

"Population pyramids for Malaysia" and "Population pyramids for Indonesia" (Washington, D.C.: U.S. Census Bureau, International Data Base, at http://www.census.gov/ipc/www/idb/pyramids.html).]

government-sponsored campaigns encouraging contraception have combined with rapid economic change and urbanization to make most couples feel that smaller families are best. As they have moved to cities, women have had more chances to work and study outside the home, opportunities that tend to reduce fertility rates in most societies. High literacy rates for both men and women and Buddhist attitudes that accept the use of contraception have also been credited for the decline in Thailand's fertility rate.

Two extremely poor and rural countries in the region show the usual correlation between poverty and high fertility. In Cambodia and Laos, fertility rates average between 3.4 and 4.8 children. Infant mortality rates are 71 per 1000 births for

Cambodia and 85 per 1000 for Laos. In Vietnam, where people are only slightly more prosperous and urbanized, the fertility rate (2.1 children per adult woman) and infant mortality rate (18 per 1000 births) are much lower. Vietnam's lower rates are explained by the fact that this socialist state provides basic education and health care to all its people, regardless of income. In Vietnam, literacy rates are more than 94 percent for men and 87 percent for women, whereas only 60 percent of women in Cambodia and only 61 percent in Laos can read. In addition, Vietnam's rapidly growing and urbanizing economy is attracting foreign investment, which provides more employment for women. Hence, careers are replacing child rearing as the central focus of many women's lives.

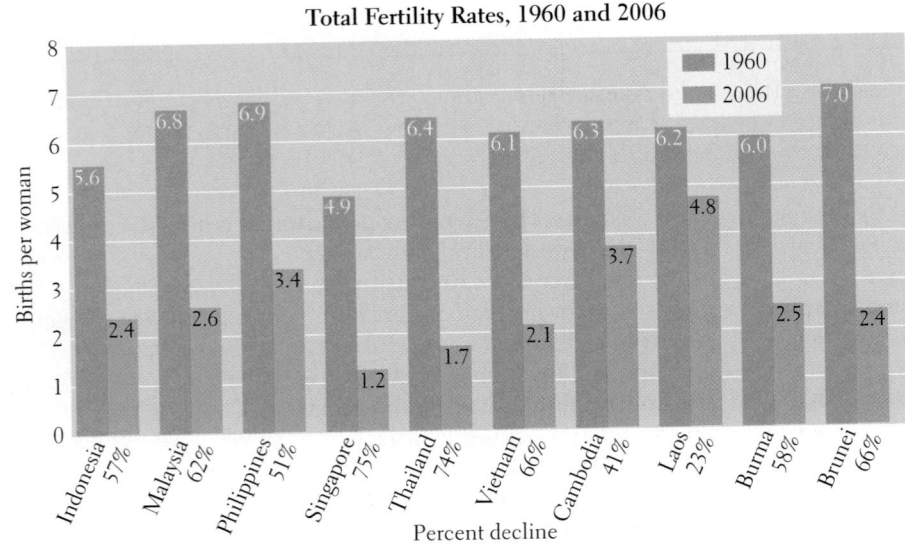

Total Fertility Rates, 1960 and 2006

Figure 10.17 Total fertility rates, 1960 and 2006. Total fertility rates declined during this period for nearly all Southeast Asian countries. Timor-Leste is the exception, but current data for that country are not available. [Data from *A Demographic Portrait of South and Southeast Asia* (Washington, D.C.: Population Reference Bureau, 1994), p. 9; Globalis, at http://globalis.gvu.unu.edu/; *2006 World Population Data Sheet* (Washington, D.C.: Population Reference Bureau).]

Urbanization

Southeast Asia as a whole is just 39 percent urban, but the rural-urban balance is shifting steadily in response to declining agricultural employment and booming urban industries. The forces driving farmers into the cities are called the **push factors** in rural-to-urban migration. They include the rising cost of farming due to the use of new technologies. **Pull factors,** in contrast, are those that attract people to the city, such as abundant manufacturing jobs. These factors have come together to create steady urbanization. Malaysia is already 62 percent urban; the Philippines, 48 percent; and Singapore, 100 percent.

Throughout Southeast Asia, employment in agriculture has been declining since the introduction of new production methods that increase costs and reduce the need for human labor. The use of farm machinery for planting and harvesting is spreading throughout the region, drastically reducing the number of people needed to perform these tasks. Meanwhile, the use of chemical pesticides and fertilizers has also spread. While such additives can increase harvests dramatically, they also drive the cost of production beyond what most farmers can afford. Many farmers have sold their land to more prosperous farmers or to corporations and moved to the cities.

Labor-intensive manufacturing industries, such as garment and shoe making, are expanding in cities and towns of the poorer countries, such as Cambodia, Vietnam, and parts of Indonesia. In the urban and suburban areas of the wealthier countries—Singapore, Malaysia, Thailand, and parts of Indonesia and the Philippines—technologically sophisticated manufacturing industries are also growing. This includes automobile assembly, chemical and petroleum refining, and assembly of computers and other electronic equipment.

In response to both push and pull factors, rural migrants are streaming into cities like Jakarta, Manila, and Bangkok, which are among the most rapidly growing metropolitan areas on earth (Table 10.1). Such a city, often the capital of a country, may become a *primate city*—one that is at least two times the size of the second largest city. Bangkok is nearly 10 times larger than Thailand's next largest metropolitan area, Chiang Mai, and Manila (Figure 10.18) is 9 times larger than Cebu, the Philippines' second-largest city. Thanks to their strong industrial base, political power, and the massive immigration they attract, primate cities can dominate whole countries.

> Often the focus of migration is the capital of a country, which may become a *primate city*—one that is at least two times the size of the second largest city.

Rarely, however, can any of these cities provide sufficient housing, water, sanitation, or even decent jobs for all the new arrivals. Of all the cities in Southeast Asia, only Singapore provides well for nearly all of its citizens. Even there, however, a significant illegal, non-citizen population lives in poverty. More typical is the experience of rural-to-urban migrants who go to

Table 10.1 Recent growth in Southeast Asia's metropolitan areas with 5 million or more inhabitants

Rank, 2007	City/ Country	Population 2000*	Population 2007 (est.)
1	Manila, Philippines	9,906,000	18,491,000
2	Jakarta, Indonesia	15,961,000	18,267,000
3	Bangkok, Thailand	6,320,200	10,230,000
4	Kuala Lumpur, Malaysia	4,428,800	6,933,000
5	Bandung, Indonesia	3,416,000 (1995)	5,980,000
6	Singapore– Johor Baharu, (Malaysia)	4,500,000 (2001)	5,490,000
7	Ho Chi Minh City, Vietnam	3,316,500	5,183,000

*Unless otherwise specified.

Source: World Gazetteer, http://www.world-gazetteer.com.

Bangkok, many of whom reside in temporary, usually self-built slums and squatter settlements such as those described in the vignette below.

Vignette Mak (age 33) and Lin (age 27), husband and wife, left their two sons in the care of Lin's parents in a village north of Bangkok. They could not afford the school fees for even one of their sons on their rural wages, yet they hoped to educate both boys, even though educating only the eldest is the custom. Therefore, Mak and Lin traveled by bus for 10 hours to reach Bangkok. After several anxious days, they both found grueling work unloading bags of flour from ships in the harbor.

Each day when they finished their work, they walked miles to their quarters in one part of a tiny houseboat anchored, with thousands of others, on the Chao Phraya River. That river is Bangkok's low-income housing site, its source of water, its primary transport artery, and its sewer. Mak and Lin had to step gingerly across dozens of boats to get to theirs, intruding repeatedly on the privacy of their fellow river dwellers. They bathed in the dangerously polluted river, washed their sweaty, flour-covered work clothes by hand, and cooked their dinner of rice over a Coleman stove. Their only entertainment was provided by the private lives of their too numerous and too near-at-hand neighbors, who were often drunk on cheap liquor.

Figure 10.18 Manila's crowded streets. [David Paul Morris.]

For two years, Mak and Lin sent remittances to their family and managed to save enough to buy a bicycle, to pay their sons' school fees, and to cover basic necessities for their parents. Eventually, they returned home. Although they were glad to be out of Bangkok, both nevertheless agreed that they would eventually go back if they could not find work near their village.

Source: Alex Pulsipher's field notes, 2000. ∎

Migration: Globalization, Conflict, and Natural Disaster

Millions of Southeast Asians are moving to places other than the cities of their home countries. Flows of workers from this world region to others are an important aspect of globalization. Many other Southeast Asians are refugees forced to migrate by violent conflict or natural disasters.

Migration and globalization. The same push and pull factors driving urbanization are also driving some people to migrate out of Southeast Asia. These migrants are a major force of globalization as they supply much of the world's growing demand for low-wage workers who are willing to travel or live temporarily in foreign countries. They are also a globalizing force within their home countries as their remittances (monies sent home) boost family incomes and supply governments with badly needed **foreign exchange** (foreign currency) that countries need to purchase imports. For example, Filipinos working abroad are that country's largest source of foreign exchange, sending home over U.S.$6 billion, and increasing household income by an average of 40 percent.

Recently, women have constituted well over 50 percent of the more than 8 million migrants from this region. Many skilled nurses and technicians from the Philippines work in European, North American, and Southwest Asian cities. About 3 million participate in the global "maid trade" (Figure 10.19). This involves primarily uneducated women from the Philippines and Indonesia who work under 2- to 4-year contracts in wealthy homes throughout Asia. The maid trade has become notorious for abusive working conditions and employers who often do not pay what they promise. The Philippines went so far as to ban the maid trade in 1988, but re-established it in 1995 after better pay and working conditions were negotiated with countries receiving the workers.

The maid trade increased exponentially after the economic crisis of the late 1990s, when many Southeast Asian women lost their factory jobs and had to seek alternative employment. Whereas about 10,000 Indonesian women migrated to Saudi Arabia each year in the 1980s, this number had risen to over 380,000 per year by 1998. An estimated 1 to 3 million Indonesian maids now work outside of the country, mostly in the Persian Gulf.

Conditions are much better for the many skilled male workers from Southeast Asia who work in the international merchant marine. They typically work onboard international freighters as seamen, cooks, and engine mechanics, though a few are now officers. They generally work for 6 months or more at a stretch, after which they return home to their families. Their earnings are usually sufficient to provide a middle-class lifestyle and education for their families.

Refugees from conflict and natural disaster. In this region, violence and natural disasters force many people to migrate. During the last half of the twentieth century, millions of mainland Southeast Asians fled into neighboring states to escape protracted conflict. Thailand, Laos, and Cambodia, in partic-

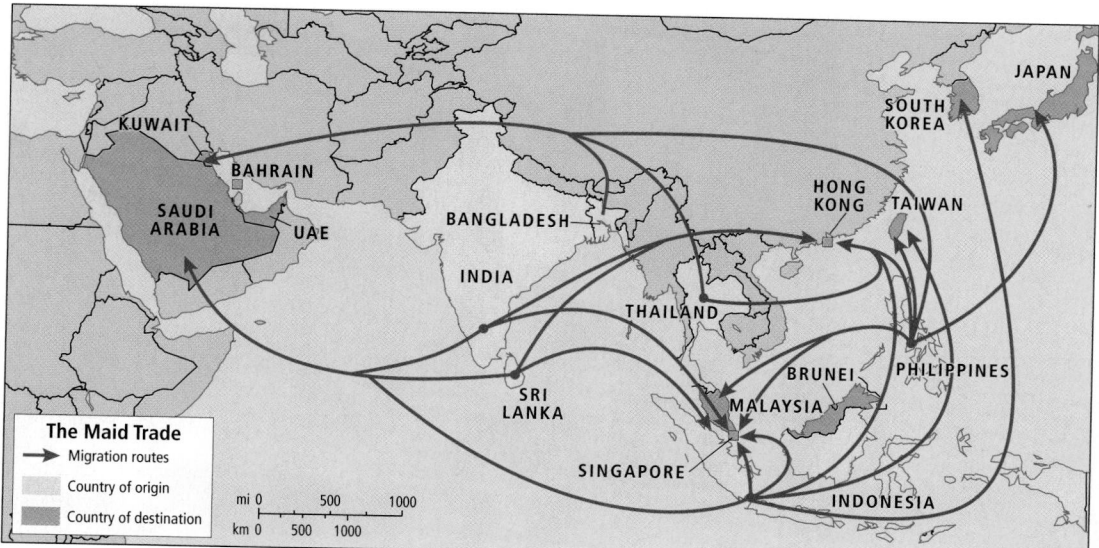

Figure 10.19 Globalization: The "maid trade." In the 1990s, between 1 million and 1.5 million Southeast Asian women were working elsewhere in Asia (including the Arab states) as domestic servants. It is estimated that by 2005 the number had more than doubled, with the majority coming from the Philippines, Indonesia, and Sri Lanka. [Adapted from Joni Seager, *The Penguin Atlas of Women in the World* (New York: Penguin Books, 2003), p. 73; with updated information from the Migration Policy Institute, at http://www.migrationinformation.org/Profiles/display.cfm?ID=364.]

ular, received many refugees during and after the Vietnam War. Currently, repression by Burma's military government has resulted in at least 680,000 refugees, with 500,000 displaced within Burma and 180,000 having fled to Thailand (Figure 10.20a), Bangladesh, India, and Malaysia. ◼ The tsunami of December 2004 complicated the refugee picture when it displaced well over 130,000 people in Sumatra alone

(in addition to killing an equal number of people on that island), plus several thousand in Thailand and Burma. Most of the refugees went to Malaysia or to another part of their home country, and many have now returned home. Figure 10.20b shows the numbers of refugees (from all causes) reported by each country.

Current Geographic Issues

Like Middle and South America, Africa, and South Asia, Southeast Asia is expanding its links to the global economy. However, it has had greater success than other areas in achieving prosperity, largely by following the example of some East Asian countries. In some cases, rapid economic change has brought financial crisis and conflict among the region's wide array of culture groups and religions. Further, gender roles are changing as populations urbanize and as women work more outside the home. In this section, we will look at these and other major economic, political, and sociocultural issues.

ECONOMIC AND POLITICAL ISSUES

The economic and political situation in the region has changed dramatically in recent years. From the mid-1980s to the late 1990s, Southeast Asian countries had some of the highest economic growth rates in the world. Their success was largely due to an economic strategy that emphasized the export of manufactured goods. For a few years beginning in 1997, however, economic growth stagnated and political instability increased as foreign investors withdrew their money. The high level of corruption revealed by the crisis strengthened calls for democracy, which expanded in some countries in the years afterward.

Strategic Globalization: State Aid and Export-led Growth

From the 1960s to the 1990s, national governments in Southeast Asia achieved strong and sustained economic expansion by emulating two strategies for economic growth pioneered earlier by Japan, Taiwan, and South Korea (see Chapter 9, pages 347–348). One was the formation of *state-aided market economies*. National governments intervened strategically in the financial sector to make sure that certain economic sectors developed. Investment by foreigners was limited so that the governments could have more control over the direction of the economy. The other strategy was *export-led growth*, which focused investment on industries that manufactured products for export, primarily to developed countries. These strategies amounted to a limited and selective embrace of globalization in that, while foreign sources of capital were not sought after, global markets for the region's products were.

> From the 1960s to the 1990s, national governments in Southeast Asia restricted and controlled foreign investment in the region.

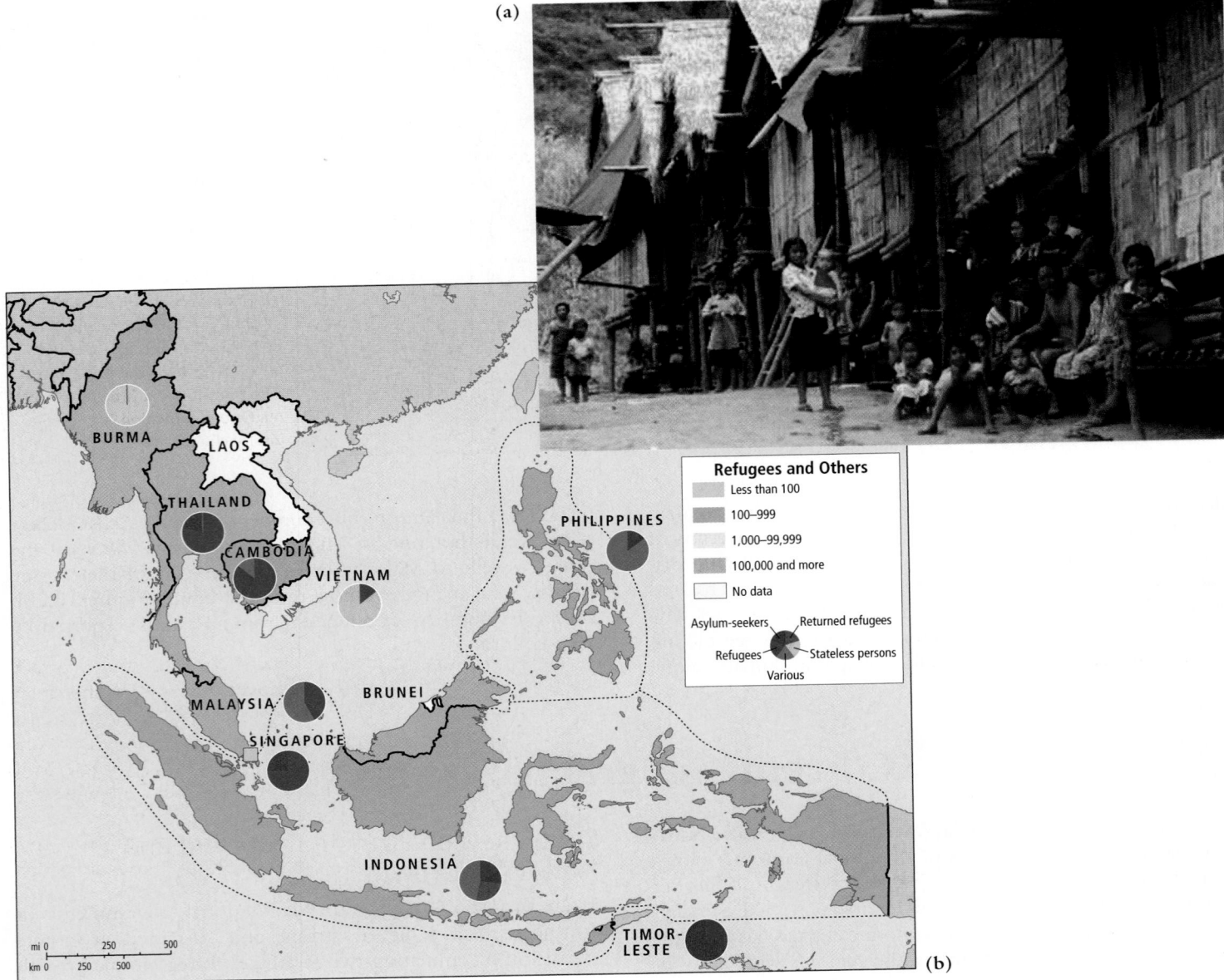

Figure 10.20 Refugees and other "persons of concern" at the end of 2005. **(a)** Karen refugees from Burma at Tham Hin camp near Bangkok. **(b)** Nearly every country in the region has refugees. The data on these persons are generally provided by each government based on its own definitions and collection methods and are therefore provisional. [(a) Sukree Sukplang/Reuters/Landov. (b) Data from "2005 global refugee trends: Statistical overview of populations of refugees, asylum-seekers, internally displaced persons, stateless persons, and other persons of concern to UNHDR," Table 1, June 2006, at http://www.unhcr.org/cgi-bin/texis/vtx/events/opendoc.pdf?tbl=STATISTICS&id=4486ceb12.]

These strategies were a dramatic departure from those used in other developing areas. In Middle and South America and parts of Africa, governments relied on import substitution industries that produced manufactured goods mainly for local use. By contrast, export-led growth allowed Southeast Asia's industries to earn much more money in the vastly larger markets of the developed world. Standards of living increased markedly, especially in Malaysia, Singapore, Indonesia, and Thailand. Another important result of Southeast Asia's success was relatively less extreme disparities in wealth (Table 10.2).

EPZs. In the 1970s, some governments in the region began adopting an additional strategy for encouraging economic development. This time foreign sources of capital were sought after, but the places they could invest were limited to specially designated free trade areas. Such **export processing zones**

Table 10.2 Disparities of wealth for selected countries in Southeast Asia and Middle and South America, as shown through income spread ratios[a]

Country	Income spread ratio, 2003	Income spread ratio, 2005
Indonesia	5.2	5.2
Malaysia	12.4	12.4
Singapore	9.7	9.7
Thailand	8.3	8.3
Vietnam	5.6	5.5
Brazil	29.7	26.4
Chile	19.3	18.7
Costa Rica	11.5	12.3
Mexico	17.0	19.3
Peru	11.7	18.4

[a]Ratio of wealthiest 20 percent to poorest 20 percent. The higher the number, the greater the discrepancy between wealthy and poor.

Sources: United Nations Human Development Report 2003, Table 13; *United Nations Human Development Report 2005,* Table 15 (New York: United Nations Human Development Programme).

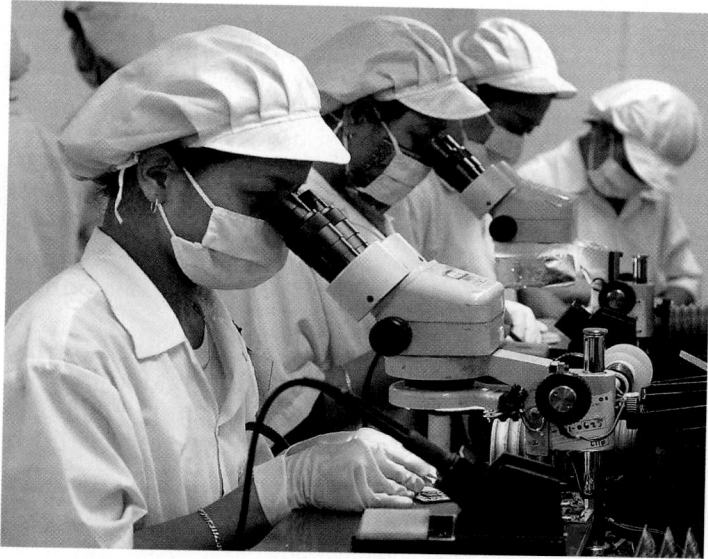

Figure 10.21 Patterns of industrialization. Employees inspect printed circuit boards at Integrated Microelectronic, Inc. (IMI), in Santa Rosa, a suburban city south of Manila. The company has operations in the Philippines, California, and Singapore, and serves clients in Japan, Europe, and the United States. [AFP Photo/Joel Nito/Newscom.]

(EPZs) are places in which foreign companies can set up industries using inexpensive, locally available labor to produce items only for export. Taxes are eliminated or greatly reduced. Since the 1970s, EPZs have expanded economic development in Malaysia, Indonesia, Vietnam, and the Philippines.

The feminization of labor. Between 80 and 90 percent of the workers in the EPZs are women, not only in Southeast Asia but in other world regions as well (Figure 10.21). The **feminization of labor** has been a distinct characteristic of globalization over the past three decades (see Chapter 3, page 121, and Chapter 9, pages 350–351). Employers prefer to hire young, single women because they are perceived as the cheapest, least troublesome employees. Statistics do show that, generally, women will work for lower wages than men, will not complain about poor and unsafe working conditions, will accept being restricted to certain jobs on the basis of sex, and are not as likely as men to agitate for promotions.

Working conditions. In general, the benefits of Southeast Asia's "economic miracle" have been unequally apportioned. In the region's new factories and other enterprises, it is not unusual for assembly-line employees to work 10 to 12 hours per day, 7 days per week, for less than the legal minimum wage, and without benefits. Labor unions that would address working conditions and wage grievances are frequently repressed by governments, and international consumer pressure to improve working conditions has been only partially effective. For example, Asian employees of the U.S. shoe manufacturing firm Nike were frequently exposed to hazardous chemicals, as well as to physical abuse and psychological cruelty on the job. Nike's labor tactics became widely known to U.S. university students when student activists revealed that many campus stores were selling Nike equipment and garments made under these substandard conditions in Southeast Asian factories. Although the students' exposure of abuses resulted in some improvements in working conditions, those improvements were relatively minor.

A much more powerful force that is driving up pay and improving working conditions is the service sector, which is growing throughout the region. In Singapore, the Philippines, and Thailand, the service sector already dominates, and in Malaysia and Indonesia, it is rapidly catching up to the manufacturing sector as a contributor to GDP. Many service sector jobs require at least a high school education, and competition for the smaller number of educated workers means that wages and working conditions are already better than in manufacturing and are likely to improve faster.

Economic Crisis and Recovery: The Perils of Globalization

The economic crisis that swept through Southeast Asia in the late 1990s forced millions of people into poverty and changed the political order in some countries. A major cause of the crisis was the lifting of controls on Southeast Asia's once highly regulated financial sector.

Deregulating investment. As part of a general move to open national economies to the free market, Southeast Asian governments relaxed controls on the financial sector in the 1990s. Previously, governments had regulated the kinds of investments that banks could make and had restricted their ability to accept money from foreign investors. By relaxing these controls, governments hoped that banks might uncover new economic opportunities. By allowing more foreign investment, governments also hoped to tap into new funds from abroad that could finance economic expansion. These hopes proved largely unfounded.

Soon, Southeast Asian banks were flooded with money from investors in the rich countries of the world who hoped to profit from the region's growing economies. Flush with cash and newfound freedoms, the banks often made reckless decisions. For example, bankers made many risky loans to real estate developers, often for high-rise office building construction. As a result, many Southeast Asian cities soon had a glut of office space. In Bangkok alone, there was U.S.$20 billion worth of unused office space by 1996.

Lifting the controls on investments was also made problematic by a kind of corruption known as **crony capitalism.** In most Southeast Asian countries, as elsewhere, corruption is encouraged by the close personal and family relationships between high-level politicians, bankers, and wealthy business owners. In Indonesia, for example, the most lucrative government contracts and business opportunities were reserved for the children of former president Suharto, who ruled the country from 1967 to 1997. His children became some of the wealthiest people in Southeast Asia. This kind of corruption expanded considerably with the new foreign investment money, much of which was diverted to bribery or unnecessary projects that brought prestige to political leaders.

The cumulative effect of crony capitalism and the lifting of controls on banks was that many ventures failed to produce any profits at all. In response, foreign investors panicked, withdrawing their money on a massive scale. In 1996, before the crisis, there was a net inflow of U.S.$94 billion to Southeast Asia's leading economies. In 1997, inflows had ceased and there was a net outflow of U.S.$12 billion.

The IMF bailout. The International Monetary Fund (IMF) made a major effort to keep the region from sliding deeper into recession. In 1998, the IMF brought in U.S.$65 billion, much of it used to rescue Southeast Asian banks from bankruptcy. This money was tied to specific reforms designed to make banks more responsible in their lending practices. The IMF also required *structural adjustment policies,* (see Chapter 3, pages 119–122), which required countries to cut government spending (especially on social services) and abandon policies intended to protect domestic industries.

After several years of economic chaos, and much debate over whether the IMF bailout helped or hurt a majority of Southeast Asians, economies began to recover. Many economies are again growing, though much more slowly than before the crisis. Older strategies, such as the establishment of EPZs, have been expanded to attract additional multinational corporations to Southeast Asia. By 2006, the crisis, while still serving as an ominous reminder of the risks of globalization, had been more or less overcome. 📹

The impact of China's growth. The main long-term effect of the Southeast Asian financial crisis was to reduce the region's ability to compete with China in attracting new industries and foreign investors. China now attracts more than twice as much foreign direct investment as Southeast Asia. Southeast Asia's wealthier countries, such as Singapore, Malaysia, and Thailand, have lost investment to China. All three are now struggling to position themselves as locations offering more highly skilled labor and more high-tech infrastructure than China.

However, China's growth is also an opportunity for Southeast Asia. Both Singapore and Malaysia are trying to "piggyback" on China's growth by winning large contracts to upgrade China's infrastructure in areas such as wastewater treatment and gas distribution. Meanwhile, in Southeast Asia's poorer countries, such as Vietnam, wages are considerably lower than in China. Vietnam has thus been able to attract investment in low-skill manufacturing that had been going to China.

Regional Trade and ASEAN

Southeast Asian countries trade more with China and the rich countries of the world than they do with one another (Figure 10.22). Trade among countries within the region has been inhibited by the fact that they all export similar goods and traditionally have imposed tariffs against one another. They export food, raw materials, and manufactured goods to the developed countries, and import consumer products, industrial materials, machinery, and fossil fuels.

These issues of economic and political cooperation are being addressed by the **Association of Southeast Asian Nations (ASEAN).** Though it started in 1967 as an anticommunist, anti-China association, ASEAN now focuses on agreements that strengthen regional cooperation. One example is the Southeast Asian Nuclear Weapons–Free Zone Treaty signed in December 1995 by all ten Southeast Asian nations. Another is the ASEAN Free Trade Association (AFTA), patterned after the North American Free Trade Agreement and the European Union. Tariffs between countries are being reduced in order to lower production costs and make ASEAN's manufacturing industries more efficient. It is hoped that Southeast Asia's products can thereby be priced lower, making them more competitive in the global market.

An even greater shift in Southeast Asia's economies will occur if a free trade agreement is reached between ASEAN and China, which is expected to occur by 2010. Tariffs have already been reduced on a range of goods since talks on the potential free trade agreement began in 2002, and trade between ASEAN and China has grown faster than trade among

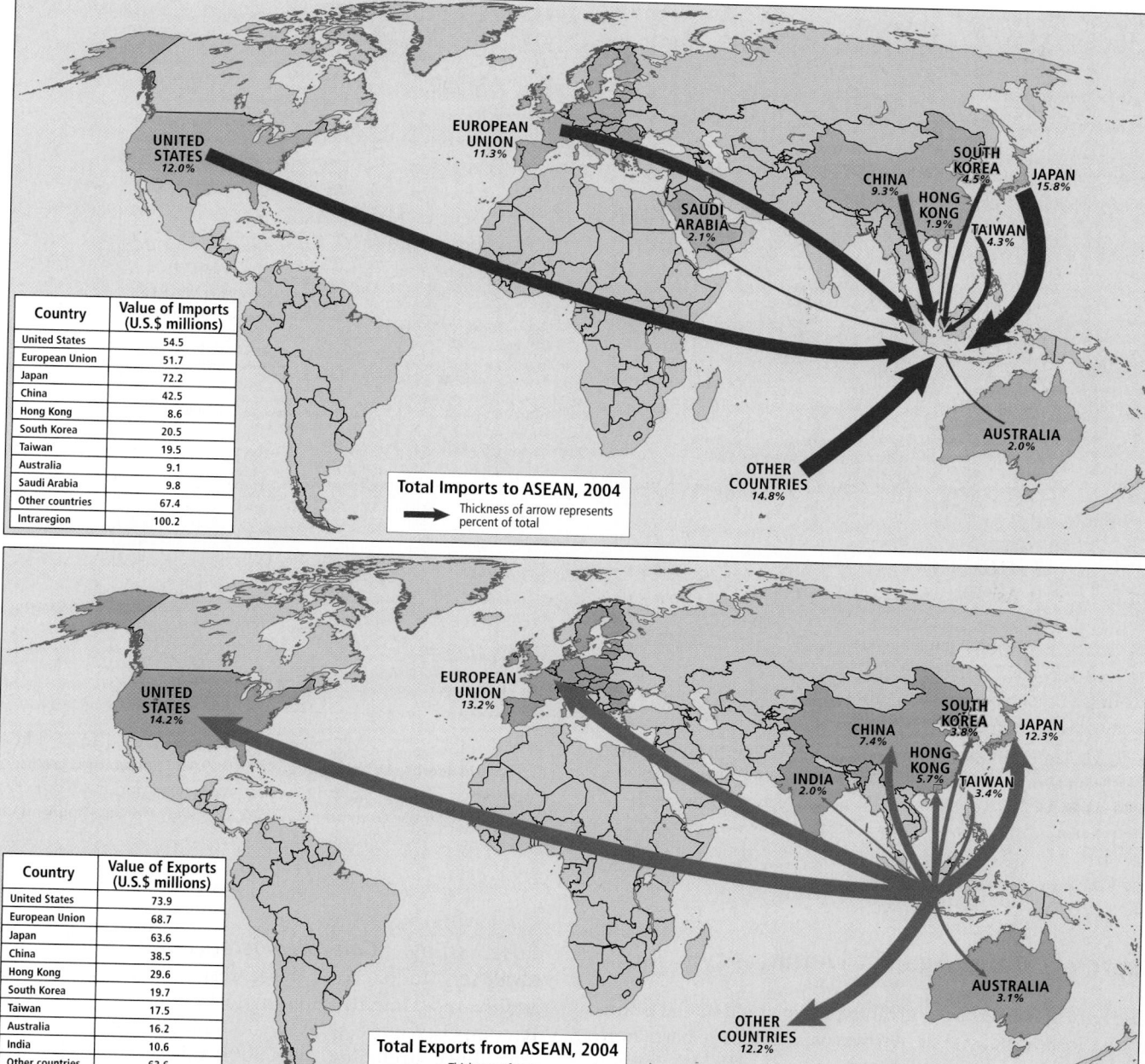

Figure 10.22 ASEAN imports and exports. The ASEAN countries are very active in world trade as both importers (top) and exporters (bottom), but trade within the region (intraregional) is also very important. [Data from *ASEAN Statistical Yearbook 2005*, Chapter 5, at http://www.aseansec.org/SYB2005/chapter-5.pdf.]

ASEAN nations. China is now the region's fourth-largest trade partner (well behind the EU, the United States, and Japan), buying primarily natural resources.

ASEAN has also seen increased cooperation among its members over issues of tourism, the fastest-growing industry in Southeast Asia. Between 1991 and 2001, the number of international visitors to the region doubled to more than 40 million, which accounted for nearly 6 percent of the world's

total. In response, ASEAN members have been working to improve the region's transport infrastructure. One such project is the Asian Highway, a web of standardized roads looping through the mainland and connecting it with Malaysia, Singapore, and Indonesia (the latter via ferry) (Figure 10.23). Eventually, the Asian Highway will facilitate ground travel through 32 Eurasian countries from Moscow to Indonesia and from Turkey to Japan.

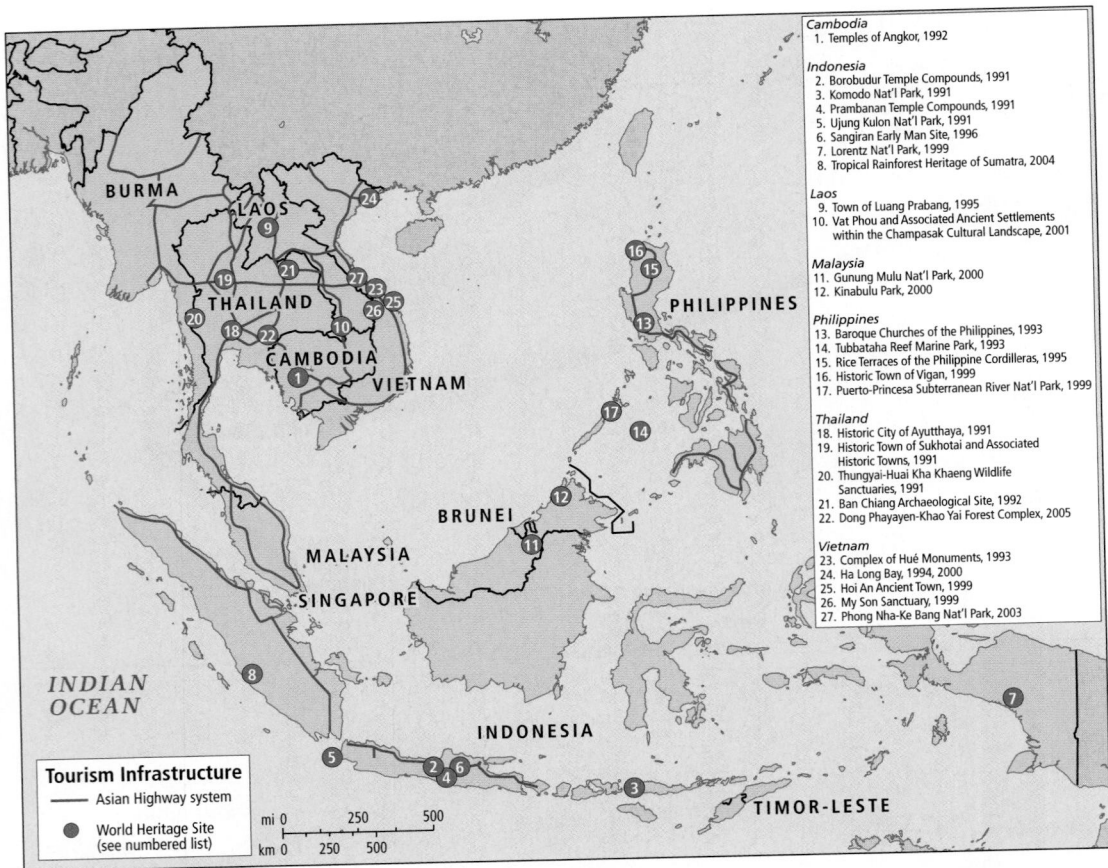

Cambodia
1. Temples of Angkor, 1992

Indonesia
2. Borobudur Temple Compounds, 1991
3. Komodo Nat'l Park, 1991
4. Prambanan Temple Compounds, 1991
5. Ujung Kulon Nat'l Park, 1991
6. Sangiran Early Man Site, 1996
7. Lorentz Nat'l Park, 1999
8. Tropical Rainforest Heritage of Sumatra, 2004

Laos
9. Town of Luang Prabang, 1995
10. Vat Phou and Associated Ancient Settlements within the Champasak Cultural Landscape, 2001

Malaysia
11. Gunung Mulu Nat'l Park, 2000
12. Kinabulu Park, 2000

Philippines
13. Baroque Churches of the Philippines, 1993
14. Tubbataha Reef Marine Park, 1993
15. Rice Terraces of the Philippine Cordilleras, 1995
16. Historic Town of Vigan, 1999
17. Puerto-Princesa Subterranean River Nat'l Park, 1999

Thailand
18. Historic City of Ayutthaya, 1991
19. Historic Town of Sukhotai and Associated Historic Towns, 1991
20. Thungyai-Huai Kha Khaeng Wildlife Sanctuaries, 1991
21. Ban Chiang Archaeological Site, 1992
22. Dong Phayayen-Khao Yai Forest Complex, 2005

Vietnam
23. Complex of Hué Monuments, 1993
24. Ha Long Bay, 1994, 2000
25. Hoi An Ancient Town, 1999
26. My Son Sanctuary, 1999
27. Phong Nha-Ke Bang Nat'l Park, 2003

Tourism Infrastructure
— Asian Highway system
● World Heritage Site (see numbered list)

Figure 10.23 Transportation infrastructure for tourism. Many of the Southeast Asian World Heritage Sites are located along the partially completed Asian Highway, which will eventually connect Europe to Indonesia. A map of the entire system is at http://www.unescap.org/ ttdw/common/TIS/AH/maps/AHMapApr04.gif. [Adapted from UN World Heritage Convention, at http://whc.unesco.org/en/list/; http://www.unescap.org/ttdw/common/tis/ah/tourism%20attractions.asp; http://www.atimes.com/atimes/Asian_Economy/images/highways.html.]

Pressures For and Against Democracy

Demands are growing for a greater public voice in the political process in Southeast Asia. However, significant barriers to democratic participation still exist (Figure 10.24). Several countries are plagued with violent conflict, and roughly one-quarter of the region's people live under undemocratic regimes that are notorious for corruption.

The greatest recent shift toward democracy occurred in the wake of the economic crisis of the late 1990s. During the crisis, massive public protests expanded democracy in both Thailand and Indonesia. In Thailand, protesters successfully agitated for constitutional provisions that reduced corruption and provided for more regular elections. Change was more dramatic in Indonesia, where after three decades of semi-dictatorial rule by President Suharto, the economic crisis spurred massive demonstrations that forced Suharto to resign. Since then, democratic parliamentary and presidential elections have initiated a new political era in the country.

Case study: Can democratization hold Indonesia together? Indonesia is the largest country in Southeast Asia and the most fragmented, physically, culturally, and politically. It comprises more than 17,000 islands (3000 of which are inhabited), stretching over 3000 miles (8000 kilometers) of ocean. It is also the most culturally diverse, with dozens of ethnic groups and multiple religions. But instability in Indonesia has many people wondering whether this multi-island country of 220 million might be headed for disintegration.

Until the end of World War II, Indonesia was not a nation at all but rather a loose assemblage of distinct island cultures, which Dutch colonists managed to hold together as the "Netherlands East Indies." When Indonesia became an independent country in 1945, its first president, Sukarno, hoped to forge a new nation out of these many parts. To that end, he articulated a national philosophy known as *Pancasila*, which was based on tolerance, particularly in matters of religion.

In recent years, separatist movements have sprouted in four distinct areas. The only one to succeed was in Timor-

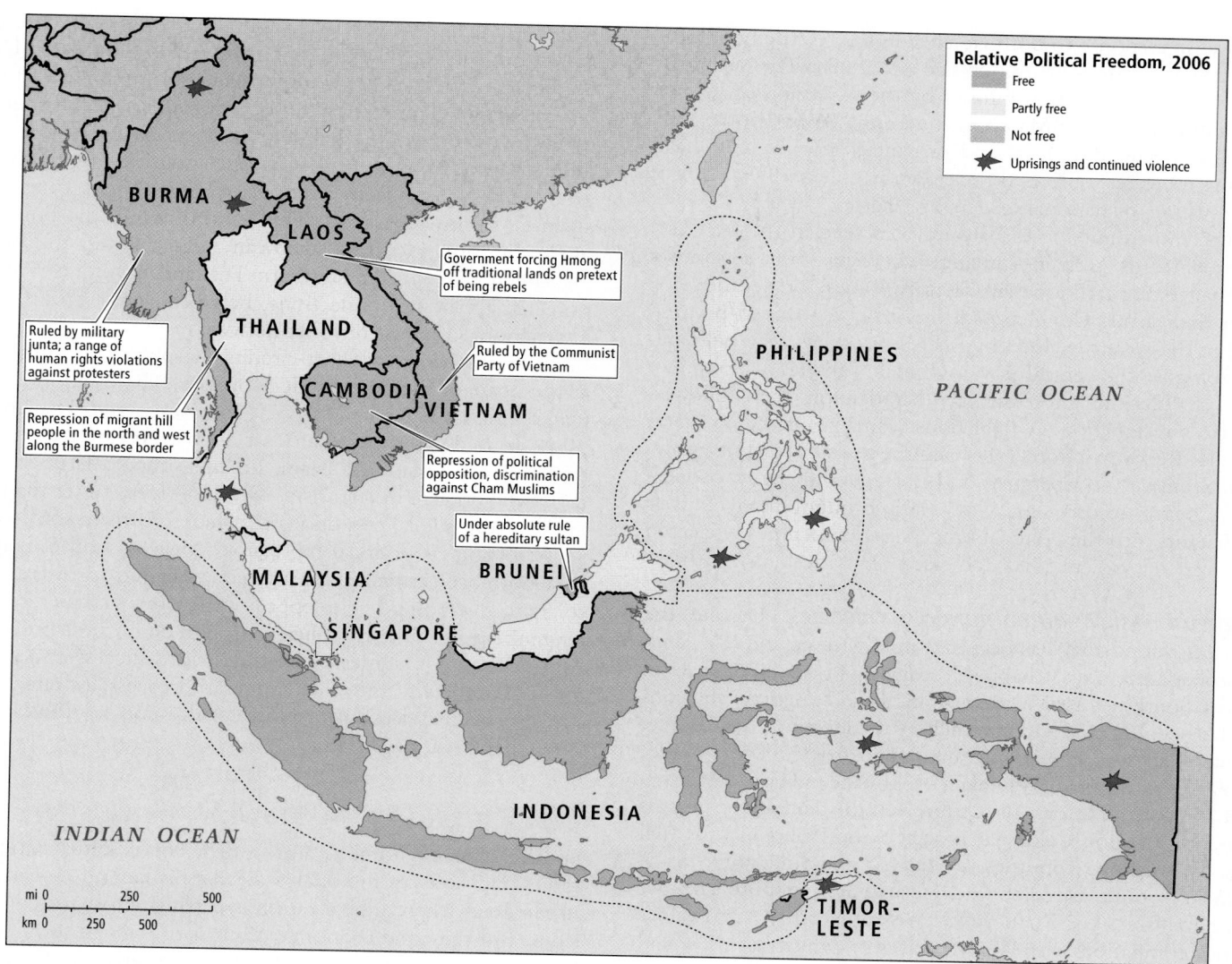

Figure 10.24 Relative political freedom in the region. The state of political freedom in Southeast Asia is debatable, with most countries experiencing political violence or election fraud of some sort over the last decade. The map here comes from the nongovernmental organization Freedom House, which assigns countries to the three categories using surveys based on foreign and domestic news reports, academic analyses, consultation (with nongovernmental organizations, think tanks, and individuals), and visits to the region. Though Freedom House admits that an element of subjectivity is inherent in the survey findings, their process emphasizes intellectual rigor and balanced and unbiased judgments. [Adapted from "Map of Freedom 2006," at http://www.freedomhouse.org/template.cfm?page= 363&year=2006.]

Leste, which became an independent country in 2002. However, its case is special in that this area was under Portuguese control until 1975, when it was forcibly integrated into Indonesia. Two other separatists movements have grown largely in response to Indonesia's **resettlement schemes.** Also known as *transmigration schemes,* these programs have relocated approximately 5 million people from crowded islands such as Java to less densely settled islands. The policies were originally initiated under the Dutch in 1905 to relieve crowding and provide agricultural labor for plantations in thinly populated areas.

> More than 5 million people have been relocated by Indonesia's resettlement schemes, which are some of the largest ever attempted by a government.

After independence, Indonesia used resettlement schemes for the same purposes, but also to bring outlying areas under closer control of the central government in Jakarta. However, the government lost rather than gained control as the newcomers inspired separatist movements in two main areas of resettlement: in the Indonesian half of the Island of New Guinea (the Indonesian provinces of Papua and West Papua) and in the Malucca islands. In 2001, three years after the expansion of Indonesian democracy following Suharto's resignation in 1988, the resettlement schemes were cancelled.

This was in response to both violence in the resettlement areas and lack of funds to continue the programs. The separatist movements were further diffused by the extension of greater local autonomy to the Maluccas and Papua/West Papua. Violence has declined as local and provincial governments have become more responsive to their own people and less subject to the whims of national leaders in Jakarta.

The most dramatic turnaround was seen in the far western province of Aceh in Sumatra, where in 2005, separatists who had battled Indonesian security forces for decades laid down their arms. Conflicts had originally developed because most of the wealth yielded by Aceh's resources, especially oil, was going to the central government in Jakarta. The conflict began to decline after the democratization that followed Suharto's resignation. A final boost came from the Tsunami of 2004. Recovery efforts following the disaster created a powerful incentive for separatists and the government to cooperate. A peace accord signed in 2005 brought many former combatants into the political process as democratically elected local leaders. Violence has decreased dramatically since then.

Southeast Asia's authoritarian tendencies. Despite the democratization that has occurred in Indonesia and Thailand, authoritarianism (see Chapter 7, page 274) is still a powerful force in Southeast Asia. Undemocratic socialist regimes still control Laos and Vietnam, and a military dictatorship runs Burma. Brunei is an authoritarian sultanate, and from 2006 to 2008, Thailand's government was taken over by the military. ▪️ Powerful and corrupt leaders have subverted the democratic process even in the region's oldest democracy, the Philippines, as well as in the wealthier countries of Malaysia and Singapore.

Some Southeast Asian leaders, such as Singapore's former prime minister, Lee Kuan Yew, have argued that Asian values are not compatible with Western ideas of democracy. Yew and other leaders assert that Asian values are grounded in the Confucian view that individuals should be submissive to authority. Hence, Asian countries should avoid the highly contentious public debate of electoral politics. Nevertheless, when confronted with governments that abuse their power, people throughout Southeast Asia have repeatedly rebelled, often in the form of pro-democracy movements.

Some pro-democracy movements have resulted in real change, as in Thailand and Indonesia in the late 1990s, but others have not. People in Burma have been futilely protesting the rule of a corrupt and undemocratic military regime for more than two decades. The regime refused to step aside when the people elected Aung San Suu Kyi to lead a civilian reformist government in 1990. Suu Kyi has been under house arrest ever since, despite having won the Nobel Peace Prize in 1991. Widespread pro-democracy protests throughout the country in 2007 were brutally repressed. ▪️

Terrorism and democracy. Like authoritarianism, terrorism has long loomed as a counterforce to democracy in this region. During the late 1990s, a series of bombs exploded in the Philippines and across Indonesia, and small terrorist cells were discovered in Malaysia, southern Thailand, and Singapore. Until the bombing of the Sari Hotel in Bali (Indonesia) in October 2002, which killed nearly 200 foreign tourists, terrorist activity in the region was local in nature. It was carried out by local militant groups pursuing domestic political agendas and grievances—most often revenge for government campaigns against Muslim separatists. The Bali bombing, the bombing of the Marriott Hotel in Jakarta in August 2003, and ongoing violence in Muslim southern Thailand have drawn attention to apparent connections between local groups and international Islamist terrorist networks.

Terrorist violence short-circuits the public debate that is at the heart of the democratic processes. The election of secular political parties in Malaysia, Indonesia, and Thailand in 2004 showed that most Southeast Asians, who are accustomed to a tolerant version of Islam, do not support Islamist militants who engage in terrorism. ▪️ However, rather than trying to address local conditions that fuel terrorism, some states are resorting to political repression. Malaysia and Singapore have both been criticized for using counter-terrorism as a cover to crack down on peaceful opposition groups that merely challenge the government's policies through democratic processes. Another route is suggested by Indonesia's success at bringing former combatants into the democratic process, a strategy that has also been successful in North Africa and Southwest Asia (see Chapter 6, page 232).

SOCIOCULTURAL ISSUES

Southeast Asia is home to a great diversity of cultures and religious traditions. Some of this diversity is fading, especially in urban areas where globalization is cutting across societal divisions. But the rapid economic changes of the past few decades have increased tensions between some groups. Economic change and urbanization are also transforming gender roles and have occasionally led to sexual exploitation.

Religious Pluralism

The major religious traditions of Southeast Asia include Hinduism, Buddhism, Confucianism and Taoism, Islam, Christianity, and animism (Figure 10.25). All originated outside the region, with the exception of the animist belief systems of the indigenous peoples. Animism takes many different forms in this region, but in general such natural features as trees, rivers, crop plants, and the rains all carry spiritual meaning. Such objects are the focus of festivals and rituals to give thanks for bounty and to mark the passing of the seasons.

> Animism takes many different forms, but in general such natural features as trees, rivers, crop plants, and the rains all carry spiritual meaning.

All of Southeast Asia's religions have undergone change as a result of exposure to one another. Hindus and Christians

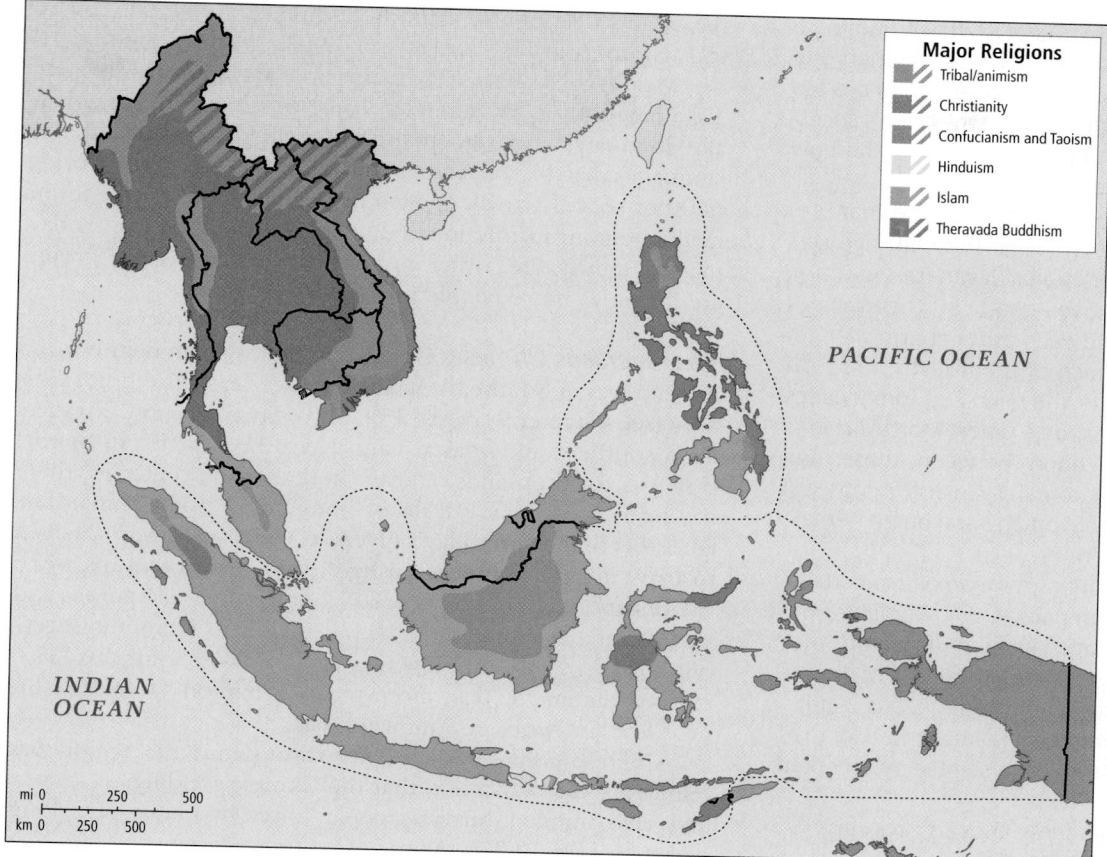

Figure 10.25 Religions of Southeast Asia. Southeast Asia is religiously very diverse: five of the world's six major religions are practiced there. Animism, the oldest belief system, is found in both island and mainland locations. [Adapted from *Oxford Atlas of the World* (New York: Oxford University Press, 1996), p. 27.]

in Indonesia, surrounded as they are by Muslims, have absorbed ideas from Islam, such as the seclusion of women. Many Muslims and Christians believe in spirits and practice rituals that have their roots in animism. Many Muslims have also absorbed ideas about kinship and marriage from indigenous belief systems, as illustrated in the following vignette.

Vignette Marta is a *Pemaes,* a woman who prepares a bride for a traditional Javanese wedding (Figure 10.26). Nearly all Javanese are Muslims, but Islam does not have elaborate marriage ceremonies, so colorful animist rituals have survived. Usually, the bride and groom have never met, their families having worked out the match. Hence, their first meeting at the wedding ceremony is

Figure 10.26 The Dahar Klimah (Dahar Kembul). In this phase of a traditional Javanese wedding, the *Pemaes* gives the bride a bundle of food (yellow rice, fried eggs, soybeans, and meat). Then the bridegroom makes three small balls of the food and feeds them to the bride; she then does the same for him. The ritual reminds them that they should share their belongings joyfully together. [Courtesy of Dirk Vranken, http:users.skynet.be/sky86158.]

surrounded by a great deal of mystery. Marta's job is to reinforce this mystery and to prepare the couple for a life together. She bathes and perfumes the bride, puts on her makeup, and dresses her, all the while making offerings to the spirits of the bride's ancestors and counseling her about how to behave as a wife and how to avoid being dominated by her husband.

The groom, meanwhile, is undergoing ceremonies that prepare him for marriage as well. Each will sign the marriage certificate before they meet. In the traditional Javanese view, it is best that a young couple not be in love—that way, they will not fall out of love. Rather, they will hold each other at arm's length from the start, not investing too much of themselves in a relationship that is bound to change over the course of the decades as they mature and as their family grows older. Marta's role is to create a magically reinforced bond between these two strangers. After marriage, as before, a man's closest friends will be his male age-mates and kin, and a woman's will be her female friends and kin.

Despite these elaborate preparations for marriage, divorce in Malaysia and Indonesia is fairly common among Muslims, who often go through one or two marriages early in life before they settle into a stable relationship. Although the prevalence of divorce is lamented by society, it is not considered outrageous. Apparently, ancient indigenous customs predating Islam allowed for mating flexibility early in life, and this attitude is still tacitly accepted in Java.

Adapted from Walter Williams, Javanese Lives *(Piscataway, N.J.: Rutgers University Press, 1991), pp. 128–134.* ∎

Cultural Pluralism

Southeast Asia is a place of **cultural pluralism** in that it is inhabited by groups of people from many different backgrounds. Over the past 40,000 years, migrants have come to the region from India, China, Southwest Asia, Japan and Korea. Many of these groups have remained distinct, partly because they live in isolated pockets separated by rugged topography or seas. However, urbanization is now bringing many of these groups together, resulting in much peaceful cultural exchange but also some violent conflict.

The globalizing urban and the traditional rural. Southeast Asia's cities are some of the most diverse parts of the region, but in some ways this diversity is decreasing as many groups are exposed to each other and to global culture. Globalization has brought a wide array of new ideas, pastimes, and products from around the world. Free market capitalism, communism, nationalism, consumerism, and environmentalism have all modified life and landscapes. This is especially true in urban areas, where their appeal often cuts across societal divisions. For example, Malaysian teens of many different ethnicities (Chinese, Tamil, Malay, Bangladeshi) spend much of their spare time following the same European soccer teams, playing the same video games, eating the same fast food, and talking to each other in English.

Meanwhile, rural areas retain more traditional influences. Take language as an example. While one main language usually dominates trade and politics in a city, dozens of different languages may be spoken in rural areas. Indeed, of the world's 6000 or so still actively spoken languages, 1000 can be found in this region, mainly in rural areas and small towns. Today, many linguistic and other cultural barriers that divide groups are falling as more people move into the cities.

The Overseas Chinese. One group that is prominent beyond its numbers in Southeast Asia is the Overseas (or ethnic) Chinese (see Chapter 9, pages 359–360). Small groups of traders from southern and coastal China have been active in Southeast Asia for thousands of years, and over the centuries, there has been a constant trickle of immigrants from China. The ancestors of most of today's Overseas Chinese, however, began to arrive in large numbers during the nineteenth century, when the European colonizers needed labor for their plantations and mines. Later, many of those who fled China's Communist revolution after 1949 sought permanent homes in Southeast Asian trading centers. Today, more than 26 million Overseas Chinese live and work in Southeast Asia.

Chinese commercial activity throughout the region has reinforced the perception that the Chinese are diligent, clever, and civic-minded businesspeople, often working very long hours (Figure 10.27). At the same time, despite the fact that most have modest incomes, the Chinese in Southeast Asia have the reputation of being rich and influential in government and commerce. The Overseas Chinese are indeed industrious, and with their region-wide connections and access to start-up money, they have been well positioned to take advantage of the new growth sectors of modernizing and globalizing economies. Sometimes new Chinese-owned enterprises have put out of business older, more traditional establishments that many local people depended on. For example, in the town of Klaten in central Java, an open-air bazaar where local people once sold an array of goods has been shut down to make way for a government-sponsored, air-conditioned shopping mall. Only the Chinese can afford the expensive store rents in the new mall.

Many low- and middle-income Southeast Asians who were hurt by the financial crisis of the late 1990s blamed their sorrows on the Overseas Chinese. A wave of violence resulted, with Chinese people assaulted, their temples desecrated, and their homes and businesses destroyed. Conflicts involving the Overseas Chinese have occurred in Vietnam, Malaysia, and many places in Indonesia (Sumatra, Java, Kalimantan, and Sulawesi) as well. Some Overseas Chinese are attempting to diffuse tensions through public education about Chinese culture and historical contributions to the Southeast Asian countries they live in. Others are financing economic and social aid projects to help their poorer, usually non-Chinese neighbors.

Figure 10.27 Overseas Chinese in Malaysia. Malay women sell breakfast from a portable stall in front of a store operated by a Malaysian of Chinese ethnicity in Lenggeng. Most businesses in Lenggeng and across Malaysia are run by ethnic Chinese (who are Malaysian citizens), and most of their customers are ethnic Malays. [Reuters/Bazuki Muhammad/Newscom.]

At the same time, some are also creating global business and financial networks that are designed in part to provide a way out of any particular area should anti-Chinese sentiment again result in violence.

Globalization and Gender: The Sex Industry

Southeast Asia has become a global center for the sex industry, which is supported in large part by sex tourism. Geared toward visitors who pay for sex, **sex tourism** in Southeast Asia grew out of the sexual entertainment industry that served foreign military troops stationed in Asia during World War II, the Korean War, and the Vietnam War. Now primarily civilian men arrive from around the globe to live out their fantasies during a few weeks of vacation. The industry is found throughout the region but is most prominent in Thailand. In 2008, 14 million tourists visited Thailand alone, up from 250,000 in 1965, and some observers estimate that as many as 70 percent were looking for sex. In recent years, some Thai government officials have encouraged sex tourism to create jobs, even though it is illegal, and the sector was praised for its role in helping the country weather the economic crisis of 1997. Some corrupt officials also support sex tourism because it provides them with a source of untaxed income from bribes.

One result of the "success" of sex tourism, as well as high local demand for sex workers, is that organized crime has entered the field, and girls and women are being coerced into sex work. Research indicates that some girls may have been sold by their families to pay off family debt. Others have been kidnapped, often at a very young age. Demographers estimate that 20,000 to 30,000 Burmese girls taken against their will—some as young as 12—are working in Thai brothels. Their wages are too low to make buying their own freedom possible. In the course of their work, they must service more than 10 clients per day, and they are routinely exposed to physical abuse and sexually transmitted diseases, especially HIV.

Vignette Twenty-five-year-old Watsanah K. (not her real name) awakens at 11:00 every morning, attends afternoon classes in English and secretarial skills, and then goes to work at 4:00 P.M. in a bar in Patpong, Bangkok's red light district. There she will meet men from Europe, North America, Japan, Taiwan, Australia, Saudi Arabia, and elsewhere, who will pay to have sex with her. She leaves work at about 2:00 A.M., studies for a while, and then goes to sleep.

Watsanah was born in northern Thailand to an ethnic minority group whose members, like many others in the area, are poor subsistence farmers. Many have recently become involved in the global drug trade by growing opium poppies. Watsanah married at 15 and had two children shortly thereafter. Several years later, her husband developed an opium addiction. She divorced him and left for Bangkok with her children. There, she found work at a factory that produced seat belts for a nearby automobile plant. In 1997, Watsanah lost her job as the result of the economic crisis that ripped through Southeast Asia. To feed her children, she became a sex worker.

Although the pay, between U.S.$400 and U.S.$800 a month, is much better than the U.S.$100 a month she earned in the factory, the work is dangerous and demeaning. Sex work, though widely practiced and generally accepted in Thailand, is illegal, and the women who do it are looked down on. As a result, Watsanah must live in constant fear of going to jail and losing her children. Moreover, she cannot always make her clients use condoms, which puts her at high risk of contracting AIDS or other sexually transmitted diseases. "I don't want my children to grow up and learn that their mother is a prostitute," says Watsanah. "That's why I am studying. Maybe by the time they are old enough to know, I will have a respectable job."

Sources: Adapted from the field notes of Alex Pulsipher and Debbi Hempel, 2000; "Sex industry assuming massive proportions in Southeast Asia," International Labor Organization News (August 19, 1998); coverage of the HIV-AIDS conference in Thailand, July 11–16, 2004, by the Kaiser Family Foundation, http: //www.kaisernetwork.org/ aids2004/kffsyndication.asp?show=guide.html ∎

Gender, Urbanization, and Empowerment

Gender roles are being transformed throughout Southeast Asia by urbanization and the changes it brings to family organization and employment. Moving to the city shifts people away from extended families and toward the nuclear family. Women have made significant gains in political empowerment and in education, but they still lag well behind men.

Family patterns. Throughout the region, it is common for a newly married couple to reside with the wife's parents. Along with this custom is a range of behavioral rules that empower the woman in a marriage, despite some basic patriarchal attitudes. For example, a family is headed by the oldest living male, usually the wife's father. When he dies, he passes on his wealth and power to the husband of his oldest daughter, not to his own son. (A son goes to live with his wife's parents and inherits from them.) Hence, a husband may live for many years as a subordinate in his father-in-law's home. Instead of the wife being the outsider, subject to the demands of her mother-in-law—as is the case, for example, in South Asia—it is the husband who must kowtow. The inevitable tension between the wife's father and the son-in-law is resolved by the custom of ritual avoidance. The wife manages communication between the two men by passing messages and even money back and forth. Consequently, she has access to a wealth of information crucial to the family and has the opportunity to influence each of the two men.

> Throughout the region, it is common for a newly married couple to reside with the wife's parents.

Urbanization has brought a shift to the nuclear family that has transformed these traditional relationships. Young couples now frequently live apart from the extended family, an arrangement that takes the pressure off the husband in daily life. Because this nuclear family unit is often dependent entirely on itself for support, wives usually work for wages outside the home. Though married women lose the power they would have if they lived among their close kin, they are empowered by the opportunity to have a career. The main drawback of this compact family structure, as many young families have discovered in Europe and the United States, is that there is no pool of relatives available to help working parents with child care and housework. Further, no one is left to help elderly parents maintain the rural family home. As a consequence, tiny urban apartments have to accommodate a young family and one or more aged grandparents in need of care. On the brighter side, working grandmothers in their vigorous middle years can often help their adult children financially.

Vignette Buaphet Khuenkaew, 35, lives in Ban Muang Wa, a village near the northern Thai city of Chiang Mai. She married at 18 and has two children: a son, 10, and a daughter, 17 (Figure 10.28). A Buddhist with a sixth-grade education, Buaphet is both a homemaker and a seamstress. Six days a week she drives the family motor scooter 30 minutes to her job in Chiang Mai, where she sews buttonholes in men's shirts for 2800 baht (U.S.$118) per month. The children perform weekday household chores when they return from school.

Buaphet's husband, Boontham, is a farmer who is about 5 years older than she. The couple knew each other before they were married. He would visit her at her parents' home, and eventually they fell in love. One day, he and his parents came to the house with a bride price of 10,000 baht (U.S.$420) in gold, and he asked her to marry him. She accepted. They used the gold to build their house on land owned by her mother, across the street from where Buaphet was born. They have electricity and a small television set. Drinking water comes from a well, is filtered through stones, and is stored in ceramic jars. Many household activities, including bathing, washing clothes, and washing dishes, take place in small shelters outside, but meals are eaten inside.

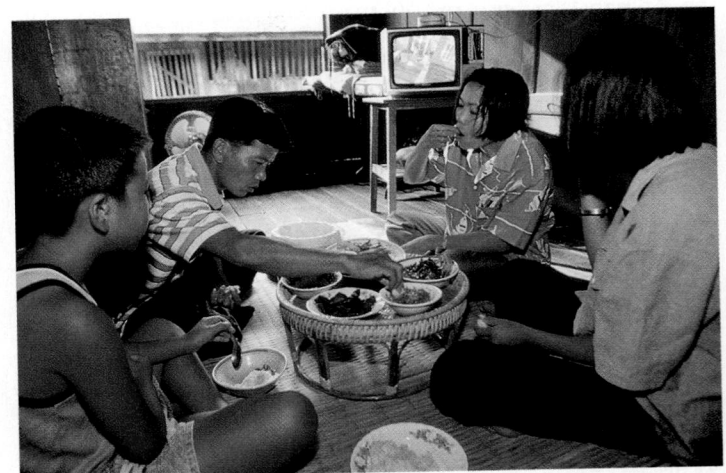

Figure 10.28 Buaphet sharing the morning meal with her family before setting off to work. [Joanna Pinneo/Material World.]

Although her husband feels that men are rightly regarded as superior in Thai society, Buaphet reports that she and her husband have an egalitarian marriage in which all decisions are made jointly. As is common for married couples in the region, they do not spend much of their leisure time together. She regularly spends time with her female friends and relatives, and he with his male friends and family. Buaphet says she is happy with her life, but also regularly complains about not having the appliances and up-to-date furnishings her friends have.

Adapted from Faith D'Alusio and Peter Menzel, Women in the Material World *(San Francisco: Sierra Club Books, 1996), pp. 228–39.* ∎

Political and economic empowerment of women. Women have made some impressive gains in politics in Southeast Asia. Economically, they still earn less money than men and work less outside the home, but this will likely change as their level of education increases.

Southeast Asia has had several prominent female leaders over the years, most of whom have risen to power in times of crisis as the leaders of movements opposing corrupt or undemocratic regimes. In the Phillipines, Corazon Aquino became president in 1986 after leading the opposition to Ferdinand Marcos, whose 21 year reign was infamous for its corruption and authoritarianism. Gloria Macapagal-Arroyo became president in 2001 after opposing a similarly corrupt president.

In Indonesia, Megawati Sukarnoputri became president in 2001 after decades of leading the opposition to Suharto's notoriously corrupt 31 year reign. And in Burma, opposition to the military dictatorship has been led by a woman, Aung San Syu Ki, for more than two decades. All of these women were wives or daughters of powerful political leaders, which raises some questions of nepotism. However, such criticisms cannot account for the several countries where the percentage of female national legislators is well above the world average of 18 percent: Vietnam (26 percent), Laos (25 percent), Singapore (24 percent), the Philippines (20 percent), and Cambodia (19 percent).

Despite these successes in politics, women still lag well behind men in economic empowerment. Men have a higher rate of employment outside the home than women throughout the region (Table 10.3, column IV). It should be noted, however, that these data do not include all the work that women do in the home, in the informal economy, and in

Table 10.3 Gender comparisons for Southeast Asia and the United States: income, education level, and labor force participation

Country (I)	Estimated earned income (in U.S.$, adjusted for PPP), 2003 (II)		Eligible students enrolled inpost-secondary education (percent), 2002/2003 (III)		Labor force participation outside the home (percent), 2003 (IV)	
	Female	Male	Female	Male	Female	Male
Brunei	11,716	26,122	17	10	51	80
Burma	1,011[a]	1,389	15	9	66	88
Cambodia	1,807	2,368	2	5	80	83
Indonesia	2,289	4,434	15	19	56	82
Laos	1,391	2,129	4	9	75	82
Malaysia	6,075	12,869	33	26	49	88
Philippines	3,213	5,409	34	27	50	79
Singapore	16,489	32,089	ND[b]	ND	50	81
Thailand	5,784	9,452	42	36	73	78
Vietnam	2,026	2,964	9	11	73	86
United States	29,017	46,456	96	70	60	72

[a]Data from *CIA World Factbook*, 2005.

[b]ND = no data

Source: United Nations Human Development Report 2005 (New York: United Nations Development Programme), Table 25, Table 27, and Table 28.

traditional agriculture. When women do work in the formal economy, their wages average only about one-half to two-thirds those of men.

Changes may be on the way. In Brunei, Burma, Malaysia, the Philippines, and Thailand, significantly more women than men are completing training beyond secondary school (Table 10.3, column III). Hence, if training qualifications were the sole consideration for employment, women would appear to have an advantage over men. This advantage may be significant if service sector economies, which generally require more education, become dominant in more countries. The service economy already dominates in the Philippines and Singapore.

Reflections on Southeast Asia

Southeast Asia is often held up as a model for other developing regions for its "miraculous" economic development. While many crises have accompanied this success, there are hopeful signs that at least some problems are being resolved. Indeed, the strategic globalization that provided the foundation for Southeast Asia's economic growth may provide some guidance as this region navigates an uncertain future.

While the economic success of this region is indeed remarkable, so are the environmental and social crises that have accompanied it. The enormous amounts of carbon released into the atmosphere by deforestation have put this region at the center of efforts to deal with global warming. Many cities are also facing crises as changes in rural food production systems, combined with the growth of employment in urban-based, export-oriented manufacturing have created a massive tide of rural-to-urban migrants. The region's largest cities have found their infrastructure swamped by ever increasing demands for water, housing, and sanitation.

Yet in the face of these massive problems, there have been many positive changes. Population growth is slowing due to urbanization, economic growth, and the success of some government policies to reduce birth rates. Despite the continuation of military dictatorships or undemocratic rule in a number of countries, democracy is strengthening throughout the region. The bright side of the economic crisis of the 1990s was that it exposed corruption and brought reform to both Thailand and Indonesia. Even on issues of gender, there are signs of change. Women's access to education and their political empowerment are increasing, and greater attention is being paid to the abuse of women in the region's huge sex industry.

Looking toward the future, China looms ever larger in trading relationships of this region. When the China–Southeast Asian free trade agreement comes into effect in 2010, this region will be under greater pressure to strengthen the areas where it can have an edge over China—high-value, technologically sophisticated industries. Otherwise, Southeast Asia will be stuck playing second fiddle to its much larger neighbor, able to attract investment only when wages and other productions costs in China rise beyond those in Southeast Asia. And yet, trade with China also offers many opportunities, and Southeast Asian companies are already profiting from some of them. Whether the majority of Southeast Asians can similarly benefit from China's growth is a key question for this region's future.

Chapter Key Terms

archipelago 369
Association of Southeast Asian
  Nations (ASEAN) 388
Australo-Melanesians 377
Austronesians 378
coral bleaching 375

crony capitalism 388
cultural pluralism 394
detritus 371
export processing zones (EPZs) 386
feminization of labor 387
foreign exchange 384

pull factors 383
push factors 383
resettlement schemes 391
sex tourism 395
wet (paddy) rice cultivation 374

Critical Thinking Questions

1. What do you think are the most serious threats to Southeast Asia posed by global warming? What do you think are the most promising responses to global warming emerging from this region?

2. What countries in the region have achieved the greatest success in controlling population growth? How do population issues differ among Southeast Asian countries? How do the population issues of the region as a whole compare to those of Europe or Africa?

3. How is urbanization influencing gender roles in families?

4. What changes in agriculture have produced migration toward cities?

5. In what ways did state aid to market economies and export-led growth amount to a strategic and limited embrace of globalization?

6. How did the economic crisis of the 1990s result in expanded democracy in some countries? What countries in this region are the least democratic? Why do you think that is the case?

7. Describe the spatial distribution of major religious traditions of this region. How have these religious traditions influenced each other over time?

8. What factors have produced major flows of refugees in Southeast Asia?

9. Gender roles and mating relationships in traditional Southeast Asian families are not necessarily what an outsider might expect. What are some of the characteristics that interested you most? Why? Discuss some of the ways in which gender roles vary from those in South Asia or East Asia.

10. Why are the Overseas Chinese resented by some people in this region?

Map labels (left page):

Kauai
Oahu
Molokai
Honolulu · Maui
Hawaii
(U.S.) · Hawaii

Line
Islands

Kiritimati

P o l y n e s i a

Marquesas
Islands

Cook Islands
(New Zealand) · Papeete · Tahiti
French Polynesia
(France)

Avarua
Rarotonga

Pitcairn
Islands
(U.K.)

SOUTH PACIFIC
OCEAN

30°N
20°N
10°N
0° Equator
10°S
20°S
30°S
40°S
50°S

Land Elevations		Ocean Depths	
meters	feet	meters	feet
		0	0
4877	16,000	300	984
3353	11,000	3500	11,483
2134	7000	5000	16,404
914	3000		
305	1000		
152	500		
0	0		

mi 0   200   400   600   800
km 0   200   400   600   800   1000   1200

1:42,000,000
Mercator Projection

CHAPTER 11

Oceania: Australia, New Zealand, and the Pacific

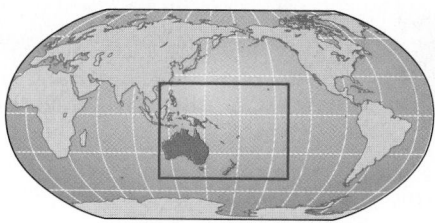

Global Patterns, Local Lives In October 2003 in Brisbane, Australia, a crowd of 47,000 waits eagerly as New Zealand's national rugby team takes the field opposite the team from Tonga, a Polynesian archipelago in the South Pacific (Figure 11.1). The famous Haka is about to begin.

The Haka is a highly emotional and physical dance traditionally performed by the **Maori,** the indigenous people of New Zealand, to motivate fellow warriors and intimidate opponents before entering battle. Dances like this have long been a part of many cultures in the islands of Oceania, but the Haka has now become an integral part of rugby, the region's most popular sport. Before almost every international match for the past century, the New Zealand team, The All Blacks, has performed the Haka, chanting, screaming, jumping, stomping their feet, poking out their tongues, widening their eyes to show the whites, and beating their thighs, arms, and chests in unison (Figure 11.2). Until now, only the New Zealand team performed the Haka, but tonight is a different story. Halfway through New Zealand's Haka, the Tongan team responds with its own Haka. The crowd roars its approval of this scene, a revival of the traditional prelude to battle throughout Oceania's long history.

To see videos of a Haka, go to http://www.Youtube.com and type in "haka."

Adapted from an article by Phil Wilkins, "Tonga can only match the Kiwis in the Haka," Brisbane, Australia, October 25, 2003.

The Haka is an example of how indigenous culture in Oceania is being revived and celebrated. At the same time, the fact that a Maori war dance is being performed before a game

Figure 11.1 Regional map of Oceania.

Figure 11.2 The Haka, a Maori tradition. The fierceness of the Haka challenge is visible on the faces of the New Zeland All Blacks rugby team as they face the Australian Wallabies in Christchurch, New Zealand, in 2006. [AP Photo/NZPA, Pool.]

of rugby, which was brought to Oceania by the British (Figure 11.3) shows how deeply globalization has penetrated this region. Oceania, which comprises Australia, New Zealand, Papua New Guinea, and the myriad Pacific islands (Figure 11.4), was dominated politically and economically by people of European descent for more than 400 years. Now, Oceania finds itself subject to a new wave of globalization emanating from Asia, which has far-reaching economic implications.

And yet, the Haka itself seems to be a globalizing phenomenon. After Tonga took on the New Zealanders in 2003, the Haka spread quickly to other sports and places, some of them far beyond the South Pacific. The University of Hawaii's football team regularly performs a Haka before its games, as does Brigham Young University in Utah and a number of college football teams in Mexico. Even a high school football team in Trinity, Texas, which has many players of Tongan

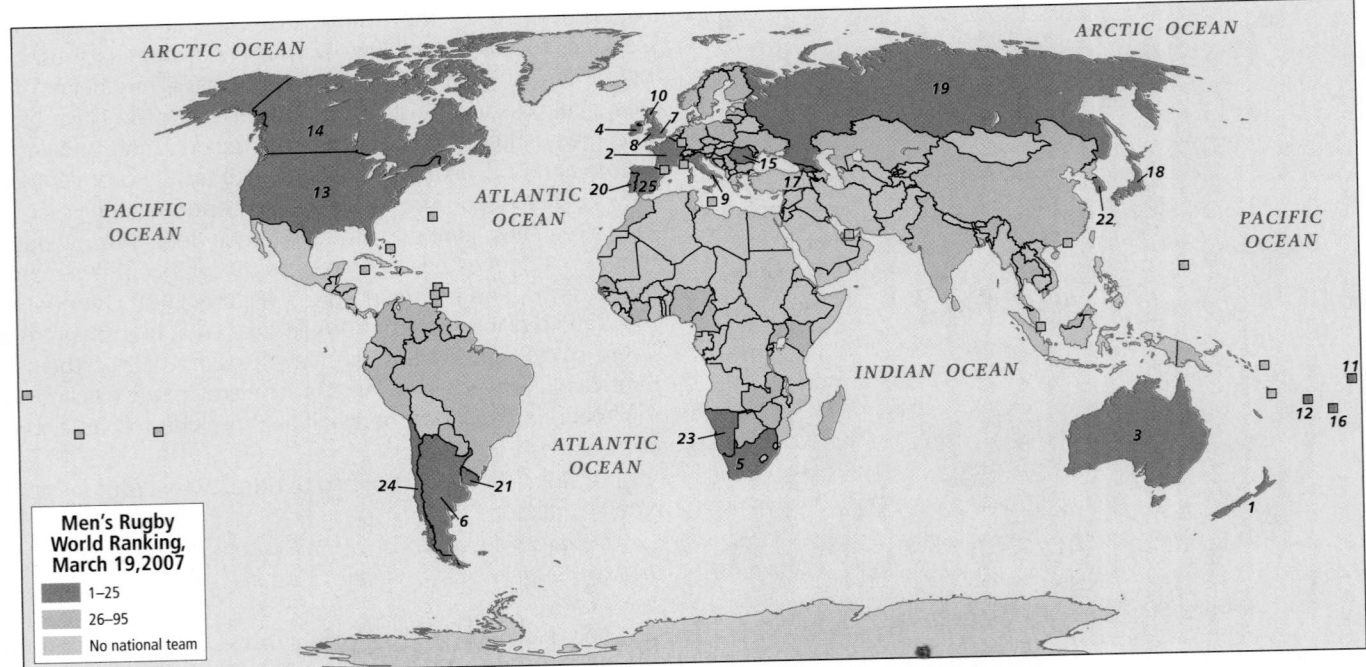

Figure 11.3 Rugby around the world. Rugby is played by women, men, boys, and girls in over 100 countries. The Men's World Cup rugby competition began in 1987, when it was won by New Zealand. Rankings are based on a system of points for match results, relative team strength, and margin of victory, and there is an allowance for home field advantage. Rankings can change weekly. [For additional information, see http://www.irb.com/EN/IRB+Organisation/.]

Northern Mariana
Islands
(U.S.)

Hawai
(U.S.)

Guam
(U.S.)

MARSHALL
ISLANDS

PALAU

FEDERATED STATES OF MICRONESIA

KIRIBATI

PAPUA NEW GUINEA

NAURU

TUVALU

SOLOMON
ISLANDS

Tokelau
(New Zealand)

Coral Sea
Islands
(Australia)

Wallis &
Futuna
(France)

SAMOA

VANUATU

American
Samoa
(U.S.)

Cook Islands
(New Zealand)

French Polynesia
(France)

FIJI

New
Caledonia
(France)

TONGA

AUSTRALIA

Norfolk
Island
(Australia)

Pitcairn
Islands
(U.K.)

NEW ZEALAND

Figure 11.4 Political map of Oceania. Thirteen of these entities are independent countries; the others are territories of or are otherwise affiliated with other nations. Not depicted on the map are the Pacific Island Wildlife Refuges, a widely scattered, essentially uninhabited group of northern Pacific islands that constitute a U.S. territory.

descent, performs a Haka before its games. Recently, the global sports company Adidas made the All Blacks' Haka the center of a worldwide ad campaign, prompting at least one Maori community to sue them for using their cultural traditions for commercial gain.

Questions to consider as you study Oceania. (Also see Views of Oceania, pages 404–405.)

1. How has globalization transformed Oceania? Europe and the United States have both exerted a powerful influence on Oceania. Now, trade and social interaction with Asia are increasing. What changes are resulting from Oceania's reorientation toward Asia?

2. What impact will global warming have on Oceania, particularly in relation to water? Sea level rise threatens some Pacific islands with submersion, and water supplies throughout the region may be strained as rainfall patterns change. What other major changes are likely to occur with global warming?

3. How is urbanization changing Oceania's population? In some parts of this region the shift to the cities has been accompanied by broad-based affluence as expanding urban service economies bring decent jobs to more people. Where has urbanization resulted in less positive change?

4. In what ways have European systems for producing food and fiber changed environments in Oceania? Oceania has been transformed by plants and animals brought to the region as part of European farming and herding systems. How have these new systems impacted the native species and landscapes of Oceania?

5. How are issues of gender, democracy, and economic empowerment developing differently across Oceania? Women's access to jobs and elected office varies dramatically across this region. Where have women gained the most political and economic empowerment?

Views of Oceania

For a tour of these places, visit www.whfreeman.com/pulsipher

[Background image: Trent Schindler NASA/GSFC Code 610.3/587 Scientific Visualization Studio]

1 & 2. Population. Two population patterns exist in Oceania. Australia, New Zealand, and Hawaii have wealthy, highly urbanized, and slower growing populations, while Papua New Guinea and many Pacific islands are much poorer, rural, and fast growing. Shown here are Melbourne, Australia (1), and the Solomon Islands (2). See pages 419–421 for more. Lat 37°46'48"S, Lon 144°59'56"E and Lat 8°06'19"S, Lon 156°49'26"E. [Batterbu: http://commons.wikimedia.org/wiki/File:Cafes.jpg; DOD]

Solomon Islands

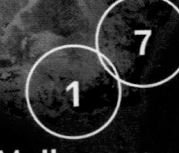

Sydney

7

1

Melbourne

7. Urbanization. Vibrant urban-based service economies have expanded to employ three quarters of the population in Australia and New Zealand. Shown here is Sydney, Australia. See page 419–421 for more.
Lat 33°51'20"S, Lon 151°13'E.
[Niagara Falls Chief Engineer Kevin Serbello]

3. Globalization and the environment. Hundreds of nuclear weapons have been tested by the United States and France in Oceania. Health impacts on the environment and Pacific islanders have been significant. Shown here is a 1954 nuclear test in the Marshall Islands See page 414–416 for more. Lat 11°36'N, Lon 165°23'E. [National Nuclear Security Administration/Nevada Site Office]

4. Food production. Agricultural expansion and associated deforestation have resulted in the extinction of many species found only in Oceania. Agriculture has transformed Hawaii (shown here), giving it more threatened or endangered species than any other U.S. state. See page 413–414 for more. Lat 22°12'26"N, Lon 159°28'22"W. [gh5046]

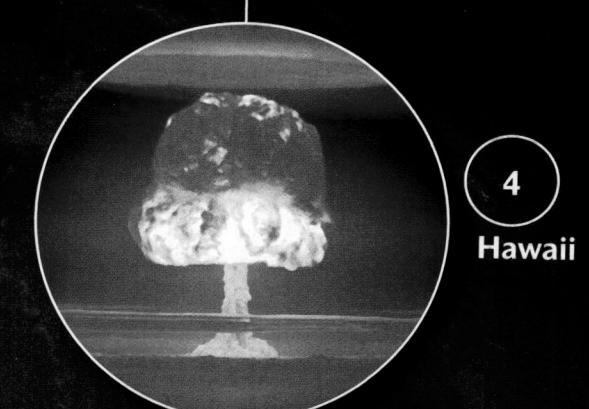

**3
Marshall
Islands**

**4
Hawaii**

**5
American
Samoa**

5. Global warming and water. Global warming is causing the oceans to warm, threatening coral reefs. Fish that live on coral reefs are a major food source for Pacific islanders. Shown here is a reef off of American Samoa. See page 410–413 for more. Lat 14°10'21"S, Lon 169°39'33"W. [Karen Koltes, PhD—Office of Insular Affairs]

**6
Wellington**

6. Gender and democracy. New Zealand was among the first countries to grant women the right to vote, and has elected two female prime ministers. Shown here is former prime minister Helen Clark in the capitol, Wellington. See page 429–431 for more. Lat 41°16'42"S, Lon 174°46'36"E. [DoD photo by R. D. Ward]

The Geographic Setting

Terms to Be Aware Of

You may see different place names in similar locations on some maps in this chapter. This reflects the political evolution of this region, where some islands together form a single country but are also sometimes identified by their individual names on maps. For example, the Caroline Islands, located north of New Guinea, still go by that name on maps and charts, but they have been divided into two countries: Palau (a small group of islands at the western end of the Caroline Islands) and the Federated States of Micronesia, which extends over 2000 miles west to east, from Yap to Kosrae. In addition, there are the three larger island groupings—Micronesia, Melanesia, and Polynesia—that are not political units, but are based on ethnic and cultural links.

PHYSICAL PATTERNS

The huge expanses of the Pacific Ocean in Oceania make it the largest world region in terms of total area. The Pacific is both a link and a barrier in this region. For some living things, the Pacific links widely separated lands. Plants and animals have found their way from island to island by floating on the water or swimming, and humans have long used the ocean as a way to make contact with other peoples by visiting, trading, and raiding. Sea life is a source of food, and today it is also a source of income as catches are sold in the global market. The movement of the water and its varying temperatures influence climates on all the region's landmasses. But the wide expanses of water have also profoundly limited the natural diffusion of plant and animal species and kept Pacific Islanders relatively isolated from one another. The vast ocean has imposed solitude and fostered self-sufficiency and subsistence economies well into the modern era.

Continent Formation

The largest landmass in Oceania is the ancient continent of Australia at the southwestern perimeter of the region (see Figure 11.1). The Australian continent is partially composed of some of the oldest rock on earth and has been relatively stable for more than 200 million years, with very little volcanic activity and only an occasional mild earthquake. Australia was once a part of the great landmass, called **Gondwana,** that formed the southern part of the ancient supercontinent Pangaea (see Figure 1.21 on page 25). What became present-day Australia broke free from Gondwana and drifted until it eventually collided with the part of the Eurasian Plate on which Southeast Asia sits. That impact created the mountainous island of New Guinea to the north of Australia.

Australia is shaped roughly like a dinner plate with a lumpy, irregular rim and two bites taken out of it: one in the north (the Gulf of Carpentaria) and one in the south (the Great Australian Bight). The center of the plate is the great lowland Australian desert. The lumpy rim is composed of uplands; the Eastern Highlands (labeled "Great Dividing Range" in Figure 11.1) are the highest and most complex of these. Over millennia, the forces of erosion—both wind and water—have worn Australia's landforms into low, rounded formations, some quite spectacular (Figure 11.5).

Off the northeastern coast of the continent lies the **Great Barrier Reef,** the largest coral reef in the world and a World Heritage Site since 1981. It stretches in an irregular arc for more than 1250 miles (2000 kilometers) along the coast of Queensland, covering 135,000 square miles (350,000 square kilometers). The Great Barrier Reef is so large that it influences Australia's climate. It interrupts the westward-flowing ocean currents in the mid–South Pacific circulation pattern,

Figure 11.5 Uluru (also known as Ayers Rock). This land formation is a smooth remnant of ancient mountains. The site is held sacred by central Australian Aborigines. It is also among Australia's most popular tourist destinations. [Torsten Blackwood/AFP/Getty Images.]

shunting warm water to the south, where it warms the southeastern coast of Australia. Threats to the health of the Great Barrier Reef are discussed on page 412.

Island Formation

The islands of the Pacific were created (and are being created still) by a variety of processes related to the movement of tectonic plates. The islands found in the western reaches of Oceania—including New Guinea, New Caledonia, and the main islands of Fiji—are remnants of the Gondwana landmass; they are large, mountainous, and geologically complex. Other islands in the region are volcanic in origin and form part of the Ring of Fire (see Figure 1.22 on page 26). Many of this latter group are situated in boundary zones where tectonic plates are either colliding or pulling apart. For example, the Mariana Islands east of the Philippines are volcanoes that were formed when the Pacific Plate plunged beneath the Philippine Plate. The two much larger islands of New Zealand were created when the eastern edge of the Indian-Australian Plate was thrust upward by its convergence with the Pacific Plate.

The Hawaiian Islands were produced through another form of volcanic activity associated with **hot spots,** places where particularly hot magma moving upward from Earth's core breaches the crust in tall plumes. Over the past 80 million years, the Pacific Plate has moved across one of these hot spots, creating a string of volcanoes 3600 miles (5800 kilometers) long. The youngest volcanoes, only a few of which are active, are on or near the islands known as Hawaii.

Volcanic islands exist in three forms: volcanic high islands, low coral atolls, and raised or uplifted coral platforms known as *makatea*. *High islands* are usually volcanoes that rise above the sea into mountainous, rocky formations that contain a rich variety of environments. New Zealand, the Hawaiian Islands, Tahiti, and Easter Island are examples of high islands. An **atoll** is a low-lying island, or chain of islets, formed of coral reefs that have built up on the circular or oval rim of a submerged volcano (Figure 11.6). These reefs are arranged around a central lagoon that was once the volcano's crater. As a consequence of their low elevation, atoll islands tend to have only a small range of environments and very limited supplies of fresh water.

Climate

Although the Pacific Ocean stretches nearly from pole to pole, most of Oceania is situated within the tropical and subtropical latitudes of that ocean. The tepid water temperatures of the central Pacific bring mild climates year-round to nearly all the inhabited parts of the region (Figure 11.7). The seasonal variation in temperature is greatest in the southernmost reaches of Australia and New Zealand.

Moisture and rainfall. With the exception of the vast arid interior of Australia, much of Oceania is warm and humid nearly all the time. New Zealand and the high islands of the

Figure 11.6 An atoll in the Tuamotu Archipelago of French Polynesia. As is the case here, the land area of an atoll is usually not continuous, but instead forms a sort of necklace of flat islets around a central lagoon. Often, the necklace surrounds one or more islands that are the remnants of the old volcanic core. [David Doubilet.]

Pacific receive copious rainfall and once supported dense forest vegetation, although much of that forest is gone after 1000 years of human impact (see Figure 11.10b on page 411). Travelers approaching New Zealand, either by air or by sea, sometimes notice a distinctive long, white cloud that stretches above the two islands. A thousand years ago, the Maori settlers also noticed this phenomenon, and they named the place *Aotearoa*, "land of the long white cloud."

The distinctive mass of moisture is brought in by the legendary **roaring forties** (named for the 40th parallel south), powerful air and ocean currents that speed around the far Southern Hemisphere virtually unimpeded by landmasses. These westerly winds (blowing west to east) deposit a drenching 130 inches (330 centimeters) of rain per year in the New Zealand highlands and more than 30 inches (76 centimeters) per year on the coastal lowlands. At the southern tip of New Zealand's North Island, the wind averages more than 40 miles per hour (64 kilometers per hour) about 118 days a year. Cabbages have to be staked to the ground or they will blow away.

By contrast, two-thirds of the continent of Australia is overwhelmingly dry. The Great Dividing Range blocks the movement of moist easterly winds (blowing east to west), so rain does not reach the interior. As a result, a large portion of Australia receives less than 20 inches (50 centimeters) of rain per year, and humans have found rather limited uses for this territory. But the eastern (windward) slopes of the highlands receive more abundant moisture. This relatively moist eastern rim of Australia was favored as a habitat by both the indigenous people and the Europeans who displaced them after 1800. During the southern summer, the fringes of the monsoon that passes over Southeast Asia and Eurasia (see Figure 10.7 on page 371) bring moisture across Australia's northern coast. There, annual rainfall varies from 20 to 80 inches (50 to 200 centimeters).

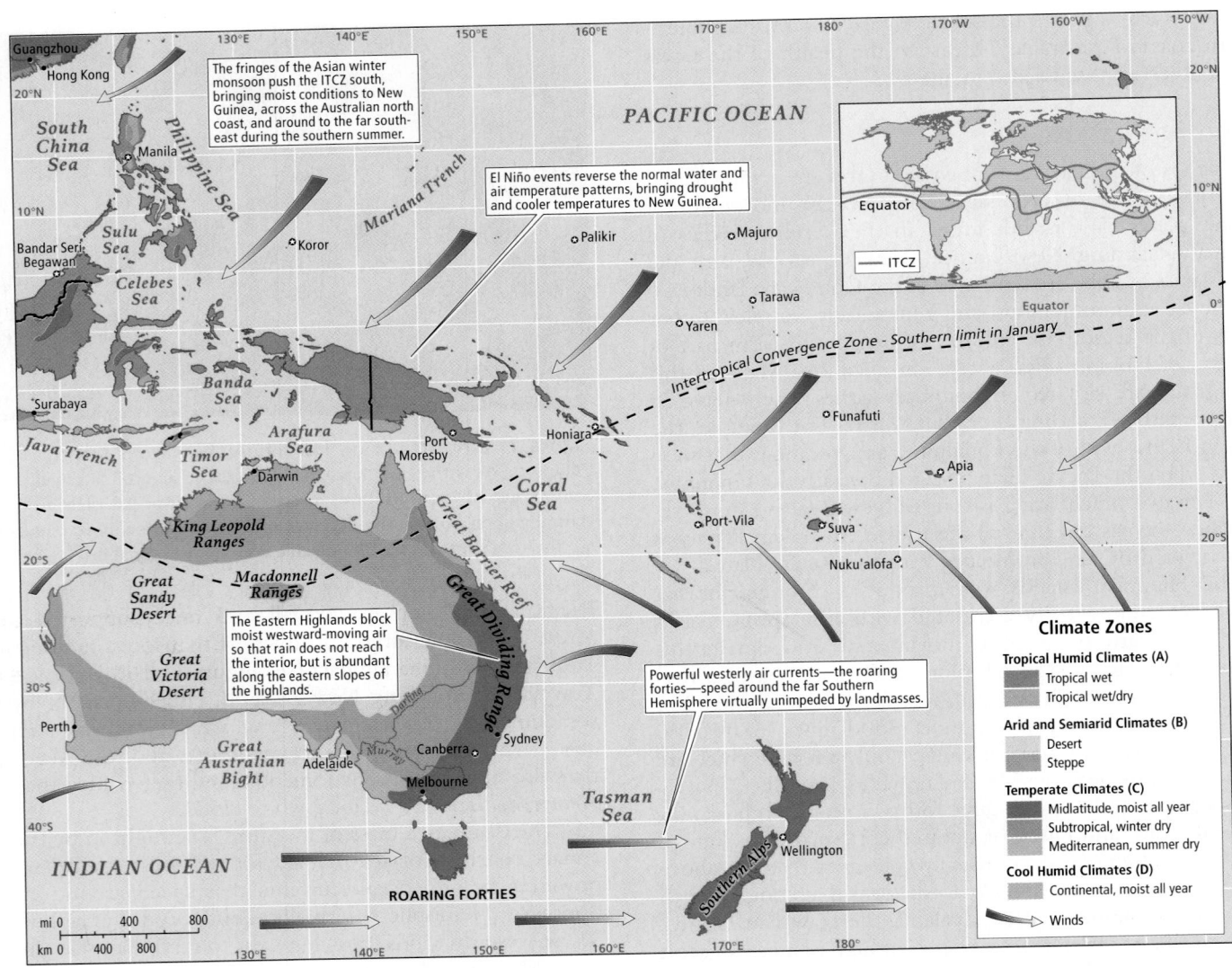

Figure 11.7 Climates of Oceania.

Overall, Australia is so arid that it has only one major river system, which is in the temperate southeast where most Australians live. There, the Darling and Murray rivers drain one-seventh of the continent, flowing into the Indian Ocean at Adelaide. One measure of the overall dryness of Australia is that the entire average *annual* flow of the Murray-Darling river system is equal to just one day's average flow of the Amazon in Brazil.

Many of the mountainous high islands in the Pacific exhibit orographic rainfall patterns, with a wet windward side and a dry leeward side (see Figure 1.24 on page 28). Low-lying islands across the region vary considerably in the amount of rainfall they receive. Some of these islands lie directly in the path of trade winds, which deliver between 60 and 120 inches of rain per year on average. These islands support a remarkable variety of plants and animals. Others, particularly those

on the equator, receive considerably less rainfall and are dominated by grasslands that support little animal life.

El Niño. In Chapter 3, we introduced the **El Niño** phenomenon, a pattern of changes in the circulation of air and water in the Pacific that occurs irregularly every 2 to 7 years. Although these cyclical changes, or oscillations, are not yet well understood, scientists have worked out a model of how the oscillations may occur (Figure 11.8).

The El Niño event of 1997–1998 illustrates the effects of this phenomenon. By December of 1997, the island of New Guinea (above Australia; see Figure 11.1) had received very little rainfall for almost a year. Crops failed, springs and streams dried up, and fires broke out in tinder-dry forests. The cloudless sky allowed heat to radiate up and away from elevations above 7200 feet (2200 meters), so temperatures at high ele-

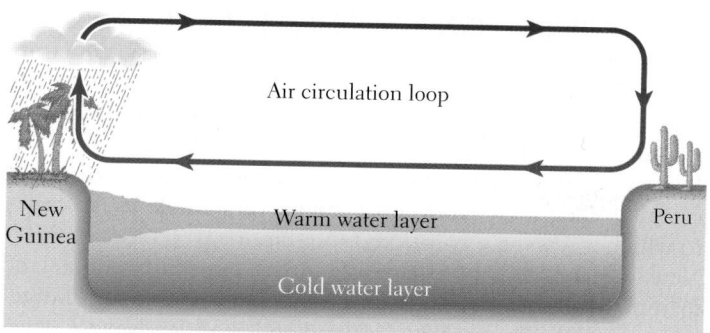

(a) Normal equatorial conditions. Water in the equatorial western Pacific (the New Guinea/Australia side) is warmer than water in the eastern Pacific (the Peru side). Due to prevailing wind patterns, the warm water piles up in the west. Warm air rises above this warm-water bulge in the western Pacific and forms rain clouds. The rising air cools and, once in the higher atmosphere, moves in an easterly direction. In the east, the dry cool air descends, bringing little rainfall to Peru.

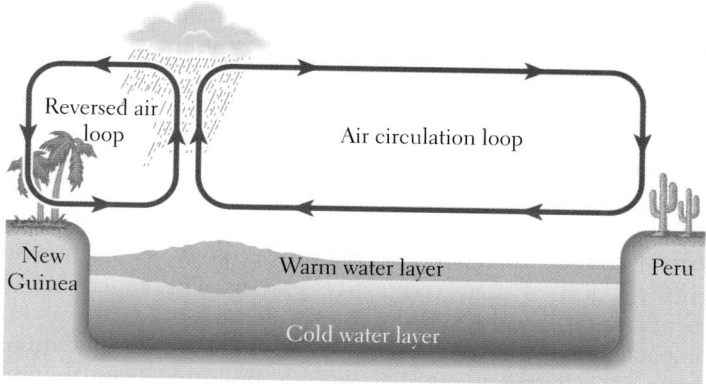

(b) Developing El Niño conditions. As an El Niño event develops, the ocean surface's warm-water bulge (orange) begins to move east. The air rising above it splits into two formations, one circulating east to west in the upper atmosphere and one west to east.

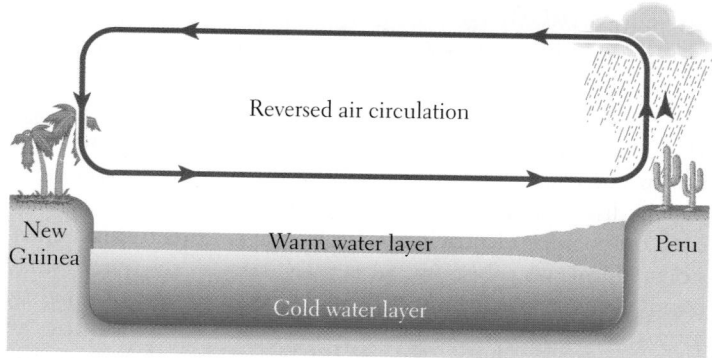

(c) Fully developed El Niño. Slowly, as the bulge of warm water at the surface of the ocean moves east, it forces the whole system into the fully developed El Niño with air at the water surface and in the upper atmosphere flowing in reverse of normal (**a**). Instead of warm, wet air rising over the mountains of New Guinea and condensing as rainfall, cool, dry, cloudless air descends to sit at the earth's surface. Meanwhile, in the east, the normally dry, clear coast of Peru experiences clouds and rainfall.

Figure 11.8 A model of the El Niño phenomenon. [Adapted from Environmental Dynamics Research, Inc., 1998; Ivan Cheung, George Washington University, Geography 137, Lecture 16, October 29, 2001, at http://www.gwu.edu/~/geog137/download/lecture16.ppt.]

vations dipped below freezing at night for stretches of a week or more. Tropical plants died, and people unaccustomed to chilly weather sickened. Meanwhile, along the Pacific coasts of North, Central, and South America, the warmer-than-usual weather brought unusually strong storms, high ocean surges, and damaging wind and rainfall. Recently, an opposite pattern, in which normal conditions become unusually strong, has been identified and named La Niña, though scientists have barely begun to study it.

Fauna and Flora

The fact that Oceania comprises an isolated continent and numerous islands has had a special effect on its animal life (*fauna*) and plant life (*flora*). Many of its species are **endemic,** meaning they exist in a particular place and nowhere else on earth. This is especially true of Australia, but many Pacific islands also have endemic species.

Plant and animal life in Australia. The uniqueness of Australia's plant and animal life is the result of the continent's long physical isolation, large size, relatively homogeneous landforms, and arid climate. Since Australia broke away from Gondwana more than 65 million years ago, its plant and animal species have evolved in isolation. One spectacular result of this long isolation is the presence of more than 144 living species of endemic marsupial animals. **Marsupials** are mammals that give birth to their young at a very immature stage and then nurture them in a pouch equipped with nipples. The best-known marsupials are the kangaroos; other species include wombats, the koala, and bandicoots. The various marsupials fill ecological niches that in other regions of the world are occupied by rats, badgers, moles, cats, wolves, ungulates (grazers), and bears. The **monotremes,** egg-laying mammals that include the duck-billed platypus and the spiny anteater, are endemic to Australia and New Guinea. Some of the 750 species of birds known in Australia migrate in and out, but more than 325 species are endemic.

> The uniqueness of Australia's plant and animal life is the result of the continent's long physical isolation, large size, relatively homogeneous landforms, and arid climate.

Most of Australia's endemic plant species are adapted to dry conditions. Many of the plants have deep taproots to draw moisture from groundwater and small, hard, shiny leaves to reflect heat and to hold moisture. Much of the continent is grassland and scrubland with bits of open woodland; there

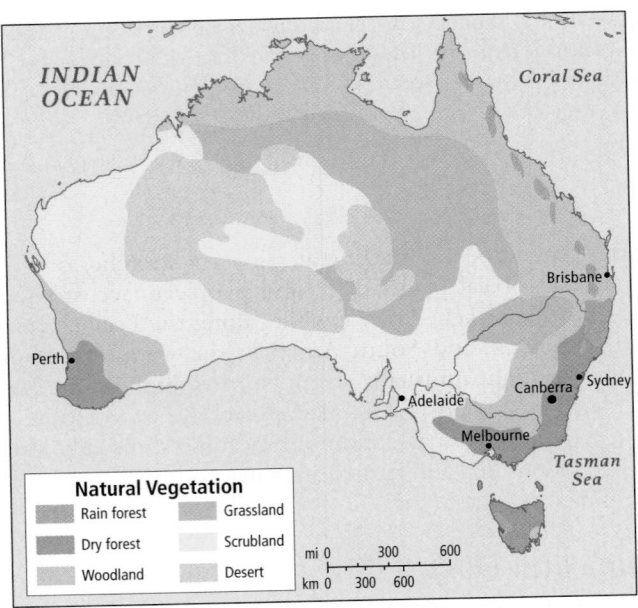

Figure 11.9 Australia's natural vegetation. Much of Australia is grassland and scrubland; a few forests can be found in the Eastern Highlands, in the far southwest, and in Tasmania. [Adapted from Tom L. McKnight, *Oceania* (Englewood Cliffs, N.J.: Prentice Hall, 1995), p. 28.]

are only a few true forests, found in pockets along the Eastern Highlands, the southwestern tip, and in Tasmania (Figure 11.9). Two plant genera account for nearly all the forest and woodland plants: *Eucalyptus* (450 species, often called "gum trees") and *Acacia* (900 species, often called "wattles").

Plant and animal life in New Zealand and the Pacific islands. Naturalists and evolutionary biologists have long been interested in the species that inhabited the Pacific islands before humans arrived. Charles Darwin formulated many of his ideas about evolution after visiting the Galàpagos Islands of the eastern Pacific (see Figure 3.1 on pages 94–95) and the islands of Oceania.

Islands gain plant and animal populations from the sea and air around them as organisms are carried from larger islands and continents by birds, storms, or ocean currents. Once these organisms "colonize" their new home, they may evolve over time into new species that are unique to one island. High, wet islands generally contain more varied species because their more complex environments provide niches for a wider range of wayfarers and thus greater opportunities for evolutionary change.

The flora and fauna of islands are also modified by human inhabitants once they arrive. In prehistoric times, Asian explorers in oceangoing sailing canoes brought plants such as bananas and breadfruit ▆, and animals such as pigs, chickens, and dogs. Today, human activities from tourism to military exercises to urbanization continue to change the flora and fauna of Oceania.

Generally, the diversity of land animals and plants is richest in the western Pacific, near the larger landmasses. It thins out to the east, where the islands are smaller and farther apart. The natural rain forest flora is rich and abundant on New Zealand, New Guinea, and the high islands of the Pacific. However, the natural fauna is much more limited on these islands. While New Guinea has fauna comparable to Australia, to which it was once connected (see Figure 10.6 on page 370), New Zealand and the Pacific islands have no indigenous land mammals, almost no indigenous reptiles, and only a few indigenous species of frogs. New Zealand and the islands were never connected to Australia and New Guinea by a land bridge that land animals could cross. On the other hand, indigenous birds have been numerous and varied. Two examples are New Zealand's kiwi and the huge moa (a bird that grew up to 12 feet [3.7 meters] tall) and was a major source of food for the Maori people until they hunted it to extinction before Europeans arrived. Today, New Zealand may well be the country with the most introduced species of mammals, fish, and fowl, nearly all brought in by European settlers.

ENVIRONMENTAL ISSUES

Despite Oceania's relatively small population, there is great public awareness of environmental issues. Oceania faces a host of environmental problems. Global warming has brought a broad array of threats to the region's ecology, as has human introduction of many nonnative species. The expansion of herding, agriculture, and the human population itself has also impacted much of the region (Figure 11.10). Even remote areas have been subjected to intense pollution from mining and the testing of nuclear weapons, especially during the cold war era. Nevertheless, solutions to many of Oceania's environmental problems are emerging.

Global Warming

Oceania is a minor but significant contributor of greenhouse gases. Australia has some of the world's highest greenhouse gas emissions on a per capita basis (see Figure 1.16 on page 19). Much like the United States, its emissions result from the use of automobile-based transportation systems to connect a widely dispersed network of cities and towns. Heavy dependence on coal for electricity generation also leads to high emissions. However, because Australia has a relatively small population (21 million in 2007), it accounts for only slightly more than 1 percent of global emissions. Other countries in Oceania also contribute negligible amounts of greenhouse gases. Nevertheless, the region is highly vulnerable to the effects of global warming.

Sea level rise. As we have seen, global warming may raise sea levels by melting glaciers and ice caps (see Chapter 1, page 18). Obviously, this issue is of great concern to residents of islands that already barely rise above the waves (Figure 11.11).

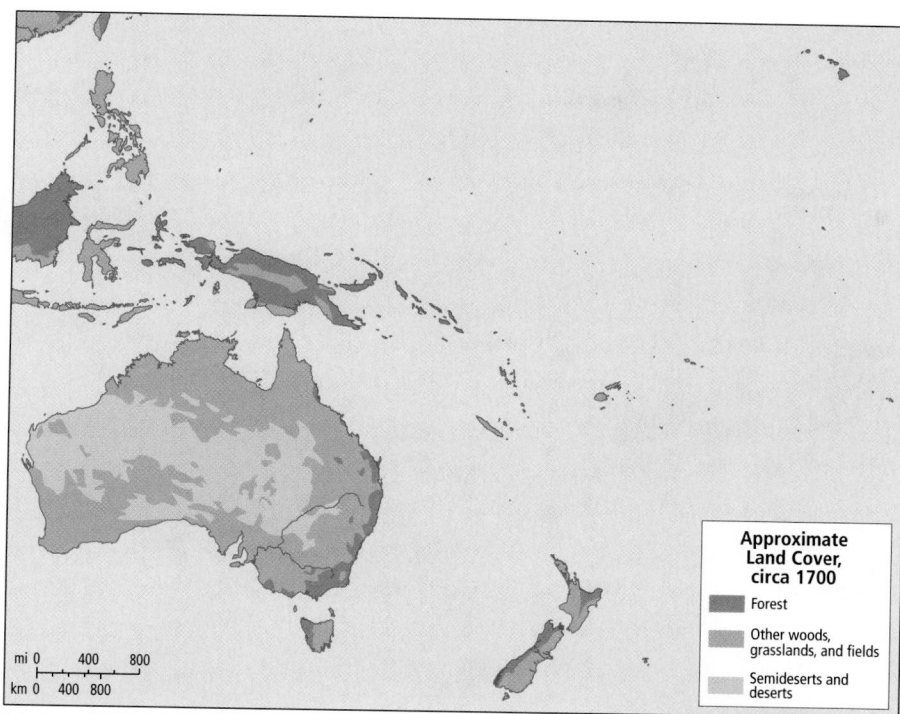

(a) circa 1700

Approximate Land Cover, circa 1700

- Forest
- Other woods, grasslands, and fields
- Semideserts and deserts

mi 0   400   800
km 0   400   800

Figure 11.10 Land cover (pre-European) and human impact (2002). Australia has been occupied by humans for at least 60,000 years, so its land cover was no doubt modified by 1700, before Europeans arrived **(a).** By 2002 **(b),** human impact was evident across Australia, but it was especially intense in the southeast. New Zealand was occupied by humans only a few thousand years ago, but human impact is now intense and the same can be said for many of the Pacific islands. [Adapted from *United Nations Environment Programme, 2002, 2003, 2004, 2005, 2006* (New York: United Nations Development Programme), at http://maps.grida.no/go/graphic/human_impact_year_1700_approximately and http://maps.grida.no/go/graphic/human_impact_year_2002.]

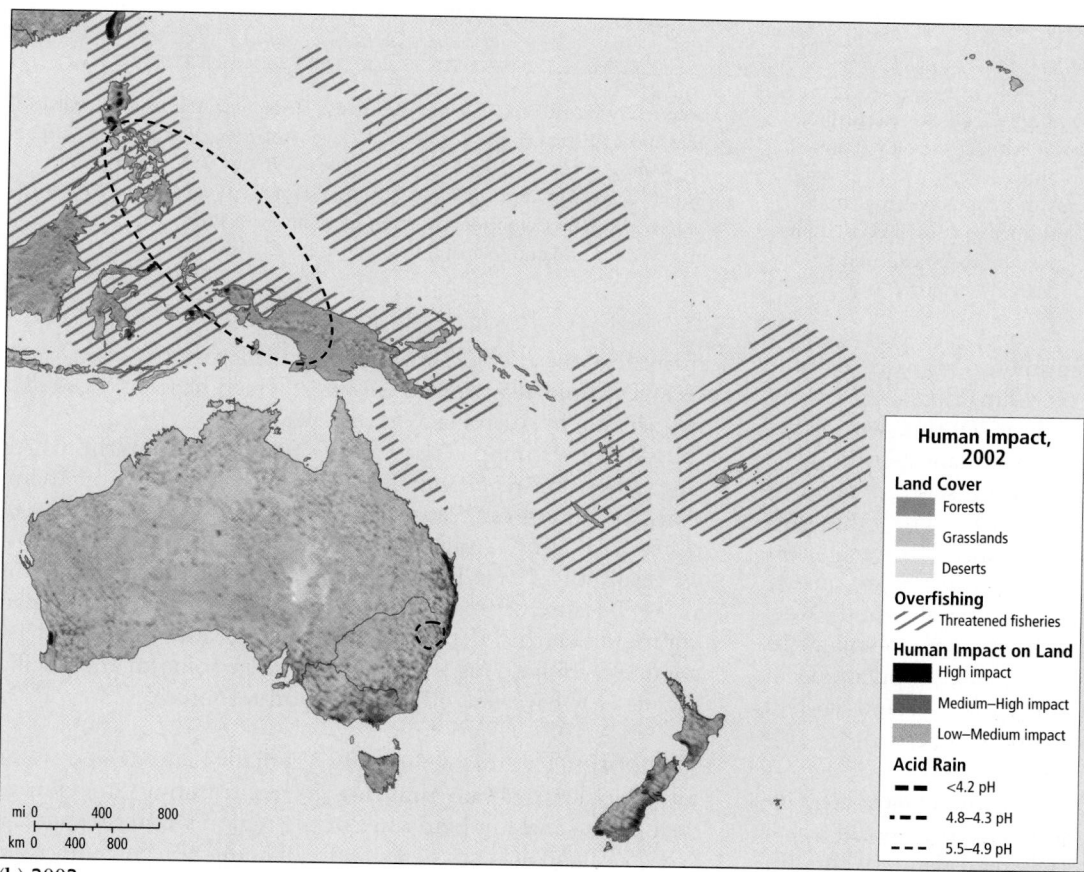

(b) 2002

Human Impact, 2002

Land Cover
- Forests
- Grasslands
- Deserts

Overfishing
- Threatened fisheries

Human Impact on Land
- High impact
- Medium–High impact
- Low–Medium impact

Acid Rain
- <4.2 pH
- 4.8–4.3 pH
- 5.5–4.9 pH

mi 0   400   800
km 0   400   800

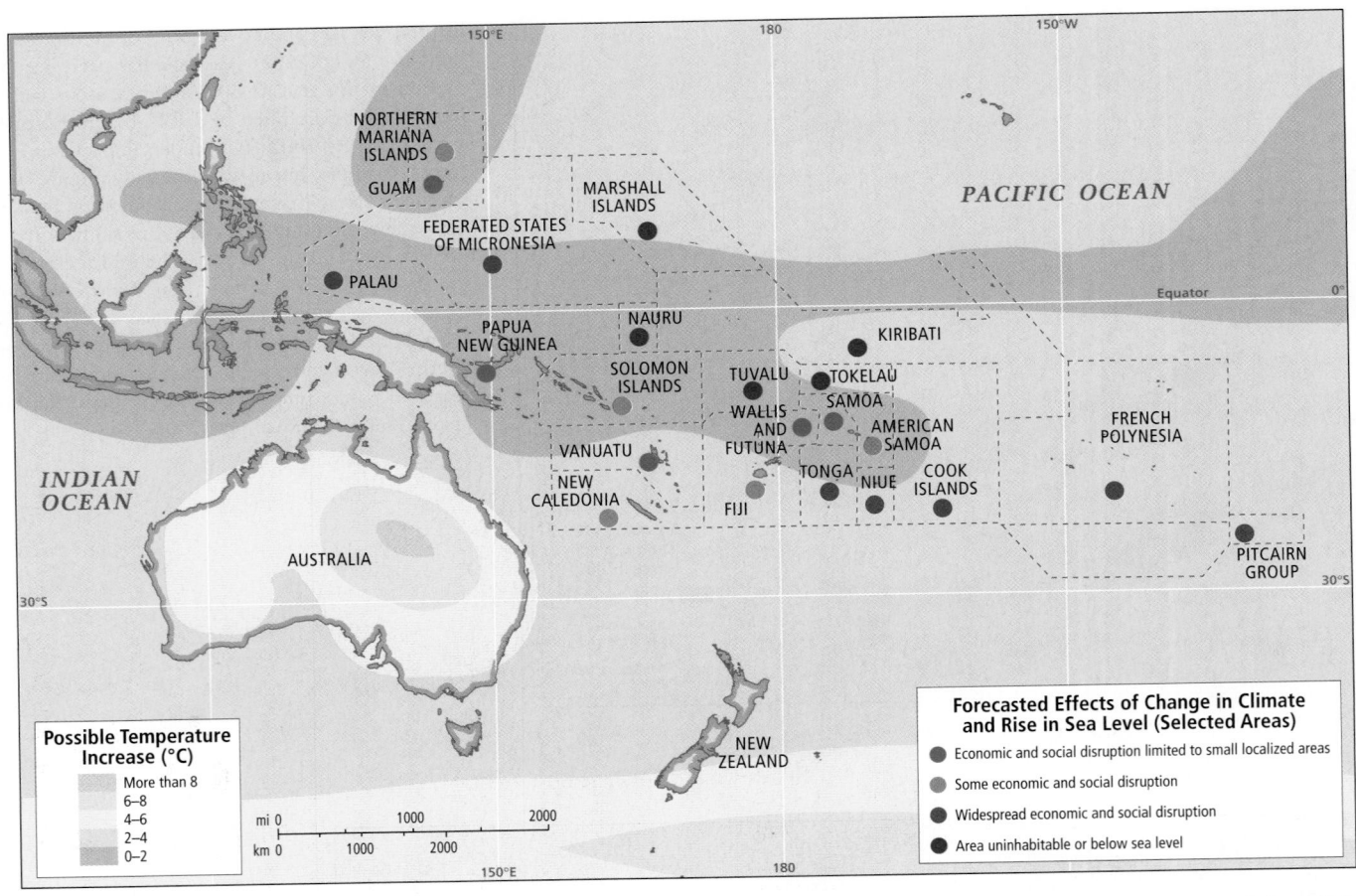

Figure 11.11 Global warming as it may affect Oceania. According to the Intergovernmental Panel on Climate Change, "Sea-level rise . . . has contributed to erosion of . . . beaches and barriers; loss of coastal dunes and wetlands; and drainage problems in many low-lying, mid-latitude coastal areas. Highly diverse and productive coastal ecosystems, coastal settlements, and island states will continue to be exposed to pressures whose impacts are expected to be largely negative and potentially disastrous in some instances." [Map adapted from Richard Nile and Christian Clerk, *Cultural Atlas of Australia, New Zealand, and the South Pacific* (New York: Facts on File, 1996), p. 223; quote from Robert T. Watson and the Core Writing Team, *Climate Change Report: Synthesis Report* (Cambridge: Cambridge University Press, 2001), at http://www.grida.no/climate/ipcc_tar/vol4/english/index.htm.]

If sea levels rise the 4 inches (10 centimeters) per decade predicted by the International Panel on Climate Change, many of the lowest-lying Pacific atolls, such as Tuvalu, will disappear under water within 50 years. Other islands, some with already very crowded coastal zones, will be severely reduced in area and will become more vulnerable to storm surges and cyclones. Some islanders are quietly migrating to Australia and New Zealand. Others are lobbying to resettle their entire populations in Australia or New Zealand should their islands be submerged.

> If sea levels rise the predicted 4 inches (10 centimeters) per decade, many of the lowest-lying atolls, such as Tuvalu, will disappear under water.

Other water-related vulnerabilities. Much of Oceania is vulnerable to changes in the region's climate that could result from global warming. Parts of Australia and many of the low Pacific islands are already experiencing water shortages, and many fear these could worsen if rainfall patterns alter. Such changes could also worsen wildfires, which have emerged as a major issue in Australia in recent years.

Global warming is also causing the oceans to warm, which threatens coral reefs and the fisheries that depend on them. Warmer waters can cause *coral bleaching* (see Chapter 10, pages 375–376), a phenomenon all reefs in this region, but especially the Great Barrier Reef, have experienced in recent years. Because so many fish depend on reefs, many human communities that depend on fishing are also threatened by coral bleaching. This is especially true in some of the Pacific islands that have few other local food resources.

Responses. Oceania is pursuing alternative energy and water technologies that are reducing its contributions to climate change and making better use of the region's water resources. New Zealand has set a goal of obtaining 95 percent of its energy from renewable sources by 2025. Much of this will come from wind power, for which this region has excellent potential, especially in areas near the "roaring forties." Geothermal energy (see Chapter 10, page 376) is either already being used or planned for future use throughout Oceania, with the exception of some of the low non-volcanic Pacific islands. Solar energy is widely used in many remote Pacific islands where the cost of importing fuel is prohibitive.

Oceania is also a world leader in the implementation of water technologies. Some are age-old methods that are simple but effective, such as harvesting rainwater from roofs and the ground itself for household use. Most buildings in rural Australia, New Zealand, and many Pacific islands get at least part of their water this way, taking pressure off surface and groundwater resources. Australia is now stretching its water resources further with extensive use of highly efficient drip irrigation technologies in agriculture (see Chapter 6, page 213) and new low-cost water-filtration techniques. 📹

Invasive Species and Food Production

The many unique endemic plants and animals of Oceania have been displaced by **invasive species,** organisms that spread into regions outside of their native range, adversely affecting economies or environments. Many exotic plants and animals were brought to this region by Europeans to support their food production systems. Ironically, many of these same species are now major threats to food production.

Australia. When Europeans first settled the continent, they brought many new animals and plants with them, sometimes on purpose, sometimes unintentionally. Nonnative species from Europe, Asia, Africa, and the Americas have displaced or even driven to extinction at least 41 native Australian bird and mammal species and more than 100 plant species. In all likelihood, many more disappeared before they were biologically classified.

European rabbits are among the most destructive of the introduced species. Rabbits were brought to Australia by early British settlers who enjoyed eating them. Many were released for hunting, and with no natural predators, they multiplied quickly, consuming so much of the native vegetation that many indigenous animal species starved. Moreover, rabbits became a major source of agricultural crop loss and reduced the capacity of many grasslands to support herds of introduced sheep and cattle. Attempts to control the rabbit population by introducing European foxes and cats backfired as these animals became major invasive species themselves. Foxes and cats have driven several native Australian predator species to extinction without having much effect on the rabbit population. Intentionally introduced diseases have proven more effective at controlling the rabbit population, though rabbits have repeatedly developed resistances to them.

Herding has also had a huge impact on Australian ecosystems. Because the climate is arid and soils in many areas are relatively infertile, the dominant land use in Australia is the

Figure 11.12 The Dingo Fence. The world's longest fence snakes 3200 miles (5800 kilometers) between Yalata in South Australia and Jandowae, near the coast of Queensland. It attempts to separate the dingo, Australia's indigenous dog, from sheep herds. A single dingo can kill as many as 50 sheep in one night of marauding. The fence requires constant patrolling and mending, but it has dramatically reduced the dingoes' killing of sheep. [Medford Taylor/National Geographic Image Collection.]

grazing of introduced domesticated animals—primarily sheep, but also cattle. More than 15 percent of the land has been given over to grazing, and Australia leads the world in exports of sheep and cattle.

Dingoes, the indigenous wild dogs of Australia, prey on the introduced sheep and young cattle. To separate the wild dogs from the herds, the Dingo Fence—the world's longest fence, extending 3200 miles (5800 kilometers)—was built (Figure 11.12). The fence is also a major ecological barrier to other wild species, though kangaroos, the natural prey of dingoes, have learned to live on the sheep side of the fence, where their population has boomed beyond sustainable levels.

New Zealand. The environment of New Zealand has been transformed even more extensively by introduced species and food production systems than Australia's environment. No humans lived in New Zealand until about 1000 years ago,

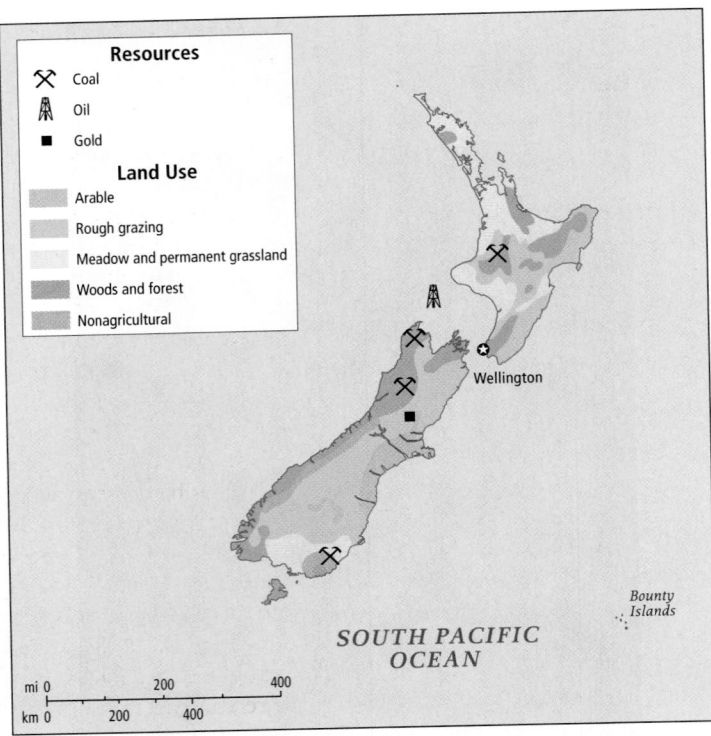

Figure 11.13 Land uses and natural resources of New Zealand. As a result of European settlement and the clearing of land for farming, only 23 percent of New Zealand remains forested. [Adapted from Richard Nile and Christian Clerk, *Cultural Atlas of Australia, New Zealand, and the South Pacific* (New York: Facts on File, 1996), p. 194.]

when the Polynesian Maori people settled there. When they arrived, dense midlatitude rain forest covered 85 percent of the land. The Maori were cultivators who brought in yams and taro as well as other nonnative plants and birds. By the time of European contact, forest clearing and over hunting by the Maori had already degraded many environments and driven several species of birds to extinction.

European settlement in New Zealand dramatically intensified environmental degradation associated with food production (Figure 11.13) and invasive species. Attempts to re-create European farming and herding systems in New Zealand resulted in environments that are actually hostile to many native species, a growing number of which are now extinct. Today, only 23 percent of the country remains forested, with ranches, farms, roads, and urban areas claiming more than 90 percent of the lowland area.

> Settlers tried to re-create European farming and herding systems in New Zealand, resulting in an environment that has become hostile to many native species.

Most of the cleared land is used for export-oriented farming and ranching. Grazing has become so widespread that today there are 15 times as many sheep as people, and 3 times as many cattle. Both of these activities have severely degraded environments. Soils exposed by the clearing of forests proved infertile, forcing farmers and ranchers to augment them with agricultural chemicals. The chemicals, along with feces from sheep and cattle, have severely polluted many waterways, causing the extinction of some aquatic species.

As in Australia, Europeans brought many new animals, such as brushtail possums, which are native to Australia and were introduced to New Zealand as a source of meat and pelts. They have been particularly damaging to native tree species, whose leaves they eat, and to the cattle industry, which is threatened by diseases carried by the possums. Rabbits have damaged both crop agriculture and herding areas, and cats and other predators have been introduced to control booming rabbit populations. The result has been the extinction of 50 percent of New Zealand's bird species.

Pacific islands. In the Pacific islands, many unique species of plants and animals have been driven to extinction as islands were deforested and converted to agriculture. Many other species died as new pests, plants, and animals were introduced accidentally or on purpose. In Hawaii, for example, extensive conversion of forests to land used for agriculture caused the extinction of numerous plant, bird, and snail species. Hawaii is home to more threatened or endangered species than any other U.S. state despite having less than 1 percent of the U.S. landmass. ▰

Globalization and the Environment in the Pacific Islands

As the Pacific islands have become more connected to the global economy over the years, flows of both resources and pollutants have increased dramatically. Mining, nuclear pollution, and tourism are all examples of how globalization has transformed environments in the Pacific islands. Recent global-scale initiatives to better manage oceanic resources highlight the extent to which Pacific island environments have become globalized.

Mining. Two mines have had particularly devastating effects on the environment. Both were operated by foreign-owned mining companies that took advantage of poorly enforced or nonexistent environmental laws. In the case of the Ok Tedi Mine on Papua New Guinea, huge amounts of mine waste devastated river systems. Tens of thousands of indigenous subsistence cultivators were forced into new mining market towns, where their skills were of little use and where they needed cash to buy food and pay rent.

Eventually, 30,000 of these people sued the Australian parent mining company, BHP, for U.S.$4 billion. Two villagers, Rex Dagi and Alex Maun, traveled to Europe and the United States to explain their cause and meet with international environmental groups. They and their supporters convinced U.S. and German partners in the Ok Tedi mine to divest their shares. In 1996, the parties reached an out-of-court settlement that set several important precedents: (1) the Papuans won U.S.$125 million in trust funds for damage mitigation;

(2) the Papuans won a 10 percent interest in the mine; and (3) the settlement provided that any further disputes would be heard in Australian rather than Papua New Guinea courts, making it less likely that the mining company could manipulate the justice system.

The most extreme case of environmental disaster due to mining took place on the once densely forested island of Nauru (one-third the size of Manhattan, located northeast of the Solomon Islands). During most of the last half of the twentieth century, the people of Nauru lived in prosperity. In fact for a few years, the country had the highest per capita income on earth, thus attracting many immigrants. Nauru's wealth was based on the proceeds from the strip-mining of high-grade phosphates used in the manufacture of fertilizer, derived from eons of bird droppings (guano). The mining companies were owned first by Germany, then by Japan, and finally by Australia.

The phosphate reserves are now depleted, the proceeds ill-spent, and the environment destroyed (Figure 11.14). Junked mining equipment sits on miles of bleached white sand where forest once stood. To avoid financial disaster, the government has been exploring shady ways of making money, including money laundering for the Russian mafia (see Chapter 5, page 193), and off-the-record offshore banking services for illegal rain forest loggers in Africa, South America, and Southeast Asia.

Nuclear pollution. The geopolitical aspects of globalization have hit this region especially hard. From the 1940s to the 1960s, the United States exploded 106 nuclear bombs in tests, primarily over the Marshall Islands. Similarly, Mururoa in French Polynesia has been the site of 180 nuclear weapons tests and the recipient of numerous shipments of nuclear waste from France. The weapons tests and imported waste have become major environmental issues for the Pacific islands. Can-

cer, infertility, birth defects, and miscarriages have been widespread among the populations of many islands.

One response to this pollution was the 1985 Treaty of Rarotonga, which established the South Pacific Nuclear Free Zone. Most independent countries in Oceania have signed this treaty, which bans nuclear weapons testing and nuclear waste dumping on their lands. As a result of political pressure from France and the United States, however, French Polynesia and U.S. territories such as the Marshall Islands have not signed the treaty. Japan, North Korea, South Korea, France, and the United States have all explored the possibility of depositing their radioactive waste in the Marshall Islands.

Tourism. Even tourism, which until recently was considered a "clean" industry, has been shown to create environmental problems. Foreign-owned tourism enterprises have often accelerated the loss of wetlands and worsened beach erosion by clearing coastal vegetation. Tourism has also strained many island water resources, and inadequate methods of disposing of sewage and trash from resorts has polluted many once pristine areas. Ecotourism (see Chapter 3, pages 107–108) is now a common element of development throughout the Pacific, but environmental impacts from tourism are still generally high.

The United Nations Convention on the Law of the Sea. Implementation of the 1994 UN Convention on the Law of the Sea (UNCLOS) has revealed how the globalization of Pacific island economies has thwarted environmental protection. UNCLOS is based on the idea that all problems of the world's oceans are interrelated and need to be addressed as a whole. It establishes rules governing all uses of the world's oceans and seas and has been ratified by 157 countries. (The United States is not one of them.)

The treaty allows islands to claim rights to ocean resources 200 miles (320 kilometers) out from the shore. Island countries

Figure 11.14 Environmental consequences of mining. Intensive mining of phosphate on the Pacific island of Nauru has left a landscape of jagged limestone outcroppings. [Courtesy U.S. Department of Energy's Atmospheric Radiation Measurement Program.]

can now make money by licensing privately owned fleets from Japan, South Korea, Russia, and the United States to fish within these offshore limits. However, there is no overarching enforcement agency, and protecting the fisheries from overfishing by these rich and powerful licensees has turned out to be an enforcement nightmare for tiny island governments with few resources. Similarly, it has proved difficult to monitor and control the exploitation of seafloor mineral deposits by foreign companies. Even mining operations conducted legally outside the 200-mile limit have the potential to pollute fisheries and other ocean resources. ▰

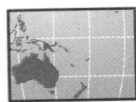

Ten Key Aspects of Oceania

- Oceania has only 34 million people living on a total land area slightly larger than the contiguous United States but spread out in bits and pieces across an ocean larger than the Eurasian landmass.
- Asia is now the major destination for Oceania's exports and is an increasing source of Oceania's imports.
- The trend toward urbanization is strongest in Australia, New Zealand, and some of the wealthier Pacific islands. It is weakest in Papua New Guinea and many smaller Pacific Islands.
- As part of its efforts to combat global warming, New Zealand has set a goal of obtaining 95 percent of its energy from renewable sources by 2025.
- Throughout Oceania, the introduction of food production systems from elsewhere has resulted in the spread of ecologically and economically damaging invasive species.
- From the 1920s to the 1960s, a "whites only" immigration policy barred Asians, Africans, and the Pacific islanders from legally migrating to either Australia or New Zealand.
- In both Australia and New Zealand, indigenous peoples have begun to win court cases that have granted them control over some of the land taken from them by Europeans.
- Recent decades have seen divisions emerge in Oceania over democratization and traditional notions of power and problem solving embodied in the Pacific Way.
- In many Pacific island groups, the number of tourists far exceeds the island population.
- On issues of gender, a striking disparity is emerging between Australia and New Zealand, where women are gaining political and economic empowerment, and Papua New Guinea and the Pacific islands, where change is much slower.

HUMAN PATTERNS OVER TIME

Vignette *With courage, you can travel anywhere in the world and never be lost. Because I have faith in the words of my ancestors, I'm a navigator.*

—Mau Piailug

In 1976, Mau Piailug made history by sailing a traditional Pacific island voyaging canoe across the 2400 miles (3860 kilometers) of deep ocean between Hawaii and Tahiti (149°E, 17°S). He did so without a compass, charts, or other modern instruments, using only methods passed down through his family. He relied mainly on observations of the stars, the sun, and the moon to find his way. When clouds covered the sky, he used the patterns of ocean waves and swells, as well as the presence of seabirds, to tell him of distant islands over the horizon.

Piailug reached Tahiti 33 days after leaving Hawaii and made the return trip in 22 days. His voyage settled a major scholarly debate over how people settled the many remote islands of the Pacific without navigational instruments, thousands of years before the arrival of Europeans. Some thought that navigation without instruments was impossible and argued that would-be settlers simply drifted about on their canoes at the mercy of the winds, most of them starving to death on the seas, with a few happening on new islands by chance. It was hard to refute this argument because local navigational methods had died out almost everywhere. However, in isolated Micronesia, where Piailug is from, indigenous navigational traditions had survived.

Piailug learned his methods from his grandfather in secret because the Germans who first colonized Micronesia, and later the Japanese, banned long-distance navigation to prevent their Micronesian forced laborers from sailing away. After World War II, however, Piailug was free to sail, and he began making small voyages of several hundred miles. Eventually he attracted the attention of some Hawaiians who were building a traditional voyaging

Figure 11.15 A traditional Pacific island voyaging canoe. In 1999, the *Hokule'a* sailed from Hawaii to Easter Island and back—a round trip of about 14,000 miles (22,530 kilometers)—guided only by traditional navigational techniques, such as the reading of wave patterns. [Cary Wolinsky.]

canoe, *Hokule'a* (Figure 11.15), with the aim of proving that traditional technologies were adequate for long-distance Pacific travel. Since the successful 1976 voyage, Piailug has trained several students in traditional navigational techniques, which have become a symbol of cultural rebirth and a source of pride throughout the Pacific.

Source: Richard Nile and Christian Clerk, Cultural Atlas of Australia, New Zealand, and the South Pacific *(New York: Facts on File, 1996), pp. 63–65.* ∎

The Peopling of Oceania

The longest-surviving inhabitants of Oceania are Australia's **Aborigines,** who migrated from Southeast Asia 50,000–70,000 years ago (Figure 11.16). Amazingly, some memory of this ancient journey may be preserved in Aboriginal oral traditions, which recall mountains and other geographic features that are now submerged under water. At about the same time that the Aborigines were settling Australia, related groups were

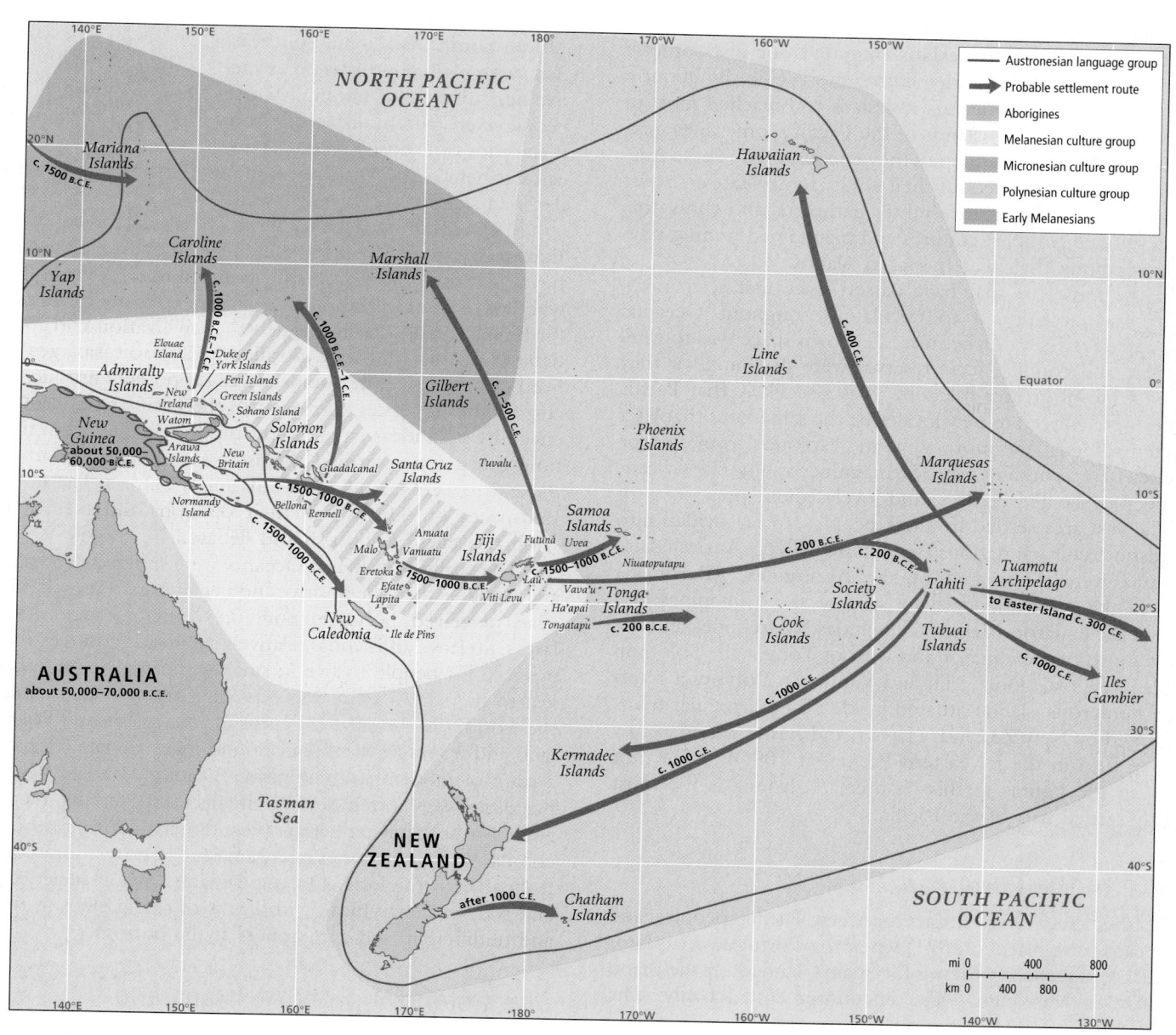

Figure 11.16 Primary culture groups in the island Pacific. By about 25,000 years ago, people were spread across a large part of New Guinea and had begun moving across the ocean to nearby Pacific islands. Movement into the more distant Pacific islands apparently began with the arrival of the Austronesians, who went on to inhabit the farthest reaches of Oceania. [Adapted from Richard Nile and Christian Clerk, *Cultural Atlas of Australia, New Zealand, and the South Pacific* (New York: Facts on File, 1996), pp. 58–59.]

settling nearby areas. These were the **Melanesians,** so named for their relatively dark skin tones, a result of high levels of the protective pigment melanin.

The Melanesians spread throughout New Guinea and other nearby islands, giving this area its name, Melanesia. They lived in isolated pockets. One indication of the great age and isolation of the Melanesian settlement of New Guinea is the existence of hundreds of distinct yet related languages on that island. Like the Aborigines, the Melanesians survived mostly by hunting, gathering, and fishing, although some groups—especially those inhabiting the New Guinea highlands—also practiced agriculture.

Much later, about 5000 to 6000 years ago, groups of linguistically related **Austronesians** migrated out of Southeast Asia and continued the settlement of the Pacific. By about a thousand years ago, some Austronesians had reached most of the remaining far-flung islands of the Pacific, sometimes mixing with the Melanesian peoples they encountered. These Austronesians were renowned for their ability to navigate over vast distances. They were fishers, hunter-gatherers, and cultivators who developed complex cultures and maintained trading relationships among their widely spaced islands.

In the millennia that have passed since first settlement, humans have continued to circulate throughout Oceania. Some apparently set out because their own space was too full of people and conflict, food reserves were declining, or they wanted a life of greater freedom. It is also likely that Pacific peoples were enticed to new locales by the same lures that later attracted some of the more romantic explorers from Europe and elsewhere: sparkling beaches, magnificent blue skies, beautiful people, scented breezes, and lovely landscapes.

The vast area settled by these people can be divided into four distinct cultural regions (see Figure 11.16). Australia and Tasmania were dominated by the Aborigines. **Micronesia** refers to the small islands lying east of the Philippines and north of the equator. **Melanesia** includes New Guinea and the islands south of the equator and west of Tonga (the Solomon Islands, New Caledonia, Fiji, and Vanuatu). **Polynesia** refers to the numerous islands situated inside a large irregular triangle formed by New Zealand, Hawaii, and Easter Island (a tiny speck of land in the far eastern Pacific, at 109°W, 27°S, not shown in the figures in this chapter). Polynesia is the most recently settled part of the Pacific.

Arrival of the Europeans

The earliest recorded contact between Pacific peoples and Europeans took place in 1521, when the Portuguese explorer Ferdinand Magellan (working for Spain), landed on the island of Guam in Micronesia. The encounter ended badly. The islanders, intrigued by European vessels, tried to take a small skiff. For this crime, Magellan had his men kill the offenders and burn their village to the ground. A few months later, Magellan was himself killed by islanders in what became the Philippines, which he had claimed for Spain. Nevertheless, by the

1560s, the Spanish had set up a lucrative trade route between Manila in the Philippines and Acapulco in Mexico. Explorers from other European states followed, first taking an interest mainly in the region's valuable spices. The British and French explored extensively in the eighteenth century.

The Pacific was not formally divided among the colonial powers until the nineteenth century, and by that time, the United States, Germany, and even Japan had joined France and Britain in taking control of various island groups. European colonization of Oceania proceeded according to the models developed in Latin America, Africa, South Asia, and Southeast Asia, with major emphasis on extractive agriculture and mining. Native people were often displaced from their lands or exposed to exotic diseases to which they had no immunity, and their populations declined sharply.

> Europe colonized Oceania according to the models developed in Latin America, Africa, South Asia, and Southeast Asia, with major emphasis on extractive agriculture and mining.

Many enduring notions about the Pacific arose from the European explorations of the eighteenth and nineteenth centuries. During this time, European thinkers were debating whether or not civilization actually improves the quality of life for human beings. Some argued that civilization corrupts and debases people, and they glorified the **"noble savages"** who lived primitive lives in distant places supposedly untouched by corrupting influences. Explorers of the Pacific who were influenced by such ideas were caught off guard when from time to time the islanders armed themselves and rebelled, attacking those who were taking their lands and resources. The surprised Europeans quickly revised their opinions and relabeled the "noble savages" as brutish and debased.

The realities of life in Oceania were much more balanced and complex than the positive and negative extremes that Europeans perceived. In Australia and on New Guinea and the other larger Melanesian islands, relatively plentiful resources made it possible for people to live in small, simple societies, less subject to the stratification and class tensions seen in so much of the world. On the smaller islands of Micronesia and Polynesia, land and resources were scarcer, and many people coexisted in a state of moderate antagonism. Although warfare occurred, hostilities were often settled ritualistically and by means of annual tribute-paying ceremonies, rather than by resorting to combat. Individual rulers rarely amassed large territories or controlled them for long. On these islands, many societies were and still are hierarchical, with layers of ruling elites at the top and undifferentiated commoners at the bottom.

The Colonization of Australia and New Zealand

Although all of Oceania has experienced European or American rule at some point, the most Westernized parts of the region are Australia and New Zealand. The colonization of these two countries by the British has resulted in many parallels with North America. In fact, the American Revolution was a major impetus for "settling" Australia because once the North American colonies were independent, the British needed somewhere else to send their convicts. In early nineteenth-century Britain, a relatively minor theft—for example, of a piglet—might be punished with seven years hard labor in Australia.

A steady flow of English and Irish convicts arrived in Australia until 1868, and most chose to stay in the colony after their sentences were served. They are given credit for Australia's rustic self-image and egalitarian spirit. They were joined by a much larger group of voluntary immigrants from the British Isles who were attracted by the availability of inexpensive farmland. Waves of these immigrants arrived until World War II. New Zealand was settled somewhat later, in the mid-1800s. Although its population also derives primarily from British immigrants, New Zealand was never a penal colony.

Another similarity among Australia, New Zealand, and North America was the treatment of indigenous peoples by European settlers. In both Australia and New Zealand, native peoples were killed outright, annihilated by infectious diseases, or shifted to the margins of society. The few who lived on territory the Europeans thought undesirable were able to maintain their traditional way of life. However, the vast majority who survived lived and worked in grinding poverty, either in urban slums or on cattle and sheep ranches. Today, native peoples still suffer from pervasive discrimination and maladies such as alcoholism and malnutrition. Even so, some progress is being made toward improving their lives, as is described on pages 426–428.

Oceania's Shifting Global Relationships

During the twentieth century, Oceania's relationship with the rest of the world changed at least three times: from a predominantly European focus to identification with the United States and Canada to the currently emerging linkage with Asia. Up until roughly World War II, the colonial system gave the region a European orientation. In most places, the economy depended largely on the export of raw materials to Europe. Thus, even when a colony gained independence from Britain, as Australia did in 1901 and New Zealand did in 1907, people remained strongly tied to their mother countries. Even today, the Queen of England remains the titular head of state in both countries. During World War II, however, the European powers provided only token resistance to Japan's invasion of much of the Pacific and its bombing of northern Australia. ▄▀

After the war, the United States became the dominant power in the Pacific, and U.S. investment became increasingly important to the economies of Oceania. Australia and New Zealand joined the United States in a cold war military alliance, and both fought alongside the United States in Korea and Vietnam, suffering considerable casualties and experiencing significant antiwar activity at home. U.S. cultural influences were strong, too, as North American products, technologies, movies, and pop music penetrated much of Oceania.

By the 1970s, another shift was taking place as many of the island groups were granted self-rule by their European colonizers and Oceania became steadily drawn into the growing economies of Asia. Since the 1960s, Australia's thriving mineral export sector has become increasingly geared toward supplying Asian manufacturing industries (first Japan in the 1960s, and increasingly China since the 1990s). Similarly, since the 1970s, New Zealand's wool and dairy exports have gone mostly to Asian markets. Despite occasional backlashes against "Asianization," Australia, New Zealand, and the rest of Oceania are becoming increasingly transformed by Asian influences. Many Pacific islands have significant Chinese, Japanese, Filipino, and Indian minorities, and the small Asian minorities of Australia and New Zealand are increasing. On some Pacific islands, such as Hawaii, Asians are now a majority of the population.

POPULATION PATTERNS

Oceania occupies a huge portion of the planet, yet it has only 34 million people, fewer than in the state of California (Figure 11.17). They live on a total land area slightly larger than the contiguous United States but spread out in bits and pieces across an ocean larger than the Eurasian landmass. The Pacific islands, including Hawaii, have slightly more than 4 million people; Australia has 20.4 million; Papua New Guinea, 5.9 million; and New Zealand, 4 million.

Two different population patterns exist in Oceania. Australia, New Zealand, and Hawaii, like many wealthy countries, have highly urbanized, relatively older, and more slowly growing populations with life expectancies around 80 years. The Pacific islands and Papua New Guinea, like many developing countries, have much more rural, younger, and rapidly growing populations with life expectancies in the sixties or low seventies. The overall trend throughout the region, however, is toward smaller families, aging populations, and urbanization.

Population densities remain relatively low in Australia, at 7 people per square mile (3 per square kilometer), and New Zealand, at 39 per square mile (15 per square kilometer). Densities vary widely in the Pacific islands; some are sparsely settled or uninhabited while others—including some of the smallest, such as the Marshall Islands and Tuvalu—have 800 to 1000 people per square mile (307 to 386 per square kilometer).

Urbanization in Oceania

The global trend of migration from the countryside to cities is highly visible in Oceania, where 73 percent of the population lives in urban areas (Table 11.1). The shift from agricultural and resource-based economies toward service economies is a major driver of urbanization, especially in Australia, New

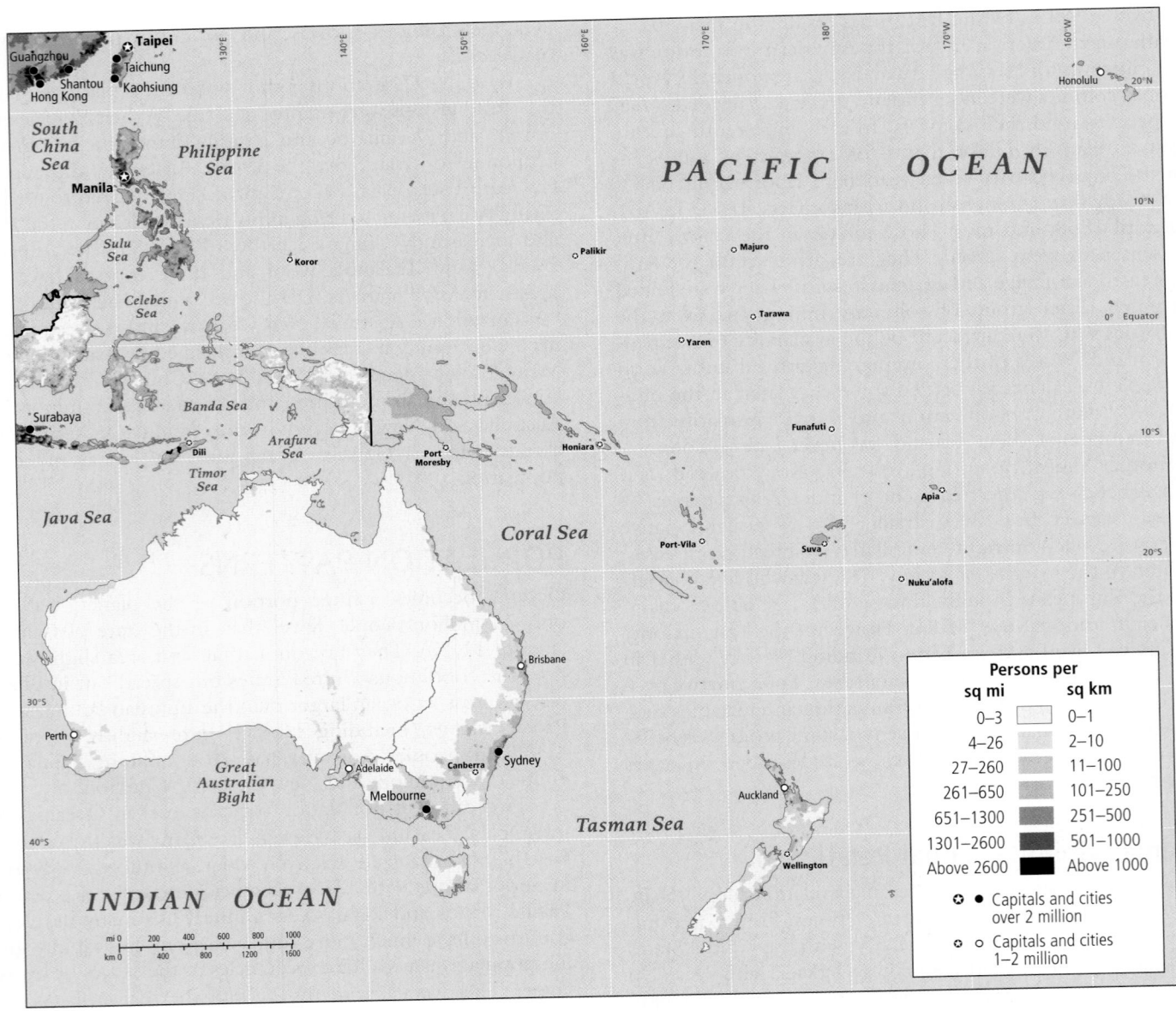

Figure 11.17 Population density in Oceania.

Zealand, and some of the wealthier Pacific islands such as Fiji. These trends are weakest in Papua New Guinea and many smaller Pacific Islands.

Australia and New Zealand. Australia and New Zealand have among the highest percentages of city dwellers outside Europe. More than 90 percent of Australians live in a string of cities along the country's well-watered and relatively fertile eastern and southeastern coasts. Similarly, 86 percent of New Zealanders live in Auckland, the country's one large city, and several medium-sized cities on both the North Island and South Island. The vast majority in these two countries live in modern comfort and work in a range of occupations typical of highly industrialized societies. Vibrant urban-based service

economies have expanded to employ about three-quarters of the population in both countries. Declining employment in mining and agriculture, where mechanization has dramatically reduced the number of workers needed, has also contributed to urbanization.

Pacific islands. Throughout the Pacific, urban centers have transformed natural landscapes, and in some small states, such as Palau and the Marshall Islands, they have become the dominant landscape. Although cities are places of opportunity, they can also be sites of both cultural change and conflict.

The great majority of Pacific island towns, and all the capital cities, are located in ecologically fragile coastal settings. Many of these towns were established during the colonial era

Table 11.1 Percentage of the population living in cities and towns in selected places in Oceania

Country	Total population	Urban population	Percent urban
Australia	20,600,000	18,746,000	91
Federated States of Micronesia	100,000	22,000	22
Fiji	800,000	368,000	46
French Polynesia	300,000	159,000	53
Guam	200,000	186,000	93
Hawaii	1,300,000	1,183,000	91
Kiribati	100,000	43,000	43
Marshall Islands	100,000	68,000	68
Nauru	10,000	10,000	100
New Caledonia	200,000	142,000	71
New Zealand	4,100,000	3,649,000	89
Palau	20,000	15,400	77
Papua New Guinea	6,000,000	780,000	13
Samoa	200,000	44,000	22
Solomon Islands	500,000	80,000	16
Tonga	100,000	23,000	23
Tuvalu	10,000	4,700	47
Vanuatu	200,000	42,000	21
Totals	34,840,000	25,565,100	73.4

Source: 2006 World Population Data Sheet (Washington, D.C.: Population Reference Bureau).

> The great majority of Pacific island towns, and all the capital cities, are located in ecologically fragile coastal settings.

for access to shipping and were situated in places suitable for only limited numbers of people. Consequently, little land is available for development, and access to housing is limited. Squatter settlements have been a visible feature of the region's urban areas for several decades. The discharge of untreated sewage and other wastes into coastal waters and lagoons has damaged marine environments, reduced the productivity of subsistence fisheries, and periodically led to the outbreak of diseases such as cholera. Air pollution is a new phenomenon in the Pacific islands, as is noise pollution.

Cultural changes, too, have resulted from urbanization. Although many urban residents were born outside the towns and maintain close connections to their rural home communities, some urban islanders have disavowed rural life, ethnic identity, and cultural commitments. Increasingly, people are marrying in town and across language divisions, creating new patterns of social alliances and networks. These changes, along with the adoption of urban lifestyles, are creating new social tensions and changing the very nature of social life in the island Pacific. Urban unemployment and crime are on the rise, and low economic growth restricts the revenue available to governments to manage urban development.

Current Geographic Issues

Many current geographic issues in Oceania are related to the transition under way from European to Asian and inter-Pacific cultural influences. This changing cultural orientation is part of the process of globalization occurring everywhere. Oceania's old relationships were built on historical factors, such as the settlement of Australia and New Zealand by Europeans.

Its new relationships, in contrast, are influenced by economic and geographic considerations, particularly its physical proximity to Asia.

New ways of interacting, such as e-mail, the Internet, and rapid air travel are giving the people of Oceania greater access to information about unfamiliar ways of life and to the chance to experience distant places personally. As a consequence, cultural sensitivity is growing and old prejudices are being modified. For instance, New Zealand and the Pacific islands are finding philosophical grounds for a closer mutual identity, including acceptance of their common Polynesian as well as European cultural heritage. They are also joining in public awareness of environmental issues. Australia participates in this new closeness too. However, some regional policing and administrative duties that it has inherited from the British occasionally place it in a dominant role in Oceania that causes resentment in other parts of the region.

ECONOMIC AND POLITICAL ISSUES

Oceania has been powerfully transformed by global relationships over the past two hundred years. Now, new forces of globalization, driven largely by Asia's growing affluence and enormous demand for resources, are shifting trade, migration, and tourism within Oceania.

Globalization and Oceania's New Asian Orientation

Globalization has long been a force in this region where so many people are of European descent and where Europe, and more recently the United States, have exerted a powerful influence on trade and politics. For the past several decades, however, globalization has been reorienting this region toward Asia, which buys 71 percent of Australia's exports (mainly coal, iron ore, and other minerals) and 44 percent of New Zealand's exports (mainly meat, wool, and dairy products; Figure 11.18).

Asia is also an increasing source of the region's imports. Because there is little manufacturing in Oceania, most manufactured goods are imported from Asia, with China, Japan, South Korea, Singapore, and Thailand the leading trading partners. Clothing, computers, electronics, machinery, and vehicles are the top Asian imports for both Australia and New Zealand. Both Australia and New Zealand have free trade relationships either completed or in negotiation with Asia's two largest economies, China and Japan.

> Both Australia and New Zealand have free trade relationships with Asia's two largest economies, China and Japan.

The Pacific islands are even further along in their reorientation toward Asia. Not only are coconut, forest, and fish products from the Pacific islands sold to Asian markets, but Asian companies increasingly own these industries. Fishing fleets from Asia regularly ply the offshore waters of Pacific island nations. Asians also dominate the Pacific island tourist trade, both as tourists and as investors. And increasing numbers of Asians are taking up residence in the Pacific islands. In Hawaii, for example, Asians now make up 57 percent of the population and have widespread economic influence.

The Stresses of Asia's economic rise for Australia and New Zealand. For Australia and New Zealand, Asia's global economic rise has meant not only increased trade but also increased competition with Asian economies in foreign markets. Throughout the region, local industries used to enjoy protected or "preferential" trade with Europe. They have lost that advantage because new European Union regulations stemming from the EU's membership in the World Trade Organization prohibit such arrangements. Now, these industries face stiff competition in their trade with Europe from larger companies in Asia that benefit from much cheaper labor.

Since the 1970s, competition from Asian companies has meant that increasing numbers of workers in Australia and New Zealand have lost jobs and seen their hard-won benefits scaled back or eliminated. This has been especially traumatic for the labor movements of these countries, which historically have been among the world's strongest. Australian coal miners' unions successfully agitated for the world's first 35-hour work week. Other labor unions won a minimum wage, pensions, and aid to families with children long before such programs were enacted in many other industrialized countries. For decades, these arrangements were highly successful. Both Australia and New Zealand enjoyed living standards comparable to those in North America but with a more egalitarian distribution of income.

Competition from Asian companies also led to lower corporate profits. As corporate profits fell, so did government tax revenues, which necessitated cuts to previously high rates of social spending on welfare, health care, and education. The loss of social support, especially for those who have lost jobs, has contributed to rising poverty in recent years. Australia now has the second-highest poverty rate in the industrialized world, after that of the United States.

The future: a mixed Asian and European orientation? Despite the powerful forces pushing Oceania toward Asia, important factors still favor strong ties with Europe and North America. The Asian economic recession of the late 1990s made clear that Oceania must maintain broad contacts with economies outside Asia. Another factor supporting Western connections is the lingering fear of Chinese aggression, justified to some extent by China's expansionist moves toward Taiwan. In spite of increasing trade links and a recent move by China to expand diplomatic and cultural relations with Australia, both Australia and New Zealand remain staunch military allies of the United States. Over the years, both have participated in U.S.-led wars in Korea, Vietnam, Afghanistan, and Iraq.

In parts of the Pacific islands, strong links to Europe and North America are also upheld by continuing political domination. In Micronesia, the United States controls Guam and the Northern Mariana Islands, and in Polynesia, American Samoa. The Hawaiian Islands are a state within the United States. Similarly, the 120 islands of French Polynesia—including Tahiti and the rest of the Society Islands, the Marquesas Islands, and the Tuamotu Archipelago—are considered Overseas Lands of France. Any desire for independence in these possessions has not been sufficient to override the financial benefits of aid, subsidies, and investment money provided by France and the United States.

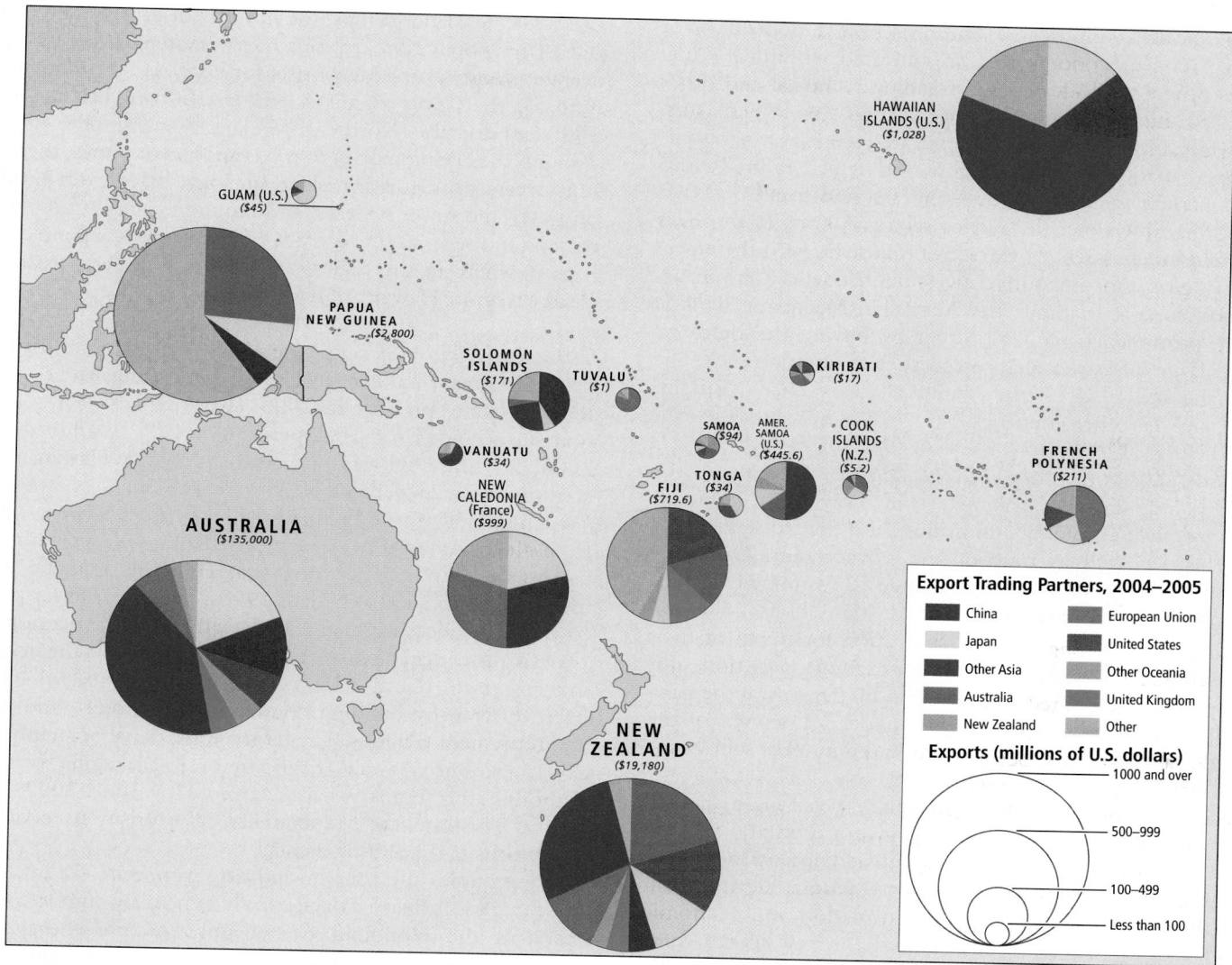

Figure 11.18 Exports from Oceania. The colors of each pie chart indicate a country's export trading partners. The "Other" sections can include trade with Canada, Mexico, the Caribbean, non-EU Europe, sub-Saharan Africa, and other locales, some of them new trade partners. The volume of trade is indicated in millions of U.S. dollars. (Figures for Hawaii do not include exports to other parts of the United States.) [Major sources: https://www.cia.gov/cia/publications/factbook/geos/Australia, https://www.cia.gov/library/publications/the-world-factbook/geos/nz.html#Econ, http://www.stats.govt.nz/economy/exports-and-imports/default.htm, and http://www.ita.doc.gov/td/industry/otea/state_reports/hawaii.html#Markets.]

From Raw Materials Exports to Service Economies

For decades, natural resources and agricultural products supported the two dominant economies of this region, Australia and New Zealand. Australia, for example, is the world's largest exporter of coal, bauxite, and a number of other minerals and metals. It also supplies about 50 percent of the world's wool used for clothing. New Zealand specializes in dairy products, meat, fish, wool, and timber products. Neither country has been a major supplier of more profitable manufactured goods to the world market.

Although their manufacturing and materials processing capacity has grown over time, unprocessed raw materials still compose the bulk of exports from the two countries. The shift toward trading more with Asia than with Europe has had little effect on this pattern because most Asian economies have a much greater need to import raw materials than manufactured goods. Nevertheless, raw materials export industries are of decreasing importance in the economies of Australia and New Zealand. One reason is that they now employ fewer people because of mechanization. This shift has been essential for these industries to stay globally competitive with other countries that have much cheaper labor.

Today, both economies are dominated by diverse and growing service sectors. However, their service economies have links to their export sectors. The extraction of minerals and management of herds and agriculture have become technologically sophisticated enterprises that depend on a dynamic service economy and an educated workforce. Australia is now a world leader in providing technical and other services to mining companies, sheep farms, and winemakers. Meanwhile, New Zealand's well-educated workforce and well-developed marketing infrastructure has helped it break into luxury markets for dairy products, meats, and fruits.

At the same time, the service sector in both countries is becoming independent of extractive industries with the emergence of globally competitive investment, finance, and insurance sectors. For example, the Australian Macquarie Bank is gaining prominence on Wall Street by buying the rights to operate large infrastructure projects such as ports, tunnels, airports, and toll roads in the United States, Canada, Britain, China, and 15 other countries.

Economic change in the Pacific islands.

In general, the Pacific islands are also making a shift away from extractive industries, such as mining and fishing, and toward service sector industries such as tourism and government. On many islands, the stress of economic change is cushioned by self-sufficiency and resources from abroad. Many households still rely on fishing and subsistence cultivation for much of their food supply. On the islands of Fiji, for example, part-time subsistence agriculture engages more than 60 percent of the population, though it accounts for just under 17 percent of the economy. Statistics also rarely include remittances sent home from Pacific islanders working abroad.

Islanders who can be self-sufficient while saving extra cash for travel and occasional purchases of manufactured goods are sometimes said to have achieved **subsistence affluence.** If there is poverty, it is often related to geographic isolation, which means a lack of access to information and economic opportunity. Although computers and the new global communication networks are not yet widely available in the Pacific islands, they have the potential to alleviate some of this isolation.

In the relatively poor and undereducated nations (the Solomon Islands, Tuvalu, and parts of Papua New Guinea, for example), conditions typify what has been termed a **MIRAB economy**—one based on <u>mi</u>gration, <u>r</u>emittance, <u>ai</u>d, and <u>b</u>ureaucracy. Foreign aid from former or present colonial powers supports government bureaucracies that supply employment for the educated and semiskilled. This type of economy has little potential for growth.

> Conditions on many of the smaller Pacific islands typify a MIRAB economy—one based on migration, remittance, aid, and bureaucracy.

Tourism

Tourism is a growing part of the economy throughout Oceania, with tourists coming largely from Japan, Korea, Taiwan, Southeast Asia, the Americas, and Europe (Figure 11.19). In many Pacific island groups, the number of tourists far exceeds the island population. In Guam, for example, there are, on average, more than 650 tourists for every 100 residents. The ratio for the Cook Islands is 351 to 100; for Palau, 304 to 100; and for the Northern Mariana Islands, 603 to 100. Though they bring money to the islands' economies, these visitors create problems for island ecology, place extra burdens on water and sewer systems, and require expensive accommodations and services that are out of reach for local people. Perhaps nowhere in the region are the issues raised by tourism clearer than in Hawaii.

Case study: Hawaii.

Since the 1950s, travel and tourism has been the largest industry in Hawaii, producing 22 percent of the gross state product in 2005. (By comparison, travel and tourism accounts for 10.4 percent of GDP worldwide.) In 2005, tourism employed one out of every five Hawaiians and accounted for 23 percent of state tax revenues.

The point of origin of Hawaii's visitors has been shifting from North America to Asia, and by 1995, 40.3 percent of all visitors to Hawaii came from Asia. Thus the dramatic slump in Asian economies in the late 1990s, which led fewer people to travel for pleasure, had a major effect on the economy of Hawaii. The decline in tourism hurt not only the tourism industry itself, but also the construction industry, which had been thriving by building condominiums, hotels, and resort and retirement facilities. The terrorist attacks of September 11, 2001, also hurt Hawaii's economy by discouraging air travel. Although the industry had recovered by June 2002, these slumps illustrate the vulnerability of tourism to economic downturns and political events.

Sometimes the tourism industry can seem like an invading force to ordinary citizens. For example, an important segment of the Honolulu tourist infrastructure—hotels, golf courses, specialty shopping centers, import shops, and nightclubs,—is geared to visitors from Japan, and many such facilities are owned by Japanese investors. Hawaiian citizens and other non-Japanese shoppers can be made to feel out of place. Across Hawaii, the demand for golf courses by Asian tourists has brought about what Native (indigenous) Hawaiians view as desecration of sacred sites. Land that was once communally owned and cultivated, and then confiscated by the colonial government, has been sold to Asian golf course developers. Now the only people with access to the sacred sites are fee-paying tourist golfers. Relocation by retired people looking for a sunny spot—often called "residential tourism"—has also had an impact.

> **Vignette** In the 1990s, the officials of a major U.S. mainland Protestant Christian denomination voted to build a retirement home for their church members on the Hawaiian island of Oahu. The idea was to acquire land and build a multilevel care facility to which church members from the mainland could retire, living independently until they needed nursing home care. The church officials proceeded to look for affordable land close to Honolulu, yet

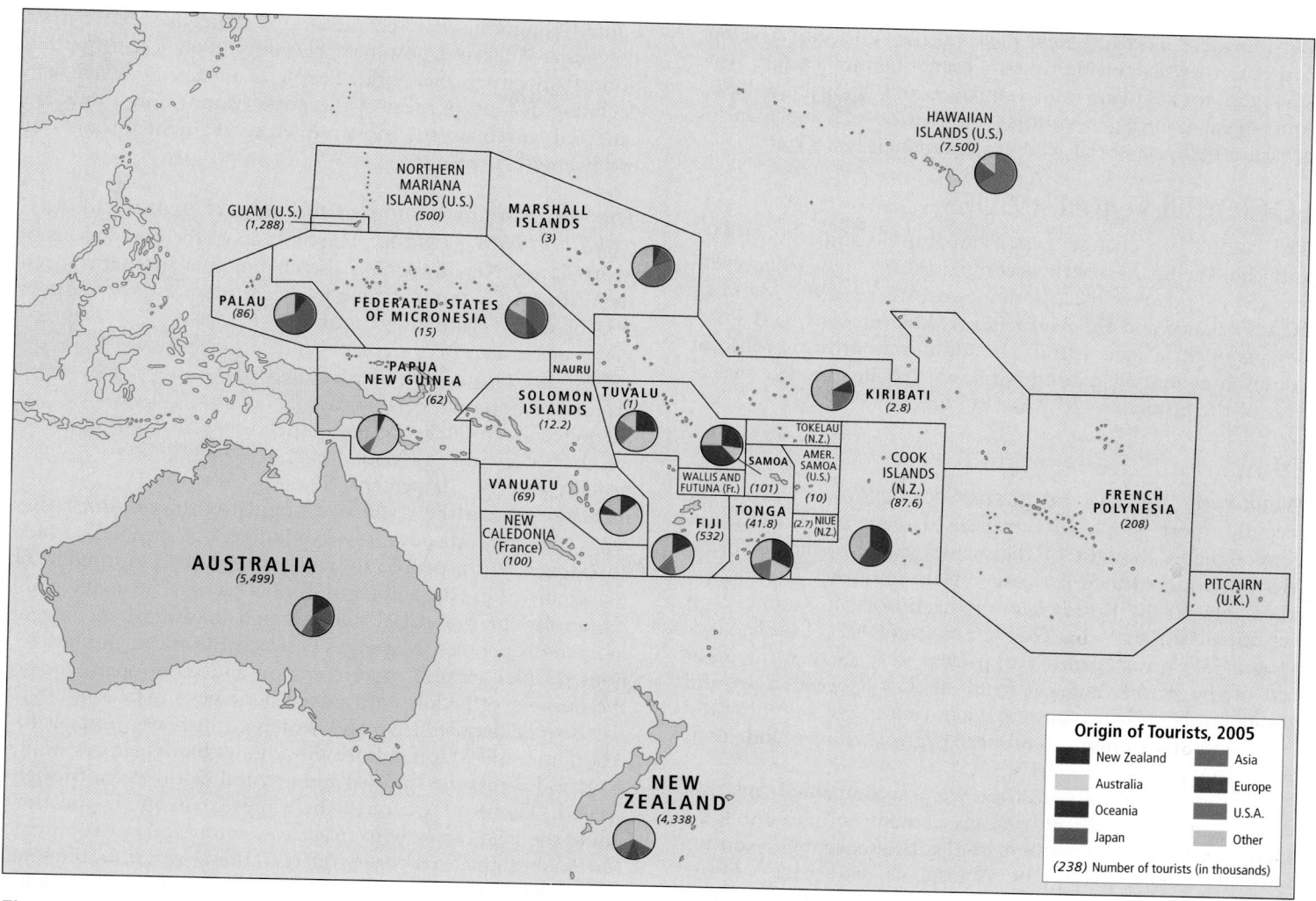

Figure 11.19 Tourism in Oceania. Tourism plays a major role in the economies of all countries in Oceania. The 10.5 million visitors to the region in 2005 contributed some U.S. $24.5 billion to local economies. The origins of the tourists reflect changing trade patterns in the region, with most by far coming from Asia. [Adapted from World Tourism Organization, at http://www.world-tourism.org/facts/tmt.html; *Financial Times World Desk Reference* (New York: Dorling Kindersley), 2004; Cook Islands, at http://166.122.164.43/archive/2004/March/tcp-ck.htm; Guam, at http://www.travelweeklyeast.com/articles/standard.asp?isArticle=1924&pCat=6&rmenu=articles.]

with landscapes of rural tropical beauty. They found a suitable tract in Pauoa Valley, one of the last valleys near Honolulu where rural people of Native Hawaiian origins still live in extended family compounds and grow traditional gardens.

If the Native Hawaiians were the only inhabitants of Pauoa Valley, their solidarity and Hawaiian laws would have protected the land from being sold. However, new immigrants from North America and Asia share the valley, and some were eager to sell. If the retirement home were built, life would change irreversibly, even for those residents who chose not to sell. The ecology and ambience of the valley would be transformed by the introduction of a large complex with 141 apartments, a 106-bed nursing home, large concrete parking lots, and manicured grounds. Once the church officials understood the issues raised by the Native Hawai-

ians, they decided to build the retirement home in the mainland United States rather than in Pauoa Valley.

Source: Conrad "Mac" Goodwin's research notes, Pauoa Valley, Oahu, Hawaii. ■

Sustainable tourism. Some islands have attempted to deal with the pressures of tourism by adopting the principle of sustainable tourism, which aims to decrease the imprint of tourism and minimize disparities between hosts and visitors. Samoa, for example, has created the Samoa Tourism Authority in conjunction with the South Pacific Regional Environment Programme. (The name *Samoa* refers to the independent country that was formerly known as Western Samoa; that country is politically distinct from American Samoa, a U.S. territory.)

With financial aid from New Zealand, the Authority develops and monitors sustainable tourism components: beaches, wetland and forest island environments, and knowledge-based tourism experiences for visitors (information-rich explanations of island political, social, and environmental issues).

SOCIOCULTURAL ISSUES

The cultural sea change away from Europe and toward Asia and the Pacific has been accompanied by new respect for indigenous peoples: the Aborigines of Australia; the Maori of New Zealand; and the Melanesians, Micronesians, and Polynesians of the Pacific islands. In addition, a growing sense of common economic ground with Asia has heightened awareness of the attractions of Asian culture.

Ethnic Roots Reexamined

Weakening of the European connection. Until very recently, most people of European descent in Australia and New Zealand thought of themselves as Europeans in exile. Many considered their lives incomplete until they had made a pilgrimage to the British Isles or the European continent. In her book *An Australian Girl in London* (1902), Louise Mack wrote: "[We] Australians [are] packed away there at the other end of the world, shut off from all that is great in art and music, but born with a passionate craving to see, and hear and come close to these [European] great things and their home[land]s."

These longings for Europe were accompanied by racist attitudes toward both indigenous peoples and Asians. Most histories of Australia written in the early twentieth century failed to even mention the Aborigines, and later writings described them as amoral. From the 1920s to the 1960s, a "whites only" immigration policy barred Asians, Africans, and the Pacific islanders from legally migrating to either Australia or New Zealand. And, as we have seen, trading patterns further reinforced connections to Europe.

When migration from the British Isles slowed after World War II, both Australia and New Zealand began to lure immigrants from southern and eastern Europe, many of whom had been displaced by the war. Hundreds of thousands came from Greece, the former Yugoslavia, and Italy. The arrival of these non-English-speaking people began a shift toward a more multicultural society. With the demise of the whites-only immigration policies, there was an influx of Vietnamese refugees in the early 1970s following the United States' withdrawal from Vietnam. More recently, growing demand for information technology specialists throughout the service sector has been met by recruiting skilled workers from India (8135 workers in 2003–2004 alone).

Asians remain a small percentage of the total population in both Australia and New Zealand, but new immigration policies are increasing the numbers of Asian immigrants, especially from China, Vietnam, and India (Figure 11.20). Although most immigrants to New Zealand continue to come from the United Kingdom, the second largest influx has come from Polynesia. Auckland now has the largest Polynesian population of any city in the world. People of European descent are declining as a proportion of the population, though they are expected to remain the most numerous segment throughout the twenty-first century.

The social repositioning of indigenous peoples in Australia and New Zealand. The makeup of the populations of Australia and New Zealand is also changing in another respect. In Australia, for the first time in recent memory, the number of people who claim indigenous origins is increasing. Between 1991 and 1996, the number of Australians claiming Aboriginal origins rose by 33 percent. In New Zealand, the number claiming a Maori background rose by 20 percent. These increases are not the result of a population boom, but rather more positive attitudes toward indigenous peoples, which encourages more people to acknowledge their Aborigine or Maori ancestry. Discrimination is now recognized as the main reason for the low social standing and impoverished state of indigenous peoples. Marriages between European and indigenous peoples are also more common and more open, and so the number of people with mixed heritage is increasing.

> Between 1991 and 1996, the number of Australians claiming Aboriginal origins rose by 33 percent.

Recent decades have also witnessed increased respect for Aboriginal and Maori culture. Aborigines base their way of life on the idea that the spiritual and physical worlds are intricately related. The dead are everywhere present in spirit, and they guide the living in how to relate to the physical environment. Much Aboriginal spirituality refers to the *Dreamtime,* the time of creation when the human spiritual connections to rocks, rivers, deserts, plants, and animals were made clear. Unfortunately, very few Aboriginal people practice their own cultural traditions, and many live in impoverished conditions.

Aboriginal land claims. In 1988, during a bicentennial celebration of the founding of white Australia, a contingent of some 15,000 Aborigines protested that they had little reason to celebrate. During the same 200 years, they had been excluded from their ancestral lands, had lost basic civil rights, and had effectively been erased from Australian national consciousness. Into the 1960s, Aborigines had only limited rights of citizenship, and it was even illegal for them to drink alcohol.

Until 1993, Aborigines were assumed to have no prior claim to any land in Australia. British documents indicate that during colonial settlement, all Australian lands were deemed to be available for British use. The Aborigines were thought to be too primitive to have concepts of land ownership since their nomadic cultures had "no fixed abodes, fields or flocks, nor any internal hierarchical differentiation." The Australian High Court declared this position void in 1993. After that, Aboriginal groups began to win some land claims, mostly for

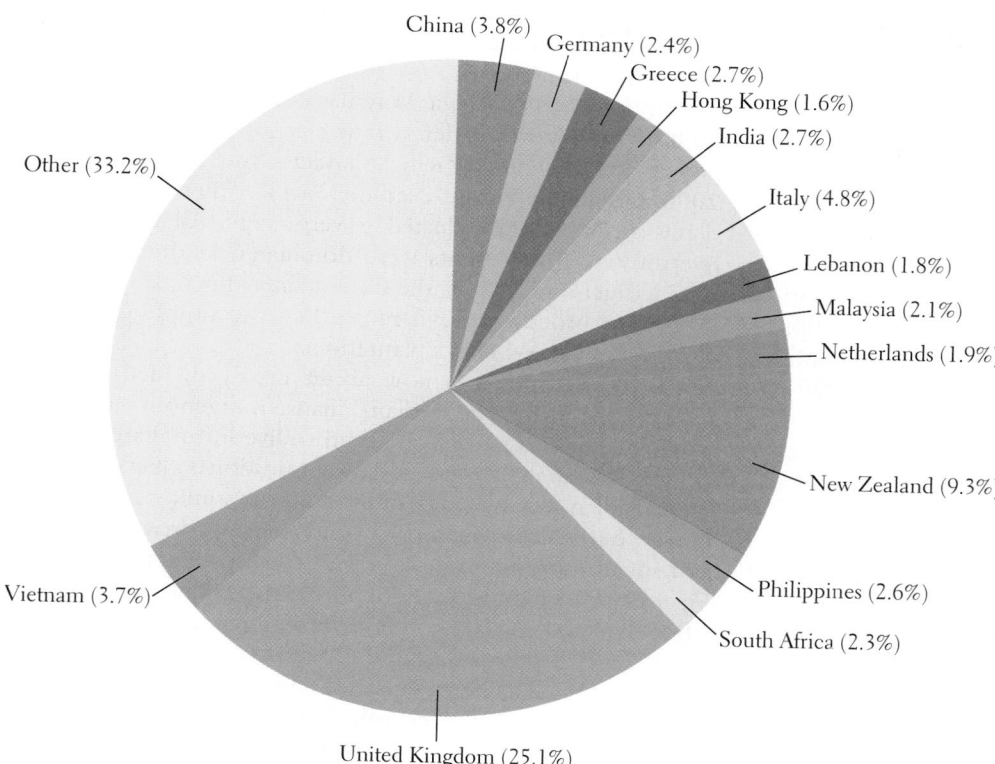

Australia's Foreign-Born Population, 2004
(percent by country of origin)

China (3.8%)
Germany (2.4%)
Greece (2.7%)
Hong Kong (1.6%)
India (2.7%)
Italy (4.8%)
Lebanon (1.8%)
Malaysia (2.1%)
Netherlands (1.9%)
New Zealand (9.3%)
Philippines (2.6%)
South Africa (2.3%)
United Kingdom (25.1%)
Vietnam (3.7%)
Other (33.2%)

Figure 11.20 Australia's cultural diversity in 2004. About 24 percent (4.75 million) of Australia's people were born in other places, making Australia one of the world's most ethnically diverse nations. [Data from Year Book Australia, 2006, at http://www. abs.gov.au/Ausstats/abs@.nsf/bb8db737e2af8 4b8ca2571780015701e/0D98319CE458B36 4CA2570DE0006A39B?opendocument.]

land in the arid interior previously claimed only by the Australian government. Court cases to restore Aboriginal rights and lands continue. Figure 11.21 shows the Aboriginal Embassy, a permanent installation in Canberra that advocates for Aborigines.

Maori land claims. In New Zealand, relations between the majority European-derived population and the indigenous Maori have proceeded only somewhat more amicably. In 1840, the Maori signed the Waitangi Treaty with the British, assuming they were granting only rights of land usage, not

Figure 11.21 The Australian Aboriginal Embassy. The Embassy is a collection of wooden structures in front of the Old Parliament House in Canberra. It was erected in 1972 as a protest against the seizure of Aboriginal lands by European settlers and the continuing refusal of the Australian government to acknowledge that injustice. Nearby is a permanent encampment of Aboriginal people who staff the facility and welcome the many visitors. [Mac Goodwin.]

ownership. The Maori did not regard land as a tradable commodity, but rather as an asset of the people as a whole, used by families and larger kin groups to fulfill their needs. Geographer Eric Pawson writes: "To the Maori the land was sacred . . . [and] the features of land and water bodies were woven through with spiritual meaning and the Maori creation myth." The British, on the other hand, assumed that the treaty had transferred Maori lands to them, giving them *exclusive* rights to settle the land with British migrants and to extract wealth through farming, mining, and forestry.

By 1950, the Maori had lost all but 6.6 percent of their former lands to European settlers and the government. Maori numbers shrank from a probable 120,000 in the early 1800s to 42,000 in 1900, and they came to occupy the lowest and most impoverished rung of New Zealand society. In the 1990s, however, the Maori began to reclaim their culture, and they established a tribunal that forcefully advances Maori interests and land claims through the courts. Since then, nearly half a million acres of land and several major fisheries have been transferred back to Maori control.

Maori numbers rebounded during the twentieth century, reaching an estimated 650,000 by 2006. This number includes the many New Zealanders who previously hid their Maori origins but are now proud to claim them. Overall, New Zealand may lead the world in addressing past mistreatment of indigenous peoples. Nonetheless, the Maori still have notably higher unemployment, lower educational attainments, and poorer health than the New Zealand population as a whole.

Politics and Culture: Democracy and the Pacific Way

Recent decades have seen stark divisions emerge in Oceania between democracy—the system of government that dominates in New Zealand and Australia—and the **Pacific Way,** a political and cultural philosophy that influences governments throughout the rest of the region. Based on traditional notions of power and problem solving, the Pacific Way refers to a way of settling issues based in the traditional culture of many Pacific islands. It favors consensus and mutual understanding over open confrontation, and respect for traditional leadership (especially the usually patriarchal leadership of families and villages) over free speech, personal freedom, and democracy.

The Pacific Way as a political and cultural philosophy originally developed in Fiji around the time of its independence from the United Kingdom in 1970. Subsequently, it spread throughout the Pacific islands, most of which gained independence in the 1970s and 1980s. Since then, the Pacific Way has often been invoked to uphold the notion of a regional identity shared by Pacific islands that grows out of their own particular history and social experience. It was particularly important to academics given the task of writing new textbooks to replace those used by their former colonial masters. The new texts focused students on their own cultures and places before they studied Britain, France, or the United States.

Appeals to the Pacific Way have also been used to uphold attempts by Pacific island governments to control their own economic development and solve their own political and social problems, even if their methods are criticized by outsiders and foreign governments.

In politics, the Pacific Way has emerged as a philosophical basis for overriding democratic elections that challenge the power of indigenous Pacific islanders. In 1987, 2000, and 2006, indigenous Fijians used the Pacific Way to justify coups d'état against legally elected governments. All three of the overthrown governments were dominated by Indian (rather than indigenous) Fijians, the descendants of people from India who were brought to Fiji more than a century ago by the British to work on sugar plantations.

Fiji's population is now about evenly divided between indigenous Fijians and Indian Fijians. Indigenous Fijians are generally less prosperous and tend to live in rural areas where community affairs are governed by traditional chiefs. By contrast, Indian Fijians hold significant economic and political power, especially in the urban centers and in areas of tourism and sugar cultivation. In response to the coups, many Indian Fijians left the islands, resulting in a loss of badly needed skilled labor that has slowed economic development.

Political responses around Oceania to the coups has been divided. Australia, New Zealand, and the United States (via the state government of Hawaii) have demanded that the election results stand and the Indian Fijians be returned to office. The rest of Oceania has appealed to the Pacific Way in arguments supporting the coup leaders. Like Fiji, most islands are governed by leaders of indigenous descent who have not always had the strongest respect for democracy, especially when it could threaten their hold on power.

Some experts on the region suggest that the Pacific Way may be weakening in the face of democracy as a new guiding political principle. They point to the fact that Fiji's last coup was harshly criticized by many Pacific island leaders. However, others argue that challenges to the coup leaders actually derived from the critics' own lack of respect for traditional leadership enshrined in the Pacific Way.

Regardless of its political status, the Pacific Way is likely to endure, especially as a concept that upholds regional identity and traditional culture. Further, some organizations now use the Pacific Way as the basis of an integrated approach to economic development and environmental issues. For example, the South Pacific Regional Environmental Programme builds on traditional Pacific island economic activities, such as fishing, and local traditions of environmental knowledge and awareness, to promote grassroots economic development and environmental sustainability.

Forging Unity in Oceania

A sense of unity is growing throughout Oceania as people develop more appreciation for the region's cultural complexity. Though the great diversity of languages in the region may sometimes make communication difficult, travel and sports are two forces that help to bring people closer together.

Inter-island travel. One way in which unity is manifested in Oceania is through inter-island travel. Today, people travel in small planes from the outlying islands to hubs such as Fiji, where jumbo jets can be boarded for Auckland, Melbourne, or Honolulu. Cook Islanders call these little planes "the canoes of the modern age." New Zealanders can migrate to Australia to teach or train, a businessman from Kiribati in Micronesia can fly to Fiji to take a short course at the University of the South Pacific, or a Cook Islands teacher can take graduate training in Hawaii.

Languages in Oceania. The Pacific islands—most notably Melanesia—have a rich variety of languages. In some cases, the islands in a single chain have several different languages. A case in point is Vanuatu, a chain of 80 mostly high volcanic islands to the east of northern Australia. At least 108 languages are spoken by a population of just 180,000—an average of one language for every 1600 people! Another example is New Guinea, the largest, most populous, and most ethnically diverse island in the Pacific. No fewer than 800 languages are spoken on New Guinea by a populace of 5.5 million.

Languages are both an important part of a community's cultural identity and a hindrance to cross-cultural understanding. In Melanesia and elsewhere in the Pacific, the need for communication with the wider world is served by a number of **pidgin** languages that are sufficiently similar to be mutually intelligible (Figure 11.22). Pidgins are made up of words borrowed from several languages of people involved in trading relationships. Over time, they can grow into fairly com-

plete languages, capable of fine nuances of expression. When a particular pidgin is in such common use that mothers talk to their children in it, then it can literally be called a "mother tongue." In Papua New Guinea, a version of pidgin English is the official language.

Sports as a unifying force. Sports and games are a major feature of daily life throughout Oceania, and the region has both shared them with and borrowed them from cultures around the world. Long-distance sailing, now a world-class sport, was an early skill in this region. Surfing evolved in Hawaii from ancient navigational customs that matched human wits against the power of the ocean. On hundreds of Pacific islands and in Australia and New Zealand, the rugby field, the volleyball court, the soccer field, and the cricket pitch (Figure 11.23) are important centers of community activity. Baseball is a favorite in the parts of Micronesia that were U.S. trust territories. Women compete in the popular sport of netball, which is similar to basketball but without a backboard.

Sports competitions (including native dances such as the Haka described in this chapter's opening vignette) are the single most common and resilient link among the countries of Oceania. Such competitions encourage regional identity, and they provide opportunities for ordinary citizens to travel extensively around the region and to sports venues in other parts of the world. The South Pacific Games—featuring soccer, boxing, tennis, golf, and netball, among other sports—are held every four years. The Micronesians hold periodic games that incorporate tests of many traditional skills, such as spearfishing, climbing coconut trees, and racing outrigger canoes.

Gender Roles in Oceania

Perceptions of Oceania are colored by many myths about how men and women are and should be. As always, the realities are more complex.

Gender myths and realities. Because of Oceania's cultural diversity, many different roles for men and women exist. In the Pacific islands, men traditionally were cultivators, deepwater fishers, and masters of seafaring. In Polynesia, they also were responsible for many aspects of food preparation, including cooking. Men fill many positions in the modern world, but idealized male images continue to be associated with vigorous activities.

In Australia and New Zealand, the supermasculine, white, working-class settler has long had prominence in the national mythologies. In New Zealand, he was a farmer and herdsman. In Australia, he was more often a many-skilled laborer—a stockman, sheep shearer, cane cutter, or digger (miner)—who possessed a laconic, laid-back sense of humor. He went from station (large farm) to station or mine to mine, working hard but sporadically, gambling, and then working again until he had enough money or experience to make it in the city. There, he often felt ill at ease and chafed to return to the wilds. Now

Figure 11.22 A poster written in pidgin English. *Mi No Poret Mi Gat Banis* is pidgin for "I'm not afraid, I have protection." The poster is part of Papua New Guinea's program to combat an alarming increase in HIV-AIDS. [Torsten Blackwood/AFP Photo.]

Figure 11.23 Sports as a unifying force. In the Trobriand Islands, near New Guinea, the British game of cricket has been reformulated to include local traditions. Village teams of as many as 60 men dress in traditional garments and decorate their bodies in ways that are reminiscent of British cricket uniforms, such as the painting-on of white shin pads, yet also include elements of magical decoration. Chants and dances are part of Trobriand cricket matches, which can go on for weeks. [Robert Harding Picture Library, London.]

immortalized in songs, novels, films such as the *Crocodile Dundee* series, and U.S. TV advertisements, these men are portrayed as a rough and nomadic tribe whose social life was dominated by male camaraderie and frequent brawls. No small part of this characterization derived from the fact that many of Australia's first immigrants were convicts.

Today, as part of larger efforts to recognize the diversity of Australian society, new ways of life for men are emerging and are breaking down the national image of the tough male loner. For example, the Australian city of Sydney annually hosts one of the world's largest gay pride parades. Nonetheless, the old model persists and remains prominent in the public images of Australian businessmen, politicians, and movie stars.

Perhaps the most enduring myth Europeans created regarding Oceania was their characterization of the women of the Pacific islands as gentle, simple, compliant love objects. (Tourist brochures still promote this notion.) There is ample evidence to suggest that Pacific Islanders did have more sexual partners in a lifetime than Europeans did. However, the reports of unrestrained sexuality related by European sailors were no doubt influenced by the exaggerated fantasies one might expect from all-male crews living at sea for months at a time. The notes of Captain James Cook are typical: "No women I ever met were less reserved. Indeed, it appeared to me, that they visited us with no other view, than to make a surrender of their persons." Over the years, such notions about Pacific island women have been encouraged by the paintings and prints of Paul Gauguin, the writings of novelist Herman Melville (*Typee*), and the studies of anthropologist Margaret Mead (*Coming of Age in Samoa*), as well as by movies and musicals such as *Mutiny on the Bounty* and *South Pacific*.

In reality, women's roles in the Pacific islands varied considerably from those in Europe, but not in the ways European explorers imagined. Women often exercised a good bit of power in family and clan, and their power increased with motherhood and advancing age. In Polynesia, a woman could achieve the rank of ruling chief in her own right, not just as the consort of a male chief. Women were primarily craftspeople, but they also contributed to subsistence by gathering fruits and nuts and by fishing. And in some places—Micronesia, for example—lineage was established through women, not men.

Gender, democracy, and economic empowerment. Today, there is a trend toward equality across gender throughout Oceania, but there is persistent inequality as well. A striking disparity is emerging between Australia and New Zealand (where women are gaining political and economic empowerment) and Papua New Guinea and the Pacific islands (where change is much slower).

In Australia and New Zealand, women's access to jobs and policy-making positions in government has improved, particularly over the last few decades. New Zealand and the Australian province of South Australia were among the first places in the world to grant women full voting rights (1893 and 1895, respectively). New Zealand has elected two female prime ministers, and Australia, one deputy prime minister (similar to the vice president in the United States). Moreover, in both countries, the proportion of females in national legislatures (33 percent in New Zealand, 27 percent in Australia) is well above the global average of 18 percent.

In both Australia and New Zealand, young women are pursuing higher education and professional careers and postponing marriage and childbearing until their thirties. Nonetheless, both societies continue to reinforce the housewife role for women in a variety of ways. For example, the expectation is that women, not men, will interrupt their careers to stay home to care for young or elderly family members. In both countries women receive on average only about 86 percent of the pay that men receive for equivalent work. This is, however, a smaller gender pay gap than in many developed countries.

In Papua New Guinea and the Pacific islands women are generally less empowered politically and economically. No woman has yet been elected to a top-level national office, and women are a tiny minority in national legislatures, if present at all. Few countries report enough statistics to determine what the gap in pay between men and women is.

> In Papua New Guinea and the Pacific islands women are less empowered politically and economically than in Australia or New Zealand.

Throughout Papua New Guinea and the Pacific islands, gender roles and relationships vary greatly over the course of a lifetime. Today, many young women fulfill traditional roles as mates and mothers and practice a wide range of domestic crafts, such as weaving and basketry. Then in middle age, they may return to school and take up careers. Some women, with the aid of government scholarships, pursue higher education or job training that takes them far from the villages where they raised their children. Accumulating age and experience may boost women into positions of considerable power in their communities.

Throughout their lives, women in Papua New Guinea and the Pacific islands contribute significantly to family assets through the formal and informal economies. Most traders in marketplaces are women, and the items they sell are usually made and transported by women. Yet like women everywhere, they have trouble obtaining credit to expand their businesses. Of the 2039 loans approved by the Agricultural Bank of Papua New Guinea in January 1991, only 4 percent went to women. Hence, while women in Papua New Guinea and the Pacific islands often have considerable economic and political power, it is often not recognized by formal institutions like banks or representative political bodies.

Reflections on Oceania

In this region that has been so powerfully shaped by the globalizing influence of outsiders, several ongoing waves of transformation will define much of what happens in the near future. The economic reorientation away from Europe and toward Asia is likely to continue. The shift in employment away from mining and agriculture and toward service economies is equally profound, and in the case of Australia and New Zealand may moderate the shift toward Asia. The boom in Asian tourism in the Pacific islands, however, will strengthen the shift toward Asia. Thriving service economies are likely to sustain high levels of urbanization, and hence living standards will probably continue to rise. Population growth will likely continue to slow, and populations will age.

Several ongoing transformations set Oceania on a course toward weathering new challenges. Democracy is well established in Australia, New Zealand, and Hawaii, as is the political and economic empowerment of women. However, both of these trends are less well established in Papua New Guinea and the smaller Pacific islands. Nevertheless, these areas also have traditions of cooperation and consensus that may help societies navigate future stresses.

It is difficult to make detailed predictions about the future of Oceania because of the uncertainties brought by global warming. While some low-lying islands are clearly threatened with submersion, the precise nature of the impacts facing Australia and New Zealand are less certain. Water stress could constrain agricultural expansion, and fire could become much more of a concern. Changes in Oceania's regional climate could further threaten the many unique ecosystems of this region, much as the importation of European food systems did. However, advances in water technologies could reduce the vulnerability of many areas to water shortage, drought, and food shortage. Further, New Zealand's ambitious goals for renewable energy are among the most admirable in the world. However, the future of Oceania, much like its past, will likely remain powerfully linked to events in distant places. Perhaps more than any other world region, Oceania is not the master of its own destiny.

Chapter Key Terms

Aborigines 417
atoll 407
Austronesians 418
El Niño 408
endemic 409
Gondwana 406
Great Barrier Reef 406
hot spots 407

invasive species 413
Maori 401
marsupials 409
Melanesia 418
Melanesians 418
Micronesia 418
MIRAB economy 424
monotremes 409

noble savage 418
Pacific Way 428
pidgin 429
Polynesia 418
roaring forties 407
subsistence affluence 424

Critical Thinking Questions

1. As Australia and New Zealand move away from intense cultural and economic involvement with Europe, new policies and attitudes have had to evolve to facilitate increased involvement with Asia. If you were a college student in Australia or New Zealand, how might you be experiencing these changes? Think about fellow students, career choices, language learning, and travel choices.

2. Discuss the emerging cultural identity of the Pacific islands, taking note of the extent to which Australia and New Zealand share or do not share in this identity. What factors are helping to forge a sense of unity across Polynesia and beyond? (First, review the spatial extent of Polynesia.)

3. Discuss the many ways in which Asia has historic, and now increasingly economic, ties to Oceania. Include in your discussion patterns of population distribution, mineral exports and imports, technological interactions, and tourism.

4. Describe the main concerns in Oceania related to global warming. Which parts of the region are likely to be most affected? To what extent can the countries of Oceania exercise control over their likely future as the climate warms?

5. Australia and New Zealand differ from each other physically. Compare and contrast the two countries in relation to water, vegetation, and prehistoric and modern animal populations.

6. Indigenous peoples worldwide are beginning to speak out on their own behalf. Discuss how the indigenous peoples of Australia, New Zealand, and the Pacific islands are serving as leaders in this movement and what measures they are taking to reconstitute a sense of cultural heritage.

7. Oceania is one of the most urbanized regions on earth. Discuss how and why this fact varies from popular impressions of the region.

8. How is tourism both boosting economies and straining environments and societies throughout the Pacific islands? What solutions are being proposed to reduce the negative impacts of tourism?

9. Compare how women have been empowered, or not, politically and economically in Australia, New Zealand, Papua New Guinea, and the Pacific islands.

10. Compared with other regions, Australia and New Zealand are somewhat unusual in having achieved broad prosperity on the basis of raw material exports. How would you account for this achievement?

Glossary

Aborigines (p. 417) the longest surviving inhabitants of Oceania, whose ancestors, the Australoids, migrated from Southeast Asia possibly as early as 50,000 years ago, over the Sundaland landmass that was exposed during the ice ages

acculturation (p. 129) adaptation of a minority culture to the host culture enough to function effectively and be self-supporting; cultural borrowing

acid rain (p. 60) acidic precipitation that has formed through the interaction of rainwater or moisture in the air with sulfur dioxide and nitrogen oxides emitted during the burning of fossil fuels

age distribution or **age structure** (p. 42) the proportion of the total population in each age group

agribusiness (p. 75) the business of farming conducted by large-scale operations that produce, package, and distribute agricultural products

agroecology (p. 313) the practice of traditional, nonchemical methods of crop fertilization and the use of natural predators to control pests

agroforestry (p. 248) the raising of economically useful trees

Ainu (p. 359) an indigenous cultural minority group in Japan characterized by their light skin, heavy beards, and thick, wavy hair who are thought to have migrated thousands of years ago from the northern Asian steppes

air pressure (p. 27) the force exerted by a column of air on a square foot of surface

animism (p. 275) a belief system in which natural features carry spiritual meaning

apartheid (p. 256) a system of laws mandating racial segregation in South Africa, in effect from 1948 until 1994

aquifer (p. 61) a natural underground water reservoir

archipelago (p. 369) a group, often a chain, of islands

assimilation (p. 129) the loss of old ways of life and the adoption of the lifestyle of another culture

Association of Southeast Asian Nations (ASEAN) (p. 388) an organization of Southeast Asian governments established to further economic growth and political cooperation

atoll (p. 407) a low-lying island, formed of coral reefs that have built up on the circular or oval rims of submerged volcanoes

Australo-Melanesians (p. 377) a group of hunters and gatherers who moved from the present northern Indian and Burman parts of southern Eurasia into the exposed landmass of Sundaland about 60,000 to 40,000 years ago

Austronesians (p. 418) a Mongoloid group of skilled farmers and seafarers from southern China who migrated south to various parts of Southeast Asia between 10,000 and 5000 years ago

average population density (p. 41) the average number of people per unit area (for example, per square mile or square kilometer)

Aztecs (p. 109) native people of central Mexico noted for advanced civilization before the Spanish conquest

baby boomer (p. 69) a member of the largest age group in North America, the generation born in the years after World War II, from 1947 to 1964, in which a marked jump in the birth rate occurred

biodiversity (p. 101) the variety of life forms to be found in a given area, such as the Amazon Basin, or the entire planet

birth rate (p. 42) the number of births per 1000 people in a given population, per unit of time, usually per year

Bolsheviks (p. 186) a faction of communists who came to power during the Russian Revolution

brain drain (p. 115) the migration of educated and ambitious young adults to cities or foreign countries, depriving the sending communities of talented youth in whom they have invested years of nurturing and education

brownfields (p. 85) old industrial sites whose degraded conditions pose obstacles to redevelopment

Buddhism (p. 308) a religion of Asia that originated in northern India in the sixth century B.C.E. as a reinterpretation of Hinduism; it emphasizes modest living and peaceful self-reflection leading to enlightenment

capital (p. 14) wealth in the form of money or property used to produce more wealth

capitalists (p. 186) usually a wealthy minority that owns the majority of factories, farms, businesses, and other means of production

carbon sequestration (p. 253) the removal and storage of carbon taken from the atmosphere

carrying capacity (p. 22) the maximum number of people that a given territory can support sustainably with food, water, and other essential resources

cartel (p. 227) a group that is able to control production and set prices for its products

cash economy (p. 45) an economic system in which the necessities of life are purchased with monetary currency

caste (p. 307) an ancient Hindu system for dividing society into hereditary hierarchical classes

central planning (p. 153) a central bureaucracy dictating prices and output with the stated aim of allocating goods equitably across society according to need

centrally planned economy (p. 186) economic system in which the state owns all real estate and means of production, while government bureaucrats direct all economic activity including the locating of factories, residences, and transport infrastructure

Christianity (p. 218) a monotheistic religion based on belief in the teachings of Jesus of Nazareth, a Jew, who described God's relationship to humans as primarily one of love and support, as exemplified by the Ten Commandments

civil disobedience (p. 299) the breaking of discriminatory laws by peaceful protesters

clear-cutting (p. 58) the cutting down of all trees on a given plot of land, regardless of age, health, or species

climate (p. 26) the long-term balance of temperature and precipitation that characteristically prevails in a particular region

cold war (p. 153) the contest that pitted the United States and Western Europe, espousing free market capitalism and democracy, against the USSR and its allies, promoting a centrally planned economy and a socialist state

G-1

commodities (p. 265) raw materials that are exported to wealthier countries that process them into more valuable manufactured goods

communal conflict (p. 309) a euphemism for religion-based violence in South Asia

Communism (p. 186) an ideology, based largely on the writings of the German revolutionary Karl Marx, that calls on workers to unite to overthrow capitalists and establish an egalitarian society where workers share what they produce

Communist Party (p. 186) the political organization that ruled the USSR from 1917 to 1991; other communist countries, such as China, Mongolia, North Korea, and Cuba, also have communist parties

Confucianism (p. 338) a Chinese philosophy that teaches that the best organizational model for the state and society is a hierarchy based on the patriarchal family

contested space (p. 123) any area that several groups claim or want to use in different and often conflicting ways, such as the Amazon or Palestine

coral bleaching (p. 375) color loss, which results when photosynthetic algae that live in the corals are expelled

country (p. 47) a political division of territory that has control over its own affairs

coup d'état (p. 125) a military- or civilian-led forceful takeover of a government

Creole (p. 112) a European person, usually of Spanish descent, born in the Americas

crony capitalism (p. 388) a type of corruption in which politicians, bankers, and entrepreneurs, sometimes members of the same family, have close personal as well as business relationships

cultural homogeneity (p. 32) uniformity of ideas, values, technologies, and institutions among culture groups in a particular area

cultural homogenization (p. 136) the tendency toward uniformity of ideas, values, technologies, and institutions among associated culture groups

cultural pluralism (p. 394) the cultural identity characteristic of a region where groups of people from many different backgrounds have lived together for a long time, yet have remained distinct

Cultural Revolution (p. 341) a series of highly politicized and destructive mass campaigns launched in 1966 to force the entire population of China to support the continuing revolution

culture (p. 29) all the ideas, materials, and institutions that people have invented to use to live on earth that are not directly part of our biological inheritance

culture group (p. 29) a group of people who share a particular set of beliefs, a way of life, a technology, and usually a place

currency devaluation (p. 269) the lowering of a currency's value relative to the U.S. dollar, the Japanese yen, the European euro, or other currency of global trade

czar (p. 186) title of the ruler of the Russian empire; derived from the word "caesar," the title of the Roman emperors

death rate (p. 42) the ratio of total deaths to total population in a specified community, usually expressed in numbers per 1000 or in percentages

deindustrialization (p. 159) reduction of industrial capacity

delta (p. 26) the triangle-shaped plain of sediment that forms where a river meets the sea

democratization (p. 46) the transition toward political systems guided by competitive elections

demographic transition (p. 45) the change from high birth and death rates to low birth and death rates that usually accompanies a cluster of other changes, such as change from a subsistence to a cash economy, increasing education rates, and urbanization

demography (p. 40) the study of population patterns and changes

desertification (p. 215) a set of ecological changes that converts nondesert lands into deserts

detritus (p. 371) dead organic material (such as plants and insects) that collects on the ground

diaspora (p. 218) the dispersion of Jews around the globe after they were expelled from the eastern Mediterranean by the Roman Empire beginning in 73 C.E.; can now refer to other dispersed culture groups

dictator (p. 125) a ruler who claims absolute authority, governing with little respect for the law or the rights of citizens

digital divide (p. 10) the discrepancy in access to information technology between small, rural, and poor areas and large, wealthy cities that contain major government research laboratories and universities

divide and rule (p. 271) the deliberate intensification of divisions and conflicts between communities by European colonial powers

double day (p. 167) the longer workday of women with jobs outside the home who also work as caretakers, housekeepers, and cooks for their families

dowry (p. 311) a price paid by the family of a bride to the groom (opposite of bride price), formerly a custom practiced only by the rich

dumping (p. 162) the cheap sale on the world market of overproduced commodities, lowering global prices and hurting producers of these same commodities elsewhere in the world

early extractive phase (p. 117) a phase in Central and South American history, beginning with the Spanish conquest and lasting until the early twentieth century, characterized by a dependence on trade in raw materials

ecological footprint (p. 24) a method of estimating the amount of biologically productive land and sea area needed to sustain a human population at its current standard of living

economic core (p. 66) the dominant economic region within a larger region; in nineteenth-century North America the core included southern Ontario and the north-central part of the United States (chiefly Illinois, Indiana, Ohio, New York, New Jersey, and Pennsylvania)

economic diversification (p. 228) the expansion of an economy to include a wider array of economic activities

economies of scale (p. 158) reductions in the unit costs of production that occur when goods or services are efficiently mass produced, resulting in a rise in profits per unit

ecotourism (p. 107) nature-oriented vacations often taken in endangered and remote landscapes, usually by travelers from industrialized nations

El Niño (p. 104) periodic climate-altering changes, especially in the circulation of the Pacific Ocean, now understood to operate on a global scale

endemic (p. 409) belonging or restricted to a particular place

erosion (p. 26) the process by which fragmented rock and soil are moved over a distance, primarily by wind and water

ethnic cleansing (p. 159) the systematic removal of an ethnic group or people from a region or country by deportation or genocide ethnic group (p. 29) a group of people who share a set of beliefs, a way of life, a technology, and usually a geographic location

ethnicity (p. 88) the quality of belonging to a particular culture group

euro (p. 159) the official (but not required) currency of the European Union as of January 1, 1999

European colonialism (p. 9) the practice of taking over the human and natural resources of often distant places to produce wealth for Europe

European Union (EU) (p. 136) a supranational institution including most of West, South, North, and Central Europe, established to bring economic integration to member countries

evangelical Protestantism (p. 131) a Christian movement that focuses on personal salvation and empowerment of the individual through miraculous healing and transformation; some practitioners preach the "gospel of success" to the poor—that a life dedicated to Christ will result in prosperity for the believer

exchange or service sector (p. 13) the part of the economy based on the bartering and trading of resources, products, and services

Export Processing Zones (EPZs) or free trade zones (p. 121) specially created legal spaces or industrial parks within a country where, to attract foreign-owned factories, duties and taxes are not charged

extended family (p. 130) a family consisting of related individuals beyond the nuclear family of parents and children

external debts (p. 119) the debts a country owes to other countries or international financial institutions

extractive resource (p. 13) a resource such as mineral ores, timber, or plants that must be mined from the earth's surface or grown from its soil

fair trade (p. 15) trade that values equity throughout the international trade system; now proposed as an alternative to free trade

favelas (p. 115) Brazilian urban slums and shantytowns built by the poor; called colonias, barrios, or barriadas in other countries

female circumcision (p. 278) the removal of the labia and the clitoris and sometimes the stitching nearly shut of the vulva

female seclusion (p. 225) the requirement that women stay out of public view

feminization of labor (p. 387) the increasing representation of women in both the formal and informal labor force

Fertile Crescent (p. 216) an arc of lush, fertile land formed by the uplands of the Tigris and Euphrates river systems and the Zagros Mountains, where nomadic peoples began the earliest known agricultural communities

feudalism (p. 149) a social system once prevalent in Europe and Asia and elsewhere in which a class of professional fighting men, or knights, defended the monarch and the peasants or serfs, who cultivated the lands of their protectors

floating population (p. 327) jobless or underemployed people who have left economically depressed rural areas for the cities and move from place to place looking for work

floodplain (p. 26) the flat land around a river where sediment is deposited during flooding

food security (p. 334) the ability of a state to supply a sufficient amount of basic food to the entire population consistently

foreign direct investment (FDI) (p. 119) the amount of money invested in a country's businesses by citizens, corporations, or governments of other countries

foreign exchange (p. 384) foreign currency that countries need to purchase imports

formal economy (p. 13) all aspects of the economy that take place in official channels

fossil fuels (p. 207) sources of energy formed from the remains of dead plants and animals

free trade (p. 14) the movement of goods and capital without government restrictions

frontal precipitation (p. 29) rainfall caused by the confrontation and interaction of large air masses of different temperatures and densities

Gazprom (p. 194) in Russia, the state-owned energy company; it is the tenth-largest oil and gas entity in the world

gender (p. 37) the culturally and biologically defined sexual category of a person

gender structure (p. 42) the proportion of males and females in each age group of a population

genetically modified organism (GMO) (p. 76) a living thing whose DNA has been modified by humans to produce desired characteristics, such as larger edible parts or resistance to pests and diseases

Geneva Conventions (p. 73) treaties concerning the conduct of war, some of which protect the rights of prisoners of war

genocide (p. 46) the deliberate destruction of an ethnic, racial, or political group

gentrification (p. 85) the renovation of old urban districts by middle-class investment, a process that often displaces poorer residents

geopolitics (p. 46) the use of strategies by countries to ensure that their best interests are served

glasnost (p. 188) literally, "openness"; the opening up of public discussion of social and economic problems that occurred in the Soviet Union under Mikhail Gorbachev in the late 1980s

global economy (p. 11) the worldwide system in which goods, services, and labor are exchanged

global warming (p. 17) the observed warming of the earth's surface and climate in recent decades as atmospheric levels of greenhouse gases increase

globalization (p. 10) the changes brought about by many types of interregional linkages and flows

Gold Rush (p. 67) when gold was discovered in California in 1849, thousands were drawn to the state with the prospect of getting rich quick

Gondwana (p. 406) the great landmass that formed the southern part of the ancient supercontinent Pangaea

grassroots economic development (p. 271) economic development projects designed to help individuals and their families achieve sustainable livelihoods

Great Barrier Reef (p. 406) the longest coral reef in the world, located off the northeastern coast of Australia

Great Leap Forward (p. 341) an economic reform program under Mao Zedong intended to quickly raise China to the industrial level of Britain and the United States

Green (p. 142) environmentally conscious

green revolution (p. 312) increases in food production brought about through the use of new seeds, fertilizers, mechanized equipment, irrigation, pesticides, and herbicides

gross domestic product (GDP) (p. 13) the market value of all goods and services produced by workers and capital within a particular country's borders and within a given year

gross domestic product (GDP) per capita (p. 39) the market value of all goods and services produced by workers and capital within a particular country's borders and within a given year divided by the number of people in the country

Group of Eight (G8) (p. 194) an organization of highly industrialized countries: France, the United States, Britain, Germany, Japan, Italy, Canada, and Russia

growth poles (p. 351) zones of development whose success draws more investment and migration to a region

guest workers (p. 164) immigrants from outside Europe, often from former colonies, who come to Europe (often temporarily) to fill labor shortages; the expectation is that they will return home when no longer needed

hacienda (p. 118) a large agricultural estate in Middle or South America, more common in the past; usually not specialized by crop and not focused on market production

hajj (p. 224) the pilgrimage to the city of Makkah (Mecca) that all Muslims are encouraged to undertake at least once in a lifetime

Harappa culture (p. 296) see Indus Valley civilization

hazardous waste (p. 63) nuclear, chemical, or industrial wastes that can have damaging environmental consequences

Hinduism (p. 306) a major world religion practiced by approximately 900 million people, 800 million of whom live in India

Hispanic (p. 53) a loose ethnic term that refers to all Spanish-speaking people from Latin America and Spain; equivalent to Latino

Holocaust (p. 153) a massive ethnic cleansing of 6 million Jews (and several million others: Roma, Slavs, the infirm, and political dissidents) perpetrated primarily by the Nazi government in Germany and the Fascist government in Italy

Horn of Africa (p. 247) the part of Africa that juts out from East Africa and wraps around the southern tip of the Arabian Peninsula

hot spots (p. 407) individual sites of upwelling material (magma) originating deep in the mantle of the earth and surfacing in a tall plume; hot spots tend to remain fixed relative to migrating tectonic plates

hub-and-spoke network (p. 78) the organization of air service in North America around hubs, strategically located airports used as collection and transfer points for passengers and cargo traveling from one place to another

hukou system (p. 326) the system in China by which citizens' permanent residence is registered

human geography (p. 5) the study of various aspects of human life that create the distinctive landscapes and regions of the world

human well-being (p. 39) various measures of the extent to which people are able to obtain a healthy life in a community of their choosing

humanism (p. 150) a philosophy and value system that emphasizes the dignity and worth of the individual

humid continental climate (p. 141) a mid-latitude climate pattern in which summers are fairly hot and moist, and winters are longer and colder the deeper into the interior of the continent one goes

import quota (p. 15) a limit on the amount of a given item that may be imported into a country over a given period of time

import substitution industrialization (ISI) (p. 119) a form of industrialization involving the use of public funds to set up factories to produce goods that previously had been imported

Incas (p. 109) Native American people who ruled the largest pre-Columbian state in the Americas, with a domain stretching from southern Colombia to northern Chile and Argentina

income disparity (p. 117) the gap in wealth and resources between the richest 10 (or 20) percent and the poorest 10 (or 20) percent of a country's population

indigenous (p. 100) native to a particular place or region

Indus Valley civilization (p. 296) the first substantial settled agricultural communities in South Asia, which appeared about 4500 years ago along the Indus River in modern-day Pakistan and along the Saraswati River in modern-day India

Industrial Revolution (p. 13) a series of inventions, innovations, and ideas that allowed manufacturing to be mechanized

informal economy (p. 13) all aspects of the economy that take place outside official channels

information technology (IT) (p. 79) the part of the service sector that relies on the use of computers and the Internet to process and transport information; includes banks, software companies, medical technology companies, and publishing houses

International Monetary Fund (IMF) (p. 15) a financial institution funded by the developed nations to help developing countries reorganize, formalize, and develop their economies

interregional linkages (p. 9) economic, political, or social connections between regions, whether contiguous or widely separated

intertropical convergence zone (ITCZ) (p. 246) a band of atmospheric currents circling the globe roughly at the equator; warm winds from both north and south converge at the ITCZ, pushing air upward and causing copious rainfall

intifada (p. 234) a prolonged Palestinian uprising against Israel

iron curtain (p. 153) a fortified border zone that separated Western Europe from Eastern Europe during the cold war

Islam (p. 211) a monotheistic religion that emerged in the seventh century c.e. when, according to tradition, the archangel Gabriel revealed the tenets of the religion to the Prophet Muhammad

Jainism (p. 308) originally a reformist movement within Hinduism, Jainism is a faith tradition that is more than 2000 years old; found mainly in western India and in large urban centers throughout the region, Jains are known for their educational achievements, nonviolence, and strict vegetarianism

jati (p. 307) in Hindu India, the subcaste into which a person is born, which largely defines the individual's experience for a lifetime

Judaism (p. 218) a monotheistic religion characterized by the belief in one God, Yaweh, a strong ethical code summarized in the Ten Commandments, and an enduring ethnic identity

kaizen (p. 348) the "continuous improvement system" pioneered in Japanese manufacturing; it ensures that fewer defective parts are produced because production lines are constantly surveyed for errors

kanban system (p. 347) the "just-in-time" system pioneered in Japanese manufacturing that clusters companies that are part of the same production process close together so that they can deliver parts to each other precisely when they are needed

knowledge economy (p. 79) markets based on the management of information, such as finance, media, and research and development

land reform (p. 119) a policy that breaks up large landholdings for redistribution among landless farmers

landforms (p. 24) physical features such as mountain ranges, river valleys, basins, and cliffs

latitude (p. 5) the distance in degrees north or south of the equator; lines of latitude run parallel to the equator, and are also called parallels

liberation theology (p. 131) a movement within the Roman Catholic Church that uses the teachings of Christ to encourage the poor to organize to change their own lives and the rich to promote social and economic equity

lingua franca (p. 279) a common language used to communicate by people who do not speak one another's native languages; often a language of trade

living wages (p. 15) minimum wages high enough to support a healthy life

longitude (p. 5) the distance in degrees east and west of Greenwich, England; lines of longitude, also called meridians, run from pole to pole (the line of longitude at Greenwich is 0° and is known as the prime meridian)

machismo (p. 130) a set of values that defines manliness in Middle and South America

maquiladoras (p. 121) foreign-owned tax exempt factories, often located in Mexican towns just across the U.S. border from U.S. towns, that hire workers at low wages to assemble manufactured goods that are then exported for sale

marianismo (p. 130) a set of values based on the life of the Virgin Mary that defines the proper social roles for women in Middle and South America

marsupials (p. 409) mammals that give birth to their young at a very immature stage and nurture them in a pouch equipped with nipples

material culture (p. 36) all the things, living or not, that humans use

medieval period (p. 149) the period in Europe circa 450–1300 C.E., during which civil society declined and commerce ceased as the Roman Empire collapsed; by 1250, town life, trade, and commerce began to revive and diffusion from outside influences and European innovation encouraged the flourishing of the arts, philosophy, and architecture

Mediterranean climate (p. 30) a climate pattern of warm, dry summers and mild, rainy winters

megalopolis (p. 84) an area formed when several cities expand so that their edges meet and coalesce

Melanesia (p. 418) New Guinea and the islands south of the equator and west of Tonga (the Solomon Islands, New Caledonia, Fiji, Vanuatu)

Melanesians (p. 418) a group of Australoids named for their relatively dark skin tones, a result of high levels of the protective pigment melanin; they settled throughout New Guinea and other nearby islands

mercantilism (p. 117) the policy by which the rulers of Spain and Portugal, and later of England and Holland, sought to increase the power and wealth of their realms by managing all aspects of production, transport, and trade in their colonies

Mercosur (p. 123) a free trade zone created in 1991 that links the economies of Brazil, Argentina, Uruguay, and Paraguay to create a common market

mestizo (p. 112) a person of mixed European and Native American descent

metropolitan areas (p. 84) cities with population of 50,000 or more and their surrounding suburbs

microcredit (p. 313) a program based on peer support that makes very small loans available to very low income entrepreneurs

Micronesia (p. 418) the small islands lying east of the Philippines and north of the equator

Middle America (p. 100) in this book, a region including Mexico, Central America, and the islands of the Caribbean

MIRAB economy (p. 424) an economy based on migration, remittance, aid, and bureaucracy

mixed agriculture (p. 250) the raising of a variety of crops and animals on a single farm, often to take advantage of several environmental niches

Mongols (p. 185) a loose confederation of nomadic pastoral people centered in eastern Central Asia, who by the thirteenth century had established by conquest an empire stretching from Europe to the Pacific

monotheisitic (p. 218) pertaining to the belief that there is only one god

monotremes (p. 409) egg-laying mammals, such as the duck-billed platypus and the spiny anteater

monsoon (p. 29) a wind pattern in which in summer months, warm, wet air coming from the ocean brings copious rainfall, and in winter, cool, dry air moves from the continental interior toward the ocean

Mughals (p. 296) a dynasty of Central Asian origin that ruled India from the sixteenth to the nineteenth century

multiculturalism (p. 33) the state of relating to, reflecting, or being adapted to diverse cultures

multinational corporation (p. 14) a business organization that operates extraction, production, and/or distribution facilities in multiple countries

multiplier effect (p. 66) economic development and diversification, which results when enterprises "spin-off" from a main industry

Muslims (p. 218) followers of Islam

nation (p. 46) a group of people who share a language, culture, political philosophy, and usually a territory

nationalism (p. 46) devotion to the interests or culture of a particular country, nation, or cultural group; the idea that a group of people living in a specific territory and sharing cultural traits should be united in a single country to which they are loyal and obedient

nationalize (p. 122) to invest control of an industry or enterprise in the government

nation-state (p. 46) a political unit, or country, formed by people who share a language, a culture, and a political philosophy

Neolithic Revolution (p. 20) a period from 20,000 to 8000 years ago characterized by the expansion of agriculture and the making of polished stone tools

New Urbanism (p. 85) the growing popularity of urban areas

noble savage (p. 418) a term coined by European Romanticists to describe what they termed the "primitive" peoples of the Pacific, who lived in distant places supposedly untouched by corrupting influences

nomadic pastoralists (p. 184) peoples whose way of life and economy are centered on the tending of grazing animals who are moved seasonally to gain access to the best grasses

nongovernmental organization (NGO) (p. 47) an association outside the formal institutions of government, in which individuals, often from widely differing backgrounds and locations, share views and activism on political, economic, social, or environmental issues

nonpoint sources of pollution (p. 180) diffuse sources of environmental contamination, such as untreated automobile exhaust, raw sewage, or agricultural chemicals that drain from fields into water supplies

North American Free Trade Agreement (NAFTA) (p. 79) a free trade agreement made in 1994 that added Mexico to the 1989 economic arrangement between the United States and Canada

North Atlantic Treaty Organization (NATO) (p. 159) a military alliance between European and North American countries developed during the cold war to counter the influence of the Soviet Union; since the breakup of the Soviet Union, NATO has expanded membership to include much of Eastern Europe and Turkey, and is now focused mainly on providing the international security and cooperation needed to expand the European Union

nuclear family (p. 90) a family consisting of a father and mother and their children

occupied territories (p. 235) Palestinian lands occupied by Israel in 1967

offshore outsourcing (p. 82) the shifting of jobs from a relatively wealthy country to one where labor or other production costs are lower

Ogallala aquifer (p. 61) the largest North American natural aquifer, which underlies the Great Plains

OPEC (Organization of Petroleum Exporting Countries) (p. 227) a cartel of oil-producing countries—including Algeria, Angola, Indonesia, Iran, Iraq, Kuwait, Libya, Nigeria, Qatar, Saudi Arabia, the United Arab Emirates, and Venezuela—that was established to regulate the production, and hence the price, of oil and natural gas

orographic rainfall (p. 29) rainfall produced when a moving moist air mass encounters a mountain range, rises, cools, and releases condensed moisture that falls as rain

Pacific Rim (Basin) (p. 69) the countries that border the Pacific Ocean on the west and east

Pacific Way (p. 428) the idea that Pacific Islanders have a regional identity and a way of handling conflicts peacefully, which grows out of their particular social experience

Pangaea hypothesis (p. 24) the proposal based on scientific evidence that about 200 million years ago all continents were joined in a single vast continent, called by geologists Pangaea

Partition (p. 299) in this context, the breakup following Indian independence that established Hindu India and Muslim Pakistan

pastoralism (p. 252) a way of life based on herding; practiced primarily in savannas, on desert margins, or in the mixture of grass and shrubs called open bush

perestroika (p. 188) literally, "restructuring"; the restructuring of the Soviet economic system in the late 1980s in an attempt to revitalize the economy

permafrost (p. 178) permanently frozen soil a few inches or feet beneath the surface

physical geography (p. 5) the study of the earth's physical processes: how they work, how they affect humans, and how they are affected by humans

pidgin (p. 429) a language used for trading; made up of words borrowed from the several languages of people involved in trading relationships

plantation (p. 118) a large estate or farm on which a large group of resident laborers grow (and partially process) a single cash crop

plate tectonics (p. 24) the scientific theory that the earth's surface is composed of large plates that float on top of an underlying layer of molten rock; the movement and interaction of the plates create many of the large features of the earth's surface, particularly mountains

pluralistic state (p. 46) a country in which political power is shared among groups, each defined by common language, ethnicity, culture, or other characteristics

political ecologist (p. 22) someone who studies power allocation in the context of the environment, development, politics, and human well-being

polygyny (p. 227) the taking by a man of more than one wife at a time

Polynesia (p. 418) the numerous islands situated inside an irregular triangle formed by New Zealand, Hawaii, and Easter Island

population pyramid (p. 42) a graph that depicts the age and gender structures of a country

populist movements (p. 130) popularly-based efforts, often seeking relief for the poor

primate city (p. 113) a city that is vastly larger than all others in a country and in which economic and political activity is centered

privatization (p. 119) the sale of industries formerly owned and operated by the government to private companies or individuals

Protestant Reformation (p. 150) a European reform (or "protest") movement that challenged Roman Catholic practices in the sixteenth century and led to the establishment of Protestant churches

pull factors (p. 383) positive features of a place that attract people to move there

purchasing power parity (PPP) (p. 39) the amount that the local currency equivalent of U.S.$1 will purchase in a given country

purdah (p. 310) the practice in South Asia of concealing women, especially during their reproductive years, from the eyes of nonfamily men

push factors (p. 383) negative features of the place where people are living that impel them to migrate

Québecois (p. 53) the French-Canadian ethnic group or members of that group; also, all citizens of Québec, regardless of ethnicity

Qur'an (or Koran) (p. 218) the holy book of Islam, believed by Muslims to contain the words Allah revealed to Muhammad through the archangel Gabriel

race (p. 38) a social or political construct based on apparent characteristics such as skin color, hair texture, and face and body shape, but of no biological significance

rain shadow (p. 29) the dry side of a mountain range, facing away from the prevailing winds

rate of natural increase (growth rate) (p. 41) the rate of population growth measured as the excess of births over deaths per 1000 individuals per year without regard for the effects of migration

recession (p. 119) a slowing of economic activity

region (p. 8) a unit of the earth's surface that contains distinct patterns of physical features and/or of human development

regional conflict (p. 319) especially in South Asia, a conflict created by the resistance of a regional ethnic or religious minority to the authority of a national or state government

regional self-sufficiency (p. 349) an economic policy in communist China that encouraged each region to develop independently in the hope of evening out the national distribution of production and income

regional trade bloc (p. 15) an association of neighboring countries that have agreed to lower trade barriers for one another

religion (p. 33) formal and informal institutions that embody value systems; most have deep roots in history, and many include a spiritual belief in a higher power (God, Yahweh, Allah) as the underpinning for their value system

religious nationalism (p. 317) the belief that a certain religion is strongly connected to a particular territory and that adherents should have political power in that territory

remittances (p. 10) earnings sent home by immigrant workers

resettlement schemes (p. 391) government plans to move large numbers of people from one part of a country to another to relieve urban congestion, disperse political dissidents, or accomplish other social purposes

Ring of Fire (p. 25) the tectonic plate junctures around the edges of the Pacific Ocean; characterized by volcanoes and earthquakes

roaring forties (p. 407) powerful westerly air and ocean currents at about 40° south latitude that speed around the far Southern Hemisphere virtually unimpeded by landmasses

Roma (p. 153) the now preferred term in Europe for Gypsy

Russian Federation (p. 178) Russia and its political subunits, which include 30 internal republics and more than 10 so-called autonomous regions

Russian Mafia (p. 193) a highly organized criminal network dominated by former KGB and military personnel

Russification (p. 191) assimilation of all minorities to Russian (Slavic) ways

Sahel (p. 246) a band of arid grassland that runs east–west along the southern edge of the Sahara

salinization (p. 212) the impregnation of the soil by salts and other minerals left by the evaporation of water, damaging soil fertility

scale (p. 4) the proportion that relates the dimensions of a map to the dimensions of the area it represents; also variable-sized units of geographical analysis from the local scale to the regional scale to the global scale

Scandinavia (p. 140) Iceland, Denmark (including Greenland and the Faroe Islands), Sweden, Norway, and Finland

Schengen Accord (p. 164) an agreement signed in the 1990s by the European Union and many of its neighbors that called for free movement across common borders

seawater desalination (p. 213) the removal of salt from seawater, usually accomplished through the use of expensive and energy-intensive technologies

secular states (p. 231) countries that have no state religion and in which religion has no direct influence on affairs of state or civil law

secularism (p. 35) a way of life informed by ethics and values that are not necessarily derived from a religious tradition

self-reliant development (p. 271) small-scale development schemes in rural areas that focus on developing local skills, creating local jobs, producing products or services for local consumption, and maintaining local control so that participants retain a sense of ownership

sex tourism (p. 395) the sexual entertainment industry that services primarily men who travel for the purpose of living out their fantasies during a few weeks of vacation

shari'a (p. 225) literally, "the correct path"; Islamic religious law that guides daily life according to the interpretations of the Qur'an

sheiks (p. 230) patriarchal leaders of tribal groups on the Arabian Peninsula

Shi'ite (or Shi'a) (p. 225) the smaller of two major groups of Muslims with different interpretations of shari'a; Shi'ites are found primarily in Iran and southern Iraq

shifting cultivation (p. 109) a productive system of agriculture in which small plots are cleared in forestlands, the dried brush is burned to release nutrients, and the clearings are planted with multiple species; each plot is used for only two or three years and then abandoned for many years of regrowth

Sikhism (p. 308) a religion of South Asia that combines beliefs of Islam and Hinduism

silt (p. 102) fine soil particles

Slavs (p. 185) a group of farmers who originated between the Dnieper and Vistula rivers in modern-day Poland, Ukraine, and Belarus

smog (p. 60) a combination of industrial air pollution and car exhaust (smoke + fog)

social safety net (p. 82) the services provided by the government—such as welfare, unemployment benefits, and healthcare—that prevent people from falling into extreme poverty

social welfare (p. 168) in Europe, elaborate tax-supported systems that serve all citizens in one way or another

soft power (p. 356) as described by some analysts, the attempt by China to soften its image in places it would like to invest

South America (p. 100) refers to the vast continent south of Central America

sovereignty (p. 46) the capability of a country to manage its own affairs

Soviet Union (p. 174) see Union of Soviet Socialist Republics

special economic zones (SEZs) (p. 350) free trade zones within China

state-aided market economy (p. 347) an economic system based on market principles such as private enterprise, profit incentives, and supply and demand, but with strong government guidance; in contrast to the free market (limited government) economic system of the United States and Europe

steppes (p. 30) semiarid, grass-covered plains

structural adjustment policies (SAPs) (p. 15) policies that require economic reorganization toward less government involvement in industry, agriculture, and social services; sometimes imposed by the World Bank and the International Monetary Fund as conditions for receiving loans

subcontinent (p. 288) often used to refer to the entire Indian peninsula, including Nepal, Bhutan, India, Pakistan, and Bangladesh

subduction (p. 25) the sliding of one lithospheric (tectonic) plate under another

subduction zone (p. 101) the zone where one tectonic plate slides under another

subsidies (p. 162) monetary assistance granted by a government to an individual or group in support of an activity, such as farming or housing construction, that is viewed as being in the public interest

subsistence affluence (p. 424) refers to a lifestyle whereby self-sufficiency is achieved for most necessities, while some opportunities to earn cash allow for travel and occasional purchases of manufactured goods

subsistence agriculture (p. 250) A practice which provides food for only the farmer's family, and is usually done on small farms

subsistence economy (p. 45) circumstances in which a family produces most of its own food, clothing, and shelter

Sunni (p. 225) the larger of two major groups of Muslims with different interpretations of shari'a

sustainable agriculture (p. 22) farming that meets human needs without poisoning the environment or using up water and soil resources

sustainable development (p. 21) improvement of standards of living in ways that will not jeopardize those of future generations

taiga (p. 178) subarctic forests

Taliban (p. 309) an archconservative Islamist movement that gained control of the government of Afghanistan in the mid-1990s

tariff (p. 14) a tax imposed by a country on imported goods, usually intended to protect industries within that country

temperate midlatitude climate (p. 140) as in south central North America, China, and Europe, a climate that is moist all year with relatively mild winters and long hot summers

temperature-altitude zones (p. 103) regions of the same latitude that vary in climate according to altitude

theocratic states (p. 231) countries that require all government leaders to subscribe to a state religion and all citizens to follow rules decreed by that religion

thermal inversion (p. 60) a warm mass of stagnant air that is temporarily trapped beneath heavy cooler air

trade deficit (p. 80) the extent to which the money earned by exports is exceeded by the money spent on imports

trade winds (p. 104) winds that blow in a generally westerly direction across the Atlantic

tsunami (p. 330) a large sea wave caused by an earthquake

tundra (p. 178) a treeless area, between the ice cap and the tree line of arctic regions, where the subsoil is permanently frozen

typhoon (p. 332) tropical cyclone or hurricane

underemployment (p. 192) the condition in which people are working too few hours to make a decent living or are working at menial jobs even though highly trained

United Nations (UN) (p. 47) an assembly of 185 member states that sponsors programs and agencies that focus on scientific research, humanitarian aid, planning for development, fostering general health, and peacekeeping assistance

urban sprawl (p. 60) the encroachment of suburbs on agricultural land

urbanization (p. 22) the process whereby cities, towns, and suburbs grow as populations shift from rural to urban livelihoods

varna (p. 307) the four hierarchically ordered divisions of society in Hindu India underlying the caste system: Brahmins (priests), Kshatriyas (warriors/kings), Vaishyas (merchants/landowners), and Sudras (laborers/artisans)

veiling (p. 225) the custom of covering the body with a loose dress and the head—and in some places, the face—with a scarf

weather (p. 26) the short-term (day-to-day) expression of climate

weathering (p. 26) the physical or chemical decomposition of rocks by sun, rain, snow, wind, ice, and the effects of life-forms

welfare state (p. 152) a social system in which the state accepts responsibility for the well-being of its citizens

West Bank barrier (p. 235) a 25-foot-high concrete wall in some places and a fence in others that now surrounds much of the West Bank and encompasses many of the remaining Jewish settlements there

wet rice (or paddy) cultivation (p. 374) a prolific type of rice production that requires the submersion of the plant roots in water for part of the growing season

World Bank (p. 15) a global lending institution that makes loans to countries that need money to pay for development projects

world cities (p. 151) cities of worldwide economic and/or cultural influence

World Trade Organization (WTO) (p. 15) a global institution made up of member countries whose stated mission is the lowering of trade barriers and the establishment of ground rules for international trade

zionists (p. 233) those who have worked, and continue to work, to create a Jewish homeland (Zion) in Palestine

Index

Note: **Boldface** indicates glossary terms; *italics* indicate foreign words or book titles; and *i* and *t* indicate illustrations and tables, respectively.